# 1st International Conference on Mechanical Electronic and Biosystem Engineering (MEBSE 2018)



IOP Conference Series: Materials Science and Engineering
Volume 557

Bogor, Indonesia
15 - 16 December 2018

**Editor:**

**Zuradzman Mohamad Razlan**

ISBN: 978-1-5108-9091-6
ISSN: 1757-8981

**Printed from e-media with permission by:**

Curran Associates, Inc.
57 Morehouse Lane
Red Hook, NY 12571

**Some format issues inherent in the e-media version may also appear in this print version.**



No changes have been made to the content of these proceedings. There may be changes to pagination and minor adjustments for aesthetics.

Printed by Curran Associates, Inc. (2019)

For permission requests, please contact the Institute of Physics at the address below.

Institute of Physics
Dirac House, Temple Back
Bristol BS1 6BE UK

Phone: 44 1 17 929 7481
Fax: 44 1 17 920 0979

techtracking@iop.org

**Additional copies of this publication are available from:**

Curran Associates, Inc.
57 Morehouse Lane
Red Hook, NY 12571 USA
Phone: 845-758-0400
Fax: 845-758-2633
Email: curran@proceedings.com
Web: www.proceedings.com

# Table of contents

## Volume 557

**The 1st International Conference on Mechanical Electronic and Biosystem Engineering (MEBSE 2018)**

**15–16 December 2018, Bogor, Indonesia**

**Accepted papers received: 17 May 2019**
**Published online: 28 June 2019**

### Preface



### Papers

**Mechatronics Engineering**

**Mechanical Engineering**

IOP Conf. Series: Materials Science and Engineering **557** (2019) 011001 doi:10.1088/1757-899X/557/1/011001

# Preface

The 1st International Conference on Mechanical Electronic and Biosystem Engineering (MEBSE 2018) was held on 15-16 December 2018 in Bogor, Indonesia, one of the most popular tourist destinations in Indonesia. MEBSE 2018 was organized by Malaysian Society of Instrumentation and Control Technology (MSICT), School of Mechatronic Engineering, Universiti Malaysia Perlis (UniMAP), Bogor Agricultural University (IPB), and Technopreneur At UniMAP Sdn. Bhd (TUSB).

The MEBSE 2018 aims to provide a platform for participants from the diverse field of Mechanical, Mechatronics and Biosystem Engineering backgrounds to exchange ideas and findings to promote cooperation, networking as well as knowledge sharing in both engineering practice and research in higher education. The conference includes the fields of Mechatronics, Ergonomics and Biomedical Engineering, Fluid Thermal and Energy Conversion, Postharvest, Food and Biosystem Engineering, Network, Data and Automation, Non-Destructive Evaluation and Remote Sensing. The conference received more than 100 papers from Malaysia, Indonesia and other countries over the world. After a rigorous peer-review process, 89 papers were accepted. The accepted papers that focused on theory, applications, design and experimental studies relevant to mechanical, mechatronics, and biosystem engineering field were invited to present in this conference.

I wish to record my gratitude to all reviewers, both local and international, for their time and effort in reviewing the papers. Their feedback to the authors has been excellent and useful to ensure the high quality of papers for the conferences. My sincere thanks also go to all sponsors for their generosity. Last but not least, I would like to thank also the organizing committee members for their hard work, cooperation and suggestions in ensuring the success of this conference.

**Ir. Dr. Zuradzman Mohamad Razlan**
**Head Editor**
**MEBSE 2018**

IOP Conf. Series: Materials Science and Engineering **557** (2019) 011001 doi:10.1088/1757-899X/557/1/011001

IOP Conf. Series: Materials Science and Engineering **557** (2019) 011002 doi:10.1088/1757-899X/557/1/011002

# Peer review statement

All papers published in this volume of *IOP Conference Series: Materials Science and Engineering* have been peer reviewed through processes administered by the proceedings Editors. Reviews were conducted by expert referees to the professional and scientific standards expected of a proceedings journal published by IOP Publishing.

# Development of Automated Spray-Painting System for Anti-Static Coating Process

**Rudzuan M N [1,4], Wan Khairunizam[1], I Zunaidi[3], Z M Razlan[1], Shahriman A B[1], Rozman A R[2] and Shaharizal A[4]**

[1]School of Mechatronic Engineering, Universiti Malaysia Perlis, Arau, Perlis, MALAYSIA

[2]NS Aries Sdn. Bhd.

[3]Faculty of Technology, University of Sunderland, St Peter's Campus, Sunderland, SR6 0DD, United Kingdom

[4]Aerospace Composites Malaysia (ACM) Sdn. Bhd.

Corresponding author: rudzuan@unimap.edu.my

**Abstract.** This paper is presented to document a development process of automated system for the spray-painting process and for the purpose of advantage of implementation of automation to the production line instead of manual spray process and know-how of thickness control when applying three spray guns compared to single gun as normally implemented in automated spraying system. Three spraying could reduce the process time to spray the workpieces, in this case, is airplane wing's parts built from composite material. But these three guns automations system comes with challenging tasks in order to find an even thickness of overlapping spray pattern coming from these three separated guns. Ultimately, various studies on atomization parameters and other factors resulted in successful mass production.

## 1. Introduction

Spray painting process has been widely used in painting and coating applications, where the paint or coating liquid is atomized and flown to deposit on the target surface (workpiece). In this study, the workpiece is a composite part uses in making airplane wings. From manual spraying process, the gantry type automation system is developed to counter inconsistency in painting thickness, to increase productivity and for the environmental and health issue of manual painting to the painters. This Automation Spray Painting System or in-short, ASPS is consists of three spray guns located on three-axis gantry type carrier and moved in X, Y and Z-axis powered by three electrical motor. Spray gun is type HVMP (High Volume Medium Pressure) by De VILBISS. Mixture solution supplies to the spray gun are transferred by a pressurized tank to ensure a constant supply. A separate pressurized tank is used to fill with thinner for flushing out balance paint solution from transfer tube.

This ASPS system is designed for a company X located in northern region of Peninsular Malaysia. The development of ASPS was started in December 2017 and successfully installed and tested at Paint Booth No.1 in August 2018. This paper is organized as follow subsequent: Section II encompasses literature review of the related researches of the manual and automated spray painting. Section III presents the structure of the system and main parts and components use in the developed system. Section IV describes the detail about Spray Gun position and setting parameters. Section V expresses the conclusion over the developed system.

IOP Conf. Series: Materials Science and Engineering **557** (2019) 012001 doi:10.1088/1757-899X/557/1/012001

## 2. Literature Review

The spray thickness and the evenness of coating layer are the major issues in the painting process. A lot of studies are undertaken in the analyses of spraying path, spray overlap, spray gun orientations and spray flow rate as reported by Wang et.al [1], From et.al. [2] and Kout et.al. [3]. Controlling spraying path is an important parameter in achieve uniformity or evenness of coating layer thickness [5]. Real time simulation and CAD-guided tool planning also contributed much to the uniformity of spray pattern distribution and thickness deposition [4,6]. Radfar et al. utilized ANFIS to predict paint thickness of the automobile paint process using a robot [7].

Company X is facing quality problems concerning paint thickness and color unevenness with existing manual painting line. Manual painting is carried out using a single gravity feed spray gun as in Figure 1, held by the painter based on the individual experience and expertise to spray the workpiece. The company plans to upgrade existing manual systems to automatic paint systems, control thickness, maintain quality, and improve productivity and efficiency. Currently the manual spray is painting based on the spraying parameter stated in Table 1.

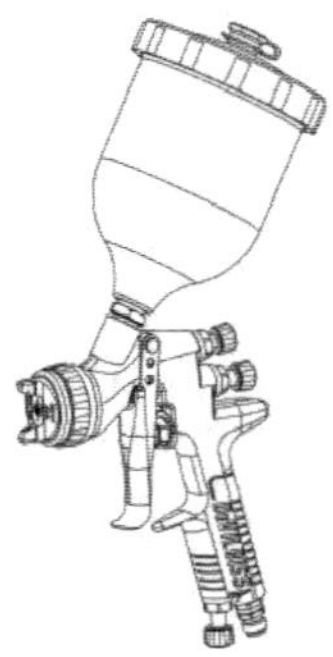

Figure 1. Manual Gravity Feed Spray Gun

Table 1. Manual Painting Parameter

| Control Items | Specification |
|---|---|
| Manual spray gun type | DeVilbiss GTIPRO |
| Paint Thickness | 0.4-0.8 mils (10µ~20 µm) |
| Resistivity (Anti-Static) | < 300kOhm |
| Distance nozzle to Panel | 8-10 inch |
| Overlap between paint | 4 inch (X-direction) and 50% (Y-direction) |
| Flash-Off time* | > 15 minutes |

Flash-off time is the necessary waiting time taken to warm room for heating process once a coat is applied on a material. If the flash-off time is short, irregularities such as uneven thickness in the paint job can be expected. With the implementation of automated system ASPS, productivity and effectiveness are expected to be improved as stated in Table 2.

IOP Conf. Series: Materials Science and Engineering **557** (2019) 012001 doi:10.1088/1757-899X/557/1/012001

Table 2. Comparison Cycle Time

| | Cycle Time [minutes] |
|---|---|
| Manual | 35 |
| ASPS (estimated) | 25 (29% shorter time) |

Increased and consistent product quality is the most widely regarded advantages gained by automation system of all types. For painting or coating, film thickness tolerance and appearance (visual) are the most important point to tackle. An industry standard assumption is that a paint savings of 15% to 30% are achieved when manual painters are replaced by automation (reduce overspray). Flexible to change, expand and adapt to a different kind of workpiece size and paint thickness. Savings achieved through film thickness tolerance control and trigger accuracy will be related directly to many other savings :

Reduces stack VOC (volatile organic compounds) at Exhaust system, Reduce rework jobs thus increase customer's satisfactory, Reduce overspray thus reduce filter/chemical usage. Other automation advantages would be less reliable on expert painters thus reduce Human Resources cost for hiring & training, and a safer and more effective workplace and the work process are ergonomically optimized.

## 3. Development of ASPS

Technically, the system consists of gantry spray gun (see Figure 2) with movement by an explosion-proof type servo motor, automatically adjust spraying distance by height sensor. Controlled by PLC, this automation spray painting system provides all necessary data for managing the process optimally, including the selection of workpiece type, pressure control, paint volume, XYZ spray direction & positioning, alarm buzzer etc.

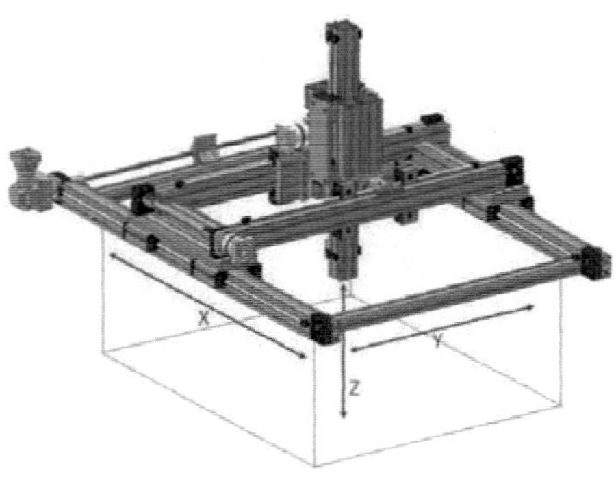


Figure 2. Gantry Spray Gun

With the implementation of this automation process, painter (who did manual painting) is not necessary to be inside painting booth and this is good for the safety and health of the worker. Furthermore, space for painter inside the booth for manual spraying process becomes unnecessary thus new cart can be designed to increase the number of the workpiece (panel to be painted) that can be allocated on the cart. Figure 3 shows the painted workpieces on the new designed cart. This new cart design could increase the number of the panel based on the increasing painting area, up to 70%.

IOP Conf. Series: Materials Science and Engineering **557** (2019) 012001 doi:10.1088/1757-899X/557/1/012001

The overall proposed system is consists of the following items.

- Main structure with nozzle bar and height sensor
- Automated Spray Gun (Nozzle) c/w pressure tank, flow counter, tubing etc
- Explosion-proof motor for XY control c/w cable
- Main Control Panel c/w Push button and Tower light

Figure 3. Painted workpieces on the new cart

### A. Spraying Paint Booth

Dimension of the paint booth is 7000mm x 5500mm with entering door in front of the booth, blower on the top roof and vacuum mechanism on the floor. There is an exit door to warm room on the rear site of the booth. Painted workpieces have to be transferred to warm room for 45 minutes after 15 minutes flashing time inside the paint booth. Flashing process is a partial evaporation process or settle down process when a liquid changes its phase and becomes partly vapour and partly liquid.

### B. Automated Gantry System

To fit with the booth dimension, the gantry system (Figure 4) is designed at L6950mm x W4800mm and Height 2380mm built from mild steel 3" x 2" square pipe. The guns can be moved freely in X and Y-axis supported by cable chain covered by flexible bellow and powered by a set of AC servomotors. For gun height adjustment, the gun also can be moved up and down and controlled by another servomotor. All servomotor movement is controlled by PLC equipped with Touch screen panel for automation mode and Teach pendant for manual control mode.

### C. Spray Gun

LVMP gun with high transfer efficiency that provides superior cost performance DA-300 type is a product of De VILBISS. It is used in this development for the reasons of the compact and high performance general-purpose automatic spray gun with superior fine atomization and transfer efficiency.

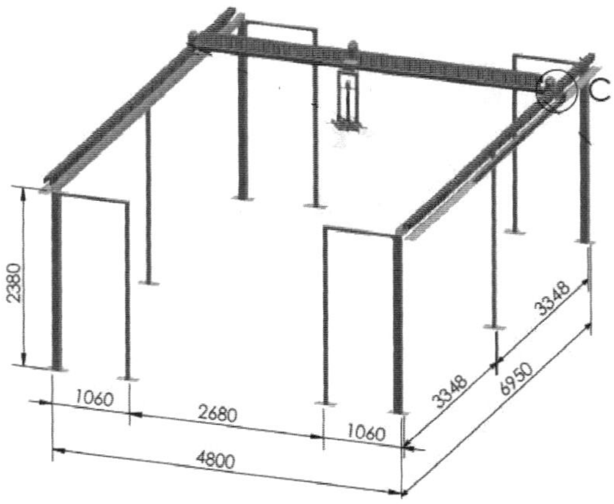


Figure 4. Automated Gantry System

There are four main characteristics highlighted by the manufacture as follow [8].

■ *Compact size and High Performance*
Featuring superior atomization, the gun is designed to be suitable for spraying metallic and pearl type paints.
■ *Air cap*
Superior atomization is made possible by using the DeVilbiss air cap which is already highly regarded in the market.
■ *LVMP*
Compared to conventional guns and HVLP guns, the LVMP gun achieves finer atomization and higher transfer efficiency with much less air consumption.
■ *Waterborne*
Has stainless steel passage for waterborne paint.

Table 3. LVMP Gun Specification

| SPECIFICATIONS | |
|---|---|
| Atomization Air Pressure (Gun Inlet) | 0.25 MPa |
| Fluid Pressure (Max.) | 0.7 MPa |
| Working Pressure (CYL) | 0.35 ~ 0.5 MPa |
| Weight | 507 g |
| Fluid Inlet | G 1/4 |
| Cylinder Air Inlet (CYL) | G 1/4 |
| Atomization Air Inlet | G 1/4 |
| Measurement | L150 × W38 × H76 mm |

Table 3 shows the specification of the spray gun sets by DeVILBISS.

**D. Pressure Tank**

There are two pressure tank provided in this system development. To reduce flushing time or cleaning time, separate pressure tank is used. One pressure tank is use to fill mixture solution paint and another one is use for solvent or thinner for flushing or cleaning purpose. After finished spraying process, pressure tank with the solvent will connect to the hose (tube) for flushing process. With this versatile PT-10M pressure tank (Figure 5), almost professional painting result can be easily carried out. Special materials such as enamel paint, chemical liquid, solvent-based coating, can be transported as smoothly as required [9].

IOP Conf. Series: Materials Science and Engineering **557** (2019) 012001 doi:10.1088/1757-899X/557/1/012001

Figure 5. PT-10M Pressure Tank

**E. AC Servomotor**

There are three AC servomotors (Figure 6) are used in this development – HG-SR52, HG-SR102 and HG-KR23B, all are from Mitsubishi Electric Co. Each of the servomotors is installed for controlling the spray gun mechanism in X-axis, Y-axis and Z-axis, respectively.

Figure 6. HG-SR52 AC Servomotor

The selection of the servomotor can be decided by using Motor sizing software provided by the manufacturer. It can be downloaded online and the servomotor selection can be done by setting all necessary parameters such as load, gear type etc based on actual application.

**F. New Cart**

New spray cart in Figure 4 is designed with new dimension L2370mm x W820mm and H920mm to maximize the number of cart that can be enter to the paint booth and more workpieces can be put on the cart for spraying process.

Figure 7. New Cart

## 4. Spray Gun

### A. Paint Material and Mixing liquid

There are two paint material are used in this spray painting process named as Liquid A and Liquid B where both of them has to mix in ratio of 1:1. With the spray gun flow-rate of 200cc/min, mixture paint of 1 can 3.79L (1 gallon) of Liquid A and 1 can 3.79L of Liquid B theoretically can be used for 37.9 minutes of painting based on consumed solution volume as calculated in (1).

$$painting\ time, t\ [\mathrm{min}] = \frac{solution\ vol\ [\mathrm{cc}]}{flow\ rate\ [^{\mathrm{cc}}/_{\mathrm{min}}]} \quad (1)$$

One complete cycle of painting takes 2.4 minutes painting time make 1 mixture liquid can be use for 15 to 16 lots of painting products.

### B. Spray Gun Position

In this development, 3 spray guns are used and they are positioned 100mm each other and one in the middle is shifted 100mm as can be seen from the bottom as shown in Figure 5. The guns are positioned exactly right angle to the workpiece as seen from the side view in Figure 6.

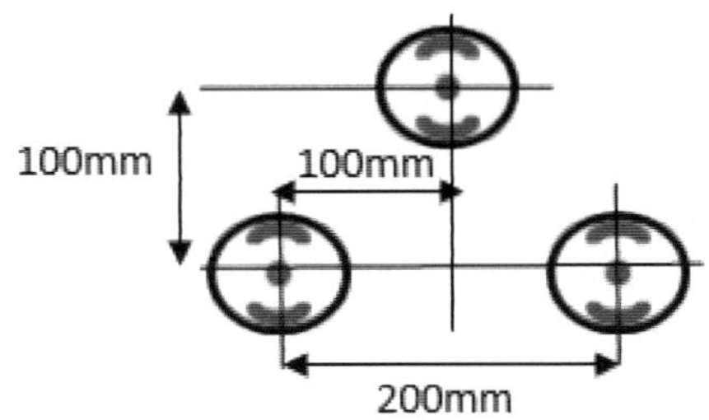


Figure 8. Gun Position (bottom view)

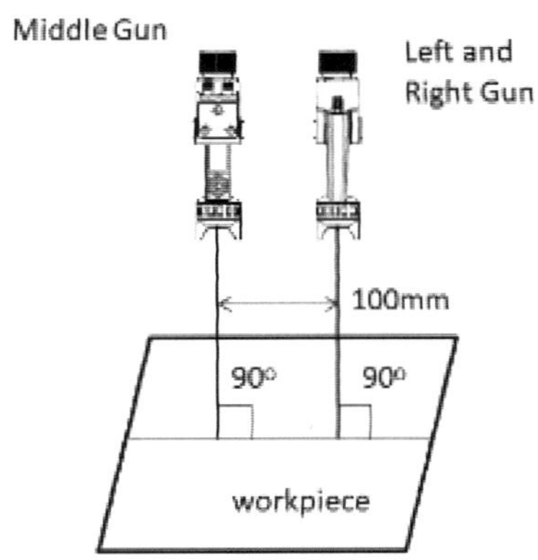


Figure 9. Gun angle to the Workpiece

The guns are designed in such position so that when the paint is sprayed out from the designated height with the correct opening valve (gun knob setting), the desired spray pattern of each gun is not overlapped each other as shown in Figure 7.

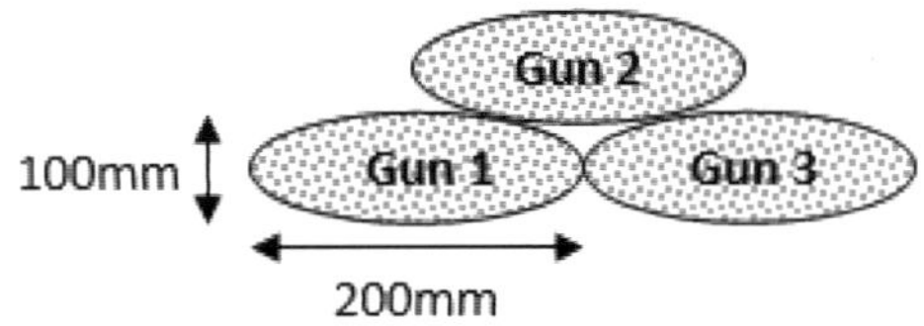


Figure 10. Spray Pattern from 3 Guns

**C. Fluid pressure and Gun (air-cap) pressure setting**

Fluid pressure is needed to move-out spray paint from the tank and transfers it through-out long tube to the spray gun. While air-cap pressure is the pressure needed to atomization the paint liquid during the spraying process.

On the design stage, fluid pressure and air-cap pressure were under-went several testing and the optimum condition was as shown in Table 4.

Table 4. Fluid pressure and Air-cap pressure

| | Parameter range | Optimum |
|---|---|---|
| Fluid Pressure | 0 ~ 1.0 bar | 0.8 bar |
| Air-cap Pressure | 0 ~ 25 bar | 2.5 bar |

**D. Nozzle distance and Paint volume setting**

Nozzle to work-piece (as shown in Figure 8) is tested for range of 8 to 10 inches while for the Gun knob valve opening (as shown in Figure 9) is tested for 0.5 turn to 2.5 turns. The turn is measured based on the fully close as 0 turn and complete 1 turn as 1.0 turn. In the other word, the turn also can be translate in degree unit as 0 turn as 0°, 1 complete turn as 360°, 1.5 turns as 440° and so on.

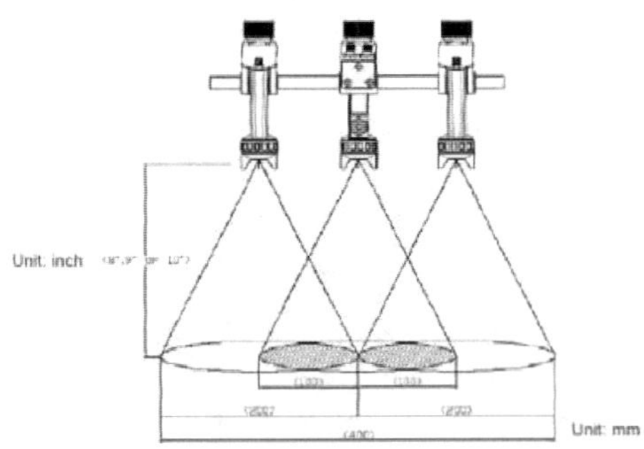


Figure 11. Nozzle to work-piece height setting

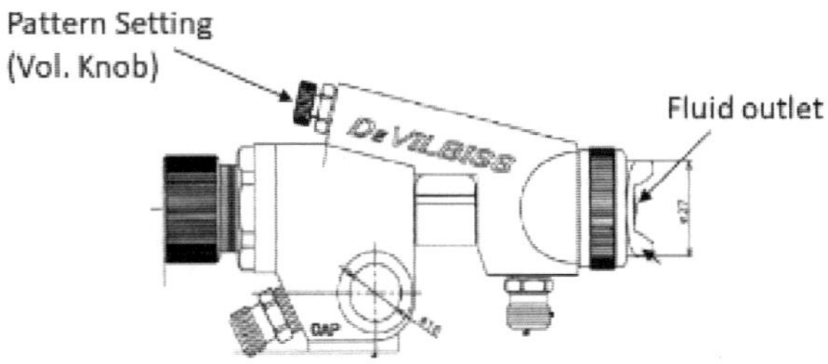


Figure 12. Spray Gun Volume Knob and Fluid outlet

The detail of spray pattern test and result of film thickness is reported in the next papers.

## 5. Conclusion

This report stated the overall process of development of Automated Spray Painting Process (ASPS) and some of the important parameters setting in achieving best spray pattern with an optimum overlapped area between the three guns in to achieve target thickness. This film thickness is controlled at 0.4 to 0.8mils.

## References

[1] Wang Z, Liu C, Cheng L, Fan X. Optimization of Spraying Path Overlap Rate Based on MATLAB. Proc. IEEE Consumer Eletronics, Communications and Networks(CECNet) 2012; 2:2731-2734.

[2] From PJ, Gunnar J, Gravdahl JT. Optimal paint gun orientation in spray paint applications-exprimental results. Proc. IEEE Transaction on Automation Science and Engineering 2011;8:438-442.

[3] Kout A, Muller H. Parameter optimization for spray coating. Proc. of The Advances in Engineering Software 2009;40(10):1078-1086.

[4] Kim D, Yoon Y, Hwang S, Lee G, Park J. Visualizing spray paint deposition in VR training. Proc. IEEE Virtual Reality Conference

[5] 2007:307-308.

[6] Sheng W, Chen H, Xi N, Chen Y. Tool path planning for compound surfaces in spray forming processes. Proc. IEEE Transaction on Automation Science and Engineering 2005;2:240-249.

[7] Chen H, Xi N, Sheng W, Dahl J. Optimizing material distribution for tool trajectory generation in surface manufacturing. Proc. the

[8] IEEE/ASME International Conference on Advanced Intelligent Mechatronics Monterey 2005;1389-1394.

[9] Radfar R, Jassbi J, Ghoreishi F, Khanmohammadi S, Alborzi M. Proactive Quality paint thickness control using ANFIS. Proc. IEEE Systems Man and Cybernetics (SMC) 2010;5-10.

[10] Catalog DeVILBISS DA-300. http://www.carlisleft.co.jp/

[11] PT-10M Pressure Tank (Manual Agitator). http://www.prowin-tools.com.tw/en/product-82354

[12] AC Servomotor. http://www.mitsubishielectric.co.jp/fa/products/

[13] Luangkularb S, Prombanpong S, Tangwarodomnukuk V. Material consumption and dry film thickness in spray coating process. Proc. CIRP 2014 : 789-794

[14] Madhuraghava P, Fakruddin B. D, Subhash R. V, Sunil N. Modelling and Structural, Analysis of a 6-DOF Robot Spray Coating Manipulator. Proc. The International Journal of Engineering and Science (IJES) Vol. 7 2018 : 48-56

[15] Choikhrue M, Prombanpong S, Sriyotha P. An Effect of Coating Parameter to Dry Film Thickness in Spray Coating Process, Proc. the International Conference on Materials Engineering and Nanotechnology 2016 : 95-98

**Acknowledgment**
Special thanks to all members of ACM, TUSB, NS Aries and School of Mechatronics Engineering, of Universiti Malaysia Perlis (UniMAP) for providing the research equipment's and internal foundations for this project. This work is supported by the research project funded by ACM under the contract research agreement signed between ACM and TUSB.

# Ultra-Wideband Confocal Microwave Imaging for Brain Tumor Detection

**M A Jamlos[1], W A Mustafa[1], Nurasma Husna[1], S Z Syed Idrus[2], Wan Khairunizam[3], I Zunaidi[4], Z M Razlan[3] and Shahriman A B[3]**

[1]Faculty of Engineering Technology, Kampus Sg. Chuchuh, Universiti Malaysia Perlis, 02100 Padang Besar, Perlis, Malaysia.

[2]Center of Excellence Geopolymer and Green Technology, Universiti Malaysia Perlis, 01000 Kangar, Perlis, Malaysia.

[3]Advanced Intelligent Computing and Sustainability (AICoS) Research Group, School of Mechatronics Engineering, Universiti Malaysia Perlis, 02600 Arau, Malaysia.

[4]Faculty of Technology, University of Sunderland, St Peter's Campus, Sunderland, SR6 0DD, United Kingdom

**Abstract.** A high-performance UWB array antenna to identify a tumor in a human brain phantom using confocal microwave imaging technique is presented. The antenna operated within the ultra-wideband (UWB) from 1.2 GHz to 10.8 GHz. Its high gains of between 5.2 dB and 14.5 dB and compact 90 mm × 45 mm size enabled its functionality and eased its integration into the microwave imaging system. Besides developing UWB array antenna as the sensor, a low-cost human head phantom represented by tissue simulating liquid as layers of skin, fat, skull, and brain are also developed. The antenna transmits the signal through the head phantom from 9 different points and receives the scattered waves under two different conditions; in the presence and absence of the tumor. The recorded scattering parameters are further processed in an improved delay and sum imaging algorithm using Matrix Laboratory software to generate an accurate image. The produced image indicated the presence of the tumor with an error of less than 2 cm.

**Keywords**: Ultrawide Band, Microwave Imaging, Brain Tumor, Array Antenna, Head Phantom

## 1. Introduction

In 2030, it is estimated that around 13.2 million people worldwide will suffer and die due to cancer since cancer is one of the most complex diseases in the world [1]. Early cancer detection allows early cancer treatment and consequently improves the survival rate. Currently, conventional X-ray, magnetic resonance imaging (MRI), computed tomography (CT Scan) and ultrasound technique are the common imaging modalities utilized to detect cancer [2]. Microwave cancer imaging provides several substantial benefits including being fast, safe, low-cost, and non-invasive with good detection accuracy [3–5]. A brain tumor is defined as a mass of tissue formed by an accumulation of abnormal cells within the brain. It is arguably the most fatal of all cancers [6,7]. On the contrary, most researchers are still concentrating research efforts on breast cancer detection and imaging. Microwave imaging is one of the potential technologies in visualizing internal human structure by exposing it to electromagnetic fields at microwave frequencies between 300MHz and 30GHz [8]. Techniques for microwave imaging are generally divided into three major types: active, passive or hybrid technique. As for active microwave imaging, several antennas transmit microwave signals into tissues and the reflected (backscattered) signals are used to generate microwave images.

IOP Conf. Series: Materials Science and Engineering **557** (2019) 012002 doi:10.1088/1757-899X/557/1/012002

Active microwave imaging can be further separated into the ultra-wideband (UWB) radar imaging and tomography imaging for detection purpose. UWB radar imaging technique was initially introduced solely for breast tumor detection [9,10]. Similar operating principles were then implemented for stroke and tumor detection in the brain [11,12]. Radar-based techniques are less complex as it only needs to identify strong scattering point caused by inclusions with high frequent occasion [13]. Image processing algorithms such as the delay and sum confocal algorithm can then be used to detect the tumor. In microwave imaging, having high gain and good directional beam-width in a wide frequency band antennas are required for target detection, localization systems, and cancer screening applications. Tumor imaging using microwave energy for early treatment is currently gaining attention from the research community. Antenna arrays functioned to enhance the focus of the EM energy into the locality of the tumors. The antenna arrays are very efficient to capture the scattered signal that can be analyzed to solve the inverse problem and obtain the constructed images of the preferred tissues. Furthermore, increasing the number of element implemented could enhance the performance of antenna arrays in imaging and treatment [14]. Compared to advanced biomedical imaging systems such as the magnetic resonance imaging (MRI) and computed tomography scan (CT scan), current microwave imaging technology is far behind in term of capabilities. However, microwave imaging systems do have the great potential of being simple, portable and cost-effective [15].

High gain and wide bandwidth UWB antenna are very essential in successfully penetrating different material with high details of the image. Antenna with a significant size which has a length of several wavelengths required to achieve such high performance. Hence, it is a great challenge to reduce the physical dimensions of the antenna so that it can be incorporated in a compact microwave imaging detection system and at the same time maintain its broadband, high-gain, and distortionless performance [16–18]. While it is desirable to produce compact-size antennas for biomedical applications, it is very important to ensure that antennas demonstrate a good directivity and high power gain as well. However, it seems that the existing antennas exhibit poor directivity, moderate gain too large to be integrated into a portable technology, very expensive and difficult to manufacture [19,20]. In this paper, the ultra-wideband antenna applied with confocal microwave imaging for brain tumor detection is proposed. This paper organized as follow, introduction, ultra-wideband confocal microwave imaging technique, results, and discussion and conclusion.

## 2. Ultra-Wideband Confocal Microwave Imaging Technique

### *2.1 Sensor Design*

Figure 1 demonstrated the simulated design of UWB reflector array antenna using Taconic (TLY-5) with a dielectric constant of $\varepsilon r = 2.2$, a thickness of $t = 1.5748{+/-}0.02$ and tangent loss of $\tan \delta = 0.0009$ as the substrate. The antenna is printed with $4\times1$ copper radiating patch array properly connected with quarter wave transformer transmission line associated with a copper parasitic element for the front side as shown in figure 1. The patches comprise of four identical circulars with a diameter of 15 mm. Parasitic element is placed on very close to feeding line with the gap only 0.2 mm. As shown in figure 1, each quarter-wave transmission line has its own specific wide dimension for 50Ω, 70.71Ω and 100Ω to ensure equal current distribution towards all four patches could be realized. Quarter-wave transformers of 70.71 Ω are used to have an ideal match between the 100 Ω lines and the 50 Ω lines [21].

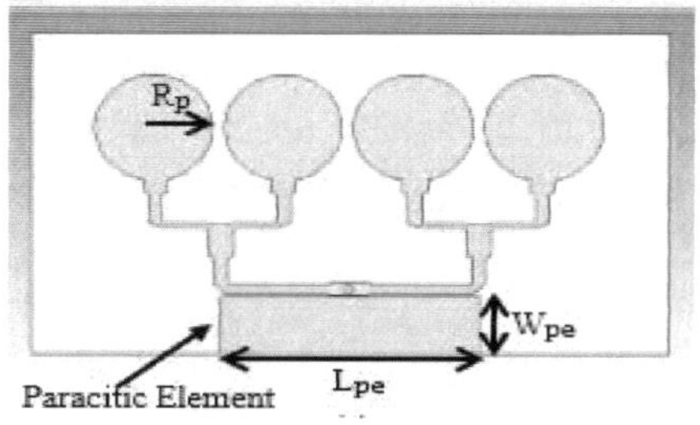


**Figure 1.** Quarter wave transformer

Some important parameters of the designed antenna are optimized to obtain the best result in term of compact size, high gain and wide bandwidth. The optimized dimensions for the antenna are tabulated in Table 1.

**Table 1.** Antenna Optimized Dimensions

| Parameter | Dimension |
|---|---|
| $R_p$ | 7.50 mm |
| $W_{pe}$ | 8.00 mm |
| $L_{pe}$ | 32.4 mm |

### 2.2 *Sensor Performances*

The simulation results of the reflection coefficient, S11 less than -10dB is shown in figure 2. From the figure, it clearly shows the proposed antenna recorded a wide range of operating frequency started from 2.6 GHz until 13.1 GHz which fulfilled the characteristic of the UWB antenna. Reflection coefficient less than -10 dB is selected due to the condition where 90% of the signals are successfully transmitted while only the left 10% is reflected back [22]. Partial ground technique and additional of parasitic element assure lower reflection coefficient achieved.

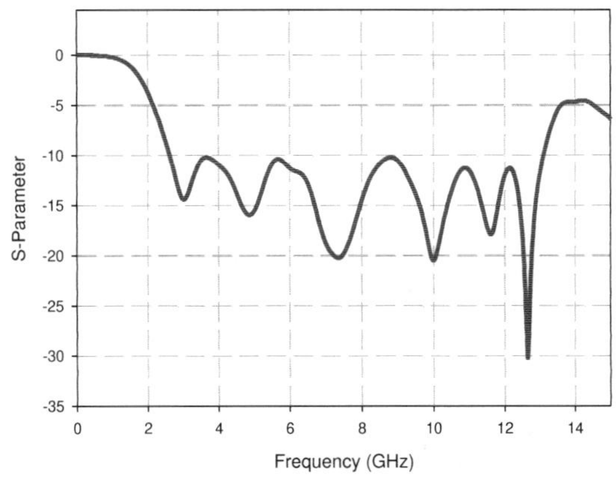


**Figure 2.** Graph of reflection coefficient (S11)

Figure 3 shows a graph for frequency domain and time domain for one same cycle of signals. Frequency domain represented by negative data values meanwhile time domain recorded positive data value. The data in the frequency domain need to be transformed to the time domain since confocal algorithm depends on round trip time for image generation.

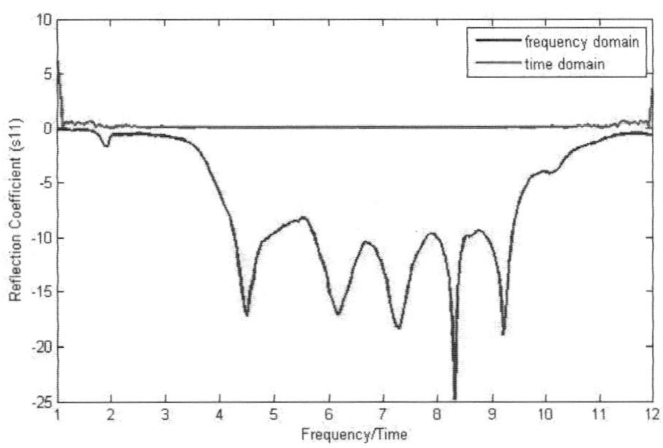


**Figure 3.** Frequency domain and time domain for one same cycle of signals

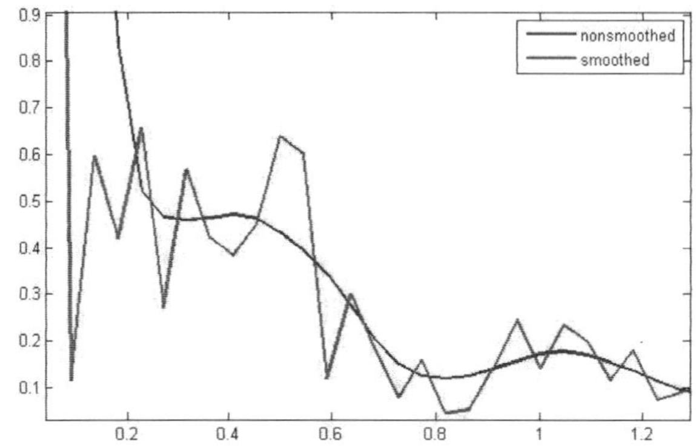


**Figure 4.** Smooth signal of time domain using IFFT

Meanwhile, figure 4 shows a smooth signal of time domain after undergoing the IFFT process towards the nonsmoothed original signal. The nonsmoothed signal associated with high and low spike form meanwhile smoothed signal represented by spline curve line. The smoothed signal is vital in generating high spatial resolution image for microwave brain tumor detection.

### *2.3 Microwave Imaging Measurement System*

Figure 5 illustrated the measurement setup for brain tumor detection using microwave radar imaging technique. It consists of VNA, UWB sensor and human head phantom with the presence of a tumor. The measurements take place in the anechoic chamber to eliminate the unwanted signal that could interfere with the desired signal throughout the whole process of obtaining the data.

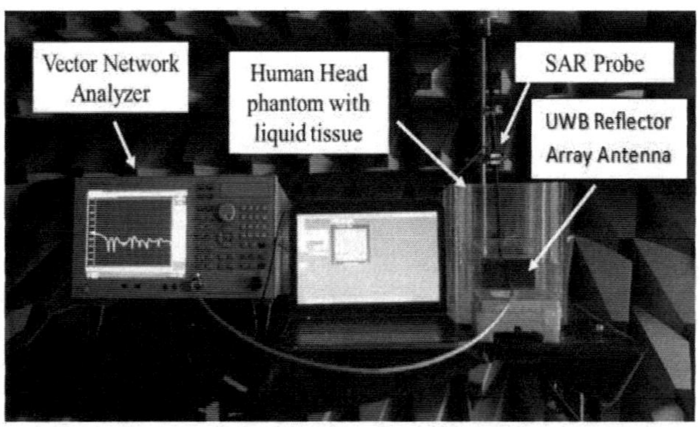


**Figure 5.** Measurement setup for microwave imaging

The UWB sensor is perfectly placed 10 mm away from the multilayer human head phantom to have maximum signal penetration. The sensor radiated the signal towards human head phantom with two different conditions which are one with tumor present meanwhile another one is without the tumor present to obtain the difference of the scattering parameter value. The sensor radiated the signal towards nine different areas to cover the whole one-sided area of the phantom as shown in figure 6 where the tumor located at position 5. Both data with and without tumor are subtracted from each other and the difference obtained is then processed in order to produce the desired image. Figure 7 shows the flow chart of the UWB confocal algorithm for data generation and image construction. The MATLAB software is used to process the data obtained using confocal microwave imaging algorithm.

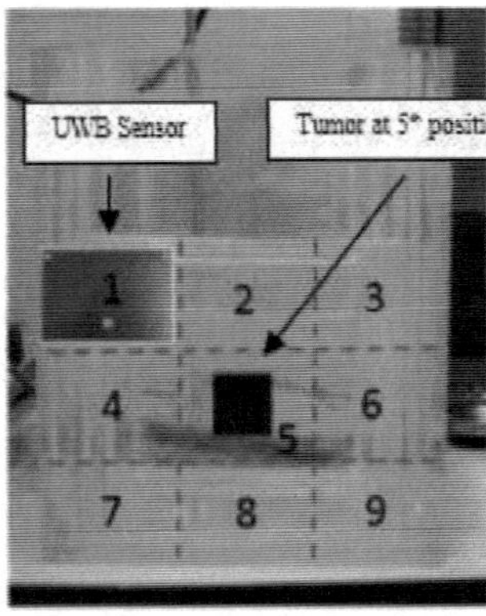


**Figure 6.** Scanning Area for Human Head Phantom

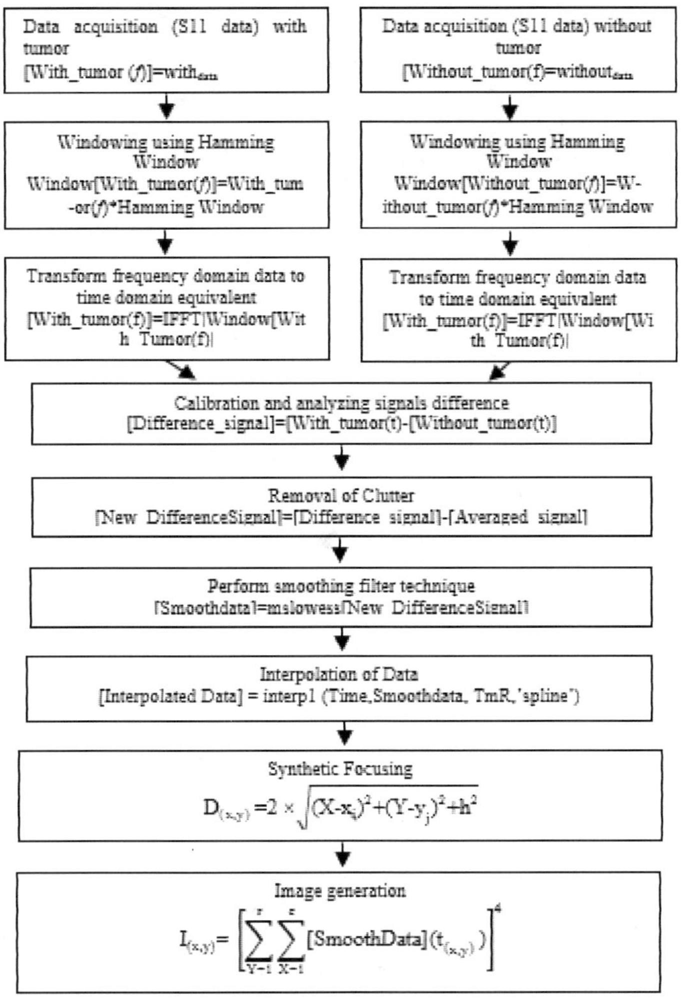


**Figure 7.** UWB Confocal Microwave Imaging Algorithm for image construction

IOP Conf. Series: Materials Science and Engineering **557** (2019) 012002 doi:10.1088/1757-899X/557/1/012002

For the algorithm, averaging and interpolation step has been applied in order to have a better and clearer image. Averaging functioned to eliminate the effects of clutters originated from dominating reflections of the environment, equipment, and the antenna. On the other hand, theoretically, interpolation is a method of constructing new data points within the range of a discrete set of known data points. Meanwhile specifically for imaging, interpolation assists in filling the empty element between the adjacent pixels by assuming based on that particular adjacent pixels value [23–25].

## 3. Results and Discussions

Figure 8 shows the images processed using the original algorithm and the UWB confocal algorithm without tumor and with the tumor present, respectively. The images demonstrated the efficiency of the improved delay and sum algorithm with clearer and sharper images and areas with higher pixel intensity. Tumor in the image is represented by the highest color intensity (reddish orange spot), resulting from the strongest scattering region. This corresponds to tumor existence in the human head phantom.

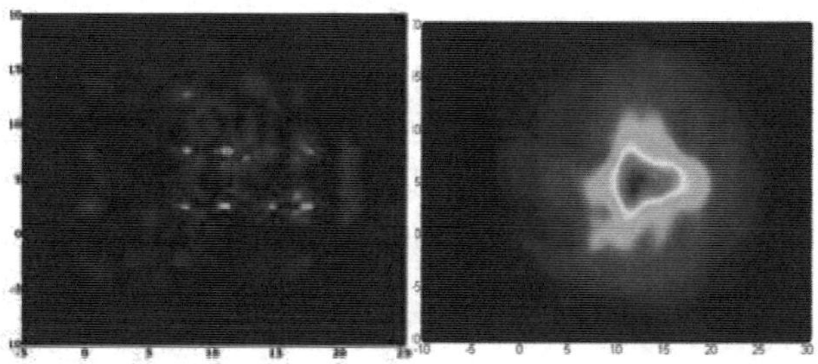

**Figure 8.** Resulted image without tumor; (a) with tumor (b)

Finally, an analysis of the localization error is performed to evaluate the effectiveness of the improved confocal microwave imaging algorithm. Since no minimum acceptable values are provided in literature for these errors, they are performed based on their relative distance from the real location of the tumor. Table 2 summarizes this for both UWB confocal and the original algorithm. The improved algorithm recorded a lesser error of 1.5 cm compared with the original algorithm, which recorded 1.8 cm.

**Table 2.** Tumor Localization Error

| | Actual position | | Position in image | | |
|---|---|---|---|---|---|
| Axis | X' (cm) | Y' (cm) | X (cm) | Y (cm) | L.E (cm) |
| UWB confocal | 10 | 5 | 10.5 | 5.5 | 1.5 |
| Original | 10 | 5 | 11.0 | 6.2 | 1.8 |

## 4. Conclusion

A high-performance antenna to identify a tumor in a human brain phantom using confocal microwave imaging technique is presented. It operates within the ultra-wideband (UWB) from 1.2 GHz to 10.8 GHz and produced gains of up to 14.5 dB in a compact 90 mm × 45 mm form. An in-house developed head phantom is used to experimentally validate the operation of this system. A UWB confocal microwave imaging algorithm which is used for image processing successfully improved the quality of the resulting image. Proposed future work includes investigating this system on the larger experimental region and with the presence of multiple tumors.

## References

[1] Mohammed B J, Abbosh a. M, Ireland D and Bialkowski M E 2012 Compact wideband antenna immerssed in optimum coupling liquid for microwave imaging of brain stroke *Prog. Electromagn. Res. C* **27** 27–39

[2] Poplack S P, Fanning M W, Dun Li, Meaney P M and Paulsen K D 2002 A clinical prototype for active microwave imaging of the breast *IEEE Trans. Microw. Theory Tech.* **48** 1841–53

[3] Barnes F S U O C-B and Greenebaum B U O W-P 2006 *Handbook of Biological Effects of Electromagnetic Fields (Third Edition)* vol 10032

[4] Töpfer F and Oberhammer J 2016 Microwave cancer diagnosis *Principles and Applications of RF/Microwave in Healthcare and Biosensing* pp 103–49

[5] Pollacco D 2016 *Microwave Medical Imaging*

[6] Association A B T 2010 *Brain Tumor Primer: A Comprehensive Introduction to Brain Tumors*

[7] Dotter P and Wade E 2006 Brain Tumor *Am. J. Nurs.* **51** 300

[8] Sudirman R, Chew K M, Mahmood N H, Seman N and Yong C Y 2013 Human Brain Microwave Imaging Signal Processing: Frequency Domain (S-parameters) to Time Domain Conversion *Engineering* **05** 31–6

[9] Hagness S C, Taflove A and Bridges J E 1998 Two-dimensional FDTD analysis of a pulsed microwave confocal system for breast cancer detection: Fixed-focus and antenna-array sensors *IEEE Trans. Biomed. Eng.* **45** 1470–9

[10] Hagness S C, Taflove A and Bridges J E 2002 FDTD analysis of a pulsed microwave confocal system for breast cancer detection pp 2506–8

[11] Ireland D and Bialkowski M E 2011 Microwave Head Imaging For Stroke Detection *Prog. Electromagn. Res. M* **21** 163–75

[12] Guo L and Abbosh A M 2014 Compressive sensing for stroke detection in microwave-based head imaging *IEEE Antennas and Propagation Society, AP-S International Symposium (Digest)* pp 1911–2

[13] Camacho L M and Tjuatja S 2005 FDTD Simulation of microwave scattering from a lung tumor *IEEE Antennas and Propagation Society, AP-S International Symposium (Digest)* vol 3 A pp 815–8

[14] Moon J S and Gaskill D K 2011 Graphene: Its fundamentals to future applications *IEEE Trans. Microw. Theory Tech.* **59** 2702–8

[15] Eesuola A, Chen Y and Tian G Y 2011 Novel ultra-wideband directional antennas for microwave breast cancer detection *IEEE Antennas and Propagation Society, AP-S International Symposium (Digest)* pp 90–3

[16] Abbosh A M 2008 Directive Antenna for Ultrawideband Medical Imaging Systems *Int. J. Antennas Propag.* **2008** 1–6

[17] Mobashsher A T and Abbosh A M 2016 Performance of directional and omnidirectional antennas in wideband head imaging *IEEE Antennas Wirel. Propag. Lett.* **15** 1618–21

[18] Abbosh A M 2008 Compact antenna for microwave imaging systems *2008 Cairo International Biomedical Engineering Conference, CIBEC 2008*

[19] Anon 1988 CHALLENGE OF THE PATCH: THE MICROSTRIP PATCH ANTENNA. *Microwaves & RF* **27**

[20] Skrivervik A K 2013 Implantable antennas: The challenge of efficiency *2013 7th European Conference on Antennas and Propagation, (EuCAP), Gothenburg* pp 3627–31

[21] Wahab N A, Maslan Z Bin, Muhamad W N W and Hamzah N 2010 Microstrip rectangular 4×1 patch array antenna at 2.5GHz for WiMax application *Proceedings - 2nd International Conference on Computational Intelligence, Communication Systems and Networks, CICSyN 2010* pp 164–8

[22] Jais M I, Jamlos M F, Jusoh M, Sabapathy T, Nayan M K A, Shahadah A, Rahim H A, Ismail I and Fuad F A A 2013 1.575 GHz dual-polarization textile antenna (DPTA) for GPS application *IEEE Symposium on Wireless Technology and Applications, ISWTA* pp 376–9

[23] Shao W and Adams R S 2011 UWB microwave imaging for early breast cancer detection: A novel confocal imaging algorithm *IEEE Antennas and Propagation Society, AP-S International Symposium (Digest)* pp 707–9

[24] Li X and Hagness S C 2001 A Confocal Microwave Imaging Algorithm for Breast *IEEE Microw. Wirel. COMPONENTS Lett.* **11** 130–2

[25] Xiao X and Kikkawa T 2015 UWB Imaging for Early Breast Cancer Detection by Confocal Algorithm

# Development of predictive models for cervical cancer based on gene expression profiling data

**A A Abdullah[1], N K Abu Sabri[1], Wan Khairunizam[1], I Zunaidi[2], Z M Razlan[1] and Shahriman A B[1]**

[1]School of Mechatronic Engineering, Universiti Malaysia Perlis, Pauh Putra Campus, 02600 Arau, Perlis, Malaysia

[2]Faculty of Technology, University of Sunderland, St Peter's Campus, Sunderland, SR6 0DD, United Kingdom

**Abstract.** Cervical cancer and the prediction of clinical outcome are among the most important emerging applications of gene expression microarray technology with feature sequencing of microRNA. By using reliable and dependable classification of machine learning algorithms available for microarray gene expression profiling data is the key in order to develop the most suitable and possible predictive model to be used by patient. In this paper, two-machine learning algorithms have been used which are Support Vector Machine (SVM) and Random Forests (RF) for the predictive models of cervical cancer. We identify and evaluate the performance of these two algorithms in order to know which algorithm has better performance. In this study, 714 features and 58 samples are used to develop predictive model for cervical cancer and our computational results show that RF algorithm outperform SVM algorithm with the accuracy of 94.21%. Our data also underline the importance of variables, which give the significant role in predicting the occurrence of cervical cancer.

## 1. Introduction

Cervical cancer is the second most common cancer among women. It may cause mortality and morbidity [1], [2]. The high mortality rate of the cervical cancer is because of the lack of awareness of women for early detection of cervical cancer [3]. Even though cervical cancer is dangerous which may lead to life threatening but it is potentially curable. Cervical cancer occurs in a woman's cervix [4]. The cervix is the lower part of the uterus. It connects the uterus to the vagina [4]. There are two main types of cervical cancer, which are squamous cell carcinoma and adenocarcinoma. The squamous cell carcinoma cancer type usually occurred in the epithelium lining of the cervix [5]. As for adenocarcinoma cancer type, it is developed from gland cells. The adenocarcinoma of cervical cancer is a premalignant glandular condition [6]. Squamous cell carcinoma has the highest percentage of cervical cancer cases which are 90% and 10% of cervical cancer cases are adenocarcinoma [1]. Hence, adenocarcinoma neoplasia of the cervical cancer is less common than squamous cell carcinoma neoplasia of cervical cancer when it is being diagnosed [6].

Cervical cancer is caused by a virus named human papillomavirus (HPV). When the infection of the human papillomavirus (HPV) at the cervix left untreated, cervical cancer is developed [7]. In cervical cancer, human papillomavirus (HPV) is the most important infectious agent because it contributes to neoplastic progression. Neoplastic progression is the progression of the abnormal growth of the cervical cancerous cells and proliferation of the abnormal cells due to a malignant process [8]. However, most scientific studies have found that human papillomavirus (HPV) infection alone is inadequate to induce the malignant of cervical cancer. Other host genetic variations also play important role in the development of cervical cancer in women [1]. Commonly, microarray gene expression profiling is used to identify and differentiate the expressed gene expression in a pre-cancerous and cancerous cell in the cervix [1]. A microarray or also known as DNA chip is an arrangement of DNA molecules that have been chemically bonded to a fine grid of surfaces. The purpose of the microarray gene expression is to interpret and analyze the genes expression state in complementary DNA prepared from mRNA in which the hybridization is taking place on the array [9]–[11]. Genome-wide expression profiling has the potential to get more accurately predict outcome at the end of the research since it allows selection of the genes that associate most strongly with the outcome by screening a large number of these genes simultaneously. Expression profiling studies show that genes are statistically significant differences under changed experimental conditions [12].

In this study, the predictive model for determining whether the cells are cancerous or pre-cancerous for cervical cancer can be developed by just looking at their gene expression profiles. In order to develop the predictive model, machine learning algorithms approaches have been implemented. Artificial intelligence and machine learning techniques have been applied in various medical diagnostic systems [13]–[16] and there are many researchers utilized the techniques in the image processing of Pap smear images [17]–[23]. However, there are fewer studies on implementing machine learning methods on the cervical cancer based on gene expression profiling data. Hence, in order to overcome this problem, we implemented unsupervised and supervised machine learning techniques, which are Principal Components Analysis (PCA), Support Vector Machine (SVM) and Random Forest (RF), respectively. The Principal Components Analysis (PCA) is used to do the data screening and data preprocessing for the model. The data that have been preliminaries in the Principal Components Analysis (PCA) will be used in Support Vector Machine (SVM). As for support machine vector (SVM) and Random Forest (RF), they are used to predict the accuracy of the predictive model for cervical cancer whether the cell is pre-cancerous or cancerous cells.

## 2. Proposed Methods

In this study, we used dataset, which is obtained from the Gynecologic Oncology Group Tissue Bank (PA, USA) for classification of cervical cancer based on the gene expression profiling data [24]. First, the dataset will be undergoing data extraction process in order to get and retrieving the relevant data or information in the dataset. The data pre-processing then applied on the dataset for eliminating the irrelevant and redundant data containing in the dataset. In order to perform classification of the gene expression profiling data of microarray dataset into their cluster, a tree-like structure is constructed by using hierarchical clustering. By implemented machine learning algorithms of Support Vector Machine (SVM) and Random Forests (RF) on the dataset, predictive model for cervical cancer can be developed. In Support Vector Machine (SVM) model, Principal Components Analysis (PCA) algorithm is used as for the feature selection technique. It is very useful in order to reduce the high dimensionality of dataset as to facilitate the algorithm to produce a good performance. The new dataset constructed will be executed by using Support Vector Machine (SVM) model in order to get the prediction accuracy. As a comparison, Random Forests (RF) is also applied to get the accuracy of predictive model for cervical cancer.

### *2.1. Data extraction*

Data extraction is a process of retrieving relevant data or information out of data sources, which is dataset containing microarray data analysis. In this stage, the data will be extracted and loaded into the staging area of the relational database of the cervical cancer. When data extraction is implemented, the extraction logic is applied in order to extract the data from the dataset. The data column of the dataset is the data regarding the normal and tumor of samples data. And as for the data in the row is about the features of the sequencing of microRNAs. The data matrix produce from the data extraction is changed into a data frame because it may ease us to interpret the data for further processing.

### *2.2. Data pre-processing*

Data pre-processing is an important step in the data mining process. It is implemented on the dataset because the dataset obtained from microarray analysis contains irrelevant, unreliable and redundant of data or noise present in the dataset. Data pre-processing is performed by undergoing the process of data cleaning, data scaling and normalization of data. In data cleaning process, the corrupt or irrelevant data from the datasets are detected. Then, the data that have been identified as incomplete and irrelevant will be filtered out of the dataset. After cleaning the corrupted data, a final dataset should be consistent with other similar dataset in the system and can be used for further processing. Other than data cleaning process, data scaling is also implemented on the dataset to pre-processing the dataset. In data scaling, the data will be centering and scaling whenever there are data with different units and ranges. By scaling the data, the final dataset will be easy to be interpreted. The mean value of sample data from the datasets is reduced by using the centering transformation method. As for scaling transformation, it will divide the value of each sample data by the standard deviation of the dataset. Thus, the numerical data of the dataset stability is improved. The normalization of the data is important to organize the columns and tables of the database to reduce the redundancy of the data. It is used to improve the integrity of the dataset in this study.

### *2.3. Hierarchical clustering*

The subgroups are defined based on the genes expression profiling data of the dataset. In order to classify the gene expression profiles of data into their own clusters, hierarchical clustering is implemented to create the hierarchical, which is the tree-like structure of the data. It is also referred as a dendrogram. The hierarchical clustering is implemented by measuring the feature similarities of the gene expression profiles. Hierarchical clustering is implemented by measuring the maximum distance matrix between the samples data. In every stage of hierarchical clustering, the two nearest clusters are merged together into a new cluster or subtypes. The clustering process is repeated until all of the samples data is agglomerated into a single cluster.

### *2.4. Principle component analysis (PCA)*

In order to develop the predictive model by using the support vector machines (SVM), principal components analysis (PCA) is used as dimensional reduction technique because it is useful to collapse the features into a smaller set of principal components. The principal components analysis (PCA) is used in this study because the dataset contains hundreds of features from the sequencing of microRNA. The principal components analysis (PCA) is needed in order to reduce the dimensionality of the hundreds of features in the dataset. By implemented the machine learning algorithm on the dataset, it helps to collapse the hundreds of features into a smaller set of principal components.

### *2.5. Support vector machine (SVM)*

Support vector machine (SVM) algorithm is implemented by performing a classification on the final dataset. The classification of the algorithm is implemented by constructing a multidimensional hyperplane to the selective samples data. It will maximize the margin between the two-data clusters, which are normal (N) and tumor (T). The type of the SVM parameter used in this study is set to the C-classification. As for the kernel, linear of kernel type is selected.

In order to evaluate the predictive model performance, classification of accuracy, sensitivity and specificity need to be calculated, as shown in (1), (2) and (3).

$$\text{Accuracy} = (TP+TN)/(TP+FP+FN+TN) \tag{1}$$

$$\text{Sensitivity} = TN/(TP+FN) \tag{2}$$

$$\text{Specificity} = TP/(FP+TP) \tag{3}$$

TN is the number of true negative, FP is defined as the number of false positive, TP is defined as number of true positive and FN is the number of false negative. These coefficients are defined and illustrated as a confusion matrix in the Table 1.

**Table 1.** Confusion matrix.

| | | Reference | |
|---|---|---|---|
| | | Negative (0) | Positive (1) |
| Prediction | Negative (0) | TN | FP |
| | Positive (1) | FN | TP |

*2.6. Random forest (RF)*

In order to make sure the result obtained is valid and can be generated for generating the prediction model, the dataset is randomly partitioned into training and testing sets. In this project, 70% of the samples data are used for training and the rest of the 30% for testing. The type of the RF parameter is set to the classification. As for the number of trees in the RF (ntree), it is set to 300. ntree is the number of trees to grow in the Random Forest (RF). The ntree should not be set to a small number because we need to ensure that every input row gets predicted at least a few times. The last RF parameter that needs to be considered is number of variables tried at each split (mtry). mtry is a number of variables that randomly sampled as candidates at each split. The number of variables tried at each split can be calculated by using (4), where p is number of variables in dataset. After the calculation is done, it is set to 27.

$$\sqrt{p} = \sqrt{714} = 26.72 \approx 27 \tag{4}$$

In order to evaluate the accuracy of random forest algorithm, OOB estimate of error rate from RF module and predictive model performance by calculating accuracy, sensitivity and specificity are considered. For both model performances, confusion matrix as illustrated in Table 1 is constructed.

## 3. Results and Discussion

*3.1. Hierarchical clustering*

The result of the clustering is presented using a heat map, as shown in Figure 1. Hierarchical clustering is mapped in the heatmap is using one-channel microarray data, which contain the gene expression profiles data. As we can see from Fig. 1, it contains several colors with two dendrogram. The tree-like structures presented above and the left of the heatmap are known as dendrogram. It works by calculating the pairwise distance between all of the data in the dataset. The data with least distant apart will be merged together until rest of the data have been agglomerated to represent as the dendrogram as illustrated in Figure 1. The samples data, which contain normal and tumor, are shown as rows by one row per data sample. The columns represent the features of the sequencing microRNAs below the heatmap figure. The features of sequencing microRNA levels are represented by the color scale which ranging from orange to white according to heat color. Orange indicates a low level of features of sequencing microRNA level while white indicates high level of features of sequencing microRNA level. The tree-like structure (dendrograms also provide some qualitative means of assessing the similarity between the features of microRNA and data samples.

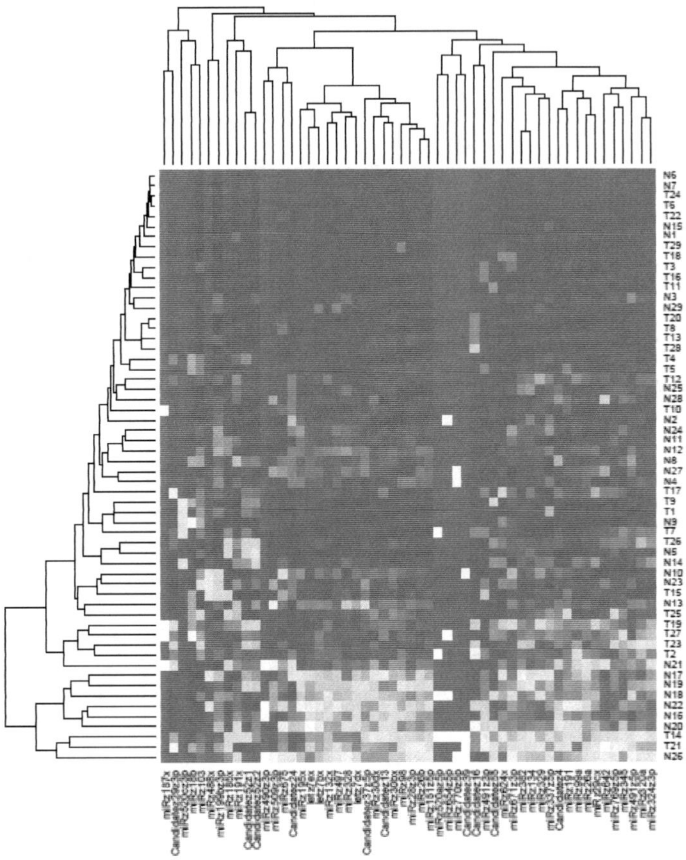

**Figure 1.** Hierarchical clustering by using heatmap.

### *3.2. Support vector machine (SVM)*

According to the test data, there are 16 sample data that have been used for testing in order to develop the predictive model are shown in Table 2. In order to perform the prediction on the sample data, confusion matrix needs to be evaluated. The confusion matrix is demonstrated in the Table 3. Based on the confusion matrix in Table 3, we can see the performance of the support vector machine (SVM) algorithm for prediction. For the normal sample data, the model has predicted that 7 out of 8 sample data are normal while the rest of the sample data, which is only 1 sample data, has been predicted as tumor. As for the tumor sample data, 7 out of 8 samples have been predicted as tumor by the model and only 1 sample data has been predicted as normal. From the confusion matrix, the performance of support vector machine (SVM) algorithm model can be evaluated and demonstrated in Figure 2.

**Table 2.** Selective samples data for SVM.

| | Normal | Tumor |
|---|---|---|
| Selective Samples Data | N12 | T2 |
| | N17 | T4 |
| | N18 | T9 |
| | N20 | T10 |
| | N24 | T12 |
| | N25 | T13 |
| | N27 | T14 |
| | N29 | T22 |

**Table 3.** Confusion matrix for SVM (1 indicates as tumor while 0 indicates as normal).

| | | Reference | |
|---|---|---|---|
| | | Negative (0) | Positive (1) |
| Prediction | Negative (0) | 7 | 1 |
| | Positive (1) | 1 | 7 |

```
            Accuracy : 0.875
              95% CI : (0.6165, 0.9845)
 No Information Rate : 0.5
 P-Value [Acc > NIR] : 0.00209

               Kappa : 0.75
Mcnemar's Test P-Value : 1.00000

         Sensitivity : 0.8750
         Specificity : 0.8750
      Pos Pred Value : 0.8750
      Neg Pred Value : 0.8750
          Prevalence : 0.5000
      Detection Rate : 0.4375
Detection Prevalence : 0.5000
   Balanced Accuracy : 0.8750

    'Positive' Class : 0
```

**Figure 2.** Prediction result for cervical cancer using SVM algorithm.

Based on Figure 2, the accuracy of the predictive model is 0.8750. The Kappa statistic is calculated for the overall accuracy of the predictive model. Kappa statistic value of the predictive model is 0.75, which indicates a good arrangement. The value for both sensitivity and specificity of the predictive model are 0.8750. Hence, we can conclude that the developed predictive model of cervical cancer based on the gene expression profiles data is a good prediction. Since, the accuracy of the predictive model is up to the 80%. So, it is proved that SVM algorithm can be used to develop the predictive model of cervical cancer based on the gene expression profiling data.

### *3.3. Random forest (RF)*

The reliability and accuracy of the RF prediction model performance for cervical cancer of gene expression profiling data is shown in the Figure 3. According to the training data used in this study, there are 41 sample data that have been selected in order to train the data. Thus, the accuracy of predictive model produced at the end of this study is high which is greater than 80% as desired outcome. The testing data used throughout the process is shown in Table 4. For RF algorithm, cross-validation is not required in order to get the unbiased estimate of the test set error. In training data stage of Random Forests (RF) model, out-of-bag (OOB) estimate of error rate is determined. OOB is the error of mean prediction on each training sample. In order to estimate the prediction error, the data used in the estimation is not in the bootstrap sample for each bootstrap iteration and related tree. Generally, OOB error rate is estimated internally during the program run. The OOB estimate of error rate and the confusion matrix for the RF model is demonstrated in Figure 3.

**Table 4.** Selective samples data for RF.

| | Normal | Tumor |
|---|---|---|
| Selective Samples Data | N2 | T2 |
| | N4 | T3 |
| | N5 | T5 |
| | N8 | T8 |
| | N11 | T21 |
| | N16 | T24 |
| | N20 | T29 |
| | N21 | |
| | N24 | |
| | N26 | |

```
            Type of random forest: classification
                  Number of trees: 300
No. of variables tried at each split: 27

        OOB estimate of  error rate: 14.63%
Confusion matrix:
   0  1 class.error
0 15  4  0.21052632
1  2 20  0.09090909
```

**Figure 3.** Out-of-bag (OOB) estimate of error rate and confusion matrix in RF model.

As shown in the Figure 3, out-of-bag (OOB) estimate of error rate in RF model is 14.63%. This means that the accuracy of the RF model is quite good because it is up to 85%. In order to assess the performance of predictive model by using Random Forest (RF) algorithm for cervical cancer, confusion matrix needs to be evaluated by using testing data. The confusion matrix for RF is demonstrated in Table 5.

**Table 5.** Confusion matrix for RF (1 indicates as tumor while 0 indicates as normal).

| | | Reference | |
|---|---|---|---|
| | | Negative (0) | Positive (1) |
| Prediction | Negative (0) | 10 | 1 |
| | Positive (1) | 0 | 6 |

According to the confusion matrix in Table 5, the visualization of prediction performance by using RF algorithms can be analyzed. The model has predicted that 10 out of 10 are the normal sample data. None of normal samples is predicted as tumor. As for the tumor samples data, 6 out of 7 samples data are predicted as tumor and only 1 sample data is predicted as normal. From the confusion matrix in Table 5, the performance of prediction by Random Forests (RF) algorithms can be demonstrated in Figure 4.

```
               Accuracy : 0.9412
                 95% CI : (0.7131, 0.9985)
    No Information Rate : 0.5882
    P-Value [Acc > NIR] : 0.001559

                  Kappa : 0.8759
 Mcnemar's Test P-Value : 1.000000

            Sensitivity : 1.0000
            Specificity : 0.8571
         Pos Pred Value : 0.9091
         Neg Pred Value : 1.0000
             Prevalence : 0.5882
         Detection Rate : 0.5882
   Detection Prevalence : 0.6471
      Balanced Accuracy : 0.9286

       'Positive' Class : 0
```

**Figure 4.** Prediction result for cervical cancer using RF algorithm.

As illustrated in the Figure 4, the accuracy of the predictive model is 0.9412, which in percentage is approximately 94.12%. This prediction accuracy by using RF algorithm is higher than SVM model. Besides that, kappa statistic value of the predictive model is 0.8759, which indicates a substantial arrangement. The sensitivity of this predictive model is perfect which 1.0. This can be seen in the confusion matrix in Table 5 where there is no error in predicting the normal sample data. On the other hand, the specificity of predictive model is 0.8571, which show that it is a good predictive model. In RF model, variables that play importance role in interpretation and prediction can be evaluated. The variables importance with Mean Decrease Accuracy (MDA) and Mean Decrease Gini (MDG) is shown in Figure 5.

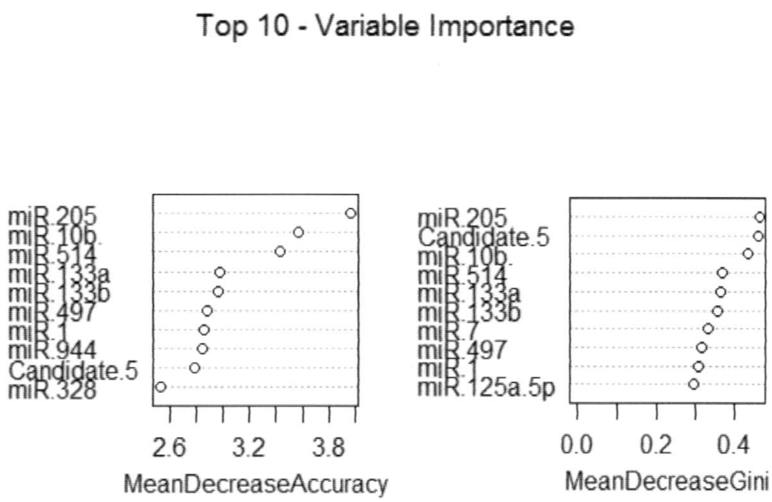


**Figure 5.** The top 10 of variables importance in RF model.

As illustrated in the Figure 5, each variable plotted in the graph indicates how important the variables are in the classification of RF model. On the x-axis, the importance of the variables is plotted and the variables importance involved in the classification on y-axis. The variables are plotted by plotting the most important variables at the top to the least important variables by making a dot-plot on x-axis. Based on Figure 5, miR. 205 candidates is the most important variable for both MDA and MDG with the value of 3.96919433 and 0.462774856 respectively. According to the Zhang Yue et al, they stated that miR. 205 is a tumor-suppressing onco-miRNA [25]. In addition, Hong Xie et al. have concluded that miR. 205 plays an important role in pathogenesis of cervical cancer by promoting cell proliferation and migration [26]. miR. 10b is also one of the variables importance in predicting cervical cancer with the value of 3.57406978 for MDA and 0.434711805 for MDG. Besides that, in

Zhang Yue et al. studies, they concluded that miR. 10b is able to promote the cervical cancer tumor invasion [25]. Ren Hou et al. reported that miR. 10b is one of the tumor suppressor microRNA which downregulated in cancer tissues. But it will inhibit cervical cancer growth and metastasis when it is in overexpression state [27]. Dongling Zou et al. stated that the anomalistic expression of miR. 10b in cervical cancer shows its play role in determining cervical carcinogenesis [28]. Thus, depending on the data recorded as shown in Figure 5, we are able to predict that the sample data of gene expression profile with a high value of this top 10 of the important variables are tumor. The investigation of miRNA levels in cervical cancer cells certainly can open new possibilities for studying molecular markers in the context of early diagnosis and prognosis programs [29].

## 4. Conclusion

In this study, we used dataset, which is obtained from the Gynecologic Oncology Group Tissue Bank (PA, USA) [24] for classification of cervical cancer based on the gene expression profiling data. We conducted the comprehensive benchmarking of Support Vector Machine (SVM) and Random Forests (RF) to evaluate the performance of the predictive models based on the accuracy, Kappa value, sensitivity and specificity. Based on our computational result, we can conclude and prove that Random Forests (RF) machine learning algorithm can be successfully used for predicting cervical cancer based on the gene expression profiling data with the microarray dataset. The accuracy obtained in cervical cancer prediction is better than previous researches results. The average Random Forests (RF) model's accuracy obtained is 94.21%, which may be acceptable in many applications. Besides that, the investigation for a high overall performance classifier with microarray gene expression and other "omics" data is still in progress. Random Forests (RF) algorithms exhibit one of the best classification performances in predicting cervical cancer.

## References

[1] C. M. Martin, K. Astbury, L. McEvoy, S. O'Toole, O. Sheils, and J. J. O'Leary, "Gene expression profiling in cervical cancer: identification of novel markers for disease diagnosis and therapy.," *Methods Mol. Biol.*, vol. 511, pp. 333–359, 2009.

[2] M. Lupu, C. Caruntu, M. A. Ghita, V. Voiculescu, S. Voiculescu, A. E. Rosca, A. Caruntu, L. Moraru, I. M. Popa, B. Calenic, M. Greabu, and D. E. Costea, "Gene Expression and Proteome Analysis as Sources of Biomarkers in Basal Cell Carcinoma," *Disease Markers*, vol. 2016. 2016.

[3] S. W. Purnami, P. M. Khasanah, S. H. Sumartini, V. Chosuvivatwong, and H. Sriplung, "Cervical cancer survival prediction using hybrid of SMOTE, CART and smooth support vector machine," in *AIP Conference Proceedings*, 2016, vol. 1723.

[4] D. R. Tobergte and S. Curtis, *Principles of Biochemistry 5th ed*, vol. 53, no. 9. 2013.

[5] M. Volpacchio, J. C. Vilanova, and A. Luna, "Cervix and vagina," *Learn. Imaging*, vol. 1, no. 6, pp. 165–186, 2012.

[6] M. P. Hopkins, J. a Roberts, and R. W. Schmidt, "Cervical adenocarcinoma in situ.," *Obstetrics and gynecology*, vol. 71, no. 6 Pt 1. pp. 842–844, 1988.

[7] A. Zayed, "Cervical Cancer: Types, Symptoms, Causes, Treatments and Medications," *ConsumerHealthDigest*, 2017. [Online]. Available: https://www.consumerhealthdigest.com/health-conditions/cervical-cancer.html.

[8] A. Deverakonda and N. Gupta, "Diagnosis and treatment of cervical cancer: a review," *J. Nurs. Heal. Sci.*, vol. 2, no. 3, pp. 1–11, 2016.

[9] "DNA Microarray microarray," in *Encyclopedia of Public Health*, 2008, pp. 298–299.

[10] M. Debnath, G. B. K. S. Prasad, and P. S. Bisen, "Microarray," in *Molecular Diagnostics: Promises and Possibilities*, 2010, pp. 193–208.

[11] A. M. Charpe, "DNA microarray," in *Advances in Biotechnology*, vol. 9788132215547, 2013, pp. 71–104.

[12] A. H. Klopp and P. J. Eifel, "Gene expression profiling in cervical cancer: State of the art and

future directions," *Cancer Journal*, vol. 12, no. 3. pp. 170–174, 2006.

[13] A. A. Abdullah, B. S. Chize, and Y. Nishio, "Implementation of an improved cellular neural network algorithm for brain tumor detection," in *2012 International Conference on Biomedical Engineering (ICoBE)*, 2012, pp. 611–615.

[14] A. A. Abdullah and S. Kanaya, "Prediction of Biological Activities of Volatile Metabolites Using Molecular Fingerprints and Machine Learning Methods," *J. Telecommun. Electron. Comput. Eng.*, vol. 10, no. 1–17, pp. 91–96, 2018.

[15] A. A. Abdullah, N. S. Fadil, and W. Khairunizam, "Development of Fuzzy Expert System for Diagnosis of Diabetes," in *2018 International Conference on Computational Approach in Smart Systems Design and Applications, ICASSDA 2018*, 2018.

[16] A. A. Abdullah, A. Yaakob, and Z. Ibrahim, "Prediction of Spinal Abnormalities Using Machine Learning Techniques," in *2018 International Conference on Computational Approach in Smart Systems Design and Applications, ICASSDA 2018*, 2018.

[17] A. A. Abdullah, A. Fonetta, D. Giong, N. Adilah, and H. Zahri, "Cervical cancer detection method using an improved cellular neural network ( CNN ) algorithm," *Indones. J. Electr. Eng. Comput. Sci.*, vol. 14, no. 1, pp. 210–218, 2019.

[18] N. A. Obukhova, A. A. Motyko, U. Kang, S.-J. Bae, and D.-S. Lee, "Automated image analysis in multispectral system for cervical cancer diagnostic," *Conf. Open Innov. Assoc. Fruct*, vol. 2017–April, 2017.

[19] P. Sukumar and R. K. Gnanamurthy, "Computer Aided Detection of Cervical Cancer Using Pap Smear Images Based on Adaptive Neuro Fuzzy Inference System Classifier," *J. Med. Imaging Heal. Informatics*, vol. 6, no. 2, pp. 312–319, 2016.

[20] W. William, A. Ware, A. H. Basaza-Ejiri, and J. Obungoloch, "A review of image analysis and machine learning techniques for automated cervical cancer screening from pap-smear images," *Computer Methods and Programs in Biomedicine*, vol. 164. pp. 15–22, 2018.

[21] M. Orozco-Monteagudo, A. Taboada-Crispi, and H. Sahli, "Biologically inspired anomaly detection in pap-smear images," in *Lecture Notes in Computer Science (including subseries Lecture Notes in Artificial Intelligence and Lecture Notes in Bioinformatics)*, 2013, vol. 8259 LNCS, no. PART 2, pp. 17–24.

[22] B. Taha, J. Dias, and N. Werghi, "Classification of cervical-cancer using pap-smear images: A convolutional neural network approach," in *Communications in Computer and Information Science*, 2017, vol. 723, pp. 261–272.

[23] Abid Sarwar, "Analysis of Machine Learning and Statistics Tool Box (Matlab R2016) over Novel Benchmark Cervical Cancer Database," *Int. J. Trend Sci. Res. Dev.*, vol. 2, no. 1, pp. 619–622, 2018.

[24] D. Witten, R. Tibshirani, S. G. Gu, A. Fire, and W. O. Lui, "Ultra-high throughput sequencing-based small RNA discovery and discrete statistical biomarker analysis in a collection of cervical tumours and matched controls," *BMC Biol.*, vol. 8, 2010.

[25] Z. Yue, Z. Yun-Shan, and X. Feng-Xia, "miR-205 mediates the inhibition of cervical cancer cell proliferation using olmesartan," *JRAAS - J. Renin-Angiotensin-Aldosterone Syst.*, vol. 17, no. 3, 2016.

[26] H. Xie, Y. Zhao, S. Caramuta, C. Larsson, and W. O. Lui, "miR-205 Expression Promotes Cell Proliferation and Migration of Human Cervical Cancer Cells," *PLoS One*, vol. 7, no. 10, 2012.

[27] R. Hou, D. Wang, and J. Lu, "MicroRNA-10b inhibits proliferation, migration and invasion in cervical cancer cells via direct targeting of insulin-like growth factor-1 receptor," *Oncol. Lett.*, vol. 13, no. 6, pp. 5009–5015, 2017.

[28] D. Zou, Q. Zhou, D. Wang, L. Guan, L. Yuan, and S. Li, "The Downregulation of MicroRNA-10b and its Role in Cervical Cancer," *Oncol. Res. Featur. Preclin. Clin. Cancer Ther.*, vol. 24, no. 2, pp. 99–108, 2016.

[29] B. Pardini, D. De Maria, A. Francavilla, C. Di Gaetano, G. Ronco, and A. Naccarati, "MicroRNAs as markers of progression in cervical cancer: A systematic review," *BMC Cancer*, vol. 18, no. 1, 2018.

# Muscle Fatigue Detections During Arm Movement using EMG Signal

**B N Cahyadi[1], Wan Khairunizam[1], I Zunaidi[2], Lee Hui Ling[1], Shahriman A B[1], Zuradzman M R[1], W A Mustafa[1] and N Z Noriman[3]**

[1]School of Mechatronic Engineering, University Malaysia Perlis (UniMAP), Malaysia

[2]Faculty of Technology, University of Sunderland, St Peter's Campus, Sunderland, SR6 0DD, United Kingdom

[3]Faculty of Engineering Technology, Kampus Sg. Chuchuh, Universiti Malaysia Perlis, 02100 Padang Besar, Perlis, Malaysia.

*Corresponding author: basrinoorc@gmail.com*

***Abstract*** — Electromyography is a study of muscle function through the analysis of electrical signals emanated during muscular and muscle contractions. In the post-stroke rehabilitation process, monitoring of muscle activity is very important to know the developments of muscle strength. EMG signals, which produced by muscle activity have information such as muscle contractions, muscle strength and muscle weakness. Rehabilitation process that takes a long period of time can cause muscle fatigue, and the rehabilitation becomes inefficient. The objective of this research is to analyze the muscle fatigue during arm movements by using EMG signals. In this study, deltoid and biceps are monitored by using EMG and the signal are analyzed by using MATLAB. Five healthy subjects are selected to perform the rehabilitation in the experiments. Functional and fundamental movements are used in the data collections. The mean as a feature from a frequency domain is proposed to be used in the analysis. The results show that the signal contractions from deltoid and biceps muscles decreased constantly by the time. In the process of rehabilitation, the stroke sufferers should not do the exercise in a long period of time.

## 1. Introduction

A stroke rehabilitation by using repetitive movements is a way to recover muscle disability. Usually, the side effect of stroke are hemiplegia and hemiparesis but more than 80% of stroke patients suffer from hemiparesis. Hemiparesis is a condition of which one side of the body experiencing weakness [1]. To recover muscle function, a stroke patient has to do physical treatment as fast as possible. The physical treatment including flexion and abduction of the arms are two important rehabilitation exercises for the recovery of muscle weakness [2, 12, 18, 19].

A researcher claims that's a weakening of muscle strength can be treated by intents use of active movement involving repetitive task, task-oriented and task-variegated [3], but the others claim that's the repetitive task produces muscle fatigue [4-5, 15]. The relationship between a repetitive task and muscle fatigue has been proved in numerous studies, in which individual authors discussed various indices of muscle fatigue [5-6].

Muscle fatigue is an exercise-induced decrease in the ability to produce force. Muscle fatigue can occur when someone does exercise without rest. This condition also has an impact on stroke patients if they do the repetitive task in a long period [11]. The existence of muscle fatigue during arm movements for stroke rehabilitation can be determined by using EMG signal [5]. The muscle fatigue can be found by decreasing the power spectrum density of the raw EMG signal. Power spectrum density (PSD) is a measure of a signal's power intensity in the frequency domain. In practice, the PSD is computed from the Fast Fourier Transform (FFT) spectrum of a signal. In the EMG signal processing, the PSD is usually applied as indices to characterize of EMG signals, especially for muscle contraction [7-8]. There are authors used frequency feature extraction to detect the muscle fatigue in the raw EMG signal [4, 6].

The aim of this study is to analyse muscle fatigue during repetitive arm movement by using frequency domain. The outline of this research paper consists of four sections, section 1 discusses an introduction and related researches, section 2 describes the proposed research methodologies and section 3 discusses the finding of the research. The conclusions of this research are discussed in section 4.

## 2. Methodologies

Figure 1 shows the workflow for this research from the preparation of the hardware and software until discussion and conclusion.

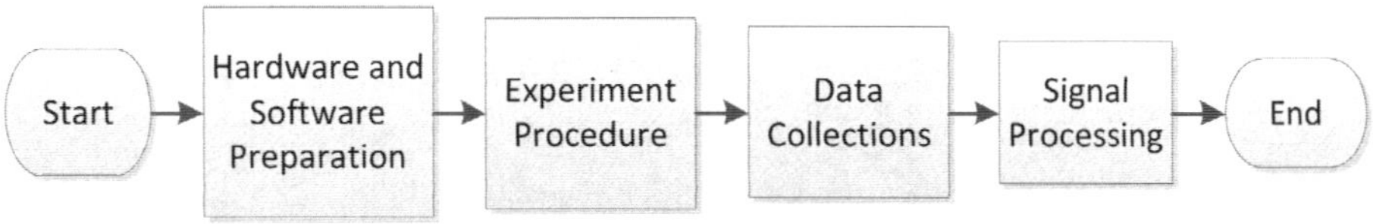


**Figure 1.** The flow of the experiments

### *2.1. Hardware and software*

The systems used for data collecting in this experiment were DELSYS Bagnoli EMG 8 channels and the surface electromyography sensor type single differential. The software used for interfacing was EMG works acquisitions 4.3.2 and MATLAB r2016a for the signal analysis. A raw EMG signal was sampled at 1 kHz and a band-pass filter 10-500 Hz was used to remove DC-artifacts.

### *2.2. Experiment procedure*

The participant had to complete three arm movements on three sessions on different days. In each session they need to do six repetitions for each movement. Three designs of arm-movements are shown in figure 3. The procedure of the movement is briefed as follows:

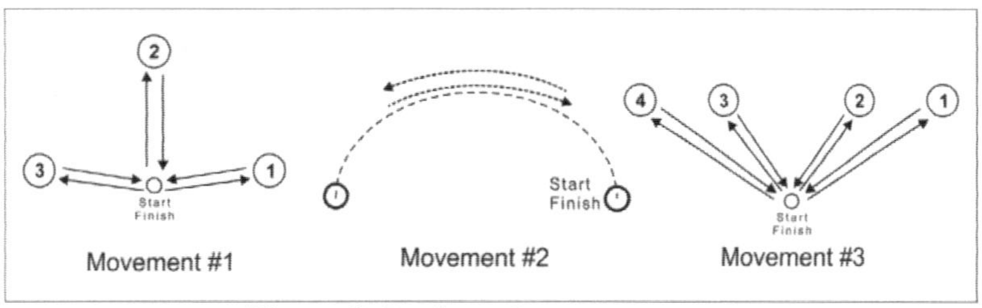


**Figure 2.** Design of arm movements

1. Movement #1
The participant needs to reach all the target with a right hand. The first step is to put the hand in the start position within 5 s - reach target #1 – back to the start position. The second step is to reach target #2 – back to the start position. The last step is to reach target #3 – back to the start position and hold for 5 s.
2. Movement #2
The participant needs to move the hand from right to left positions with parabola movement and repeat it three times.
3. Movement #3
Movement #3 has the same rule with movement #1 but it has four targets and the participant needs to reach all the targets.

*2.3. Data collections*

The raw EMG signal was acquired from the deltoid lateral and biceps muscles. At deltoid muscle, the surface electrode needs to be placed at the distance of one finger width distal and anterior. At biceps muscle, the electrode needs to be placed between the medial acromion and the fossa cubit as shown in figure 2. The EMG electrode placement was referred to as the SENIAM group muscle position [16]. Before electrode placement, the surface must have to be cleaned to reduce the resistance of the skin.

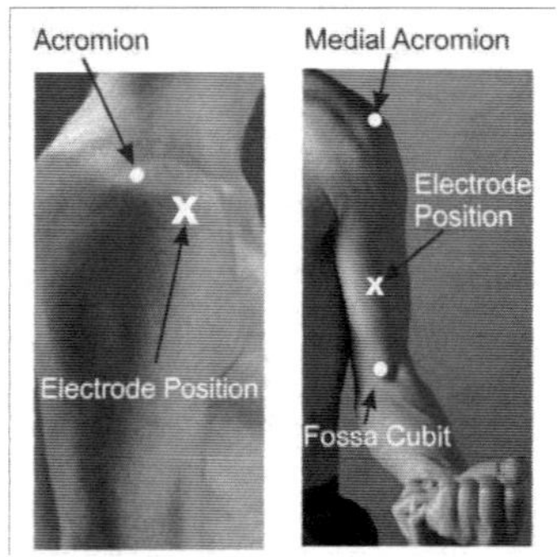


**Figure 3.** Sensor placement for deltoid and biceps [16]

Five healthy male and female with an average age of 25 years old were selected as participants for the experiments. Before data collections, the participants had been briefed all the procedures of the movements. The participants must be free from muscle abnormalities and they were not allowed to do muscle exercise within 24 hours prior to the data collections.

*2.4. Signal processing*

The data analysis was done by using MATLAB r2016a. The raw EMG signals were filtered by using Butterworth filter fifth order with the cut off frequency of 20 Hz [9, 13]. A frequency domain feature was chosen for muscle fatigue analysis because the frequency or spectral analysis were mostly be used in the assessment of muscle fatigue and motor unit analysis. On other hands, mean frequency domain (MNF) of the power spectrum was usually applied for muscle fatigue analysis [7-8].

Power Spectral Density (PSD) is defined as the amount of power per frequency interval. A Fourier Transform was used to transform the EMG signal from time to frequency domains. The periodogram method from the PSD estimator was used to find the frequency signal components in the raw EMG signal. Most of the previous researchers used the Welch method for PSD calculation in MATLAB [10, 14, 17] as shown in (1).

$$P_{XX}(w) = \frac{S\,(e^{fw})}{2p} \quad (1)$$

Where,

$P_{XX}$ (w) = Power spectrum density at a window
$S\,(e^{fw})$ = Power spectrum using FFT
$p$ = Estimation of spectrum value using periodogram
$w$ = Sampling window
$f$ = frame window

The mean frequency of a spectrum was calculated as the sum of the product of the spectrogram intensity and the frequency, divided by the total sum of spectrogram intensity [7]. The definition of MNF is given by:

IOP Conf. Series: Materials Science and Engineering **557** (2019) 012004 doi:10.1088/1757-899X/557/1/012004

$$\mathrm{MNF} = \frac{\sum_{j=1}^{m} f_j P_j}{\sum_{j=1}^{m} P_j} \quad (2)$$

Where,

$f_j$ = frequency value at a data sampling $j$
$Pj$ = EMG power spectrum at a data sampling $j$
m = length of data sampling $j$
$j$ = Number of sampling data

## 3. Result and Discussion

The data collection consists of four sessions. Each session consists of three movements and in each movement there are six repetitions. In these experiments, the PSD signals are generated from each trial and the results are shown in figure 4.

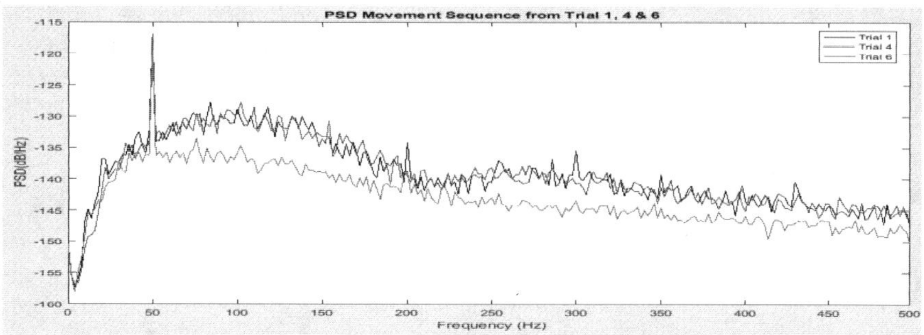


**Figure 4.** PSD signal from first, fourth and sixth trials

PSD signal changes occur in all movements and almost all signals decrease constantly, as shown in table 1. Table 1 shows the MNF values from session-1 from movement #1 until movement #3. The mean frequency value from movement #1 shows that's the MNF value in trial-6 lower than trial-1. This condition occurs in deltoid and biceps muscles (deltoid; trial-1=124.03 Hz – trial-6=122.32 Hz and Biceps; trial-1=128.10 Hz – trial-6=122.27 Hz). The MNF value for movement #2 decreased constantly (deltoid; trial-1=122.24 Hz – trial-6=114.80 Hz and Biceps; trial-1=145.58 Hz – trial-6=122.58 Hz). There is a difference for movement #3 that the value of the deltoid muscle does not decrease (deltoid; trial-1=89.51.24 Hz – trial-6=100.17 Hz and Biceps; trial-1=126.01 Hz – trial-6=103.7 Hz). The reason is the deltoid muscle is exhausted. There is two main cause of muscle fatigue: the limitations of a nerve's ability to generate a sustained signal and the reduced ability of the muscle fibre to contract.

**Table 1.** MNF value from session-1

| **Trial** | **Movement #1** | | **Movement #2** | | **Movement #3** | |
|---|---|---|---|---|---|---|
| | **D (Hz)** | **B (Hz)** | **D (Hz)** | **B (Hz)** | **D (Hz)** | **B (Hz)** |
| 1 | 124.03 | 128.10 | 122.24 | 145.58 | 89.51 | 126.01 |
| 2 | 124.77 | 134.49 | 117.56 | 148.53 | 91.97 | 129.94 |
| 3 | 122.66 | 138.90 | 122.87 | 136.27 | 100.17 | 103.73 |
| 4 | 125.06 | 124.48 | 124.50 | 136.81 | 83.62 | 104.55 |
| 5 | 122.54 | 128.73 | 119.29 | 121.35 | 89.70 | 106.23 |
| 6 | 122.32 | 122.27 | 114.80 | 122.58 | 100.17 | 103.73 |

*D= Deltoid muscle, B= Biceps muscle*

The deltoid and biceps muscles contractions during arm movement are shown in figure 5. When completing all movements, the contractions from deltoid and biceps muscles decrease constantly. The contraction movement #3 is smaller than #2 and #1. Because during the movement #3 the muscle has fatigue.

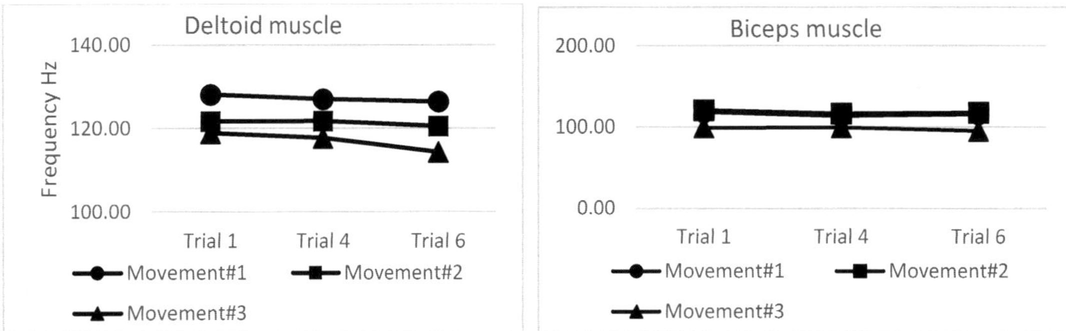


(a) (b)

**Figure 5.** MNF value for deltoid muscle

The results from session 1 to 4 are shown in table 2. The MNF values from deltoid and biceps in trial-6 smaller than trial 1. The reason is the muscle is exhausted when reaching trial #6.

**Table 2.** MNF value from all sessions

| Trial | Session-1 | | Session-2 | | Session-3 | | Session-4 | |
|---|---|---|---|---|---|---|---|---|
| | D(Hz) | B(Hz) | D(Hz) | B(Hz) | D(Hz) | B(Hz) | D(Hz) | B(Hz) |
| 1 | 112 | 133 | 115 | 117 | 156 | 140 | 123 | 112 |
| 6 | 112 | 116 | 113 | 117 | 154 | 134 | 120 | 109 |

*D= Deltoid muscle, B= Biceps muscle*

The decreasing in frequency value during the arm movement is caused by reducing the amount of spectrum in the raw EMG signal and it can be interpreted as muscle fatigue, as shown in figure 4.

## 4. Conclusions

The main objective of this research is to detect muscle fatigue during the arm movement by using raw EMG signals. The deltoid and biceps muscles have been analyzed in this experiment. The experimental results show that the mean frequency can be used for muscle fatigue detections during arm movements. Decreasing of MNF values from trial #1 to #6 caused by reducing the number of the power spectrum in each trial. The repetitive task for post-stroke rehabilitations is a suitable method for muscle weakness recovery, but the monitoring of muscle activity is too important. The muscle fatigue can occur when the stroke survivor exercises with the long period of time without rest.

**Acknowledgments**

TUSB Research UniMAP supports this research under the research grant 2017/08/0006

**Reference**

[1] Mayo Clinic Staff 2017, May 24. *Stroke rehabilitation: What to expect as you recover*. Retrieved from http://www.mayoclinic.org/stroke-rehabilitation/art-20045172
[2] J McIntosh and D Webberley 2016 *Stroke: Causes, Symptoms, Diagnosis and Treatment*, Medical News Today. [Online]. Available: http://www.medicalnewstoday.com/articles/7624.php.

[3] Afsar A I, Mirzayev I, Yemisci O U, 2018 *Virtual Reality in Upper Extremity Rehabilitation of Stroke Patients: A Randomized Controlled Trial* The Journal of Stroke and Crerbrovascular Diseases **27**(12): 1-6.

[4] Roman-Liu D, Tokarski T and Wojcik K 2004 *Quantitive assessment of upper-limb muscle fatigue depending on the conditions of repatitive task load.* The journal of Electromyography Kinesiol, 14:671-682.

[5] A Subasi and M Kemal Kiymik 2010 *Muscle Fatigue detections in EMG using Time-frequency Methods, ICA and Neural Network* The journal of medical system DOI: 10.1007/s10916-0099292-7

[6] Hostens I, Seghers J, Spaepen A and Ramon H 2004 *Validation of the wavelet spectral estimation technique in biceps brachii and brachioradialis fatigue assessment during prolonged low-level static and dynamic contractions* The journal of Electromyography kinesiol **14**:205-215.

[7] S Thongpanja, A Phinyomark, P Phukpattaranont and C Limsakul 2013 *Mean and Median Frequency of EMG Signal to Determine muscle Force based on Time Dependent Power Spectrum.* The Journal of Electronika IR Electrotecnika **19**(3), 51-56.

[8] N Nazmi, M Azizi A R, S Yamamoto, S Anom Ahmad, H Zamzuri and S A Mazlan 2016 *A review of classification technique of EMG signal during isometric and isotonic contractions* The Journal of Sensor **16**(8) 1-28.

[9] Zahari T, Abdelhakim D, Nizam Uddin A, Norhafizan A, Raja Ariffin, Raja Ghazilla, Anwa A M. and Yashim Wong 2015 *IIR filter Order and Cut-off Frequency Influences on EMG Signal Smothing* The Journal of Biomedical Research **26**(4): 616-620.

[10] Krauss T P, L Shure and J N Little 1995 *Signal Processing Toolbox: For Use with Matlab* The Mathworks, Inc.

[11] Jing-jing, Zhen Qin, Peng-yuan Wang, Yang Sun and Xia Liu 2017 *Muscle Fatigue: General Understanding and Treatment* The Journal of the Experimental and Molecular Medicine **49.** DOI: 10.1038/emm.2017.194

[12] Gopura 2010 *A study on Human Upper-Limb Muscles Activities during Daily Upper-Limb Motions* The Journal of Bioelectromagnetism **12**(2): 54-60.

[13] M H Jali, I M Ibrahim, M F Sulaima, T A Izzudin 2014 *EMG signal feature extraction of different arm movement for rehabilitation device* The Journal of applied engineering research, **9** (21),11151-11162.

[14] S Angelova, Simeon Ribagin, Rositsa Raikova and Ivanka V 2018 *Power Frequency Spectrum Analysis of Surface EMG Signals of Upper-limb Muscles during Elbow Flexion A Comparison Between Healthy Subjects and Stroke Survivors* The Journal of Electromyography and Kinesiology **38**: 7-16.

[15] Daniel R, Rogers, Dawn T, Maclsaac 2013 *A Comparison of EMG Based Muscle Fatigue Assessments during Dynamic Contractions* The Journal of Electromyography and Kinesiology **23**(5), 1004-1011

[16] Recommendation for sesor locations on individual muscles. Retrieved from http://seniam.org/sensor_location.htm

[17] B N Cahyadi, Wan Khairunizam, I Zunaidi and S H Majid 2018 *Analysis of EMG based Arm-Movement Sequence Using Mean and Median Frequency.* 5th International Conference on Electrical Engineering, Computer Science and Informatics (EECSI). Malang, Indonesia: IEEE

[18] B N Cahyadi, Wan Khairunizam, I Zunaidi and S H Majid 2018 *Upper-limb Muscle Strength Analysis for Movement Sequence Based on Maximum Voluntary Contraction using EMG signal*. International Conference on Computational Approach in Smart Systems Design and Applications (ICASSDA). Kuching, Malaysia: IEEE.

[19] R Suhaimi, Wan khairunizam, Aswad A R, Nazrul H 2014 *Analysis of EMG based Muscle Activity for Stroke Rehabilitaion* International Conference on Electronics Design (ICED). Penang, Malaysia: IEEE

# Navigation Mobile Apps Utilization and K–MACS Algorithm for VRPTW Model of Fruits Distribution

**S Hartati[1], K B Seminar[2], Sutrisno[3]**

[1]Department of MIT for Natural Resources Management, Bogor Agricultural University, Indonesia
[2]Faculty of Agriculture Technology, Bogor Agricultural University, Indonesia
[3]Department of Mechanical and Biosystem Engineering, Bogor Agricultural University, Indonesia

**Abstract.** This study proposed an extended multiple ant-colony system for vehicle routing problem with time windows (K-MACS VRPTW) in delivery cases considering by the temperature sensitive and perishable products (TSPPs) of strawberries. K-MACS VRPTW is organize with K-Means to partitioning, and MACS, a hierarchy of two artificial ant colonies with objective function to minimizes vehicles used and the transportation times. The designed model is concerning to obtain optimal shipment of box reefer vehicle both Colt-L300 and colt diesel-double (CDD) delivery routes and the unviolated trips time of distribution-center about 8 and 12 working hours. The parameters was the calculation shortest distances of clustering in *Euclidean* and optimal routing in *Haversine*. Experimental results was better than the single MACS-VRPTW solution, that is savings costs as diminish the vehicles used by 5 instead 4 vehicles with shorter total distance and quicker travel time. The trials of the assigned feasible routes is using Voyager, the free navigation mobile apps with API Google direction service for optimization, was found a better nodes sequences of routes mapping.

***Keywords***: *ant colony system, K–MACS, K–Means, perishable products, VRPTW*

## 1. Introduction

The challenging strategy in the logistics and distribution field is optimizing the costs and time delivery from suppliers to customers thus satisfying constraints. In the distribution of temperature sensitive and perishable products (TSPPs), the strawberries fruits particularly, customers's satisfaction are defined from the two aspects of freshness and time window (Wang *et al*, 2017).One of the most critical issues that challenge many fresh fruits logistics is the transportation. Any delay on transportation would negatively impact the price of fresh fruits and lead to a huge financial consequences (Bahebshi, 2017).Hence, the determination of delivery schedule and the routes from sources to destinations is an important thing in distribution management. Such problems known as vehicle routing problems (VRPs), the task of designing delivery or pickup routes, that originates and terminates at a central depot, for a not violated capacitated vehicles that serve a given set of geographically dispersed customers with known demand in the field of transportation and supply chain (Marinakis and Migdalas, 2002; Oyana and Scott, 2008).

The distributions main problem in Indonesia lies on infrastructure and geographical separated locations complexity. The muddy or bumpy dessert roads of village and the obvious circumstances of crowded cities roads deal with congestion and the traffic-conditions change minutes by minute are leading on that the shortest route does not necessarily mean it will be the quickest and then spend most of the cost on the fuel.The vehicle routing problems with time windows (VRPTW) is an extension of the VRPs could be a preferred solution to find a set of shortest vehicle routes or quickest travel time (Baran and Scaerer, 2003).

Most previous studies on the VRPs have focused on seeking for the shortest path to service a number of customers with a fleet of capacitated vehicles without consideration of the direction of next visits destination or the pattern of resulted routes. VRPs have contribution to find the shortest path, but not have many affect on the time constraints. VRPTW is one of its extensions of time constraints to visiting all of customer without violating the stipulation that (i) every route starts and ends at the central depot (ii) all customers must be visited only once by one vehicle, and (iii) ensure that the vehicle capacity and total trip time constraint is not violated.

This study constructed a *stochastic* VRPTW model with the *K-MACS metaheuristic* approach, to obtain optimal delivery routes and capacitated vehicle loads for a perishable horticulture produce distribution, as well as fleet dispatching and departure times. While, the *MACS* is extension of *ant colony system* (ACS) with two objectives of artificial ant colonies which one the most efficient of *ant colony optimization* (ACO) based implementations. Unfortunately, in VRPTW, an insertion of customers violates the system of the ACS and cause of its time constraints, it's rather difficult and not effective to solve the unfeasible situation that some customers with their time windows and demands are not visited. The novelty of the presented approach consists in combination of *K-Means* clustering to identifying and grouping each customers location of fleet services, and the *multiple ant colony system* (MACS) for customer sequences priority routing and optimization processes. Furthermore, the model execution was applied of *navigation mobile apps,* which has equipment of *API Google direction* service to compare and analyze the result.

## 2. The VRPTW Algorithm

Most solution techniques for VRPTW fall in three categories: *exact methods, heuristics and metaheuristics* (Castro and Gutierrez, 2012). This study is *heuristic* VRPs with hard time windows, which is the time constraint is known through perishable produce self-life and the depot distance diversified conditions of traffic at different timesand greatly affect to the travel time.

Constructive heuristics build feasible routes steps by step. Two-phase methods are based on the decomposition of the VRPTW solution process into the two separate sub-problems (Cordeau *et al.*, 2001):

(1). Clustering: determine a partition of the customers into subsets, each corresponding to a route, and
(2). Routing: determine the sequence of customers on each route.

### 2.1 K-Means Clustering

The K-Means clustering is one of the simplest and flexible heuristics clustering technique, which provides efficient analysis of large geospatial data that can minimize sum of squares of the distance from all ***n*** samples emerging in clustering domain to clustering centers to seek for the minimum k clustering on the basis of objective function (kumar and Ramaswami, 2007).

The $k$ data objects as original clustering centers (initial centroid) randomly selected from clustering domain by K-Means algorithm, the centers of the clusters are iteratively recalculated as the means of the coordinates $(x_j, y_j)$ of the customer points assigned to the cluster. The algorithm clusters observations into k groups, where k is provided as an input parameter. It then assigns each observation to clusters based upon the observation's proximity to the mean of the cluster. The cluster's mean is then recomputed and the process begins again (Shafeeq and Hareesha, 2012).

2.2 Ant Colony Optimization (ACO)

Ant colony optimization (ACO) is metaheuristic approach that imitates ants' behavior to construct and optimize feasible routes, where ants cooperate in their search for food by depositing chemical traces (pheromones). In analogy to the biological example, ACO base on indirect communication within a colony of simple agents, called (artificial) ants, mediated by (artificial) pheromone trails. The pheromone trails in ACO serve as distributed, numerical information, which the ants used as probabilistically construct solutions to the problem solved and which the ants adapt during the algorithm's execution to reflect their search experience (Dorigo and Stutzle, 2009). Artificial ants "walk" on the solution space, adding a new decision to the solution at each step. When an ant completes its walk, its solution evaluated. After the evaluation, artificial pheromones are lefts on the path that chosen by the ant. The amount of these artificial pheromones depends on the result of the evaluation: more pheromones are puts for the better solutions (Toklu *et al.*, 2013).

2.3 Multiple Ant Colony System (MACS)

Ant colony system (ACS) is an elitist ACO algorithm, the most important component of ACS for constructing new solutions is the management of pheromone trails, it`s the constructive procedure of *new_active_ant* for ant *m*. The MACS-VRPTW algorithm initializes and coordinates the two ACS based colonies objective functions. The first goal colony, denoted as ACS-VEI, diminish the number of vehicles used, while the second colony, as ACS-TIME, optimizes the travel time distribution ($\mathbf{t}_{ij}$) as feasible solutions found by ACS-VEI. The two colonies collaborate sharing a global best solution and exchanging information through pheromone updating (Gambardella, 1999).

The pheromone informationis given by the visibility, denoted as below equation (1), where $L_{\Psi}^{h}$ representing an initial estimation of the total delivery time, while and $J_{\Psi}^{h}$ is represent an initial estimation of the total traveling time, and n is the initial number of nodes (depots + customers). The formula is (Barán and Schaerer, 2003):

$$\tau_0 = \frac{1}{n\,x\,(L_{\Psi}^{h} J_{\Psi}^{h})} \tag{1}$$

The visibility of the ACS-TIME objective which consisting of the travelling time $\boldsymbol{\eta}_J$***(ij)*** and the delivery time $\boldsymbol{\eta}_L$ ***(ij)*** calculation as adopted from (Donati *et al.*, 2008). To choose the next node $C_j$ to be visited by an ant m of cluster k is chosen in$N_i^m$ by the following procedure (Barán and Schaerer, 2003):

| **Algorithm 1.** Chooses nextnode |
|---|
| **Procedure** ChooseNext_Node |
| Chooseto do exploitation with probability r |
| or exploration otherwise; |
| **if**(exploitation) |
| Choose $C_j$with the highest attractiveness value |
| **else** exploration |
| Choose the $C_j$ with larger probability$r_0$ |

The set of available nodes, in this case the ant is not located in other cluster, not located in a duplicated depot, includes not yet visited duplicated depots. λ = *m/k,* and β weights the relative importance of the objectives with respect to the pheromone trail, given by$\tau_0$.The two ways probability, determined by a fixed cut-off parameter $r_0 \epsilon$ (0, 1), and a random number r for each step,r $\epsilon$ (0, 1):

(a) Exploitation: pick the $\mathbf{C}_j$which maximizes p(**j**), if$\mathbf{r}_0 \leq \mathbf{r}$;
(b) Exploration: pick the $\mathbf{C}_j$distributed as p(**j**), if$\mathbf{r} \geq \mathbf{r}_0$.

The visibility of the ACS-TIME objective which consisting of the travelling time and the delivery time calculated in following equation (2) and equation (3) as:

$$\eta_J(i,j) = \frac{1}{\sum \tilde{t}_{ij}} \tag{2}$$

The visibility of delivery time related to the service time and the width of the time windows. It is calculated as adapted from [9]:

$$Dt_j = \max(At_i + t_{ij}, a_j)$$
$$\Delta t_{ij} = Dt_j - At_i$$
$$d_{ij} = \max(1, (\Delta t_{ij} . (bj - At_i))$$
$$\eta_L(i,j) = \frac{1}{d_{ij}} \tag{3}$$

During the execution of the ACS, the pheromones be updated in two ways: local update using equation (4) and global update using equation (5). Local updating is performed during solutions construction, while global updating is performed at the end of the constructive phase. This update done according to the formulation:

$$\tau_{ij} = (1 - \rho).\tau_{ij}(t) + \rho.\tau_0 \tag{4}$$
$$\tau_{ij} = (1 - \rho).\tau_{ij}(t) + \rho/(L_\psi^h.J_\psi^h) \tag{5}$$

Where parameter $\boldsymbol{\rho}$ configures the amount of pheromone decrease imposed by the local update, and $\boldsymbol{\tau_0}$is calculated initially. The effect of local updating is to change dynamically the desirability of edges: every time an ant uses an edge the quantity of pheromone associated to this edge is decreased and the edge becomes less attractive. On the other side, global updating is used to intensify the search in the neighborhood of the best solution computed

## 3. The Design Model and Algorithm

This K – MACS VRPTW composed of vary different problem types(customer $C_n$ clustered in vehicle used $V_n$, customer demand, vehicle load, route distance $\boldsymbol{d_{ij}}$, travel time $\boldsymbol{t_{ij}}$, starts time, service time, ends time, and also depot time window/time agent).

This research was using quantitative data which has been taken in CV. Yan's fruits and vegetables of Lembang, Bandung. The major types of transportation used was the road systems with vary *caroseries*-reeferbox of vehicle as main modes such as 4 units *ColtL300* measuring 237cm x 155cm x 129cm which in a single transport can load the crates to 85 approximately, and 2 units *Colt Diesel Double* (CDD) measuring 560cmx200cmx220cm which can load more than 240 crates approximately. Each data set contains 4 to 35 customers use 6 unit vehicles to serve all the customers in JABODETABEK (Jakarta, Bogor, Depok, Tangerang, Bekasi) area, including 4 is *L300 Colt* and 2 is *CDD* was the 5 units for operational vehicles and one between for backup.

IOP Conf. Series: Materials Science and Engineering **557** (2019) 012005 doi:10.1088/1757-899X/557/1/012005

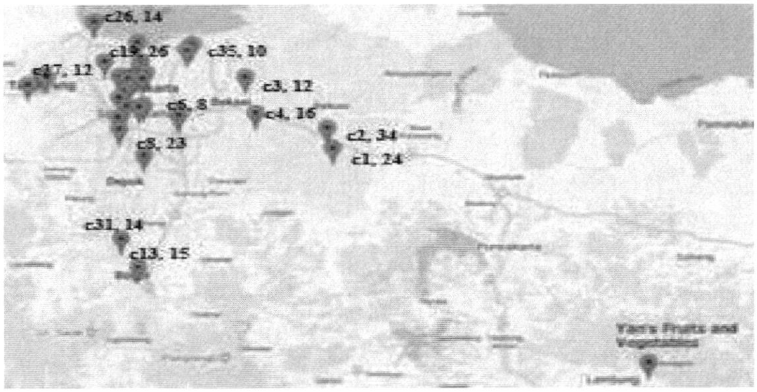


**Figure 1.** Customer distribution map.

The time widows of this study is that company's agreement with all of customers is enforce the rules asthe arrival goods hours should not than be later10:00 am, and also the company has time windows that the travel time such 480 minutes (8 working hours per-day) for vehicle type of *L300 Colt* and about 720 minutes *(*12working hours per-day) for *CDD*.Hence, the departing times of dispatching vehicles from Bandung to JABODETABEK area should be adjusted as the length of travel times from depot to the first customer and then to another, also the total service time of all customer destination.

*In* this study, the constructive procedureof *new_active_ant*was *using pseudorandom rule* (the probabilistic rule processing where select node *j* uses probability smaller$q_0$ node with the highest attractiveness value and *exploitation* mechanism) and global pheromone trail is updated during solutions construction.

### 3.1 Distances

The most important variable in VRP problems is the distance (*d*) of delivery points and the region as input basic data in order to determine and estimating time travel and the total distance covered. In this study, we used *Haversine* distance and *x, y Cartesian* coordinate of geographic latitude and longitude as following equation:

$$d = 2r \arcsin\left(\sqrt{\sin^2\left(\frac{\varphi_2-\varphi_1}{2}\right) + \cos\varphi_1 \cos\varphi_2 \sin^2\left(\frac{\lambda_2-\lambda_1}{2}\right)}\right) \quad (6)$$

$$x = r' \cos\lambda = r\cos\varphi\cos\lambda \quad (6a)$$

$$y = r' \sin\lambda = r\cos\varphi\cos\lambda \quad (6b)$$

Where $r' = r\cos\varphi$ and $r = \sqrt{x^2 + y^2}$

### 3.2 Distances Stochastic Hard Time VRPTW

In the classic VRP with hard time windows (VRPTW), vary capacitated vehicles (Q) isscheduled to visit the given set of N customers $C_i$,each characterized by ademand ($q_i$), a time window ($tw_j = [a_j, b_j]$) and a service time ($s_i$), with routes starting and ending at a depot with time constraint[$t_i$, $e_j$] is specified. One arc existing oriented arc a$\epsilon$A, information about the travel time must be given to deduce the time necessary to traverse the arc when starting the trip at a time ***t***. The location of each customer address have been known also the distance from the depot to each customer or the customer distance between, then the estimated travel time (t) to each customer address can be calculated by dividing the distance (S) with the average speed of the vehicle (V) as following equation:

$$\mathbf{t} = \frac{\mathbf{S}}{\mathbf{V}} \tag{7}$$

*t = time (hours)*
*S = space or distance (km)*
*V= velocity or speed (km/hours)*

The time constraint is denoted by a predefined local markets time. The travel time is stochastic, vehicle travel speed may be affected by factors such as traffic volume, the weather and accidents, which are characterized as random in nature. Let $\tilde{t}_{ij}^{1}$ denote the new travel time on the arc ($\mathbf{C}_i$, $\mathbf{C}_j$) for vehicle *V*, *A* denote the sets of arc with travelling barriers and the weight respectively, can be expressed as $\tilde{t}_{ij}^{1} = w.t_{ij}$, where ***w*** is weight of travelling barriers (traffic congestion and damaged roads). Assume some arc of *A* have the weight ***w*** of being travelled barriers. Then, expected travel time on arc ($\mathbf{C}_i$, $\mathbf{C}_j$) for vehicle *V*, $\tilde{t}_{ij}^{1}$, is given by:

$$\tilde{t}_{ij} = \begin{cases} w.t_{ij}, & if\ (V_i, V_j) \in A \\ 0, & Otherwise \end{cases}$$

$$i=1, \ldots n;\ j=1, \ldots n;\ V=1, \ldots m \tag{8}$$

Where $t_{ij}$ represent travel time on arc ($\mathbf{C}_i$, $\mathbf{C}_j$), because of the randomness of travel time on arc, arrival time at each customer location is characterized as a random variable. Since the real-time traffic conditions for every arc is unknown before departure from the distribution center, arrival time at each customer is difficult to predict. Hence, in this study, the visibility of travel time is consisting of travelling barriers weight includes of traffic congestion and damage roads rated as ***w*** (0.5 and 0.25).

3.3 Distances K-MACS VRPTW Approach

Stage 1: *K–Means* Clustering

The *K–Means* for the VRPs consist of the following steps:

Step1. Determine the *K* clusters, which is the clusters number are equal to *r*vehicles (*K*= *r*) or less (max *K* = *r* – 1). This step, applied for finding solutions of vehicles number required with the constraint of capacitated vehicle and the customer time window.

Step2. Randomly choose each centroids of K *cluster* such as x, y coordinates of customers location points as initial cluster center

Step3. Denote the cluster center as the beginning of a route to determine the customer closest in order to assigning customer location of the cluster construction.

Step4. Calculates the mean value of customers distance to depots and the customers distance to their neighbors are calculated by *Euclidean-distance*.

$$d_{ij}(x_{ik}, c_{jk}) = \left[ \sum_{k=1}^{n} |\, x_{ik} - c_{jk}\, |^2 \right]^{1/2} \tag{9}$$

Then, constructs cluster for the minimum customer distance to the centroid and set of each depot.

Step5. Repeat Step 4 until no change.

Stage 2: Routing and Optimization

The second stage is the routing and optimization using MACS-VRPTW algorithm. The best feasible Insertion is forth process to improve routing as following algorithm works:

1. Initial value of MACS, $\Psi^{gb}$ is a feasible VRPTW solution found with *K-Means* clustering and others heuristic routing method to identify customer (pheromone) service priority.
2. The next step, $\Psi^{gb}$ is simultaneously looked for an improved and feasible solution, that is: (i) a solution that has a smaller number vehicle using; (ii) it has the same number of tours and has a faster travel time.
3. Denote the duplicated depot as the beginning ofa route, and ACS parameters to choose next visit nodes contains (probability ρ, attractiveness α, visibility β, vehicle capacity Q, ants number $\boldsymbol{m}$ (in this case equal to number cluster $\boldsymbol{k}$), weightparameter $w_j$ of travel time which is contain of traffic congestion and damage roads rated as $\boldsymbol{w}$ (0.5 and 0.25) as given in equation (8), iterations number, and vehicle network, G = (v, A)
4. Set the queue order route of locations that satisfies all of constraint with nearest neighbor heuristic
5. When a new best solution is found, it is then used to perform a global pheromone update, so that both colonies can make use of the updated information about the performance of the new solution.
6. Activate ACS-VEI and ACS-TIME.
7. Decision processes.

Once each ant has built a complete solution, and then tentatively improved using a *local search* procedure. Next, the best solution found from the beginning of the trial is used to update the pheromone trails. Then, the process iterated by starting again $m$ ants until a termination condition is met.

## 4. Results and Analysis

Refers to figure1 i.e the mapof nodes data which is containing thenumber of customers, real time data usagecorresponding longitude & latitude oflocations and then converted into a corresponding(X,Y) coordinates, and also the'Great Circle Distance' between customer's location which is calculated use the equation (6a) and equation (6b), customers demand and the time constraint. The best K-Means result of customer distances clustering to **k** number clusteras much as available vehicle transportation can made for 5 cluster (K= V1,V2, V3, V4, V5) as the following table shows the clustered customers of K-Means clustering result:

Table 1.K-Means clustering result

| Cluster Name | Customer Members | | | Route Distance (km) | Vehicles | | Time Windows (minutes) | | |
|---|---|---|---|---|---|---|---|---|---|
| | Points | Total Cust | Total Demand (crates) | | Max Load (crates) | Vehicle Type | Depot Time | Travel Time | Service Time |
| Cluster1 (V1) | C1, C2, C4, C8, C9, C14, C24, C25, C34 | 9 | 214 | 239.65 | 240 | CDD | 720 | 361 | 235 |
| Cluster2 (V2) | C10, C20, C22, C29, C31 | 5 | 75 | 283.37 | 85 | Colt L300 | 480 | 425 | 85 |
| Cluster3 (V3) | C5, C19, C21, C35 | 4 | 70 | 201.98 | 85 | Colt L300 | 480 | 303 | 80 |
| Cluster4 (V4) | C7, C13, C26, C27 | 4 | 64 | 245.15 | 85 | Colt L300 | 480 | 368 | 70 |
| Cluster5 (V5) | C3, C6, C11, C12, C15, C16, C17, C18, C23, C28, C30, C32, C33 | 13 | 221 | 285.41 | 240 | CDD | 720 | 420 | 260 |

The K-Means result as above table1 show, the routes then rearranges inter-route or intra-route using local search optimizations and LINGO 17.0 solver to decrease the total travel distance (or total travel time) by the vehicles. The processes is resulting initials routes solution of $\psi^{gb}$as:**D** - C25 - C14 - C9 - C8 - C24 - C4 - C1 - C2 - C34 - **D** - C20 - C10 - C29 - C22 - C31 - **D** - C5 - C35 - C19 - C21 - **D** - C27 - C7 - C26 - C13 - **D** - C30 - C12 - C32 - C16 - C18 - C17 - C15 - C23 - C3 - C11 - C6 - C28 - C33 – **D.** Where, the agent needs 5 units of vehicle transportation with total distances 1255.54 km, and the total transportation time 2616,34 minutes including travel time and delivery time.

Considering to the constructive procedure of *new_active_ant*isaiming to find a feasiblesolution with one vehicle less than the number of vehicles used in time windows by maximizing number of visited customers. Hence, we drowned out the initials solution clusters and operated vehicle to being 4 units for serves 30 customers with total distances 972,17km; total travel time 1461 minutes; average of delivery time 70 minutes; and the ACS_VEI initials pheromone $\tau_0$ : 3,54 x $10^{-8}$.

In *exploitation* mechanism of ACS_VEI, an agent of ant *m*select node $C_j$depend on the attractiveness value of pheromone information, the information indicates how well seems to visit customer $C_j$after $C_i$considering the solutions already found. The highest attractiveness and the visibility node then its will be selected to visit destination by the agent. Thesolution computed since the beginning of the trial with the highest number of visited customers is stored in the variable$\psi^{ACS_VEI}$. Asolution is better than $\psi^{ACS\_VEI}$ only when the number of visited customers is increased, then the *pheromone intensity* updated in *global-update*. Experimentshave been done with the following parameter settings: $m$=4 ants; $q_0$=0.9; β=1 and ρ=0.1.

IOP Conf. Series: Materials Science and Engineering **557** (2019) 012005 doi:10.1088/1757-899X/557/1/012005

The next step is ACS_TIME activation to finding the $\psi^{gb}$ with implement a local search procedure to improve the quality of the ACS_VEI feasible solutions. In this procedure, customer is searched for the best feasible insertion (shortest travel time) until no further feasible insertion is possible.The$\psi^{gb}$ feasible routes solution of ACS_TIME as [**D** - C19 - C14 - C25 - C17 - C15 - C23 - C6 - C11 - C13 - C9 - C24 - C4 - C26 - C34 - **D** - C16 - C32 - C20 - C2 - C7 - C8 - C10 - C28 - C3 - C33 - C35 - C5 - **D** - C27 - C29 - C1 - C18 - **D** - C21 - C35 - C12 - C31 - C30 – **D**]. With the total distances 1019.68 kmand total transportation times 2261 minutes.

The following table is the result of last processes of K-MACS algorithm routing and the optimization using *inter-route* and *intra-route* of *local search* method:

Table 2. Feasible route solution of K-MACS and the Local search optimization

| Cluster Name | Customer Members: Points | Customer Members: Tot Cust | Customer Members: Total Demand (crates) | Route Distance (km) | Vehicles: Max Load (crates) | Vehicles: Type | Time Windows (minutes): Travel Time | Time Windows (minutes): Service Time | Time Windows (minutes): Depot Time |
|---|---|---|---|---|---|---|---|---|---|
| C1 (Vehicle 1) | D-C19-C14-C25-C17-C15-C23-C6-C11-C13-C9-C24-C4-C26-C34-D | 14 | 238 | 256,686 | 240 | CDD | 385,03 | 260 | 645 |
| C2 (Vehicle 2) | D-C16-C32-C20-C2-C7-C8-C10-C28-C3-C33-C35-C5-D | 12 | 237 | 271,054 | 240 | CDD | 406,58 | 280 | 686,6 |
| C3 (Vehicle 3) | D-C27-C29-C1-C18-D | 4 | 84 | 263,175 | 85 | Colt L300 | 394,76 | 85 | 479,8 |
| C4 (Vehicle 4) | D-C27-C29 -C1-C18-D | 5 | 82 | 243,135 | 85 | Colt L300 | 364,7 | 85 | 449,7 |

To analyze the methods was used in this research, we compares the method result steps by step on the along processing as well as merging both K-Means and MACS algorithms between with the *local search* optimization result for each algorithms. In the following table is presented the performance metrics calculation for every feasible solution run of each algorithm result concerned of total distances and total travel time, where it is easy to note that theunification of MAC-VRPTW is improve the optimal K-Means result with or without *local search* optimization.

Table 3. Comparison results of algorithms used

| M[a] | K-MEANS | | | K-MEANS & LOCAL SEARCH | | | K-MEANS & MACS (K-MACS) | | | K-MACS & LOCAL SEARCH | | |
|---|---|---|---|---|---|---|---|---|---|---|---|---|
| | V[b] | RD[c] | TT[d] | V[b] | RD[c] | TT[d] | V[b] | RD[c] | TT[d] | V[b] | RD[c] | TT[d] |
| V1 | 9 | 250,6 | 376 | 9 | 239,6 | 361 | 14 | 283,8 | 426 | 14 | 256,69 | 385 |
| V2 | 13 | 420,7 | 631 | 13 | 285,4 | 429 | 11 | 359,5 | 539 | 12 | 271,05 | 407 |
| V3 | 4 | 226,3 | 339 | 4 | 202 | 303 | 4 | 283,4 | 425 | 4 | 263,17 | 395 |
| V4 | 4 | 247,2 | 371 | 4 | 245,2 | 368 | 6 | 340,6 | 511 | 5 | 243,13 | 365 |
| V5 | 5 | 297,6 | 446 | 5 | 283,4 | 425 | - | - | - | - | - | - |
| Tot | 35 | 1442,4 | 2164 | 35 | 1255,6 | 1886 | 35 | 1267,3 | 1901 | 35 | 1034,05 | 1551 |

Notes :
[a]Methods
[b]Visits
[c]Route Distance (km)
[d]Travel time (minutes)

When the feasible routes result was tried to applying into *Google Maps* to navigate, the figure of map result are out-of sync neither by *Google Maps* calculation (shortest distance and the travel time estimation) nor the continuity of directions travel lines. Then, its' tried into routes optimizations mobile Appwith *API Google direction service* commonly available applications (we used *Voyager* mobile App) in *play store* of *Android* smart phone to gets the optimize routes comparison as the following example presented of cluster-4 (Vehicle 4) feasible result.

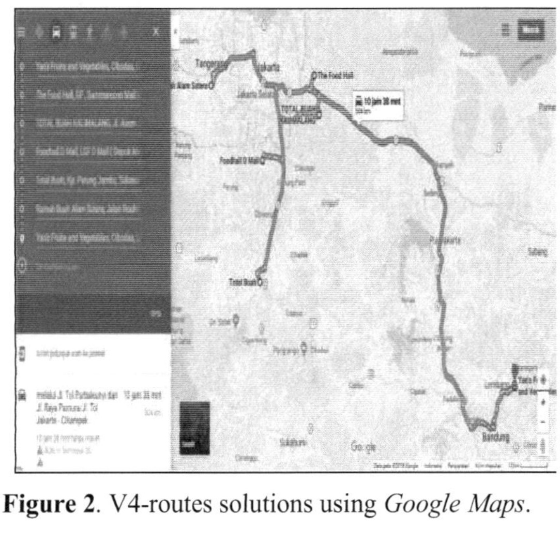

**Figure 2**. V4-routes solutions using *Google Maps*.

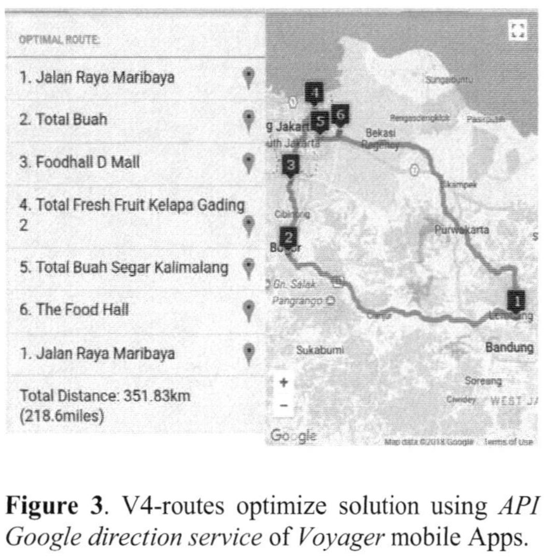


**Figure 3**. V4-routes optimize solution using *API Google direction service* of *Voyager* mobile Apps.

## 5. Conclusions

The present research modifies the MACS-VRPTW algorithmwith merge of *heuristics* K-Means clustering and *metaheuristics ACS* supported byenhanced local search proceduresto propose. The model is developed to solvethevehicle routing problem with time windows (VRPTW)ofthe temperature sensitive and perishable products (TSPPs) logistics transportation,which is provides a combined approach for both minimizing the number of vehicles used and the total transport time. From the results and analysis it is concluded that, the model built can be help savings distribution costs by diminish the vehicles used from 5 to 4 units as 2 *Colt*-L300 and 2 *CDD Box*-reefer with total trip time less than 8*working hours* and 12 *working hours* approximately constraint. According to optimizations assigning trials of each feasible routes in *Voyagernavigationmobile App*s with *API Google direction* service of *Android* smart-phone, it's found the better result of visit nodes sequences of routes due to direction of traffic network consideration. Therefore, in additions ofthe shortest mileage parameter, the *API direction service* is important to optimization routes of geographical visiting point and the obtained result substantiates the effectiveness of the application of the proposed method. This work can also be refined further for different applications.

**References**

[1]. Carlisle D 2014 The impact of food perishability issues in the vehicle routing problem (graphicx: Enhanced support for graphics. Retrieved from Amorim P, Almada-Lobo B (2010, April) *Computers and Industrial Engineering* **vol. 67**, no. 1: 223–233.

[2]. Bahebshi R O 2017 Maroco Dynamical Modeling of Fresh Vegetables while Considering Degradation Effect on Pricing. *Proceedings of the International Conference on Industrial Engineering and Operations Management Rabat* **April** 11-13: 1817-1825

[3]. Barán B, Schaerer M 2003 Austria A Multiobjective Ant Colony System For Vehicle Routing Problem With Time Windows *Proceedings of the 21st IASTED International Conference Applied Informatics* 97-102

[4]. Bogataj M, Bogataj L, Vodopivec R 2005 Stability of perishable goods in cold logistics chains. *International Journal of Production Economics* 93/94(8): 345–356.

[5]. Chen X, Wan W S, Xu X H 2006 China The Real-Time Time-Dependent Vehicle Routing Problem *Transportation Research Part E: Logistics and Transportation review* **vol. 42** No. 5: 383-408.

[6]. Cordeau J F, Gendreau M, Laporte G 1997 A tabu search heuristic for periodic and multi-depot vehicle routing problems. *Networks* **vol. 30**:105–119.

[7]. Cordeau J F, Gendreau M, Laporte G, Mercier A 2001 A Unified Tabu Search heuristic for vehicle routing problems with time windows. *Journal of the Operational Research Societ* **vol.52**: 928–936.

[8]. Cordeau J F, Desaulniers G, Desrosiers J, Solomon M.M, Soumis F 2001 Philadelphia The VRP with Time Windows, The Vehicle Routing Problem *SIAM Monographs on Discrete Mathematics and Applications* 157-194

[9]. Donati A V, Montemanni R, Casagrande N, Rizzoli A E, Gambardella L M 2008 Time dependent vehicle routing problem with a multi ant colony system *European Journal of Operational Research*. **vol.185**:1174–1191

[10]. Dorigo M, Gambardella L M 1997 Ant colony system: Acooperative learning approach to the travelling salesman problem *IEEE Transactions on Evolutionary Computation* **vol.1**: 53–66.

[11]. Dorigo M, Gambardella L M 1997 Ant colonies for the travelling salesman problem *BioSystems* **vol. 43** no. 2: 73–81

[12]. Dorigo M, Di Caro G 1999 London The Ant Colony Optimization metaheuristic *New Ideas in Optimization* 11–32

[13]. Dorigo M, Stutzle T 2010 Handbook of Metaheuristics In: M. Gendreau, J.Y. Potvin Editors. *International Series in Operation Research and Management Science 146* (Brussels:Springer Science + Business Media, LLC) 227-263

[14]. El-Sherbeny N A 2010 Cairo Vehicle routing with time windows: An overview of exact, heuristic and metaheuristic methods. *Journal of King Saud University (Science)*

[15]. Gambardella L M, Taillard E D, Dorigo M 1999 Ant colonies for the quadratic assignment problem *Journal of theOperational Research Society* **vol.50**: 167–176.

[16]. Gambardella L M, Taillard E D, Agazzi A 1999 London MACS-VRPTW: A multiple ant colony system for vehicle routing problems with time windows *In New Ideas in Optimization* 63–76

[17]. Gambardella L M, Dorigo M 1996 New York Solving symmetric and asymmetric TSPs by ant colonies *In Proceedings of the 1996 IEEE International Conference on Evolutionary Computation (ICEC'96)* 622–627

[18]. Gen M, Cheng R 1997 Genetic Algorithms and Engineering Design (New York:Wiley)

[19]. Guiterrez C J 2012 UK Multi Objective Tools for the Vehicle Routing Problem with Time Windows *PhD Thesis* (United Kingdom: Postgraduate University of Nottingham)

[20]. Potvin J Y 1996 Genetic algorithms for the travelling salesman problem *Annals of Operations Research* **vol.63**: 339–370

[21]. Iredi S, Merkle D, Middendorf M 2001 Zurich Bi-Criterion Optimization with Multi Colony Ant Algorithms *Lecture Notes in Computer Science* **March**: 360-370.

[22]. Jeon G, Leep H R, Shim J Y 2007 A vehicle routing problem solved by using a hybrid genetic algorithm *Computers and Industrial Engineering* **vol. 53**: 680-692

[23]. Kumar S P, Ramaswami K S 2014 Fuzzy K-Means Cluster Validation for Institutional Quality Assesment *Green Computing Communication and Electrical Engineering (ICGCCEE) International Conference* 1- 8.

[24]. Marinakis Y, Migdalas A 2002 Heuristic solutions of vehicle routing problems in supply chain management *Combinatorial and Global Optimization* 205–236.

[25]. Merkle D, Middendorf M 2002 Modelling the dynamics of ant colony optimization algorithms *Evolutionary Computation* **vol. 10** no. 3: 235–262

[26]. Oyana T J, Scott K E 2008 A Geospatial Implementation of a Novel Delineation Clustering Algorithm Employing the K-means *The European Information Society*

[27]. Renaud J, Laporte G, Boctor F F 1996 A tabu search heuristic for the multi-depot vehicle routing problem *Computers and Operations Research* **vol. 23** 229-235

[28]. Shafeeq B M, Hareesha K S 2012 Dynamic Clustering of Data with Modified K-Means Algorithm *International Conference on Information and Computer Networks (ICICN) IPCSIT* **vol. 27**

[29]. Surekha, Sumathi S 2011 India Solution To Multi-Depot Vehicle Routing Problem Using Genetic Algorithms *World Applied Programming* **vol.1** no.3

[30]. Toklu N E, Montemanni R, Gambardella L M 2013 A robust multiple ant colony system for the capacitated vehicle routing problem *IEEE International Conference on Systems, Man, and Cybernetics*

[31]. Wang X, Sun X, Dong J, Wang M, Ruan J 2017 Optimizing Terminal Delivery of Perishable Products considering Customer Satisfaction. *Hindawi Publishing Corporation Mathematical Problems in Engineering*

[32]. Xuping W, Xiaoyu S, Jie D, Meng W, Junhu R 2017 Optimizing Terminal Delivery of Perishable Products considering Customer Satisfaction. *Mathematical Problems in Engineering* Article ID 8696910: 1-12

# A Comprehensive Review on Document Image (DIBCO) Database

**W A Mustafa[1], Wan Khairunizam[2], I Zunaidi[3], Z M Razlan[2] and Shahriman A B[2]**

[1]Faculty of Engineering Technology, Kampus Sg. Chuchuh, Universiti Malaysia Perlis, 02100 Padang Besar, Perlis, Malaysia.
[2]School of Mechatronic Engineering, Universiti Malaysia Perlis, Arau, Perlis, MALAYSIA
[3]Faculty of Technology, University of Sunderland, St Peter's Campus, Sunderland, SR6 0DD, United Kingdom

*wanazani@unimap.edu.my*

**Abstract.** One of the most significant current discussions in image processing is a document analysis. Now, many types of document database were established in order to address the issue of binarization effectiveness. In this paper, a comprehensive review of the document database was presented. Review based on an image from Document Image Binarization Contest (DIBCO) from 2013 to 2017 which consists of handwritten and printed image. The best algorithm for each year is discussed and analyzed. Based on the results, the technique using background estimation and stroke edges is better performance for the overall database. Besides, the method using the combination of Laplacian operator and canny edge detection also shows the successful result, especially in the printed image. Implications of the review give the direction for future binarization approach developments.

## 1. Introduction

In the recent decades, the binarization technique has been one of the major interesting research subjects due to detect the text images. However, there has been little discussion and challenges about the degradation of input document images caused by large background ink stains, contrast variation and illumination problem [1–3]. Binarization is one of the important process in order to separate between the object (text) and background by removing all unwanted noised [4,5]. A large and growing body of literature has investigated a new method in order to segment the backgorund and foreground document images [6,7]. However, selecting the most optimum method for binarization is a difficult task due to the presence of a variety of degradation in document images.

There are many algorithms that can be used for binarization process but the earliest and most enduring image binarization methods is Otsu's method [12,13]. This method works by finding a threshold value that differentiates pixels into two components which is lighter and darker pixels. The lighter pixel will be classified as white meanwhile for the darker pixel it is black. The method has its own weakness, it cannot binarize image with almost the same pixels color. Besides, Ntirogiannis et al. [14] have done a combined approach for the binarization of handwritten document images. The research was inspired by the low detection rate of faint characters in binarization of handwritten document images. Firstly, image normalization based on background compensation and estimation is applied. For the second step, global thresholding is applied to the image. On the binarized images, very small components are discarded and representative characteristics of a document image such as the stroke width and the contrast are computed. Furthermore, local adaptive binarization is performed on the normalized image taking into account the aforementioned characteristics. Finally, the two binarization outputs are combined at connecting component level. For DIBCO 2011 dataset, the average F-measure and PSNR are 94.05% and 21.65 respectively.

In this paper, a comprehensive review of the document image database from 2012 to 2017 was presented. The database consists of two types of image which is handwritten and printed image. The specification in terms of size and resolution also was discussed. Furthermore, the best binarization method for each year from the first rank to the third rank was explained. The rest of this paper is organized as follows: Section 2 describes an overview of the document image. Sub-section 2.1 to 2.5 described the database image. Finally, Section 3 explained the conclusion of this work.

## 2. Overview Database

Recently, researchers have shown an increased interest publication in document binarization techniques [15–17]. Every year, a new document database was provided by a specialist in order to increase the challenging for researchers. Based on the review and publication number, many research used a document database from http://www.iit.demokritos.gr/~bgat and http://utopia.duth.gr/~ipratika/ website [18,19]. The contest of binarization using a popular document database was organized called as International Document Image Binarization Contest (DIBCO). In this paper, a comprehensive review of the document database from DIBCO 2013 to DIBCO 2017 was analyzed.

### *2.1 DIBCO 2013*

Database 2013 presents are 8 handwritten images and 8 printed images (shown in Figure 1) and can be download at http://utopia.duth.gr/~ipratika/DIBCO2013/benchmark. Overall, the database images show the different size and same resolution as show in Table 1. However, two (2) image from the handwritten group consists of different resolution which is 96 ppi (HW06 and HW08). Based on image size, database 2013 is largest compared to the previous database (2009-2012) [20]. Example, 'HW01' image size is 4161 x 1094 pixel.

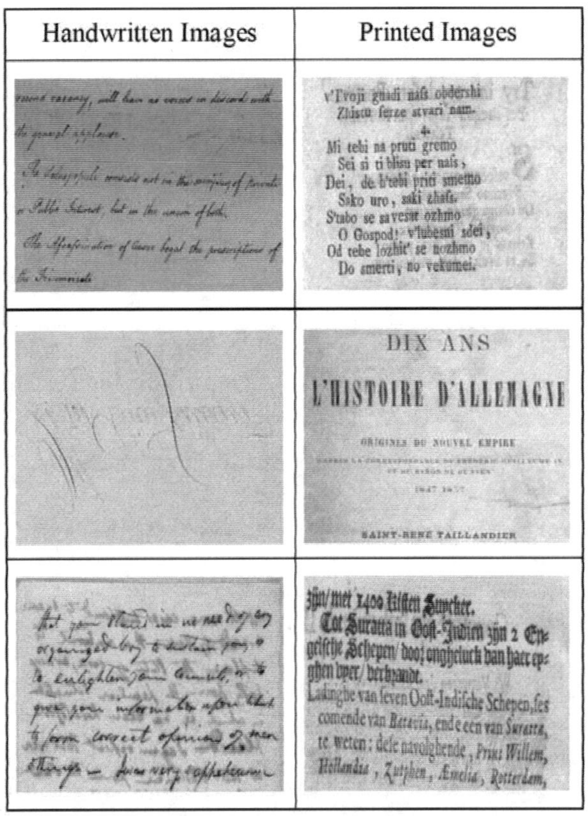

Figure 1: Sample image from DIBCO Database 2013

Table 1: Specification of the image database

| Images | Size (pixels) | Resolution (ppi) |
|---|---|---|
| HW01 | 4161 x 1094 | 72 |
| HW02 | 1136 x 559 | 72 |
| HW03 | 2290 x 504 | 72 |
| HW04 | 3297 x 1097 | 72 |
| HW05 | 1493 x 1613 | 72 |
| HW06 | 3088 x 1152 | 96 |
| HW07 | 1903 x 807 | 72 |
| HW08 | 2292 x 1082 | 96 |
| PR01 | 1557 x 965 | 72 |
| PR02 | 1485 x 847 | 72 |
| PR03 | 1192 x 956 | 72 |
| PR04 | 2287 x 870 | 72 |
| PR05 | 2251 x 429 | 72 |
| PR06 | 2275 x 737 | 72 |
| PR07 | 871 x 369 | 72 |
| PR08 | 1661 x 479 | 72 |

In 2013, Pratikakis et al. [21] organized Document Image Binarization Contest at 12th International Conference on Document Analysis and Recognition. The objective is to select the best algorithm based on 4 measurements calculation such as F-measure and PSNR. Eighteen (18) research groups have participated in the competition with twenty-three (23) distinct algorithms. Again, B.Su et al. National University of Singapore and the Institute for Infocomm Research, Singapore achieved the first rank Therefore, this method is an improvement from the method was proposed in 2009. They used the local maximum and minimum and local image gradient to obtained the local contrast [22,23]. Second rank belonging by Howe from Smith College, MA, USA. The main idea is using the combination of Laplacian operator and Canny edge detection. The technique is described in previous section [24,25]. Moghaddam et al. University of Quebec, Canada and Smith College, MA, USA proposed combination techniques between Laplacian energy binarization method introduced in [25] and the Ensemble-of-Expert (EoE) framework [26].

### *2.2 DIBCO 2014*

Refer on the http://users.iit.demokritos.gr/~bgat/HDIBCO2014/benchmark, the document database 2014 consists ten (10) handwritten image without any printed images. All images show the different pixel size but same resolution (96ppi). Figure 2 illustrated a few sample images from database 2014. Based on observation, the major problem of this database caused by ink fading and background artifact. The details specification was described in Table 2. The largest image is 'H03' (2675 x 1255) and the smallest image is 'H06' (775 x 460). All the image labeled from 'H01' to 'H10'.

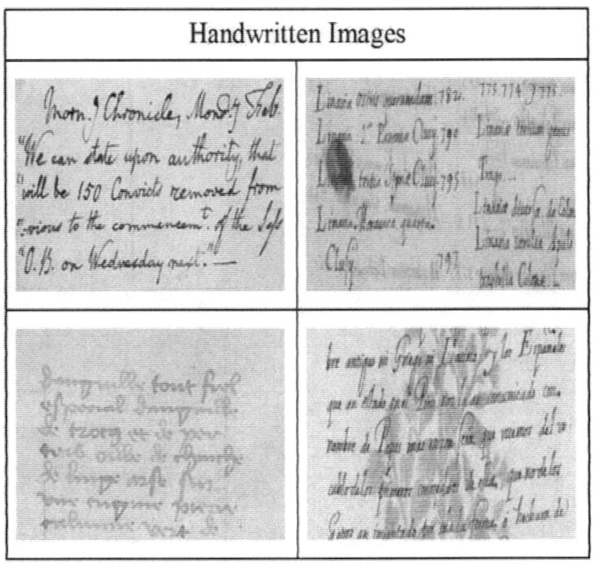


Figure 2. Image from DIBCO Database 2014

Table 2. Specification of the image database

| Images | Size (pixels) | Resolution (ppi) |
|---|---|---|
| H01 | 1761 x 707 | 96 |
| H02 | 1881 x 455 | 96 |
| H03 | 2675 x 1255 | 96 |
| H04 | 1105 x 339 | 96 |
| H05 | 1317 x 288 | 96 |
| H06 | 775 x 460 | 96 |
| H07 | 1449 x 463 | 96 |
| H08 | 1925 x 688 | 96 |
| H09 | 1474 x 482 | 96 |
| H10 | 1498 x 407 | 96 |

According to the Competition on Handwritten Document Image Binarization (H-DIBCO 2014), seven (7) distinct research groups have participated in the competition with eight (8) different algorithms since one participating research group has submitted two algorithms. Based on four evaluations such as F-measure, pseudo F-measure, Peak Signal Noise Ratio (PSNR) and Distance Reciprocal Distortion Metric (DRD), the winner is the method by Mesquita et al. [16] from Centro de Informática, Universidade Federal de Pernambuco, Brazil. This method based on human visual system perceives

IOP Conf. Series: Materials Science and Engineering **557** (2019) 012006 doi:10.1088/1757-899X/557/1/012006

and improved the Howe's binarization method [25], where Howe's method may fail in some situations in which its parametric choice is optimal for only a part of the image, as in cases of degradation generated by smudges or uneven illumination. Howe from Smith College, Department of Computer Science, Northampton (MA), USA achieved the second rank [25]. The main part is the canny threshold to separate the foreground and the background. The details technique can be referred to section 2.4. The third rank is Nafchi et al. [27] from Synchromedia Lab, Icole de technologies suprieure, University of Quebec, Montreal, Quebec, Canada. This method involved a few steps which are; (1) Denoised image with phase preserved, (2) Maximum moment of phase congruency covariance and (3) Locally weighted mean phase angle.

*2.3 DIBCO 2015*

Based on two established link for document database is http://vc.ee.duth.gr/ and http://users.iit.demokritos.gr/~bgat/ referred that no any document database provided in 2015. However, a few papers were published 2015 used an Indian document dataset [28,29]. In International Conference on Document Analysis and Recognition (ICDAR 2015) only used the datasets are generated from a set of historical manuscripts written between the 17th and 20th centuries, collected by Archives of Quebec (Canada) [30].

*2.4 DIBCO 2016*

Latest, the database 2016 can be accessed at http://vc.ee.duth.gr/h-dibco2016/benchmark. Database 2016 provided 10 handwritten images with different size and resolution. The image '7' show the lowest resolution which is 96ppi and the rest image is same resolution (300ppi). The problem of the images is the low quality background and fade ink as show in Figure 3. The specification details of the database are present in Table 3. Many researchers agreed that difficulties and challenges task in order to detect the text region in a non-uniform background condition [31,32].

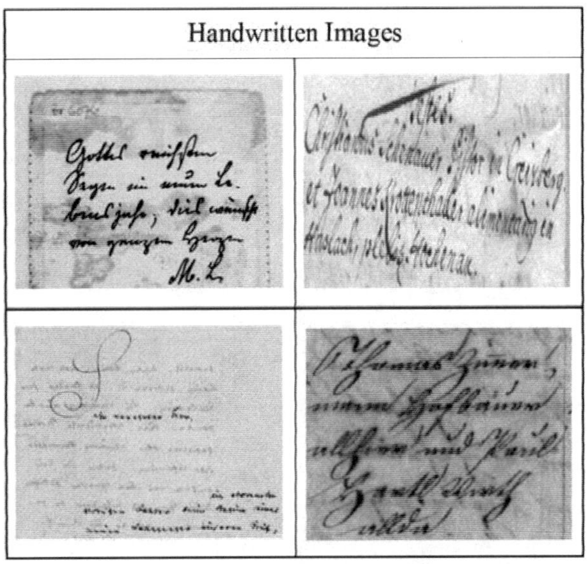


Figure 3. Image from DIBCO Database 2016

Table 3. Specification of the image database

| Images | Size (pixels) | Resolution (ppi) |
|---|---|---|
| 1 | 1510 x 1067 | 300 |
| 2 | 2259 x 1023 | 300 |
| 3 | 2417 x 1064 | 300 |
| 4 | 2363 x 615 | 300 |
| 5 | 2628 x 867 | 300 |
| 6 | 1364 x 788 | 300 |
| 7 | 963 x 656 | 96 |
| 8 | 1782 x 334 | 300 |
| 9 | 1339 x 302 | 300 |
| 10 | 378 x 315 | 300 |

In Handwritten Document Image Binarization Contest (H-DIBCO 2016), Pratikakis et al. [33] listed the ranking from nine (9) research groups with twelve (12) distinct algorithms. The winner is Nati Kligler and Ayellet Tal from Israel Institute of Technology, Israel. This method involved three stages; (1) pre-processing that used Creating the Visibility Score Map Hidden Point Removal (HPR) operator [34]. (2) Binarization – for the database 2016, they suggest to applied the binarization method presented by [25] and (3) Post-processing- use standard denoising. In this competition, Hassaine et al. from the University of Bordeaux, France and Qatar University submit are three methods where the 'Method 3' get the second place and 'Method 2' get the third place [35]. Method 2 based on Otsu binarization method [36], while Method 3 is the combination of Method 2 and the restoration of optical soundtracks [37].

*2.5 DIBCO 2017*

International Competition on Document Image Binarization (ICDAR 2017), eighteen (18) research groups have participated in the competition with twenty-six (26) different algorithms. The database of DIBCO 2017 contains two types which are; ten (10) handwritten images and ten (10) printed image. This database image can be reached at http://vc.ee.duth.gr/dibco2017/benchmark/. Figure 4 shows the example of an image in the DIBCO 2017 database. The database image present different of resolution (Pixels Per Inch) and the size of the pixel as shown in Table 4. The handwritten image labeled 'H' and the printed image 'P'.

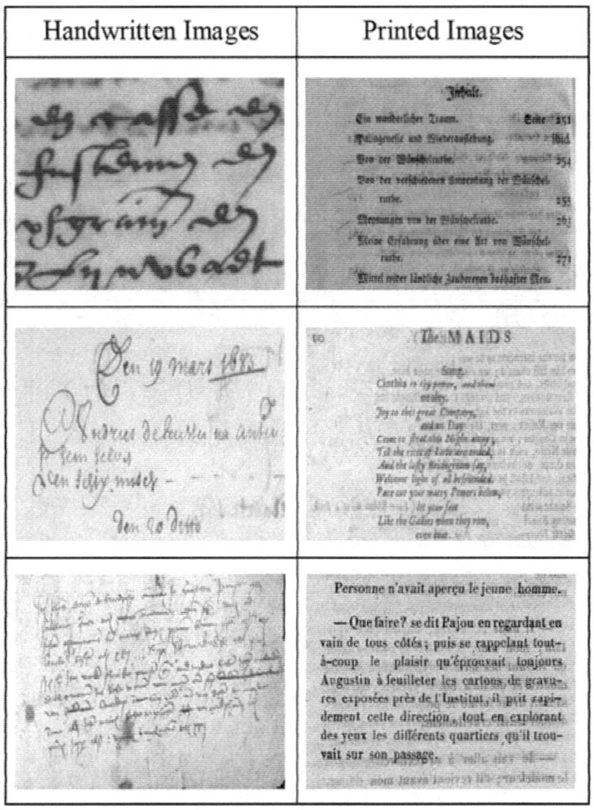


Figure 4. Image from DIBCO Database 2016

Table 4. Specification of the image database

| Images | Size (pixels) | Resolution (ppi) |
|---|---|---|
| H1 | 2397 x 1350 | 300 |
| H2 | 2095 x 1781 | 96 |
| H3 | 2379 x 1308 | 96 |
| H4 | 2439 x 1229 | 300 |
| H5 | 2291 x 874 | 300 |
| H6 | 351 x 292 | 300 |
| H7 | 593x 376 | 300 |
| H8 | 1303 x 594 | 300 |
| H9 | 1744 x 809 | 300 |
| H10 | 1721 x 924 | 300 |
| P1 | 2092 x 951 | 72 |
| P2 | 1083 x 877 | 300 |
| P3 | 1139 x 433 | 72 |
| P4 | 1225 x 739 | 300 |
| P5 | 1179 x 516 | 72 |
| P6 | 1553 x 2073 | 72 |
| P7 | 1233 x 2206 | 72 |
| P8 | 1050 x 608 | 300 |
| P9 | 1320 x 768 | 96 |
| P10 | 1075 x 775 | 300 |

IOP Conf. Series: Materials Science and Engineering **557** (2019) 012006 doi:10.1088/1757-899X/557/1/012006

Based on the contest at 14th IAPR International Conference on Document Analysis and Recognition [38], the first ranks winner comes from the Institute for Infocomm Research, Singapore. They proposed binarization techniques based on four steps; (1) document background extraction, (2) stroke edge detection, (3) local thresholding, and (4) post-processing [22]. Fabrizio and Marcotegui from Université Pierre, France achieved second place with finding a new algorithm using toggle mapping operator [9]. They focused on the morphological operator such as erosion and dilation in order to divide the input image into three classes; foreground, background and homogeneous. The third rank is a method proposed by Henault et al. based on local probabilistic models and the calculus of variation [39,40].

## 3. Conclusion

Binarization plays a vital role in the document recognition and analysis. The low quality image such as degraded and non-uniform background will effect on the post-processing stage. This study sets out to review in detail the available information on the document image database from the established website. The review focuses on DIBCO 2013 to DIBCO 2017 based on quantity, pixel size and resolution. The best binarization method for each database image also was discussed. The results of this review indicate that the binarization technique using background estimation and stroke edges is best and efficient for the overall database. The method using the combination of Laplacian operator and canny edge detection also successful result especially for the printed image. Further research might explore the comparison effect between handwritten and printed image on binarization field.

## 4. References

[1] Singh B M, Sharma R, Ghosh D and Mittal A 2014 Adaptive binarization of severely degraded and non-uniformly illuminated documents *Int. J. Doc. Anal. Recognit.* 393–412

[2] Mustafa W A and Yazid H 2016 Illumination and Contrast Correction Strategy using Bilateral Filtering and Binarization Comparison *J. Telecommun. Electron. Comput. Eng.* **8** 67–73

[3] Mustafa W A and Yazid H 2017 Image Enhancement Technique on Contrast Variation : A Comprehensive Review *J. Telecommun. Electron. Comput. Eng.* **9** 199–204

[4] Rivest-Hénault D, Farrahi Moghaddam R and Cheriet M 2012 A local linear level set method for the binarization of degraded historical document images *Int. J. Doc. Anal. Recognit.* **15** 101–24

[5] Mustafa W A and Kader M M M A 2018 Binarization of Document Images: A Comprehensive Review *J. Phys. Conf. Ser.* **1019** 1–9

[6] Mustafa W A, Khairunizam W, Ibrahim Z, Ab S and M.Razlan Z 2018 Improved Feng Binarization Based on Max-Mean Technique on Document Image *IEEE International Conference on Computational Approach in Smart Systems Design and Applications (ICASSDA)* (IEEE) pp 1–6

[7] Mustafa W A, Aziz H, Khairunizam W, Ibrahim Z, Ab S and M.Razlan Z 2018 Review of Different Binarization Approaches on Degraded Document Images *IEEE International Conference on Computational Approach in Smart Systems Design and Applications (ICASSDA)* (IEEE) pp 1–8

[8] Moghaddam R F and Cheriet M 2009 Low quality document image modeling and enhancement *Int. J. Doc. Anal. Recognit.* **11** 183–201

[9] Nina O, Morse B and Barrett W 2010 A Recursive Otsu Thresholding Method for Scanned Document Binarization *IEEE Workshop on Applications of Computer Vision (WACV)* pp 307–14

[10] Zhang Y and Wu L 2011 Fast Document Image Binarization Based on an Improved Adaptive Otsu ' s Method and Destination Word Accumulation *J. Comput. Inf. Syst.* **6** 1886–92

[11] Reza Farrahi Moghaddamn and Cheriet M 2012 AdOtsu: An adaptive and parameterless generalization of Otsu's method for document image binarization *Pattern Recognit.* **45** 2419–

31
[12] Ntirogiannis K, Gatos B and Pratikakis I 2013 Performance evaluation methodology for historical document image binarization *IEEE Trans. Image Process.* **22** 595–609
[13] Lokhande S S and Dawande N A 2015 A survey on document image binarization techniques *Proceedings - 1st International Conference on Computing, Communication, Control and Automation, ICCUBEA 2015* pp 742–6
[14] Ntirogiannis K, Gatos B and Pratikakis I 2014 A combined approach for the binarization of handwritten document images *Pattern Recognit. Lett.* **35** 3–15
[15] Wagdy M, Faye I and Rohaya D 2013 Fast and efficient document image clean up and binarization based on retinex theory *Proceedings - 2013 IEEE 9th International Colloquium on Signal Processing and its Applications, CSPA 2013* pp 58–62
[16] Mesquita R G, Mello C A B and Almeida L H E V 2014 A new thresholding algorithm for document images based on the perception of objects by distance *Integr. Comput. Aided. Eng.* **21** 133–46
[17] Soua M, Kachouri R and Akil M 2015 Improved Hybrid Binarization based on Kmeans for Heterogeneous document processing *9th International Symposium on Image and Signal Processing and Analysis, ISPA 2015* pp 210–5
[18] Mustafa W A and Kader M M M A 2018 Binarization of Document Image Using Optimum Threshold Modification *J. Phys. Conf. Ser.* 1**019** 1–8
[19] Mustafa W A, Yazid H and Jaafar M 2018 An Improved Sauvola Approach on Document Images Binarization *J. Telecommun. Electron. Comput. Eng.* **10** 43–50
[20] Mustafa W A and Kader M M M A 2018 Document Image Database ( 2009 - 2012 ): A Systematic Review *J. Phys. Conf. Ser.* **1019** 1–7
[21] Pratikakis I, Gatos B and Ntirogiannis K 2013 ICDAR 2013 document image binarization contest (DIBCO 2013) *Proceedings of the International Conference on Document Analysis and Recognition, ICDAR* pp 1471–6
[22] Lu S, Su B and Tan C L 2010 Document image binarization using background estimation and stroke edges *Int. J. Doc. Anal. Recognit.* **13** 303–14
[23] Su B, Lu S and Tan C L 2013 Robust document image binarization technique for degraded document images *IEEE Trans. Image Process.* **22** 1408–17
[24] Howe N R 2011 A Laplacian energy for document binarization *Proceedings of the International Conference on Document Analysis and Recognition, ICDAR* pp 6–10
[25] Howe N R 2013 Document binarization with automatic parameter tuning *Int. J. Doc. Anal. Recognit.* **16** 247–58
[26] Reza Farrahi Moghaddam, Fereydoun Farrahi Moghaddam M C 2013 Unsupervised ensemble of experts (EoE) framework for automatic binarization of document images *Comput. Vis. Pattern Recognit.* 703–7
[27] Ziaei Nafchi H, Farrahi Moghaddam R and Cheriet M 2013 Historical document binarization based on phase information of images *Lecture Notes in Computer Science (including subseries Lecture Notes in Artificial Intelligence and Lecture Notes in Bioinformatics)* vol 7729 LNCS pp 1–12
[28] Kavitha S, Shivakumara P, Hemantha Kumar G and Tan C L 2015 A robust script identification system for historical Indian document images *Malaysian J. Comput. Sci.* **28** 283–300
[29] Ahmadi E, Azimifar Z, Shams M, Famouri M and Shafiee M J 2015 Document image binarization using a discriminative structural classifier *Pattern Recognit. Lett.* **63** 36–42
[30] Hedjam R, Nafchi H Z, Moghaddam R F, Kalacska M and Cheriet M 2015 ICDAR 2015 MultiSpectral Text Extraction Contest *13th Int. Confrence Doc. Anal. Recognit. - ICDAR'15* 1181–5
[31] Gatos B, Pratikakis I and Perantonis S J 2004 An Adaptive Binarization Technique for Low Quality Historical Documents *Doc. Anal. Syst. VI* 102–13
[32] Pai Y-T, Chang Y-F and Ruan S-J 2010 Adaptive thresholding algorithm: Efficient computation technique based on intelligent block detection for degraded document images

*Pattern Recognit.* **43** 3177–87

[33] Pratikakis I, Zagoris K, Barlas G and Gatos B 2017 ICFHR 2016 handwritten document image binarization contest (H-DIBCO 2016) *Proc. Int. Conf. Front. Handwrit. Recognition, ICFHR* 619–23

[34] Katz S, Tal A and Basri R 2007 Direct visibility of point sets *ACM Trans. Graph.* **26** 24

[35] Maadeed S Al, Ayouby W, Hassaïne A and Aljaam J M 2012 QUWI: An Arabic and English handwriting dataset for offline writer identification *Proceedings - International Workshop on Frontiers in Handwriting Recognition, IWFHR* pp 746–51

[36] Hassaïne A, Al-Maadeed S and Bouridane A 2012 A set of geometrical features for writer identification *Lecture Notes in Computer Science (including subseries Lecture Notes in Artificial Intelligence and Lecture Notes in Bioinformatics)* vol 7667 LNCS pp 584–91

[37] Hassaïne A, Decencière E and Besserer B 2009 Efficient restoration of variable area soundtracks *Image Anal. Stereol.* **28** 113–9

[38] Pratikakis I, Zagoris K, Barlas G and Gatos B 2018 ICDAR2017 Competition on Document Image Binarization (DIBCO 2017) *Proc. Int. Conf. Doc. Anal. Recognition, ICDAR* **1** 1395–403

[39] Farrahi Moghaddam R and Cheriet M 2009 RSLDI: Restoration of single-sided low-quality document images *Pattern Recognit.* **42** 3355–64

[40] Rivest-Hénault D and Cheriet M 2007 Image Segmentation Using Level Set and Local Linear Approximations *Image Analysis and Recognition: 4th International Conference, ICIAR 2007, Montreal, Canada, August 22-24, 2007. Proceedings* ed M Kamel and A Campilho (Berlin, Heidelberg: Springer Berlin Heidelberg) pp 234–45

IOP Conf. Series: Materials Science and Engineering **557** (2019) 012007 doi:10.1088/1757-899X/557/1/012007

# Colour Component Analysis Approach for Malaria Parasites Detection Based on Thick Blood Smear Images

**T A Aris[1], A S A Nasir[1], Z Mohamed[2], H Jaafar[1], W A Mustafa[1], Wan Khairunizam[3], M A Jamlos[1], I Zunaidi[4], Z M Razlan[3] and Shahriman A B[3]**

[1]Faculty of Engineering Technology, Universiti Malaysia Perlis, UniCITI Alam Campus, Sungai Chuchuh, 02100 Padang Besar, Perlis, Malaysia

[2]Department of Microbiology & Parasitology, School of Medical Sciences, Health Campus, Universiti Sains Malaysia, 16150 Kubang Kerian, Kelantan, Malaysia

[3]Advanced Intelligent Computing and Sustainability (AICoS) Research Group, School of Mechatronics Engineering, Universiti Malaysia Perlis, 02600 Arau, Perlis, Malaysia

[4]Faculty of Technology, University of Sunderland, St Peter's Campus, Sunderland, SR6 0DD, United Kingdom

**Abstract.** Malaria is a plasmodium parasite disease that affects millions of people in the world every year. Hence, early detection tests are needed to prevent the malaria parasites spread throughout the body. For centuries, manual microscopic blood test remains as the most common method that still being used for malaria detection. However, this procedure contains the probability of miscalculation of parasites due to human error. Computerized system is recognized as a quick and easy ways to analyze a lot of blood samples images by providing direct visualization on the computer screen without the need to examine under the microscope. Therefore, this paper aims to analyze different colour components for improving the parasites counting performance based on thick blood smear images. In this study, five different colour spaces namely YCbCr, RGB, CMY, HSV and HSL have been analyzed and eight colour components which are Y, Cb, R, G, C, M, S and L have been extracted in order to determine which colour component is the best for malaria parasites counting. Overall, experimental results indicate that segmentation using Y component of YCbCr proved to be the best with average counting accuracy of 98.48% for 100 images of malaria thick blood smear.

## 1. Introduction

Malaria is one of the severe global contagious diseases caused by plasmodium parasites that transmitted through infected female Anopheles mosquito. In 2016, based on statistic by World Health Organization (WHO), there was a total of 216 million of malaria cases worldwide which causes 445 000 deaths [1]. Over time, malaria parasites can distort normal blood cells and normally, one to two weeks after a person is infected, the first symptoms of malaria appear such as fever, headache and vomiting. Microscopy blood test is well-known as the most common method used for malaria detection and consists of thick and thin blood smears test [2]. The thick blood smear test is used to find the level density of malaria parasites infection.

Digital image processing has been applied on medical field as it is not only the most efficient method, but also the cheapest. One of the examples is in application of image segmentation for malaria detection. The quality of segmentation is the importance element for the success of image analysis process. A number of researchers have proposed variety of image segmentation approach to improve malaria detection [3]. For example using the watershed distance transform [4], enhance *k*-means clustering [5], cascaded enhanced *k*-means and fuzzy *c*-means clustering [6], Poisson distribution thresholding [7], Otsu's thresholding [8] and mathematical morphology [9].

Arco *et al.* [10] worked on thick blood smear images and suggested adaptive thresholding method for classification based on pixel size. The system will classify either the pixel belongs to the background or to the parasites and white blood cells. Furthermore, it is only established the pixels around it and analyzed its neighbourhood. Then, morphological methods are applied to evaluate the area of connected components, labelling and counting the parasites. May *et al.* [11] worked on counting infected red blood cells in thin blood smears for malaria detection. First, the images were converted to CIELAB colour space. Next, Otsu's thresholding was used for conversion into binary image by calculated threshold value for each image. Dilation and erosion were performed for noise removal. Based on the previous studies, segmentation plays a major role in improving counting performance for malaria detection. Therefore, this study will utilize the potential of various colour components for segmentation process in order to improve counting performance of malaria parasites.

## 2. Methodology

In this study several image processing techniques have been implemented to obtain the segmented malaria parasites on thick blood smear images. The proposed procedures are summarized as follows:

**Step 1 :** Capture the malaria thick blood smear images.
**Step 2 :** Apply image enhancement technique namely modified global contrast stretching.
**Step 3 :** Apply colour conversion on enhanced images using YCbCr (luma, blue-difference chroma, red-difference chroma), RGB (red, green, blue), CMY (cyan, magenta, yellow), HSV (hue, saturation, value) and HSL (hue, saturation, luminance) colour spaces.
**Step 4 :** Apply Otsu's thresholding on Y, Cb, R, G, C, M, S and L component.
**Step 5 :** Apply watershed segmentation to separate the clumping parasites.
**Step 6 :** Apply noise removal to remove smaller and larger pixels than malaria parasites pixels.

### *2.1. Image acquisition*

The malaria thick blood smear samples were retrieved from Department of Microbiology and Parasitology, Hospital Universiti Sains Malaysia (HUSM). The blood smear samples have been examined using 100X magnification lens of computerized Leica DLMA microscope. In this study, 100 malaria images were captured with 800 x 600 resolutions and saved in bitmap (*.bmp) format.

### *2.2. Contrast enhancement technique using modified global contrast stretching (MGCS)*

In this study, the images are captured under low light condition, which is resulted as a low quality images. Therefore, image enhancement is needed to improve the quality of images for visual awareness. The images are enhanced by using MGCS technique to emphasize and sharpen the images quality. This technique use particular minimum and maximum values that lie in a specified percentage of pixels from the total number of pixels in the RGB image. The enhancement of malaria images depends directly on the minimum and maximum values that will be used during the contrast stretching process. Detail descriptions of the MGCS enhancement method can be found in [12].

### *2.3. Colour conversion on enhanced image using YCbCr, RGB, CMY, HSV and HSL colour spaces*

The malaria thick blood smear images are captured in RGB which are difficult for the segmentation process. Therefore, colour conversion is applied on enhanced images for easing the segmentation process and also as they can describe colour equivalent as how the human eye interpreted colour. The enhanced images are converted into five colour spaces namely YCbCr, RGB, CMY, HSV and HSL. Figure 1 shows the conversion of colour spaces into colour component.

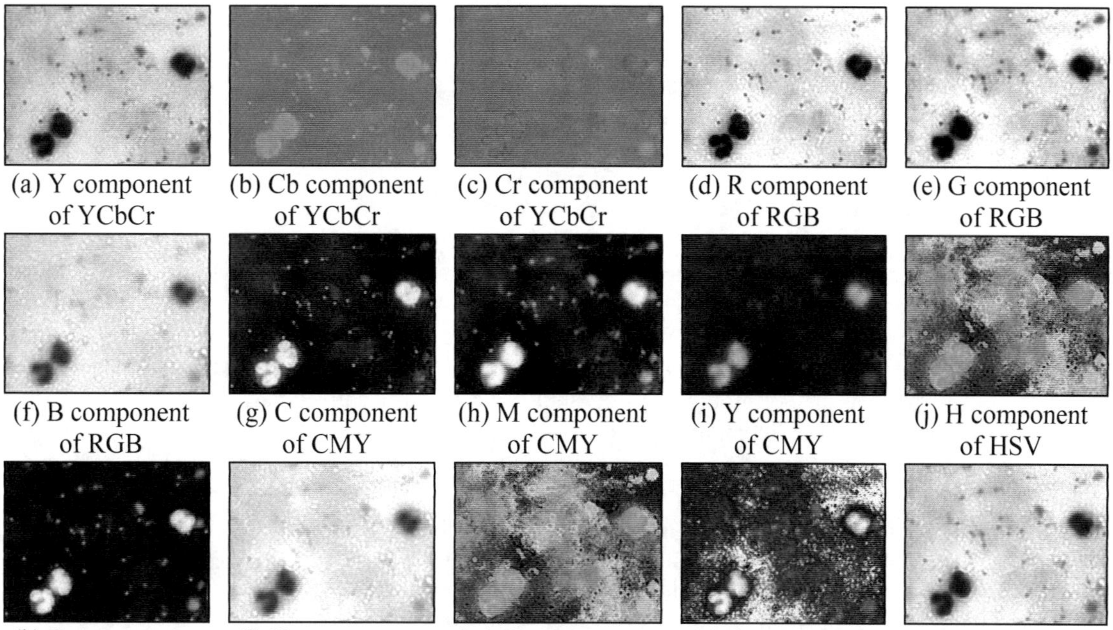

(a) Y component of YCbCr (b) Cb component of YCbCr (c) Cr component of YCbCr (d) R component of RGB (e) G component of RGB
(f) B component of RGB (g) C component of CMY (h) M component of CMY (i) Y component of CMY (j) H component of HSV
(k) S component of HSV (l) V component of HSV (m) H component of HSL (n) S component of HSL (o) L component of HSL

**Figure 1.** Conversion of YCbCr, RGB, CMY, HSV and HSL colour spaces into colour component

Apparently eight colour components have been chosen to be segmented, which are Y and Cb components from YCbCr colour spaces, R and G components from RGB colour spaces, C and M components from CMY colour spaces, S component from HSV colour spaces and L component from HSL colour spaces. These colour components are chosen as they shows better contrast between foreground and the background as can be seen in figure 1. The conversion from RGB images to five different colour spaces will be interpreted as the following equation [13]:

$$Y = 0.299R + 0.587G + 0.114B \quad (1)$$

$$Cb = 128 - 0.169R - 0.331G + 0.511B \quad (2)$$

$$C = 1 - R \quad (3)$$

$$M = 1 - G \quad (4)$$

$$S = 1 - \frac{3}{R+G+B}\min(R,G,B) \quad (5)$$

$$L = \frac{(max_{(R,G,B)} + min_{(R,G,B)})}{2} \quad (6)$$

*2.4. Otsu's thresholding on Y, Cb, R, G, C, M, S and L colour component*

After applying the colour conversion on enhanced images, Otsu's thresholding method is proceeding on the eight selected colour components. Otsu's thresholding is used to conceive a binary image from the grey level image by finding the maximum variance between the classes [14]. Basic principle of Otsu's thresholding is to split the image into two classes in term of foreground and background is called as bi-level thresholding (BT). However, Otsu's thresholding can also split the images into more than two classes, and it is called as multilevel thresholding (MT). Multilevel thresholding segments a gray level image pixel into several distinct classes [14]. In this study, Y, Cb, R, G, C and M component are segmented into three classes, which are for malaria parasites, white blood cells (WBCs) and the background regions. Meanwhile, S and L component are segmented into two classes, which are for malaria parasites and the background regions. Nonetheless, the thresholding value is differs due to different colour component is applied.

*2.5. Separation of clumping parasites by using watershed segmentation with distance transform*

The malaria parasites are close to each other and some of them are overlapping. These overlapping malaria parasites might affect the performance of accuracy counting. Therefore watershed with distance transform is used to separate the clumped parasites. Euclidean distance transform calculates the difference between the pixel and the nearest non-zero pixel to separate each connected parasites. The detail descriptions of this method can be found in [15].

*2.6. Removing the lesser and greater pixels than malaria parasites pixels size*

After the segmentation process is done, some noise appears on the image. An arithmetic operation namely area opening is used to remove the noise. This operation allows the eliminating of the unwanted object that is smaller and larger than the structuring element of the desired object. Thus, any regions which are lesser or greater than malaria parasites pixels are considered as non-malarias and will be get rid from the image through area opening process. After all, the size of malaria parasites pixels also differ due to different colour component is used.

*2.7. Evaluation of malaria parasites counting*

Next, malaria parasites counting process is carried out to verify the accuracy of counting performance by using difference colour component. The parasite counting is performed by using the region growing algorithm. This algorithm labels the malaria parasites according to their order in the malaria image and then determines the total number of the parasites based on the labelled image. The accuracy of the counted malaria parasites is determined based on the average counting percentage in order to evaluate the performance of each colour component to produce a good counting analysis by using the following equation:

$$Accuracy = \left(1 - \frac{Manual\_Count - System\_Count}{Manual\_Count}\right) x100\% \quad (7)$$

## 3. Result and Discussions

In this study, the proposed image processing procedure have been employed and analyzed on 100 malaria thick blood smear images. Firstly, the malaria images are enhanced by using MGCS since they are low quality images. Figure 2 shows the original malaria images and the enhanced images after applying the MGCS technique.

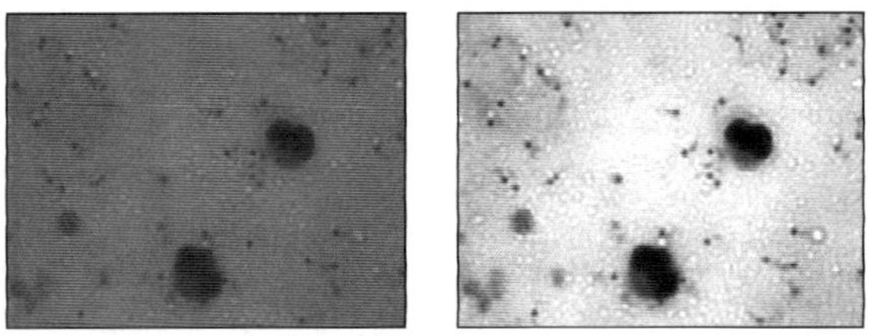

(a) Original image (b) Enhanced image

**Figure 2.** Original images of malaria thick blood smear and enhanced images after applying MGCS technique.

From the enhanced image, colour conversion has been applied using YCbCr, RGB, CMY, HSV and HSL colour spaces for easing segmentation process. Y, Cb, R, G, C, M, S and L colour component have been selected to be segment since its shows a good distinct between malaria parasites on foreground and the background of the images. Figure 3 shows the differences between colour components.

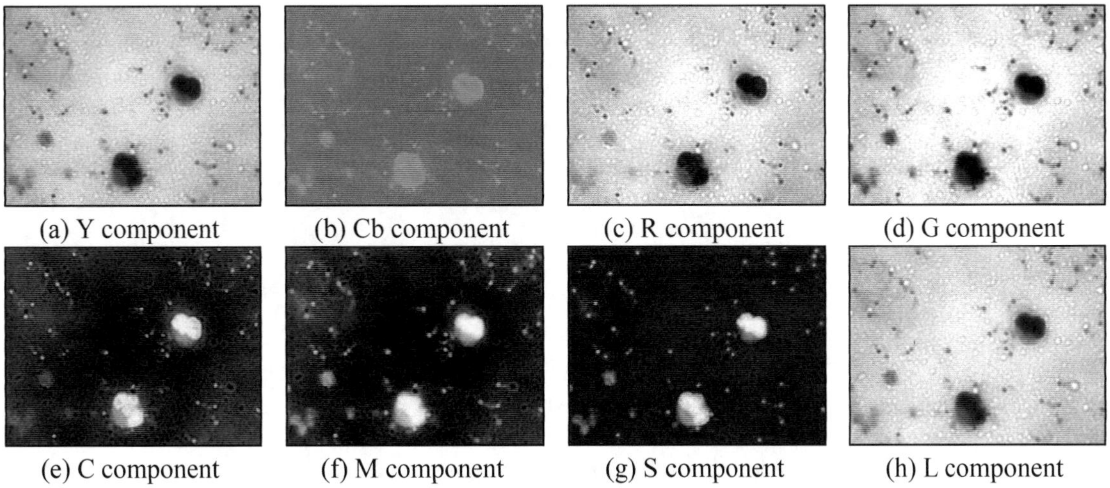

(a) Y component (b) Cb component (c) R component (d) G component

(e) C component (f) M component (g) S component (h) L component

**Figure 3.** The differences between colour components.

Next, Otsu's thresholding has been applied to segment the malaria parasites from the background region. The clumping malaria parasites have been separated by using watershed segmentation with euclidean distance transform to obtain accurate result during counting process. Finally, the smaller and larger pixels that lie in between malaria parasite pixels have been removed through area opening technique from arithmetic operation and noise removal of image border has been done to obtain a clean segmented image and improve counting accuracy. The final results obtain after performing Otsu's thresholding, watershed segmentation, and noise removal process are presented in figure 4.

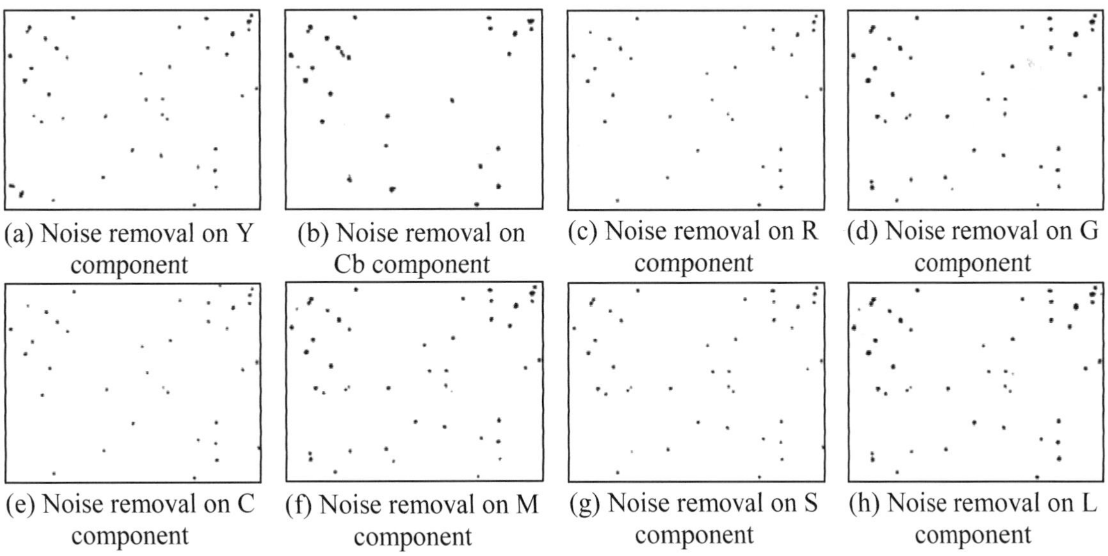

(a) Noise removal on Y component (b) Noise removal on Cb component (c) Noise removal on R component (d) Noise removal on G component

(e) Noise removal on C component (f) Noise removal on M component (g) Noise removal on S component (h) Noise removal on L component

**Figure 4.** Final results obtain after performing Otsu's thresholding, watershed segmentation, and noise removal processes.

The counting process of malaria parasite has been carried on to certify the performance of different colour component. Based on average of malaria parasites count obtained from table 1, the Y component of YCbCr colour spaces has provide a better counting performance with average accuracy of 98.48% for 100 malaria images as compared to the other colour components.

**Table 1.** Average malaria parasites count for 100 images.

| Details | Manual Counting | YCbCr [Y] | YCbCr [Cb] | RGB [R] | RGB [G] | CMY [C] | CMY [M] | HSV [S] | HSL [L] |
|---|---|---|---|---|---|---|---|---|---|
| Total Parasites | 3936 | 3876 | 3761 | 4199 | 3674 | 4243 | 3997 | 4112 | 3751 |
| Accuracy (%) | - | **98.48** | 95.55 | 93.32 | 93.34 | 92.20 | 98.45 | 95.53 | 95.30 |

## 4. Conclusion

In this paper, the results of average counting of malaria parasites by applying eight types of colour components have been displayed. Comparisons between Y, Cb, R, G, C, M, S and L colour components extracted from YCbCr, RGB, CMY, HSV and HSL colour spaces have been operated as input images to Otsu's thresholding in order to identify the best colour component for segmentation of malaria image for counting of malaria parasites process. The proposed segmentation techniques have been applied on 100 malaria thick blood smear images. Based on the quantitative findings, the results indicate that segmentation using Y component of YCbCr colour spaces has proven to be the best colour component for segmenting malaria image with average counting of 98.48%.

## 5. References

[1] World Health Organization 2017 World Malaria Report 2017

[2] Mathison B A and Pritt B S 2017. Update on malaria diagnostics and test utilization *Journal of clinical microbiology* **55**

[3] Poostchi M, Silamut K, Maude R, Jaeger S and Thoma 2018 Image analysis and machine learning for detecting malaria *Translational Research* **194**

[4] Salamah U, Sarno R, Arifin A Z, Sarimuddin S, Nugroho A S, Rozi I E and Asih P B S 2018 Segmentation of Malaria Parasite Candidate from Thickblood Smear Microscopic Images using Watershed and Adaptive Thresholding *Journal of Telecommunication, Electronic and Computer Engineering (JTEC)* **10**

[5] Nasir A S A, Mashor M Y and Mohamed Z 2018 Enhanced k-means clustering algorithm for malaria image segmentation *Journal of Advanced Research in Fluid Mechanics and Thermal Sciences* **42**

[6] Abdul Nasir A S, Jaafar H, Wan Mustafa W A and Mohamed Z The Cascaded 2018 Enhanced k-Means and Fuzzy c-Means Clustering Algorithms for Automated Segmentation of Malaria Parasites *MATEC Web of Conferences* **150**

[7] Suryawanshi M S, Dixit V 2013 Improved technique for detection of malaria parasites within the blood cell images *Int J Sci Eng Res* **4**

[8] Sulur K M, Nasir A S A, Mustafa W A, Jaafar H and Mohamed Z 2017 Analysis of Colour Constancy Algorithms for Improving Segmentation of Malaria Images *Journal of Telecommunication, Electronic and Computer Engineering* **10**

[9] Loddo A, Di Ruberto C and Kocher M (2018) Recent advances of malaria parasites detection systems based on mathematical morphology *Sensors* **18**

[10] Arco J, Gorriz J, Ramirez J, Alvarez I and Puntonet C 2014 Digital image analysis for automatic enumeration of malaria parasites using morphological operations *Expert Syst Appl* **42**

[11] May Z, Aziz S A M and Salamat R A 2013 Automated Quantification and Classification of Malaria Parasites in Thin Blood Smears *International Conference on Signal and Image Processing Applications (ICSIPA)*

[12] Abdul-Nasir A S, Mashor M Y and Mohamed Z 2012 Modified global and modified linear contrast stretching algorithms: New colour contrast enhancement techniques for microscopic analysis of malaria slide images *Computational and Mathematical Methods in Medicine*

[13] Nishad P M 2013 Various colour spaces and colour space conversion *Journal of Global Research in Computer Science* **4**
[14] Kotte S, Kumar P R and Injeti S K 2016 An efficient approach for optimal multilevel thresholding selection for gray scale images based on improved differential search algorithm *Ain Shams Engineering Journal*
[15] Aris T A, Nasir A S A and Mustafa W A 2017 Analysis of Distance Transforms for Watershed Segmentation on Chronic Leukaemia Images *Journal of Telecommunication, Electronic and Computer Engineering* **10**

**Acknowledgments**
The authors would like to express their gratitude to Hospital Universiti Sains Malaysia (HUSM) for providing the malaria thick blood smear samples.

# Image Segmentation Approach for Acute and Chronic Leukaemia Based on Blood Sample Images

**N A A Khairudin[1], F N M Ariff[1], A S A Nasir[1], W A Mustafa[1], Wan Khairunizam[2], M A Jamlos[1], I Zunaidi[3], Z M Razlan[2] and Shahriman A B[2]**

[1]Faculty of Engineering Technology, Universiti Malaysia Perlis, UniCITI Alam Campus, Sungai Chuchuh, 02100 Padang Besar, Perlis, Malaysia

[2]Advanced Intelligent Computing and Sustainability (AICoS) Research Group, School of Mechatronics Engineering, Universiti Malaysia Perlis, 02600 Arau, Perlis, Malaysia

[3]Faculty of Technology, University of Sunderland, St Peter's Campus, Sunderland, SR6 0DD, United Kingdom

Corresponding author: hanisayunie@gmail.com

**Abstract.** The uncontrolled development of abnormal white blood cells (blast) in bone marrow is the starting point where leukaemia cancer begins. It does not die like it should. Instead, it goes on dividing and crowding out the healthy blood cells, making it difficult for these healthy bloods to function normally. The diagnosis process from haematologist consumes a lot of time. Therefore, a good segmentation procedure is necessary in order to successfully identify and diagnose leukaemia cells automatically from blood sample images. This paper proposes segmentation procedures which consist of contrast enhancement, extraction of hue component information, as well as segmentation based on Otsu's thresholding and watershed technique. The experimental results shows that the proposed segmentation procedure has successfully segmented 200 images consisting of acute and chronic leukaemia with average accuracy, sensitivity and specificity obtained of 98.90%, 82.14% and 99.49%, respectively. The result of segmentation performance achieved shows the significant of this approach. Based on results obtained, this segmentation technique is applicable to segment both acute and chronic leukaemia images with good segmentation performance.

## 1. Introduction

Leukaemia can be classified into acute and chronic leukaemia based on how quickly the disease develops. Acute leukaemia occurs in a short period of time while chronic leukaemia takes time to fully develop. Besides, leukaemia cells can spread to other body part, such as the spleen, lymph nodes and brain. Acute leukaemia can be divided into two classes which are acute lymphocytic leukaemia (ALL) and acute myeloid leukaemia (AML), while chronic leukaemia can be divided into chronic lymphocytic leukaemia (CLL) and chronic myeloid leukaemia (CML). Leukaemia recognition starts by a complete blood count (CBC) process. If any abnormality appears, a study of morphological bone marrow smear analysis is arranged to check the present of a leukaemia cell [1]. Problem starts when some haematologists still conduct the complete blood count process manually using light microscope which may cause their eyes get tired when concentrating at the blood sample too long [2]. As a result, the procedure to obtain the outcome is slow and the accuracy obtained is not standardized [3].

These limitations trigger the development of digital image processing for blood cells recognition through computer algorithms and several processes [4]-[9]. Chen *et al.* [4] have measured the difference between the performances of watershed segmentation with different distance transforms for binary image. A successful segmentation result obtained from combination of watershed and chessboard distance transforms. Meanwhile, Mohammed *et al.* [5] have suggested watershed algorithm and optimal threshold to segment CLL and the proposed method has produced 99.92% for nucleus segmentation performance.

Image segmentation on white blood cells (WBCs) based on global Otsu's thresholding has been used for acute leukaemia detection based on blood sample images [6]. The proposed method has produced accuracy of 93% when applied onto 108 images. Prasad *et al.* [7] generated an automatic segmentation by using watershed segmentation, threshold and chessboard distance measurement on blood stream images. A minimum counting error reached +0.5% for 200 objects in the dataset to count the blood cells by automatic segmentation that consist three colour images of blood cells.

Image segmentation is an important technique that is required in accomplishes leukaemia diagnosis through digital image processing. This paper aims to improvise the segmentation technique in order to increase the segmentation performance of leukaemia diagnosis process.

## 2. Methodology

In this paper, image processing techniques are separated into two parts which are pre-processing and image segmentation parts. In pre-processing, contrast enhancement and colour component from HSV (Hue, Saturation, Value) colour space are applied on acute and chronic leukaemia images. As for the image segmentation part, Otsu's thresholding and watershed algorithms are used to segment the abnormal white blood cells (blast) in leukaemia images. Preparations are taken to separate the blast from other components in blood image such as platelets, red blood cells (RBCs) and background region. Starting with preparation of blood slide, it will undergo a few processes until the blast are segmented. Then, the segmentation performance for each type of leukaemia will be analysed.

### *2. 1. Image acquisition*

The blood samples of ALL, AML, CLL and CML were provided by Hospital Universiti Sains Malaysia (HUSM). These blood samples were prepared on slides and being analysed at 40X magnification by using a Leica DLMA digital microscope. The staining process helps to get a clear vision of the blast cell. In this study, a total of 200 leukaemia images were captured by a resolution setting of 808x608 and saved in bitmap (*.bmp) format.

### *2. 2. Image processing*

In this study, several image processing techniques have been implemented to obtain the segmented blast on leukaemia images. The proposed procedures are summarized as follows:

**Step 1 :** Original image undergoes enhancement by using linear contrast stretching (LCS) technique to increase the quality of the image.
**Step 2 :** Extract the hue component information of HSV colour space.
**Step 3 :** Apply the image segmentation technique by using Otsu's thresholding to segment the blast.
**Step 4 :** Remove noises and unwanted region from the target image in order to clean the image.
**Step 5 :** Separate the touching cells by using watershed segmentation.

#### *2. 2. 1. Contrast enhancement using LCS technique for leukaemia image*

Linear contrast stretching (LCS) technique is carry out in order to improve the brightness level and contrast level of the image because some images are blur. This technique is based on the original brightness and the contrast level of images to be adjusted [10]. The algorithm of LCS is defined in equation (1).

$$outRGB(x,y) = 255 * \left[\frac{(inRGB(x,y) - aRGB)}{bRGB - aRGB}\right] \quad (1)$$

Where $in_{RGB}(x, y)$ is the pixel value of original RGB, $out_{RGB}(x, y)$ is the pixel value of a new RGB, $a_{RGB}$ is minimum RGB value and $b_{RGB}$ is maximum RGB value.

#### *2. 2. 2. HSV colour space on the enhanced image*

HSV colour space mostly used to improve the RGB colour space. Hue corresponds directly to the concept of hue in the colour basics section [11]. The advantage of HSV is it has a simple conceptual which each of its element attributes directly correspond to the basic colour models. The disadvantage is saturation attributes which correspond to the mixture of a colour with white (tinting), so desaturation colours increase the total intensity. Based on analysis of leukaemia images, it has been found that hue component in HSV colour space covers most information of the blast. Thus, the procedure is developed by using hue component according to the following equations [12]:

$$Hue = \begin{cases} \theta & if\ B \leq G \\ 360° - \theta & if\ B > G \end{cases} \tag{2}$$

$$\theta = cos^{-1}\left\{\frac{1/2\ [(R-G)+(R-B)])}{[(R-G)^\wedge 2+\ (R-B)(G-B)]^\wedge(1/2)}\right\} \tag{3}$$

*2. 2. 3 Segmentation of blast using Otsu's Thresholding*

Otsu's thresholding [13] helps minimize variance within class and perform the image in a binary image. The optimum threshold is calculated to separate the image that contains two classes of pixels so that their combined spread become equivalents or minimal in order for inter-class variance is maximal. Equations (4) and (5) show the algorithms of class probability. Weights $\omega_0$ and $\omega_1$ are the possibilities of the two classes differentiate by a threshold *t* and $\sigma_0^2$ and $\sigma_1^2$ are variances of these two classes. The class possibility $\omega_{0,1}$ (t) is computed from the histograms.

$$\omega_0(t) = \sum_{i=0}^{t=1} p(i) \tag{4}$$

$$\omega_1(t) = \sum_{i=1}^{L=1} p(i) \tag{5}$$

*2. 2. 4 Separating overlapping cells using watershed segmentation process*

In order to separate a joint blast on the image, watershed segmentation is used. The contours are calculated correspond to the blast to the catchment basins and watershed lines by them. The gradient makeover of the original image often suits this condition. The morphological gradient based on 3x3 square structuring element of the initial grey-level image is defined as the difference between the dilation and the erosion of the grey value image [14]. The detail descriptions of this method can be found in [15].

*2.3 Segmentation performance for the proposed method*

Segmentation performance acts as a bench mark for the competency of the proposed method. It can be classified based on accuracy, specificity and sensitivity of the segmented region. The quality of the segmented image is performed based on similarity of pixels by comparing the results of segmented image against the manual segmented image. Accuracy, specificity and sensitivity [16] are obtained based on the following equations:

$$Accuracy = \frac{TP+TN}{TP+TN+\ FP+FN}\ X\ 100 \tag{6}$$

$$Specificity = \frac{TN}{TN+FP}\ X\ 100 \tag{7}$$

$$Sensitivity = \frac{TP}{TP+FN}\ X\ 100 \tag{8}$$

Accuracy is the percentage of pixels that is correctly classified to the entire area of the cell. Sensitivity is a true positive measure in which refers to the amount of images that contain a cancerous masses which have been classified correctly. Specificity is a true negative measure that refers to the proportion of images containing a cancerous masses that have been incorrectly classified. TP is for true positive, TN is for true negative, FP is for false positive and FN is for false negative [16].

## 3. Results and Discussions

In this research, the proposed segmentation procedures for both acute and chronic leukaemia images have been tested on 50 images of ALL, AML, CLL and CML. The results of pre-processing process are shown in figure 1. Starting with original image for each class which states by images (a)-(d), the brightness level and contrast level of blast are increased through LCS process as shown in images (e)-(h). LCS is reflected to each range of RGB colour space that appears in the image. The range of each colour space is used for contrast stretching process to represent each range of colour and set the minimum and maximum values of each colour space.

Hue component covers most of the information of blast cell. After the hue component has been applied on LCS image, the blast appeared as a brighter part and the unwanted region appeared as a darker part and almost similar with the background image. As a result, the blast can easily be seen after applying hue component as presented in images (i)-(l). Based on the observation, the shapes of the blast in AML images are almost unchanged which may enhance in next segmentation process. Meanwhile, the hue component is only able to highlight the nucleus area of the blast in ALL, CLL and CML images, which results on the missing cytoplasm area of the blast.

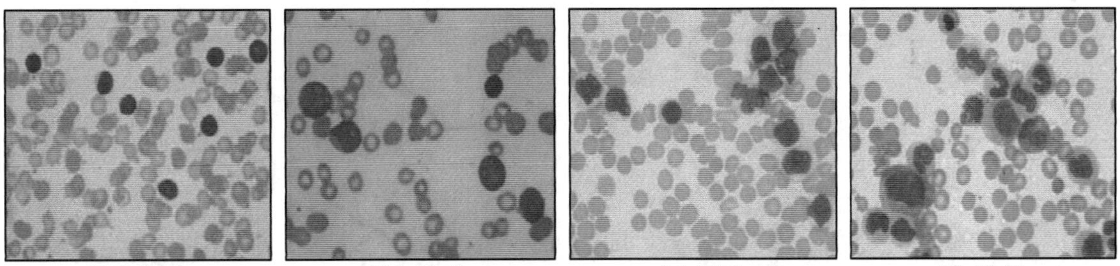

(a) ALL (b) AML (c) CLL (d) CML

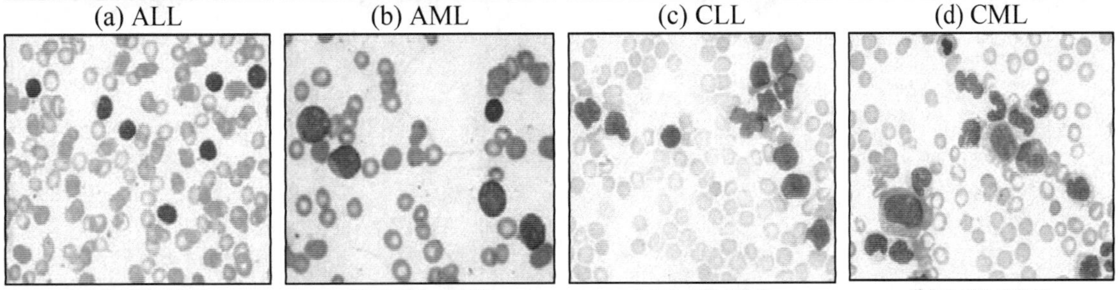

(e) LCS ALL (f) LCS AML (g) LCS CLL (h) LCS CML

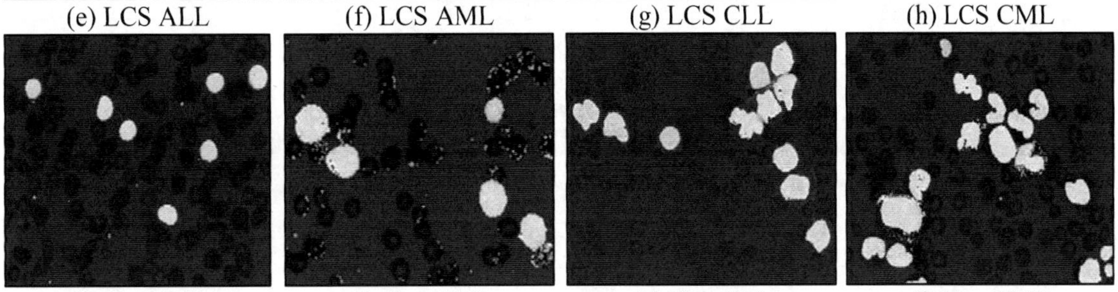

(i) Hue ALL (j) Hue AML (k) Hue CLL (l) Hue CML

**Figure 1.** Original images and results of images after the pre-processing process.

Later, the blast images have been segmented using Otsu's thresholding technique. By using Otsu's thresholding technique, the blasts have been segmented from the RBCs and background regions which results on the form of binary image as shown in figure 2. From these images, the desired segmented objects are represented by white colour, while the rest has been represented by black. The unwanted regions are still appeared on the segmented image but in the forms of small pixels region. Therefore, image filtering and object remover have been applied to smoother the image and to remove the unwanted regions from the images. Finally, the targeted images have been segmented through watershed process, in order to separate the touching cells. The final segmented images obtained after separating the touching cells are presented in figure 3. Qualitatively, the proposed segmentation approach has successfully produced fully segmented blasts for AML images. However, this approaches only suitable for segmenting the nucleus in ALL, CLL and CML images.

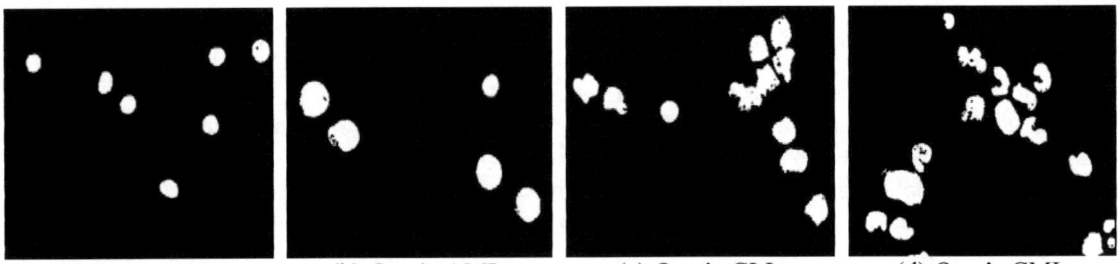

(a) Otsu's ALL (b) Otsu's AML (c) Otsu's CLL (d) Otsu's CML

**Figure 2**. Results of segmentation using Otsu's thresholding on hue images.

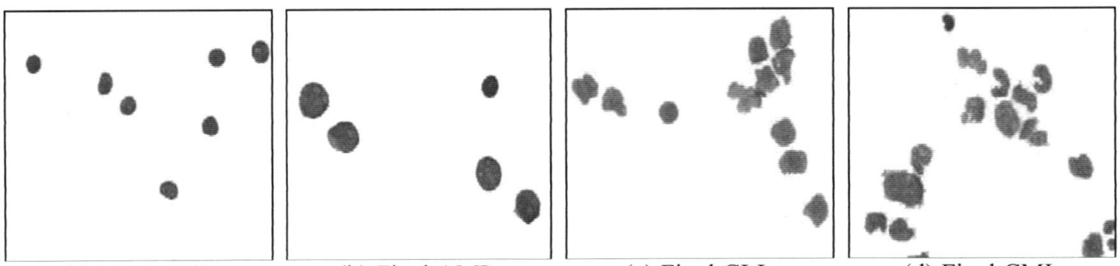

(a) Final ALL (b) Final AML (c) Final CLL (d) Final CML

**Figure 3**. The final segmented images after separation of touching cells using watershed.

**Table 1**. Leukaemia segmentation performance.

| Class of Leukaemia | Accuracy | Sensitivity | Specificity |
|---|---|---|---|
| ALL | 99.54% | 78.60% | 98.63% |
| AML | 99.37% | 83.55% | 100% |
| CLL | 98.98% | 81.40% | 99.99% |
| CML | 97.71% | 85.00% | 99.33% |
| Total 200 images | 98.90% | 82.14% | 99.49% |

Segmentation performance is obtained from the comparison result of segmented image with manual segmented image. The results of segmentation performance based on accuracy, sensitivity and specificity percentage is tabulated in table 1. The results show that the proposed segmentation approach gave a satisfied result with the highest value on specificity with 99.49%, followed by accuracy of 98.90% and sensitivity of 82.14%.

## 4. Conclusion

The results show accomplishment in improvising segmentation technique for leukaemia images through the segmentation performance. Linear contrast enhancement and hue colour component are able to highlight the targeted image. Meanwhile, each class of leukaemia cell is able to fully segment and separated by using Otsu's thresholding technique combined with watershed segmentation. Overall, the proposed segmentation procedure has produced segmentation performance with average accuracy, sensitivity and specificity obtained of 98.90%, 82.14% and 99.49%, respectively. The percentage results of segmentation performance achieved concluded that proposed segmentation procedure is applicable for segmentation both acute and chronic leukaemia images.

**References**

[1] Abdul Nasir A S, Mashor M Y and Rosline H 2011 Detection of acute leukaemia cells using variety of features and neural networks *The 5$^{th}$ Kuala Lumpur International Conference on Biomedical Engineering (BioMed 2011)*
[2] Chung J, Ou X, Kulkarni R P and Yang C 2015 Counting white blood cells from a blood smear using fourier ptychographic microscopy *PloS One **10***
[3] Meera V and Mathew S A 2014 Fuzzy Local Information C Means Clustering For Acute Myelogenous Leukaemia Image Segmentation *Proceeding of the International Conference on Innovation and Advances in Science Engineering and Technology*
[4] Chen Q, Yang X and Petriu E M 2004 Watershed Segmentation for Binary Images with Different Distance Transforms *Proceedings of the 3rd IEEE International Workshop on Haptic, Audio and Visual Environments and Their Applications*
[5] Mohammed E A, Mohamed M M A, Naugler C and Far B H 2013 Chronic lymphocytic leukaemia cell segmentation from microscopic blood images using watershed algorithm and optimal thresholding *2013 26th Annual IEEE Canadian Conference on Electrical and Computer Engineering (CCECE)*
[6] Joshi M D, Karode A H and Suralkar S R 2013 White Blood Cells Segmentation and Classification to Detect Acute Leukaemia *International Journal of Emerging Trends & Technology in Computer Science (IJETTCS)* **2**
[7] Prasad A S, Latha K S and Rao S K 2013 Separation and counting of blood cells using geometrical features and distance transformed watershed *International Journal of Engineering and Innovative Technology (IJEIT)* **3**
[8] Halim N H A, Mashor M Y, Nasir A S A and Hassan R 2018 The Effectiveness of Colour Constancy on Segmenting Leukemia Cells Using Unsupervised Clustering Technique *Journal of Physics: Conference Series* **1019**
[9] Rodellar J, Alférez S, Acevedo A, Molina A and Merino A 2018 Image processing and machine learning in the morphological analysis of blood cells *International journal of laboratory hematology* **40**
[10] Abdul-Nasir A S, Mashor M Y and Mohamed Z 2012 Modified global and modified linear contrast stretching algorithms: New colour contrast enhancement techniques for microscopic analysis of malaria slide images *Computational and Mathematical Methods in Medicine*
[11] Sharma B and Nayyer R 2016 Use and Analysis of Color Models in Image Processing *Journal of Food Processing & Technology*
[12] Gonzalez R C and Woods R E 2007 *Digital Image Processing* (Pearson)
[13] Otsu N 1979 A threshold selection method from gray-level histograms *IEEE transactions on systems, man, and cybernetics 9*
[14] Malpica N, de Solorzano C O, Vaquero J J, Santos A, Vallcorba I, García- Sagredo J M and Del Pozo F 1997 Applying watershed algorithms to the segmentation of clustered nuclei *Cytometry* **28**
[15] Aris T A, Nasir A S A and Mustafa W A 2017 Analysis of Distance Transforms for Watershed Segmentation on Chronic Leukaemia Images *Journal of Telecommunication, Electronic and Computer Engineering* **10**
[16] Kourou K, Exarchos T P, Exarchos K P, Karamouzis M V and Fotiadis D I 2015 Machine learning applications in cancer prognosis and prediction *Computational and structural biotechnology journal **13***

**Acknowledgements**
The authors would like to acknowledge Hospital Universiti Sains Malaysia (HUSM) for their helps in providing the leukaemia blood samples.

# Blood Vessel Extraction Using Combination of Kirsch's Templates and Fuzzy C-Means (FCM) on Retinal Images

**Wan Azani Mustafa[1,*], Ahmad Syauqi Mahmud[1], Wan Khairunizam[2], Z M Razlan[2], A B Shahriman[2], and I Zunaidi[3]**

[1] Department of Electrical Technology Engineering, Faculty of Engineering Technology, Universiti Malaysia Perlis, UniCITI Alam Campus, Sungai Chuchuh, 02100 Padang Besar, Perlis, Malaysia

[2] School of Mechatronics Engineering, Universiti Malaysia Perlis, 02600 Arau, Malaysia

[3]Faculty of Technology, University of Sunderland, St Peter's Campus, Sunderland, SR6 0DD, United Kingdom

*E-mail:wanazani@unimap.edu.my*

**Abstract.** Disease diagnosis based on retinal image analysis is very popular in order to detect a few critical diseases such as diabetic retinopathy, high blood pressure, cancer and glaucoma. The important part in the retinal is a blood vessel. Besides, the blood vessel study plays an important part in different medical areas such as ophthalmology, oncology, and neurosurgery. The significance of the vessel analysis was helped by the continuous overview in clinical studies of new medical technologies intended for improving the visualization of vessels. In this paper, a new blood vessel detection based on a combination of Kirsch's templates and Fuzzy C-Means (FCM) was proposed. The main objective of this study is to improve the detection result of FCM and achieved more effective performance compared to the Kirsch's templates result. The proposed method experimented on 20 images is utilized namely from Digital Retina Images for Vessel Extraction (DRIVE) dataset. The resulting images are compared with the benchmark images based on a few image quality assessment (IQA) such as accuracy, sensitivity and specificity. The total average of accuracy is 92.64%, while sensitivity and specificity got 95.73% and 60.45% respectively.

## 1. Introduction

The retinal blood vessel is recognized as a crucial part in both cardiovascular disease diagnosis and ophthalmological such as diabetic retinopathy and glaucoma [1,2]. Diabetic retinopathy is a diabetes complication that affects the eyes [3–5]. The statistic of this disease increases in community health and it's also the reason for the loss of sight. Hence, precise recognition of retinal blood vessel is vital. Manual diagnosis is usually performed by analyzing the images from a patient, as not all images show signs of diabetic retinopathy [6]. It raises the time and tips to an incorrect diagnostic decision for ophthalmologists. Hence, an automatic segmentation of the vasculature might preserve the work of the ophthalmologists and will support in portraying spotted injuries. The characteristics of retinal vasculature together with tortuosity, width, length, angles, and branching pattern can play a part in the diagnostic result. Nevertheless, even though promising, manual segmentation of retinal blood vessels is a repetitive work and time consuming, and it involves specialized expertise for even though the finest vessel could contribute to the differential diagnosis list [7]. The demand for faster and automatic study of the retinal vessel images must upraise for supporting ophthalmologists with this unpredictable and monotonous work.

## 2. Literature Review

In the retinal image, the blood vessel is one vital part and it acts as milestones for registration of retinal images of the similar patient collected from dissimilar sources. Over the previous era, blood vessel studies enable to determine several eye diseases. The extraction of blood vessels and vascular intersections in retinal images may assist physicians to diagnose eye disease on behalf of patient screening and clinical study [8–11]. The presence of a blood vessel may deliver data about the pathology of diseases, including diabetes and high blood. In recent developments, many approaches to automated retinal blood vessel segmentation were suggested. The summary of a few research and study about blood vessel diagnosis was shown on table 1.

**Table 1.** Summary the previous research about blood vessel detection

| No. | Author | Research focus | Method | Performance |
|---|---|---|---|---|
| 1. | Elbalaoui *et al.* [12] | Detection of Blood Vessel | -Adaptive thresholding<br>- Hessian multiscale enhancement filter | Accuracy = 93.43% |
| 2. | Wilfred Franklin *et al.* [13] | Blood vessel segmentation | -Multilayer perceptron neural network. | - |
| 3. | Fan *et al.* [14] | Blood Vessel Segmentation | - Hierarchical Image Matting Model | Accuracy = 94.1% |
| 4. | Wan Azani [15] | Blood Vessel Extraction | - Morphological Operation | Sensitivity = 99.6%<br>Specificity = 47.9% |
| 5. | Roychowdhury *et al.* [16] | Blood Vessel Segmentation | - High-pass filtering<br>- Gaussian Mixture Model | Accuracy = 92.3% |
| 6. | Salazar-Gonzalez *et al.* [17] | Blood Vessel Segmentation | - Graph cut technique | Accuracy = 94.12% |
| 7. | Bandara *et al.*[18] | Blood Vessel Segmentation | -Hough line transformation | Accuracy = 93.11% |
| 8. | Barua *et al.* [19] | Segmentation of Blood Vessels | -Artificial Neural Networks (ANN) classifier | Accuracy = 92% |
| 9. | Lili Xu and Shuqian Luo [8] | Segmentation of Blood Vessels | - Adaptive local thresholding<br>- Support Vector Machine (SVM) | Sensitivity = 77%<br>Accuracy = 93.2%. |

IOP Conf. Series: Materials Science and Engineering **557** (2019) 012009 doi:10.1088/1757-899X/557/1/012009

## 3. Research Methodology

The proposed system consists of three stages, in the pre-processing part, the aim is to covert the retinal color image to the grayscale image and the second stage is the feature extraction of the image. Next, the third stage is a blood vessel segmentation using Fuzzy C-Means. The proposed method for blood vessel detection is illustrated in Figure 1.

### *3.1. Pre-Processing*

Many researchers agree that the pre-processing process is very important, especially in the medical image [20,21]. Pre-processing able to eliminate noise and reduce the illumination and contrast effect [22,23].

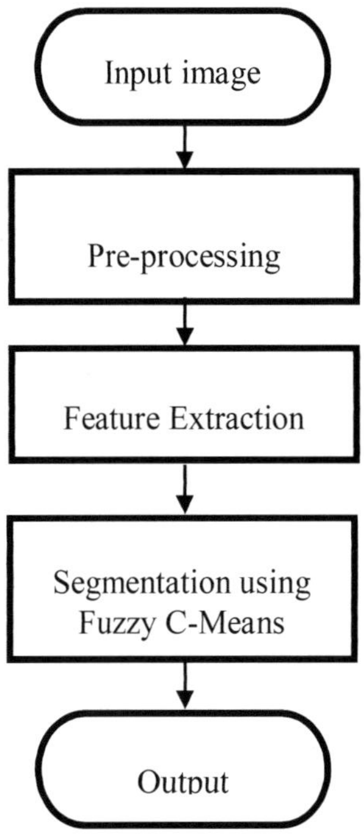


**Figure 1:** Flowchart blood vessel extraction

### *3.2. Feature Extraction*

Image feature attributes or distinctive aspect was important in image processing. Features extracted from the image useful especially for classification and image recognition. Characteristics of extracted during this phase helps to classify pixels whether it is vessel or not. In this paper, the extraction using threshold, elements of the restructuring, closing morphology and Kirsch template used.

### *3.3. Kirsch's templates*

Kirsch template is one of the first order derivative versions of discrete for enhancement and detection. This technique was used in order to detect the edge of blood vessel by using the eight direction of template which rotated fairly by 45 °. From the templates result, the greater will be considered for the output of products and then extracted. Kirsch template can set and reset thresholds to find the perfect image [24]. Figure 2 shows the arrays of Kirsch's templates.

| | | | | | | |
|---|---|---|---|---|---|---|
| 5 | -3 | -3 | | -3 | -3 | 5 |
| 5 | 0 | -3 | | -3 | 0 | 5 |
| 5 | -3 | -3 | | -3 | -3 | 5 |
| | 0° | | | | 180° | |
| -3 | -3 | -3 | | -3 | 5 | 5 |
| 5 | 0 | -3 | | -3 | 0 | 5 |
| 5 | 5 | -3 | | -3 | -3 | -3 |
| | 45° | | | | 225° | |
| -3 | -3 | -3 | | 5 | 5 | 5 |
| -3 | 0 | -3 | | -3 | 0 | -3 |
| 5 | 5 | 5 | | -3 | -3 | -3 |
| | 90° | | | | 270° | |
| -3 | -3 | -3 | | 5 | 5 | -3 |
| -3 | 0 | 5 | | 5 | 0 | -3 |
| -3 | 5 | 5 | | -3 | -3 | -3 |
| | 135° | | | | 315° | |

**Figure 2**: Arrays of Kirsch's method

### *3.4. Segmentation using Fuzzy C-Means (FCM)*

Fuzzy C-Means (FCM) is widely used in pattern recognition as a clustering method. By updating the cluster centers and the membership grades for each unique pixel, FCM shifts the cluster centers to the "true" location within a set of pixels. To accommodate the introduction of fuzzy partitioning involved the membership matrix is randomly and membership of function of the data point [25]. FCM consist a few mathematical equation and complicated, however the segmentation performance is good.

## 4. Result And Discussion

In this research, the programs were run in MATLAB R2017b from an HP laptop with Intel® Core™ i7-45000 CPU @2.40GHz and 8.00GB RAM. The method experimented with the 20 retinal images from DRIVE online database and can be download at https://www.isi.uu.nl/Research/Databases/DRIVE/. A DRIVE image is an established database and specific for the blood vessel detection. The size image of each is 565 x 584 pixels with 24-bit depth and 96 dpi. The original image is shown in Figure 3 (a). Different channels namely red, green and blue will be extracted. The vessels are visible in the red channel. In the pre-processing stage, the input image is resized and the green channel image is separate because the blood vessel appears brighter in the green channel image will be employed as shown in Figure 3 (b). In this paper, a comparison between ground truths vessel extractions from DRIVE Database as a benchmark is compared with the Kirsch Template method and is shown in Figure 3 (c) and Figure 3 (d). Then, a comparison is made between ground truths with our proposed method as shown in Figure 3 (e). Table 2 shows the result in term of accuracy, sensitivity and specificity between the two methods and the benchmark images.

IOP Conf. Series: Materials Science and Engineering **557** (2019) 012009 doi:10.1088/1757-899X/557/1/012009

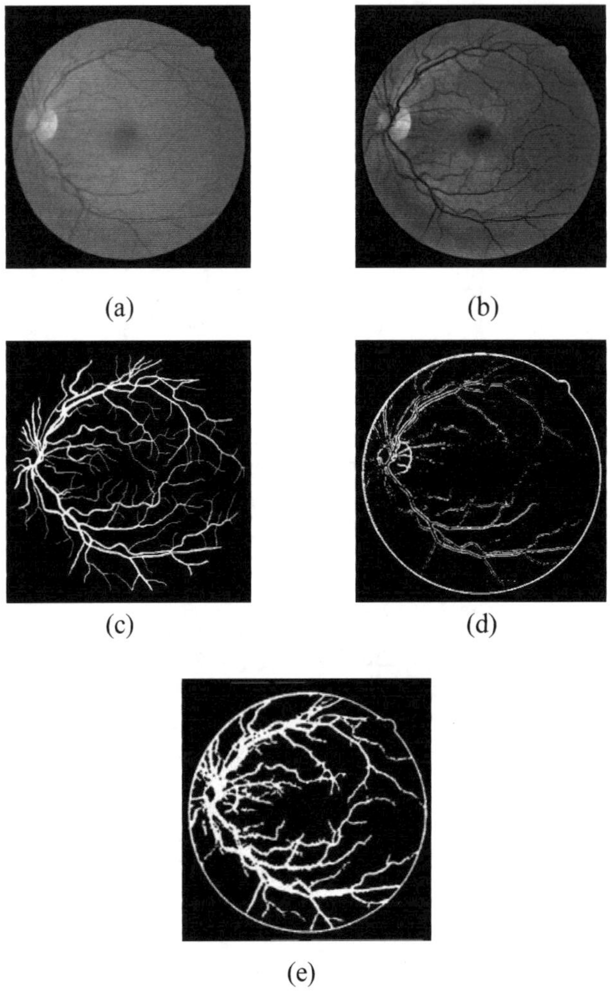

**Figure 3:** Resulting of the retinal image; (a) original image, (b) image after applying pre-processing, (c) ground truth image, (d) image after applying the Kirsch's technique and (e) image after applying the proposed method.

**Table 2.** Comparison of resulting performance after applying proposed method

| Image | Ground truth vs Kirsch Templates | | | Ground truth vs Proposed method | | |
|---|---|---|---|---|---|---|
| | Accuracy % | Sensitivity % | Specificity % | Accuracy % | Sensitivity % | Specificity % |
| 1 | 90.25 | 96.35 | 28.04 | 92.08 | 94.25 | 69.96 |
| 2 | 90.13 | 96.83 | 31.41 | 93.02 | 94.52 | 79.87 |
| 3 | 89.36 | 97.59 | 14.91 | 91.69 | 96.80 | 45.49 |
| 4 | 90.92 | 96.66 | 34.26 | 93.48 | 96.33 | 65.44 |
| 5 | 90.28 | 97.52 | 20.25 | 93.23 | 96.80 | 58.64 |
| 6 | 89.68 | 97.31 | 18.84 | 92.29 | 96.81 | 50.33 |
| 7 | 90.97 | 97.11 | 29.96 | 92.74 | 95.76 | 62.63 |

| | | | | | | |
|---|---|---|---|---|---|---|
| 8 | 90.21 | 96.97 | 18.41 | 92.32 | 96.79 | 44.80 |
| 9 | 90.77 | 97.33 | 16.35 | 93.00 | 96.83 | 49.59 |
| 10 | 91.10 | 97.34 | 21.62 | 93.28 | 96.46 | 57.88 |
| 11 | 91.27 | 96.53 | 37.76 | 92.17 | 93.92 | 74.33 |
| 12 | 90.53 | 96.99 | 22.13 | 92.32 | 95.61 | 57.48 |
| 13 | 90.07 | 96.97 | 26.35 | 92.60 | 96.14 | 59.95 |
| 14 | 91.15 | 96.85 | 26.37 | 93.06 | 95.28 | 67.90 |
| 15 | 92.69 | 97.08 | 35.82 | 93.76 | 95.37 | 72.98 |
| 16 | 90.49 | 96.56 | 29.36 | 92.23 | 95.95 | 64.85 |
| 17 | 90.09 | 96.48 | 20.93 | 92.06 | 95.75 | 52.10 |
| 18 | 90.10 | 96.33 | 17.68 | 91.65 | 94.61 | 57.28 |
| 19 | 90.75 | 96.77 | 24.19 | 92.96 | 96.37 | 66.37 |
| 20 | 91.04 | 97.00 | 15.89 | 92.92 | 96.24 | 51.07 |
| Average | 90.59% | 96.93% | 24.53% | 92.64 % | 97.73 % | 60.45 % |

In order to demonstrate the effectiveness, three parameters are used namely the accuracy, sensitivity and specificity. The equation for accuracy, sensitivity and specificity can refer to [26,27]. In this work, the proposed method is compared with Kirsch Templates method in term of accuracy, sensitivity and specificity. Based on the result, a proposed method obtained slightly higher in term of accuracy which is 92.64 % compare than 90.59%. The specificity result also shown improvement from 24.53% to 60.45 %. Lastly, performance based on sensitivity give the highest result which is 97.73 % compare than 96.93%.

## 5. Conclusion

Diabetic retinopathy is a diabetes complication that affects the eyes. The statistic of this disease increases in community health and it also the reason of loss of sight. Hence, precise recognition of retinal blood vessel is vital. In this paper, new approaches based on a combination of Kirsch Templates and FCM were proposed. The aim is to improve the detection technique of Kirsch Templates. Based on the result, the proposed method successful to overcome the Kirsch Templates performance shown by Accuracy = 92.64%, Sensitivity = 97.73% and Specificity = 60.45. Although our proposed method can detect blood vessel effectively but still fail to detect small vessels. The future way of segmentation research will be to develop quicker and more accurate more automated methods.

## 6. Acknowledgment

This work was supported by Ministry of Higher Education Malaysia under the Fundamental Research Grant Scheme (FRGS/1/2018/SKK13/UNIMAP/02/1)

## References

[1] Adalarasan R and Malathi R 2018 Automatic Detection of Blood Vessels in Digital Retinal Images using Soft Computing Technique *Mater. Today Proc.* **5** 1950–9
[2] Mustafa W A, Yazid H and Yaacob S 2014 A Review : Comparison Between Different Type of Filtering Methods on the Contrast Variation Retinal Images *IEEE International Conference on Control System, Computing and Engineering* pp 542–6
[3] Gardner T W 2017 Diabetic Retinopathy: A Position Statement by the American Diabetes Association *Diabetes Care* **40** 412–8
[4] Antonetti D A, Klein R and Gardner T W 2012 Diabetic retinopathy. *N. Engl. J. Med.* **366** 1227–39
[5] Mustafa W A, Abdul-nasir A S and Yazid H 2018 Diabetic Retinopathy ( DR ) on Retinal

Image : A Pilot Study *J. Phys. Conf. Ser. Pap.* **1019** 1–6

[6] Rodrigues L C and Marengoni M 2017 Segmentation of optic disc and blood vessels in retinal images using wavelets, mathematical morphology and Hessian-based multi-scale filtering *Biomed. Signal Process. Control* **36** 39–49

[7] Singh D, Dharmveer and Singh B 2015 A new morphology based approach for blood vessel segmentation in retinal images *11th IEEE India Conf. Emerg. Trends Innov. Technol. INDICON 2014*

[8] Xu L and Luo S 2010 A novel method for blood vessel detection from retinal images *Biomed. Eng. Online* **9** 14

[9] Mustafa W A and Kader M M M A 2018 Automatic Blood Vessel Detection on Retinal Image Using Hybrid Combination Techniques *Malaysian Appl. Biol.* **47** 47–52

[10] Mustafa W A, Yazid H and Kamaruddin W 2017 Combination of Gray-Level and Moment Invariant for Automatic Blood Vessel Detection on Retinal Image *J. Biomimetics, Biomater. Biomed. Eng.* **34** 10–9

[11] Mustafa W A, Yazid H, Yaacob S and Basah S 2014 Blood vessel extraction using morphological operation for diabetic retinopathy *IEEE Reg. 10 Symp.* 208–12

[12] Elbalaoui A, Fakir M, Taifi K and Merbouha A 2016 Automatic Detection of Blood Vessel in Retinal Images *Proc. - Comput. Graph. Imaging Vis. New Tech. Trends, CGiV 2016* 324–32

[13] Franklin S W and Rajan S E 2014 Computerized screening of diabetic retinopathy employing blood vessel segmentation in retinal images *Biocybern. Biomed. Eng.* **34** 117–24

[14] Fan Z, Lu J, Li W, Wei C, Huang H, Cai X and Chen X 2017 A Hierarchical Image Matting Model for Blood Vessel Segmentation in Fundus images *Comput. Vis. Pattern Recognit.* 1–10

[15] Azani W and Mustafa B W 2014 Blood Vessel Extraction using Morphological Operation for Diabetic Retinopathy 208–12

[16] Roychowdhury S, Koozekanani D D and Parhi K K 2015 Blood vessel segmentation of fundus images by major vessel extraction and subimage classification *IEEE J. Biomed. Heal. Informatics* **19** 1118–28

[17] Salazar-Gonzalez A, Kaba D, Li Y and Liu X 2014 Segmentation of the blood vessels and optic disk in retinal images *IEEE J. Biomed. Heal. Informatics* **18** 1874–86

[18] Bandara A M R R and Giragama P W G R M P B 2018 A Retinal Image Enhancement Technique for Blood Vessel Segmentation Algorithm 1–5

[19] Barua B and Hasan M 2016 A New Approach of Detection and Segmentation of Blood Vessels for the Classification of Healthy and Diseased Retinal Images

[20] Mustafa W A and Yazid H 2018 Conversion of the Retinal Image Using Gray World Technique *J. Biomimetics, Biomater. Biomed. Eng.* **36** 70–7

[21] Antal B and András Hajdu 2012 Improving microaneurysm detection using an optimally selected subset of candidate extractors and preprocessing methods *Pattern Recognit.* **45** 264–70

[22] Mustafa W A, Yazid H and Kader M M M A 2018 Luminosity Correction Using Statistical Features on Retinal Images *J. Biomimetics, Biomater. Biomed. Eng.* **37** 74–84

[23] Mustafa W A and Yazid H 2017 Image Enhancement Technique on Contrast Variation : A Comprehensive Review *J. Telecommun. Electron. Comput. Eng.* **9** 199–204

[24] H.S. Bhadauria 2013 Vessels Extraction from Retinal Images *IOSR J. Electron. Commun. Eng.* **6** 79–82

[25] Akhavan R and Faez K 2014 A novel retinal blood vessel segmentation algorithm using fuzzy segmentation *Int. J. Electr. Comput. Eng.* **4** 561–72

[26] Mustafa W A, Abdul-Nasir A S and Mohamed Z 2018 Malaria Parasites Segmentation Based on Sauvola Algorithm Modification *Malaysian Appl. Biol.* **47** 71–6

[27] Mustafa W A, Abdul-nasir A S, Mohamed Z and Yazid H 2018 Segmentation Based on Morphological Approach for Enhanced Malaria Parasites Detection *J. Telecommun. Electron. Comput. Eng. segmentation* **10** 15–20

IOP Conf. Series: Materials Science and Engineering **557** (2019) 012010 doi:10.1088/1757-899X/557/1/012010

# Growth Monitoring of Harumanis Mango Leaves *(Mangifera Indica)* at Vegetative Stage Using SPAD Meter and Leaf Area Meter

**N Roslan[1], A A Aznan[1,a)], R Ruslan[1], M N Jaafar[2], F A Azizan[1]**

[1] School of Bioprocess Engineering, Universiti Malaysia Perlis, Kompleks Pusat Pengajian Jejawi 3, 02600 Arau, Perlis

[2] Institute of Sustainable Agrotechnology, Universiti Malaysia Perlis, 02100 Padang Besar, Perlis

[a)] Corresponding author: aimiathirah@unimap.edu.my

**Abstract.** Current advancement in orchard farming has results in ultra-high density planting of harumanis mango trees under greenhouse production system. The innovative approach in monitoring the nutrient status is deemed necessary to manage the growth of the trees. Sufficient application of nutrient especially nitrogen will exhibit healthy green color on the leaf while deficiency in nitrogen will result in poor leaf color. The level of greenness in the leaf will indicate the overall healthiness of the plant. There are very limited studies that indicate the growth of harumanis leaves at discrete time interval during it developmental stage. The objective of this research is to study the growth development of harumanis leaves by using SPAD meter and area meter at vegetative stage. SPAD reading were taken from the leaves at the age of 2$^{nd}$ to 7$^{th}$ week by using a Minolta SPAD-502 meter. Leaves were sampled from 32 trees and statistical analysis was conducted to determine the relationship of age to the growth of the leaves for its area expansion and the SPAD reading. The result for SPAD reading of the leaves shows significant increase with the growing period while leaves area shows less significance different during the growing period. This study concluded that SPAD meter can be used as a reliable tool to indicate the growth of the leaves instead of the leaf area measurement.

## 1. Introduction

Harumanis is the iconic products from Perlis due to the aroma, texture and the sweetness. Lately, it has risen in demand and the selling price. However, very limited study that has been done for Harumanis grown under high density planting system (HDPS) in greenhouse especially on the nutrient management aspect. Nutrient management is very important to monitor each individual tree requirement of Harumanis trees planted under this advance orchard farming.

Generally, good management and monitoring of the growth development of the trees will result in good quality produce. The quality of the produce can be predicted from the growth performance shown by the trees. Various methods have been used by researcher to determine plant growth especially via plant tissue analysis. Plant tissue sampling is essential to determine nutrient requirement for plant growth. Visually, healthiness of the trees can be indicated by the greenness of the leaves.

Leaves that exhibit healthy green color will indicate the sufficiency in the nutrient availability in the soil while leaves with poor green color results from nutrient deficiency.

The growth of leaves and important physiological events of the trees greatly depends on the age. As example Ramírez et al.[7] agreed that the age of the vegetative stems is important to induce flowering. Leaf growth was characterized into several stages. The study from Ramírez et al.[8] has described the vegetative growth of 'Keitt' and 'Tommy Atkins' variety. According to study, vegetative growth will undergo several stages from resting period, bud break, bud elongation, early leaf elongation and elongation of leaf green stage before another flush take place. It is a repetitive and continuous process throughout the growing period. A typical leaf growth displays sigmoidal curve for the log, linear and stationary phase. In normal development, leaf increase in size during its active growth period. Abnormal growth pattern indicates the state of health of the tree.

Leaf area measurement is vital in the plant and nutritional study. Ghoreishi et al.[1] has use Li-3100 area meter to measure leaf area. In the study, simple model was established for leaf area prediction of mango and significant result was obtained for leaf growth estimation. Mokhtarpour et al.[4] claimed that some method to measure plant leaf area is time consuming. The leaf area measurement using area meter was expected to overcome the constraints of other measuring tools. In the study, leaf growth model for maize were established and significant relationships were obtained between leaf area and the other growth parameter.

On the other hand, SPAD-502 has been recognized as one of the reliable tools to determine plant nutrient status[12,13]. It provides data in numerical values that indicated the level of greenness in the leaf. Theoretically, the level of greenness is directly proportional to the amount of the chlorophyll present in the leaves. It is also directly correlated to their nitrogen concentration. Difference in SPAD reading may cause by different chlorophyll concentration present in the leaves. Higher SPAD reading will indicate healthier plants with sufficient nutrient. Netto et al.[6] claimed that SPAD meter has ability to detect nitrogen deficiency, thus it can be useful to improve nitrogen management. However, the values may vary depends to the varieties, cultivar and different development stage[10]. In Malaysia, the correlation between SPAD meter and leaf nitrogen concentration has been widely studied in paddy.

The aim of this research was to study the vegetative growth of Harumanis leaves by comparing two measuring tools using leaf area meter and SPAD meter. Correlation analysis between the SPAD reading and leaves area was carried out to determine if there is correlation between both parameters. Then, the best tools indicating leaves growth was selected based on the healthiness exhibit by the leaves.

## 2. Material and method

### *2.1. Study area*

The study was carried out in a greenhouse (GH03) located at the Institute of Sustainable Agriculture (INSAT), Universiti Malaysia Perlis (UniMAP), Padang Besar, Perlis. The GPS position is N6.654321, E100.265557, at 53m above mean sea level. INSAT were located at the northern area, zone 1 of Malaysia characterized with drought in January to March followed by rainy season in September to December with maximum temperature recorded up to 40°C. The soils generally are loamy clay of lateritic in origin. The planting follows the standard plant husbandry practice for high density planting except for its pruning and micro-climate controlled. The age of the trees is one year and the trees size is almost similar due to uniform pruning. The planted areas are supplied with drip irrigation system.

*2.2. Plant samples and leaf measurement*

A total of 32 trees per greenhouse were selected based on completely randomized design of staggered grid sampling. Leaves were selected from a healthy stem at four different sides of the trees. Each week, five trees of the same leaves age were chosen for data collection. The age of the leaves was monitored from the date of bud emergence. For non-destructive measurement of leaves greenness, Minolta SPAD-502 meter were used. The reading was recorded in SPAD unit. Measurement using SPAD meter was taken at four point per leaf at about halfway from the leaf tip and collar; about middle point between the leaf midrib and margin. The measurement within one tree was grouped as a single sample. Then, leaves from each branch were collected for destructive analysis. The collected fresh leaves were transported to laboratory in insulated box and washed with distilled water to remove dirt. The samples were measured on the area basis by using area meter (Li-3100C, Li-Cor, Inc., Lincoln, Nebraska, USA).

*2.3. Data analysis*

The statistical analysis was carried out using SPSS Statistics software. In the analysis, the data collected for each tree of the same age was grouped together as a single sample. T-test paired two samples per means for p-value<0.05 were conducted to distinguish significance different between the age of the leaves for both the SPAD value and area of the leaves. The relationship of the SPAD value and the area to the age were evaluated and lastly, correlation between the age, SPAD reading and area were established using correlation coefficient analysis.

## 3. Result and discussion

*3.1. Relationship between area expansion and the age of the leaves*

The result for the area expansion affected by the age of the leaves was shown in Figure 1. The graph showed that the growth of the leaves increase with increasing leaves age. Overall growth trend shows sigmoidal curve (double curve). These observations are in line with the result obtained by Gonzalez et al.[2] for leaves growth. From the graph, the early growth of young and in matured leaves showed a slower growth compared to the expansion period. The leaves grow actively during expansion period. During early growth from the age of 2$^{nd}$ to 3$^{rd}$ week, the growth mainly focused on the quantitative increase in cell number and less on elongation of cell. During the expansion period from the 3$^{rd}$ to 6$^{th}$ week, the leaves exhibit rapid growth in the leaves area size. At this state there are rapid increase in cell division and cell size. There was no significant different from the age of 4$^{th}$ to 5$^{th}$ weeks. In this period, the leaves may rest to store enough nutrients to promote further elongation and expansion of the leaves. The matured leaves at the age of 6$^{th}$ to 7$^{th}$ week do not show any significant changes in leaf area due to deceleration in the cell division and elongation. The growth is almost absence. In this study, more than 80% correlation was obtained between leaves area and leaves age. Thus, it is proved that the leaves area increased in parallel with the increasing leaves age.

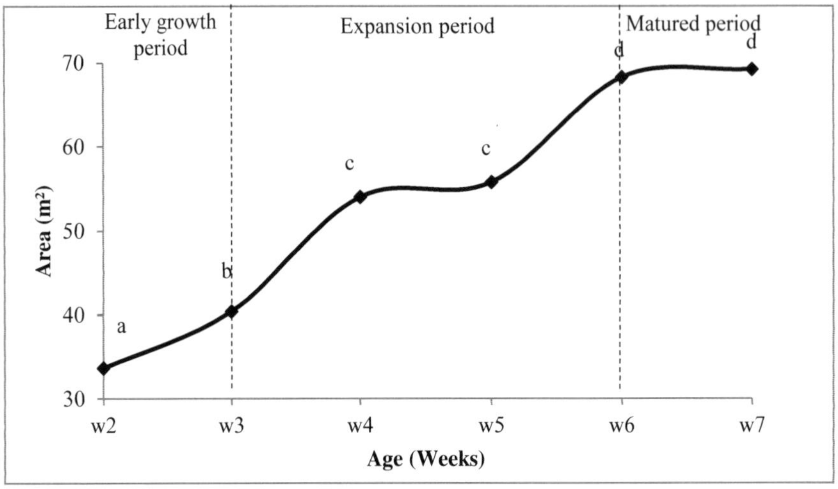


**Figure 1.** Area expansion of leaves affected by the age of the leaves of Harumanis mango grown in greenhouse at Sg Chuchuh, Perlis.

*3.2. Relationship between SPAD reading and the age of the leaves*

The result for the SPAD reading related to the age of the leaves was shown in Figure 2. The graph showed that the SPAD reading of the leaves increase with increasing leaves age. From the graph, there was significant increase in SPAD reading at age $2^{nd}$ week (average 17.3 SPAD units) until reach peak at $6^{th}$ week (average 40.36 SPAD units). However, the value of SPAD reading dropped at $7^{th}$ week to average of 34.34 SPAD units. The greenness of leaves measured by using SPAD meter was positively correlated with chlorophyll content in the leaves. Thus, increase in SPAD reading results from increase in chlorophyll content. Decrease in SPAD reading during $7^{th}$ week in matured leaves may relate to several factors that may disturb the SPAD reading prior to measurement. When too high chlorophyll concentration exists in the leaves, it could disturb the accuracy of the instruments[9]. On the other hand, the light intensity during the time of measurement may give influence to the measurement of the SPAD reading as agreed by Martínez and Guiamet[3]. According to Nauš et al.[5], light will affect the chlorophyll distribution in the leaves due to the chloroplast movement. The non-uniform chlorophyll distribution in the leaves also may disturb the reading of SPAD meter[11]. Technically, SPAD meter needs to be controlled by user to eliminate error in the reading during the field measurement.

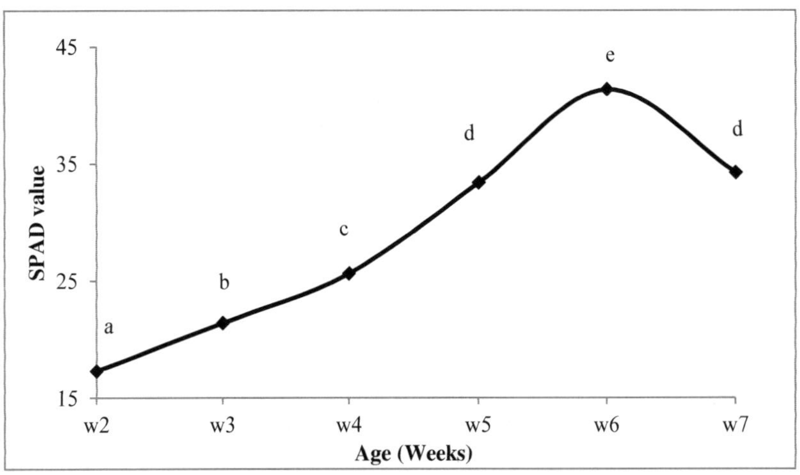


**Figure 2.** Changes in SPAD reading affected by the age of the leaves of Harumanis mango grown in greenhouse at Sg Chuchuh, Perlis.

### *3.3. Correlation between area and SPAD reading*

The correlation between SPAD reading and area of the leaves were shown in Figure 3. The graph shows positive gradient between both parameters. A moderate positive correlation of 62.89% was obtained between SPAD reading and leaf area measurement. Between the parameter itself, both of the leaves area and SPAD reading shows significant different since the p-value is less than 0.05. On the other hand, interaction effect between the SPAD reading and leaf area measurement do not give significant different. However, due to the higher F value in the SPAD reading compared to leaf area measurement, it was concluded that the SPAD reading able to give more accurate value indicating the healthiness of the Harumanis mango plant.

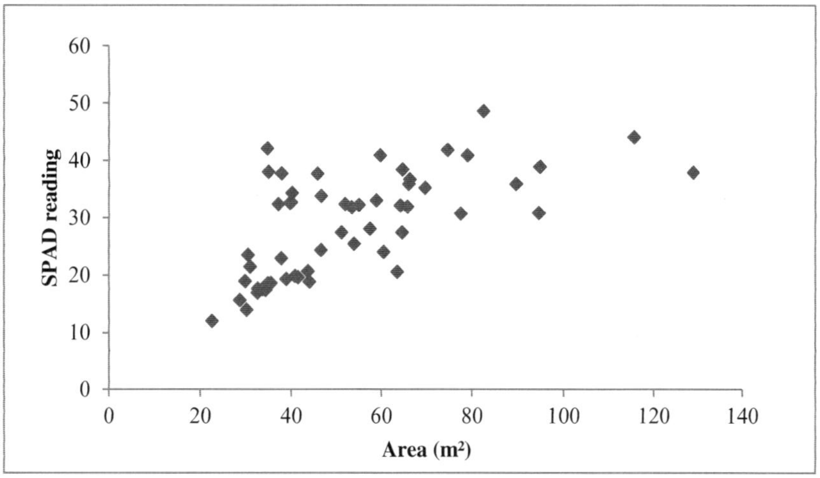


**Figure 3.** Correlation between SPAD reading and area of the leaves.

## 4. Conclusion

As conclusion, it was proved that both leaf area and SPAD measurement could give significant different to indicate plant healthiness. However, the value from SPAD meter were found to give higher significant different between the ages of the leaves compared to the leaf area measurement. Based on the result and consideration to select the best method, SPAD meter was chosen as a reliable and effective tool to aid in the evaluation of growth performance in Harumanis plants. These result consistent with the claim from previous findings that strongly support the use of SPAD meter for determining plant nutrient status. Thus, it can be elucidated that the SPAD data may be used as the ground truth for remote sensing of mango farm. This study expected to act as benchmark to indicate healthiness of Harumanis leaves at specific development period. However, other important information such as chlorophyll and total nitrogen content is still lacking. Therefore, further study is required to fill the gap in Harumanis mango farming.

## References

[1] Ghoreishi M, Hossini Y and Maftoon M 2012 Simple models for predicting leaf area of mango (Mangifera indica L.) *Journal of Biology and Earth Sciences* **2** B45-B53

[2] Gonzalez N, De Bodt S, Sulpice R, Jikumaru Y, Chae E, Dhondt S, Van Daele T, De Milde L, Weigel D, Kamiya Y and Stitt M 2010 Increased leaf size: different means to an end *Plant Physiology* **153** 1261-1279

[3] Martínez D and Guiamet J 2004 Distortion of the SPAD 502 chlorophyll meter readings by changes in irradiance and leaf water status *Agronomie* **24** 41-46.

[4] Mokhtarpour H, Teh C B, Saleh G, Selamat A B, Asadi M E and Kamkar B 2010 Non-destructive estimation of maize leaf area, fresh weight, and dry weight using leaf length and leaf width *Communications in Biometry and Crop Science* **5** 19-26.

[5] Nauš J, Prokopová J, Řebíček J and Špundová M 2010 SPAD chlorophyll meter reading can be

pronouncedly affected by chloroplast movement *Photosynthesis Research* **105** 265-271
[6] Netto A T, Campostrini E de Oliveira J G and Bressan-Smith R E 2005 Photosynthetic pigments, nitrogen, chlorophyll a fluorescence and SPAD-502 readings in coffee leaves *Scientia Horticulturae* **104** 199-209
[7] Ramírez F, Davenport T L, Fischer G and Pinzón J C A 2010 The stem age required for floral induction of synchronized mango trees in the tropics *Scientia Horticulturae* **123** 443-453
[8] Ramírez F, Davenport T L, Fischer G, Pinzón J C A and Ulrichs, C 2014 Mango trees have no distinct phenology: the case of mangoes in the tropics *Scientia Horticulturae* **168** 258-266
[9] Uddling J, Gelang-Alfredsson J, Piikki K and Pleijel H 2007 Evaluating the relationship between leaf chlorophyll concentration and SPAD-502 chlorophyll meter readings. *Photosynthesis research* **91** 37-46
[10] Wang S, Zhu Y, Jiang H and Cao W 2006 Positional differences in nitrogen and sugar concentrations of upper leaves relate to plant N status in rice under different N rates *Field Crops Research* **96** 224-234
[11] Xiong D, Chen J, Yu T, Gao W, Ling X, Li Y, Peng S and Huang J 2015 SPAD-based leaf nitrogen estimation is impacted by environmental factors and crop leaf characteristics *Scientific reports* **5** 13389
[12] Yang H, Yang J, Lv Y and He J 2014 SPAD values and nitrogen nutrition index for the evaluation of rice nitrogen status *Plant Production Science* **9** e88421
[13] Yuan Z, Cao Q, Zhang K, Ata-Ul-Karim S T, Tian Y, Zhu Y, Cao W and Liu X 2016 Optimal leaf positions for SPAD meter measurement in rice *Frontiers in Plant Science* **7** 719

**Acknowledgement**
The author would like to acknowledge the support from the Fundamental Research Grant Scheme (FRGS) under a grant number of FRGS/1/2016/WAB01/UNIMAP/03/6 from the Ministry of Higher Education Malaysia.

# Analysis of Human Behaviour while Controlling the Steering Wheel of a Buggy Electric Vehicle (EV)

**Hafiz Halin[1], Wan Khairunizam[1], K Ikram[1], Hasri Haris[1], I Zunaidi[2], S A Bakar[1], Z M Razlan[1], and N Z Noriman[3]**

[1]Advanced Intelligent Computing and Sustainable Research Group (AICOS), [2]School of Mechatronic Engineering, Universiti Malaysia Perlis (UniMAP), Pauh Putra Campus, 02600 Arau,Perlis, Malaysia .
[2]Faculty of Technology, University of Sunderland, St Peter's Campus, Sunderland, SR6 0DD,United Kingdom
[3]Faculty of Engineering Technology, Universiti Malaysia Perlis, UniCITI Alam Campus, Sungai Chuchuh, 02100 Padang Besar, Perlis, Malaysia

**Abstract.** Human ability to make a decision and recognize pattern are based on their intelligence level and experience. This ability makes the driver alert and knows what they need to do at certain situations. This paper aims to study human behavior while driving the vehicle at the desired path. Certain criteria subjects are selected with the purpose to study their driving pattern. The subjects must own a driving license and driving experience less than 10 years. The subject then drives the vehicle at the designed paths, which is straight and right turn. The drive pattern for the subjects shows a similar pattern but different characteristics. The experiments used to create a fuzzy controller that imitate human behavior while controlling the steering wheel. The steering wheel angle, position, heading and speed of the vehicle are recorded during the experiments. This paper only focus on the analysis of human behaviour based on steering wheel angle recorded.

## 1. Introduction

Steering wheel control need for an autonomous vehicle navigation system. The steering wheel usually controls by using a motor [1]–[3] or other suitable actuator. The controller use to control the actuators for navigation. However, the navigation by controller alone make the autonomous vehicle lack of comfort and human-like factor. The future passengers of the autonomous vehicle needs to feel safe and comfort while using the autonomous vehicles. Other researchs related to steering wheel control which are focusses on the comfort and safety of the passenger still in the development phase [4], [5]. Fuzzy controller [6] and Neural Network [7] developed to imitate human ability of the decision making. Development of autonomous car by Google, Toyota, BMW, and others multinational company still lacks the fundamental experiments to support the autonomous vehicle success. Steering rate control [8], steering by wire [3], yaw stability [5], and zero-radius steering [9] are the methods developed in order to improve the steering control of an autonomous vehicle.

The data collection and analysis used for future reference in developing the controller for the steering wheel. The experiment uses 5 subjects to gather data on steering wheel control [10]. The data use to set the baseline or benchmark in order to create a controller that imitate human driving pattern. The controller needs to retain the passengers' confort for an autonomous drive in the future. The driving experience helps the driver to determine the appropriate action based on the path taken [11].

This paper consists of an abstract, introduction, methodology, result and discussion, and a conclusion. Abstract explain the brief description of this paper. The introduction consists of research background and inspiration for research. The methodology explains the method of experiments. Result and discussion present experiments result and explanation from the result. Finally, the conclusion concludes the finding from the experiments.

## 2. Methodology

### *2.1. Flow Chart*

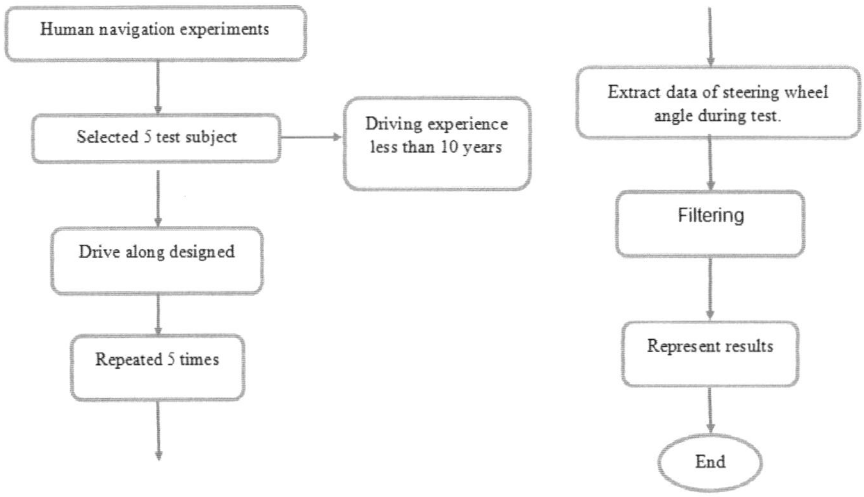


**Figure 1**. Flow chart for Human navigation test

Figure 1 shows the process flow for the human navigation experiments. The selected subjects must have a valid driving license with driving experience less than 10 years. There are 2 path selected for data collection, which are straight and right turn. The experiments repeats 5 times for each path. Then, the steering wheel are extracted and filtered. The selected result represent in Figure 3 and Figure 4.

### *2.2. Test setup*

The experiments consist of right turn and straight path maneuver with 5 subjects selected in order to study the characteristics of human while driving. The path for experiments shown in Figure 2. The experiments done at Institut Kemahiran Mara (IKM), Beseri, Perlis Malaysia. Distance of the path is 40m include the corner. The subject start at point (1) and stop at point (2). The subjects instruct to maintain a constant speed during the test. The subject must own a class D driving license with driving experience less than 10 years [12]. Data of steering wheel angle taken while the subjects drive on the paths. The result selected after 5 trials. The vehicle used for this experiment is a Yamaha electric vehicle. The specification of the vehicle show in table 1.

**Figure 2.** (a) Full satellite image of IKM Beseri, Perlis; (b) zoom in image and path for experiments

**Table 1.** Vehicle specification

| | |
|---|---|
| Length | 231cm |
| Width | 111cm |
| Weight | 249kg |
| Wheel base | 163cm |
| Top speed | 30km/h |
| Drive motor | 2kW |
| Battery | 48V DC |
| Steering | Rack and pinion gear |
| Brake | Mechanical brake cable system to drum brake |

## 3. Result and Discussion

The result presented in Figure 3 and Figure 4. The positive value of steering wheel angle indicate the rotation on clockwise direction while negative value is rotation to counter clockwise direction. In Figure 3, the transition from straight line into the curve path can be seen at time 11.7 second by subject #1. The steering angle change smoothly while take the corner. At 16.3 second, the sudden change in steering angle when the subject try to drive straight again. Inefficiency of rack and pinion steering system can be seen as the driver turn the steering. Subject #2 drive more smoothly. The straight drive and drive at curve can be distinguish by the different of steering wheel angle. At 10.5 second, the subject #1 enter the curve with maximum steering angle of 157°. The orientation of the vehicle straight again at 20.9 second. The drive pattern for both subject different in term of smoothness. The transition from straight path to curve are more smother for subject #1. However subject #4 use less effort in order to take the same curve. This is because the maximum steering angle by subject #4 less than subject #1 even they taking the same curve.

Figure 4 show the steering wheel angle for the straight path for subjects #3, #4, and #5. The steering wheel angle recorded for the straight path are varied from -16° to 15°. The uneven road and road gradation are the main cause for the angle variation for the straight path navigation. The variation of the steering wheel angle are consider small if compared to the maximum steering wheel angle recorded, which is 408°. Subject #5 recorded the highest steering wheel angle (absolute value of 16.5°) at 18.3 sec.

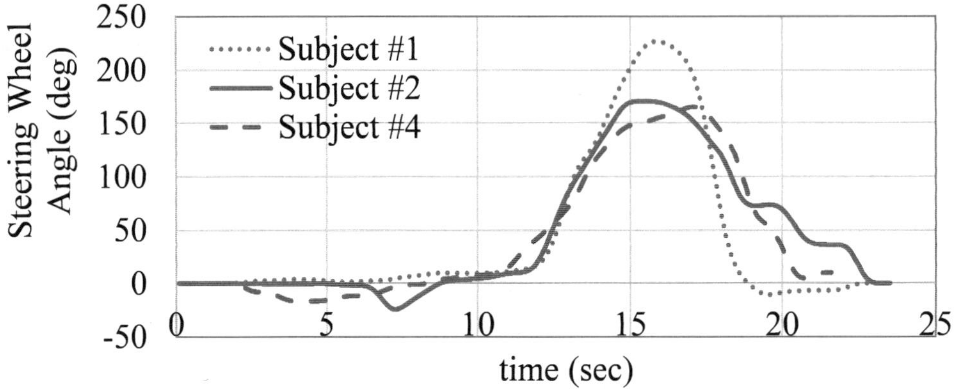


**Figure 3.** Steering wheel angle for the selected subjects on right turn path

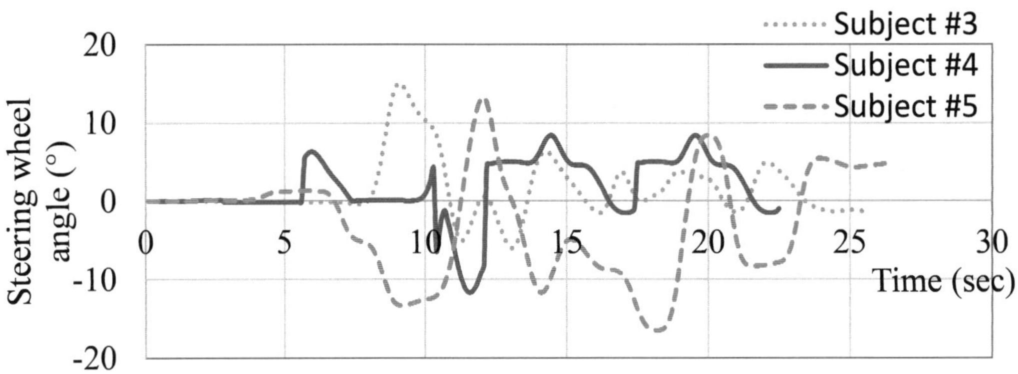


**Figure 4.** Steering Wheel Angle for the straight path for selected subjects.

## 4. Conclusion

Drive pattern for different subject show different characteristic. The bell shape graph for steering wheel angle show same pattern but different characteristic for each subject. The behaviour when subject take the curve based on their experience and attitude. The speed taken for the experiments almost constant throughout the experiments. From the experiments, the extracted data use to develop a Fuzzy controller. The average and standard deviation use to develop the membership function for the input and output of the Fuzzy controller.

## 5. References

[1] E. P. Ping and S. K. Swee, "Simulation and experiment of automatic steering control for lane keeping manoeuvre," in *2012 4th International Conference on Intelligent and Advanced Systems (ICIAS2012)*, 2012, vol. 1, pp. 105–110.

[2] K. R. S. Kodagoda, W. S. Wijesoma, and E. K. Teoh, "Fuzzy speed and steering control of an AGV," *IEEE Trans. Control Syst. Technol.*, vol. 10, no. 1, pp. 112–120, 2002.

[3] Jae-Sung Im, Fuminori Ozaki, Nobutomo Matsunaga, and Shigeyasu Kawaji, "Control of steering-by-wire system using bilateral control scheme with passivity approach," in *2007 International Conference on Control, Automation and Systems*, 2007, pp. 1488–1493.

[4] J. Guldner, W. Sienel, Han-Shue Tan, J. Ackermann, S. Patwardhan, and T. Bunte, "Robust automatic steering control for look-down reference systems with front and rear sensors," *IEEE Trans. Control Syst. Technol.*, vol. 7, no. 1, pp. 2–11, 1999.

[5] N. E. Kahveci and P. A. Ioannou, "Parameter identification and adaptive control for improved vehicle yaw stability performance," in *49th IEEE Conference on Decision and Control (CDC)*, 2010, pp. 2904–2909.

[6] Lin Cai, A. B. Rad, W. L. Chan, and Kai-Yuan Cai, "A robust fuzzy PD controller for automatic steering control of autonomous vehicles," in *The 12th IEEE International Conference on Fuzzy Systems, 2003. FUZZ '03.*, 2003, vol. 1, pp. 549–554.

[7] K. M. Junaid, K. M. Usman, K. Attaullah, and J. A. Raza, "A Neural Network Based Adaptive Autopilot for Marine Applications," pp. 1–6, 2006.

[8] I. Bae, J. H. Kim, and S. Kim, "Steering rate controller based on curvature of trajectory for autonomous driving vehicles," in *2013 IEEE Intelligent Vehicles Symposium (IV)*, 2013, no. Iv, pp. 1381–1386.

[9] YuHon Tee, Y. Tan, BoonYew Teoh, EngBeng Tan, and ZhenYang Wong, "A compact design of zero-radius steering autonomous amphibious vehicle with direct differential directional drive - UTAR-AAV," in *2010 IEEE Conference on Robotics, Automation and Mechatronics*, 2010, pp. 176–181.

[10] W. Soehren and W. Hawkinson, "Prototype personal navigation system," *IEEE Aerosp. Electron. Syst. Mag.*, vol. 23, no. 4, pp. 10–18, 2008.

[11] D. Fernandez Llorca *et al.*, "Autonomous pedestrian collision avoidance using a fuzzy steering controller," in *IEEE Transactions on Intelligent Transportation Systems*, 2011.

[12] T. Lajunen and H. Summala, "Driving experience, personality, and skill and safety-motive dimensions in drivers' self-assessments," *Pers. Individ. Dif.*, vol. 19, no. 3, pp. 307–318, Sep. 1995.

# Document Images Binarization Using Hybrid Combination of Fuzzy C-Means and Deghost Method

**W A Mustafa[1*], Hairy Aziz[1], Wan Khairunizam[2], I Zunaidi[3], Z M Razlan[2] and Shahriman A B[2]**

[1]Department of Electrical Technology Engineering, Faculty of Engineering Technology, Universiti Malaysia Perlis, UniCITI Alam Campus, Sungai Chuchuh, 02100 Padang Besar, Perlis, Malaysia

[2]School of Mechatronics Engineering, Universiti Malaysia Perlis, 02600 Arau, Malaysia

[3]Faculty of Technology, University of Sunderland, St Peter's Campus, Sunderland, SR6 0DD, United Kingdom

*E-mail:wanazani@unimap.edu.my*

**Abstract.** This paper presents a document binarization approach to the document image. Thousands of historical documents usually hold important information within them. It is usually stored in the national archives and library around the globe waiting to be scanned to retrieve the content it holds. However, many environmental factors, improper handling, and the poor quality of the materials used in the document creation cause it to suffer a high degree of degradation of various types. In order to retrieve the content of the degraded historical document, a binarization approach on document images must be applied. The processes of the binarization separate pixel value of an input image into two values which is white as the background and black as foreground text. The proposed system consists of three parts. In the first part, the image pre-processing operation is done before the binarization process to enhance image quality. In this part, Contrast Stretching and Mean Filter is applied onto the image to remove noise on the image. The second part will be to apply the binarization algorithm on the document image that has undergone an image pre-processing operation. By applying the Fuzzy C-Means algorithm to the document images, the images will be converted to a binary image and divided into two components, which is text and background. The last step of the proposed method will be performing the Deghost operation to remove "ghost" entities that may have appeared on the document image. The method will undergo imaging quality analysis, such as PSNR, Accuracy, and F-measure to determine the effectiveness of the proposed method. The experimental results on H-DIBCO 2013 dataset show the robustness, reliability, and efficiency in the proposed approach.

## 1. Introduction

Thousands of historical document images usually hold important information within them. Document images are usually badly degraded and suffer from degradation as it began to age over time. Some type of degradation that the document image suffers is bleed through, smear and uneven illumination image contrast variation [1–6]. It is important to retrieve text from a document image as it may hold important information. The major problem faced is degradation, when it occurs, the document image is in a very poor quality which in most cases is hard to decipher [7–11].

In these past years, an increasing amount of literature on discussing the automated image processing method that has been performed solely purpose document images has shown an increasing pattern [12,13]. The proposed image processing approaches are basically based on a new algorithm and methodology [14,15]. Mostly, researchers have agreed that the difficulties of document image binarization are badly degradation of the old and historical manuscript of historical document images [16–21]. In 2014, an improved degraded document image binarization using a guided image filter was proposed by Kaur et al [22]. The proposed method provides in-depth analysis of the work of a new technique which has the ability to binarized documents in a more efficient manner. It consists of three parts, the first part will be responsible for applying the guided image filter to smooth and restore a degraded document which will improve the visibility of the degraded image. Now second part comes into action, in this part adaptive image contrast enhancement are going to be applied. At last, the third part is under action which can do the true binarization. The contrast map is going to be binarized using adaptive thresholding thus it will be integrated with a Hybrid edge map to recognize the text stroke edge pixels. The comparative analysis has shown that the proposed algorithm provides quite a significant improvement over the available algorithms. Image Quality Analysis (IQA) acquired the average of 0.9896 for Accuracy, 95.84% for F-measure, and 22.89 for PSNR.

Ntirogiannis et al. [23] have done a combined approach for the binarization of handwritten document images. The research was inspired by the low detection rate of faint characters in binarization of handwritten document images. Initially, background estimation is applied along with image normalization based on background compensation. Afterward, global binarization is performed on the normalized image. In the binarized images, very small components are discarded and representative characteristics of a document image such as the stroke width and the contrast are computed. Furthermore, local adaptive binarization is performed on the normalized image taking into account the aforementioned characteristics. Finally, the two binarization outputs are combined at connecting component level. For DIBCO 2011 dataset, the average F-measure and PSNR are 94.05% and 21.65 respectively.

In the year 2017, Peng et al. [24] have reported Document Image binarization using Convolutional Encoder-Decoder. In the proposed method, mid-level document image representations are learned by a stack of convolutional layers that compose the encoder in this architecture. Then, by mapping low-resolution representations of the original size through the decoder, the binarization image is obtained, which is then composed by a series of transposed convolutional layers. The method compares the proposed method with other binarization algorithms both qualitatively and quantitatively on the public dataset. The experimental results show that the proposed method has comparable performance to the other hand-crafted binarization approaches and has more generalization capabilities with limited in-domain training data. For H-DIBCO 2016 dataset, the average F-measure and PSNR are 88.07% and 18.13 respectively.

Document image binarization using a discriminative structural classifier was written by author Ahmadi et al. [25]. The study done was to propose a statistical framework for binarizing degraded document images based on the concept of conditional random fields (CRFs). The CRFs are discriminative graphical models which display conditional distributions that are used in structural classifications. The proposed graphical framework enables the depending labeling of all the sites of the image despite the independent pixel-by-pixel binarization of other methods. The performance of the proposed system is evaluated on different document image datasets and is compared with several well-known binarization methods. Experimental results show comparable performance with respect to other state-of-the-art methods. By using DIBCO 2009 dataset, the average F-measure and PSNR are 91.12% and 18.63 respectively.

In this paper, a binarization method based on Fuzzy C-Means and Deghost technique is proposed. The proposed method came from a combination of the Fuzzy C-Means clustering algorithm and Deghost method. In order to prove the effectiveness of the proposed method, three image quality assessment (IQA) which are Accuracy, F-measure, and Peak Signal Noise Ratio (PSNR) have been calculated. This paper is organized into four sections. Section 2 and 3 explains the Fuzzy C-Means and Deghost method, Section 4 shows the results performance in terms of the IQA. Section 5 describes the summary of the work.

## 2. Fuzzy C-Means Clustering Algorithm

The first step is applying Fuzzy C-means algorithm into the processed image. The process is needed to separate text from the background. The standard FCM objective function for partitioning $\{x_k\}_{k=1}^{N}$ into $c$ clusters is given by Equation [26].

$$J = \sum_{i=1}^{c} \sum_{k=1}^{N} u_{ik}^{p} \, ||x_k - v_i||^2 \tag{1}$$

where $\{v_i\}_{i=1}^{c}$ are the prototypes of the clusters and the array $[u_{ik}] = U$ represents a partition matrix, $U \in U$ , namely in Equation 2 [26] .

$$U \{ u_{ik} \in [0,1] | \sum_{i=1}^{c} u_{ik} = 1 \, \forall \, k \; and \; 0 < \sum_{k=1}^{N} u_{ik} < N \, \forall \, i \} \tag{2}$$

The parameter $p$ is a weighting exponent on each fuzzy membership and determines the amount of fuzziness of the resulting class friction. The FCM objective function is minimized when high membership values are assigned to voxels whose intensities are close to the centroid of its particular class, and low membership values are assigned when the voxel data is far from the centroid. The proposed technique, two clusters value has been used.

## 3. Deghost Method

The next processed is applying the Deghost method to the processed image. The Deghost method removes "ghost" entities in the processed image.

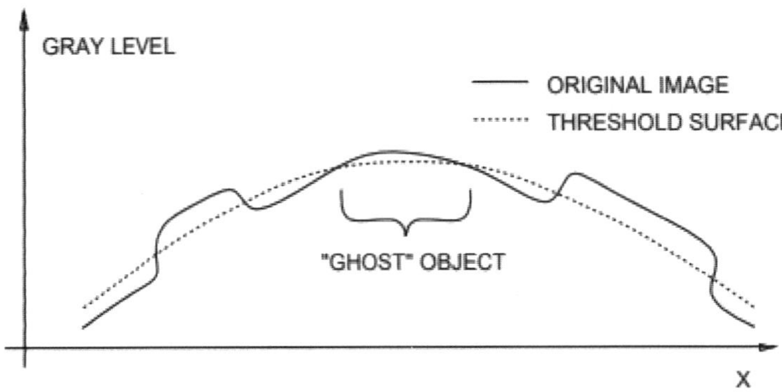


**Figure 1:** Cross section through an image, illustrating how "ghost" objects might occur

The main steps of the algorithm:

1. The gradient magnitude image G of the smoothed image is calculated, using Sobel's edge operator.
2. A value for Objects having an average gradient below a threshold $T_p$ is selected. There is no automatic method to specify $T_p$, so it is specified by trial and error.
3. For all four-connected print components, the average gradient of the edge pixels is calculated.

Based on figure 2 after applying the Deghost method on the image, some unwanted noise after the binarization process is successfully removed or reduced. Figure 2 (top) image shows the resulting after applied segmentation process using FCM and figure 2 (bottom) shows the result of a combination between FCM and enhanced using the Deghost technique.

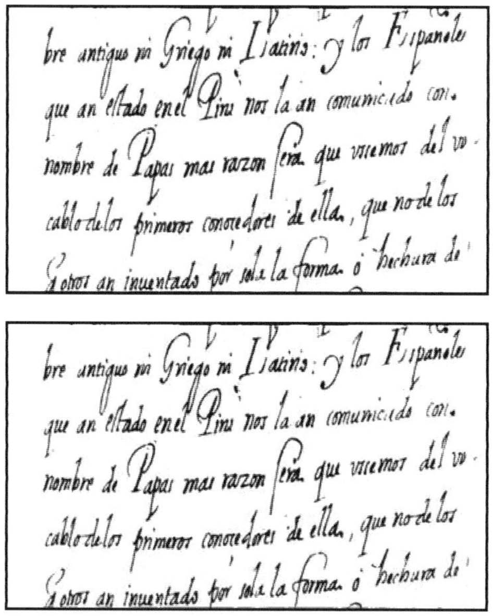

**Figure 2:** Image before (top) and after applying the Deghost method (bottom).

### 4. Results And Discussion

In this study, 16 document images from H-DIBCO 2013 dataset which have non-uniform image background is used as a test subject. The result of document image after the binarization process is shown in Figure 3. According to Figure 3, it is clear that the proposed method is better than the comparative method as it contains less noise and its accuracy is comparable to ground truth. Therefore, it is hard to compare all the method in Figure 3 using only human naked eyes. Thus, Image Quality Analysis (IQA) must be referred to find the most effective one.

| Method | Image 1 | Image 2 | Image 3 |
|---|---|---|---|
| Original | 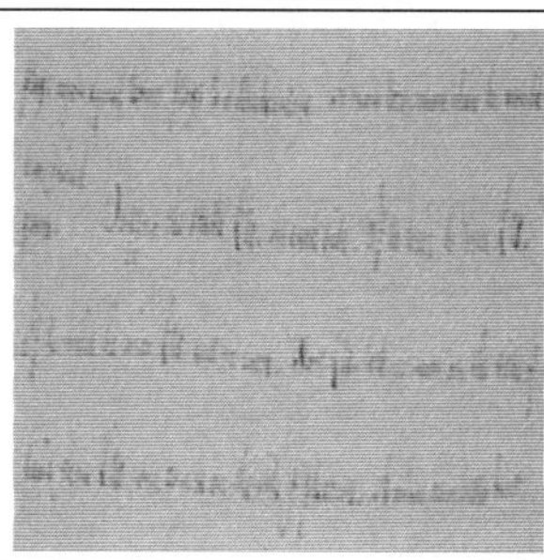 | 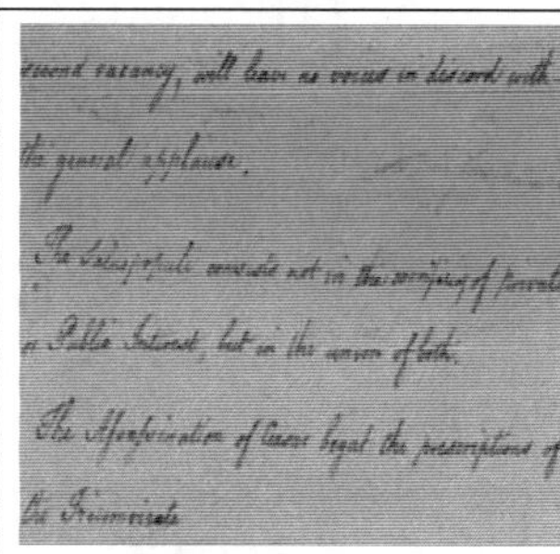 | 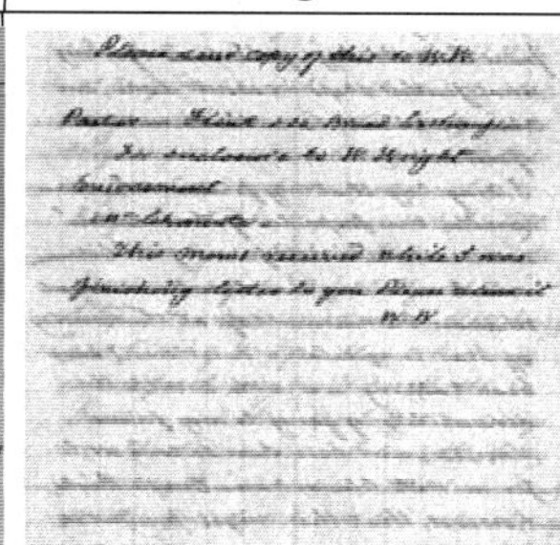 |
| Ground truth | 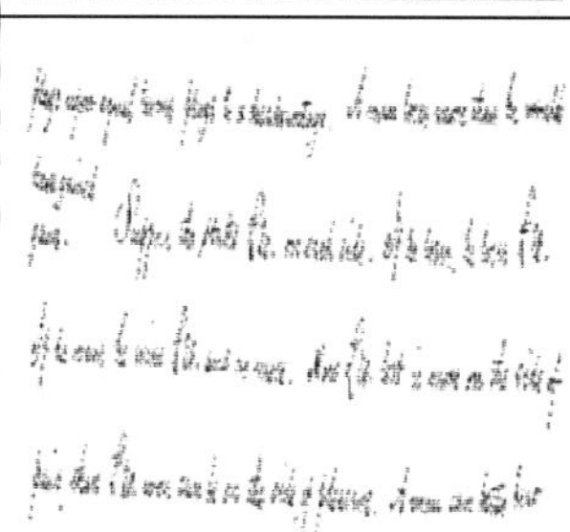 | 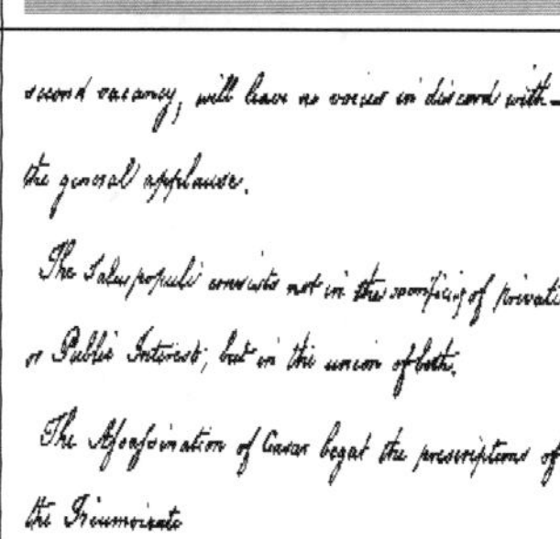 | 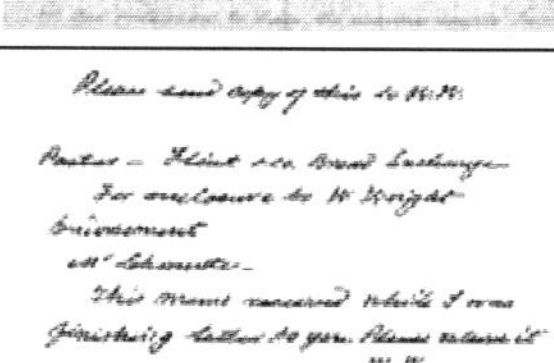 |

IOP Conf. Series: Materials Science and Engineering **557** (2019) 012012 doi:10.1088/1757-899X/557/1/012012

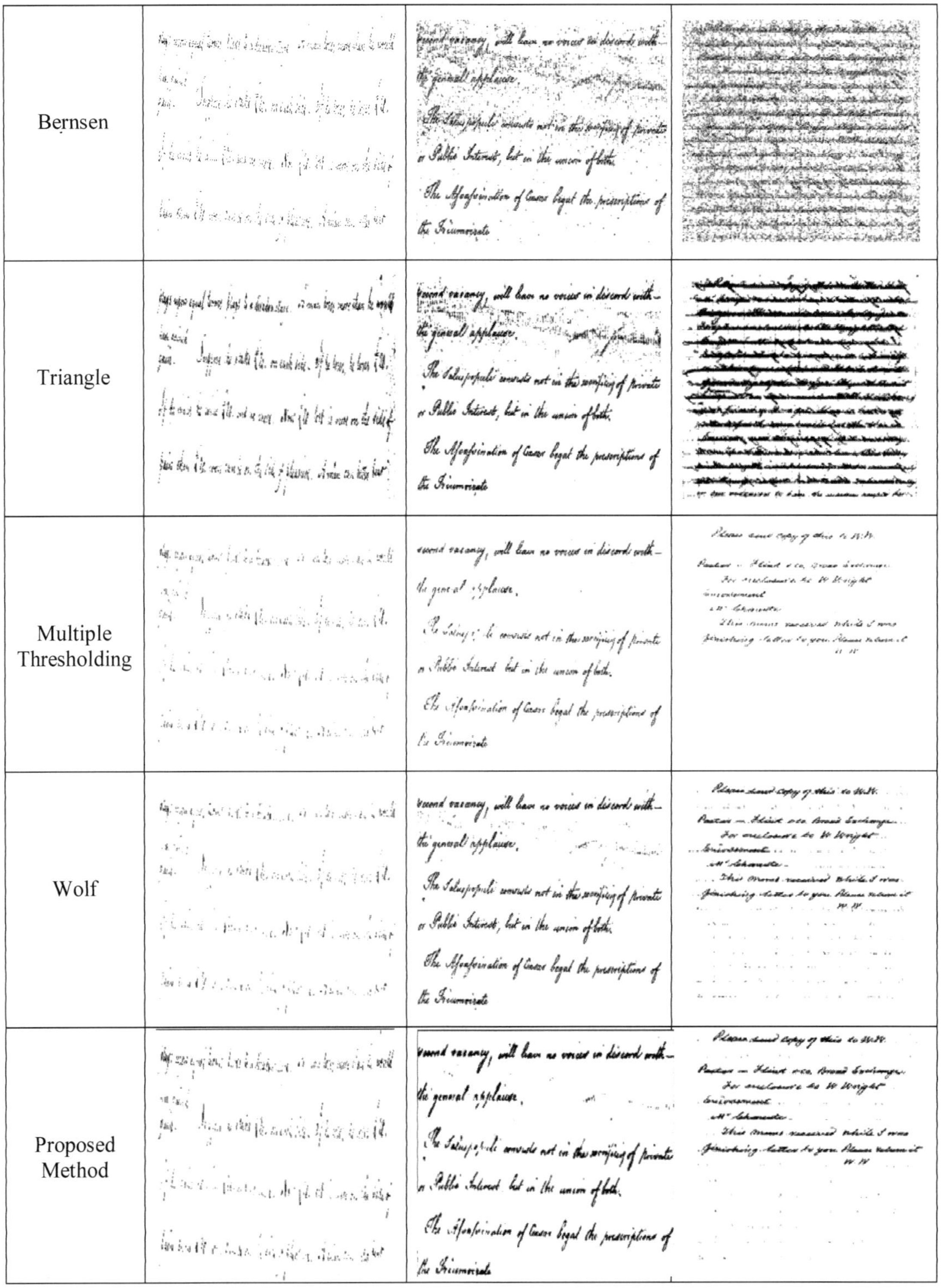


**Figure 3:** Image Binarization using H-DIBCO 2013 for selected binarization method

Based on Figure 3, the proposed method images are comparable to the ground truth image as both images are almost identical from the comparison below. The proposed method image also has a low amount of noise compared to another method in Figure 3. It can be stated from proposed method image, that it is better than the other comparison method. In order to assess the overall performance of each selected binarization method, a quantitative analysis is conducted. The performances of different binarization techniques are evaluated by using three objective indices. These indices are F-measure, Accuracy, and PSNR. The pixels similarity of the resultant segmented image and the segmented image that use the selected method determines the quality of the segmented image. All calculation is based on true positive (TP), true negative (TN), false positive (FP), precision (P), recall (R), the maximum pixel value ($Max_I$) mean squared error (MSE), and false negative (FN). The segmentation accuracy can be obtained by calculating the percentage of pixels that are correctly segmented and background in the image. Accuracy score reaches its best value at 100 and worst at 0. The accuracy equation can be referred in Equation 3 [27,28].

$$Accuracy = \left(\frac{TP+TN}{TP+TN+FP+FN}\right) \tag{3}$$

The result of IQA for Accuracy is presented in Table 1. Bernsen method gives out the lowest value of 85.68% for Accuracy compare to other methods. This is due to the processed images that still have much noise after the binarization process. However, the proposed method achieves the highest result with 96.06%. From the data in Table 1, it is apparent that all selected method gave the satisfied result.

**Table 1:** Accuracy performance against H-DIBCO 2013 dataset for the selected method

| Accuracy (%) | | | | | |
|---|---|---|---|---|---|
| **Image** | **Bernsen** | **Triangle** | **Multiple Threshold** | **Wolf** | **Proposed Method** |
| 1 | 99.17 | 98.58 | 97.91 | 96.29 | 98.26 |
| 2 | 93.10 | 96.90 | 97.15 | 93.38 | 98.02 |
| 3 | 96.84 | 95.85 | 95.43 | 93.48 | 95.72 |
| 4 | 68.33 | 97.73 | 99.23 | 96.97 | 99.50 |
| 5 | 70.24 | 77.43 | 98.93 | 97.47 | 99.39 |
| 6 | 91.82 | 6.69 | 95.63 | 94.15 | 98.09 |
| 7 | 94.36 | 98.47 | 97.95 | 97.73 | 97.89 |
| 8 | 68.36 | 94.91 | 98.96 | 95.18 | 99.13 |
| 9 | 92.79 | 97.61 | 97.55 | 95.04 | 95.87 |
| 10 | 80.79 | 95.53 | 97.82 | 92.41 | 98.92 |
| 11 | 96.56 | 98.74 | 98.33 | 94.76 | 97.88 |
| 12 | 89.42 | 66.91 | 77.81 | 88.90 | 80.09 |
| 13 | 90.13 | 87.41 | 54.51 | 84.23 | 93.59 |
| 14 | 89.30 | 84.89 | 59.33 | 90.81 | 96.24 |
| 15 | 87.96 | 93.50 | 95.34 | 80.31 | 92.80 |
| 16 | 61.77 | 89.21 | 96.16 | 89.47 | 95.60 |
| **Average** | 85.68 | 86.27 | 91.13 | 92.54 | **96.06** |

Besides, F-measure also was calculated based on Equation 4 [27].

$$F - measure = 2 \times \left(\frac{P \times R}{P + R}\right) \times 100 \quad (4)$$

The F-measure score is the harmonic average of the precision and recall, where the F-measure score reaches its best value at 1 and worst at 0. The result of F-Measure should be higher to show a well-segmented image. The result of IQA for F-measure is presented in Table 2. From the data in Table 2, it is apparent that all selected method gave the satisfied result. The proposed method achieves the highest result with 74.96%. This is due to the processed images that have been well binarized. However, the Wolf method gives out the lowest value of 23.32% for the F-Measure compare to other methods.

**Table 2:** F-Measure performance against H-DIBCO 2013 dataset for the selected method

| F-Measure (%) | | | | | |
|---|---|---|---|---|---|
| **Image** | **Bernsen** | **Triangle** | **Multiple Threshold** | **Wolf** | **Proposed Method** |
| 1 | 88.53 | 84.33 | 63.50 | 9.60 | 74.02 |
| 2 | 60.53 | 80.30 | 73.11 | 2.54 | 84.82 |
| 3 | 75.65 | 75.20 | 47.14 | 2.37 | 56.95 |
| 4 | 19.44 | 80.65 | 91.28 | 53.73 | 94.51 |
| 5 | 14.56 | 21.58 | 79.33 | 32.32 | 90.67 |
| 6 | 56.62 | 12.53 | 51.52 | 22.49 | 84.07 |
| 7 | 33.61 | 69.86 | 19.16 | 2.31 | 34.81 |
| 8 | 19.76 | 69.87 | 90.64 | 33.83 | 92.54 |
| 9 | 52.05 | 79.72 | 68.23 | 4.75 | 38.44 |
| 10 | 42.59 | 79.95 | 86.10 | 26.06 | 93.74 |
| 11 | 76.85 | 91.45 | 87.12 | 44.22 | 83.86 |
| 12 | 64.73 | 42.86 | 51.18 | 41.54 | 50.47 |
| 13 | 74.91 | 74.49 | 44.70 | 42.70 | 83.29 |
| 14 | 62.34 | 60.04 | 35.82 | 32.03 | 81.79 |
| 15 | 73.47 | 86.64 | 87.68 | 13.14 | 79.60 |
| 16 | 25.82 | 60.44 | 76.03 | 9.47 | 75.73 |
| **Average** | 52.59 | 66.87 | 65.78 | 23.32 | **74.96** |

PSNR is most commonly used to measure the quality of reconstruction of lost compression code. The equation of PSNR as follows in Equation 5 [27].

$$PSNR = 10 \log_{10}\left(\frac{Max_I^2}{MSE}\right) \quad (5)$$

PSNR is most commonly used to measure the quality of reconstruction of lost compression code. The result of IQA for PSNR is presented in Table 3. Bernsen method gives out the lowest value of 10.13 dB for PSNR out of all the selected methods. This is due to the processed images that have not well reconstructed. However, the proposed method achieves the highest result with 16.03 dB. From the data in Table 3, it can be said that all selected method gave the satisfied result.

**Table 3:** PSNR performance against H-DIBCO 2013 dataset for the selected method

| PSNR (dB) | | | | | |
|---|---|---|---|---|---|
| **Image** | **Bernsen** | **Triangle** | **Multiple Threshold** | **Wolf** | **Proposed Method** |
| 1 | 20.79 | 18.49 | 16.79 | 14.30 | 17.60 |
| 2 | 11.61 | 15.09 | 15.46 | 11.79 | 17.04 |
| 3 | 15.00 | 13.82 | 13.40 | 11.86 | 13.68 |
| 4 | 4.99 | 16.43 | 21.15 | 15.19 | 22.98 |
| 5 | 5.26 | 6.46 | 19.72 | 15.97 | 22.15 |
| 6 | 10.87 | 0.30 | 13.60 | 12.33 | 17.18 |
| 7 | 12.49 | 18.17 | 16.88 | 16.44 | 16.75 |
| 8 | 5.00 | 12.93 | 19.83 | 13.17 | 20.63 |
| 9 | 11.42 | 16.21 | 16.10 | 13.05 | 13.84 |
| 10 | 7.17 | 13.50 | 16.61 | 11.20 | 19.68 |
| 11 | 14.63 | 18.99 | 17.77 | 12.81 | 16.73 |
| 12 | 9.75 | 4.80 | 6.54 | 9.55 | 7.01 |
| 13 | 10.06 | 9.00 | 3.42 | 8.02 | 11.93 |
| 14 | 9.71 | 8.21 | 3.91 | 10.36 | 14.25 |
| 15 | 9.19 | 11.87 | 13.32 | 7.06 | 11.43 |
| 16 | 4.18 | 9.67 | 14.16 | 9.78 | 13.56 |
| **Average** | 10.13 | 12.12 | 14.29 | 12.05 | **16.03** |

## 5. Conclusion

Today, there is a strong move toward digitalization to preserve the content of the historical document for future generations. The huge amount of digital data produced requires automatic processing, enhancement, and recognition. There is a need for binarization approach on the document image to retrieve information on a document. In this paper, a combination of the Fuzzy C-Means clustering algorithm and Deghost method are used to create a binarization technique. The target is to suggest a maximum threshold value in order to detect the text image. Besides, the goal of this study is to acquire result in term of reducing blur image and eliminate the noise. In order to analyze the result performance, a few IQA such as F-measure, and PSNR was calculated. The proposed method successfully to improve the image quality in term of achieved highest result.

## 6. References

[1] Mustafa W A and Yazid H 2017 Image Enhancement Technique on Contrast Variation : A Comprehensive Review *J. Telecommun. Electron. Comput. Eng.* **9** 199–204

[2] Pardhi S and Kharat G U 2017 An Improved Binarization Method for Degraded Document *Int. J. Res. Advent Technol.* 1–5

[3] Lu D, Huang X, Liu C, Lin X, Zhang H and Yan J 2016 Binarization of degraded document image based on contrast enhancement *Chinese Control Conf. CCC* **2016-Augus** 4894–9

[4] Mysore S, Gupta M K and Belhe S 2016 Complex and degraded color document image binarization *3rd Int. Conf. Signal Process. Integr. Networks, SPIN 2016* 157–62

[5] Mustafa W A and Kader M M M A 2018 Binarization of Document Images: A Comprehensive Review *J. Phys. Conf. Ser.* **1019** 1–9

[6] Mustafa W A and Kader M M M A 2018 Document Image Database ( 2009 - 2012 ): A Systematic Review *J. Phys. Conf. Ser.* **1019** 1–7

[7] Ayyalasomayajula K R and Brun A 2017 Document Binarization Combining with Graph Cuts and Deep Neural Networks *36th Swedish Symposium on Image Analysis* pp 1–6

[8] Obafemi-Ajayi T, Agam G and Frieder O 2010 Historical document enhancement using LUT classification *Int. J. Doc. Anal. Recognit.* **13** 1–17

[9] Mustafa W A and Yazid H 2016 Illumination and Contrast Correction Strategy using Bilateral Filtering and Binarization Comparison *J. Telecommun. Electron. Comput. Eng.* **8** 67–73

[10] Kuk J G and Cho N I 2009 Feature Based Binarization of Document Images Degraded by Uneven Light Condition *2009 10th Int. Conf. Doc. Anal. Recognit.* 748–52

[11] Kaur J and Mahajan R 2014 A Review of Degraded Document Image Binarization Techniques *Int. J. Adv. Res. Comput. Commun. Eng.* **3**

[12] Mustafa W A, Khairunizam W, Ibrahim Z, Ab S and M.Razlan Z 2018 Improved Feng Binarization Based on Max-Mean Technique on Document Image *IEEE International Conference on Computational Approach in Smart Systems Design and Applications (ICASSDA)* (IEEE) pp 1–6

[13] Mustafa W A, Aziz H, Khairunizam W, Ibrahim Z, Ab S and M.Razlan Z 2018 Review of Different Binarization Approaches on Degraded Document Images *IEEE International Conference on Computational Approach in Smart Systems Design and Applications (ICASSDA)* (IEEE) pp 1–8

[14] Mustafa W A, Yazid H and Jaafar M 2018 An Improved Sauvola Approach on Document Images Binarization *J. Telecommun. Electron. Comput. Eng.* **10** 43–50

[15] Mustafa W A and Kader M M M A 2018 Binarization of Document Image Using Optimum Threshold Modification *J. Phys. Conf. Ser.* **1019** 1–8

[16] Kaviya Selvi K and Sabeenian R S 2015 Restoration of degraded documents using image binarization technique *ARPN J. Eng. Appl. Sci.* **10** 2813–7

[17] Ranganatha D and Holi G 2015 Hybrid binarization technique for degraded document images *Souvenir of the 2015 IEEE International Advance Computing Conference, IACC 2015* pp 893–8

[18] Zemouri E T, Chibani Y and Brik Y 2014 Restoration based Contourlet Transform for historical document image binarization *Int. Conf. Multimed. Comput. Syst. -Proceedings* **0** 309–13

[19] Singh B M, Sharma R, Ghosh D and Mittal A 2014 Adaptive binarization of severely degraded and non-uniformly illuminated documents *Int. J. Doc. Anal. Recognit.* 393–412

[20] Jia F, Shi C, He K, Wang C and Xiao B 2018 Degraded document image binarization using structural symmetry of strokes *Pattern Recognit.* **74** 225–40

[21] Jyoti D, Raj B, Sharma A and Kapoor K 2016 Document Image Binarization Technique for Degraded Document Images By Using Morphological Operators *Int. J. Adv. Res. Ideas Innov. Technol.* **2** 1–7

[22] Kaur E J and Mahajan R 2014 Improved Degraded Document Image Binarization Using Guided Image Filter *Int. J. Sci. Res. Educ.* **4** 242–9

[23] Ntirogiannis K, Gatos B and Pratikakis I 2014 A combined approach for the binarization of handwritten document images *Pattern Recognit. Lett.* **35** 3–15

[24] Peng X, Cao H and Natarajan P 2017 Using Convolutional Encoder-Decoder for Document Image Binarization *2017 14th IAPR Int. Conf. Doc. Anal. Recognit.* 708–13

[25] Ahmadi E, Azimifar Z, Shams M, Famouri M and Shafiee M J 2015 Document image binarization using a discriminative structural classifier *Pattern Recognit. Lett.* **63** 36–42

[26] Mesquita R G, Silva R M A, Mello C A B and Miranda P B C 2015 Parameter tuning for document image binarization using a racing algorithm *Expert Syst. Appl.* **42** 2593–603

[27] Lokhande S S and Dawande N A 2015 A survey on document image binarization techniques *Proceedings - 1st International Conference on Computing, Communication, Control and Automation, ICCUBEA 2015* pp 742–6

[28] Mustafa W A and Yazid H 2016 Background Correction using Average Filtering and Gradient Based Thresholding *J. Telecommun. Electron. Comput. Eng.* **8** 81–8

# An Overview on Overvoltage Phenomena in Power Systems

**N S Othman[1], M N K H Rohani[1], W A Mustafa[2], C L Wooi[1], A S Rosmi[1], N F M Shakur[1], R Juliangga[1], Wan Khairunizam[3], I Zunaidi[4], Z M Razlan[3] and Shahriman A B[3]**

[1]Centre of Excellence for Renewable Energy (CERE), School of Electrical System Engineering, Pauh Putra Main Campus, Universiti Malaysia Perlis, 02600 Arau, Perlis, Malaysia.

[2]Department of Electrical Technology Engineering, Faculty of Engineering Technology, Universiti Malaysia Perlis (UniMAP), UniCITI Alam Campus, Sungai Chuchuh, 02100 Padang Besar, Perlis Malaysia.

[3]School of Mechatronics Engineering, Universiti Malaysia Perlis, 02600 Arau, Malaysia

[4]Faculty of Technology, University of Sunderland, St Peter's Campus, Sunderland, SR6 0DD, United Kingdom

*E-mail: syakinaothman@gmail.com*

**Abstract.** Overvoltage happens in a condition where the voltage is increased and exceed its design limit. This situation may lead to harmful damage to machines or related equipment that connected to the system. Overvoltage can exist in a form of transient, voltage spike or permanent, depending on its duration. Types of overvoltage consist of lightning overvoltage and switching overvoltage. Overvoltage that caused by lightning is considerate as natural phenomena, while switching overvoltage exists from the system itself, either by the interruption of faults or inappropriate connection of circuit breaker contacts. This paper is discussed about overvoltage phenomenon including causes and effects of overvoltage and overvoltage protection towards power system.

## 1. Introduction

Overvoltage in power system is defined as the increase in voltage for the very short time in the power system. It is also known as the voltage transients or voltage surge [1-3]. The overvoltage in the power system can be classified into two factors which are an internal factor (temporary, switching) or external factor (usually lightning) [4-6]. Overvoltage in power systems can happen for several reasons such as lightning, faults, and disconnection with the most destructive is caused by a lightning strike to the power system [7-8].

A single lightning strike can carry up to 300 kV voltages and almost 30 kA of current which is very high and can cause device and insulation breakdown [9]. Therefore, it is vital to design the power system so that the expected overvoltage will be lower than the value that can be withstood by the system [10]. But such a system will be costly and is not practical in real life application. The best way is to design the power system so that the damage from the overvoltage can be minimized.

The protection against overvoltage is very important to ensure smooth operation of the power system and to protect the insulation of power equipment (air, oil, SF6) which is very sensitive of the high voltage [11].

## 2. Types and Causes of Overvoltage

Overvoltage phenomena in the power system can be classified into two types which are the internal overvoltage and external overvoltage. Internal overvoltage is caused by the changes in the operating conditions of the power system. There are three common types of internal overvoltage in the power system.

### 2.1 Temporary over voltage (Sustained over voltage)

According to IEEE Standard 100-2000, the temporary overvoltage is defined as an oscillatory phase-to-ground or phase-to-phase overvoltage that is at a given location of relatively long duration (seconds, even minutes) and that is undamped or only weakly damped [12]. This type of transient is typically caused by faults (earth fault, single-phase faults etc), switching operations (or sudden disconnection) or from nonlinearities (Ferro resonance effects, harmonics). Temporary overvoltage can be differentiating from other types by the amplitude, duration, the frequencies of oscillation or the fall time ([C/PE4] 1313.1-1996, C57.12.80-1978r) [13].

The magnitude of temporary overvoltage is characteristically low compared to other types of overvoltage. The magnitude usually just a few percents to 50 percent above the nominal operating voltage except for several cases such as ferro resonance which can reach up to 300-400 percent above the nominal operating voltage [14-15]. The most common occurrence of temporary overvoltage is happening between healthy phase and the earth. When a fault happens, it will cause a voltage drop at the faulted phase and may result in overvoltage at the unfaulted phases.

### 2.2 Slow-front over voltage (Switching overvoltage)

Switching overvoltage, as the name itself suggest is caused by the switching operation in the power system or sometimes can be because of fault [16-17]. This type of overvoltage is quite common since switching operations are carried out routinely in the power system as part of the testing process. Due to the Ferranti Effect, when an unloaded long line is charged, the voltage at the receiving end will increase in a significantly high amount that causes overvoltage to the system [18]. The same case happens in transformer operation where the transient can occur when the primary side of the transformer is switched on. The frequency content of switching transients depends on system parameters as illustrates in probability distribution curves of measured line switching overvoltage as shown in Figure 1 [19].

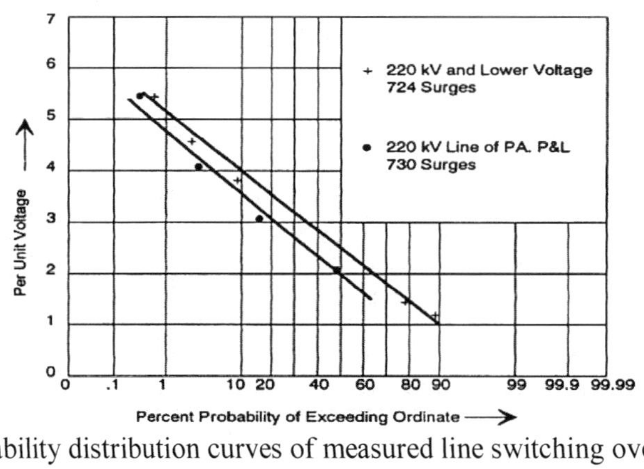


**Figure 1.** Probability distribution curves of measured line switching overvoltage [19].

### 2.3 Very-fast overvoltage

Very-fast overvoltage is happening due to the arc interruption and restriking when opening disconnector. It can be caused by switching operation, a fault in gas-insulated substation (GIS) and certain lighting condition. In GIS, the substation operation itself can cause overvoltage because of the fast breakdown of the gas gap and the unrestraint propagation of the surge in addition to the high oscillation frequency (50 kHz-100 MHz) that result from the very short distance between two adjacent transition point in GIS [20-21]. However, the transient surges are only dangerous to equipment located close to the disconnector (heats wiring, causes internal resonance).

External overvoltages originates from atmospheric phenomena, primarily due to lightning [22-23]. It can be because of direct lightning stroke, electromagnetically induced overvoltages due to lightning discharge taking place near the line (side stroke), back flashover and electrostatically induced voltages.

This type of overvoltage is also known as lightning overvoltage or fast-front overvoltage. A lightning can stroke generally anywhere on a power system or nearby object even if it tends to strike the tallest structure. Unlike other types of overvoltage, which is proportional to the system voltage, lightning overvoltages are dependent on system impedances instead of system voltage. It is almost impossible economically and practically to protect distribution or lower-kilovolt-level transmission lines to withstand this type of overvoltage.

## 3. Overvoltage Protection

Overvoltage may be of transient or persistent nature. There are two categories, internal and external. This overvoltage can source destruction to insulators and substation gear [24-25]. Thus, it is required to provide facilities to protect insulators and other gear from the damaging of overvoltage [26]. The process of protecting the electrical system from the destruction that may be triggered by overvoltage. The use of devices such as bent sirens close to the transmission line and Zener diodes for electronic circuits.

There are dissimilar methods can be used for overvoltage protection, with its specific characteristics. Performance, cost, difficulty and functioning mode all need to be considered when defining the method to use.

### *3.1 Surge Arresters*

Series of construction SiC resistors and spark gaps, which are positioned in porcelain covering, and are commonly named "spark gap arresters" [27]. Overvoltage has a different voltage value, characteristics that differentiate overvoltage protection devices is their rated surge capacity and the achievable level of protection [28].

- Level 1 lightning arrester: Defends against overvoltage and great currents caused straight or secondary lightning strikes
- Level 2 surge arrester: Defends against overvoltage caused electrical switching processes
- Level 3 surge arrester: Defends electrical loads against overvoltage

### *3.2 SCR Crowbar*

Figure 2 shows the crowbar circuit that seats a short circuit across the output if an overvoltage situation is practiced. Characteristically thyristors for example SCRs are used for as they can switch great currents and continue on pending any charge has isolated. Often the thyristor is related back to a fuse that isolates the regulator from needing any more voltage located upon it [29].

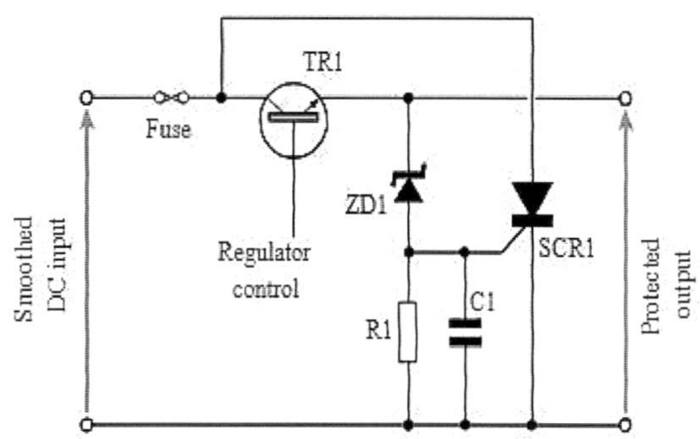


**Figure 2.** Thyristor crowbar circuit [29].

### *3.3 Voltage Clamping*

In the simplest protection system can be on condition that used a Zener diode located across the output as illustrated in Figure 3. With the Zener diode voltage selected to be slightly overhead the determined line voltage, below normal situations it determination not conduct. If the voltage increases too great, then it will start to conduct, clamping the voltage at a worth somewhat overhead the line voltage [30].

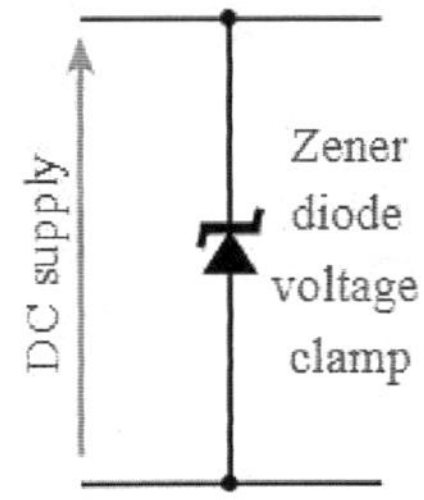


**Figure 3.** Simple Zener diode over-voltage clamp [30].

## 4. Conclusion

Overvoltage absolutely will cause harm and damage to the insulation of the distribution network equipment. Therefore, it is vital to identify the types and causes of overvoltage so that undesirable damage can be minimized. Various methods of overvoltage protection provide safety towards the performance of power systems networks such as surge arresters, SCR crowbar, and voltage clamping.

## References

[1] Sze S M 1969 *Physics of Semiconductor Devices* (New York: Wiley–Interscience)
[2] Dorman L I 1975 *Variations of Galactic Cosmic Rays* (Moscow: Moscow State University Press) p 103
[3] Caplar R and Kulisic P 1973 *Proc. Int. Conf. on Nuclear Physics (Munich)* vol 1 (Amsterdam: North-Holland/American Elsevier) p 517
[4] Szytula A and Leciejewicz J 1989 *Handbook on the Physics and Chemistry of Rare Earths* vol 12, ed K A Gschneidner Jr and L Erwin (Amsterdam: Elsevier) p 133
[5] Kuhn T 1998 Density matrix theory of coherent ultrafast dynamics *Theory of Transport Properties of Semiconductor Nanostructures (Electronic Materials* vol 4*)* ed E Schöll (London: Chapman and Hall) chapter 6 pp 173–214
[6] Xue H and Popov M 2014 Analysis of switching transient overvoltages in the power system of floating production storage and offloading vessel *Elcectric Power Syst. Res.* **115** 3–10
[7] A H S, M M and T B 2015 Impact of surge arrester number and placement on reliability and lightning overvoltage level in high voltage substations *Int. J. Electr. Power Energy Syst.* **65** 146–58
[8] Reza S and Javad G 2017 The effect of grounding system modeling on lightning-related studies of transmission lines *J. Appl. Res. Technol.* **15** 545–54
[9] Franc B, Filipović-Grčić B and Milardić V 2016 Lightning overvoltage performance of 110 kV air-insulated substation *Electr. Power Syst. Res.* **138** 78–84
[10] Yu Z, Zhu T, Wang Z, Lu G, Zeng R, Liu Y, Luo J, Wang Y, He J and Zhuang C 2016 Calculation and experiment of induced lightning overvoltage on power distribution line *Electr. Power Syst. Res.* **139** 52–9
[11] Saad H, Rault P and Dennetière S 2018 Study on transient overvoltages in converter station of MMC-HVDC links *Electr. Power Syst. Res.* **160** 397–403
[12] IEEE 2000 *100-2000 - The Authoritative Dictionary of IEEE Standards Terms*
[13] Santiago J and Tavares M C 2019 Analysis of half-wavelength transmission line under critical balanced faults: Voltage response and overvoltage mitigation procedure *Electr. Power Syst. Res.* **166** 99–111
[14] Shariatinasab R, Kermani B and Gholinezhad J 2019 Transient modeling of the wind farms in order to analysis the lightning related overvoltages *Renew. Energy* **132** 1151–66
[15] Jin E S, Chen X F, Cao D L and Bo Z Q 2012 Theoretical Analysis on High Frequency Ferroresonant Overvoltage *Energy Procedia* **17** 233–40

[16] Han R, Yang Q, Sima W, Zhang Y, Sun S, Liu T and Chen S 2016 Non-contact measurement of lightning and switching transient overvoltage based on capacitive coupling and pockels effects *Electr. Power Syst. Res.* **139** 93–100

[17] Filipović-Grčić B, Franc B, Uglešić I, Pavić I, Keitoue S, Murat I and Ivanković I 2017 Monitoring of transient overvoltages on the power transformers and shunt reactors - Field experience in the Croatian power transmission system *Procedia Eng.* **202** 29–42

[18] Pengcheng Y, Shuiming C and Jinliang H 2010 Effect of Different Arresters on Switching Overvoltages in UHV Transmission Lines *Tsinghua Sci. Technol.* **15** 325–8

[19] Santoso S and Beaty H W 2017 Power System Overvoltages *Standard Handbook for Electrical Engineers* (McGraw-Hill Education)

[20] Feng S, Zheng L, Lie J, Sun Y, Chen G and Ji X 2011 Research on arc-grounding overvoltage in the 10kV distribution system *Energy Procedia* **16** 1785–91

[21] Suo C, Xu X, Zhang W, Gui J, Tan X and Li T 2017 GIS Very Fast Transients Overvoltage Testing Using Electric Field Measurement *Proceedings of the Fifteenth China International Nanoscience and Technology Symposium* pp 23–4

[22] Trainba M, Christodoulou C A, Vita V and Ekonomou L 2017 Lightning overvoltage and protection of power substations *WSEAS Trans. Power Syst.* **12** 107–14

[23] Rodrigues R B, Mendes V M F and Catalão J P S 2012 Protection of interconnected wind turbines against lightning effects: Overvoltages and electromagnetic transients study *Renew. Energy* **46** 232–40

[24] Mokhtari G, Nourbakhsh G, Zare F and Ghosh A 2013 Overvoltage prevention in LV smart grid using customer resources coordination *Energy Build.* **61** 387–95

[25] Mestas P and Tavares M C 2010 Comparative Analysis of Techniques for Control of Switching Overvoltages During Transmission LInes Enerization *Elcectric Power Syst. Res.* **80** 115–20

[26] Soloot A H, Høidalen H K and Gustavsen B 2012 The assessment of overvoltage protection within energization of offshore wind farms *Energy Procedia* **24** 151–8

[27] Napolitano F, Borghetti A, Nucci C A, Martinez M L B, Lopes G P and Dos Santos G J G 2014 Protection against lightning overvoltages in resonant grounded power distribution networks *Electr. Power Syst. Res.* **113** 121–8

[28] Datsios Z G and Mikropoulos P N 2016 Modeling of lightning impulse behavior of long air gaps and insulators including predischarge current : Implications on insulation coordination of overhead transmission lines and substations *Electr. Power Syst. Res.* **139** 37–46

[29] Patel S S 2015 Over Voltage Protection Using Crowbar Devices for Low Voltage Loads 6703–5

[30] Wei L, Liu Z and Skibinski G L 2011 DC Bus Voltage Clamp Method to Prevent Over-Voltage Failures in Adjustable Speed Drives *IEEE Trans. Ind. Appl.* **47** 1178–785

IOP Conf. Series: Materials Science and Engineering **557** (2019) 012014 doi:10.1088/1757-899X/557/1/012014

# Selection of Morphological Features in Classifying Weedy Rice and Rice Seed Varieties using Discriminant Function Analysis

**R Ruslan[1,2,3], S K Bejo[1], I H Rukunuddin[2] and A A Aznan[2]**

[1]Department of Biological and Agricultural Engineering, Universiti Putra Malaysia, UPM 43400 Serdang, Selangor Malaysia

[2]School of Bioprocess Engineering, Universiti Malaysia Perlis, UniMAP 02600 Arau, Perlis Malaysia

[3]Center of Excellence for Advanced Sensor Technology (CEASTech) Universiti Malaysia Perlis, UniMAP 02600 Arau, Perlis Malaysia

**Abstract.** A comparison on classification accuracy and morphological features selection between stepwise discriminant function analysis (DFA) and enter all variables DFA were made for classification of Malaysian rice seed varieties namely MR297, MR263, MR284, MR219 and a group of weedy rice. Eighteen morphological features were used for the discriminant analysis. The classification accuracies of MR297 and MR263 against weedy rice group were maintained at 99.1% and 98.9% respectively, either using stepwise DFA or normal DFA. The classification accuracy for MR284 decrease from 95.0% to 93.7% using stepwise DFA with the reduction to five morphological features selected in the analysis. However for MR219, the classification accuracy increased by 0.3% using optimized features in stepwise DFA. Thus it can be concluded the optimization of morphological features through stepwise DFA classification does not necessarily increase the accuracy for discriminating weedy rice and Malaysian rice seed varieties.

## 1. Introduction

Production of rice seed in Malaysia is greatly depends on the purity of the cultivated rice seed produced under the certified rice seed program monitored by the Department of Agriculture (DOA). Under this program, the farmers are ensured to receive high quality seeds by monitoring the seed production activities starting from planting to harvesting, processing and seed sampling in the laboratory [1]. Rice seed sampling is one of the crucial part where the processed rice seeds are tested for germination rate, moisture content, and finding foreign/unwanted seed. Normally, seed producers having difficulty in getting clearance to meet the specification from DOA where they fail the test in finding foreign/unwanted seed. Thus, the purposed of this project is to replace manual process of distinguishing the unwanted seed or known as weedy rice from the cultivated rice seed samples. With the advancement of technology, a machine vision-based inspection system was developed for this project to identify the cultivated rice seeds and weedy rice seeds.

## 2. Related Research

Substantial work related to machine vision-based in agriculture especially in the rice seed industry has been carried out. Common features used in classifying the rice grains are morphological [2][3], color, textural and combination of the said features [4]. Some researchers such as [5] identify the sterile lemma traits in discriminating between rice seed cultivars. The classifiers tested for image classification were done using either parametric or non-parametric such as support vector machine (SVM), neural network, k-nearest neighbour and discriminant function analysis (DFA). The classification accuracy up to 85% were achieved in classifying rice seed cultivars. Other works involving rice seed classification were to discriminate between weedy rice and rice seed cultivars [6]. [4] extracted texture and shape feature of four Indian rice variety using vision based and image processing and evaluated the classification accuracy using neural network. The classification accuracy were highest for shape features, followed by texture-n-shape and texture features at 88%, 87.27% and 82.61% respectively. Work by [5] used microscopic imaging to identify 45 rice seed cultivar using morphological, colors and sterile lemma traits of the rice grains with overall classification accuracy reached to 85.02% by support vector machine (SVM) classifier.

## 3. Materials and Method

### *3.1. Machine Vision Setup*

A machine vision-based setup consist of a CMOS 6MP 1/1.8" GigE camera (MVCA060-10GC, HIK Vision) and 6MP 25mm focal length lens (MVL-HF2528M-6MP, HIK Vision) was setup for this project. The area scan camera resolution is 3072 x 2048 and has sensor size of 2.4 µm x 2.4 µm per pixel. A laptop with 12Gb RAM memory and 2.8 GHz (Intel i7) processing speed was attached to the camera through Ethernet cable CAT6 with speeds up to 1000 Mb/s. Image processing and analysis was programmed in LabVIEW 2016 (National Instruments, Texas USA) visual programming language. A low angled LED lighting (TMS Lite) was used and provides even and controlled lighting distribution on the seed samples. The seed sample was kept at a distance of 13.7 cm from the lens and camera.

### *3.2. Rice Seed Samples*

Cultivated rice seed samples used in this project were supplied from Rice Seed Bank (MARDI Seberang Perai, Malaysia). Four rice seed varieties which in demand every planting seasons (MR297, MR263, MR284 and MR219) and two weedy rice variants (the unwanted/off type rice seed similar to cultivated rice) gathered from plantation area were used in this project. Figure 1 displays sample images of the cultivated rice seed and weedy rice variants. The shape of the rice seed varieties is almost similar to each other and weedy rice variants used in this project is slightly smaller than cultivated rice seed varieties and have an awn.

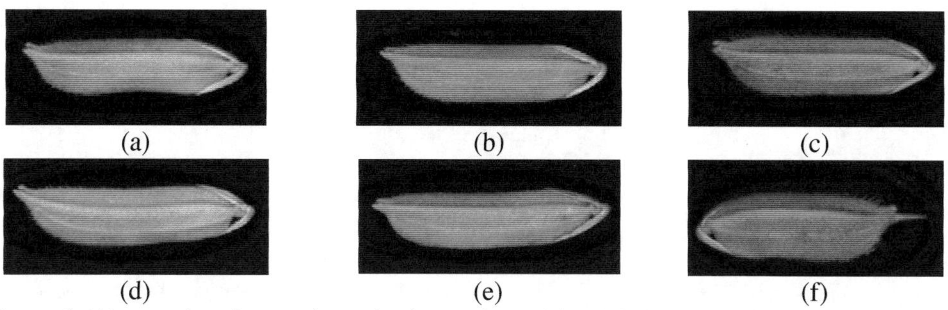


**Figure 1.** Rice seed variety and weedy rice variants (a) MR219 (b)MR284 (c) Weedy Rice Variant A (d) MR297 (e) MR263 (f) Weedy Rice Variant B

### *3.3. Morphological features*

Feature extraction was programmed in LabVIEW environment and IMAQ Particle Analysis VI was used. The following morphological features [7] as depicted in table 1 were extracted from labelled images of individual seed kernels:

**Table 1.** Morphological features explanation

| No | Features | Unit | Descriptions |
|---|---|---|---|
| 1. | Length | mm | Length of the bounding rectangular width of the seed kernel |
| 2. | Width | mm | Length of the bounding rectangular height of the seed kernel |
| 3. | Area | $mm^2$ | Number of pixels in the area of the seed kernel |
| 4. | Perimeter | mm | Length of the pixel boundary of the seed kernel |
| 5. | Convex Hull Perimeter | mm | Length of the smallest convex polygon containing all pixels in the seed kernel. |
| 6. | Seed Maximum Feret Diameter | mm | The longest length of the line segment connecting the two end points on the seed kernel |
| 7. | Major Axis Length | mm | Length of the major axis of the ellipse that has the same area and same perimeter as the kernel |
| 8. | Minor Axis Length | mm | Length of the minor axis of the ellipse that has the same area and same perimeter as the kernel |
| 9. | Aspect Ratio | - | Major axis length/ Minor axis length |
| 10. | Rectangular Aspect Ratio | - | Length/Width |
| 11. | Thinness Ratio | - | $(\text{Seed Perimeter})^2/(\text{Seed Area}*\pi)$ |
| 12. | Area ratio | - | (Length/Width)* Seed Area |
| 13. | Ellipse Ratio | - | Major Axis Length/Minor Axis Length |
| 14. | Rectangular Diagonal | mm | Length of the opposite corner of the bounding rectangular of the seed kernel |
| 15. | Hydraulic Radius | mm | Disk area/disk perimeter where the disk radius is the distance between a pixel boundary to the centroid of the seed kernel |
| 16. | Angle Orientation | radian | The angle of the line that passes through the seed centroid about which the particle has the lowest moment of inertia. |
| 17. | Moment XX | - | Moment on the X direction to the particle center of mass |
| 18. | Moment YY | - | Moment on the Y direction to the particle center of mass |

### *3.4. Image Processing and Analysis*

Simple image processing technique were employed to analysed the seed kernel images and features extraction. Figure 2 shows the flowchart for the image processing procedure run in the LabVIEW program.

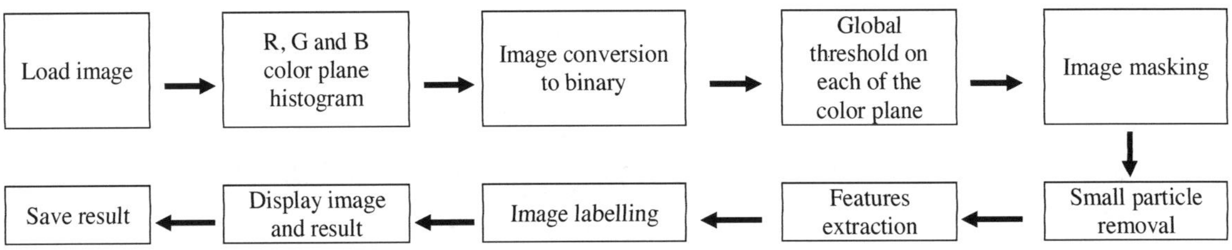


**Figure 2.** Image processing flowchart

Image analyses were performed based on non-parametric analysis. Discriminant analysis was conducted using methods (i) enter all independent variables into Discriminant Function Analysis (DFA) and (ii) stepwise DFA (SPSS). The analyses were arranged one-to-one scheme between weedy rice group and each cultivated rice seed. For both cross validation analysis, features selection were based on the minimization of the Wilks' Lambda. Morphological features that minimize the Wilks' Lambda value is selected as the predictors for the function equation. Criteria for entry and removal of the features into the predictors equation (F criteria) was set at 0.05 and 0.10, respectively.

## 4. Result and discussion

### *4.1. Morphological features selection using DFA and stepwise DFA*

Stepwise DFA technique removes or retained independent features based on the reduction of overall Wilks' Lambda in each entry while the normal DFA would enter all variables at once. In general, overall number of features selected for the discriminant function decreased from 12 to six, seven, five and eight for MR297, MR263, MR284 and MR219, respectively as indicated in table 2. In normal DFA technique, all 12 features selected for the function generation were the same in the four seed varieties. Meanwhile, the most significant features in stepwise DFA were convex hull perimeter, minor axis length and area ratio as it was selected by three rice seed varieties out of four for classification function generation.

**Table 2.** Morphological features selection using stepwise and normal discriminant function analysis.

| No | Morphological Features | Rice Seed Varieties | | | | | | | | Number of times feature appeared in stepwise DFA |
|---|---|---|---|---|---|---|---|---|---|---|
| | | MR297 | | MR263 | | MR284 | | MR219 | | |
| | | Stepwise DFA | DFA | Stepwise DFA | DFA | Stepwise DFA | DFA | Stepwise DFA | DFA | |
| 1. | Width | ✓ | ✓ | | ✓ | | ✓ | | ✓ | 1 |
| 2. | Length | | ✓ | | ✓ | | ✓ | | ✓ | 0 |
| 3. | Seed Area | | ✓ | | ✓ | | ✓ | ✓ | ✓ | 1 |
| 4. | Convex Hull Perimeter | ✓ | ✓ | ✓ | ✓ | | ✓ | ✓ | ✓ | 3 |
| 5. | Seed Perimeter | | ✓ | | ✓ | | ✓ | ✓ | ✓ | 1 |
| 6. | Seed Maximum Feret Diameter | | ✓ | | ✓ | ✓ | ✓ | ✓ | ✓ | 2 |
| 7. | Major Axis Length | ✓ | | ✓ | | | | | | 2 |
| | Minor Axis Length | | ✓ | ✓ | ✓ | ✓ | ✓ | ✓ | ✓ | 3 |
| 9. | Aspect ratio | | ✓ | ✓ | ✓ | | ✓ | | ✓ | 1 |
| 10. | Rectangular Diagonal | | | | | | | ✓ | | 1 |
| 11. | Thinness Ratio | | | | | | | ✓ | | 1 |
| 12. | Hydraulic Radius | ✓ | | | | ✓ | | | | 2 |
| 13. | Angle Orientation | | ✓ | ✓ | ✓ | ✓ | ✓ | | ✓ | 2 |
| 14. | Rectangular Aspect Ratio | | ✓ | ✓ | ✓ | | ✓ | | ✓ | 1 |
| 15. | Moment XX | | ✓ | | ✓ | | | ✓ | ✓ | 2 |
| 16. | Moment YY | | | | | | | | | 0 |
| 17. | Ellipse Ratio | ✓ | | | | | | | | 1 |
| 18. | Area Ratio | ✓ | ✓ | ✓ | ✓ | ✓ | ✓ | | ✓ | 3 |
| | Total features | 6 | 12 | 7 | 12 | 5 | 12 | 8 | 12 | |
| | Overall classification accuracy | 99.1% | 99.1% | 98.9% | 98.9% | 93.7% | 95.0% | 98.3% | 98.0% | |
| | Canonical correlation | 0.928 | 0.929 | 0.909 | 0.910 | 0.835 | 0.856 | 0.905 | 0.905 | |
| | Wilks Lambda | 0.138 | 0.138 | 0.174 | 0.172 | 0.303 | 0.268 | 0.181 | 0.180 | |
| | Eigenvalue | 6.239 | 6.262 | 4.749 | 4.804 | 2.299 | 2.738 | 4.525 | 4.551 | |

MR297 rice seed group has shown highest discriminatory ability among other seed groups by having smallest values of Wilks Lambda. High variance in the independent features selected was explained by the function indicated from the large eigenvalues. The lowest classification accuracy among the rice seed varieties groups was shown by MR284 due to lower number of features (five out of 18) selected in the stepwise DFA, thus increased the Wilk's Lambda value to 0.303.

The classification accuracies of MR297 and MR263 with weedy rice group were maintained at 99.1% and 98.9% respectively, either using stepwise DFA or not. The classification accuracy for MR284 decreased from 95.0% to 93.7% using stepwise DFA with the reduction to five morphological features selected in the function. In this case, the classifier performance decline when lower number of features were selected by stepwise DFA. However for MR219, the classification accuracy increased by 0.3% using optimized features in stepwise DFA.

## 5. Conclusion

It can be deduced from the results, either using optimized features or not, the generation of the function coefficient greatly depending on the discriminatory ability from Wilks' Lambda reduction during entry of the independent features. The most significant features selected in stepwise DFA were convex hull perimeter, minor axis length and area ratio.

## 6. References

[1] Ismail A T and Said A R 2012 Certified Paddy Seed Production and Processing, *Int. Conf. on Agricultural and Food Engineering for Life* (Cafei2012) pp1-5

[2] Gayathridevi T, Neelamegam P and Sudha S Machine Vision based Quality Analysis of Rice Grains

[3] Guzman J D and Peralta E K 2008 Classification of Philippine Rice Grains Using Machine Vision and Artificial Neural Networks *Proc. World Conf. on Agricultural and IT (*Nagatsuka T:Tokyo) Tokyo University of Agriculture pp 41- 48

[4] Chaugule A and Mali S M 2014 Evaluation of Texture and Shape Features for Classification of Four Paddy Varieties Journal of Engineering Hindawi Publishing Corporation
Volume pp 8

[5] T Y Kuo, S Y Chen, H A Lin, C L Chung, Y F Kuo 2015 Identifying Rice Grains of Various Cultivars Using Machine Vision ASABE Paper No. 152188279 St. Joseph, Mich.: ASABE

[6] A.A. Aznan, I.H. Rukunudin, A.Y.M. Shakaff, R. Ruslan, A. Zakaria and F.S.A. Saad 2014 Application of Image Processing Technique to Extract Morphological Characteristics of Weedy Rice Seeds Variants for Malaysian Seed Industry". Adv. Environ. Biol., vol 8(22), pp. 112- 115

[7] IMAQ Vision for LabVIEW User Manual 2000 National Instruments Corp, Austin, Texas

IOP Conf. Series: Materials Science and Engineering **557** (2019) 012015 doi:10.1088/1757-899X/557/1/012015

# Distributed Online Averaged One Dependence Estimator (DOAODE) Algorithm for Multi-class Classification of Network Anomaly Detection System

**Mukrimah Nawir[1], Amiza Amir[1,2], Naimah Yaakob[1], R.Badlishah Ahmad[2,1] Anuar Mat Safar[1], Mohd. Nazri Mohd. Warip[1] and I Zunaidi[1,3]**

[1]Embedded, Network, and Advanced Computing (ENAC) Research Cluster, School of Computer and Communication Engineering, Universiti Malaysia Perlis (UniMAP), Pauh Putra, 02600, Perlis

[2]Faculty of Informatics and Computing, Universiti Sultan Zainal Abidin (UniSZA), 22200, Besut, Terengganu

[3]Faculty of Technology, University of Sunderland, St Peter's Campus, Sunderland, SR6 0DD, United Kingdom

E-mail: mukrimahnawir@yahoo.com; amizaamir@unimap.edu.my; naimahyaakob@unimap.edu.my; badli@unisza.edu.my; anuarms@unimap.edu.my; nazriwarip@unimap.edu.my;

**Abstract.** Network monitoring system consists of large data streams, distributed architecture, and multiple computers that are geographically located all over the world caused a difficulty to detect abnormalities in the system. In addition, when handling network traffic, the data in network is fast incoming and requires an online learning where immediately response and predict the pattern of network traffic for classification once there is an event or request occur. Therefore, this paper aims to develop an effective and efficient network anomaly detection system by using distributed online averaged one dependence estimator (DOAODE) classification algorithm for multi-class network data to overcome these issues. The finding of DOAODE algorithm for multi-class classification is high in accuracy with average 83% and fast to train the network traffic recorded less than ten seconds and takes shorter time when the number of nodes increases.

## 1. Introduction

Network traffic can be classified into binary or multi-class classification. Binary classification is the process to classify that involved only two classes of network traffic either it is normal or attack class [1]. Whereas, multi-class classification is where the network traffic contains several classes which is a normal and several network threats present in a network system. Commonly, the network anomaly detection system is built for binary classification that insufficient to detect the network traffic's patterns in a system and motivates this paper to investigate the performance of classifier to detect anomalous data that dealing with multi-class network data [2][3].

The architecture of network system can be either a centralized or distributed computing. For distributed computing, the results of prediction are combining geographically by allocating multiple classifier at every distributed sites and the local data analyses that hold the information and prediction of network traffic. The knowledge from these all local monitoring sites being combined by using distributed aggregation for making a final prediction. Additionally, the network traffic fast incoming into the system and require a fast learning classifier (an online learning) that enable a frequently update the classifier for classification [4][5]. Therefore, this paper develops an effective and efficient distributed online averaged one dependence estimator (DOAODE) algorithm for multi-class classification of network anomaly detection system performed on the UNSW-NB15 dataset.

The structured of this paper as follows: Section 2 is the previous works that develop a network anomaly detection system by using a machine learning algorithm. Subsequent section is the experiment setting including the network dataset, classifier, and distributed online classifier of updating the classifier (Section 3) followed by the results (Section 4) of conducted experiment based on classification rate and training time for network anomaly detection of the proposed method. Last but not least, the conclusion and some recommendation to improve this present work is provided in Section 5.

## 2. Related Works

In this section, the related works of developing a network anomaly detection system by using machine learning algorithm were summarized and tabulated in Table 1. In [6], the authors developed network anomaly detection by using centralized batch AODE to classify the multi-class of network traffic in NSL-KDD dataset. An improvement made by using ensemble techniques proposed by [7] and produced an accuracy of multi-class classification is 98.73% with 38 numbers of attack types. Although, their work high in accuracy with highest number of attacks compared this current work but they do not consider the time for classification which is important to ensure the performance is efficiently without buffering the network traffic for making a prediction.

Authors by [8] proposed an online Naïve Bayes classifier to solve the issue of profile change over time due to network traffic it inspects and need to retrain with the recent traffic instances. According to their work, online learning able to handle the presence of continuous flow of data where each of new traffic instance arrived were involved in the training phase to updating the network anomaly detection system. Yet, the centralized classifier that they build bring up an issue of single point failure to occur when there are powerful attacks compromised the system cause fatal destruction toward the network system. Therefore, it is important to design a classifier in distributed architecture to overcome this issue.

The closer work to this present work was presented in paper [9], where the authors designed a distributed online Naïve Bayes (NB) algorithm instead of averaged one dependence estimator as a classifier that detect the network traffic's patterns. Also, the authors used different dataset (NSL-KDD) dataset which is outdated dataset and had been criticism by many researchers. However, results in low classification rate due to the assumption of NB algorithm where the attributes independently to each other. Hence, to solve these issues an appropriate and dependence features assumption machine learning algorithm which is averaged one dependence estimator (AODE) algorithm to increase the accuracy and fast training classification to discriminate normal and several attacks (multi-class network traffic) by using new benchmark UNSW-NB15 network dataset.

**Table 1.** Related Works of Network Anomaly Detection by using Classification Algorithm

| Ref. | Algorithm | Architecture | Learning | Dataset | Classification | Performances |
|---|---|---|---|---|---|---|
| [6] | AODE | Centralized | Batch | NSL-KDD | Multi-class | Accuracy multi-class classification (3 classes) = 96.64% |
| [7] | Ensemble | Centralized | Batch | NSL-KDD | Multi-class | Accuracy =98.73% |
| [8] | NB | Centralized | Online | KDD-CUP99 | Binary, Multi-class (measured only F1-score no accuracy result provided) | Accuracy (binary classification) = 93%, Running time = 768.28 seconds |
| [9] | NB | Distributed | Online | NSL-KDD | Binary, Multi-class | Binary classification accuracy = 97.05% time = 6.23 seconds multi-class classification accuracy = 85% time: 24.87% |
| Our | AODE | Distributed | Online | UNSW-NB15 | Multi-class | Accuracy in the range 83%, Training time = less than 10 seconds |

## 3. Experiment Setup

The proposed to develop network anomaly detection system consists of three major steps: online classification, distributing network data, and distributed aggregation as in Figure 1. Presently, this experiment develops a distributed online classifier for network anomaly detection system toward the labelled network dataset named UNSW-NB15 dataset. This dataset consists 44 features with multi-class (ten class; where one is a normal class and the rest are attacks type).

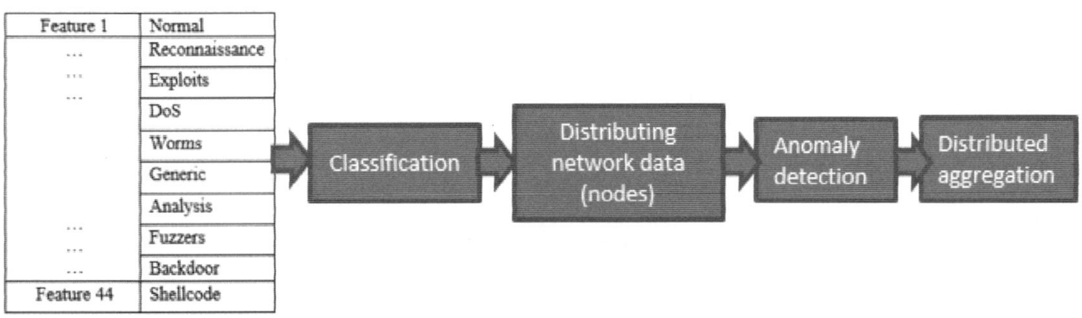


**Figure 1.** Proposed Method of NADS

### *3.1. UNSW-NB15 Dataset*

The network dataset, UNSW-NB15, is public and available at [10]. It is a newest labelled network dataset that provided two label features either conducted a binary (label feature used is *"label"*) or multi-class classification (label feature is *"attack_cat"* used). Also contains realistic normal and several type of attacks that allowing the benchmarking dataset to representing the real network system. In this present work, there only 44 features used instead of 49 features provided by the contributor due to the unnecessarily to be used and to increase the performance of network anomaly detection system. Additionally, there is 2, 540, 044 data instances of this dataset but for classification purpose the raw data in csv file converted to arff file with the amount about 257 673 data instances. For multi-class classification, there are nine types of different attacks being considered with one normal traffic. They are reconnaissance, shellcode, backdoor, analysis, generic, worms, Dos, fuzzers, and exploits [11] [12].

### *3.2. Distributed Online Averaged One Dependence Estimator (DOAODE) Algorithm*

Averaged One Dependence Estimator (AODE) algorithm [6] [13] [14] is a supervised machine learning algorithm that postulate the features/attributes are dependent one to another feature by averaging all the classifier generated. Moreover, this algorithm can be design in online manner to solve the issue of frequent updated classifier and produced a fast classifier to train the network traffic. In this work, we design the AODE algorithm within a distributed architecture by distribute the network traffic to the present nodes in network. As these nodes has its own classifier (AODE algorithm) for local prediction. The distributed online averaged one dependence estimator (DOAODE) algorithm works as in Figure 2.

For instance, when there are five nodes in a network then each of the nodes hold an AODE classifier so there are five AODE classifiers present in the system for multi-class classification the UNSW-NB15 dataset. Later, one of the nodes that act as a global classifier to aggregate all the results of prediction from local classifiers to make a final prediction of network traffic as either normal or anomalous data by using distributed majority voting. For distributed majority voting, we only consider an odd number of nodes, then when the major or highest voting determine the pattern of network traffic. For example, out of these five nodes three of them predict that the network traffic is a normal traffic and another two nodes are predicted that the patterns are anomalous and the obtained result out of these voting the network traffic is classified as a normal pattern.

IOP Conf. Series: Materials Science and Engineering **557** (2019) 012015 doi:10.1088/1757-899X/557/1/012015

**<u>DOAODE Algorithm:</u>**
**Input:** instance $\mathbf{x} =< x_1, \ldots, x_m >$
**Output:** the class estimation of AODE

**Training phase:**

**Begin**
Given an instance $\mathbf{x} =< x_1, \ldots, x_m >$
At initiator node: Find the neighbour nodes, NN
Calculate probability of y given $\mathbf{x}$: $p(y, \mathbf{x}) = p(y|\mathbf{x})p(\mathbf{x})$ then
$p(y|\mathbf{x}) = \frac{p(y,\mathbf{x})}{p(\mathbf{x})} = \frac{p(y)p(\mathbf{x}|y)}{p(\mathbf{x})}$
**End**

**Testing phase:**

**Begin**
Find the attribute values $x_i \in \mathbf{x}$ in test sample
Calculate $p(y, \mathbf{x}) = p(\mathbf{x}, x_i, y) = p(y, x_i)p(\mathbf{x}|y, x_i)$
Calculate probability of y given $\mathbf{x}$: $p(y|\mathbf{x}) = \frac{p(y,x_i)p(\mathbf{x}|y,x_i)}{p(\mathbf{x})}$
Upon receiving results from nodes NN
Output the class values
Aggregate the results by using majority voting
Update classifier
Repeat
**End**

**Figure 2.** Distributed Online Averaged One Dependence Estimator (DOAODE) Algorithm

## 4. Results

The proposed distributed online averaged one dependence estimator (DOAODE) algorithm is measured its performances based on classification rate and training time on the network dataset which is UNSW-NB15 dataset that consists 257 673 instances, 10 classes, and 44 features.

### *4.1. Classification Rate of DOAODE Algorithm for Network Anomaly Detection System*

Table 2 is the result of classification rate (accuracy) of distributed online averaged one dependence estimator (DOAODE) algorithm being compared with another different designs which are distributed online Naïve Bayes (DONB), distributed batch averaged one dependence estimator (DBAODE), and distributed batch Naïve Bayes (DBNB) algorithm for network anomaly detection of the multi-class classification. The accuracy of DOAODE and DBAODE algorithms are outperformed and higher than DONB as well as DBNB algorithm where both designs AODE algorithm achieved averagely about 83%.

Meanwhile, the Naïve Bayes algorithm only obtained in average 70% for multi-class classification UNSW-NB15 dataset for network anomaly detection system when the number of nodes is varies in the network system. The trend of accuracies results for DOADE algorithm inconsistent depending the global node that made final prediction after collecting all the results from the local nodes that present in the network system. Due to distributed classifier cause a difficulty in making decision and when the number of nodes increases the number of conflicting local prediction results. Yet, the distributed classifier avoids the single point failure of the network to be occurred where when there is a powerful network threat compromise the system cause the whole destruction of network system.

**Table 2.** Classification Rate of DOADE Algorithm for NADS

| Classifier/ | Classification Rate (%) | | | |
|---|---|---|---|---|
| **Number of Nodes** | **DOAODE** | **DONB** | **DBAODE** | **DBNB** |
| **5** | 83.87 | 70.89 | 83.83 | 70.84 |
| **10** | 83.91 | 70.84 | 83.89 | 70.87 |
| **15** | 83.70 | 70.83 | 83.79 | 70.84 |
| **20** | 83.67 | 70.73 | 83.79 | 70.79 |
| **25** | 83.51 | 70.74 | 83.81 | 70.73 |

### *4.2. Training Time of DOAODE Algorithm for Network Anomaly Detection System*

When dealing with network data in a system the importance measurement is the training time due to the fast data incoming and rapid updating data in a network required an efficient classifier that can determine the patterns of network data for classification purpose. Additionally, most of the machine learning algorithms have low classification time therefore the concern is the training time for a classifier to classify the network traffic where usually expensive and increase dramatically with an increasing of training data. Table 3 shows the experimental results for DOAODE algorithm compared with various designs of classification algorithm including the DONB, DBAODE, and DBNB algorithm for network anomaly detection based on training time.

From the experimental, the proposed method (DOAODE algorithm) shows a reasonable faster classifier to train the network traffic where with the highest number of nodes (25 nodes) produced only 1.78 seconds to build the classifier model for anomaly detection. In addition, the obtained result shows that the decreasing trend of training time with the number of nodes increasing (network size become large). Although, it is slower training classifier compared to DONB still the amount of training time of DOAODE algorithm acceptable as a fast classifier for anomaly detection which is took less than 10 seconds.

**Table 3.** Training Time of DOAODE Algorithm for NADS

| Classifier/ | Training Time (Seconds) | | | |
|---|---|---|---|---|
| **Number of Nodes** | **DOAODE** | **DONB** | **DBAODE** | **DBNB** |
| **5** | 8.76 | 0.89 | 25.25 | 1.00 |
| **10** | 4.96 | 0.46 | 28.39 | 0.49 |
| **15** | 3.73 | 0.31 | 56.54 | 0.32 |
| **20** | 2.19 | 0.25 | 63.11 | 0.24 |
| **25** | 1.78 | 0.19 | 74.23 | 0.20 |

## 5. Conclusion

In summary, this paper proposed a distributed online implementation of averaged one dependence estimator (DOAODE) classifier for network anomaly detection system. Based on the independence assumption of Naïve Bayes classifier motivate this paper to choose an AODE algorithm that is assuming the attributes' dependency by averaging all the estimation of probabilities of the present models in network traffic accurately gives an outperformed, high classification accuracy, and fast training time for anomaly detection the network dataset (UNSW-NB15) where the obtained of accuracy for multi-class classification is in the range 83% and took less than 10 seconds when the number of nodes are varies from 5 to 25 nodes.

## 6. References

[1] Nawir M, Amir A, Lynn OB, Yaakob N, Ahmad RB. Performances of Machine Learning Algorithms for Binary Classification of Network Anomaly Detection System. In *Journal of Physics: Conference Series 2018 May* (Vol. 1018, No. 1, p. 012015). IOP Publishing.

[2] Nawir M, Amir A, Yaakob N, Lynn OB. Multi-Classification of UNSW-NB15 Dataset for Network Anomaly Detection System. *Journal of Theoretical & Applied Information Technology*. 2018 Aug 15;96(15).

[3] Rettig L, Khayati M, Cudré-Mauroux P, Piórkowski M. Online anomaly detection over big data streams. *In2015 IEEE International Conference on Big Data (Big Data)* 2015 Oct 1 (pp. 1113-1122). IEEE.

[4] Ahmad S, Purdy S. Real-time anomaly detection for streaming analytics. arXiv preprint arXiv:1607.02480. 2016 Jul 8.

[5] Limthong K. Real-time computer network anomaly detection using machine learning techniques. *Journal of Advances in Computer Networks*. 2013 Mar;1(1).

[6] Sultana A, Jabbar MA. Intelligent Network Intrusion Detection System using Data Mining techniques. *In Applied and Theoretical Computing and Communication Technology (iCATccT), 2016 2nd International Conference on 2016 Jul 21 (pp. 329-333)*. IEEE.

[7] Vasudha K. Intrusion Detection System using Decision Tree-based Attribute Weighted AODE. *International Journal of Advances Research in Computer and Communication Engineering,* 2014 December (pp. 451-467).

[8] Gumus F, Sakar CO, Erdem Z, Kursun O. Online Naive Bayes classification for network intrusion detection. *InProceedings of the 2014 IEEE/ACM International Conference on Advances in Social Networks Analysis and Mining* 2014 Aug 17 (pp. 670-674). IEEE Press.

[9] Idhammad M, Afdel K, Belouch M. Distributed Intrusion Detection System for Cloud Environments based on Data Mining techniques. *Procedia Computer Science*. 2018 Dec 31;127:35-41.

[10] Weka 3 – Mining Big Data with Open Source Machine Learning Software in Java. Retrieved 2016-11-21, from http://www.cs.waikato.ac.nz.ml/weka/bigdata.html

[11] Moustafa N, Slay J. UNSW-NB15: A Comprehensive Data set for Network Intrusion Detection Systems (UNSW-NB15 Network Data set). In *Military Communications and Information Systems Conference (MilCIS),* 2015 2015 Nov 10 (pp. 1-6). IEEE.

[12] Moustafa N, Slay J. The evaluation of Network Anomaly Detection Systems: Statistical analysis of the UNSW-NB15 Data set and the Comparison with the KDD99 data set. *Information Security Journal: A Global Perspective*. 2016 Apr 4;25(1-3):18-31.

IOP Conf. Series: Materials Science and Engineering **557** (2019) 012015 doi:10.1088/1757-899X/557/1/012015

[13] Webb GI, Boughton J, Wang Z. Averaged one-dependence estimators: preliminary results. *In Proceedings of the Australasian Data Mining Workshop 2002 Dec 2 (Vol. 2002).* Language Thought, and Reality: Selected Writings of Benjamin Lee Worf. MIT Press.

[14] Webb GI, Boughton JR, Wang Z. Not so naive Bayes: aggregating one-dependence estimators. Machine learning. 2005 Jan 1;58(1):5-24.

**Acknowledgments**

The research reported in this paper is supported by Research Acculturation Grant Scheme (RAGS). The authors would also like to express gratitude to the Malaysian Ministry Higher of Education (MOHE) and University Malaysia Perlis (UniMAP) for the facilities provided.

IOP Conf. Series: Materials Science and Engineering **557** (2019) 012016 doi:10.1088/1757-899X/557/1/012016

# Enhancement of Motor Imagery Brain Computer Interface Performance Using Channel Reduction Method based on Statistical Parameters

**Mohd Shuhanaz Zanar Azalan[1], M P Paulraj[2], Abdul Hamid Adom[1]**

[1]School of Mechatronic Engineering, Universiti Malaysia Perlis, Arau, Perlis, Malaysia [2]Sri Ramakrishna Institute of Technology, Coimbatore, Tamilnadu, India

**Abstract.** In this paper, a novel method to reduce the number of EEG channels for a Motor Imagery-based Brain Computer Interfaced (BCI) system without compromising its performance is proposed. By reducing the number of EEG channels, the number of features can be reduced and this has to be achieved without sacrificing the classification accuracy and computational time of the BCI. EEG signals were recorded from 10 subjects using a 19-channel EEG amplifier. Higuchi Fractal features were extracted from the recorded signals and modelled using Neural Networks (NN). A simple statistical analysis based on standard deviation was then used for the channel reduction process. The classification accuracy of the NN model formulated with the 19 channels features were compared to that of the model with features selected using statistical method. From the results it was observed that using this approach, the number of EEG channels can be reduced up to 30% without sacrificing its classification performance.

## 1. Introduction

A BCI is a system that acts as link for the brain signal to communicate with computer system without going through a usual route of peripheral nerves and muscles [1]. Research about BCI have been explored for over two decades, and is considered as a new research area. The interest in BCI has been increasing rapidly and this is evidence in the number of publications on the subject matter. The group of researcher who had done a research about BCI also increase year to year [2].

However, there are still great opportunities in this research area. Many improvements in many aspects of BCI performance can be done for example by removing features that may act as noise to the system through channel selection process. New channel selection algorithm based on high order statistical parameters using motor imagery data is presented in this paper. The proposed channel selection algorithm was used to reduce the feature dimension in order to enhance the classification performance of the recorded motor imagery data features.

The motor imagery data are based on a novel asynchronous data collection protocol. Motor imagery can be defined as the process to carry out pretended movement of arm or other parts of human body. The concept of pretended movement of arm can be defined such as preparation for movement, passive observations of action, and mental operations of sensorimotor representations [3]. The selection of channel is used to reduce the size of data dimension and has been widely used to increase the performance of BCI systems. This refers to the process of choosing a subset of variables from the original dataset such that the optimal features are retained. Some researchers claim that the selection of variables will improve classification accuracy or decreasing the size of features without significantly decreasing the accuracy of the classifiers [4].

In order to find suitable channel, S. M. Park et al.[5] study four channel selection methods namely binary particle swarm optimization (BPSO), BPSO with a channel impact factor, a genetic algorithm (GA), and a harmony search (HS) using a support vector machine (SVM).The performance for each method are compared in term of the classification accuracy and the number of selected electrodes. The results showed that BPSO with a channel impact factor selected least number of electrodes and improved accuracy by 20.4%.

Hossain et al.[6] proposed channel selection approach using wavelet-based Maximum Entropy on the Mean (wMEM). They used the concept of sparsity assumption. The channels located around the areas with high MI-related amplitude is selected. It is observed that the number of channel and the computation time can be reduced without reducing the classification accuracy of the system.

Qiu et al.[7] proposed a channel selection method by improving the Sequential floating forward selection (SFFS) method. The result indicates that the improved SFFS yielded better performance than using all channels and support vector machine recursive feature elimination method with an average improvement of 3.8%. The number of channel is reduced with the use of only 30.8 out of 118 channels.

Shan et al.[8] proposed channel selection based on Relief algorithm. Support vector machine was used as the classifier. Three different datasets were used to validate the method. The results indicate that the proposed method yielded average classification accuracies of 85.2%, 94.1%, and 83.2 %, respectively for Dataset 1, Dataset 2, and Dataset 3. The average accuracies were improved by 31.7%, 8.0% and 19.7 % for Dataset 1, Dataset 2 and Dataset 3 respectively.

Yang et al.[9] proposed channel selection method based on Fisher's discriminant analysis for evaluating the discriminative power features for each channel. The result indicates that the method can reduce the number of channels (from 118 channels to 9 in average) without a decrease in mean classification accuracy.

The channel selection plays a major role in reducing the setup time, to minimize the time for placing the cap on the scalp and to avoid the increase in computation time due to the presence of number of channels. In spite of numerous approaches published in the literature, data dimensional reduction through variable selection is still an on-going research and there is opportunity to improve it. It is impossible to say which technique amongst all the reported is the best for channel selection. It is highly subjective in nature and very application specific. Researchers are still looking for new techniques which are relatively simple and robust learning model to improve multiclass motor imagery classification accuracy and learning time.

In this paper a channel selection methods known as Channel Selection based on Standard Deviation of raw energy signal (CSSD) is proposed to enhance the classification performance by selecting significant channel and discard a channel that may give irrelevant information. The Higuchi fractal feature from four different bands namely alpha 2 (11-12 Hz), beta 1 (13-15 Hz) and beta 2 (16-18 Hz) and beta 3 (19-25 Hz) were extracted from all the 19 channels of the EEG raw data. The extracted features were then classified into four different types of hand rotational movements using a feed-forward neural network model. The extracted features were then minimized using CSSD and a network model was subsequently developed using the minimized features. The performances of the two network model were then compared.

## 2. Methodology

### *2.1. Data Acquisition*

The EEG raw signal was recorded from ten healthy subjects. All the subjects were from University Malaysia Perlis. The experimental work was conducted in air-conditioned laboratory with normal level of noise. This condition makes it more natural since BCI systems would usually be implemented in situations with many distractions. The age range of the participated subjects was between 22 to 34 years. All the chosen subjects must fulfil the criteria such that they must be medication free, free from illness, enough slept at least six hours in the night the day before the experiment was conducted and have no neurological or psychiatric disorders. All the ten subjects who have participated in the data recording process are right handed and not familiar with the EEG recording equipment.

A non-cue-based protocol using motor imagery as a predefined task is proposed in this experimental work. All the ten subjects performed four different imaginary tasks. They were asked to go through four different imagery task sessions. Each session represents a specific hand movement task. The four different imaginary tasks employed in the experimental procedure were relax, right arm movement, left arm movement and both arm movement. The subjects were given a video demonstration showing a person moving his right forearm up and down for 10 seconds. After the video presentation, the subjects were given a relaxation period of 20 seconds. Then the subjects were requested to imagine moving that they were making up and down right forearm movements. The subjects were asked to repeat the same imagination process for 10 times; simultaneously, the EEG signals corresponding to the 10 trials were recorded. Between the trails, the subjects were given a resting period of 15 seconds. For 'relax' task, the subjects will be in a state of rest and do not perform any task. Subjects were requested as much as possible not to think of anything.

The EEG raw signals were acquired using Mindset-24 Topographic Neuro Mapping Instrument of Nolan Computer System LLC. This EEG instrument is a portable acquisition system which consists of 19 EEG bipolar channels. All the 19 electrodes were positioned as per the international 10- 20 method of electrode placement.

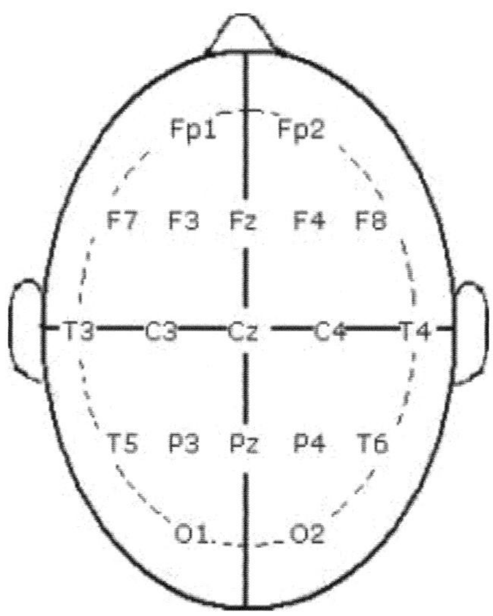


**Figure 1.** ''10–20'' system of electrode placement. F = frontal, T = temporal, C = central, O = occipital, P = parietal. Odd numbers = left hemisphere, even numbers = right hemisphere. [10]

### *2.2. Pre-processing and Feature Extraction*

The signals were recorded for 10 seconds. The EEG signals were sampled at the rate of 256 samples per seconds. The signals were then segmented in such a way that all the segmented frames have equal number of samples. Each frame has 128 sample data points (corresponding to 0.5 s) with 50% overlapping between two consecutive signal frames. A notch filter is used to remove the 50 Hz power line frequency noise from the raw EEG signal. This 50Hz power line frequency noise is due to electrical interference from the EEG recording equipment. Every signal frame is band pass filtered using a second order Chebyshev bandpass filter into four different sub-band frequencies namely alpha 2 (11-13 Hz), beta 1 (13-15 Hz), beta 2 (16-18 Hz) and beta 3 (19-30 Hz). Due to extracted frequency band lies between 8 – 30 Hz, so there's no additional step to remove ocular artefact and muscle artefact energy. It is because most of the ocular artefacts generally appear under 4 Hz frequency band and most of the muscle artefact energy is not contained in 8-30 Hz frequency band [11].

In this paper, Higuchi fractal dimension features were extracted from the recorded EEG. Fractal Dimension (FD) is a measurement process to quantify the signal self-similar characteristic based on the illustrative presentation of the signal. A single non integer value (fractional) is obtained through the process. The fractal feature values corresponding to the EEG signals lies between 1 and 2. Higuchi proposed a new mathematical process to calculate the time series average. The proposed mathematical process differentiated between Higuchi method and Burlaga and Klein method. The proposed mathematical process is suitable for short time series signals in order to obtain a solid fractal dimension value. By using Higuchi's method, a solid fractal dimension features are brought out from the Motor Imagery EEG signals. The fractal dimension value $F_d$ can be computed by using following Equation (1):

$$F_d = \frac{log(L_k)}{log(k)} \tag{1}$$

where $L_k$ is the mean length value for all the curve length of the subset time series[12].

### *2.3. Channel Selection*

In this paper, a channel selection based on statistical parameters using standard deviation of raw energy signals are proposed. The algorithm is discussed using following step:

Step 1: Segmentation of the recorded EEG signals into number of frame is lead oft in the process of reducing the channel. After the segmentation process, total of 39 frames per channel are gained. Each frame comprises of 128 number of sample with an overlapping of 50% between the successive frames. Thus for 10 trials we have F (390) frames.

Step 2: Compute and formulate raw energy signals matrix $\boldsymbol{R}^k$. $\boldsymbol{R}^k$ represents raw energy signals obtained from all the *F* frames and *C* channels for the $k^{th}$ task. It can be formulate as follow:

$$\boldsymbol{R}^k = \begin{bmatrix} r_{1,1}^k & r_{1,2}^k & r_{1,3}^k & \cdots & r_{1,j}^k & \cdots & r_{1,C}^k \\ r_{2,1}^k & r_{2,2}^k & r_{2,3}^k & \cdots & r_{2,j}^k & \cdots & r_{2,C}^k \\ r_{3,1}^k & r_{3,2}^k & r_{3,3}^k & \cdots & r_{3,j}^k & \cdots & r_{3,C}^k \\ \vdots & \vdots & \vdots & \vdots & \vdots & \vdots & \vdots \\ r_{i,1}^k & r_{i,2}^k & r_{i,3}^k & \cdots & r_{i,j}^k & \cdots & r_{i,C}^k \\ \vdots & \vdots & \vdots & \vdots & \vdots & \vdots & \vdots \\ r_{F,1}^k & r_{F,2}^k & r_{F,3}^k & \cdots & r_{F,j}^k & \cdots & r_{F,C}^k \end{bmatrix} \tag{2}$$

where $r_{i,j}^k = \sum_{q=1}^{N} \left[x_{i,j}^k(q)\right]^2$ is the raw energy value of the signals in the $i^{th}$ frame, $j^{th}$ channel and for the $k^{th}$ task. $x_{i,j}^k(q)$ is the $q^{th}$ data sample of the $i^{th}$ frame, $j^{th}$ channel and $k^{th}$ task where *N* is total number of sample in each frame.

Step 3: Compute standard deviation for column of matrix $\boldsymbol{R}^k$ for $k^{th}$ task. Matrix *D* is then formulated and written as:

$$\boldsymbol{D} = \begin{bmatrix} sd_1^1 & sd_2^1 & sd_3^1 & \cdots & sd_j^1 & \cdots & sd_C^1 \\ sd_1^2 & sd_2^2 & sd_3^2 & \cdots & sd_j^2 & \cdots & sd_C^2 \\ sd_1^3 & sd_2^3 & sd_3^3 & \cdots & sd_j^3 & \cdots & sd_C^3 \\ \vdots & \vdots & \vdots & \cdots & \vdots & \cdots & \vdots \\ sd_1^k & sd_2^k & sd_3^k & \cdots & sd_j^k & \cdots & sd_C^k \\ \vdots & \vdots & \vdots & \cdots & \vdots & \cdots & \vdots \\ sd_1^T & sd_2^T & sd_3^T & \cdots & sd_j^T & \cdots & sd_C^T \end{bmatrix} \quad (3)$$

where $sd_j^k$ is standard deviation of the $j^{th}$ channel and corresponding to the $k^{th}$ task.

Step 4: Using the standard deviation matrix *D*, sum all standard deviation values for each column. These values are then arranged in the descending order.

$$\boldsymbol{KC} = \left[ kc_1, kc_2, kc_3, \cdots, kc_j, \cdots kc_c \right] \quad (4)$$

where $kc_i < kc_{i+1}, i = 1,2,\cdots,C-1$ and the associated channels are represented as $\boldsymbol{K_S}$ :

$$\boldsymbol{K_S} = \{ks_1, \ ks_2, \ ks_3, \ \cdots \ ks_j, \ \cdots \ ks_C\} \quad (5)$$

Step 5: Build up the neural network model by utilizing the feature set extracted from all the *C* channels. The recorded EEG signal from *'B'* selected frequency band which is derived from band frequency were considered in this process. Classification accuracy gain by developed neural network model is marked as a main classification performance (MCP).

Step 6: Set channel index value, set of channel S and channel performance tolerance (CPT) as follows:
index=2

$$\boldsymbol{S} = \{ks_1, ks_2\}$$

channel performance tolerance (CPT) = 2%.

Step 7: Develop neural network model using the feature set extracted from channel sets. Obtain its classification accuracy and mark it as CP.

Step 8: If (MCP-CP > 2%)
Set index= index+1;
Reformulate the channel set as $S = S \cup ks_{index}$
Go to step 7.
Else
Terminate the process.

### *2.4. Classification*

In this paper standard feed-forward neural network with one hidden layer was developed and implemented to classify the four imagery tasks for each subject. Feed-forward neural network was used because of its simplicity and generalization capabilities. One of the advantages of using artificial neural networks as a classification method is because of their robustness to choices of parameter values and their similarity to other nonlinear regression method.

Higuchi fractal features of four different bands from the selected channel were extracted. A neural network model is developed using the extracted features. The trained neural network model was tested and the classification performance with computation time was obtained and tabulated. The hidden neurons and network learning rate for the network were chosen as 6 and 0.5 respectively. Through simulation, the number of hidden neurons and the network learning rate was chosen in such a way to give the highest classification accuracy. Both hidden and output neurons were activated using log sigmoid activation function. Training tolerance and testing tolerance were set to 0.01 and 0.1 respectively. The network iteration process was performed until the Mean Square Error (MSE) value reached below 0.0001 or a maximum epoch values of 500 has been reached. The network was trained using Levenberg-Marquardt (LM) algorithm. Binary normalization algorithm was used to normalize the training and the testing samples.

The performances of the channel selection algorithm are evaluated and its performance in each subject is observed. Set of feature from four different bands is obtained from each selected channel. This set of feature is used as an input feature and a neural network model is developed using the input feature. The trained neural network model is tested and the classification performance with its computation time is obtained and tabulated.

In the analysis, the classification performance of selected channel subset is evaluated by 10-fold cross validation. All datasets are split into ten subsets of approximately equal size. Randomly, one dataset is used for testing and the remainders are training. The same procedure is repeated ten times and the observation of performance for each network model is made. The performance is observed base on the average, maximum, minimum classification rate and their average computation time. The proposed algorithms are implemented in MATLAB 2015 on Intel(R) Core(TM) i5-2450 CPU @ 1.8 GHz; RAM: 4 GB.

## 3. Results and Discussions

The channel selection method based on the standard deviation of raw energy signal was applied to the Higuchi fractal features. The list of the best channels after the channel reduction process is obtained for each subject. Table 1 depicts the result obtained by each subject after reducing the channel.

**Table 1.** Classification Performance of Channel Selection using CSSD

| Subject | All channel | | CSSD | | |
|---|---|---|---|---|---|
| | Mean Accuracy (%) | Mean Computation time (sec) | Number selected channels | Mean Accuracy (%) | Mean Computation time (sec) |
| Subject 1 | 95.0 | 4.1 | 15 | 94.5 | 3.1 |
| Subject 2 | 96.9 | 2.8 | 13 | 97.1 | 2.0 |
| Subject 3 | 95.5 | 2.8 | 16 | 94.5 | 2.0 |
| Subject 4 | 99.2 | 2.5 | 13 | **98.4** | 2.2 |
| Subject 5 | 95.4 | 3.7 | 17 | 95.1 | 3.3 |
| Subject 6 | 97.5 | 2.9 | 16 | 97.7 | 2.1 |
| Subject 7 | 98.2 | 2.6 | 14 | 98.3 | 2.1 |
| Subject 8 | 93.8 | 2.4 | 16 | **93.7** | 2.3 |
| Subject 9 | 96.9 | 3.7 | 18 | 96.8 | 2.8 |
| Subject 10 | 97.2 | 2.9 | 14 | 97.3 | 1.8 |
| Mean | 96.5 | 3.0 | 15.2 | 96.3 | 2.3 |
| SD | 1.6 | 0.5 | 1.6 | 1.7 | 0.5 |

A channel selection method based on CSSD algorithm can reduced the number of channel between 1 to 6 channels. The number of reduce channel also depend on the subject. From Table 1, it is observed that subject #4 has the highest mean classification accuracy of 98.4% and the subject #8 has the lowest mean classification accuracy of 93.7%.

It is clear from the Table 1 that a reduction in the number of channels not only maintains the same classification accuracy but also increases the classification accuracy for some subject. The classification performances of each the subjects were being remain in the range of -1% to +0.5% when compared to the network model classification rates of the same subject obtained using 19 channels. It is observed that classification accuracy for subject #2, #6, #7 and #10 is increase in the range of 0.1% to 0.2%.

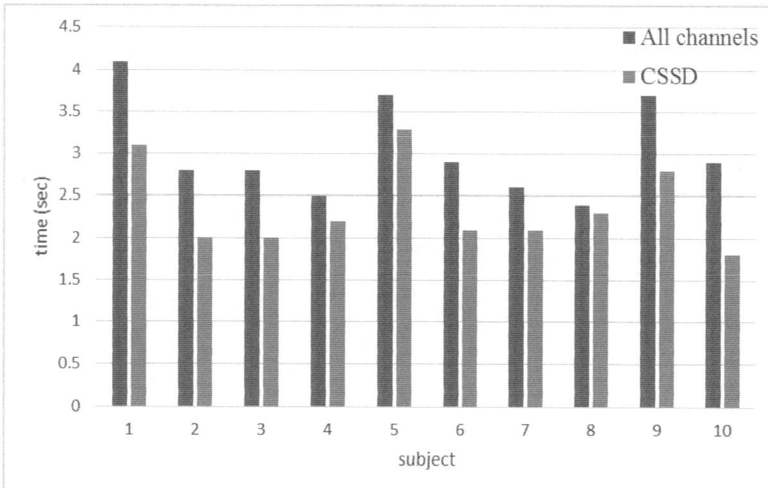


**Figure 2.** Computation time comparison between All Channels and CSSD

The comparison of computation time is shown in Figure 2. The computation time of the CSSD was significantly decrease in the range of 4% to 37% for each subject. On average, CSSD method reduced the computation time by 22%.

The percentage of channel selected by the CSSD are shown in Figure 3. Channel FP1, FP2, F7, F3, Fz, F4 and P3 were selected by all the subjects. It is observed that channel F8, Cz and P4 were selected by more than 90% of subjects. All those channels are distributed in frontal lobe and parietal lobe of the cerebral cortex. The less significant channel are located in occipital lobe and temporal lobe of the cerebral cortex. It can be seen in Figure 3 where channel in this area were selected by less than 80% of the subjects.

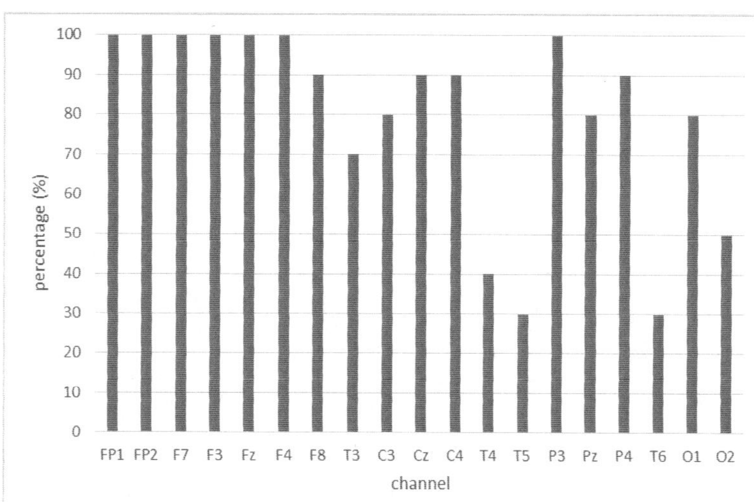


**Figure 3.** Percentage of channel selected using CSSD

Overall the number of channels is reduced in the range of 5% to 20%. The distributions of channels selected by the CSSD are shown in Figure 3. The distributed channels shows are referred to the number of times for each channels was selected by every subjects. Overall, it is observed that most channels that covers the frontal lobe are selected by 100% of subjects. The motor areas of cerebral cortex are covered in frontal lobe. Channels that related to motor imagery also located in this area. Due to this, most selected channels are cover in this area. Primary somatosensory cortex is part of the parietal lobe. The information related to the perception from external stimuli are processed in this area.

The proposed protocol involved the perception from external stimuli in order to induce the motor imagery signal. Due to this, most channel that distributed in parietal lobe of the cerebral cortex is selected by more than 90% of the subjects. Most channels that are located in the occipital and temporal lobes of the cerebral cortex were selected by less than 70% of the subjects. This is because the channels covered in this area are not related with the motor imagery and coordination movement.

The combination of selected data set channels in different area of the brain that related motor movement will give the best performance and accuracy to discriminate four different motor imagery task. The improved performances and the correct of selection channels in motor cortex area proof that the proposed channel selection method can determine the significant channel that relate to the motor imagery task.

## 4. Conclusions

The paper shows that the reduction in channels using the proposed algorithm do not result in significant loss in performance compared to that using the whole 19 channels. It also has the benefit of reduced training and testing time. It was also observed that the number of selected channel depends on the individual subjects. This may be due to the concentration levels of individuals vary.

From the above, it can be concluded that the proposed channel reduction method helps to find the reduced number of channels and indirectly can reduce the number of input features and computation time. In the future work, the research work can be extended with the fusion of different channel selection techniques to improve the performance of the classification accuracy.

## References

[1] J. R. Wolpaw *et al.*, "Brain-computer interfaces for communication and control.," *Clin. Neurophysiol.*, vol. 113, no. 6, pp. 767–91, 2002.

[2] Shih et al, "Brain-computer interfaces in medicine," *Mayo Clinic Proceedings*, vol. 87, no. 3. pp. 268–279, 2012.

[3] M. Jeannerod, "The representing Brain : Neural correlates of motor imagery and intention," *Behav. Brain Sci.*, vol. 17, pp. 187–245, 1994.

[4] D. Koller and M. Sahami, "Toward {Optimal} {Feature} {Selection}," *Int. {Conference} {Machine} {Learning}*, pp. 284–292, 1996.

[5] S. M. Park, J. Y. Kim, and K. B. Sim, "EEG electrode selection method based on BPSO with channel impact factor for acquisition of significant brain signal," *Optik (Stuttg).*, vol. 155, pp. 89–96, 2018.

[6] M. S. Hossain, S. Saha, M. A. Habib, A. Al Noman, T. Sharfuddin, and K. I. Ahmed, "Application of wavelet-based maximum entropy on the mean in channel optimization for BCI," *1st Int. Conf. Med. Eng. Heal. Informatics Technol. MediTec 2016*, 2017.

[7] Z. Qiu, J. Jin, H. K. Lam, Y. Zhang, X. Wang, and A. Cichocki, "Improved SFFS method for channel selection in motor imagery based BCI," *Neurocomputing*, vol. 207, pp. 519–527, 2016.

[8] H. Shan, H. Xu, S. Zhu, and B. He, "A novel channel selection method for optimal classification in different motor imagery BCI paradigms.," *Biomed. Eng. Online*, vol. 14, no. 1, p. 93, 2015.

[9] Y. Yang, I. Bloch, S. Chevallier, and J. Wiart, "Subject-Specific Channel Selection Using Time Information for Motor Imagery Brain–Computer Interfaces," *Cognit. Comput.*, vol. 8, no. 3, pp. 505–518, 2016.

[10] T. Thompson, T. Steffert, T. Ros, J. Leach, and J. Gruzelier, "EEG applications for sport and

performance," *Methods*, vol. 45, no. 4, pp. 279–288, 2008.
[11] E. Gysels and P. Celka, "'Phase synchronization for the recognition of mental tasks in a brain-computer interface., ' {IEEE} {Transactions} on {Neural} {Systems} and {Rehabilitation} {Engineering}," *Ieee Trans. Neural Syst. Rehabil. Eng.*, vol. 12, no. 4, pp. 406–415, 2004.
[12] T. Higuchi, "Approach to an irregular time series on the basis of the fractal theory," *Phys. D Nonlinear Phenom.*, vol. 31, no. 2, pp. 277–283, Jun. 1988.

## Acknowledgments

The authors would like to thank Dato' Professor Dr. Zul Azhar Zahid Jamal, the vice chancellor, Universiti of Malaysia Perlis for his continuous support and encouragement.

# Automated Chicken Weighing System Using Wireless Sensor Network for Poultry Farmers

**C.C. Lee[1], A.H. Adom[1], M.A. Markom [1] and E.S.M.M. Tan[2]**

[1] School of Mechatronic Engineering, Universiti Malaysia Perlis, Kampus Pauh Putra, 02600 Arau, Perlis, Malaysia.
[2] Centre of Diploma Studies, Universiti Malaysia Perlis, S2-L1-26, Kampus UniCITI Alam Sg.Chuchuh, 02100 Padang Besar, Perlis, Malaysia

**Abstract.** In recent years, in the quest for cost-effective operations, slower than expected chicken growth has become a big issue for the poultry farmers in Malaysia. To be profitable, the chickens need to reach 1.8kg in only 38 days from hatching. At present, the farmers manually check the weight of chickens twice a day, in the morning and evening, to monitor the growth rate of chickens. However, the workers in such farms do not weigh the chickens according to an established SOP, which in turn results in inaccurate growth-rate data obtained. This paper describes the development of an automated chicken weighing system using wireless sensor network (WSN) for poultry farmers to replace the manual approach. The weighing scales are equipped with wireless data transfer system to enable automatic data transfer to the Cloud using a Wi-Fi module. The data collected are then compared to the expected growth rates and used to monitor of the growth rate of chickens. The proposed growth rate monitoring system enables them the daily monitoring of the chicken development rate and enables the rectification of problems as soon as any discrepancies are detected. Problems such as insufficient or poor distribution of feed and the occurrence of disease can be rectified at an early stage.

## 1. Introduction

Poultry meat is one of the most popular and cheapest protein based-food among Malaysians as the meat is socially and religiously acceptable [1]. Malaysians consumed 51.2kg/year of poultry meat in 2016 [2] and has been steadily increasing over the years.

Poultry meat has recorded the highest self-sufficiency ratio (SSR) among the livestock commodities which is 98.0% in 2016 [2], produced 305,058,206 of chickens and 10,655,926 of ducks [3]. According to the federation of livestock famers' associations of Malaysia, there are approximately 2,676 chicken farms in 2015 which produced 725 million of broilers a year [4]. There are about 15,010 tons of chicken meat and 52.71 million of live chickens were exported a year, mostly to Singapore [4]. This shows that poultry farming is an important agricultural industry in Malaysia.

However, poultry farmers in Malaysia follow a strict schedule to maximize profits where the broiler chickens need to reach 1.8kg in 38 days from hatching and be harvested immediately. This is to allow time for the farmers to clean the poultry house and new batch of chicks to come in. Since the chicks are pre-ordered one month earlier before harvest, and will arrive on time, there is no empty room for delays. If the broiler chickens are unable to reach 1.8kg in 38 days, the farmers have to sell the livestock for a fraction of the price to other poultry farms, and hence incur losses. In addition, the margin between broiler chickens' production cost and sales is quite small which results in reduction of revenue to the farmers [1].

There are several factors which cause the retardation of broiler chickens' growth. One of the most important factors is the uneven distribution of broiler chickens themselves in the coop. The broiler chickens should be intensively reared in the coop at all times where the poultry farmers can only place one animal at every 0.80ft$^2$ to 1ft$^2$ [5]. However, some of the poultry farmers overstock the coop beyond the standard operating procedure. This results in the broiler chickens not having proper access to the food and water due to overcrowding. In addition, the broiler chickens will gorge each other when the food is available which can impact both feed conversion and body weight gain.

Another important factor is that sometimes the broiler chickens experience suboptimal temperature which is caused by the environmental control system of the farms. Heat stress in the coop can cause the broiler chickens to reduce its feed intake by around 16.4% [6, 7, 8] and increase the water consumption [7]. This could result in the body weight of the broiler chickens to be lower by 32.6% [6, 8] and increased the feed conversion ratio by 25.6% [6].

The goal of the proposed project is to enable the farmers to monitor the development of their chickens wherever and whenever using an automated weighing system which utilising wireless data transmission method or IoT. The robust weighing scale was developed using aluminium plates and 5kg load cell which able to transmit the weight data to the workstation through wireless transceiver module. The wireless transceiver module of the weighing scales will transmit the data wirelessly to the workstation every 15 seconds. The collected daily weight values will be averaged and compared to the reference weight values and used to detect any discrepancies or variations. The current average weight along with the reference will be transmitted to the Cloud through Wi-Fi in real time. These values will show the growth rate of the chickens, and any differences between the recorded and reference values to provide an indication of a possible problem.

## 2. Problem Statement

The general scenario and associated problems with poultry farmers have been introduced above. At present, to monitor the broiler chickens' growth, the farmers in Malaysia have to manually weigh the livestock twice a day, in the morning and evening by using the analogue or digital weighing scale. However, the workers in the poultry farms do not weigh the broiler chickens according to any established standard operating procedure (SOP), which in turn results in inaccurate growth-rate data obtained.

Therefore, this research proposed the development of an automated chicken weighing system to replace the manual approach with a more accurate data collection approach. The system is equipped with wireless sensor network (WSN) that enables daily monitoring of the chicken development. Equipped with growth rate reference, the system also enables any problems that may arise to be detected as soon as any discrepancies are recorded. Hence problems such as insufficient or poor distribution of feed and the occurrence of disease can be rectified at an early stage.

## 3. Proposed System

The development of the automated chicken weighing system using wireless sensor network (WSN) consists of two parts which are the weighing scales and a workstation that collects the weight data.

### *3.1. Weighing Scale*

The robust weighing scale was developed using aluminium plates and 5kg load cell where it can handle loads from 500 to 5,000 grammes. The Arduino Microcontroller was used to collect data from the load cell amplifier and transmit it to the workstation through the wireless transceiver module. Two units of 3.7vdc (3.8amp) rechargeable batteries were used as the power supply for the weighing scale. In addition, the weighing scale was designed to be water and dust resistant so that the internal electronic components are fully enclosed. The proposed system architecture of the weighing scale is as shown in the Figure 1.

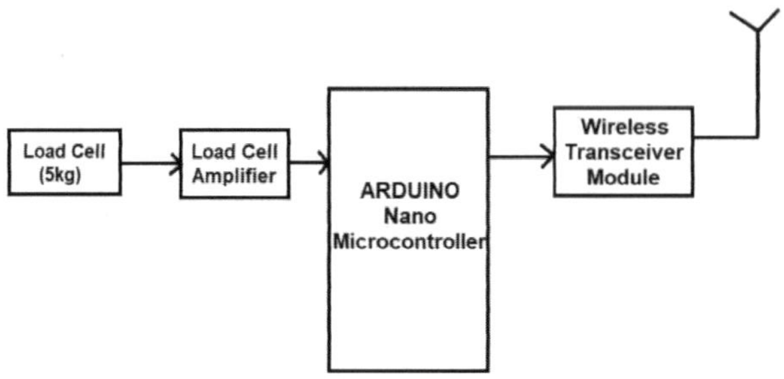


**Figure 1.** Proposed system architecture of weighing scale.

*3.2. Workstation*

The wireless transceiver module of the weighing scales will transmit the data wirelessly to the workstation every 15 seconds. The collected daily weight values will be averaged and compared to the reference weight values and used to detect any discrepancies or variations. The current average weight along with the reference will be transmitted to the Cloud. These values will show the growth rate of the chickens, and any differences between the recorded and reference values to provide an indication of a possible problem. This enables the farmers to keep track of the broiler chickens' development by access the daily weight readings and comparisons wherever and whenever. The proposed system architecture of the workstation is as shown in the Figure 2.

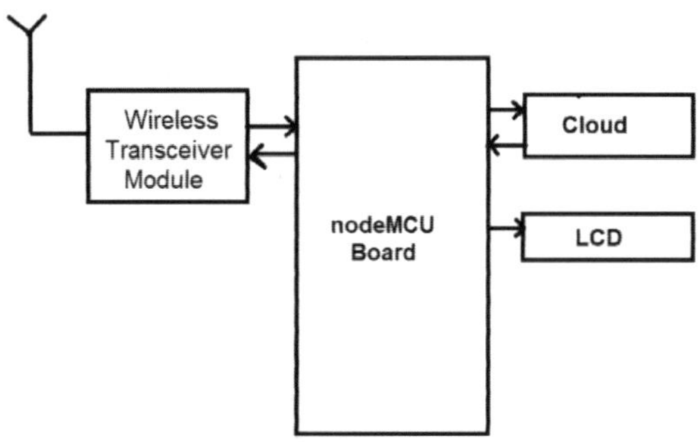


**Figure 2.** Proposed system architecture of workstation.

## 4. Real Time Data Collection in Chicken Coops

The proposed system is installed in three chicken coops with different dimensions to perform the real time data collection, designated coop A, B and C.

The chicken coop A which is 200ft long and 80ft wide needs 20 sets of weighing scales and one workstation. The chicken coops B and C, on the other hand, which are 180ft long / 80ft wide and 100ft long / 50ft wide need 18 and 12 sets of weighing scales respectively. The data from each weighing scale are transferred wirelessly to the workstation. To obtain a more accurate weight data, the weighing scale is placed at every 18ft in the coop because the movement of one group of chickens is 3 meters in radius.

These are then averaged to get the daily weight of the chickens in the specific coop. The system will compare the average daily weight of the chickens to the reference values. The differences are evaluated: insignificant deviations correspond to normal growth rate while otherwise may indicate problems. If the latter is detected, a warning is given to the farmer via the Cloud. The farmer may also view the collected data whenever the need arises through an app via mobile phone.

The weighing scale is placed within the three coops for 38 days to obtain and monitor the development of the chickens from day to day. This is to ensure the growth rate of the chickens is as expected.

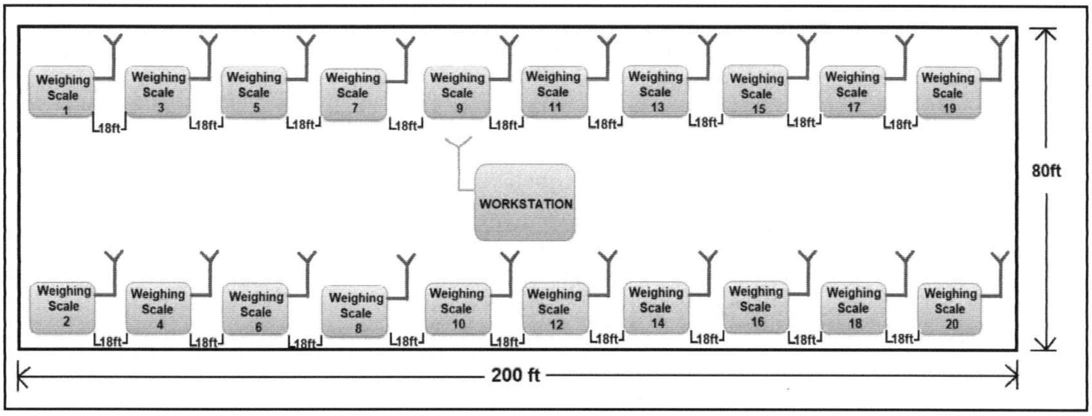


**Figure 3.** Weighing scale distribution for chicken coop A.

## 5. Result and Discussion

Based on the real time data collection in three coops as shown in Figure 4, it can be clearly seen that the growth rate of chicken is as expected from day 1 to day 12.

However, starting from day 13, the growth rate of chickens was recorded to be slower than expected. Compared to expected growth rate, the growth of chickens is 3.38% to 12.21% slower which caused the livestock take 40-42 days to reach 1.8kg. This is due to high amount of broiler chickens in one coop.

For coop A which is 200ft long and 80ft wide, the poultry farmer supposedly can only place 15,000 of broilers chickens. However, the poultry farmers placed 16,000 of broiler chickens. This results in the broiler chickens not having proper access to the food and water due to overcrowding. In addition, the broiler chickens will gorge each other when the food available which can impact both feed conversion and body weight.

The coop B and C, on the other hand, which are 180ft long / 80ft wide and 100ft long / 50ft wide supposedly can only place 14,400 and 5,000 of broiler chickens respectively. However, the poultry farmers placed 12,000 and 4,000 of broiler chickens respectively in one coop. This is why the broiler chickens in coop B and C grow well than the those in coop A.

Secondly, the unexpected growth rate of chickens might also due to the poor ventilation system in the coop. When the temperature is high in the coop, the broiler chickens drink more water than they eat the food which resulted in the body weight gain of livestock is lower.

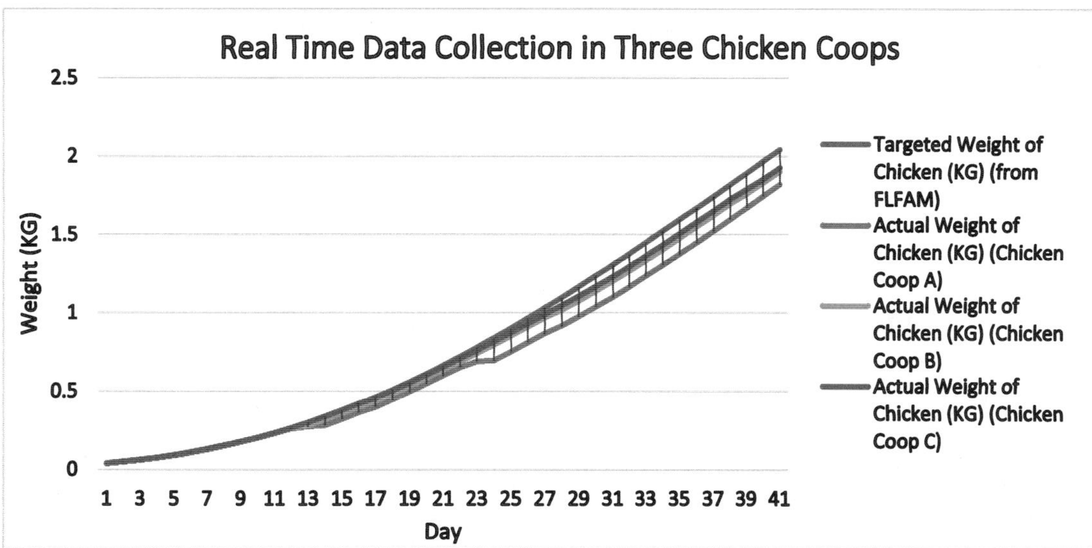


**Figure 4.** Real time data collection in three chicken coops.

## 6. Conclusion

Successful development of an automated chicken weighing system using wireless sensor network (WSN) able of collect real-time weight data of broiler chickens has been demonstrated. The proposed system aims to replace the manual approach and assist the farmers to monitor the daily development of chickens such that rapid action can be taken as soon as problems are detected. With the aid of this system, problems such as insufficient or poor distribution of feed, the occurrence of disease and poor ventilation system can be rectified at an early stage.

Further development of the system may include additional sensing modalities, to enable the temperature and humidity zoning within the coop to be monitored and its effects to the chicken growth analysed. This will enable true efficient management of the chicken coops to be implemented.

**Acknowledgments**

We would like to thank Universiti Malaysia Perlis (UniMAP) for supporting our research works.

## References

[1] Abdurofi I., Ismail, M.M., Kamal H. A. W. and Gabdo B. H.,"Economic analysis of broiler production in Peninsular Malaysia," *International Food Research Journal*, vol. 24, no. 2, pp. 761-766 , 2017.

[2] D. O. S. Malaysia, "Press Release Supply and Utilization Accounts Selected Agricultural Commodities, Malaysia 2012-2016," The Office of Chief Statistician Malaysia Department of Statistics, Department of Statistics, Malaysia., Kuala Lumpur, 2017.

[3] D.O.S. Malaysia, "Press Release Selected Agricultural Indicators, Malaysia, 2017," The Office of Chief Statistician Malaysia, Department of Statistics, Malaysia, Kuala Lumpur, 2017.

[4] FLFAM, Malaysia, "Industry Info," [Online]. Available: http://www.flfam.org.my/index.php/industry-info/the-poultry-industry/broiler-production. [Accessed 18 November 2018].

[5] DVS.Malaysia, "Guidelines of Poultry Meat Farming," June 2006. [Online]. Available: http://www.dvs.gov.my/dvs/resources/auto%20download%20images/560a42ba49275.pdf. [Accessed 18 November 2018].

[6] Lucas J. Lara, Marcos H. Rostagno, "Impact of Heat Stress on Poultry Production," *Animals,* vol. 3, pp. 356-369, 2013.

[7] Defra. Department for Environment, "Heat Stress in Poultry," Defra, London, 2005.

[8] M. U. Sohail, M. E. Hume , J. A. Byrd , D. J. Nisbet , A. Ijaz , A. Sohail ,M. Z. Shabbir , H. Rehman, " Effect of supplementation of prebiotic mannan-oligosaccharides and probiotic mixture on growth performance of broilers subjected to chronic heat stress," *Poultry Science,* vol. 91, p. 2235–2240 , 2012.

# High Gain of Directional Ultra-Wideband Array Antenna Using Flat Reflector Structure for Microwave Imaging

**Mohd Aminudin Jamlos[1], W A Mustafa[1], Wan Khairunizam[2], I Zunaidi[3], Z M Razlan[2] and Shahriman A B[2]**

[1]Faculty of Engineering Technology, Universiti Malaysia Perlis, UniCITI Alam Campus, Sungai Chuchuh, 02100 Padang Besar, Perlis, Malaysia

[2]Advanced Intelligent Computing and Sustainability (AICoS) Research Group, School of Mechatronics Engineering, Universiti Malaysia Perlis, 02600 Arau, Perlis, Malaysia

[3]Faculty of Technology, University of Sunderland, St Peter's Campus, Sunderland, SR6 0DD, United Kingdom

**Abstract.** UWB array antenna with high gain is proposed. The 4×1 array antenna consists of four identical copper circular patches are appropriately connected using quarter-wavelength transformer transmission line. The array elements and its feed network are properly optimized to have a high gain performance over wide range bandwidth frequencies. New additional flat reflector with distance of 20 mm enhances the proposed antenna gain. UWB antenna performance with reflector is better than UWB antenna without reflector structure. Antenna with reflector recorded wider operated bandwidth; 1.9GHz-10.6GHz with higher gain ranged from 3.5dB until 14.2dB compared with without reflector antenna operated from 2.2GHz to 10.6GHz with lower gain ranges of 1.9dB to 9.8dB. The reflector structure enhanced the antenna gain by 50% where the average gain throughout operated frequency for antenna with reflector and antenna without reflector is 10.2dB and 6.8dB respectively. Dimensions of 80 mm × 45 mm is considered as small UWB array antenna. The high gain and wide operator bandwidth of proposed antenna finds it very suitable to be integrated in human brain microwave imaging system due to ability of the wide signal with high energy penetrating human head structure.

## 1. Introduction

Microwave imaging could be defined as seeing the internal structure of an object by exposing electromagnetic fields at microwave frequencies of 300MHz- 30GHz [1]. It is done by producing images consist of distributed electrical property in human body called human microwave imaging. Techniques for microwave imaging generally divided by three major parts involving active, passive and hybrid techniques. As for active microwave imaging, several antennas transmit microwave signals into tissues and reflected signals known as backscattered signals are used to generate microwave images. Active microwave imaging can be further separated into ultra-wideband (UWB) radar imaging and tomography imaging. At first, UWB radar imaging technique is introduced purposely only for breast tumor detection [2] and then followed by emerging modification technique used comparable principles technique for stroke and tumor detection in the brain later [3]. Radar based techniques is preferable since it only need to identify strong scattering point caused by inclusions with high frequent occasion [4]. This kind of technique is less complicated which involve less sophisticated algorithm of delay and sum confocal microwave imaging algorithm. This algorithm is the most commonly algorithm used in ultra-wideband radar based microwave imaging technique.

Currently, microstrip UWB antennas have gaining great interested from researches due to their advantages such as simple structure, low profile, high data rate, easy integration with monolithic microwave integrated circuits (MMICs) and simple to fabricate. Thus, UWB antenna has turned to be the ideal candidate for future short-range (10 m) high-data wireless communication applications, peer-to-peer ultra-fast communications and a lot of other applications. This has inspired researchers to thoroughly investigate the design of UWB antennas [5]. Due to great benefits provided by UWB technology, various fields and domains with potential applications are deployed using this kind of technology such as in communications, imaging, radar, landmine detection, localization and biomedical systems [6, 7]. Application of microwave energy in the medical field is currently gaining interest by the research community where microwave based systems have the high potential of being simple, safe, portable and cost-effective.

Due to high gain and better return loss offered by UWB antenna that well suitable for medical applications, the utilization of UWB antenna promising significant function for detection purpose compared with most of the conventional compact antennas which have low gain and poor return loss. In microwave imaging, high gain and good directional beam-width for wide frequency band antennas are really necessary for target detection, localization systems and cancer screening applications [8]. High gain and wide operated bandwidth antenna that boost the focus of the electromagnetic energy is crucial in microwave imaging for interacting and penetrating the multilayer structures with different characteristics [9] contributes in providing images with high details. Clear and sharp images lead to accurate diagnostic and avoid false treatment given to the patient.

Currently, most of the UWB antenna structures explained in literature are consist of single patch element exhibited nearly omnidirectional radiation patterns and considerably low gain. [10]. Furthermore, single UWB antenna has been explored comprehensively before and already established worldwide. According to [11, 12], most of the compact UWB antennas demonstrated omnidirectional radiation patterns with moderately low gain and noticeable distortion within the impulse response. UWB arrays antenna can be considered to be applied once directional radiation pattern and higher value of antenna gain is required [10]. Apart from directional and high gain antenna, UWB array antenna also required to overcome the disadvantages of microstrip single antenna which is narrow bandwidth with low efficiency [13]. Partial ground technique and present of parasitic element applied in the design lead to realization of UWB array characteristic. The arrays structures are very effective in capture the scattered signal that can be analyzed to solve the inverse problem and obtain the constructed images of the preferred tissues [14].

Apart from array structure and coaxially fed, introduction of additional copper reflector contribute significantly in increasing the proposed antenna gain in order to penetrate multi-structure human head so that the signal could reach the tumor inside brain. It is done by reflecting the backwards radiation towards frontwards radiation and hence increases the gain by summing up the original frontwards radiation with reflected one. Reflector in antenna is a structure that reflects electromagnetic waves towards desired direction. Reflectors also work to stop undesirable signals from interfering either the side or back of the antenna [15]. Reflector is extremely necessary for single direction transmission and reception or maximum gain possible required. Reflectors could be either standalone device or being integrated with the antenna as a whole structure. Standalone reflector function is to redirect electro-magnetic (EM) energy where corner reflector is one of the most common standalone reflector types. This kind of reflector generally used for radar application as a passive repeater.

On the other hand, integrated reflector function to modify the radiation pattern of the antenna and increasing gain in a wanted direction. It consists of passive element located behind a radiating dipole element in order to absorb and radiates back the signal in a directional way. Among the common integrated reflector types are parabolic reflector, flat reflector, corner reflector and cylindrical reflector [16]. In this paper, flat reflector has been integrated with UWB array antenna in improving the gain and return loss.

## 2. Antenna Design and Fabrication

A 4x1 UWB array antenna has been designed in this paper. This proposed antenna consists of partial ground plane, single layer of taconic as the substrate and circular patches with parasitic element as shown in Figure 1. Parasitic element is placed on the same layer of radiating patch and very close to feeding line with the gap only 0.2 mm. The antenna has small dimensions of 80 mm x 45 mm in size. As shown in Figure 1 (b), the proposed antenna is coaxially feed using 50 Ohm SMA connector located in the middle of the lower part of the antenna back. The proposed antenna implemented the coaxial probe feed where the signal is fed directly to the radiating patch. Here, inner conductor of the coax is connected to the patch with soldering of the dielectric substrate and also its outer conductor which is directly connected to the ground plane [14]. Figure 1 demonstrated the simulated design of UWB array antenna using taconic (TLY-5) with a dielectric constant of $\varepsilon r = 2.2$, a thickness of $t = 1.5748 +/- 0.02$ and tangent loss of $\tan \delta = 0.0009$ as the substrate. The antenna is printed with 4×1 copper radiating patch array properly connected with quarter wave transformer transmission line associated with copper parasitic element for the front side as shown in Figure 1(a). The patches comprise of four identical circular with diameter of 15 mm. The parasitic element is placed on very close to feeding line with the gap only 0.2 mm. As shown in Figure 1 (a), each quarter wave transmission line has its own specific wide dimension for 50Ω, 70.71Ω and 100Ω to ensure equal current distribution towards all four patches could be realized. Quarter-wave transformers of 70.71 Ω are used to have ideal match between the 100 Ω lines and the 50 Ω lines [13].

On the other hand, Figure 1(b) shows copper partial ground plane is printed at the back side with 50Ω SMA connector coaxially fed in the middle of the lower part of the antenna back where the signal is fed directly to the radiating patch. Meanwhile Figure 1 (c) and (d) show the copper reflector with 20 mm gap functioned to reduce the side lobe and realized the directional antenna which has higher gain property by reflecting the backwards radiation towards frontwards radiation. The size of the reflector is similar with the size of the sensor. There is a hole through the reflector specially made for connector connection as shown in Figure 1(d). Small dimensions of the sensor made it suitable enough to be integrated as the sensor in microwave imaging system.

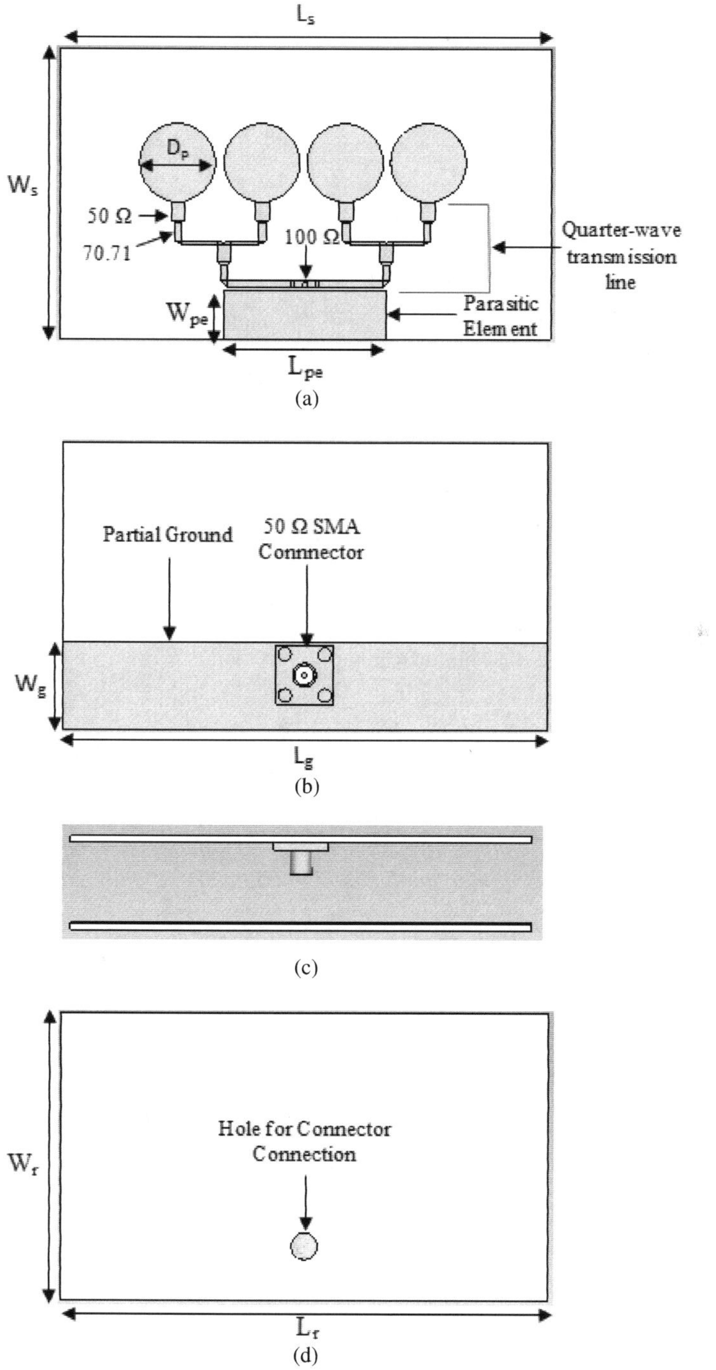


**Figure 1.** The simulated geometry of the proposed UWB array antenna, a) front view, b) transparent back view (without reflector) c) top view e) back view

All important parameters of the designed antenna are optimized to obtain the best result in term of compact size, high gain and wide bandwidth. The optimized dimensions for the antenna are listed in Table 1. Measurements of gain, patterns and s-parameter have been performed using the setup consisting of Agilent ENA 8051C and anechoic chamber. The horn antenna is used as the transmitting antenna while antenna under test (AUT), UWB array as the receiver with a distance of 0.84 m between them.

**Table 1.** Optimized Antenna Parameter

| Parameter | Quantity |
|---|---|
| $L_s$ | 90.0 mm |
| $W_s$ | 45.0 mm |
| $D_p$ | 7.50 mm |
| $W_{pe}$ | 8.00 mm |
| $L_{pe}$ | 32.0 mm |
| $R_d$ | 20.0 mm |
| $W_g$ | 18.0 mm |
| $L_g$ | 90.0mm |
| $W_r$ | 90.0 mm |
| $L_r$ | 45.0 mm |
| $L_s$ | 90.0 mm |
| $W_s$ | 45.0 mm |

Corporate feed network is utilized in realizing four elements array structure where each patch is fed parallelly using transmission lines. The transmission lines are separated into four divisions based on the number of the radiating patch. The quarter-wave transformer impedance matching technique is applied to distribute the power correspondingly toward entire patches where the feed lines of 70.71 Ω are utilized for perfect matching between the 100 Ω lines and the 50 Ω lines as shown in Figure 1 (a). Figure 2 shows Quarter-wave transmission line impedance matching for the corporate feed network.

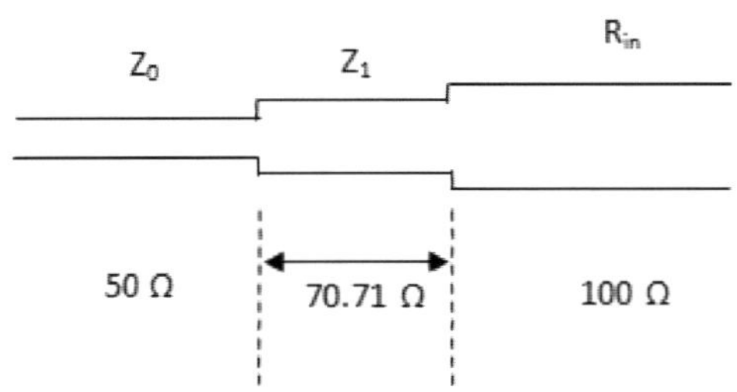


**Figure 2.** Quarter-wave transmission line impedance matching

The formula for the quarter-wave transformer is shown in following equation [13]:

$$Z_1 = \sqrt{Z_0 R_{in}} \tag{1}$$

where:

Z1 = transformer characteristic impedance
Z0 = input transmission line characteristic impedance
Rin = edge resistance at resonance.

Impedance calculation for array antenna is alike for single patch calculation. The formula to perfectly match 100 Ω and 50 Ω transmission lines is shown as followed. Via equation above where Zo= 50 Ω and Rin = 100 Ω, the transformer impedance value is:

$$Z_1 = \sqrt{50(100)}$$

$$= 70.71\ \Omega$$

The sizes for 50 Ω feedline, 70.71 Ω quarterwave transformer and 100 Ω impedance line are acquired by applying following equations [13]:

$$for \frac{W}{h} < 2$$

$$\frac{W}{h} = 8e^{A}/(e^{2A} - 2$$

$$for\, W/h > 2 \tag{2}$$

$$\frac{W}{h} = \frac{2}{\pi}[B - 1 - \ln(2B - 1) + \frac{\varepsilon_r - 1}{2\varepsilon_r}\{\ln(B - 1) + 0.39 - \left(\frac{0.61}{\varepsilon_r}\right)\}] \tag{3}$$

Where h = substrate height and

$$A = \frac{Z_0}{60}[\frac{\varepsilon_r + 1}{2}]^{\frac{1}{2}} + [\frac{\varepsilon_r - 1}{(\varepsilon_r + 1)(0.23 + \left(\frac{0.11}{\varepsilon_r}\right))}]$$

$$B = 377\pi/(2Z_0\ \sqrt{\varepsilon_r}) \tag{4}$$

**Table 2.** Optimized Antenna Parameter

| Impedance (Ω) | Length (mm) | Width (mm) |
|---|---|---|
| 50 | 8.00 | 2.90 |
| 70.71 | 8.19 | 1.50 |
| 100 | 8.40 | 0.65 |

Figure 3 shows the comparison image between the fabricated UWB antenna with reflector and UWB antenna without reflector as shown in Figure 3(a) and (b) respectively.

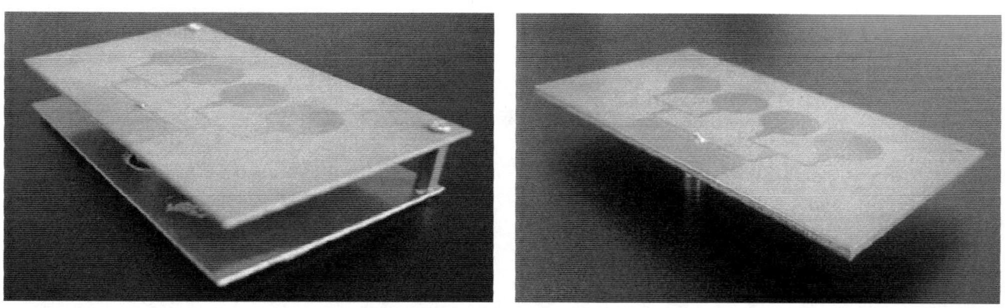

**Figure 3.** Fabricated UWB-Array antenna, a) with reflector, b) without reflector

On the other hand, Figure 4 demonstrated the geometry of the fabricated UWB array antenna with reflector in detail.

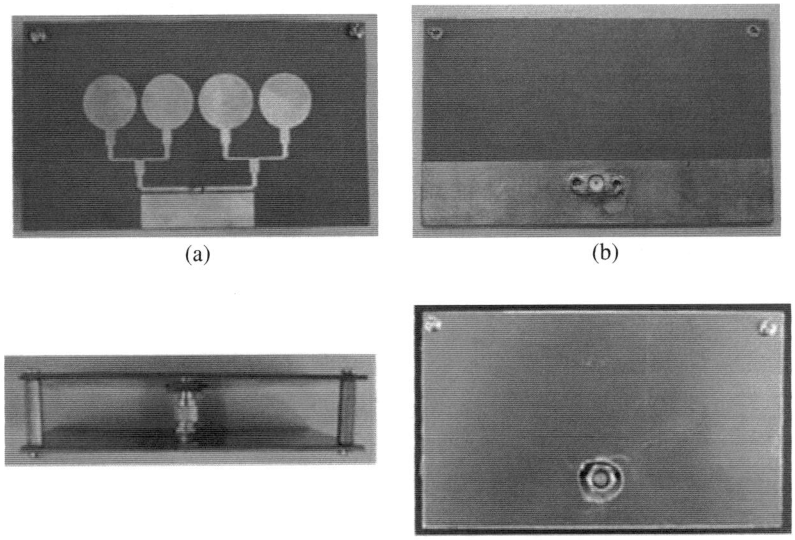

**Figure 4.** The fabricated UWB array antenna with reflector. a) front view, b) transparent back view (without reflector), c) top view, d) back view

## 3. Results and Discussion

Simulation and measurement results for both antennas with reflector and without reflector in term of reflection coefficient are shown in Figure 5. Both simulation and measurement results for each antenna fulfilling the requirement for UWB characteristic; 3.1 GHz until 10.6 GHz. The antenna with reflector performance is better than antenna without reflector performance. From the figure, it shows the antenna with reflector recorded wider range bandwidth of operated frequency compared with antenna without reflector for both simulated and measured results. For simulated, antenna with reflector recorded UWB operated frequency started from 1.8 GHz until 10.8 GHz while antenna without reflector started from 2.0 GHz until 10.8 GHz. On the other hand, antenna with reflector and without reflector recorded 1.9 GHz until 10.6 GHz and 2.2 GHz until 10.6 GHz respectively for measured results. Reflection coefficient less than -10 dB is selected due to the condition where 90% of the signals are successfully transmitted while only the left 10% is reflected back [17]. Partial ground technique and additional of parasitic element assure lower reflection coefficient achieved.

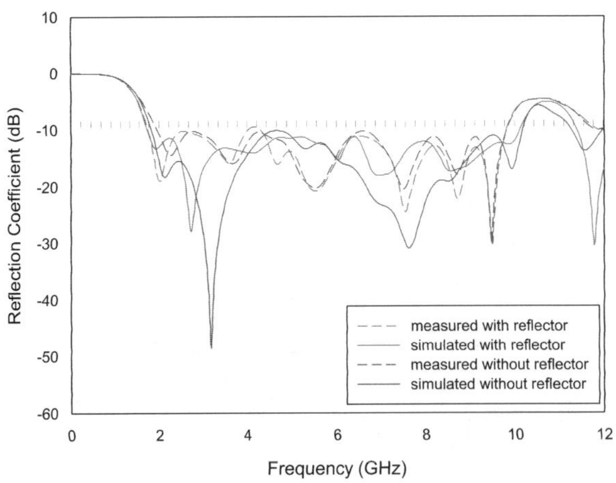


**Figure 5.** Simulated and measured reflection coefficient for UWB array antenna with reflector and without reflector

The antenna with reflector exhibits higher gain than antenna without reflector over the whole operated frequency both for simulated and measured as depicted in Figure 6. Simulated and measured results recorded almost similar readings and considered as adequate agreement between them. The measured gain for antenna with reflector and without reflector ranged from 3.5 dB until 14.2 dB and 1.9 dB until 9.8 dB respectively. The reflector structure enhanced the antenna gain by 50% where the average gain throughout operated frequency for antenna with reflector and antenna without reflector is 10.2 dB and 6.8 respectively. The significant gain enhancement is done with the function of reflector in reflecting the backwards radiation towards frontwards radiation and executed as a unidirectional beam associated with high gain.

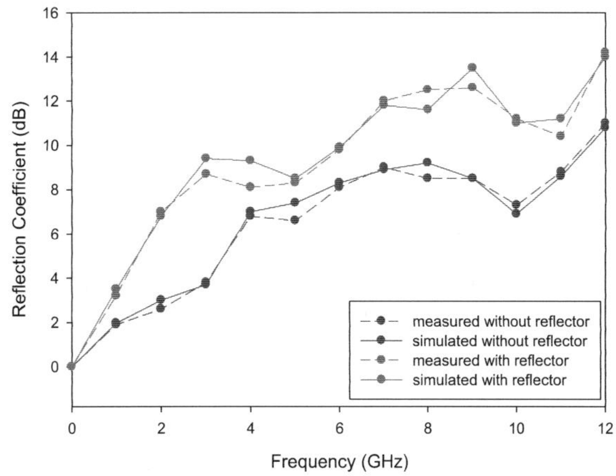


**Figure 6.** Simulated and measured gain for UWB array antenna with reflector and without reflector

In this design, one of the efficient method to obtain directional sensor and optimize the gain while maintaining the UWB frequency is by adjusting the reflector distance (Rd) as shown in Figure 7. Meanwhile, Figure 8 illustrates that the proposed UWB array's reflection coefficient differ significantly among the various Rd. Compared to other Rd, which is 5 mm, 10 mm, 15 mm, 25 mm and 30 mm, only Rd = 20 mm successfully achieves the reflection coefficient within the targeted operating frequencies. Hence, the reflector with specific distance of Rd = 20 mm has been chosen in the UWB sensor design with a dimension of 90 mm × 45 mm.

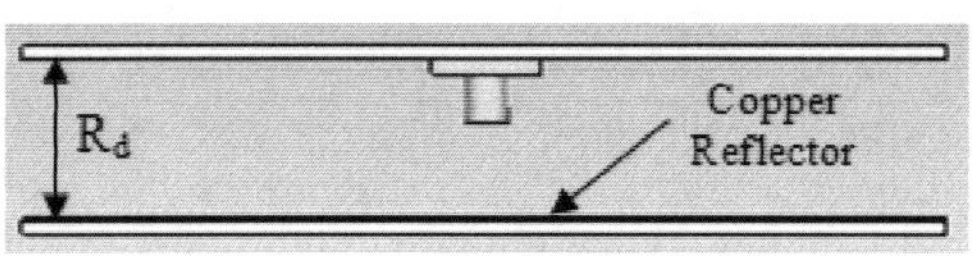


**Figure 7.** Reflector distance for UWB array antenna (top view)

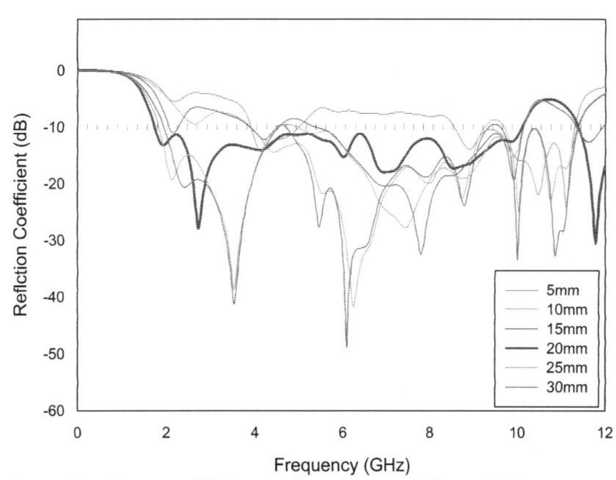


**Figure 8.** Reflection coefficient of various Rd for UWB array antenna

Instead of bandwidth and gain, radiation pattern is the other essential parameter to evaluate the proposed antenna. The measurement and simulated radiation pattern results of Azimuth-Plane for both antennas are shown in Figure 9. The figure demonstrates the polar radiation pattern for the proposed antenna at the frequency of 2 GHz and 3 GHz that are essential frequencies for brain microwave imaging application [1]. The radiation pattern indicates both antenna with reflector and without reflector does radiate over a wide frequency band [12] where antenna with reflector recorded better radiation pattern indicated by wider area covered especially the main lobe. Antenna with reflector also could be categorized under unidirectional antenna since the radiated wave radiates more on main lobe compared to the side and back lobe. In addition, simulation and measurement results indicate acceptable agreement between them. The sensor exhibits the averaged total radiation efficiency of 95% and 87% for antenna with reflector and antenna without reflector respectively. The total radiation efficiency of the antenna promising the good probe or sensor for human brain microwave imaging. The pattern, gain and frequency ranges of the antenna are the parameters as a good probe or sensor for human brain microwave imaging.

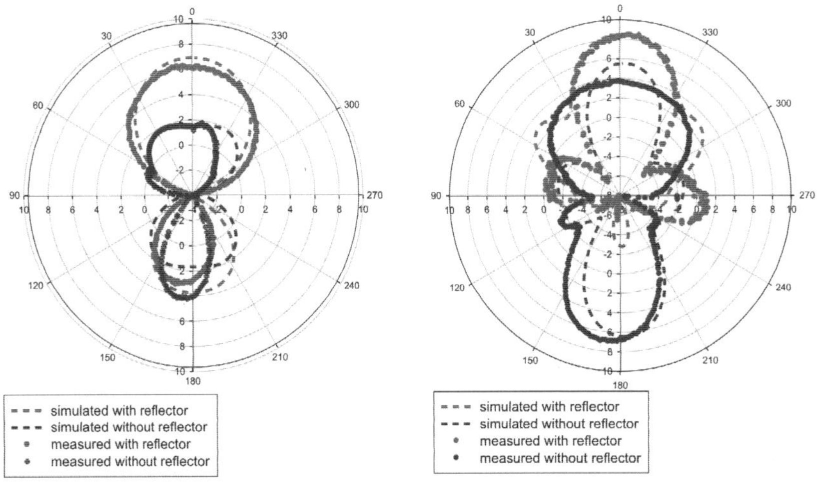


**Figure 9.** Measured and simulated polar radiation pattern in Azimuth plane for antenna with reflector and without reflector; (a) 2 GHz and (b) 3 GHz

## 4. Conclusion

A UWB array antenna with high enhancement of gain for brain microwave imaging is proposed. Additional flat reflector structure contributes significantly in improving the overall antenna performance especially the gain and radiation pattern. The antenna with reflector demonstrated wider reflection coefficient of less than -10dB started from 1.8 GHz until 10.6 GHz with the ranges of gain between 3.5 dB to 14.2 dB as compared with antenna without reflector that recorded 2.2 GHz to 10.6 GHz and 1.9 dB to 9.8 dB for the reflection coefficient and gain respectively. Moreover, measurement results have identical behavior compared to simulations. Thus, in a nutshell, the proposed antenna is very suitable to be implemented in brain microwave imaging applications due to high gain and wide bandwidth.

## 5. Acknowledgment

This work was supported by Excellent Grant Scheme (Grant No. 9009-0049) from University Malaysia Perlis, Malaysia

## 6. References

[1] K. M. Chew, R. Sudirman, N. H. Mahmood, N. Seman, and C. Y. Yong, "Human Brain Microwave Imaging Signal Processing: Frequency Domain (S-parameters) to Time Domain Conversion", Engineering, vol. 5, pp. 31-36, 2013.

[2] S. C. Hagness, A. Taflove, and J. E. Bridges, "Two-dimensional FDTD analysis of a pulsed microwave confocal system for breast cancer detection: fixed-focus and antenna-array sensors", Biomedical Engineering, IEEE Transactions, vol. 45, pp. 1470-1479, Dec. 1998.

[3] D. Ireland, and M. E. Bialkowski, "Microwave head imaging for stroke detection", Progress In Electromagnetics Research M, vol. 21, 163-175, 2011.

[4] L. M. Camacho and S. Tjuatja, "FDTD Simulation of microwave scattering from a lung tumor"; in IEEE Antennas and Propagation Society, AP-S International Symposium (Digest), Washington, Vol. 3 A, 2005, pp. 815-818.

[5] A. K. Gautam, S. Yadav, and Binod Kr Kanaujia, "A CPW-Fed Compact UWB Microstrip Antenna", IEEE Antennas and Wireless Propagation Letters, Vol. 12, pp. 151-154. 2013.
[6] X. N. Low, Z. Chen, and S. P. Terence, "A UWB Dipole Antenna with Enhanced Impedance and Gain Performance", IEEE Transactions on Antenna and Propagations, vol. 57, pp. 2959-2966, Oct. 2009.
[7] A. Eusuola, Y. Chen, and G. Y. Tian, "Novel Ultra-Wideband Directional Antenna for Microwave Breast Cancer Detection", in Antennas and Propagation (APSURSI), IEEE International Symposium, Washington, 2011, pp. 90-93.
[8] A. J. Jasmine and J. M. T. Anita, "Design of Vivaldi Antenna for Brain Cancer Detection", in International Conference on Electronics and Communication Systems (ICECS), Coimbatore, INDIA, 2014, pp. 1-4.
[9] F. S. G. B. Barnes, Handbook of biological effects of electromagnetic fields, Boca Raton: CRC/Taylor & Francis, 2007.
[10] M. GARBARUK, "Analysis of ultra-wideband linear antenna arrays", Electrical Review, pp. 75-76, 2012.
[11] A. M. Abbosh, "Directive Antenna for Ultrawideband Medical Imaging Systems", International Journal of Antennas and Propagation, pp. 1-6, 2008.
[12] N. Ojaroudi, M. Ojaroudi, F. Geran, and Sh. Amiri, "Omnidirectional Antenna with Multi-Resonance Performance for Microwave Imaging Systems", in 20th Telecommunications Forum (TELFOR), Belgrade, 2013, pp. 1162-1164.
[13] N. A. Wahab, Z. Maslan, W. Norsyafizan W. Muhamad, N. Hamzah, "Microstrip Rectangular 4x1 Patch Array Antenna at 2.5GHz for WiMax Application", in Second International Conference on Computational Intelligence, Communication Systems and Networks, Liverpool, 2010, pp. 164-168.
[14] A. F. Sheta, I. Elshafiey, A. Mohra, Z. Siddiqui, and A. R. Sebak, "A Compact Antenna for Microwave Imaging and Hyperthermia Treatment of Brain Tumor", in 15th International Symposium Antenna Technology and Applied Electromagnetics (ANTEM), Toulouse, 2012, pp. 1-4.
[15] A. Imani, Soleimani, M. Soleimani, S. Amiri, "Analysis of Dual-Reflector Antenna for Radar Applications", Applied Computational Electromagnetics Society Journal, vol. 30, issue 12, pp. 1294-1300, 2015.
[16] L. J. Zhang, S. G. Choi, B. C. Ahn, J. H. Bang, D. H. Kim, Y. T. Choi, "Simplified Feed Prime Focus Reflector Antenna", Microwave Journal, vol. 59, issue 2, pp. 108-114, 2016.
[17] M. I. Jais, M. F. Jamlos, M. Jusoh, T. Sabapathy, M. K A. Nayan, A. Shahadah, H. A. Ramli, I. Ismail, F. A. A. Fuad, "1.575 GHz Dual-Polarization Textile Antenna (DPTA) for GPS Application", IEEE Symposium on Wireless Technology & Applications (ISWTA), Kuching, 2013, pp. 376-379

IOP Conf. Series: Materials Science and Engineering **557** (2019) 012019 doi:10.1088/1757-899X/557/1/012019

# Improved KNN Scan Matching for Local Map Classification in Mobile Robot Localisation Application

**M A Markom [1], A H Adom [1], S A Abdul Shukor [1], N Abdul Rahim [1], E S Mohd Muslim Tan [2] and B Ilias[1]**

[1] School of Mechatronic Engineering, Universiti Malaysia Perlis, Kampus Ulu Pauh, 02600 Arau, Perlis, Malaysia

[2] Centre of Diploma Studies, Universiti Malaysia Perlis, Kampus Unicity, 02100 Padang Besar, Malaysia

Corresponding author: marni@unimap.edu.my

**Abstract.** Localisation is essential for autonomous mobile robot system enabling it to locate itself within its environment. One method to perform localisation is to use scan matching with iteration closest point (ICP) algorithm. However, typical ICP may be prone to inaccuracies in localisation and mapping due to problems associated with laser range data limitation such as overshoot data and blank data. This paper presents the improvement to the above problem by the inclusion of a threshold to the KNN scan matching algorithm during iteration process. The threshold is a percentage of nearest point of incoming input with respected to reference point. Threshold values of 0%, 70% and 90% were tested, and improvements of the classification performance were observed with the increase in the threshold values, with the latter achieving 100% accuracy. This work shows that the use of threshold in scan matching may improve the accuracy of local map classification.

## 1. Introduction

There has been an increase in the implementation of mobile robot systems to assist humans. The use of mobile robots has seen successful implementations in various fields, such as home appliances [1], military [2], exploration [3,4], as well as search and rescue [5,6].

In mobile robot system, mapping and localisation can be considered as priority parts. Both are usually incorporated together in order to facilitate the mobile robot to move in a particular environment autonomously. The work on localisation algorithm development typically has types of algorithm-based which are scan matching [7,8], probability method [9] and Kalman filter [10]. These algorithms are applied to calculate and estimate the mobile robot position based on surrounding information that obtained from range finder sensor such as Kinect, laser range finder and ultrasonic. However, these algorithms can only be applied depends on its application and environment [11].

This paper has worked in indoor environment with a reference map that has been established. The suitable algorithm-based is scan matching. The key idea of using scan matching is to determine a displacement vector of two 2D scans, which are taken from two different times, by rotating and shifting the two scans against each other in such a way that the two scans optimally matched. Scan matching can be categorised into two methods, Point-to-point and Feature-to-feature.

The point-to-point scan matching needs two scans compared directly. Some example of point-to-point method are Iteration Closest Point (ICP) [7] and cluster method [12]. Usually, it applies in a static environment, but, the method can be enhanced for the dynamic environment. There are two problems that should be considered when dealing with the point-to-point method;

i. Scan results of incoming input are not close to reference scans
ii. Certain point of scanning results are missing

Because of this problem, the use of point-to-point method is very limited. It is not suitable to apply for certain range finder sensor, such as ultrasonic. This sensor only provides one data sample for each scan. If the ultrasonic scan fails to give a result, then, the whole system will be useless. In order to solve the limitation, researcher used laser range finder, it provides more scanning points, or multiple sensors [13] and multiple types of sensors [14], as well as classifier to determine the matching issue [7].

Feature-to-feature method is a more flexible method to perform scan matching. It is suitable for applications in both static and dynamic environments, but still requires assistance by some sort of decision analysis, accuracy calculation and perhaps pattern recognition. This scan matching has high computational complexity compared to the point-to-point method. Also, the quality of the matching is depends on the reliability of the features. Here, some features, for example, line segments or corners, are extracted respectively from previous scan and current scan and then the two scans are matched according to the correspondences between the two feature groups [15].

Both of these point-to-point and feature-to-feature methods apply iteration process which is consists of 'rotation' and 'translation/shifting' in order to match the incoming input sample to the previous/reference scans. It will stop when the incoming input sample is matched with any of reference local maps. In order to have the stop, some studies used percentage of error, and some used number of rotation. The rotation and shifting processes are described as below:

### *1.1. The rotation*

The rotation process involves the rotation of the scanned map in terms of angle, such as 1° to 10° steps with relative to the reference map. This is to eliminate issues with the orientation of the mobile robot with respect to its environment, and hence the reference map. This rotation process will be stopped when the scan matching achieved its expected error percentage. Figure 1(a) shows illustration of the rotation process example. In this Figure, the black shape is a reference, while, the red shape is the incoming scan. The red shape is going to rotate to be matched with the red shape.

### *1.2. The shifting/translation*

The translation process is involved with a varying of x-axis and y-axis of the incoming scan to be fitted with the x-axis and y-axis of the reference. Figure 1(b) shows the translation process illustration. In this Figure, the red shape must be transformed to be matched with the black shape. The black shape is the reference and the red shape is the incoming scan. The translation process will be stopped after reaching its percentage error. Once it's stopped, the whole process of scan matching is stopped and the results are going for further classification decision.

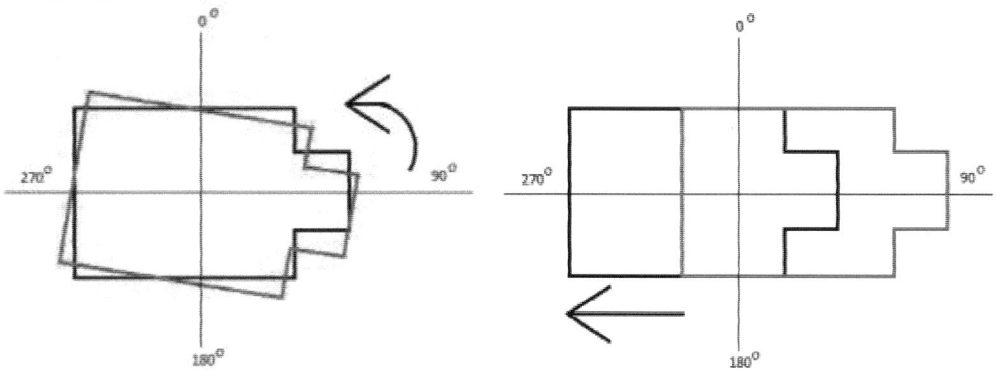


(a) (b)

**Figure 1.** (a) the rotation process, the input is altered in angles, and (b) the shifting process, the input is altered in term of horizontal and vertical.

## 2. Scan matching and threshold

For this work, scan matching is computed using Euclidean distance. When the scan matching results are unsatisfied the threshold, the iteration begins. Three values of threshold is applied which are set to be 0%, 70% and 90%. This threshold is used to represent the percentage of nearest point of incoming input with respected to the reference point. Higher the percentage means the incoming input is very similar to one of the reference local map. Else, it is otherwise. The flow of scan matching with a threshold is shown in figure 2.

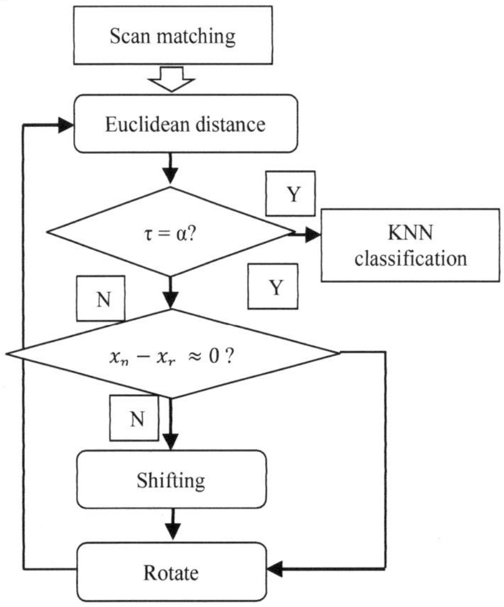


**Figure 2.** The iteration process (rotation and transformation) in the scan matching based on the Euclidean distance

## 3. Reference map

In mobile robot self-localisation, there are cases that needs reference to show and illustrate the mobile robot location. The reference can be the previous or initial map [16], location of beacon [17] and specific landmark [18].

In this study, the reference are consisted of initial local map. The initial map is obtained during the initial experiment, which is also known as exploration work. At this work, a testing location is recognised, and a few position of scanning to create local maps are established. For each location, 10 times of scanning is performed. The scanning data are analysed to be repetitive and consistent to ensure the reliability and valid. Figure 3 shows the diagram of reference local map 1 to local map 6.

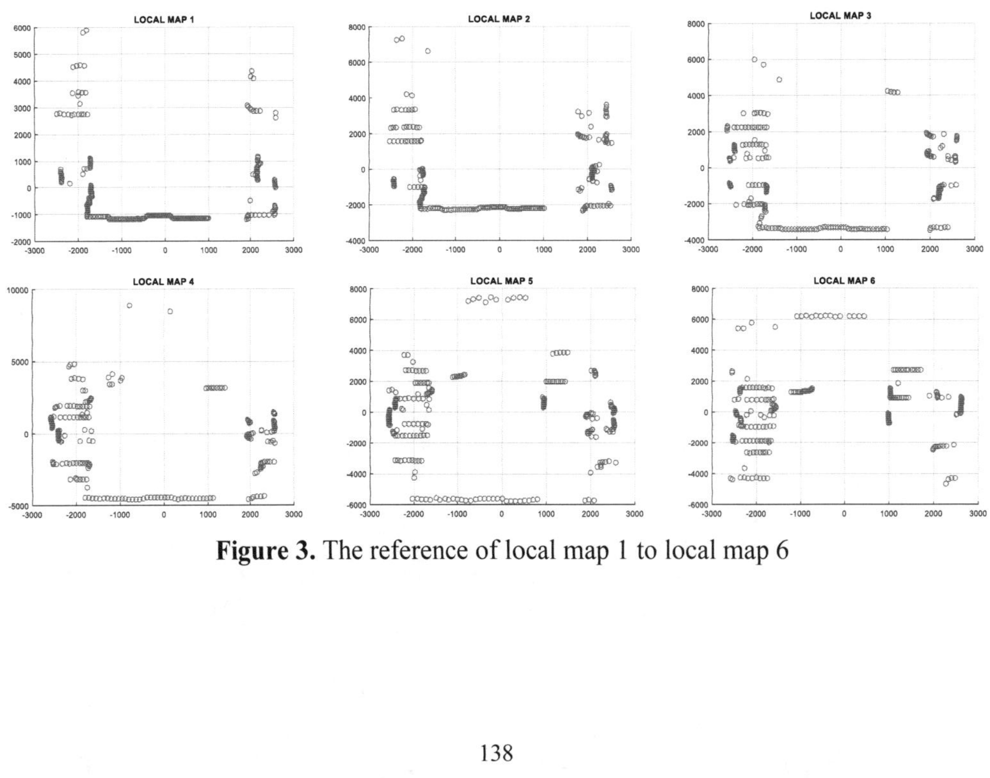


**Figure 3.** The reference of local map 1 to local map 6

## 4. The experiment

A testing is conducted using 100 sets of new scanning sample that involves the segment/area of local map 1 to local map 6 each. This testing will show the performance of the rotation and translation process in order to justify the point to stop the iteration and to get back to the main algorithm to complete the classification process. The results are shown in graphic and the percentage of accuracy. The graphic is used to represent the rotation and translation process from the beginning until it finds the match among the local maps with threshold of 90%.

## 5. Local Map Classification Using KNN

KNN is used to identify the local map that matched with the incoming input. This KNN implements Euclidean distance as its distance parameter. The flow of the scan matching with KNN and location estimation is as below;

- Ready the KNN model with sample class labels, $(s_i, \theta_i)$,
- Ready the input/unlabelled sample, $(s_j, \theta_j)$
- Convert the $\theta_i(deg)$ and $\theta_j(deg)$ into $\theta_i(rad)$ and $\theta_j(rad)$

$$\theta_{rad} = \frac{\pi}{180} \times \theta_{deg} \quad (1)$$

- Convert the point into $x$ and $y$

$$x_j = s_j + \cos\theta_{rad} \quad (2)$$

$$y_j = s_j + \sin\theta_{rad} \quad (3)$$

where $s_j$ is the distance of the scanning sample, $x_j$ and $y_j$ are the estimation results.

- Determine the *K*–parameter. Here, *K*–parameter uses is 3.
- Calculate the Euclidean distance, $d_{ij}$, for all reference samples in the KNN model, $(x_i, y_i)$, with the input, $(x_j, y_j)$

$$d_{ij} = \sqrt{(x_i - x_j)^2 + (y_i - y_j)^2} \quad (4)$$

where $i$-th and $j$-th is the number of data input sample and data reference sample.

- Repeat for all samples in the KNN model.
- Iteration process with $\tau$, $(a = 0°, 70°, 90°)$.
- Ready the Euclidean distance,$d_{ij}(m)$, get the classification results based on the lowest summation of Euclidean distance.

## 6. Results and Discussion

There will be two results of this work, i. Iteration performance results, and ii. The map classification result. All are presented as below;

### *6.1. Iteration results with and without threshold*

Table 1 shows the iteration process results for the scan matching using three thresholds. The results are showed in terms of accuracy percentage using incoming input of local map 3. Based on the table, there is no results with percentage of 0%, which is means the incoming input is able to show similarity features to the reference map. At initial iteration results, the incoming input is able to match with a reference local map which is achieved threshold of 70%. However, this result is wrong and will end up with wrong classification of local map. This is a mistake if threshold of 70% is applied. For iteration with threshold of 90%, the iteration is performed a few times until the incoming input matched with any reference local map with 90% of accuracy. Based on the table, the accuracy of the local map 3 gets higher, while, some of the local maps get lower, and, several others retain similar results. The local map 3 shows that it starts with 62% and after having its fifth rotation, it turns to 91%.

### *6.2. Map classification results*

The local map classification results are shown in Table 2 based on confusion matrix of KNN. The testing are conducted using 100 sample of incoming input for each local maps. Based on Table 2, the results of algorithm without iteration, total corrected incoming sample for local map 1 to local map 6 are 40%, 45%, 44%, 42% 38% and 32%, respectively. For threshold of 70%, the total corrected incoming sample are 78%, 80%, 76%, 74%, 70% and 69%, respectively. While, for threshold of 90%, all incoming sample are classified correctly. These show such results because there are similarity features between near local maps. For example as shown in algorithm without iteration, for local map 1, the incoming input are classified in local map 2 and local map 3, but, there are no classification results out with local map 4 to local map 6. When the threshold is increased to 90%, there are many similarity features from near local maps are eliminated, and just leave the real features for the particular class of local map. In other word, incoming input will precisely classify at its own local map.

**Table 1.** The example of iteration process result with six time of rotations with the incoming input of local map 3.

| No of iteration | ± degree (°) | Local map 1 (%) | Local map 2 (%) | Local map 3 (%) | Local map 4 (%) | Local map 5 (%) | Local map 6 (%) |
|---|---|---|---|---|---|---|---|
| - | - | 54 | 78 | 62 | 23 | 63 | 60 |
| 1 | +2 | 54 | 76 | 71 | 23 | 63 | 60 |
| 2 | +4 | 55 | 76 | 76 | 21 | 63 | 61 |
| 3 | +6 | 55 | 78 | 78 | 20 | 64 | 61 |
| 4 | +8 | 55 | 76 | 87 | 18 | 64 | 63 |
| 5 | +10 | 55 | 74 | 91 | 14 | 61 | 61 |

**Table 2**. Confusion matrix of KNN classification result

| Ref | Incoming input | | | | | | | | | | | | | | | | | |
|---|---|---|---|---|---|---|---|---|---|---|---|---|---|---|---|---|---|---|
| | No iteration | | | | | | Threshold 70% | | | | | | Threshold 90% | | | | | |
| | LM 1 | LM 2 | LM 3 | LM 4 | LM 5 | LM 6 | LM 1 | LM 2 | LM 3 | LM 4 | LM 5 | LM 6 | LM 1 | LM 2 | LM 3 | LM 4 | LM 5 | LM 6 |
| LM 1 | 26 | 10 | 0 | 0 | 0 | 0 | 70 | 2 | 0 | 0 | 0 | 0 | 100 | 0 | 0 | 0 | 0 | 0 |
| LM 2 | 10 | 23 | 12 | 2 | 0 | 0 | 8 | 75 | 4 | 0 | 0 | 0 | 0 | 100 | 0 | 0 | 0 | 0 |
| LM 3 | 4 | 12 | 30 | 6 | 1 | 0 | 0 | 3 | 70 | 4 | 0 | 0 | 0 | 0 | 100 | 0 | 0 | 0 |
| LM 4 | 0 | 0 | 2 | 25 | 12 | 2 | 0 | 0 | 2 | 70 | 5 | 0 | 0 | 0 | 0 | 100 | 0 | 0 |
| LM 5 | 0 | 0 | 0 | 5 | 17 | 12 | 0 | 0 | 0 | 0 | 62 | 4 | 0 | 0 | 0 | 0 | 100 | 0 |
| LM 6 | 0 | 0 | 0 | 4 | 8 | 18 | 0 | 0 | 0 | 0 | 3 | 65 | 0 | 0 | 0 | 0 | 0 | 100 |
| % | **40** | **45** | **44** | **42** | **38** | **32** | **78** | **80** | **76** | **74** | **70** | **69** | **100** | **100** | **100** | **100** | **100** | **100** |

## 7. Conclusion

As conclusion, the scan matching incorporated with threshold really helps localisation algorithm in term of accuracy. The work has proven by comparing the results without threshold or using low value of threshold, which is both results showed low classification accuracy compared to high value of threshold. The needs of high value of threshold, such as 90%, is due to the presence of similarity features among the near local maps. If each of local map has much different in features, then, low value of threshold can be used. The feature similarity can be analysed using statistical analysis, such as ANOVA, MANOVA and can be represented using Dendrogram.

## 8. References

[1] Do, H. M., Pham, M., Sheng, W., Yang, D., & Liu, M. (2018). RiSH: A robot-integrated smart home for elderly care. Robotics and Autonomous Systems, 101, 74-92.

[2] Siciliano, B., & Khatib, O. (Eds.). (2016). Springer handbook of robotics. Springer.

[3] Joochim, C., & Roth, H. (2010). Mobile Robot Exploration Based On Three Dimension Cameras Acquisition. IFAC Proceedings Volumes, 43(23), 116-121.

[4] Lauri, M., & Ritala, R. (2016). Planning for robotic exploration based on forward simulation. Robotics and Autonomous Systems, 83, 15-31.
[5] Bakhshipour, M., Ghadi, M. J., & Namdari, F. (2017). Swarm robotics search & rescue: A novel artificial intelligence-inspired optimization approach. Applied Soft Computing, 57, 708-726.
[6] Bai, L., Guan, J., Chen, X., Hou, J., & Duan, W. (2018). An optional passive/active transformable wheel-legged mobility concept for search and rescue robots. Robotics and Autonomous Systems, 107, 145–155
[7] Konecny, J., Prauzek, M., & Hlavica, J. (2016). ICP Algorithm in Mobile Robot Navigation: Analysis of Computational Demands in Embedded Solutions. IFAC-PapersOnLine, 49(25), 396-400.
[8] Addie, I., Marni Azira, M., Abdul Hamid, A., & Mohd Muslim Tan, E. S. (2017). Scan Matching and KNN Classification for Mobile Robot Localisation Algorithm. In Proceeding of 2017 IEEE 3rd International Symposium in Robotics and Manufacturing Automation (ROMA). 19-21 Sept. 2017. Kuala Lumpur. DOI: 10.1109/ROMA.2017.8231836
[9] Vroegindeweij, B. A., IJsselmuiden, J., & van Henten, E. J. (2016). Probabilistic localisation in repetitive environments: Estimating a robot's position in an aviary poultry house. Computers and Electronics in Agriculture, 124, 303-317.
[10] Zhu, D., Zhao, B., & Wang, S. (2018). Mobile target indoor tracking based on Multi-Direction Weight Position Kalman Filter. Computer Networks, 141, 115-127.
[11] Klancar, G., Zdesar, A., Blazic, S., & Skrjanc, I. (2017). Wheeled mobile robotics: from fundamentals towards autonomous systems. Butterworth-Heinemann.
[12] Rashid, A. T., Frasca, M., Ali, A. A., Rizzo, A., & Fortuna, L. (2015). Multi-robot localization and orientation estimation using robotic cluster matching algorithm. Robotics and Autonomous Systems, 63, 108-121.
[13] Ilias, B., Shukor, S.A.A., Adom, A.H., Ibrahim, M.F. and Yaacob, S., 2016, May. A novel indoor mobile robot mapping with USB-16 ultrasonic sensor bank and NWA optimization algorithm. In Computer Applications & Industrial Electronics (ISCAIE), 2016 IEEE Symposium on (pp. 189-194).
[14] Santos, J.M., Couceiro, M.S., Portugal, D. and Rocha, R.P., 2014, May. Fusing sonars and LRF data to perform SLAM in reduced visibility scenarios. In Autonomous Robot Systems and Competitions (ICARSC), 2014 IEEE International Conference on (pp. 116-121).
[15] Petrich, J., Brown, M. F., Pentzer, J. L., & Sustersic, J. P. (2018). Side scan sonar based self-localization for small Autonomous Underwater Vehicles. Ocean Engineering, 161, 221-226.
[16] Huang, H., Lee, D. H., Chang, K., Li, W., & Acharya, T. D. (2018). Development of mobile platform for indoor positioning reference map using geomagnetic field data. Computers & Electrical Engineering, 68, 557-569.
[17] Zhang, J., Shi, C., Sun, D., & Han, Y. (2018). High-precision, limited-beacon-aided AUV localization algorithm. Ocean Engineering, 149, 106-112.
[18] Qu, X., Soheilian, B., & Paparoditis, N. (2018). Landmark based localization in urban environment. ISPRS Journal of Photogrammetry and Remote Sensing, 140, 90-103.

# Fragmentation in MAC IEEE 802.15.4 to Improve Delay Performance in Wireless Body Area Network (WBAN)

**Wan Aida Nadia Wan Abdullah[1], Naimah Yaakob[1], R. Badlishah[2], Mohamed Elshaikh Elobaid[1], Siti Asilah Yah[1] and I. Zunaidi[3]**

[1]ENAC, School of Computer and Communication Engineering (SCCE), Universiti Malaysia Perlis (UniMAP), 02600 Arau, Perlis, Malaysia

[2]Faculty of Informatics and Computing, Universiti Sultan Zainal Abidin (UniSZA), Kuala Terengganu, Malaysia

[3]Faculty of Technology, University of Sunderland, St Peter's Campus, Sunderland, SR6 0DD, United Kingdom

**Abstract.** A Medium Access Control (MAC) protocol plays a significant role in enhancing the performances of Wireless Body Area Network (WBAN). This technology is commonly implemented IEEE 802.15.4 standard that works under CSMA/CA protocol. In addition, WBAN deals with probability of high traffic occurrence due to collection bulk of medical data that are generated from several bio-sensors through a continuous monitoring of patient's health. This phenomenon could result to congestion in the network which may lead to increment in delay and packet loss that could collapse the network. The increment in these two performances could lead to wrongly diagnosed of the diseases as well as late in medical help. Hence, the proposed mechanism aimed to reduce the delay through fragmenting the packets by decrementing the slot time in MAC 802.15.4 protocol under different number of number nodes with constant packet size and bit rate. The performances of the proposed mechanism are measured in terms of delay and packet loss by using OMNeT++ simulator tool. Based on the collected results, the proposed mechanism has shown a promising result in reducing the delay which means that it has improved the network performance of WBAN.

## 1. Introduction

A Wireless Body Area Network (WBAN) is one of the advanced technologies derived from Wireless Sensor Networks (WSN). It specifically focuses on the remote health monitoring applications by offering a continuous health monitoring over a longer period of time and real-time feedback to the user and medical staffs [1]. It is made up of several tiny of intelligent health sensors that can be wearable or implanted in the body. These sensors will be integrated and communicated through existing wireless technologies such as IEEE 802.15.4 and Internet. This has enable medical personnel to remotely monitor the condition of health of users without constraining the movement and daily activities of the user [2]. Additionally, this technology provides an earlier detection of any abnormalities in the vital signals of the body such as heart rate and blood pressure. Due to possibility of dealing with heavy traffic of data transmission, this technology is really stringent to delay and packet losses especially when there is a presence of network congestion as increment in these performances could degraded the overall performances of the system. In short, it is in need of a prompt and timely delivered of medical assistance which could save the user's life and reduce the risk of mortality. Thus, a proper choice of MAC protocol is strictly needed to gain optimum performances in terms of delay, Packet Delivery Ratio (PDR) and throughput of the system.

There are two types of Multiple Access Control (MAC) protocols that are commonly adapted in WBAN such as Carrier Sense Multiple Access (CSMA) and Time Division Multiple Access (TDMA) according to [3][4]. However, the sensors in WBAN commonly use IEEE 802.15.4 technology that implements CSMA mechanism to communicate in the system. This technology offers low-data rate of 250Kbps which is suitable to be adopted in medical application [5]. CSMA is known as a contention-based protocol where it listens first before transmitting any data to the destination node of the shared medium. It has been furthered developed into Carrier Sense Multiple Access/Collison Avoidance (CSMA/CA) and Carrier Sense Multiple Access/Collison Detection (CSMA/CD). But, CSMA faces a strict restriction in utilizing maximum bandwidth and throughput during increment number of nodes and rate of transmission. This is due to the risk of collision occurrence between packets and acknowledgement (ACK) frame in a certain time duration. Moreover, the performances of WBAN could severely impacted if there is a transmission of longer packet size which will be more elaborated in the next section. Hence, suitable selection of slot time and fragmentation of packets are highly recommended to improve the network performances in the WBAN and to guarantee its reliability as well.

To solve the above-mentioned problem, fragmentation of packet is done by exploiting a random values of slot time of MAC IEEE 802.15.4 protocol in order to measure the delay performances of WBAN under constant packet size with different number of nodes. To the best of our knowledge, this is the first attempt of this method to enhance the WBAN performances. The rest of the paper is organized as follows. Section 2 discussed on the previous and related works on IEEE 802.15.4 and its fragmentation in WBAN. Section 3 elaborates the mechanism of the proposed method and used value of parameters for the WBAN simulation purpose. Section 4 presents the detailed analysis on the output of delay performances in the WBAN. Finally, Section 5 concludes and presents some future works on the WBAN.

## 2. Related Works

### *2.1. Mechanism of IEEE 802.15.4*

As aforementioned in earlier section, IEEE 802.15.4 works on CSMA protocol which is particularly implemented CSMA/CA. This mechanism operates as if the medium is found busy, the node needs to wait for a certain duration before transmitting the data and re-listening to the shared medium whether it is free or not. The difference between CSMA/CA and CSMA/CD is the transmission is deferred for a random exponential back-off when it detects of busy medium [6]. To be exact, the deferred transmission of packet in the CSMA/CA mechanism at the end of the superframe can lead to high delay due to insufficient time to complete the transmission process since it increases the probability of the collisions. The collision of these data could be more severed if there is existence of network congestion. Also, this could be worst when it involves with the transmission of longer packet size.

This CSMA/CA can be divided into two categories such as beaconless of Unslotted CSMA/CA and beacon-enabled of Slotted CSMA/CA in the IEEE 802.15.4 [7] [8]. For the delay stringent application as medical application, a proper tuning of slot time is really important to lessen the delay which can ensure the information and medical help can be timely arrived. Hence, this paper is focused to study the mechanism of Slotted CSMA/CA in MAC IEEE 802.15.4 protocol.

The CSMA/CA initiates the number of Back-Off (NB), Back-Off Exponent (BE) and Contention Window (CW) which have default value of 0, 2 and 2 respectively. The value of BE can be varied between [2, macMinBE] which macMinBE has default value of 3. Later, node delay for a random number of uniformly generated Back-Off period (BP) that lies between [0, 2BE -1] is started to count down. The operation must begin at the boundary of BP. Once the Back-Off time expires to 0, the algorithm performs 1st Clear Channel Assessment (CCA1) to assess the channel condition. The CW is decremented by 1. The 2nd CCA which is termed as CCA2 is carried out if CW is not equal to 0. These CCAs are done to avoid any potential collisions between acknowledgement (ACK) packet. Then, the node starts to transmit data and waits for acknowledgement (ACK) packet from Coordinator.

It is worth to state that, if the channel is sensed as idle again after the successful transmission of packet, the remaining BP should be sufficient for the next transmission of packet and ACK. In contrast, if CCA1 or CCA2 spots the channel is busy, the value of BE and NB is incremented by 1 and CW is re-initialized to 2. Meanwhile, increment in BE means there is a high probability for longer delay of Back-Off. In this case, BE and NB can reach to their maximum value of 5 which is known as aMaxBE and macMaxCSMABackoff correspondingly [9]. If the maximum value of macMaxCSMABackoff is reached, a failure report will be sent to the higher layer and the packet is discarded. Else, the algorithm goes back to previous step and the Back-Off operation is re-initiated. To simplify, the MAC algorithm for Slotted CSMA/CA in IEEE 802.15.4 is depicted in the flowchart as shown in Figure 1.

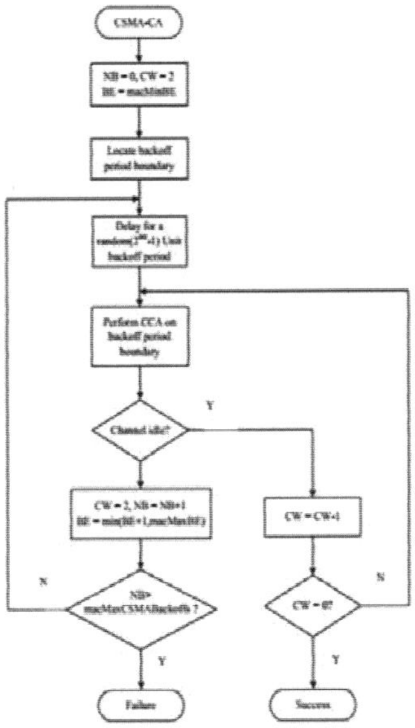


**Figure 1**. Slotted CSMA/CA algorithm [10]

### *2.2. Data Fragmentation Scheme*

The fragmentation of data is also done in the MAC protocol as well. Fragmentation is defined as a chunking a large packet into smaller one. Sending a smaller packet will shorten the time taken for the packet to arrive at the destination. In other words, there will be a reduction in delay which directly improve the performance of the system. Nevertheless, MAC and PHY overhead are added to the frame of fragmented data.

In [11], data fragmentation is done in WSN for IEEE 802.15.4 which deal with small beacon interval application. In this work, a smaller data is sent in a small Back-Off period that are not being used in IEEE 802.15.4. This means the fragmentation of data payload is depending on the remaining number of Back-Off periods. For instance, the current remaining Back-Off period are 8, then the node can only transmit its data for only 5 Back-Off period. The data is fragmented according to the value of 5 Back-Off period. This work has achieved reduction in collision and high aggregation of throughput. However, this work does not consider about delay performance which is one of the crucial measurements in the WBAN. In other work[12], the ACK frame is used when the fragmentation is not employ. On the other hand, if the fragmentation is done, the ACK frame is not used to achieve short transmission time and this mechanism followed Markov chain under a non-saturation traffic by means there is no incoming of new packet until the packet is fully served. This mechanism managed to achieve high utilization of bandwidth as well as improve the throughput and decrease the delay.

## 3. Proposed Method

A suitable packet size of WBAN is ranging from 30 to 60 bytes [13]. In this study, the packet size of 30 bytes is chosen as smaller packet size reduces the transmission delay and contains less bit error in packet. Theoretically, it requires 1 millisecond in order to have a successful packet transmission of 30 bytes under IEEE 802.15.4 which has bit rate of 250 Kbps. This can be calculated using (1).

$$Slot\ time, T\ (s) = \frac{Length\ of\ Packet\ (bits)}{Bit\ rate\ (bits\ per\ second)} \quad (1)$$

However, setting up this slot time value could result in high delay which could cause negative impact to the overall performances of WBAN. This phenomenon could lead to late in medical assistance and might increase the risk of death to the patient. This is especially true during dealing and transmitting larger packet size that can cause inevitable collision and waste the channel utilization in IEEE 802.15.4 Hence, the packets of 30 bytes should be fragmented into a smaller size under certain slot time. In short, the relationship between length of packet size is directly proportional to the slot time in the MAC protocol of the system under a constant bit rate.

The proposed method aims to reduce the transmission delay in the WBAN which could lessen the impact of network congestion by fragmenting the data through decreasing the value of slot time which ranging from 0.2 to 1.0 milliseconds. This experiment is carried out under different number of nodes from 20 to 60 and constant bit rate of 250Kbps and Contention Window (CW) of 31.

The simulation of WBAN network to test the proposed method is carried out in OMNeT++ simulator [14] using MAC protocol of IEEE 802.15.4 slotted under CSMA/CA. The details of other parameter setting is tabulated in Table 1 as follows.

**Table 1.** Settings of Parameters

| Parameters | Value |
|---|---|
| Packet Size | 30 bytes |
| Data Rate | 250 Kbps |
| Coverage Area | 50m x 50m |
| Routing Protocol | AODV |
| Simulation Time | 200 seconds |

## 4. Results and Discussions

This section highlights on output analysis from the performance of proposed method in terms of delay, percentage packet loss, throughput and Packet Delivery Ratio (PDR) in the WBAN.

### *4.1. Average Delay*

Average delay reflects the time taken for all packets to be transmitted from source to destination. This delay is measured between slot time of 0.2 to 1.0 millisecond for fix packet size of 30 bytes. The obtained results are conveyed in Figure 2.

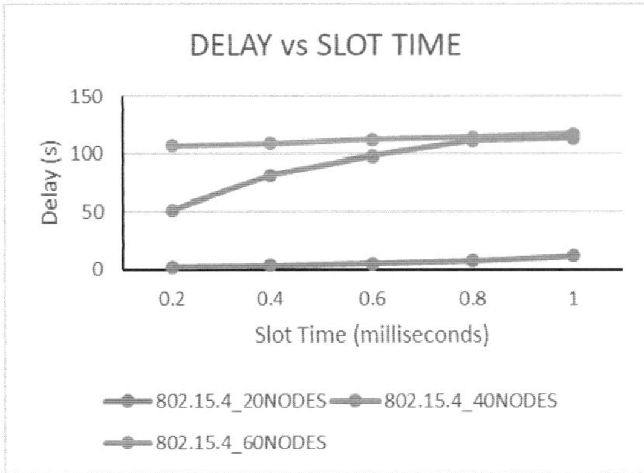


**Figure 2**. Delay performance for proposed fragmentation

In general, it can be clearly seen that there is an increment pattern of delay performances under different number of nodes through varying slot time. As slot time increases, the delay increases as well. Theoretically, more packets are generated when the number of nodes are increased. Thus, it would take longer time to transfer the packets to the destination. The packets have to wait for a longer time in order to accessing the medium during detecting the condition of the channel. This condition could prevent timely medical help and misdiagnose of the disease. From the graph, as short slot time is used, it brings to the decrement in delay and thus promise an efficient transmission of data across the network. This would increase the reliability of WBAN as well. From the plotted results, the proposed mechanism has shown a promising effects in the delay performances of the WBAN.

*4.2. Percentage Packet Loss*

Percentage packet loss indicates number of loss packets during transmitting and receiving data from the source to the destination.

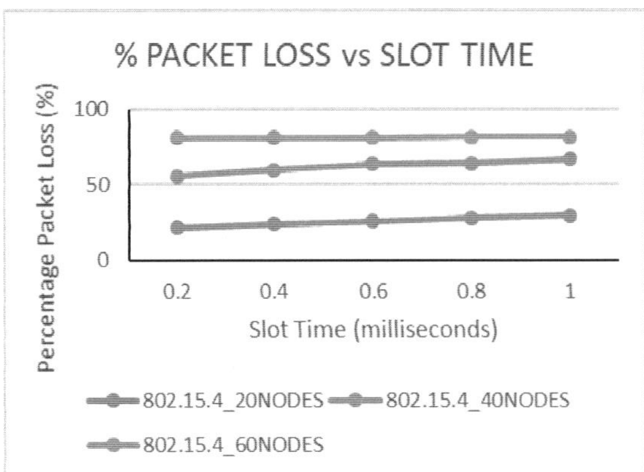


**Figure 3.** Percentage packet loss performance for proposed fragmentation

From Figure 3, the maximum rate of packet loss is noted during number of nodes are 60 compares to 20 and 40 nodes. With this major difference in loss of packets, the tendency of congestion to occur is high. Specifically, under slot time of 0.2 milliseconds, there is less number of packets loss in the WBAN compares to slot time of 1.0 milliseconds. This result has verified that implementing short slot time lead to less packet loss as more packets have been successfully accessed the medium.

*4.3. Throughput*

Throughput refers to the total number of successful bit transmission in a specific period of time.

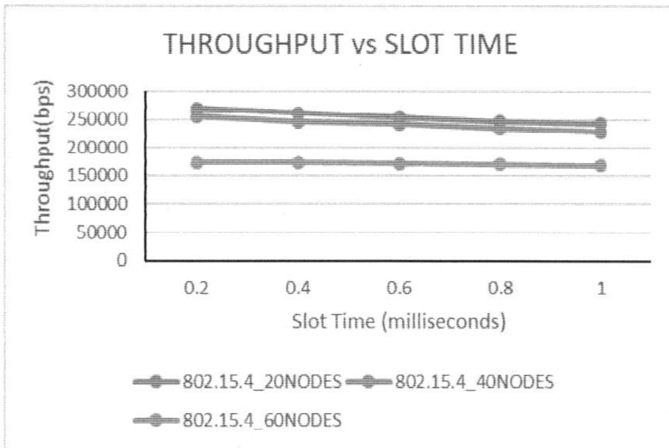


**Figure 4**: Throughput performance for proposed fragmentation

From Figure 4, low throughput is detected as the number of nodes are increased. However, in slot time of 0.2 milliseconds, the value of throughput is higher compared to 1.0 milliseconds for 20, 40 and 60 nodes. As slot time increases, throughput is slowly decreased which means many packets fail to access the channel due to longer waiting time to be served. The packets might be dropped and needed for retransmission. This has reduced the reliability of the WBAN. Hence, deploying short slot time conveys best performance in terms of throughput.

*4.4. Packet Delivery ratio (PDR)*

It reveals the ratio of number of received packets at destination to the number of transmitted packets by the source.

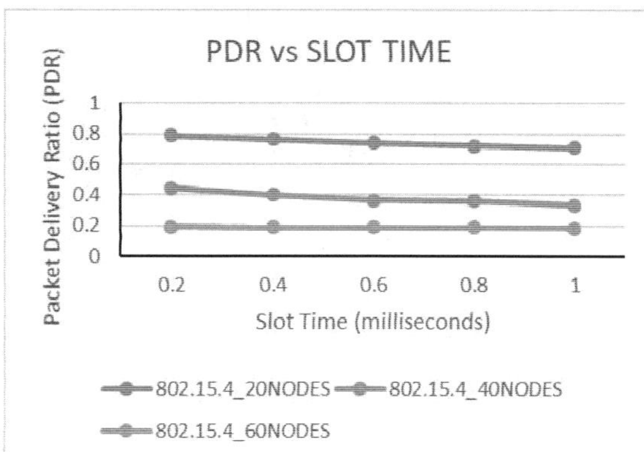


**Figure 5**: Packet Delivery Ratio (PDR) performance for proposed fragmentation

As reflects in Figure 5, the PDR is high for 20 nodes compared to 60 nodes for different slot times. This shows that a high number of packets have successfully received at the destination. Precisely, in slot time of 0.2 milliseconds, the probability of congestion phenomenon is less to present for number of nodes from 20 to 60. Thus, more packets are able to be served in the network. In short, the proposed mechanism shows better performance in PDR for high number of nodes.

## 5. Conclusion

A high delay would worsen the network performance of WBAN which could lead to packet loss especially critical packets. Based on this experiment, a fragmentation scheme to the packets through decrementing the slot time has contributed to the significant reduction in delay of WBAN. This study will be furthered boosted by modifying this scheme to be suitably adopted according to the level of importance of the packets such as normal, abnormal and critical packets in the WBAN

## 6. References

[1] X. Feng, "The Study of MAC Protocol Fairness for WBANs," in *International Conference on Computer Science and Network Technology*, 2011, pp. 795–799.

[2] S. Movassaghi, M. Abolhasan, J. Lipman, D. Smith, and A. Jamalipour, "Wireless Body Area Networks: A Survey," *Ieee Commun. Surv. Tutorials*, vol. 16, no. 3, pp. 1658–1686, 2014.

[3] R. Goyal, H. S. Bhadauria, R. B. Patel, and D. Prasad, "TDMA Based Delay Sensitive and Energy Efficient Protocol for WBAN," *J. Eng. Sci. Technol.*, vol. 12, no. 4, pp. 1067–1080, 2017.

[4] L. Filipe, F. Fdez-riverola, and N. Costa, "Wireless Body Area Networks for Healthcare Applications: Protocol Stack Review," *Int. J. Distrib. Sens. Networks*, vol. 2015, no. Special Issue on Mobile Sensing and Social Computing, p. 23 pages, 2015.

[5] S. Henna, M. Sajeel, F. Bashir, M. Asfand-e-yar, and M. Tauqir, "A Fair Contention Access Scheme for Low-Priority Traffic in Wireless Body Area Networks," *Sensors*, vol. 17, no. 9, p. 1931, 2017.

[6] U. Gupta, "Wireless with Collision Avoidance ( CSMA - CA )," *Int. J. Enhanc. Res. Sci. Technol. Eng.*, vol. 3, no. 6, pp. 229–231, 2014.

[7] B. B. Olyaei, J. Pirskanen, O. Raeesi, A. Hazmi, and M. Valkama, "Performance comparison between slotted IEEE 802.15.4 and IEEE 802.1 lah in IoT based applications," in *International Conference on Wireless and Mobile Computing, Networking and Communications*, 2013, pp. 332–337.

[8] S. Nepal, S. Dahal, and S. Shin, "Does the IEEE 802.15.4 MAC Protocol Work Well in Wireless Body Area Networks?," *J. Adv. Comput. Networks*, vol. 4, no. 1, pp. 52–58, 2016.

[9] A. Koubaa, M. Alves, and E. Tovar, "A comprehensive simulation study of slotted CSMA/CA for IEEE 802.15.4 wireless sensor networks," *2006 IEEE Int. Work. Fact. Commun. Syst.*, pp. 183–192, 2006.

[10] Z. Dahham, A. Sali, B. M. Ali, and M. S. Jahan, "An efficient CSMA-CA algorithm for IEEE 802.15.4 Wireless Sensor Networks," *2012 Int. Symp. Telecommun. Technol. ISTT 2012*, pp. 118–123, 2012.

[11] J. Yoon, H. Kim, and J. Ko, "Data Fragmentation Scheme in IEEE 802 . 15 . 4 Wireless Sensor Networks," in *IEEE 65th Vehicular Technology Conference - VTC2007-Spring*, 2007, pp. 26–30.

[12] Y. H.-A. Mouloud Atmani, Djamil Aïssani, "Modelling and analysis data fragmentation in IEEE 802.15.4 slotted CSMA / CA protocol without ACK mode," *Int. J. Crit. Comput. Syst.*, vol. 7, no. 1, pp. 4–21, 2017.

[13] H. S. Yahia and W. M. A. Monnet, "Performance of ZigBee Wireless Body Sensor Networks for Electrocardiogram Signal Transmission under Maximum Payload Size," *UKH J. Sci. Eng.*, vol. 1, no. 1, pp. 19–25, 2017.

[14] A. Varga and R. Hornig, "AN OVERVIEW OF THE OMNeT++ SIMULATION ENVIRONMENT," in *Proceedings of the First International ICST Conference on Simulation Tools and Techniques for Communications Networks and Systems*, 2008.

IOP Conf. Series: Materials Science and Engineering **557** (2019) 012021 doi:10.1088/1757-899X/557/1/012021

# Low-Complexity Distributed SDN Controller Environments on Switch Migration Testbed

**Mohd Nazri Bin Mohd Warip[1], Ng Yen Phing[1], Anuar Mat Safar[1], R. Badlishah Ahmad[1], Naimah Yaakob[1], Amiza Amir[1] and Zunaidi Ibrahim[2]**

[1]School of Computer and Communication Engineering, Universiti Malaysia Perlis, Pauh Putra Campus, 02600 Arau, Perlis, Malaysia
[2]Faculty of Technology, University of Sunderland, St Peter's Campus, Sunderland, SR6 0DD, United Kingdom

**Abstract.** Control Path Management is an important component in the Distributed SDN Controller, and plays a key role in Software Defined Networking (SDN). Testbed simulations are done with an implementation of several controllers are deployed to improve the scalability and reliability of the control plane framework by employed network clustering into several subdomains with separate controllers. This paper investigates the deployment of distributed SDN controller but logically centralized systems for decouples control plane using Distributed Traffic Engineering mechanism. The results obtained show that the proposed mechanism can achieve controller load balancing with better performance.

## 1. Introduction

The key distinction the traditional network and Software Defined Networking (SDN) is that decouples control plane from data plane [1][5]. It has been noted that to designing Software Defined Network (SDN) [1][5] are rooted in the network architecture. Low complexity distributed SDN controllers designed to improve the scalability, network flexibility management, rapid innovation and agility. In addition, the role of control plane is to enable directly programmable as well as offering centralizes network management. This paper investigates the deployment of distributed SDN controller but logically centralized systems for decoupling control plane using Load-Balanced Distributed Traffic Engineering. Figure 1 provides the typical SDN architecture and taxonomy of topology discovery.

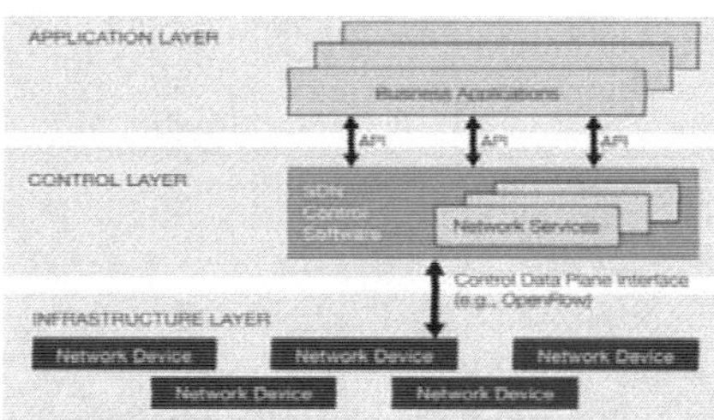


**Figure 1.** SDN Architecture[5][6]

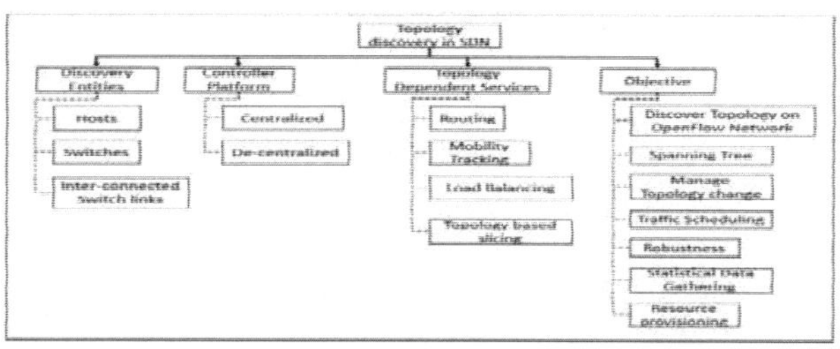


**Figure 2.** Topology Discovery in SDN [2]

**Table 1.** Comparison Between SDN and Legacy Network [3]

| | **SDN** | **Traditional Network** |
|---|---|---|
| **Features** | Decoupled control plane from data plane to provide a programmability environment | High complexity of network control and protocol problem |
| **Configuration** | Automated configuration with centralized validation | Manual configuration with error prone |
| **Performance** | Dynamic global control with cross layer information | Relative static configuration and limited information |
| **Innovation** | Can develop for new ideas by programming, easily to build virtual testbed for test environment | Difficult to implement new ideas on hardware and limited testing environment |

## 2. Simulation Model

We establish SDN simulation model setting in Mininet [4]. It is capabilities to emulate a large SDN network environment in a single machine. Mininet quickly creating a virtual network that running actual kernel, switch and software application code over a desktop. Custom network topology can run using the MiniEdit, mininet GUI editor.

## 3. SDN Controller

The SDN Controller Platforms in software-defined network (SDN) perform as the centralize network based on Ryu platform. It is able to work with the common application interface using OpenFlow and open virtual switch database (OVSDB) protocols. A distributed decision mechanism for controller load balancing based on switch migration in SDN is proposed by [6]. Type of SDN controller is shown in Table 2.

**Table 2.** Type of SDN Controller and Description [5]

| Controller | Description |
|---|---|
| POX | Communicating with SDN switches using OPenFlow or OVSDB protocols. Used as basic SDN controller. Complex controller can create through the POX components. |
| Ryu | Ryu controller was an open source product using SDN technology to increase the agility of the network and supported by NTT. Providing software components with well-defined application program interfaces (APIs). Easing creation of new network management and control applications. |
| Floodlight | Offering the ability for software adaption and development. Present of representational state transfer application program interface (REST APIs). Easing the programming interface and offering aids for developers in building products. Able to deploy in variety working environment. |
| ONOS | Offered simplified programmatic interfaces that flexible to create and deploy new network services. Ability to create new network applications without altering the data plane programs. Platform extensible and distributed controller. Grant abilities to developers to expend the network dynamically. |
| OpenDaylight | Supporting multiple SDN platforms including OpenFlow, OVSDB, NETCONF and BGP. Enable SDN controller to work with forwarding plane and network changes. |

## 4. Design of SDN Topology for Load Balancing

The design of SDN topology for load balancing is shown in Figure 3 for low complexity distributed SDN controller environments on switch migration testbed. Environment topology setting encompass of 9 hosts and 21 switch were implemented in Mininet using CLI command as below:

```
$ sudo mn -topo=custom, 4.py -mac -switch=ovsk -controller=remote
```

Each switch been renamed to show different between switch.

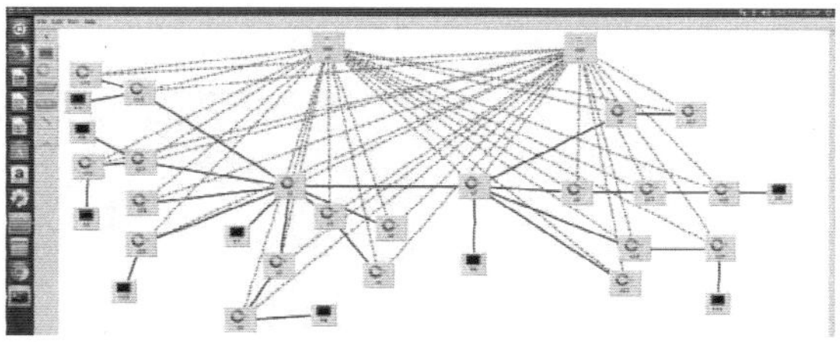

**Figure 3.** Design of SDN Topology for Load Balancing

## 5. Results and Discussions

Figure 4 provides the results that generated based on packets captured on topology network shown in Figure 3. The graph contains a lot of packets transfer delay. This might due the topology designed connecting all the switches with the controllers. The transmission took longer time might due to the noise that occur at the controller's sites.

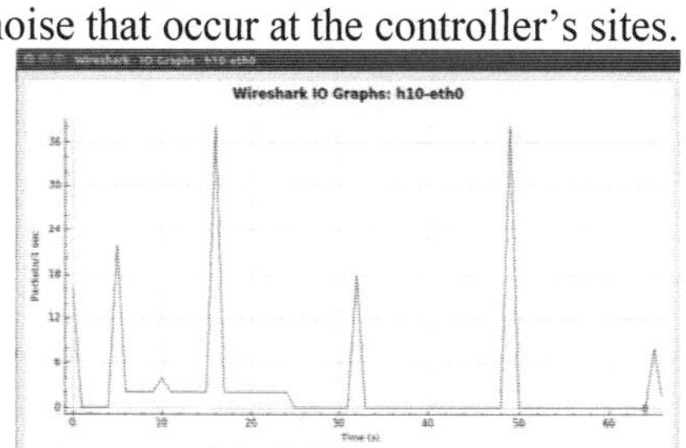


**Figure 4.** IO Graph for h10 ping with s7

**Figure 5.** Host Ping Test When Both Controllers Inactive

By using topology developed, inactive controller test being conducted to observed that the host communication. Figure 5 about the test case conducted. The connection failed for every device if both controllers being inactive. Resulting 100% packet dropped. This due to absent of centralized management. To solve the problem at least one controller needed to be activated. Second test for the design is by enable only one port of controller being activate.

**Figure 6.** Port for c0 Being Activate

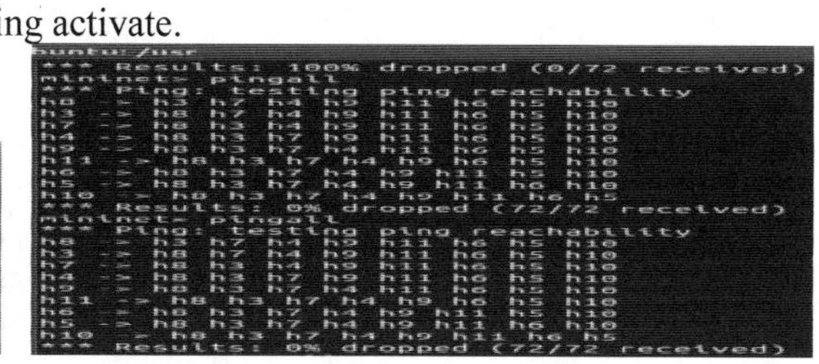

**Figure 7.** Host Ping Test After One Controller Activate

IOP Conf. Series: Materials Science and Engineering **557** (2019) 012021 doi:10.1088/1757-899X/557/1/012021

In Case 2 regarding only one controller being activated. Since the design (refer Figure 3) connecting all the devices together, it enables the controller manages the network even though only one controller being activated. This enable the devices to send packets without problem with only one controller management. In summary, event one controller fail to operate as long the backup controller present the network still can be managing.

### *5.1 UDP and TCP Test*

The UDP and TCP results is shown in this section.

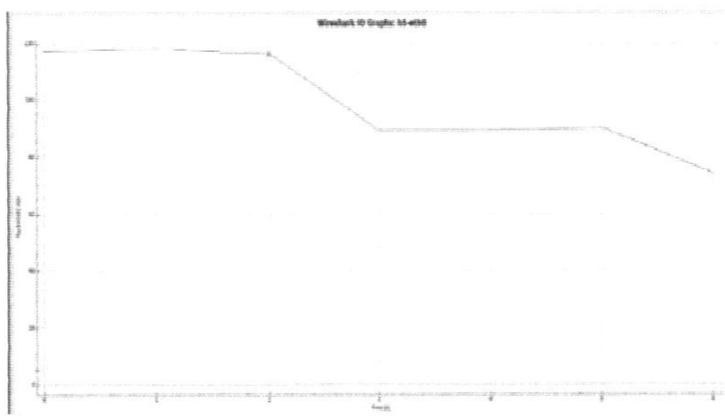

**Figure 8.** UDP Wireshark IO Graph

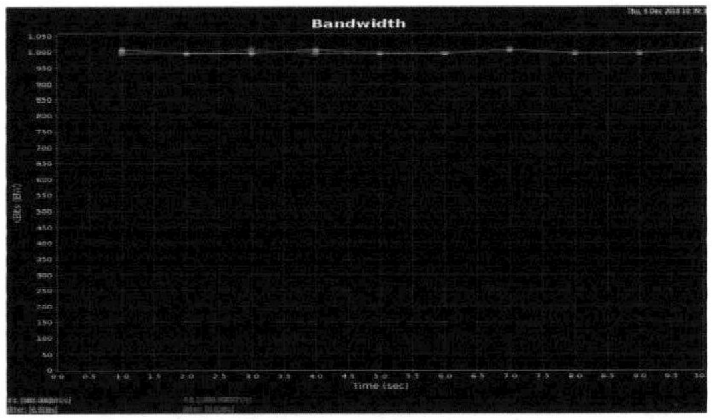


**Figure 9.** UDP Graph by Jperf

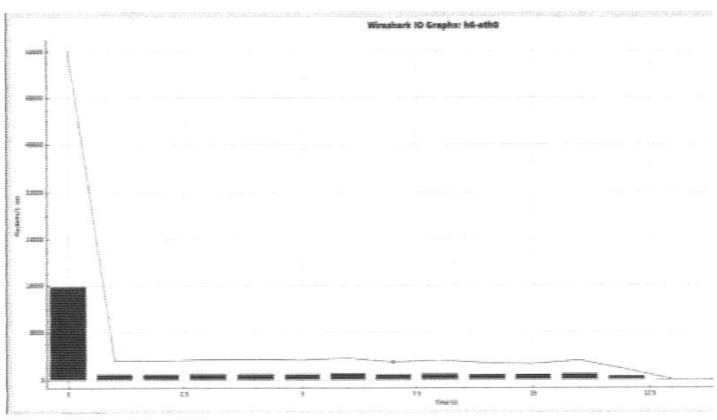

**Figure 10.** TCP Wireshark IO Graph

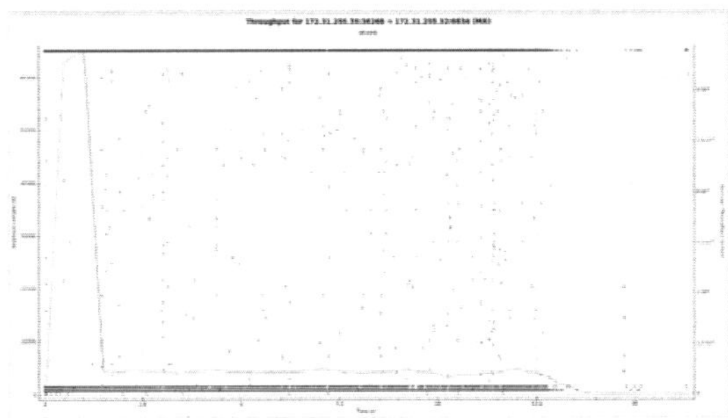

**Figure 11.** TCP Throughput Graph

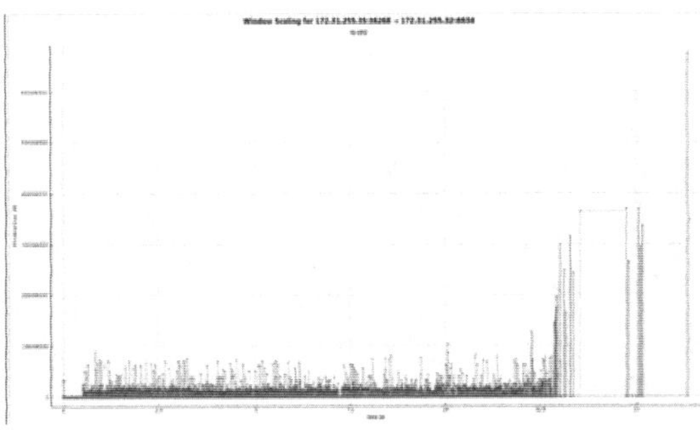

**Figure 12.** TCP Window Scaling Graph

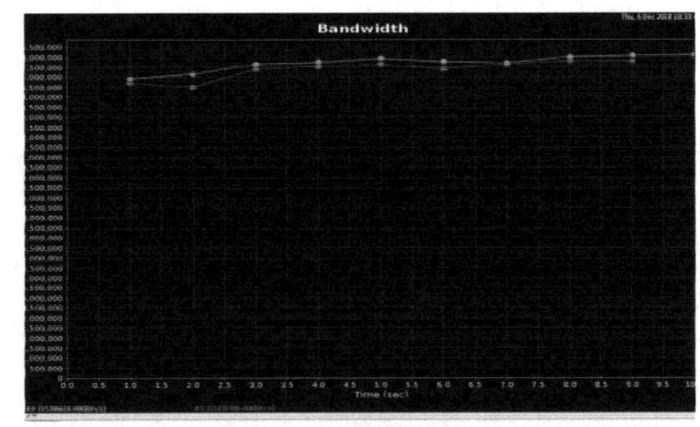


**Figure 13.** TCP Graph by Jperf

## 6. Conclusions

This study has examined the impact of the SDN controller architecture designed for increasing the network agility and performance. The separating of the network brains (controller) with the body (devices hosted on the network) also ease the network maintenance. These findings have significant implications for the understanding of how the SDN technology actually better to be deploy in bigger network enterprise due to the ability to expand and centralized management. The SDN controller functionalities can be tested over the virtual environment. The main objectives of SDN controller was to enable the fail-over system in Load balancing environment.

## 7. References

[1] B. Beach and B. Beach, "AWS Architecture Overview," *Pro Powershell Amaz. Web Serv.*, no. 1, pp. 1–6, 2014.

[2] D. Hasan and M. Othman, "Efficient Topology Discovery in Software Defined Networks: Revisited," *Procedia Comput. Sci.*, vol. 116, pp. 539–547, 2017.

[3] A. Ahurra, K. M. Kamurali, D. O. Briain, and D. Okello, "Software Defined Networking ( SDN )," no. May, 2017.

[4] K. Kaur, J. Singh, and N. S. Ghumman, "Mininet as Software Defined Networking Testing Platform," *Int. Conf. Commun. Comput. Syst.*, no. December, pp. 3–6, 2014.

[5] O. Salman, I. H. Elhajj, A. Kayssi, and A. Chehab, "SDN controllers: A comparative study," *Proc. 18th Mediterr. Electrotech. Conf. Intell. Effic. Technol. Serv. Citizen, MELECON 2016*, no. April, 2016.

[6] T. Hu, P. Yi, J. Zhang, and J. Lan, "A distributed decision mechanism for controller load balancing based on switch migration in SDN," *China Commun.*, vol. 15, no. 10, pp. 129–142, 2018.

# Geometrical and mechanical performance of multi-material fused deposition modelling

**N A Alief[1], M S Utomo[2], I Kartika[2], and Y Whulanza[1,3]**

[1] Department of Mechanical Engineering, Universitas Indonesia
[2] Research Center for Metallurgy and Materials, Indonesia Institute of Sciences
[3] Research Center for Biomedical Engineering, Universitas Indonesia

**Abstract.** One method that has been used on additive manufacturing process is fused deposition modelling (FDM). This method employs material extrusion in the form of filament to build a product. Several materials have been developed as the filament, each with its own mechanical, chemical, and thermal properties. Certain FDM machines are capable to print filaments from different materials. Although, each materials require different printing parameters. Consequently, different geometrical and mechanical properties from same design may be produced from their different printing parameters. Therefore, in order to obtain the characteristic of FDM machine output on multi-material, a comparative study is conducted. Lattice structures constructed from 3D CAD model with difference in pore configuration are printed by different filament materials on same FDM machine. We investigate the geometrical and mechanical properties of the structure, especially on the capability of the lattice structure to bear compressive load. There are 42-53% of pore dimension difference between the printed structure and the design despite material variation is used for each pore configuration.

## 1. Introduction

Fused Deposition Modelling (FDM) is currently become a popular fabrication process and commonly used for modelling, prototyping, and production application [1]. This process favour the ease of fabricating three-dimensional objects of almost any form and less waste than traditional subtractive production method [2]. As one of methods in additive manufacturing process, fused deposition modelling utilizes extrusion of material formed in filament to build structural 3D model by layering from bottom to the top. The filament is melted according to the melting temperature of certain material and extrusion occurs through the nozzle, subsequently. Materials are subjected to phase changes under relatively high temperature [3]. The material is soon after being deposited and it cools down and solidifies right after deposition [4]. The movement of the nozzle is based on the process parameters resulting the path generation. A desired structure, such as accurate dimension and splendid mechanical performance, may be achieved accurately and precisely by configuring each component of parameters.

Wide range materials, from polymers, ceramics, to metals [5], has been developed into 3D printing filaments. These materials have two main roles, which could be as the build material and support material [1]. The common material combination is Polylactic Acid (PLA) and Polyvinyl Alcohol (PVA). Besides that PLA is the most widely used material in extrusion-based 3D printing technique

[6] is easy to print and biodegradable [7]. On the other hands, PVA generally takes a role as the support material due to the ease of removal from the built structure. Both of them has intersect melting temperature value, PLA has 150-220°C [8-11] and PVA has melting temperature range of 190–220°C [12]. Therefore, this combination has been utilized in order to build three-dimensional structure.

Several comparative studies has been conducted on the FDM scope. Ceretti, et.al. [3] conducted a comparison between grain and wire FDM process. Minetola et.al. [4] presented the benchmarking analysis of Prusa i3 machine improvements. Loh, et.al [13] compared FDM with other rapid prototyping techniques. Gregor, et. al. [14] attempted to overcome technical limitations and fabricate scaffolds using current, cheap and commercially available devices and materials. Zein et.al. [15] compared the 3d printed results that have various combination of pattern. Tagami 2018 [16] conduct studies between FDM printed PLA and PVA for preparation of multi-component tablets. The characteristic of the FDM output differs due to the material variation. Therefore, a comparative study between PLA and PVA-made lattice structure with two types of pore configuration is conducted.

## 2. Methods

### *2.1. Designing the Lattice Structure*

Lattice structure formed with two types of pore configuration.in cube. One has pore dimension of 1000 μm and the other one has 1500 μm length, as shown on the Figure 1. Both has uniform length and width of 12.5 mm yet each of the configuration has different height. The 1000 microns has 14mm and the other one has 15mm. The pore itself has identical square shape. Each configuration is printed with two types of material, which are PLA and PVA.

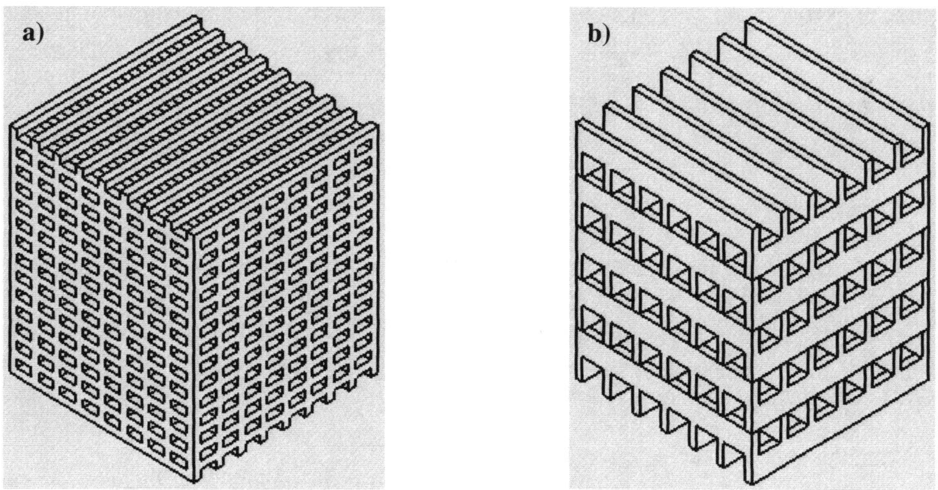


**Figure 1.** Lattice Structure Design with two types of configuration. a) 1000 μm length b) 1500μm length.

### *2.2. Printing Setup*

Before structures are fabricated, CAD models are saved on triangulated form as .stl file format. Each structure with different pore configuration are then being input to the built-in slicer software of the FDM machine. On the software, the lattice is replicated into four pieces in order to print in one batch. Afterwards, we set the material, building orientation, and process parameters. Standard mode, one of three printing options, is chosen within particular quantity of each component as seen on Table 1.

Once the process parameters are set, the nozzle path can be generated and results on .gsd file format and building estimation for each batch. The study is conducted by building two batches for each material and pore configuration.

**Table 1.** Amount of each process parameter for the standard printing mode.

| Parameter settings | Value |
|---|---|
| **Print Speed (mm/s)** | 60 |
| **Extrude Speed (mm/min)** | 45 |
| **Nozzle Temperature (°C)** | 195 |
| **Layer Thickness (mm)** | 0.1 |
| **Number of Solid Surface Layer** | 4 |
| **Number of Solid Shell Layer** | 1 |
| **Infill Line Distance (mm)** | 5 |
| **Strip Rate** | 2.6 |
| **Support Angle** | 50 |

### *2.3. Characterization*

*2.3.1. Pore Dimension Measurement.* Two lattice structures are taken as the sample from each batch of material and pore configuration, thus there are 16 pieces of lattice structure sample. The samples are then being captured by calibrated digital microscope which has measurement feature on the axial view. Each pore sample from the chosen lattice structure is measured with designated arrangement. The lattice dimension is also measured as the same view as pore measurement.

*2.3.2. Compressive Load Test Setup.* Three samples of each pore dimension and material combination were subjected to compressive load test. Since there were two material and two pore dimensions, there were twelve samples in total. Compressive load test was conducted on Shimadzu AG-50kN XPlus. The room temperature and humidity during testing were 21.5 °C and 52.3 % respectively. Each sample was subjected to tensile rate of 10 mm/min.

## 3. Result and Discussion

### *3.1. Geometrical Performance*

Based on the lattice formation, both the 1000 microns and 1500 microns have consistent square-shaped pore on the center part of lattice. Besides, the pore formation on the peripheral region has not complied the desired model yet each of the pore on this location is much alike one on another. In addition, it is observed that each corner of the lattice formed a round corner due to the cylindrical form of melted filament. The comparison of lattice formation between PLA and PVA material can be seen on the Figure 2. In contrast to the PLA lattice, PVA structure tend to have more fibrous filaments that is trapped inside the structure which need micro equipment for cleaning the fibers.

The measurement of overall lattice formation has been obtained. It is resulted in area and height of lattice structure. The 1000 microns has the average area of 155.74 $mm^2$ (standard deviation of sample [$SD_{sample}$] = 0.97) and 13.8 mm ($SD_{sample}$ = 0.11) height whereas the other pore configuration has 154.07 $mm^2$ ($SD_{sample}$ = 1.31) of the average pore area and height of 14.63 mm ($SD_{sample}$ = 0.17). The measurement also resulted in area of each pore sample. According to the pore configurations, it is obtained that the average pore area is 0.497 $mm^2$ with standard deviation of 0.08 for 1000 microns and 0.952$mm^2$ with standard deviation of 0.16 for 1500 microns. However, from the Figure 3, the lattice structures from both material has relative change with respect to the preliminary design in a range of 42% to 53%. Thus, these reveal that the layer thickness has an influence towards the proportion of printed structure. According to the machine specification, this machine has 0.4 nozzle diameter. Hence, in order to obtain the value of layer thickness, the melted filament is urged to the sides of the nozzle. This causes a reduction on the pore dimension.

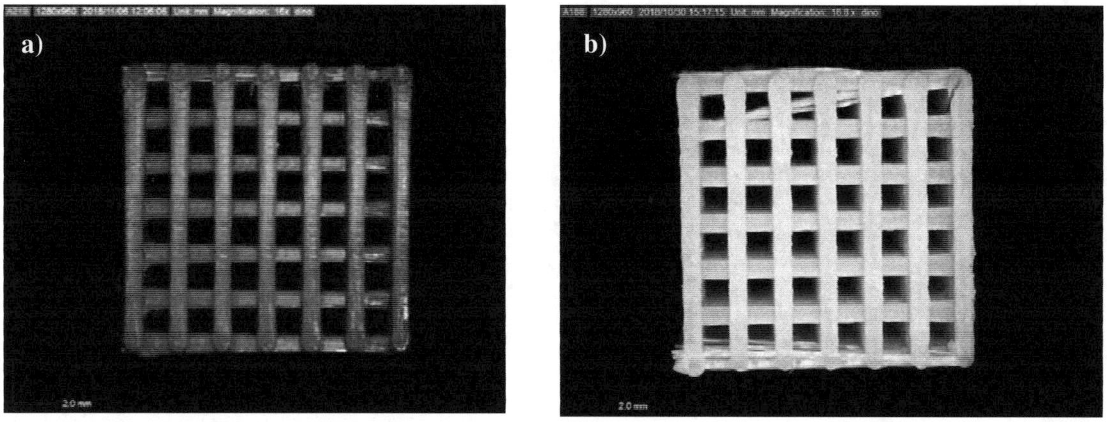


**Figure 2.** Samples of 1500 microns lattice structures with different filament materials. a) Polylactic Acid (PLA). b) Polyvinyl Alcohol (PVA).

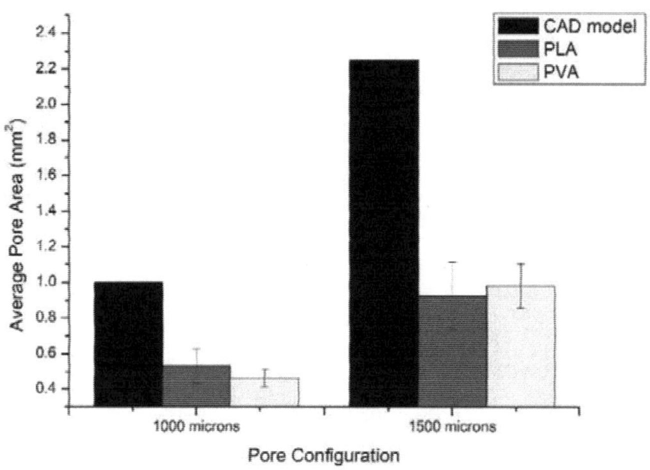


**Figure 3.** Graph of pore dimension for each pore configuration and materials.

The uniformity among pores on the central part and peripheral region is also being examined. Based on Figure 4, uniformity is occurred on each observed region regardless the material and the pore configuration. Thus, the machine can fabricate lattice precisely.

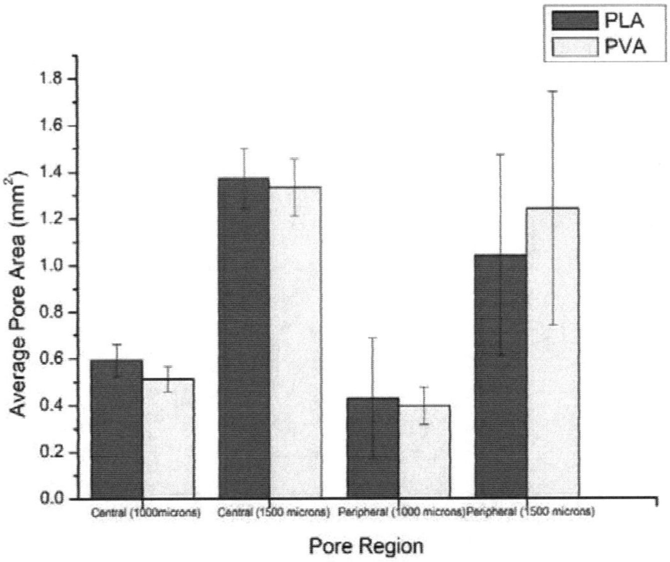


**Figure 4.** Average area in central and peripheral region for both pore configuration and materials.

*3.2. Mechanical Performance*

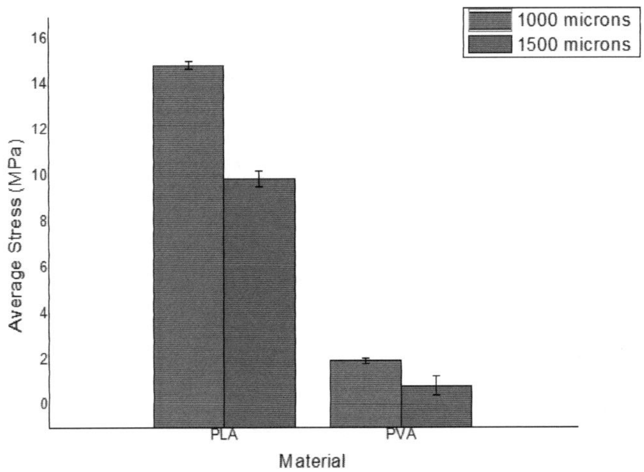


**Figure 5.** Graph of compressive load for each pore configuration and materials.

Figure 5 shows the result of compressive load test. At a glance, lattice structures made from PLA exhibit higher capability to withstand compressive load compared to lattice structures made from PVA for both pore configurations. Lattice structures made from PLA show average compressive strength of 9.87 and 14.84 MPa for pore dimension of 1500 and 1000 microns respectively. Meanwhile, lattice structures made from PVA show average compressive strength of 0.83 and 1.96 MPa for pore dimension of 1500 and 1000 microns respectively.

It can be seen from the result that both material and pore dimension affect mechanical properties of the 3D-printed polymeric lattice structures. For lattices with pore dimension of 1500 microns,

compressive strength ratio between PLA and PVA is 12:1. Meanwhile, for lattices with pore dimension of 1000 microns, compressive strength ratio between PLA and PVA is 7.6:1.
The ratio of compressive strength on pore dimension between samples made from PLA show that samples with 1500 microns has compressive strength 0.67 times lower than samples with 1000 microns. Meanwhile, the ratio of compressive strength on pore dimension between samples made from PVA show that samples with 1500 microns has compressive strength 0.43 times lower than samples with 1000 microns.

## 4. Conclusion
Geometrical evaluation of 3D-printed polymeric lattice structures by optical measurement show that using current combination of printing machine, filament material, and printing parameters, the realized products deviated 40 to 50 % from the desired model. This result should be put into consideration when 3D-printed materials are required to have precise dimension in application. Optimization is needed in order to achieve more accurate result on geometrical and mechanical performance of lattice structure. Compressive load test shows that combination of pore geometry and materials can be utilized to produce lattice structures with specific mechanical strength. This could be further characterized and utilized for specifi usage.

## Acknowledgments
This research is funded by 2018 Insinas grant scheme from Indonesian Ministry of Research and Higher Education. Authors would like to thank colleagues and technical staff for their technical and non-technical support.

## References
[1] Mohamed O A, Masood S H, and Bhowmik J L 2015 *Adv. Manuf.* **3**(1) 42-53
[2] Ramos-Lozano S, Molina-Salazar J, Rico-Pérez L, and Atayde-Campos D 2019 *Best Practices in Manufacturing Processes* (Cham: Springer) p 389-410
[3] Ceretti E, Ginestra P, Neto P I, Fiorentino A, and Da Silva J V L 2017 *Procedia CIRP* **65** 13-18
[4] Minetola P, Galati M, Iuliano L, Atzeni E, and Salmi A 2018 *Procedia CIRP* **67** 203-208
[5] Do A-V, Khorsand B, Geary S M, and Salem A K 2015 *Adv. Healthc. Mater.* **4**(12) 1742-1762
[6] Lopes L R, Silva A F, and Carneiro O S 2018 *Addit. Manuf.* **23** 45-52
[7] Bagaria V, Bhansali R, Pawar P 2018 *J. Clin. Orthop. Trauma.* **9**(3) 207-212
[8] Castro-Aguirre E, Iñiguez-Franco F, Samsudin H, Fang X, and Auras R 2016 *Adv. Drug. Deliv. Rev.* **107** 333-366
[9] Martin O and Avérous L 2001 *Polymer* **42** (14) 6209–6219
[10] Jing X, Mi H Y, Peng X F, and Turng L S 2015 *Polym. Eng. Sci.* **55** (1) 70–80
[11] Leist S K, Gao D, Chiou R, and Zhou J 2017 *Virtual. Phys. Prototyp.* **12**(4) 290-300
[12] Goyanes A, Kobayashi M, Martínez-Pacheco R, Gaisford S, and Basit A W 2016 *Int. J. Pharm.* **514**(1) 290-295
[13] Loh Q L and Choong C 2013 *Tissue. Eng. Part. B. Rev.* **19**(6) 485-502
[14] Gregor A, Filová E, Novák M, Kronek J, Chlup H, Buzgo M, Blahnová V, Lukášová V,Bartoš M, Nečas A and Hošek J 2017 *J. Biol. Eng.* **11**(1) 31
[15] Zein I, Hutmacher D W, Tan K C, and Teoh S H 2005 *Biomaterials* **23**(4) 1169-1185
[16] Tagami T, Nagata N, Hayashi N, Ogawa E, Fukushige K, Sakai N, and Ozeki T 2018 *Int. J. Pharm* **43**(1-2) 361-367

# Design of Online Application for Agricultural Machinery Service based on Android Operating System

**Desrial[1] and Prawesa Adi Kumara Indriawardhana[1]**

[1]Department of Mechanical and Biosystem Engineering, Bogor Agricultural University, Indonesia.

**Abstract.** Utilization of agricultural machinery recently becoming big issue to increase agriculture production in Indonesia. In these couple years Indonesian Ministry of Agriculture massively granted agricultural machineries to the farmers for increasing rice production. Considering that farmer's paddy field are scattered in villages area, it can be happening that agricultural machinery in one village may overutilize where the other still underutilize. The objective of this research is to design an online application for agricultural machinery service for optimizing the utilization of agricultural machinery. The application was developed under Android operating system which is commonly used for mobile phone. The method of this research was following System Development Life Cycle which include investigation, system analysis, design, implementation, installation and testing steps. There are three sub systems in this application i.e. system for operator, machinery provider and machinery user. The application contains several functions such as machinery data management, order management and machinery services scheduling. In the user application, detail data such as field location, type of machinery, order time, and supporting map of the location can be input and read by the machinery provider easily. The application was tested under several Android operating systems and the results showed that the application can run satisfactorily.

**Keywords:** Agricultural Machinery, Android, Online, Service System.

## 1. Introduction

Starting from 2013 until 2017 there were drastic increase of agricultural machineries that government grants to farmers throughout Indonesia in purpose of implementing agriculture mechanization. Despite of those massive grant from government, according to Yunus et. al. in 2017, in Kolaka Regency, usability of power tiller is only 62% which is far from fully utilization. One of the main reason is the distribution of those agricultural machinery were not well spread throughout regency as the machinery in one village may overutilize where the other still underutilize.

Indonesian Ministry of Agriculture actually already try to suppressing that problem with creating UPJA, a new form of business based on agricultural machinery services which managed by chosen farmer aided by local agriculture offices and *Brigade Alsintan*, an army task force for agricultural machinery utilization which is resulting a creation of brand new service system.

The traditional perception of service system is that of a service delivery system. The system consist of elements (resources like people with skills, competencies, and knowledge, but also things like facilities, tools, materials, computer programs) that have a structure (organization and configuration), a behavior (described as a process or mechanism), and a purpose (to deliver a service) [1].

People in Indonesia already familiar with online service system. Transportation service system such as GOJEK, Grab, and Uber are growing rapidly for past couple years. Those service system based on Android smartphone for high accessibility because Android as one of operating system for smartphone is start to dominating the market. According to StatCounter [3], in August 2017, Android smartphone user in Indonesia is 83.99% from 43% smartphone user from 173 million handphone user and projected will be growing in following years. Because of that, this research objective is to design an online application for agricultural machinery services for optimizing the utilization of agricultural machinery.

This application which only cover system for machinery services provider is a part of one system for the agricultural machinery services. The whole system main objective is to increase the usability, accessibility, and effectivity of agricultural machinery services.

## 2. Material And Methods

The method of this research was following System Development Life Cycle of O'Brien and Marakas [3] that is investigation, identifying information and data flow from current system, identifying current system problem, crucial data, weaknesses, and the new system feasibility for the sake of getting the solution; system analysis, for defining user identity, and information, functional, and non-functional requirements; design, designing new system based on investigation and analysis; implementation, creating the designed system; and installation and testing.

The application system was developed with JDK as compiler, Android Studio and Android SDK as Integrated Development Environment (IDE), Adobe Photoshop CC 2018 as UI design tools, Firebase as BaaS (Backend as Services) for authentication and database, and Google Maps API.

## 3. Results

### *3.1 System Investigation*

The current conventional system of agricultural machinery services in Indonesia is identified from survey and interview to farmers and services providers in Klaten and Sukoharjo Regency, Central Jaya Province. The whole sequence of services still counting on direct communication which is very fragile against information loss. Information accuracy is also the problem that occur frequently because of different subjectivity from farmers and services providers. The last problem identified is detail of the order often just remembered by the services providers.

The application was built in Android apps. and named as AgriMach (Jasa Tani) Apps. The application at least need database for data storage and Google Maps integration with Google Maps API for support its map-accessing function. For that purpose, the application architecture is formulated and depicted in figure 1.

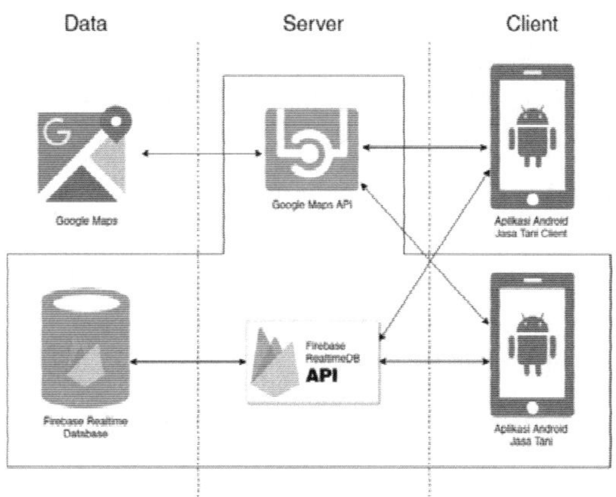


**Figure 1.** AgriMach application architecture

The newly developed application can be directly applied in the study area because the farmers are used to hire agricultural machinery service provided by UPJA or personal rental service by agricultural machinery owners. Based on data analysis from field survey, it is confirmed that the online application is organizationally feasible because that online rental service procedure was built by following the current rental system. The application is also economically feasible for the farmer because there is no additional effort and cost are needed other than installing application on their Android smartphone. Most of farmers in study area are already having smartphones with data package which has free access to Google Maps. The application is technically feasible because almost all smartphone as the application hardware also having GPS sensor which is needed for location finder as well as measurement of the map area. The application is also operationally feasible because of high rate of smartphone ownership among farmer society, several number of

IOP Conf. Series: Materials Science and Engineering **557** (2019) 012023 doi:10.1088/1757-899X/557/1/012023

internet provider with reasonable pricing, high coverage of signal at those study area, and high usability of smartphone.

*3.2 System Analysis*

*3.2.1 User Identification and Information Requirements*

The AgriMach Apps. is specially designed for agricultural machinery services in paddy cultivation which include machinery for land preparation, planting or transplanting, and harvesting that served in separate service packages. The required information for ordering machinery online service need more detail data such as location, date, type of machinery, and field conditions. Those detail information in very importance for decision making by machinery owners.

*3.2.2 Functional Requirements*

The functional requirements is obtained from system main function and scheduling. Functional requirements in this application was divided into user and owner requirement. For machinery owner the functions were managing order, selection of available machines, considering detail information of order, and ability to save and make scheduling for those orders.

*3.2.3 Non-Functional Requirements*

Non - functional requirements for the application were defined as the ease of use of the application, good user experience, easy to access, and data security. In order to perform those non functional requirements, it was decided to use Android operation system with cloud

*3.3 System Design*

System design begin with detailing the use of application by creating use case diagram with machinery services providers became main actor. Then, each of use case must be described for reference for activity diagram. Activity diagram has function as description of application flow and becoming guidelines for implementation of each stage.

System database design was made based on use case diagram. In this application three main tables of data were used i.e. table for user, table for machinery and table for order where its relationship is shown in Fig. 2. Data flow in the database occurs between users, data bases, systems and Google Maps. Almost all data is entered by users starting from personal data , technical data, and order data. However, exceptions to location data are due the data is entered by the system that gets input from Google Maps API after the user selects the location on the displayed map. The data flow of the location retrieval process is illustrated in figure 2.

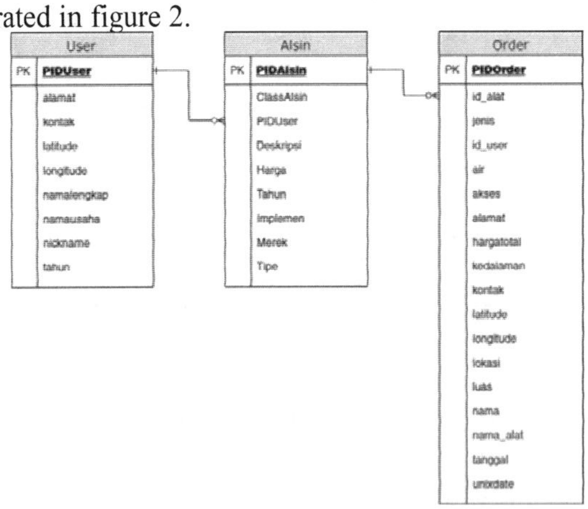


**Figure 2.** Entity relationship diagram

Design of user interface (UI) of the application was made based on anticipating what user might need to do and ensuring that interface elements are easy to be accessed, understand, and facilitate all necessary actions. The UI design is done with Adobe Photoshop CC 2018 and its sample is shown in figure 3.

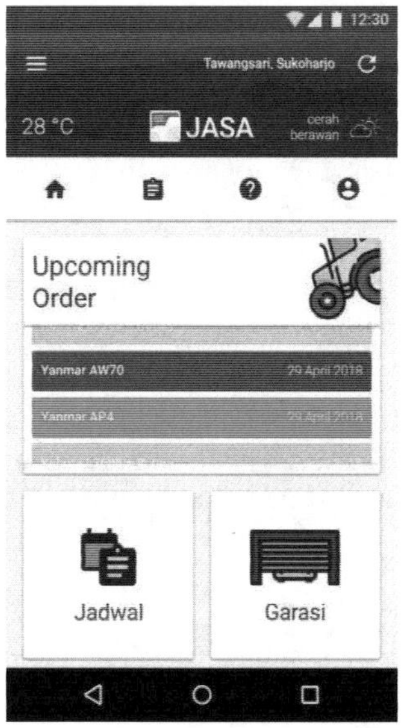


**Figure 3.** Designed UI of application

*3.4 System Implementation*

System implementation was started with database development with Firebase Realtime Database which is No-SQL system. No-SQL is all database and data storage which does not follow the principles of RDBMS and is generally related to big data which is accessed and managed on a network scale. Firebase Realtime Database is cloud-hosted database that stores data in the form of JSON documents and synchronize in real time to all connected clients. The data structure is slightly different from the general RDBMS, i.e. data of FRTDB is stored in key-value pairs. As shown in figure 4, a key can store values and have other keys that are on underneath or commonly called a node. The data structure built for the AgriMach Apps system has four nodes namely tools (alat), end order, new order, ongoing order, and user.

**Figure 4.** Root data structure

User interface (UI) for this application is implemented in Extensible Markup Language (XML) in layout resource file written in Java. Application algorithm and flow is implemented in Java classes which follow previously created activity and data flow diagram. All of those code simulated in Xiaomi Redmi Note 5 Pro with USB debugging method during the development process.

*3.5 Installation & Testing*

Finished application was compiled to android application package file (.apk) then it was installed in Android smartphone. Installation was followed by application test by using black box method [2]. This testing method ignoring the process and structure inside the system but focusing on expected input and output. The items of functional testing were user authentication, data base entry and loading, data base administration, integration and link to Google map, navigation and calender and connecting to phone dialing. Testing were carried out on four tester smartphones, namely Asus Zenfone 5 for Kitkat 4.4.2 Android OS, Samsung Galaxy Grand Prime for Android OS Lollipop 5.0, Samsung J7 for Marshmellow Android OS 6.0.1, Asus Zenfone 3 Max for Nougat Android OS 7.0, and Xiaomi Redmi Note 5 Pro for Oreo Android OS 8.1.0. Based on test results, it was proven that the application can run all its functions properly on all tested devices and all android OS. The process of requesting permission to use supporting aplication such as using GPS for finding locations and access phone application to do dialing contact numbers do not experience any problems for all devices. That function relating to the database can also run well at the time of application.

## 4. Conclusion

Online application for agricultural machinery service based on Android Operating System namely AgriMach Apps was successfully developed using Android Studio with the Firebase Realtime Database as its database. Based on the test results, this application can be used satisfactorily with the Android 5.0 (Lollipop) operating system up to Android 8.1.0 (Oreo). The application requires an internet connection for data communication with databases and GPS installed devices to access the user's location. The main functions of the application are managing personal data, owned machinery, and orders (scheduling); displaying order and machinery data; and integration with maps and navigation and calendar application. The size of the application that was successfully built was 3.98 MiB.

## References

[1] Cardoso J, Fromm H, Nickel S, Satzger G, Studer R, Weinhardt C. 2015. *Fundamentals of Service Systems*. Cham (ZG): Springer International Publishing AG.
[2] Myers GJ, Badgett T, Sandler C. 2011. *The Art of Software Testing 3rd Edition*. New Jersey (US): J Wiley & Sons.
[3] O'Brien JA, Marakas GM. 2010. *Introduction to Information Systems*. Boston (US): McGraw-Hill/Irwin.
[4] StatCounter. 2017. *Market share held by mobile operating systems in Indonesia from January 2012 to July 2017*. (https://www.statista.com/statistics/262205 /market-share-held-by-mobile-operating-systems-in-indonesia/). Accessed on December 4, 2017.
[5] Yunus L, Iswandi M, Hasan I. 2016. *Optimalisasi Kebutuhan Traktor untuk Pengolahan Tanah Sawah di Kecamatan Wundulako Kabupaten Kolaka*. JSA Vol 1, No. 1:16-31.

IOP Conf. Series: Materials Science and Engineering **557** (2019) 012024 doi:10.1088/1757-899X/557/1/012024

# Tumor Detection via Specific Absorption Rate Technique Using Ultra-Wideband Antenna

**Mohd Aminudin Jamlos[1], W A Mustafa[1], Wan Khairunizam[2], I Zunaidi[3], Z M Razlan[2] and Shahriman A B[2]**

[1]Faculty of Engineering Technology, Universiti Malaysia Perlis, UniCITI Alam Campus, Sungai Chuchuh, 02100 Padang Besar, Perlis, Malaysia

[2]Advanced Intelligent Computing and Sustainability (AICoS) Research Group, School of Mechatronics Engineering, Universiti Malaysia Perlis, 02600 Arau, Perlis, Malaysia

[3]Faculty of Technology, University of Sunderland, St Peter's Campus, Sunderland, SR6 0DD, United Kingdom

**Abstract.** This paper proposes high gain ultra-wide band (UWB) reflector array antenna for human brain tumor detection using specific absorption rate (SAR) technique in microwave detection system. Introduction of copper reflector with ideal distance contribute in increasing the proposed antenna gain. High gain antenna is essential in human microwave detection system for penetrating the multilayer structures by enhancing the focus of the electromagnetic energy into the desired structure. The antenna with reflector recorded wider bandwidth; 1.6 GHz-10.8 GHz associated with higher gain ranged from 3.2 dB until 14.1 dB compared with antenna without reflector operated from 2.2 GHz to 10.8 GHz with lower gain ranges of 2.1 dB to 10.2 dB. The antenna is successfully detecting human brain tumor based on SAR technique. Amount of energy or gain produced by the antenna which being absorbed by the human brain indicated the present of tumor inside. Human brain with tumor absorbed more energy and recorded higher SAR value compared to human brain without tumor. In this paper, human head phantom associated with tumor recorded SAR value of 2.53 W/kg and 2.51 W/kg for simulated and measured respectively meanwhile head phantom without tumor stated SAR value of 1.06 W/kg for simulated and 1.05 W/kg for measured. The biggest different in SAR value according to area precisely determine the location of the tumor in the brain which in this case position 5 (tumor location) resulted biggest SAR different value of 1.47 W/Kg for simulated and 1.46 W/Kg for measured.

## 1. Introduction

Brain tumor is defined as a mass of tissue formed by an accumulation of abnormal cells within the brain. Primary brain tumor is the condition where the tumor originated from the brain itself while secondary brain tumor occurs when the tumor originated from other parts of the body [1]. It could be the most dramatic form of human illness, and among the most rapidly fatal of all cancers. In 2030, around 13.2 million people worldwide will suffer and die cause of cancer since cancer is one of the most complex disease exist in the world [2]. Early cancer detection which allows early cancer treatment could increase the cure rates as the treatment is more efficient and effective compared with treatment done at the late stage of cancer. Application of microwave energy in the imaging and treatment for brain tumor is currently gaining interest by the research community. Microwave detection systems have the high potential of being simple, safe, portable and cost-effective [3]. In addition, almost all of current microwave imaging are concentrating on breast cancer only and have limited effort in building complete microwave brain imaging [4].

In this paper, specific absorption rate (SAR) technique to detect brain tumor for human being was proposed. Principally, the technique based on different ability of the healthy tissue and tumor tissue in absorbing energy. Tumor tissue will absorbed more energy and resulted in higher SAR value compared with healthy tissues due to present of self-blood vessel of the tumor itself. Antenna that produces energy radiated towards the human head phantom with tumor and without tumor. The result SAR values was compared between brain with tumor and without tumor in order to validate the SAR technique in detecting brain tumor. The rest of this paper is organized as follows: Section 2 describes the antenna design and performances. Methodology for the proposed technique, experimental results of comparison between SAR values of human brain with tumor and without tumor are shown in Section 3. Finally, Section 4 explained the conclusion of this work.

Conventional X-ray, magnetic resonance imaging (MRI), computed tomography (CT Scan) and ultrasound technique are the common imaging modalities utilized to detect cancer [5]. However, all those modalities come with major drawbacks such as involving ionizing radiation, too expensive, invasive, personnel dependent and long scanning time [6]. Lately, microwave frequency based device has been utilized in tumor detection for one of its medical application. Microwave cancer detection offers several significant advantages compared to others imaging technique including low costs, noninvasive, involves nonionizing radiation and have high accuracy in detecting tumor existence [7]. In Specific Absorption Rate (SAR), microwave cancer detection is realized by the interesting characteristic of the normal tissue and malignant tissue which demonstrated huge difference in term of dielectric property at the microwave frequency. Malignant tissues will record higher dielectric property due to more absorption of electric field compared to normal tissues which help the researcher to identify the tumor presence [8]. Hence, microwave based technique, SAR technique is selected as the alternative method for detection since such technique could accurately detect the presence of the tumor in the brain. SAR is used to measure the energy rate absorbed by the human body when exposed to electromagnetic field [2]. The rate could be calculated from following scientific formula (1):

$$SAR= (\sigma|E|2)/\rho \quad (1)$$

where $\sigma$ is the tissue conductivity (S/m), E is the internal electric field (V/m) and $\rho$ is tissue mass density (Kg/m3). The values of SAR can be categorized as 1g or 10g mass of tissue which equivalence to 1g or 10g spatial average SAR. According to IEEE C95.1:1999, I.6W/kg is the limit value for 1g spatial average SAR while it has been updated as 2W /kg for 10g spatial average SAR based on IEEE C95.1:2005.

Currently, utilization of UWB antennas promising significant function in cancer detection due to UWB antennas are well suited for medical applications since it offer high gain good return loss compared with most of the conventional compact UWB antennas which have low gain and poor return loss [5]. High gain and wide bandwidth antenna is essential in human microwave imaging for interacting and penetrating the multilayer structures with different characteristics by enhancing the focus of the electromagnetic energy into the desired structure [9]. Partial ground technique and present of parasitic element applied in the design lead to realization of UWB characteristic [14]. Apart from array structure [10] and coaxially fed [11], introduction of additional copper reflector contribute in increasing the proposed antenna gain in order to penetrate multi-structure human head so that the signal could reach the tumor inside brain. It is done by reflecting the backwards radiation towards frontwards radiation and hence increases the gain by summing up the original frontwards radiation with reflected one. Due to excellent electrical characteristics including high gain, low side lobe level and low cross polarization reflector antennas have been widely used in radio astronomy, microwave communication, tracking and telemetry nowadays [12].

Basically, the proposed technique used improved gain and better return loss of UWB reflector array antenna in order to detect the brain tumor using SAR technique. A human head phantom with tumor inside is realized and placed close to proposed antenna to measure the desired SAR value. It has been discovered that the SAR rate for the healthy brain model is lower compared to the model with tumor inside where the respective SAR values for both condition are tabulated to see the significant differences.

## 2. Antenna Design, Fabrication and Performances

Figure 1 demonstrated the simulated design of UWB reflector array antenna using taconic (TLY-5) with a dielectric constant of $\varepsilon r = 2.2$, a thickness of $t = 1.5748+/-0.02$ and tangent loss of $\tan \delta = 0.0009$ as the substrate. The antenna is printed with 4×1 copper radiating patch array properly connected with quarter wave transformer transmission line associated with copper parasitic element for the front side as shown in Figure 1(a). The patches comprise of four identical circular with diameter of 15 mm. Parasitic element is placed on very close to feeding line with the gap only 0.2 mm. As shown in Figure 1 (a), each quarter wave transmission line has its own specific wide dimension for 50Ω, 70.71Ω and 100Ω to ensure equal current distribution towards all four patches could be realized. Quarter-wave transformers of 70.71 Ω are used to have ideal match between the 100 Ω lines and the 50 Ω lines [13].

On the other hand, Figure 1(b) shows copper partial ground plane is printed at the back side with 50Ω SMA connector coaxially fed in the middle of the lower part of the antenna back where the signal is fed directly to the radiating patch. Meanwhile Figure 1 (c) and (d) show the copper reflector with 20 mm gap functioned to reduce the side lobe and realized the uni-directional antenna which has higher gain property by reflecting the backwards radiation towards frontwards radiation. The size of the reflector is similar with the size of the UWB antenna. There is a hole through the reflector specially made for connector connection

IOP Conf. Series: Materials Science and Engineering **557** (2019) 012024 doi:10.1088/1757-899X/557/1/012024

as shown in Figure 1(d). Small dimensions of the antenna made it suitable enough to be integrated as the signal radiator in microwave detection system.

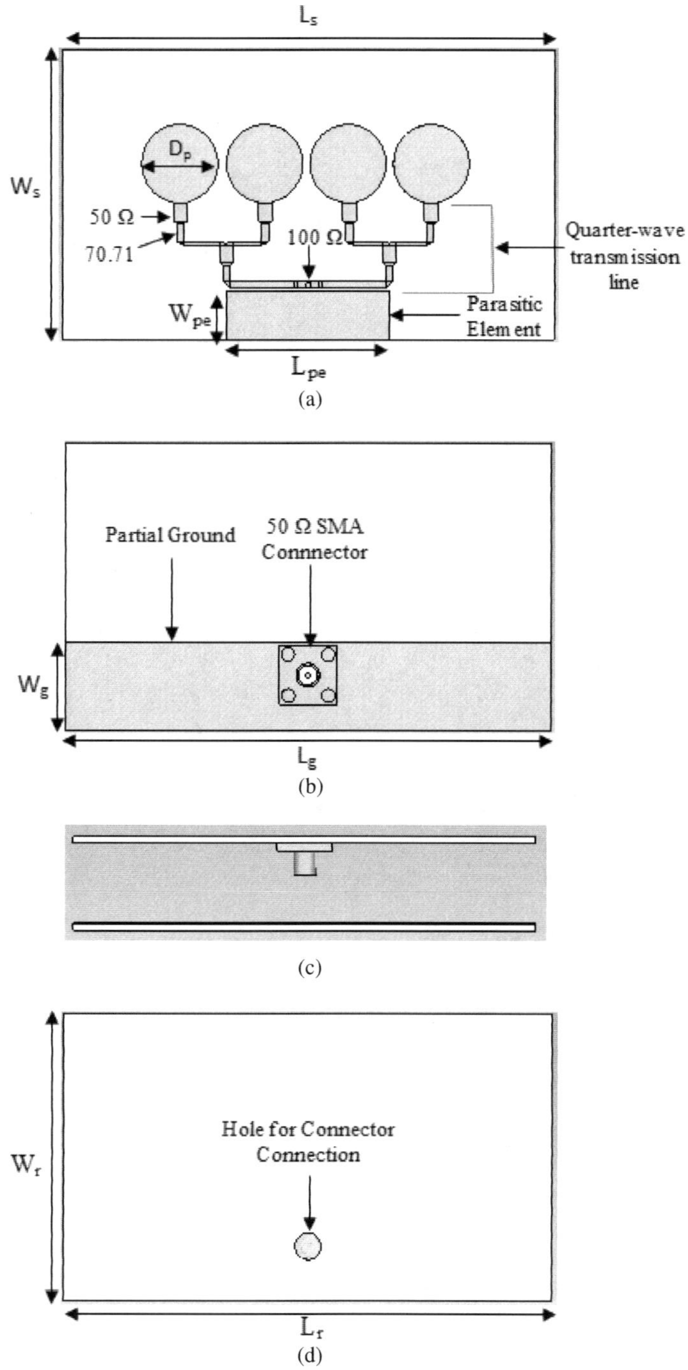


**Figure 1.** The simulated geometry of the proposed UWB array antenna, a) front view, b) transparent back view (without reflector) c) top view e) back view

IOP Conf. Series: Materials Science and Engineering **557** (2019) 012024 doi:10.1088/1757-899X/557/1/012024

Some important parameters of the designed antenna are optimized to obtain the best result in term of compact size, high gain and wide bandwidth. The optimized dimensions for the antenna are tabulated in Table 1. Measurements of gain, patterns and s-parameter have been performed using the setup consisting of Agilent ENA 8051C and Anechoic Chamber. The horn antenna is used as the transmitting antenna, whereas antenna under test (AUT), UWB reflector array as the receiver. Both antennas are placed at DAUT about 0.84 m apart.

**Table 1.** Optimized Antenna Parameter

| Parameter | Quantity |
|---|---|
| $L_s$ | 90.0 mm |
| $W_s$ | 45.0 mm |
| $D_p$ | 7.50 mm |
| $W_{pe}$ | 8.00 mm |
| $L_{pe}$ | 32.0 mm |
| $R_d$ | 20.0 mm |
| $W_g$ | 18.0 mm |
| $L_g$ | 90.0mm |
| $W_r$ | 90.0 mm |
| $L_r$ | 45.0 mm |
| $L_s$ | 90.0 mm |
| $W_s$ | 45.0 mm |

Figure 2 shows Quarter-wave transmission line impedance matching for the corporate feed network. Corporate feed network is utilized in realizing four elements array structure where each patch is fed parallelly using transmission lines. The transmission lines are separated into four divisions based on the number of the radiating patch. The quarter-wave transformer impedance matching technique is applied to distribute the power correspondingly toward entire patches where the feed lines of 70.71 Ω are utilized for perfect matching between the 100 Ω lines and the 50 Ω lines [13]. The dimensions of the feedline are shown in Table 2.

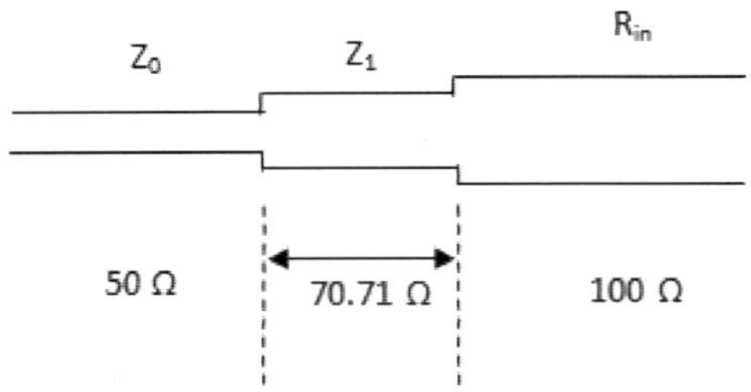


**Figure 2.** Quarter-wave transmission line impedance matching

The formula for the quarter-wave transformer is shown in following equation [13]:

$$Z_1 = \sqrt{Z_0 R_{in}}$$

where:

Z1 = transformer characteristic impedance
Z0 = input transmission line characteristic impedance
Rin = edge resistance at resonance.

Impedance calculation for array antenna is alike for single patch calculation. The formula to perfectly match 100 Ω and 50 Ω transmission lines is shown as followed. Via equation above where Zo= 50 Ω and Rin = 100 Ω, the transformer impedance value is:

$$Z_1 = \sqrt{50(100)}$$

$$= 70.71\,\Omega$$

The sizes for 50 Ω feedline, 70.71 Ω quarterwave transformer and 100 Ω impedance line are acquired by applying following equations [13]:

$$for \frac{W}{h} < 2$$

$$\frac{W}{h} = 8e^{A}/(e^{2A} - 2$$

$$for\ W/h > 2 \tag{2}$$

$$\frac{W}{h} = \frac{2}{\pi}[B - 1 - \ln(2B - 1) + \frac{\varepsilon_r - 1}{2\varepsilon_r}\left\{\ln(B - 1) + 0.39 - \left(\frac{0.61}{\varepsilon_r}\right)\right\}] \tag{3}$$

Where h = substrate height and

$$A = \frac{Z_0}{60}[\frac{\varepsilon_r + 1}{2}]^{\frac{1}{2}} + [\frac{\varepsilon_r - 1}{(\varepsilon_r + 1)(0.23 + \left(\frac{0.11}{\varepsilon_r}\right))}]$$

$$B = 377\pi/(2Z_0\sqrt{\varepsilon_r}) \tag{4}$$

**Table 2.** Optimized Antenna Parameter

| Impedance (Ω) | Length (mm) | Width (mm) |
|---|---|---|
| 50 | 8.00 | 2.90 |
| 70.71 | 8.19 | 1.50 |
| 100 | 8.40 | 0.65 |

Figure 3 shows the comparison image between the fabricated UWB antenna with reflector and UWB antenna without reflector as shown in Figure 3(a) and (b) respectively.

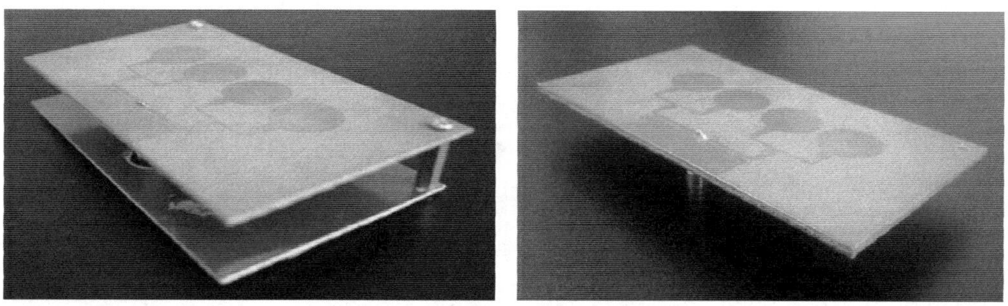

**Figure 3.** Fabricated UWB-Array antenna, a) with reflector, b) without reflector

Simulation and measurement results for both antenna; with reflector and without reflector in term of reflection coefficient are shown in Figure 4. Both simulation and measurement results for the antennas formed a good agreement fulfilling the requirement for UWB characteristic (3.1 GHz to 10.6 GHz). The UWB reflector array antenna performances are better than UWB array antenna performances. From the figure, it clearly shows the antenna with reflector recorded wider range bandwidth of operated frequency compared with antenna without reflector for both simulated and measured results. For simulated, antenna with reflector recorded UWB operated frequency started from 1.6 GHz until 11.4 GHz while antenna without reflector started from 2.1 GHz until 11.4 GHz. On the other hand, antenna with reflector and without reflector recorded 1.6 GHz until 10.8 GHz and 2.2 GHz until 10.8 GHz respectively for measured results. Reflection coefficient less than -10 dB is selected due to the condition where 90% of the signals are successfully transmitted while only the left 10% is reflected back [14]. Partial ground technique and additional of parasitic element assure lower reflection coefficient achieved.

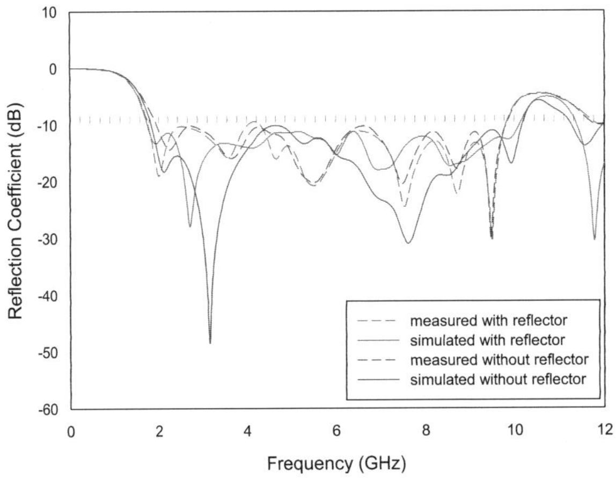


**Figure 4.** Graph of reflection coefficient (s11) for simulation and measurement

The antenna with reflector exhibits higher gain than antenna without reflector over the whole operated frequency as depicted in Figure 5. The measured gain for antenna with reflector and without reflector ranged from 3.2 dB until 14.1 dB and 2.1 dB until 10.2 dB respectively. Reflector element contributes in increasing the antenna gain by reflecting the backwards radiation towards frontwards radiation and hence increases the gain by summing up the original frontwards radiation with reflected one. Once the reflector integrated into an antenna structure, the reflector functions to modify the radiation pattern of the antenna resulted in increasing gain due to reflecting the incoming signal back to the original direction from where it came which is from front direction [12].

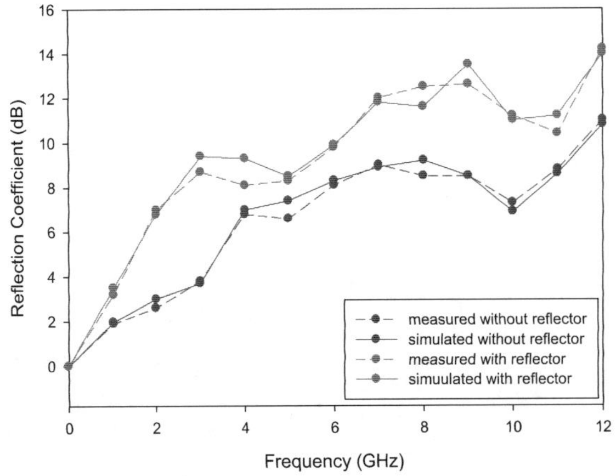


**Figure 5.** Graph of gain (dB)

IOP Conf. Series: Materials Science and Engineering **557** (2019) 012024 doi:10.1088/1757-899X/557/1/012024

Instead of bandwidth and gain, radiation pattern is the other essential parameter in evaluating the proposed antenna performance. The measurement radiation pattern results of Azimuth-Plane for both antennas are shown in Figure 6. The figure demonstrates the polar radiation pattern for the proposed antenna at the frequency of 4 GHz and 5 GHz. These two particular frequencies are vital for brain microwave imaging application. The radiation pattern indicates both antenna with reflector and without reflector does radiate over a wide frequency band [15] where reflector antenna recorded better radiation pattern indicated by wider area covered especially the main lobe. Antenna with reflector categorized under uni-directional antenna since the radiated wave radiates more on main lobe compared to the side and back lobe while antenna without reflector considered as bi-directional antenna due to mostly equal in radiated wave for front lobe and back lobe [16]. In addition, simulation and measurement results indicate high agreement between them. The antenna exhibits the averaged total radiation efficiency of 92% and 90% for antenna with reflector and without reflector respectively. The total radiation efficiency of the antenna promising the good signal radiator for human brain tumor detection. The pattern, gain and frequency ranges are the parameters indicating the antenna as an excellent microwave signal radiator for human brain tumor detection using SAR technique.

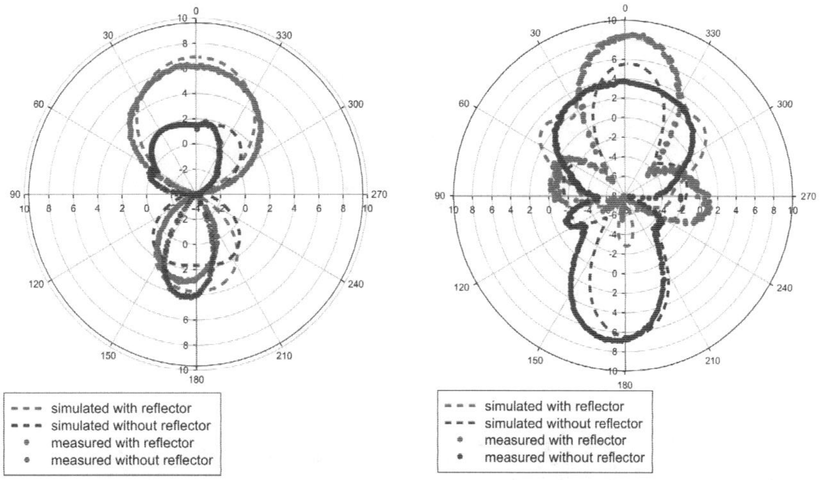


**Figure 6.** Measured and simulated polar radiation pattern in Azimuth plane for antenna with reflector and without reflector; (a) 2 GHz and (b) 3 GHz

## 3. SAR Measurement Results and Discussion

Figure 7 illustrated the measurement setup for brain tumor detection system using SAR technique. The system consist of UWB array antenna as the microwave signal radiator, Vector Network Analyzer as the power source of the antenna, multilayer human head phantom with liquid tissues that have similar dielectric property of real human head tissue structures and SAR probe to absorb energy produced by the antenna.

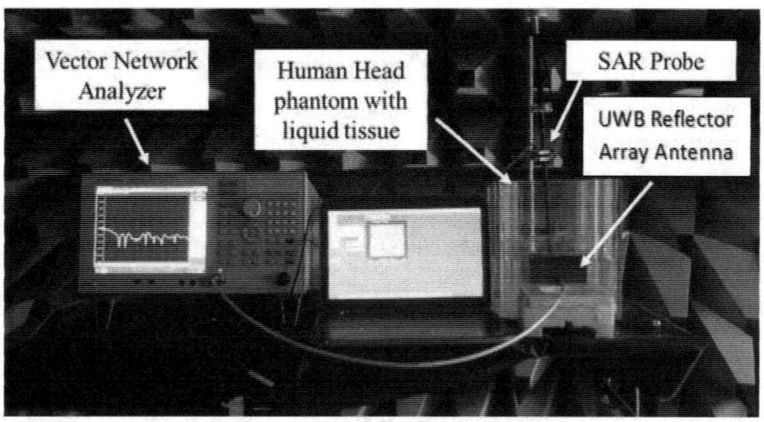


**Figure 7**. SAR measurement setup

On the other hand, Figure 8 shows the whole SAR probe (Probe Amplifier IXA – 020) used in the system which is consist of tip, processors and cable meanwhile IndexSAR software applied to generate the processed SAR values. The UWB reflector array antenna is placed 10 mm away from the multilayer human head phantom as the ideal distance to have maximum energy absorption by the probe as shown in Figure 9. Other distances would lead to less energy absorption due to air interference and high reflection of the phantom as the antenna position is too far and too close from the phantom respectively. The self-made multilayer phantom consist of 4 different layers; skin, fat, skull and brain in resembling the actual structure of human head [18]. With the brain region as the center and first surrounded by the skull layer followed by fat layer and skin layer as the outer most layers completely construct the human head phantom. The phantom is made of Plexiglas with permittivity of 2.3 (εr).

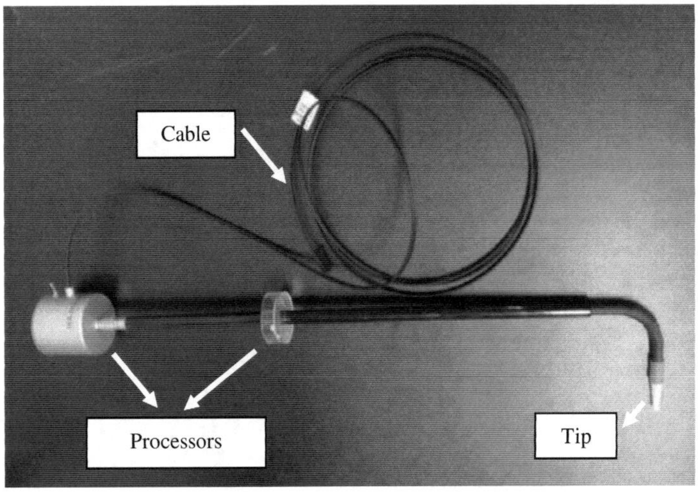


**Figure 8.** SAR probe (Probe Amplifier IXA – 020)

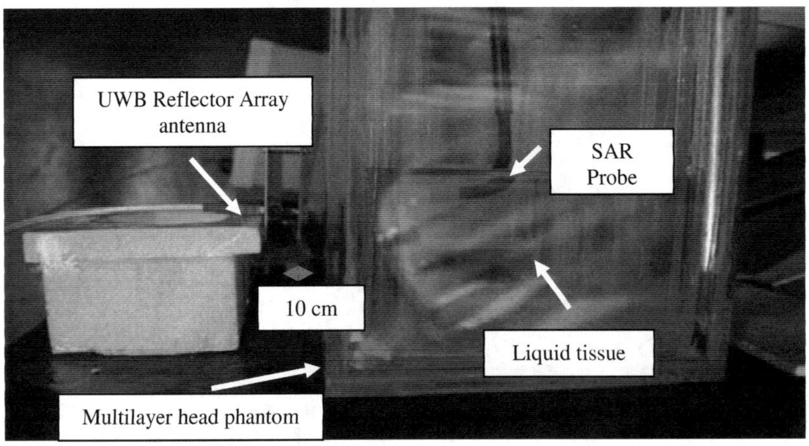


**Figure 9.** Antenna placement in SAR measurement

The whole probe tip must be immersed within the liquid tissue in order to absorb the total microwave signal/energy radiated towards the tissue by the UWB reflector antenna. The self-made phantom also provide specific compartment for tumor to be filled by liquid tumor tissue which has the similar dielectric property of actual tumor; permittivity of 63.2 at 3 GHz [17]. The major constituents of the liquids are water and Tween 20 which are mixed in different ratios to achieve the dielectric properties of the brain tumor in the frequency range of 1-5GHz. The same goes for human head structures of skin, fat, skull and brain. Ideal mixture composition of Tween 20 and water are required in order to obtain similar permittivity value for actual human head structure with human head phantom structure. The liquids are nonreactive, harmless, and safe for handling. The antenna radiated the energy towards human head phantoms within two different conditions; one with tumor present meanwhile another one is without the tumor present to obtain the different SAR value. On the other hand, in order to precisely detecting the tumor, the sensor radiated the energy towards nine different areas to cover the whole one sided area of the phantom as shown in Figure 10.

IOP Conf. Series: Materials Science and Engineering **557** (2019) 012024 doi:10.1088/1757-899X/557/1/012024

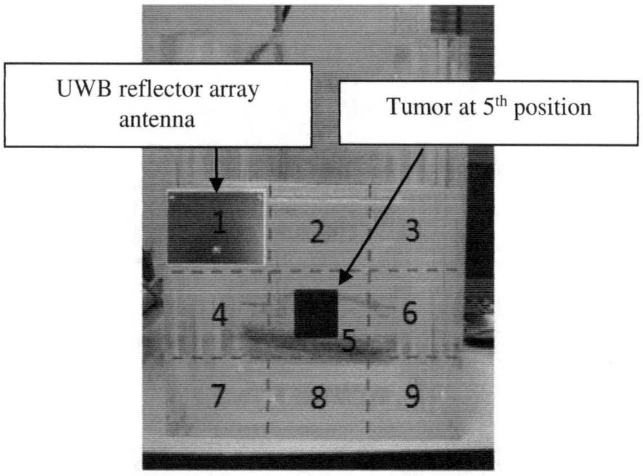


**Figure 10.** Scanning area for head phantom

As the result, the highest different in SAR value among the nine scanned areas between phantom with tumor and without tumor among indicate the tumor present at that particular position or area. The antenna precisely positioned for each area with specific coordinate values of X, Y and Z axis. The values obtained are tabulated in Table 3 for simulated and measured SAR. The results for SAR different values proposed the tumor position at area 5 that recorded the highest SAR difference values of 1.47 W/Kg and 1.46 W/Kg for simulated and measured respectively as indicated in Table 3. This is due to tumor structures consist of their own blood vessels will record higher dielectric property due to more absorption of electric field compared to normal surrounded tissues which help the researcher to identify the tumor presence.

**Table 3.** SAR difference values for simulated and measured

| Area/Coordinate | With Tumor | | Without Tumor | | SAR Difference | |
|---|---|---|---|---|---|---|
| (x,y,z) | simulated | measured | simulated | measured | simulated | measured |
| 1 (50,150,-10) | 2.07 | 1.98 | 1.04 | 1.00 | 1.03 | 0.98 |
| 2 (125,150,-10) | 2.33 | 2.27 | 1.04 | 1.01 | 1.29 | 1.26 |
| 3 (175,150,-10) | 2.00 | 2.01 | 1.05 | 1.04 | 0.95 | 0.97 |
| 4 (50,90,-10) | 2.31 | 2.34 | 1.05 | 1.02 | 1.26 | 1.32 |
| 5 (125,90,-10) | 2.53 | 2.51 | 1.06 | 1.05 | **1.47** | **1.46** |
| 6 (175,90,-10) | 2.33 | 2.29 | 1.05 | 1.02 | 1.28 | 1.27 |
| 7 (50,30,-10) | 2.02 | 1.99 | 1.04 | 1.01 | 0.98 | 0.98 |
| 8 (125,30,-10) | 2.29 | 2.26 | 1.03 | 1.01 | 1.26 | 1.25 |
| 9 (175,30,-10) | 2.04 | 2.00 | 1.02 | 1.00 | 1.02 | 1.00 |

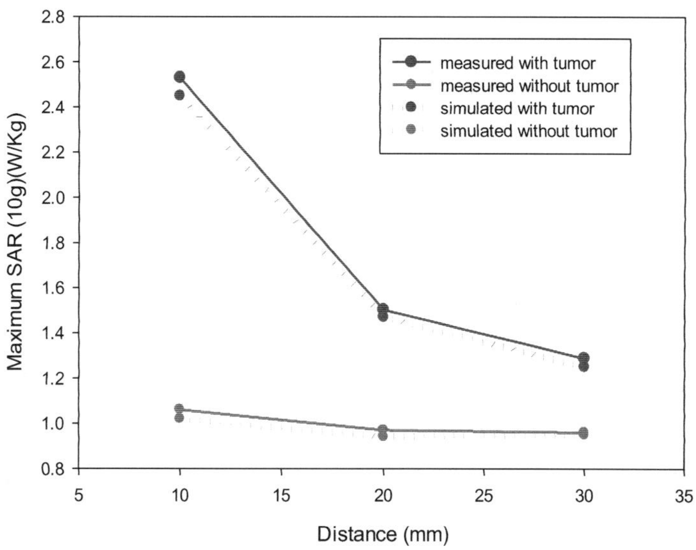


**Figure 11.** SAR difference at area 5 (tumor location); blue line & dot (with tumor), red line & dot (without tumor)

On the other hand, Figure 11 shows the graph of simulated and measured SAR values different at area 5 (tumor location) within head phantom, for with tumor and without tumor condition with variable antenna distances from the phantom; 10 mm, 20 mm and 30 mm. According to the graph, there is significant different between SAR values for with tumor and without tumor obviously at distance of 10 mm compared with two others distances indicating the 10 mm is the ideal distance between the antenna and the phantom for SAR measurement. In addition, as the distance increase, the SAR difference value become less significant due to less concentrated signal/energy manage to reach the phantom because of air resistance. Apart from that, measured values exhibit slightly less compared with simulated values due to environment effect that interfere the radiated energy such as air and humidity. However, the differences are ignorable since most of the values are less than 0.1W/kg.

## 4. Conclusion

A high gain ultra-wide band (UWB) reflector array antenna for human brain tumor detection using specific absorption rate (SAR) technique in microwave detection system is proposed. Reflector element contributes in improving the overall sensor performance especially the gain, operated bandwidth and radiation pattern leads to an excellent microwave signal radiator in SAR measurement. The UWB antenna with reflector demonstrated wider reflection coefficient of less than -10dB started from 1.6 GHz until 10.8 GHz with the ranges of gain between 3.2 dB to 14.1 dB as compared with UWB antenna without reflector that recorded 2.2 GHz to 10.8 GHz and 2.1 dB to 10.2 dB for the reflection coefficient and gain respectively. Moreover, measurements of bandwidth and reflection coefficient have identical behavior compared to simulations. The antenna is successfully detecting human brain tumor based on SAR technique. Amount of energy or gain produced by the antenna which being absorbed by the human brain indicated the present of tumor inside. Human brain with tumor absorbed more energy and recorded higher SAR value compared to human brain without tumor. In this paper, human head phantom associated with tumor recorded SAR value of 2.53 W/kg and 2.51 W/kg for simulated and measured respectively meanwhile head phantom without tumor stated SAR value of 1.06 W/kg for simulated and 1.05 W/kg for measured. The biggest different in SAR value according to area precisely determine the location of the tumor in the brain which in this case position 5 (tumor location) resulted biggest SAR different value of 1.47 W/Kg for simulated and 1.46 W/Kg for measured.

## 5. Acknowledgment

This work was supported by Excellent Grant Scheme (Grant No. 9009-0049) from University Malaysia Perlis, Malaysia

## 6. References

[1] Andrew H. Kaye, Edward R. Laws, "Brain Tumors: An Encyclopedic Approach, Expert Consult", third edition, Elsivier Limited, 2012
[2] Alzabidi, M. A., Aldhaeebi, M. A., and Elshafiey, I., "Optimization of UWB Vivaldi Antenna for Tumor Detection", First International Conference on Artificial Intelligence, Modelling & Simulation, 2013
[3] P. M. Meaney, M. W. Fanning, D. Li, S. P. Poplack, and K. D. Paulsen, "A clinical prototype for active microwave imaging of the breast," Microwave Theory and Techniques, IEEE Transactions on, vol. 48, pp. 1841-1853, 2000.
[4] Abbosh, A. M., "Directive antenna for ultrawideband medical imaging systems", International Journal of Antennas and Propagation, 2008.
[5] Jasmine Angel. J. and Anita Jones Mary. T., "Design of Vivaldi Antenna for Brain Cancer Detection", 2014 International Conference on Electronics and Communication Systems (ICECS - 2014), Coimbatore, INDIA, Feb.13 -14, 2014,
[6] F. S. G. B. Barnes, Handbook of biological effects of electromagnetic fields. Boca Raton: CRC/Taylor & Francis, 2007.
[7] Chaudhary, S., Luthra, P. K., Kumar, A., "Use of Graphene as a Patch Material in comparison to the copper and other Carbon Nanomaterials", International Journal of Emerging Technologies in Computational and Applied Sciences, 3(3), Dec.12-Feb.13, pp. 272-279, 2013
[8] Wasusathien, W., Santalunai, S., Thanaset Thosdeekoraphat, T., Thongsopa C., "Ultra Wideband Breast Cancer Detection by Using SAR for Indication the Tumor Location", International Journal of Medical, Health, Pharmaceutical and Biomedical Engineering Vol:8 No:7, 2014
[9] Xuyang Li, Yoke Leen Sit, Lukasz Zwirello, and Thomas Zwick, "A Miniaturized UWB Stepped-Slot Antenna for Medical Diagnostic Imaging", Microwave and Optical Technology Letters/Vol. 55, No. 1, 2013.
[10] Marek GARBARUK, "Analysis of ultra-wideband linear antenna arrays", (Electrical Review), ISSN 0033-2097, R. 88 NR 8/2012
[11] B.T.P.Madhav, J.ChandrasekharRao, K.Nalini, N.Durga Indira, "Analysis of Coaxial Feeding and Strip Line Feeding on the Performance of the Square Patch Antenna", Int. J. Comp. Tech. Appl., Vol 2 (5), 1352-1356, 2011

[12] Amir Moallemizadeh, R.Saraf-Shirazi, and Mohammad Bod, "Design of a Novel Compact Cup Feed for Parabolic Reflector Antennas", Progress In Electromagnetics Research Letters, Vol. 64, 81–86, 2016

[13] Norfishah Ab Wahab, Zulkifli Bin Maslan, Wan Norsyafizan W. Muhamad, Norhayati Hamzah, "Microstrip Rectangular 4x1 Patch Array Antenna at 2.5GHz for WiMax Application", 2010 Second International Conference on Computational Intelligence, Communication Systems and Networks, 2010

[14] Jamlos, M. A., Jamlos, M. F. and Ismail, A. H., "High performance of coaxial feed UWB antenna with parasitic element for microwave imaging", Microw. Opt. Technol. Lett., 57: 649–653. doi:10.1002/mop.28926, 2015.

[15] N. Ojaroudi, M. Ojaroudi, F. Geran, and Sh. Amiri, "Omnidirectional Antenna with Multi-Resonance Performance for Microwave Imaging Systems", Telfor Journal, Vol. 5, No. 2, 2013

[16] Anwer Sabah Mekki, Mohd Nizar Hamidon, Alyani Ismail, and Adam R. H. Alhawari, "Gain Enhancement of a Microstrip Patch Antenna Using a Reflecting Layer," International Journal of Antennas and Propagation, vol. 2015, Article ID 975263, 7 pages, 2015

[17] IT'IS Foundation, "Database of Tissue Properties" retrieved from http://www.itis.ethz.ch/virtual-population/tissue-properties/overview.

[18] Haoyu, A., El-Rayis, O., Haridas, N., Noordin, N. H., Erdogan, A. T., and Arslan T., "A smart antenna array for brain cancer detection", Loughborough Antennas & Propagation Conference 2012 (LAPC 2012), Loughborough, UK, 2012

# Study on Quality Parameters of Yellow Matoa Fruit (*Pometia pinnata*) Using Digital Image Processing

**A. Furay[1], U. Ahmad[2*], and S. Widodo[2]**

[1]Postharvest Technology Master Program, Graduate School, Bogor Agricultural University, Indonesia 16680

[2]Department of Mechanical Engineering and Biosisitem, Faculty of Agricultural Engineering and Technology, Bogor Agricultural University, Indonesia 16680

*Corresponding e-mail: uahmad2010@gmail.com

**Abstract**. At this time matoa fruits are sold without grading them first so high and low quality fruits are mixed. In addition to not having SNI, there is also no standard method for grading according to standardization of fresh horticultural products. Related to this issue, digital image processing can be used as an alternative method for grading. The purpose of this study were to study quality parameters and to develop quality evaluation method for yellow type matoa using digital image processing. Manual measurement using weight as quality parameter of 203 yellow matoa produced three categories of quality classes namely A, B and C. Image processing algorithm was then developed to replace the manual measurement by estimating the weight using projected area of fruits image and to measure skin color of the fruits. The overall accuracy of quality evaluation based on weight using developed algorithm was 73.89%. For quality evaluation based on visual parameter or skin color represented by red and blue color ratio (R/B), and hue and value (H/V) color ratio, it was found possible to classify the fruits into three quality classes namely class 1 (brown to yellow-brown), class 2 (yellow), and class 3 (yellow-green to green) with overall accuracy of 74.38%. Finally, using combination of those two parameters (i.e. area and skin color) nine new quality classes, namely A1, A2, A3, B1, B2, B3, C1, C2, and C3 classes were obtained with overall accuracy of 52.71%.

## 1. Introduction

Matoa (*Pometia pinnata*) has been well known exotic fruits which is widely distributed in Asia Pacific including Papua, Indonesia [1]. It has been designated as one of the important horticultural commodities in Indonesia [2]. With its unique taste and texture, matoa fruit is quite popular and has good market potential outside Papua. Its marketing has spread from local markets, supermarkets, and even export markets. Matoa fruit also has the potential for further processing into syrup, juice and other processed products that are useful and become new economic opportunities for household and industrial scale.

At this time matoa fruit does not yet have a quality standard contained in Indonesia National Standard (SNI) or in other forms that can be used as a reference for standardization. Related to this issue, image processing can be used as an alternative method for quality evaluation prior to grading operation. Camera and image processing has been widely used for automatic fruit grading and classification [3], mainly due to its capability to evaluate the variety of produces non-destructively with high accuracy and speed so it is suitable for handling large quantities. Image processing is a visual technology with camera as the sensor, used to observe, process and analyze an object without dealing directly (non-contact) with the object, leaving the object without any harm.

Appearance of products, including fruits, is typically evaluated by some characteristics such as size, shape, form, color, freshness and absence of visual defects **[4]**. One of these characteristics, color is an important sensory attribute of product quality. Therefore, color is an important criteria for grading of any product, especially fresh fruits **[5]**. Many works have been reported the use of image processing for quality evaluation of various agricultural products such as coffee bean **[6]**, melinjo (*Gnetum gnemon*) seed **[7]**, and tea leaves **[8]**; application of image processing on mango grading machine **[9]**; and constructing a vision system to monitor harvested paddy grain quality in combine harvester operation **[10]**. In the field application, color image analysis of the melons fruits in combination with discriminant analysis could be used to distinguish between harvesting ages of 46, 53, 60, and 67 days after planting with an average accuracy of 86% **[11]**. The purpose of this study was to study quality parameters and to develop an alternative quality evaluation method of yellow matoa using digital image processing. The parameters are expected to be used for building the quality standard to be applied in matoa promotion and marketing in wider areas in Indonesia and export as well. The developed image processing evaluation method is expected to be used in automatic grading of the matoa fruits in the future to guarantee the quality of the fruits especially for export.

## 2. Material and Method

The study was conducted in August 2016 - January 2017 at the Laboratory of Food and Agricultural Product Processing Engineering, Department of Mechanical and Biosystems Engineering, Bogor Agricultural University. The material used was 203 yellow matoa fruits obtained from the local farmer in Papua. Physical quality parameters were measured as follows: weight (gram) using a digital scale (Metler PM-4800); length and diameter of fruit (cm) using calipers; hardness (kgf/cm$^2$) using CR-30 model rheometer, carried out at three different points: top, middle and bottom of the fruits; total soluble solids (TSS, in % brix) using Atago PR-201 digital refractometer, carried out by slicing the fruit flesh, dropping the liquid on the refractometer and the average data were taken from three repetitions of measurement. Digital image of matoa fruits were captured using a set of machine vision system consisting of a color CCD camera (VED, OC-305 D), image frame grabber (PXC 200A), personal computer (PC) with Intel Pentium IV processor, and four 7 watts TL lamp as light source.

The obtained images were then processed using Matlab 2014 software. The algorithm include thresholding to eliminate unwanted background, noises cleaning and holes filling through a series of erosions and dilations to binary image, pixel counting to get projection area of the fruit, and color values picking and analysis from the fruit part, in RGB (red-green-blue) and HSV (hue-saturity-intensity) color models.

## 3. Result and Discussion

The results of this study will be explained in two sections which are physical parameters measurement to be used as reference values, and image processing to be used for prediction. Furthermore, the second sections will be divided into three sub-sections as to demonstrate the use of each size and color parameters individually, and combination of the two image processing paremeters for the proposed new classes in quality.

### *3.1. Physical Quality Parameters of Yellow Matoa Fruit*

Yellow-type matoa fruit typically has a light green skin color when it is unripe and turns into yellow to brown when it is ripe. Based on the color, the matoa fruit can be grouped into five color groups namely green, green-yellow, yellow, yellow-brown and brown. In terms of shape, the yellow type of matoa can be single or double fruit, either double fruit with the same or different size. Based on the form of harvest commonly traded on the market, the matoa fruit is divided into single fruits and fruit series. As a reference in compiling the standards for matoa fruit extraction, measurements of the physical quality parameters of the yellow matoa fruit were measured which included length, diameter, weight, total soluble solid, and hardness. The measurement results obtained can be seen in Table 1.

**Table 1.** Physical quality parameters of yellow matoa fruit

| | Length (cm) | Diameter (cm) | Weight (gr) | TSS (%brix) | Hardness ($kgf/cm^2$) |
|---|---|---|---|---|---|
| Means | 3.6 | 2.9 | 10.3 | 21.9 | 2.8 |
| Maximum | 8.3 | 8.3 | 20.8 | 31.2 | 6.9 |
| Minimum | 1.9 | 1.2 | 6.8 | 9.8 | 0.2 |
| Standard Dev. | 1.3 | 1.1 | 2.9 | 3.8 | 1.3 |

Because of shape of the fruit which is quite diverse, the weight are chosen among the parameters measured as a reference for grading, in addition to skin color. However, in this paper only weight of the fifth physical data measured directly from the object used for fruit classification, while the other four data were unused. Weight is very common to grade fruits based on its size, while length and diameter of the fruit were found to be less accurate to measure the size of the fruits. Meanwhile, TSS and hardness were assumed to be represented by skin color using image processing so they are not analyzed and discussed in this paper. Furthermore, considering the means, maximum, minimum and standard deviation value of the weight and practicality in grading operation, yellow matoa fruit can be grouped into three grades namely A, B and C with the criteria as shown in Table 2.

**Table 2.** Yellow matoa fruits grade based on weight

| Grade | Weight (gram) |
|---|---|
| A | >12.5 |
| B | 9.5 - 12.5 |
| C | < 9.5 |

*3.2. Quality Parameters of Yellow Matoa Fruit by Image Processing*

As discussed in the previous section, among measured parameters weight and skin color are considered to be used as main criteria for yellow matoa grading. More detail about the results of yellow matoa fruit grading using weight, skin color and combination of both parameters are discussed in the following sub-sections.

*3.2.1. Fruit grading based on weight.* In the previous section, it has been discussed that it is possible to classify yellow matoa fruit based on its weight into three grade: A, B, and C. Image processing algorithm was then developed to replace manual measurement of weight by estimating the weight using the projected area of the fruits in image. Measurements of area were made by converting color images into grayscale images and then to binary images at certain threshold values for distinguishing objects and backgrounds. The area was then determined by counting the number of white pixels of the segmented object. The counting pixels were actually the projection area which represents the size of the fruit in two-dimension. The number of pixels obtained from the area then were used as input to calculate the weight of the fruit using equation shown in Fig. 1 (weight = 0.0005 * area + 8.2741). The results of area counting from all images were then plotted again the weight of the fruits obtained from the direct measurement (weghing). Fig. 1 showed the relation between area and fruit weight, which is in linear relationship. It is shown that the fruit weight could be estimated using the developed linear mathematical model quite well with coefficient of determination 0.8499. Using this model, the criteria for matoa fruit grading based on area which is corresponding with the weight of the fruits was determined as shown in Table 3.

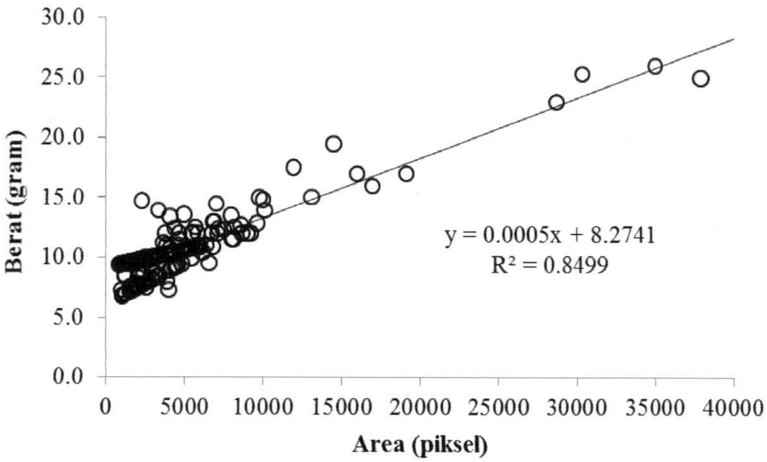


**Figure 1.** The linear relationship between the weight and the area of fruit in image.

**Table 3.** Correlation of fruit grade based on weight and projected area obtained using image processing

| Grade | Weight (gram) | Area (pixels) |
|---|---|---|
| A | > 12.5 | > 9318 |
| B | 9.5 - 12.5 | 1584 – 9318 |
| C | < 9.5 | <1584 |

Table 4 showed confusion matrix showing the comparison between manual grading and the grading by image processing algorithm based on fruit weight using the mathematical model obtained from Figure 1. The overall accuracy is 73.89% with the highest accuracy of 88.62% for grade B and the lowest accuracy of 43.55% for grade C. This result indicated that the developed model could not estimate grade C (small size fruits) well enough. It is likely due to number the sample which were not well distributed among the fruits grade, or those which grouped into A and C classes were relatively small number as compared to B class.

**Table 4.** Comparison of fruit grading based on weight by manual and image processing algorithm

| Grade (Manual) | Grade (image processing algorithm) A | B | C | Total | Accuracy (%) |
|---|---|---|---|---|---|
| A | **14** | 4 | - | 18 | 77.78 |
| B | 1 | **109** | 13 | 123 | 88.62 |
| C | - | 35 | **27** | 62 | 43.55 |
| Total | 15 | 148 | 40 | **203** | |
| Recall (%) | 93.33 | 73.65 | 67.50 | | **73.89** |

*3.2.2. Fruit grading based on skin color.* The results of measurement of skin color visual quality parameter using image processing algorithm based on maximum, minimum, and average values shows that color parameter using each RGB and HSV color models directly and individually does not show high variation values so it is difficult to use skin color as a reference for matoa fruits grading with a good result. However, visual color parameters based on red and blue ratio (R/B) and hue and value (H/V) ratio show fairly good variations that can be used as references for grading the fruits based on skin color (Fig. 2 and Fig. 3). However, it is quite difficult to distinguish all of the five classes based on the skin color (as done manually) due to high overlapping values amomg the different classes. Therefore it was decided to reduce the number of grades become three classes only, namely grade 1 (showing brown to yellow-brown skin colors), grade 2 (yellow), and grade 3 (green-yellow to green). The detail criteria used for classification the samples using color ratios from the two color models (RGB and HSV) were listed in Table 5.

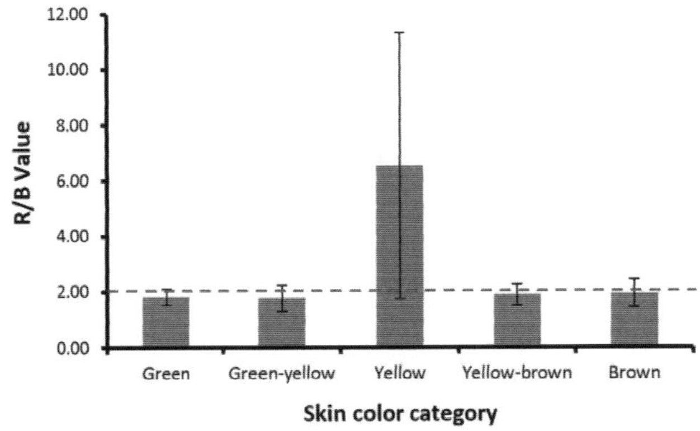


**Figure 2.** Distribution of R/B value for the skin color and selected threshold value.

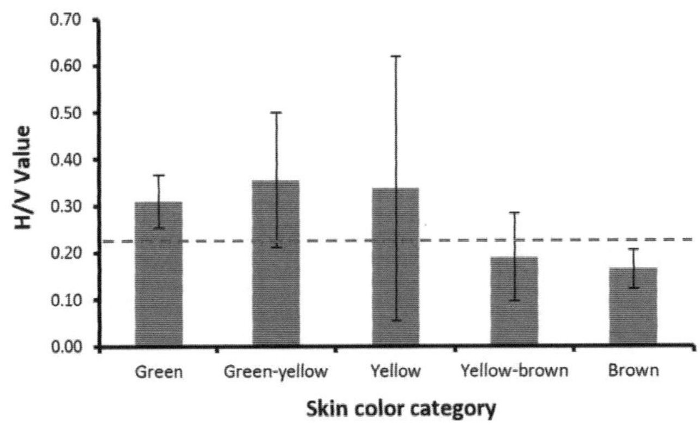


**Figure 3.** Distribution of H/V value for the skin color and selected threshold value.

**Table 5.** Criteria for fruit classification based on skin color

| Grade | Color ratio (R/B, H/V) | Skin color by |
|---|---|---|
| 1 | R/B < 2.0 & H/V < 0.25 | Yellow-brown to brown |
| 2 | R/B > 2.0 | Yellow |
| 3 | R/B < 2.0 & H/V >0.25 | Green to green-yellow |

Table 6 shows the comparison between manual grading and grading using image processing algorithm based on fruit weight. Image processing algortithm predicted the weight of the fruit by counting the number of pixels belongs to object, after a series of image processing algorithm applied to the image. Before reading the color values of the image, the background pixels were set zero so the pixels could be skipped. The overall accuracy is 74.38% with the highest accuracy of 79.13% for grade 1 and the lowest accuracy of 56.25% for grade B. Similar to previous case for grading based on weight, the poor accuracy for grade 2 is likely due to distribution of the sample where grade 2 only contributes 7.9% of the total sample as a reference (manual grading based on skin color), while the grade 1 contributes 56.7% of the total sample and resulted in the highest accuracy in classification. It can be said here that not enough number of samples causing poor classification for grades 2 and 3, while grade 1 gave a relatively good result because it had enough samples.

**Table 6.** Comparison of fruit grading based on skin color by manual and image processing algorithm

| Grade (Manual) | Grade (image processing algorithm) 1 | 2 | 3 | Total | Accuracy (%) |
|---|---|---|---|---|---|
| 1 | **91** | 14 | 10 | 115 | 79.13 |
| 2 | 1 | **9** | 6 | 16 | 56.25 |
| 3 | 20 | 1 | **51** | 72 | 70.83 |
| Total | 112 | 24 | 67 | **203** | |
| Recall (%) | 81.25 | 37.50 | 76.12 | | **74.38** |

*3.2.3. Fruit grading based on combination of weight and skin color.* Finally, using combination of area and red-blue ratio (R/B) and hue-value ratio (H/V) of skin color by image processing algorithm as grading parameters, nine new grade namely A1, A2, A3, B1, B2, B3, C1, C2, and C3 classes were proposed. The combaination means the algorithm applied two rules, first rule was to determine the size classes, which were A, B, or C using area projection. After that, the algorithm evaluated R/B and H/V values to see whether the fruit will have 1, 2, or 3 classes based on skin color. Whatever the results are, the fruit will go to one of the nine classes combination of the two parameters. Fig. 4 shows examples of the fruit images and its corresponding quality classes obtained using developed image processing algorithm that combined area and color parameter. The comparison between manual grading and grading using image processing algorithm based on combination of area and skin color was shown in Table 7. The overall accuracy was 52.71%.

**Table 7.** Comparison of fruit grading based on combination weight and color by manual and digital image processing

| Grade (Manual) | Grade (Image orocessing) A1 | A2 | A3 | B1 | B2 | B3 | C1 | C2 | C3 | Total | Accuracy (%) |
|---|---|---|---|---|---|---|---|---|---|---|---|
| A1 | **5** | - | - | 1 | - | - | - | - | - | 6 | 83.33 |
| A2 | - | **3** | 3 | - | 1 | - | - | - | - | 7 | 42.86 |
| A3 | 2 | - | **1** | - | - | 2 | - | - | - | 5 | 20.00 |
| B1 | - | - | - | **44** | 10 | 7 | 4 | 2 | - | 67 | 65.67 |
| B2 | - | - | - | 1 | **3** | 1 | - | 1 | - | 6 | 50.00 |
| B3 | - | - | 1 | 9 | - | **34** | 3 | - | 3 | 50 | 68.00 |
| C1 | - | - | - | 28 | 2 | 1 | **9** | - | 2 | 42 | 21.43 |
| C2 | - | - | - | - | - | - | - | **1** | 2 | 3 | 33.33 |
| C3 | - | - | - | 1 | - | 3 | 5 | 1 | **7** | 17 | 41.18 |
| Total | 7 | 3 | 5 | 84 | 16 | 48 | 21 | 5 | 14 | **203** | |
| Recall (%) | 71.43 | 100.00 | 20.00 | 52.38 | 18.75 | 70.83 | 42.86 | 20.00 | 50.00 | | **52.71** |

In general the accuracy of the classification is still low and unacceptable for practical use. There were several reason for poor accuracy of the developed system. Image processing algorithm used to extract feature or information of the fruits should be improved. Also, number of sample should be increase and distribution of the fruits among quality classes also should be improved to provide broader range of data and also the generalization capability of the developed algorithm.

**Figure 4.** Examples of the grading results using combination of weight and color: A1, A2, A3, B1, B2, B3, C1, C2, C3.

## 4. Conclusion

The quality parameter of yellow matoa fruits has been studied and then used as a reference to develop an alternative quality evaluation method using digital image processing. Among quality parameters studied weight and skin color of fruits was selected to classify the fruits into several classes. For quality evaluation based on weight, it was possible to classify the fruits into three classes namely A, B, and C. The developed quality evaluation based on image processing algorithm could classify the fruits based on weight with an overall accuracy of 73.89% compared to manual method. For quality evaluation based on visual parameter or skin color represented by red-blue color ratio (R/B) and hue-value color ratio (H/V), it was possible to classify the fruits into three quality classes namely class 1 (brown to yellow-brown), class 2 (yellow), and class 3 (yellow-green to green) colors with overall accuracy of 74.38%. Finally, using combination of those two parameter (i.e. area and skin color) nine new grades, namely A1, A2, A3, B1, B2, B3, C1, C2, and C3 classes were obtained with overall accuracy of 52.71%. There were several reasons for poor accuracy of the developed system. Beside image processing algorithm that should be improved to obtain more accurate feature or information of the fruits, the number and distribution of the fruits among quality classes and skin color also should be well prepared to get a balance samples number.

## References

[1] Trimedona N, Nurdin H, Darwis D, and Efdi M. 2015. Isolation of triterpenoid from stem bark of Pometia pinnata Forst & Forst. *Journal of Chemical and Pharmaceutical Research*

7(11):225–227.
[2] Ministry of Agriculture Republic of Indonesia. 2006. Decree No. 511 of 2006: Types of Commodities under Directorate General of Plantation, Directorate General of Food Crops, and Directorate General of Horticulture.
[3] Seema, A. Kumar, and G.S. Gill. 2015. Automatic Fruit Grading and Classification System Using Computer Vision: A Review. In Proceeding of the Second International Conference on Advances in Computing and Communication Engineering (ICACCE 2015). Dehradun, India, 1-2 May 2015
[4] Costa C, Antonucci F, Pallottino F, Aguzzi J, Sun DW, Menesatti P. 2011. Shape analysis of agricultural products: a review of recent research advances and potential application to computer vision. *Food and Bioprocess Technol.* 4(5): 673-692.
[5] McCaig TN. 2002. Extending the use of visible/near-infrared reflectance spectrophotometers to measure colour of food and agricultural products. *Food Research Int.* 35: 731-736.
[6] Soedibyo DW, Seminar KB, Ahmad U, Subrata IDM. 2010. The development of automatic coffee sorting system based on image processing and artificial neural network. Proc. of AFITA Int. Conf.: The Quality Information for Competitive Agricultural Based Production System and Commerce, Bogor, West Java-Indonesia: 272-275.
[7] Muqodas A.U., Widodo S., Seminar K.B., Solahudin M. 2018). Machine Vision-Based Analysis for Black Tea Quality Evaluation. In: McLellan B. (eds) Sustainable Future for Human Security. Springer, Singapore
[8] Widodo, S., M. Kalili. 2018. Quality Evaluation of Melinjo Seeds (*Gnetum gnemon* L.) Using Digital Image Processing. *Jurnal Teknik Pertanian Lampung* (Journal of Agricultural Engineering) 7(2): 106-114.
[9] Ahmad U, Mardison, Tjahjohutomo R, Nurhasanah A. 2010. Development of automatic grading machine prototype for citrus using image processing. Austral. J. of Agr. Eng. 1(5): 165-169.
[10] Mahirah J, Yamamoto K, Miyamoto M, Kondo N, Ogawa Y, Suzuki T, Habaragamuwa H, Ahmad U. 2015. Double lighting machine vision system to monitor harvested paddy grain quality during head-feeding combine harvester operation. *Machines* 3(4): 352-363.
[11] Ahmad, U., D.P. Bermani, Mardison.2018. Color distribution analysis for ripeness prediction of golden apollo melon. *TELKOMNIKA* (Telecommunication, Computing, Electronics and Control) 16(4): 1659-1666.

IOP Conf. Series: Materials Science and Engineering **557** (2019) 012026 doi:10.1088/1757-899X/557/1/012026

# Stacked Stepped-Fed Super Wideband Antenna Performance in Free Space and Liquid Medium for Biomedical Applications

**Mohd Aminudin Jamlos[1], W A Mustafa[1], Wan Khairunizam[2], I Zunaidi[3], Z M Razlan[2] and Shahriman A B[2]**

[1]Faculty of Engineering Technology, Universiti Malaysia Perlis, UniCITI Alam Campus, Sungai Chuchuh, 02100 Padang Besar, Perlis, Malaysia

[2]Advanced Intelligent Computing and Sustainability (AICoS) Research Group, School of Mechatronics Engineering, Universiti Malaysia Perlis, 02600 Arau, Perlis, Malaysia

[3]Faculty of Technology, University of Sunderland, St Peter's Campus, Sunderland, SR6 0DD, United Kingdom

**Abstract.** A new design of compact Super Wideband (SWB) antenna for performance comparison between in free space and in plain water is presented in this paper. The proposed antenna consists of basic circular patch as radiating element associated with partial ground and a zinc reflector to realize high gain and wide bandwidth. Dimension of 20 mm × 30 mm made the proposed antenna as a compact antenna. The significant feature of this design is the cautious manufacturing of the feed line with additional of two microstrip lines between the patch and the feed line known as stepped transmission line that altering the capacitance between the radiating element and the ground plane. The patch is directly connected with SMA connector using 50 Ohm mirostrip feed line through side feeding and deploy the air gap stacked. The antenna exhibit high performance in free space indicated by operated frequency ranged from 4.3 GHz until more than 15 GHz (<-10dB) with maximum gain of 9.79 dB. On the other hand, plain water medium degraded the antenna performance yet still within acceptance limit associated with resonant frequency ranged from 4.7 GHz until more than 15 GHz (<-6dB) and 2.2 dB for the maximum gain. The results reflect the reliability of the SWB antenna operated within liquid medium for various related applications especially for biomedical related applications such as microwave imaging for cancer detection.

## 1. Introduction

Current wireless communication technology of super wide band antenna (SWB) functioned to cover both the short and long range transmission for ubiquitous applications. Significant advantages offer by SWB technology especially larger channel capacity and higher time precision has made the technology more attractive and more essential toward many potential applications related with wireless communication technology. SWB technology demonstrated impedance bandwidth ratio larger than 10:1 with <-10 dB reflection coefficient [1]. SWB technology provided large bandwidth, high and fast transmission data rate as well as low power cost for many wireless communication applications [2]. Underwater sensor connection could be established through electromagnetic signals radiated in distances of few meters. Therefore, the antenna becomes an essential component in term of propagation aspects of the electromagnetic waves in liquid for such condition. Underwater sensors that deployed SWB technology have advantages of transmitting high amounts of data using extremely little power over relative short distances [3]. The main reason triggering inspiration to investigate underwater SWB antenna is the inadequate work done towards this kind of antennas. In fact, currently there are a lot of designs for SWB antennas yet all of them are function for air communication applications. Until now, all SWB devices are utilized to perfectly function for air environments. Hence, this paper introduced SWB antenna to work in plain water environments. It is compulsory for SWB technology to operate well with a high performance underwater to ensure functionality of particular applications. It is supported by the high demands for inter-medium communication systems since rapid development of trans-medium human activities. Intersect communication between air-water boundaries is gaining interest day by day as underwater communication under on-going process to established itself within the same time [4].

Apart from that, most of the electromagnetic waves studies in water environment emphasize on narrow band antennas such as dipoles or loops isolated with plastics. Loop antennas are directional antennas which have advantages of sensitive reactions towards magnetic fields realized them as the suitable candidate for seawater antenna [6]. Designing SWB antenna for underwater applications is a great challenge due to variety of potential solutions. Moreover, the device has to be compact, not bigger than 5 cm of length and wide in order for the device to be operated at the optimum level. This limitation in term of size causes additional challenge to discover a right antenna with a good performance during underwater measurement take place. Besides that, the constraint on data transmission over distance in water also have to be concentrated on since the propagation of electromagnetic waves in water is very different than in the air. High dielectric constant of water resulted in far higher attenuation compared with attenuation occurred in air.

The omnidirectionality is one of the significant features needs to be demonstrated by underwater antenna in order for the antenna to transmit in entire directions. This requirement ensures that particular antenna covered wide range for data transmission and data receiving to take place resulted in high data rates. Hence, the radiation pattern should be as much omnidirectional as possible to have high performance of underwater antenna. Antenna that contributes in equal radiation towards all directions is known as isotropic antenna. An omnidirectional antenna is one that radiates uniformly in one plane [5]. Presently, the amount of underwater applications engage with completely data exchange occurred underwater is extremely few. Among of those applications are involved with animal tracking, environmental monitoring, mainly in marine environment as well as pipeline monitoring, underwater navigation and wireless communication technologies. Current trend for underwater applications nowadays emphasize more on increasing the transmission data rate. This is due to actual acoustic or optical wireless transmission technologies do not achieve the target regarding both relative low cost and high data rates [7]. Hence, an omni-directional with high gain and high data rates of small SWB antenna operated underwater is required.

The proposed SWB antenna which demonstrated omni-directional radiation pattern with high gain for underwater (2.2 dB) and very wide bandwidth starting from 4.7 GHz until more than 15 GHz as well as compact size of 30 mm × 20 mm has met the entire requirement needed for an underwater antenna. In this study, the performance of custom-designed SWB antenna will be observed in free-space and in water. The proposed antenna applied stepped transmission lines since impedance matching is essential to obtain wider bandwidth of SWB system. In order to ensure 50 ohms characteristic impedance is received by the radiating element from the source while maintaining the SWB characteristic, the stepped impedance microstrip feeder is chosen. Theoretically, more steps present indicate better characteristic impedance yet feeder with excessive steps will increase the total electrical length and probably exceed the calculated size by resonant frequency. This might leads to additional complicated for antenna production. Two stepped impedance for transmission lines are commonly utilized for designing SWB antenna [17]. The air gap stacked using zinc as the reflector is deployed in this article. reflector plane is introduced at the back side of the antenna to reduce the back lobe. This reflector plane is used at the distance from the ground plane, which creates an air gap between the substrate and reflector [18]. The paper is organized as follows: In Section 2, single SWB antenna side feed with reflector is designed and investigated. The measurements setup for simulation and real measurement are included in Section 3 as well as the simulated and measurement results of S-parameter, gain, and pattern results. Finally, a conclusion will be drawn in Section 4 followed by acknowledgement in section 5. References would be the last section; section 6.

## 2. Antenna Design and Fabrication

Figure 1 demonstrates the structure of the proposed antenna. The dimension of the proposed antenna represented by Ls × Ws is considered as very compact in size. The radius of copper circular patch (Rp) with thickness of 0.035 mm is printed on top of the Taconic dielectric substrate. Taconic substrate (TLY-5) with a dielectric constant of $\varepsilon r = 2.2$, a thickness of $t = 1.5748+/-0.02$ and tangent loss of $\tan \delta = 0.0009$ has a relative permittivity, thickness and loss tangent of 4.7, 1.6 mm and 0.019 respectively. Meanwhile on the other side of the substrate, the copper partial ground covered the whole length of the microstrip feed line. Two additional of different dimension of microstrip lines known as stepped transmission lines are placed among the radiating patch and the feed line with different impedance characteristic as shown in Figure 1 (a). The capacitance between the ground plane and the patch could be influenced by the present of these lines in determining the impedance bandwidth and reflection coefficient values [7]. Figure 1 (b) shows a reflector consists of zinc with a thickness of 1mm and same dimension of the substrate is placed at the back of the antenna with a certain distance (D) works as the air gap stacked structure. The stacked structure functioned to reflect the back radiation towards the front. The antenna is fed by 50 Ohm SMA connector through side feeding. The feed line dimension is 3 mm in wide to perfectly match the antenna with 50 Ohm source feed. The equations to determine the width of the microstrip feed line in order to have characteristic impedance, Zo of 50Ω as follows [15]:

$$W_m = \frac{120\pi}{\sqrt{\varepsilon_r}} \quad (1)$$

wm = microstrip feed line width
h = substrate thickness
Ɛr = substrate dielectric constant,
Zo= input impedance

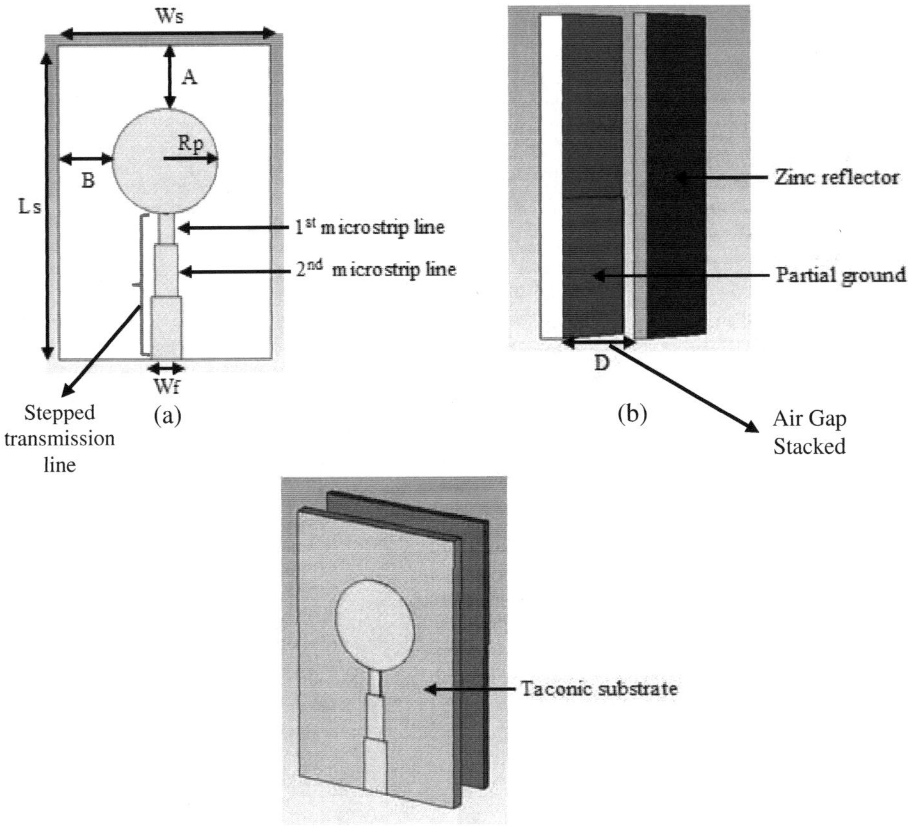


**Figure 1.** The simulated geometry of the proposed SWB array antenna, a) front view b) angled back view c) angled side view

The patch has characteristic of 7.5 mm in radius and 0.035 mm of thickness. Dimension of the circular patch is derived from the following equation [16]:

$$R = \frac{F}{\sqrt{1 + \frac{2h}{\pi \varepsilon_{eff} F[\ln(\frac{F\pi}{2h}) + 1.7726]}}} \quad (2)$$

$$F = \frac{8.791 \times 10^9}{fr\sqrt{\varepsilon_{eff}}}$$

The antenna is then fabricated on the Taconic TLY-5 with substrate thickness of 1.5748 mm, dielectric permittivity, $\varepsilon r$ of 2.2 and tangent loss, tan $\delta$ of 0.0009 as shown in Figure 2. Some important parameters of the designed antenna are optimized to obtain the best result in term of compact size, high gain and wide bandwidth. The optimized dimensions for the antenna are tabulated in Table 1. Measurements have been performed using the setup consisting of Agilent E83628 PNA Network Analyzer and Anechoic Chamber. The horn antenna is used as the transmitting antenna, whereas antenna under test (AUT), SWB array as the receiver. Both antennas are placed at DAUT about 0.84 m apart.

(a)

(b)

**Figure 2.** The fabricated SWB array antenna. a) Front view, b) Side View

**Table 1:** Antenna Parameter

| Parameter | Dimension |
|---|---|
| Substrate Length, $L_s$ | 20.0 mm |
| Substrate Width, $W_s$ | 30.0 mm |
| Patch Radius, $R_p$ | 5.0 mm |
| Stacked Air Gap, D | 5.0 mm |
| Ground Length, $L_g$ | 20.0 mm |
| Ground Width, $W_g$ | 13.0 mm |
| Feedline Width, $W_f$ | 3.0 mm |
| Patch gap with top side, A | 6.0 mm |
| Patch gap with left side, B | 5.0 mm |

## 3. Measurement Result and Discussion

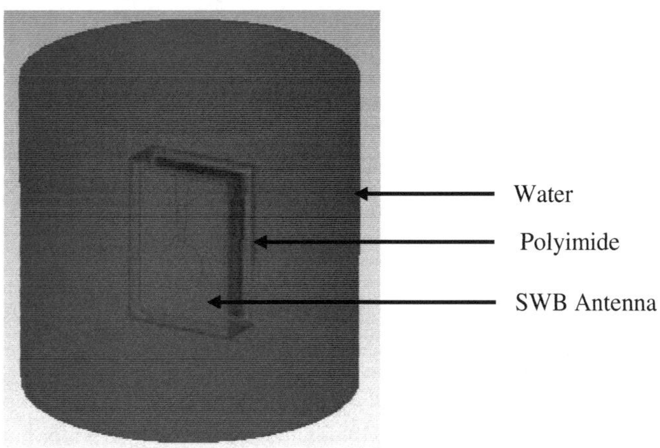


**Figure 3.** Simulation setup for SWB antenna performance in water

Figure 3 shows the simulation set up for evaluating the SWB antenna performance in water using CST software. The antenna is covered by polyimide as the casing during immersion in water. The casing has size of 30 mm x 45 mm with 15 mm thickness. The casing has hole inside in order to insert the SWB antenna. The level of the water exactly the same with the beaker used for the real measurement.

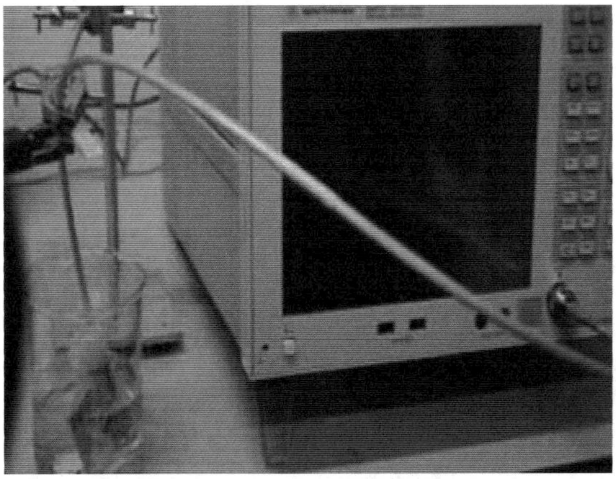

**(a)**

**(b)**

**Figure 4.** Measurement setup for a) reflection coefficient and b) radiation pattern

Figure 4 shows the measurement process was carried out in anechoic chamber with the help of Agilent Technologies ENA 8051C Network Analyzer to obtain the data for reflection coefficient and radiation pattern. The antenna covered by thin plain plastic is immersed in beaker contain 250 ml of water as shown in Figure 4(a).

In order to observe the performance of SWB antenna in different medium which is in water and in free-space, parameters that are important are reflection coefficient, gain, and radiated power. It is important also to identify the type of antenna's pattern whether directional or Omni-directional. Figure 5 shows the reflection coefficient for simulated and measured in both free space and plain water medium. Both simulated and measured in free space recorded wide bandwidth (<-10 dB) ranged from 3.7 GHz until more than 15 GHz and from 4.3 GHz until more than 15 GHz respectively. However, water medium degrades the bandwidth yet still within acceptable limit (<-6dB) [9-11]. The antenna recorded bandwidth from 4.5 GHz until more than 15GHz for simulated and from 5 GHz until more than 15 GHz for measured.

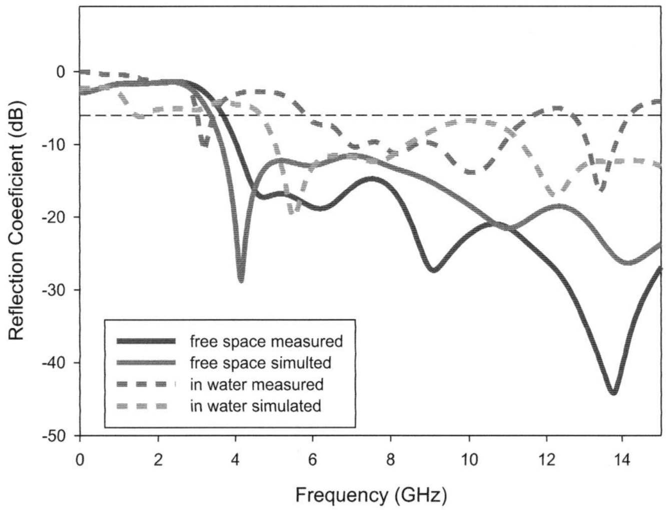


**Figure 5.** Measured and simulated reflection coefficient in free space and in plain water

For further investigation, the antenna is immersed in water with different level of 25%, 50%, 75% and 100% of immersion as shown in Figure 6. 25% of antenna immersion in water provides widest bandwidth followed by 50 %, 75% and 100% of immersion. However, fully immersion of antenna in water has been emphasize for further analysis in term of gain, reflection coefficient and radiation pattern. Electromagnetic wave propagation in water is totally different compared with propagation air since water has higher dielectric constant value. Rapid attenuation occurrence of the signals leads to degrade the SWB antenna performance as a whole [3]. Roughly, the simulated and measured recorded almost similar result in term of bandwidth wideness as shown in Figure 7.

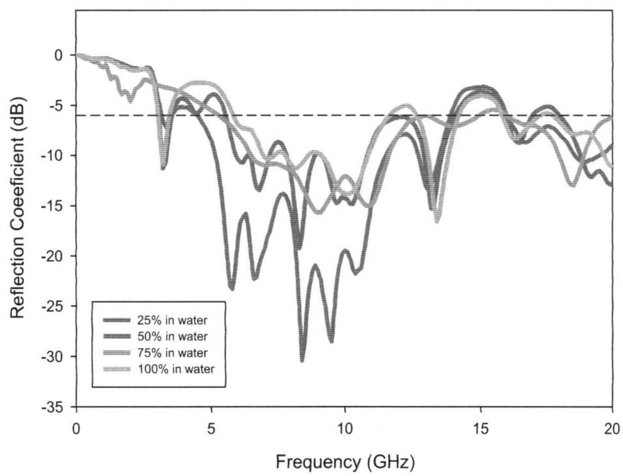


**Figure 6.** Measured reflection coefficient for immersion level

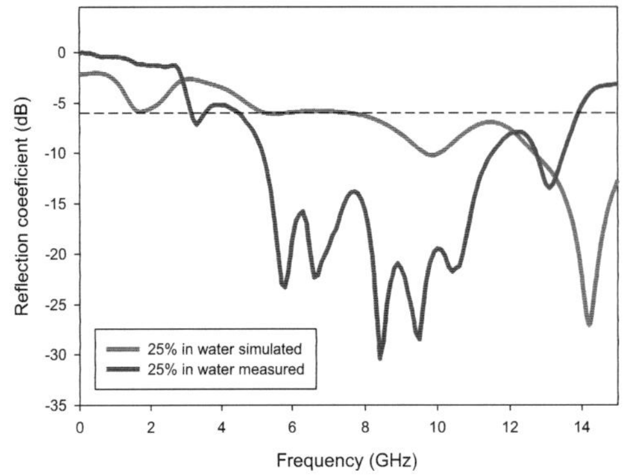


(a)

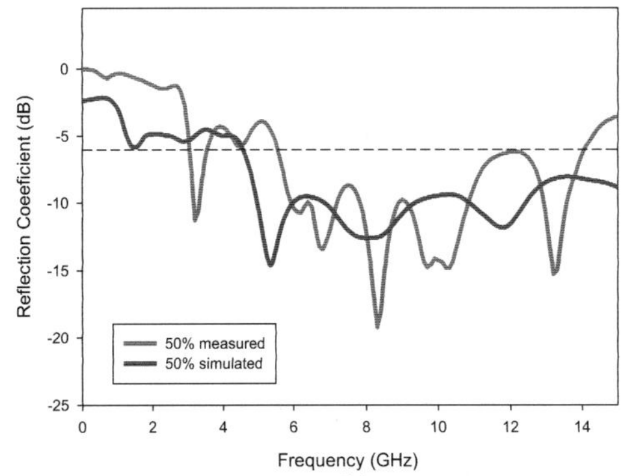


(b)

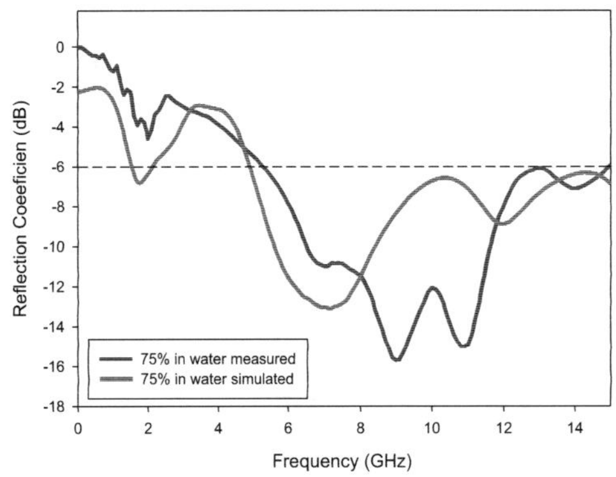


(c)

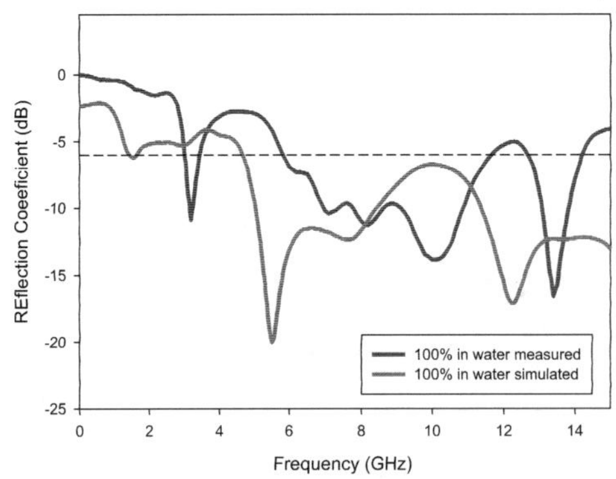


(d)

**Figure 7**. Reflection coefficient of difference immersion level; red (simulated), blue (measured) (a) 25% (b) 50% (c) 75% (d) 100%

The antenna gain in free space and plain water has been plotted in Figure 8. Both mediums show constant Increments throughout the frequency with lower gain values recorded for antenna immersed in water rather than antenna in free space. The maximum gain for antenna in free space is 6.2 dB meanwhile for antenna immersed in water, the maximum gain 2.2 dB. Table 2 summarizes the performance of the SWB antenna in term of gain and return loss.

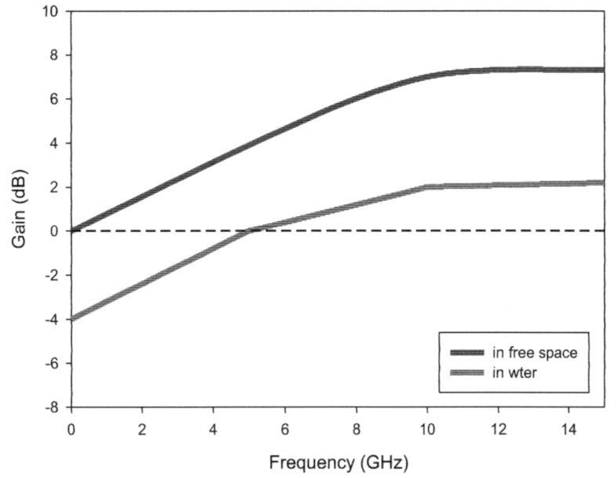


**Figure 8.** Graph of gain in free space and in water

**Table 2:** Measurement result of SWB antenna performance in free space and in water

| Frequency (GHz) | Gain (dB) | | Reflection Coefficient (dB) | |
|---|---|---|---|---|
| | Free space | Water | Free Space | Water |
| 1 | 0.5 | -3.9 | -1.66 | -2.96 |
| 5 | 4.5 | 0.1 | -16.94 | -9.6 |
| 10 | 7.2 | 2.0 | -22.38 | -6.73 |
| 15 | 7.5 | 2.2 | -26.87 | -13.06 |

Figure 9 demonstrate the radiation patterns for the proposed antenna both in free space and in water condition for selective frequencies. The figure clearly shows that the SWB antenna recorded omni-directional pattern for both medium with a little bit degradation for water medium. Omni-directional radiation pattern is very important to ensure underwater antenna radiate in all directions in order to have wider coverage for optimum data transmission.

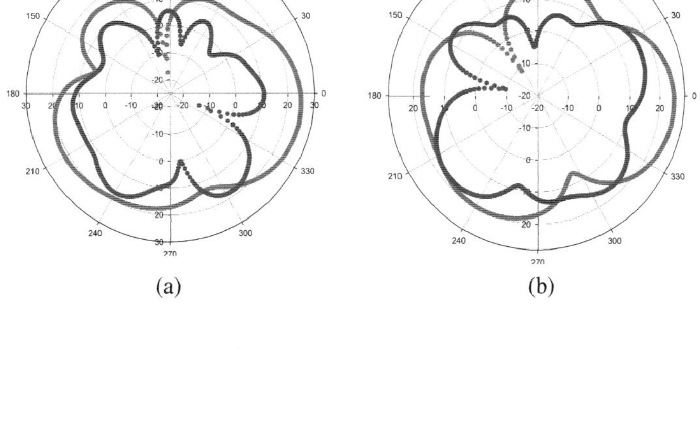

(a) (b)

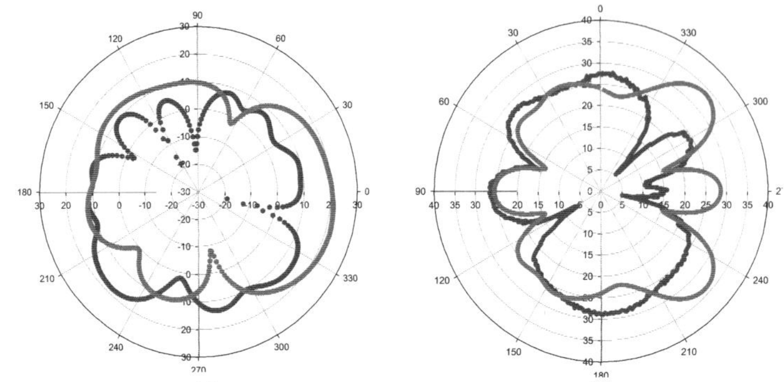

(c) (d)

**Figure 9.** In water (blue) and in free space (red) of SWB antenna radiation pattern of Azimuth-Plane a) 6GHz, b) 8GHz, c) 10GHz, d) 12GHz

The measured and simulated results of radiation pattern in water for 3 GHz, 5 GHz, 7 GHz and 9 GHz are shown in Figure 10. The radiation patterns show good performance across the whole SWB band with high degree of consistency [12]. The ripples in the radiation patterns may be due to the reflections into the field between the proposed (Rx) and horn antenna (Tx). The power loss in the RF cable that is applied in the measurement also contributes to these small scale variations [13, 14].

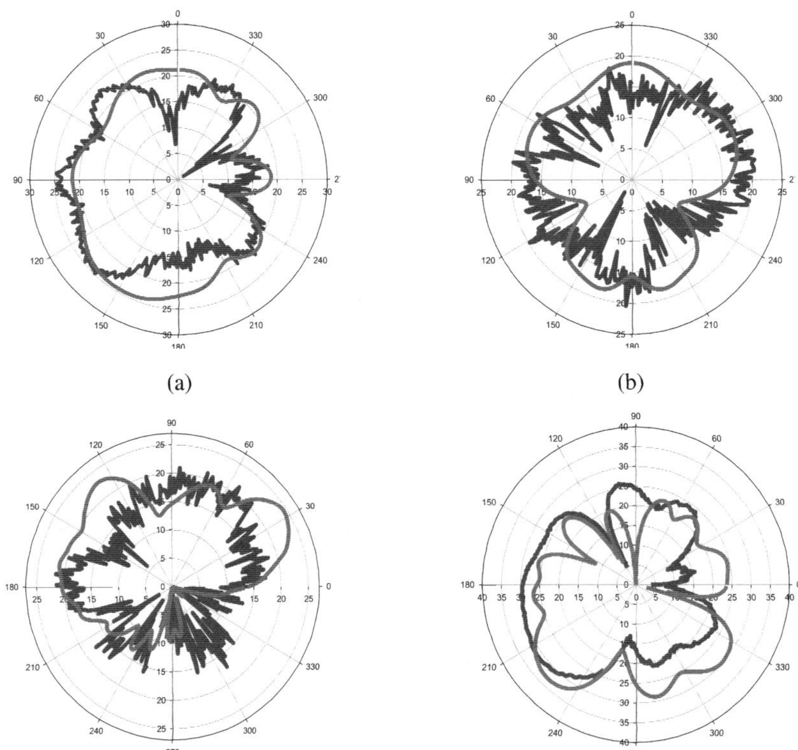

(a) (b)

(b) (d)

**Figure 10:** Comparison of Measured (blue) and Simulated (red) of SWB Array Antenna's Radiation Pattern of Azimuth-Plane in water a) 6GHz, b) 8GHz, c) 10GHz, d) 12GHz

## 4. Conclusion

A new design of compact Super Wideband (SWB) antenna for liquid application has been proposed. Eventhough the liquid medium degrades the proposed SWB antenna performance compared to free space, SWB antenna is still functional in water medium since it recorded good impedance matching (<-6dB) ranged within 4.7 GHz until more than 15 GHz with a maximum gain of 2.2 dB of omni-directional pattern. 25% of antenna immersion in water provides widest bandwidth followed by 50 %, 75% and 100% of immersion. In a nutshell, the antenna has great potential to be deployed in liquid medium for various related applications especially for biomedical related applications such as microwave imaging for cancer detection.

## 5. Acknowledgment

This work was supported by Excellent Grant Scheme (Grant No. 9009-0049) from University Malaysia Perlis, Malaysia

## 6. References

[1] Mohammad A. Dorostkar, Mohammad T. Islam and Rezaul Azim, "Design Of A Novel Super Wide Band Circular- Hexagonal Fractal Antenna", Progress In Electromagnetics Research, Vol. 139, 229-245, 2013.

[2] Xue Ni Low, ZhiNing Chen, and Terence S. P. See, "A SWB Dipole Antenna With Enhanced Impedance and Gain Performance", IEEE TRANSACTIONS ON ANTENNA AND PROPAGATIONS, VOL. 57, NO. 10, OCTOBER 2009.

[3] Giuliano Benelli and Alessandro Pozzebon, "RFID Under Water: Technical Issues and Applications", University of Siena, Department of Information Engineering, Siena, Italy.

[4] Mohd Ansor Bin Yusof and Shahid Kabir, "Underwater Communication Systems: A Review", Progress In Electromagnetics Research Symposium Proceedings, Marrakesh, Morocco, Mar. 20{23, 2011

[5] Kenneth P. Hunt, James J. Niemeier, Anton Kruger,"Short Paper: Antennas for Mussel-Based Underwater Biological Sensor Networks in Rivers", WUWNet'10, Sept. 30 – Oct 1, 2010, Woods Hole, MA, USA.

[6] A A Abdou, A Shaw1, A Mason, A Al-Shamma'a, J Cullen, S Wylie and M. Diallo," A matched Bow-tie antenna at 433MHz for use in underwater wireless sensor networks", Journal of Physics: Conference Series 450 (2013) 012048

[7] Hector Fabian GUARNIZO MENDEZ1,2, François LE PENNEC2, Christian GAC1, Christian PERSON2," High Performance Underwater UHF Radio Antenna Development", OCEANS 2011, June 6th -9th, Santander SPAIN

[8] Mohamed Nabil Srifi, "Ultra-Wideband Printed Antennas Design", Ultra Wideband Communications: Novel Trends – Antennas and Propagation, InTech, 2011.

[9] DaliaM. Elsheakh and Esmat A. Abdallah," Ultrawideband Vivaldi Antenna for DVB-T, WLAN, and WiMAX Applications", International Journal of Antennas and Propagation Volume 2014, Article ID 761634

[10] Ahmed B. Numan and Mohammad S. Sharawi, Ashley Steffes and Daniel N. Aloi," A Defected Ground Structure for Isolation Enhancement in a Printed MIMO Antenna System", 2013 7th European Conference on Antenna and Propagation (EuCAP)

[11] A. Cihangir, W. G. Whittow, C. J. Panagamuwa, F. Ferrero, G. Jacquemod, F. Gianesello, and C. Luxey, "Feasibility study of 4G cellular antennas for eyewear communicating devices," IEEE Antennas Wirel. Propag. Lett., vol. 12, no. DOI: 10.1109/LAWP.2013.2287204, pp. 1704 – 1707, 2013.

[12] N. Ojaroudi, M. Ojaroudi, F. Geran, and Sh. Amiri, "Omnidirectional Antenna with Multi-Resonance Performance for Microwave Imaging Systems", Telfor Journal, Vol. 5, No. 2, 2013.

[13] M. Jusoh1, M. F. Jamlos1, M. R. Kamarudin, F. Malek, M. H. Mat, and M. A. Jamlos, "A Novel Compact Tree-Design Antenna (Ncta) With High Gain Enhancement For SWB Application", J. of Electromagn. Waves and Appl., Vol. 25, 2474–2486, 2011

[14] SatyabrataMaiti, Amrit Mukherjee, AmlanDutta, "Design of a Novel Shaped Ultra Wideband Co-axial Feed Microstrip PatchAntenna", Fourth International Conference on Advanced Computing & Communication Technologies, 2014

[15] Amin M. Abbosh, "Directive Antenna for Ultrawideband Medical Imaging Systems", International Journal of Antennas and Propagation, 2008.

[16] K. A. Nayan, M. F. Jamlos, M. Jusoh, M. I. Jais, T. Sabapathy, "Adaptive Zero-Interference of Compact Reconfigurable Antenna for GPS Application", Wirelees Technology and Applications (ISWTA), pp 380-383, 2013.

[17] Xiaoming Zhu†, Xiaodong Yang* and Peng Chen, "Compact Band-notched SWB Antenna Design Based On Transmission Line Model", J Electr Eng Technol Vol. 9, No. 742, 2014

[18] N. S. Raghava, Asok De, "Effect of Air Gap width on the Performance of a Stacked Square Electronic Band Gap Antenna", International Journal Of Microwave And Optical Technology, Vol. 4, No. 5, September 2009

# Image Enhancement Based on Discrete Cosine Transforms (DCT) and Discrete Wavelet Transform (DWT): A Review

**Wan Azani Mustafa[1,a], Haniza Yazid[2,b], Wan Khairunizam[2,c], Mohd Aminuddin Jamlos[1,d], I Zunaidi[3], Z M Razlan[2], A B Shahriman[2]**

[1]Faculty of Engineering Technology, Universiti Malaysia Perlis, UniCITI Alam Campus, Sungai Chuchuh, 02100 Padang Besar, Perlis, Malaysia
[2] Advanced Intelligent Computing and Sustainability (AICoS) Research Group, School of Mechatronics Engineering, Universiti Malaysia Perlis, 02600 Arau, Malaysia
[3]Faculty of Technology, University of Sunderland, St Peter's Campus, Sunderland, SR6 0DD, United Kingdom

[a]*wanazani@unimap.edu.my*, [b]*hanizayazid@unimap.edu.my*, [c]*khairunizam@unimap.edu.my*, [d]*mohdaminudin@unimap.edu.my*

**Abstract:** Image enhancement is an important topic in image analysis in order to help humans and computer vision algorithms to obtain an accuracy information for analysis. The visual quality and certain image properties, such as brightness, contrast, signal to noise ratio, resolution, edge sharpness, and colour accuracy were improved through the enhancement process. The goal of image enhancement is to improve the quality of an image to become more suitable for a particular application. Till today, numerous image enhancement methods have been proposed for various applications and efforts have been directed to further increase the quality of the enhancement results and minimize the computational complexity and memory usage. In this paper, an image enhancement method based on Discrete Cosine Transforms (DCT) and Discrete Wavelet Transform (DWT) was studied. This paper presents an exhaustive review of these studies and suggests a direction for future developments of image enhancement methods. Each method shows the owned advantages and drawbacks. In future, this work will give the direction to other researchers in order to propose new advanced enhancement techniques.

## 1. Introduction

The image enhancement technique is to make the digital picture more appealing to our eyes, for example, making the images smooth or sharp [1–4]. This is an important topic in digital image processing. It can help humans and computer vision algorithms obtaining accurate information from the enhanced images. The visual quality and certain image properties, such as brightness, contrast, signal-to-noise ratio, resolution, edge sharpness, and color accuracy were improved through the enhancement process [5–9]. Recently, many image enhancement methods have been developed based on various digital image processing techniques and applications. They can be developed in the spatial domain or spatial-frequency domain. The enhanced image provides useful information for post-processing, especially in segmentation stage. Several survey papers on image normalization and segmentation methods have been presented in the literature and can be found in [10–13].

In the recent decades, Fourier Transform (Discrete Cosine Transforms (DCT) & Discrete Wavelet Transforms (DWT)) has been one of the major interesting research to enhance the images [14,15]. The discrete cosine transform (DCT) helps separate the image into parts (or spectral sub-bands) of differing importance (with respect to the image's visual quality). The DCT is similar to the discrete Fourier transform: it transforms a signal or image from the spatial domain to the frequency domain [16]. The discrete wavelet transform (DWT) is an implementation of the wavelet transform using a discrete set of the wavelet scales and translations obeying some defined rules. In other words, this transform decomposes the image into a mutually orthogonal set of wavelets [17]. In this paper, a comprehensive review of DWT and DCT was studied. The objective of this study is to analyze the advantages and drawbacks of each proposed method based on DWT and DCT.

## 2. Discrete Cosine Transforms (DCT)

One of the most significant current discussions in image enhancement is the Discrete Cosine Transforms (DCT) method. A DCT is another popular technique used because it is fast and effective. The basic function of DCT is to transform signals from the spatial representation into the frequency representation. Starting from 2006, Chen et al. [18] suggested a slight modification on Discrete cosine transforms (DCT) algorithm to improve the result performance under large illumination variations. They found that by discarding low-frequency DCT coefficients in the logarithm domain, the method capable of reducing the illumination variations effect significantly. The advantages of this method are no modelling step and bootstrap sets are required. The approach is fast and easy to be implemented in a real-time face recognition system. However, it is essential to solving a shadowing and the specularity problems because they lie in the same frequency band as some facial features.

In summary, according to an investigation by [18], the advantages using the DCT techniques are:

1) The DCT is real-valued instead of complexity (i.e., it involves magnitude and phase) such that it is easier to be implemented.
2) The DCT is more efficient for illumination variation estimation than the DWT.
3) The DCT approach is similar to the homomorphic filtering, which has been used for contrast enhancement.

Similarly, Vishwakarma et al. [19] considered the correlation of DCT in low frequency coefficients as the main parameter. This approach involved two steps: first, by using histogram equalization function. The original image was stretched and second, the low frequency of DCT coefficients was re-scaled to a lower value to compensate the illumination variations. The benefit of this technique is, it does not require recalculation of the DC coefficient for another difference illumination original image. In order to improve the previous method, [20] proposed a combination of Adaptive Histogram Equalization and DCT. Adaptive Histogram Equalization was used to stretch the contrast of the original image before applying the logarithm transform. Next, [21] again reported a new method based on a fuzzy filter that has been applied to the low frequency DCT (LFDCT) coefficient. This experiment involved two main techniques: Histogram Equalization (HE) and Contrast Limiting Adaptive Histogram Equalization (CLAHE) to find the background normalization. Based on the result, this method was efficient and successful compared to [20] and [19] methods. All the above approaches are able to perform in the real-time face recognition systems.

However, interestingly, this is contrary to a study conducted by [22] where focuses on Balanced Discrete Cosine Transfer (BDCT) to reconstruct the background image. The unbalanced background condition and low contrast were normalized before performing the segmentation process. The aim of the research is to overcome the drawback of DCT where reconstructing the background image in horizontal or vertical directions only. Based on the accuracy of segmentation result, the BDCT method achieved up to 25% better than the DCT method. Figure 1 shows the result of the BDCT method.

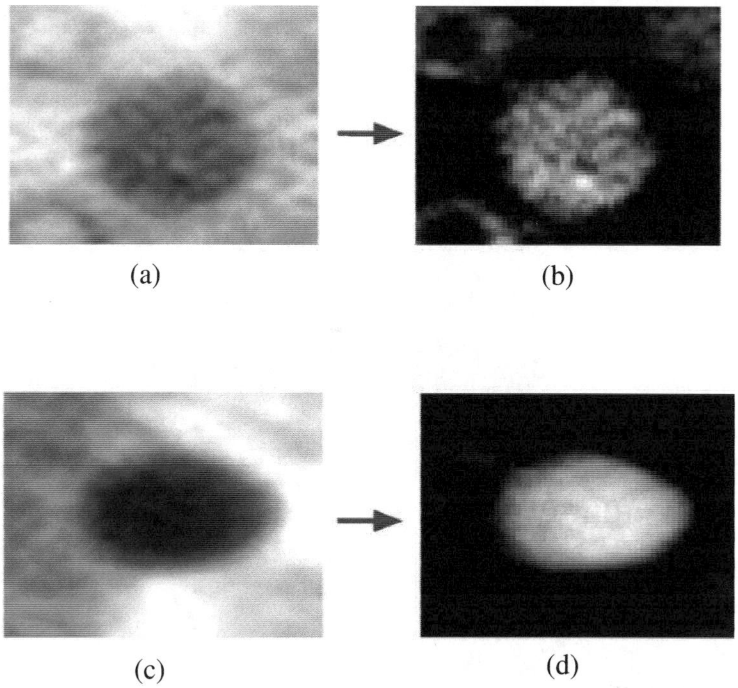

**Figure 1.** (a) and (c); The original 'Mura defect' image: (b) and (d); the resulting image after applying BDCT method [22].

In a different study, Lian, Er, & Li (2011) examined a low computational complexity illumination normalization method for face recognition application. Based on the noise estimation, the DCT coefficients were assumed as an illumination model to solve the background problem. The benefits of this approach are simple to implement and effective. However, in the future, the effect of the adaptive size selection of local area should be investigated. In a colour image, a new method for manipulating the DCT coefficients known as Multi Contrast Enhancement (MCE) was proposed by Xia, Panetta, & Agaian (2011). DCT was applied to produce an image with a few sub-bands based on the low and high frequency component. However, to improve the DCT function, alpha rooting and logarithmic enhancement function has been performed. The main advantage of this approach is simple, fast, and efficient for colour image application.

A study by Praveena & Venkatasrinu (2012) suggested a novel method of contrast enhancement for low contrast satellite images based on a combination of Singular Value Decomposition (SVD) and DCT. SVD was used to normalize the illumination and the output image was reconstructed using DCT. The technique was effective and produced a better performance compared to DWT-SVD technique [26], DWT technique, and Gamma Correction techniques. This finding is supported by [27] who proposed the technique using DCT and the Singular Value Decomposition (SVD). This technique process the image block by block. The DCT-SVD is not only robust but also effective in retaining the imperceptible property of the image.

In 2013, Goel et al. [28] presented is low frequency DCT coefficients in order to remove the illumination variation. This approach also performed contrast limiting adaptive histogram equalization (CLAHE) for optimum contrast enhancement. The above findings contradict the study by [14,28]. They argued that a combination work of the regularized histogram equalization (HE) and the DCT is more effective and comprehensive especially for remote sensing image compared to CLAHE technique. The HE was used to generate the distribution function of the original image and produced a new image with improved global contrast. Moreover, the method has a satisfactory computation time, which is suitable remote sensing and ordinary images enhancement. The comparison result of the enhancement methods is shown in Figure 2.

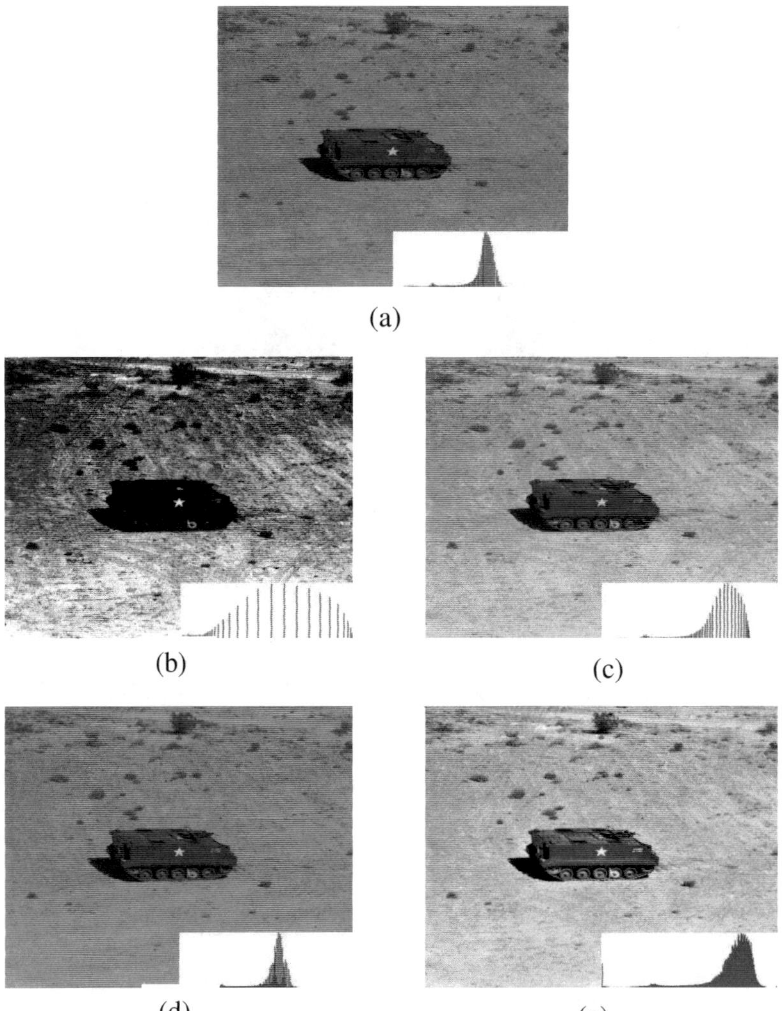


**Figure 2.** A comparisons on the Tank image: (a) Original image, (b) HE, (c) AGCWD [29], (d) DWT-SVD [26], and (e) Fu method [14].

## 3. Discrete Wavelet Transform (DWT)

Recently, wavelet transform technology has been widely applied in image processing and pattern recognition. Wavelet transforms can analyze an image in multi-resolution and detect small fluctuations at the multi-levels, they are good tools for contrast enhancement. A good contrast enhancement should be considered both a global and local information. An increasing amount of literature was discussed based on illumination normalization algorithms using wavelet-related methods were discussed, especially in face recognition application. In 2009, [30] published a paper which described a combination Multi-Scale Retinex (MSR) and DWT in order to enhance the colour images. The main purpose of MSR is to normalize the shadow effect and automatically improve the image quality. Finally, the DWT was used to fuse the sub-images into a single composite image that is more informative and suitable for visual perception.

In other research, [31] presented an improvement of wavelet algorithms. DWT was applied to get different band information and each band was processed separately. The low frequency was stretched using the HE and at the same time, the high frequency was denoised. In 2012, a 2D wavelet method was investigated by [32]. By focusing on low frequency components to find new coefficients, the result was good to normalize the shadows. In addition, this technique is simple and easy to be implemented in real-time face verification system. However, the weakness of this technique is a slow processing time.

In medical image application, [33] demonstrated a modification method of Undecimated Discrete Wavelet Transform (UDWT). The method used the hierarchical correlation of the coefficients of UDWT. The advantages of this approach are the algorithm is simple, fast, and produced better visual quality results. [34] found a significant relationship between Undecimated Wavelet Transform (UWT) and Wavelet Coefficient Mapping (WCM) technique to improve the previous method. UWT acted as a filtering to eliminate the noise in the original image and next, WCM was used to enhance the contrast. The method was tested on mammograms and chest radiograph images. Based on the assessment results, the approach is significantly superior, and faster compared the Undecimated Discrete Wavelet Transform (UDWT) [33] and sigmoid-type mapping [35]. This technique can be used not only for improving the visual quality of medical images, however also as a pre-processing module for computer-aided detection/diagnosis systems to improve the performance of screening and detecting regions of interest. Table 1 shows the comparison result of previous research.

**Table 1.** Comparison of image processing methods in terms of 4 quantitative quality metrics for mammograms [34].

| Method | Mean-To- Standard-Deviation (MSR) | Contrast Improvement Ratio (CIR) | Peak Signal-To-Noise Ratio (PSNR) |
|---|---|---|---|
| Sigmoid [35] | 6.09 | 0.29 | 36.39 |
| UDWT [33] | 5.80 | 0.28 | 38.35 |
| UWT + WCM [34] | 6.24 | 0.67 | 37.98 |

A different opinion was expressed by [36] which suggested that a combination technique of Stationary Wavelet Transform (SWT), Singular Value Decomposition (SVD), and Discrete Wavelet Transform (DWT) in order to enhance the contrast. The original image was divided into four sub-bands to process separately before reconstructing a high-resolution image using IDWT (Inverse Discrete Wavelet Transforms). Based on PSNR, the result was superiority compared to the conventional techniques such as DWT, Bilinear, and Bicubic. Similar findings were reported by [37]. They proposed enhancement technique based on DWT and

Singular Value Decomposition (SVD) for fungus images application. The procedures almost similar to the method proposed by [36], however, this approach focuses on modification of low sub-bands coefficients. This approach produced a better result compared to HE, CLAHE, and AHE. The research by [37] was supported by [38] where a contrast enhancement method based on Discrete Wavelet Transform (DWT) and Singular Value Decomposition (SVD) was proposed. Therefore, they improved the wavelet function to become productive known as Dual-Tree Complex Wavelet Transform (DTCWT). The main purpose of DTCWT is to find the high frequency sub-bands, however, it introduces the new artefacts. The Non-Local Means (NLM) filter was applied in order to eliminate the artefacts. Finally, the Inverse DTCWT was used to merge all the sub-bands image. They also described a combination technique to enhance the contrast of the satellite image. The combination involved Resolution Enhancement (RE), Dual-Tree Complex Wavelet Transform (DTCWT) and Non-Local Means (NLM) [39]. The process and flow almost similar to the above method [38], however, this approach explained more on the Resolution enhancement (RE). The Lanczos interpolator used in order to interpolate the low-resolution image and high frequency sub-bands. The resulting performance of a few techniques is shown in Figure 3.

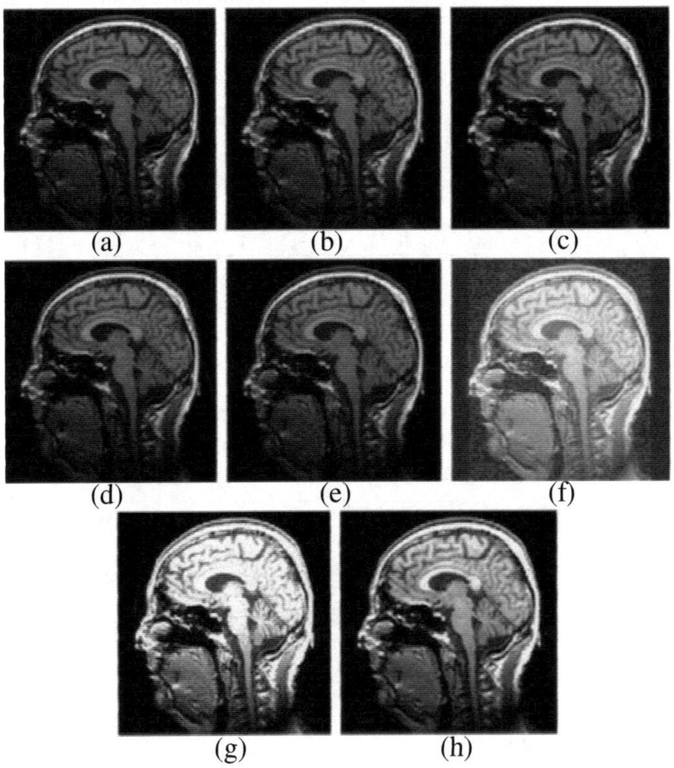


**Figure 3.** The comparison result: (a) Original image 'Brain MRI', (b) SWT-RE [40], (c) DWT-RE, (d) DTCWT-RE, (e) DTCWT-NLM-RE [39], (f) GHE, (g) SVD-based CE [41], and (h) DTCWT-NLM-SVD [38].

In 2011, [42] conducted numerical experiments on face images under different lighting conditions. A new novel method known as multi-scale dual-tree complex wavelet transform (DT-CWT) was proposed in order to reduce the halo artefacts in the non-uniform image. In a different study, [43] found a new method using knee transfer function and gamma correction based on Discrete Wavelet Transform (DWT). The procedure dealt with four frequency components which are Lowpass-Lowpass (LL), Lowpass-Highpass (LH), Highpass-Lowpass (HL), and Highpass-Highpass (HH). The outcome is effective to improve the image quality, especially for low contrast images.

In the pre-processing process, [44] demonstrated that illumination and contrast correction using DWT by applying a few steps: (1) the original image was decomposed into two components, which are low frequency and high frequency components, (2) the two components were processed separately through contrast enhancement to eliminate the effect of illumination variations and enhance the detailed of edge information, and (3) the normalized image was corrected using the Inverse Discrete Wavelet Transform (IDWT). Based on computation time results, this method performed faster compared to other methods such as Multi-Scale Retinex (MSR) [30], logarithm domain (LTV) [45], and anisotropic diffusion based smoothing technique [46].

## 4. Conclusion

One major area of digital image processing is image enhancement. The main objective of image enhancement is to improve the quality of images emphasize wanted features and make them less obscured. Generally, image enhancement can be divided into two categories which are spatial domain and frequency domain. This study discusses an overview of the background and related work in the area of image enhancement based frequency domain (DWT and DCT techniques). Based on the review, the frequency domain technique can analyze an image in multi-resolution and detect small fluctuations at the multi-levels, they are a good technique for contrast enhancement. However, the DWT technique shows a slow processing time and complex implementation. In addition, the DWT has a problem to deal with higher noise and badly non-uniform image. Besides, the DCT has a high computational cost and complicated in term of selecting suitable transfer functions or parameters. A serious drawback of this Fourier transforms approach is the algorithm does not perform perfectly to enhance the image in an extreme illumination and has a problem to deal with the high noise level.

## 5. Acknowledgment

This work was supported by Ministry of Higher Education Malaysia under the Fundamental Research Grant Scheme (FRGS/1/2018/SKK13/UNIMAP/02/1)

## 6. References

[1] Mustafa W A and Yazid H 2016 Illumination and Contrast Correction Strategy using Bilateral Filtering and Binarization Comparison *J. Telecommun. Electron. Comput. Eng.* **8** 67–73

[2] Mustafa W A and Yazid H 2017 Image Enhancement Technique on Contrast Variation : A Comprehensive Review *J. Telecommun. Electron. Comput. Eng.* **9** 199–204

[3] Mustafa W A and Kader M M M A 2018 Contrast Enhancement Based on Fusion Method : A Review *J. Phys. Conf. Ser.* **1019** 1–7

[4] Kanafiah S N A M, Mashor M Y, Mustafa W A and Mohamed Z 2018 A Novel Contrast Enhancement Technique Based on Combination of Local and Global Statistical Data on Malaria Images *J. Biomimetics, Biomater. Biomed. Eng.* **38** 23–30

[5] Mustafa W A and Yazid H 2017 Contrast and Luminosity Correction Based On Statistical Region Information *Adv. Sci. Lett.* **23** 5383–6

[6] Gonzalez R C and Woods R E 2008 *Digital Image Processing* (Upper Saddle River, NJ, USA: Prentice Hall)

[7] S. Sharma 2013 *Fundamentals of Digital Image Processing* (New Delhi: S.K. Kataria & Sons)

[8] Mustafa W A, Yazid H and Yaacob S 2014 Illumination Normalization of Non-Uniform Images Based on Double Mean Filtering *IEEE International Conference on Control Systems, Computing and Engineering* pp 366–71

[9] Mustafa W A, Yazid H and Yaacob S 2014 A Review : Comparison Between Different Type of Filtering Methods on the Contrast Variation Retinal Images *IEEE International Conference on Control System, Computing and Engineering* pp 542–6

[10] Reza Ghabousian and Nooshin Allahbakhshi 2015 Survey of Contrast Enhancement Techniques based on Histogram Equalization *Int. J. Rev. Life Sci.* **5** 901–8

[11] Jain M K and Arya I B 2014 A Survey of Contrast Enhancement Technique for Remote Sensing Images *Int. J. Electr. Electron. Comput. Eng.* **3** 1–6

[12] Singh S and Sharma S 2014 A Survey of Image Enhancement Techniques *Int. J. Comput. Sci.* **2** 1–5

[13] Mustafa W A, Yazid H and Yaacob S 2015 Illumination Correction of Retinal Images Using Superimpose Low Pass and Gaussian Filtering *International Conference on Biomedical Engineering (ICoBE)* pp 1–4

[14] Fu X, Wang J, Zeng D, Huang Y and Ding X 2015 Remote Sensing Image Enhancement Using Regularized-Histogram Equalization and DCT *IEEE Geosci. Remote Sens. Lett.* **12** 2301–5

[15] Rubel A, Lukin V, Uss M, Vozel B, Pogrebnyak O and Egiazarian K 2016 Efficiency of texture image enhancement by DCT-based filtering *Neurocomputing* **175** 948–65

[16] Tang J 2004 A contrast based image fusion technique in the DCT domain *Digit. Signal Process. A Rev. J.* **14** 218–26

[17] Amolins K, Zhang Y and Dare P 2007 Wavelet based image fusion techniques - An introduction, review and comparison *ISPRS J. Photogramm. Remote Sens.* **62** 249–63

[18] Chen W, Er M J and Wu S 2006 Illumination Compensation and Normalization for Robust Face Recognition Using Discrete Cosine Transform in Logarithm Domain *IEEE Trans. Syst. Man, Cybern. B* **36** 458–66

[19] Vishwakarma V P, Pandey S and Gupta M N 2007 A Novel Approach for Face Recognition Using DCT Coefficients Re-scaling for Illumination Normalization *International Conference on Advanced Computing and Communications.* pp 535–9

[20] Vishwakarma V P, Pandey S and Gupta M N 2009 Adaptive histogram equalization and logarithm transform with rescaled low frequency DCT coefficients for illumination normalization *Int. J. Recent Trends Eng. Technol.* **1** 318–22

[21] Vishwakarma V 2013 Illumination normalization using fuzzy filter in DCT domain for face recognition *Int. J. Mach. Learn. Cybern.* 1–18

[22] Chen L C, Chien C H and Nguyen X L 2013 An effective image segmentation method for noisy low-contrast unbalanced background in Mura defects using balanced discrete-cosine-transfer (BDCT) *Precis. Eng.* **37** 336–44

[23] Lian Z, Er M J and Li J 2011 A Novel Local Illumination Normalization Approach for Face Recognition *Adv. Neural Networks – ISNN 2011* 350–5

[24] Xia J, Panetta K and Agaian S 2011 Color image enhancement algorithms based on the DCT domain *IEEE International Conference on Systems, Man, and Cybernetics* pp 1496–501

[25] Praveena G and Venkatasrinu M 2012 A Modified SVD-DCT Method for Enhancement of Low Contrast Satellite Images *Int. J. Comput. Eng. Res.* **2** 2250–3005

[26] Demirel H, Ozcinar C and Anbarjafari G 2010 Satellite image contrast enhancement using discrete wavelet transform and singular value decomposition *Geosci. Remote Sens. Lett. IEEE* **7** 333–7

[27] Mukherjee S and Pal A K 2012 A DCT-SVD based robust watermarking scheme for grayscale image *International Conference on Advances in Computing, Communications and Informatics* pp 573–8

[28] Goel T, Nehra V and Vishwakarma V P 2013 Rescaling of low frequency DCT coefficients with Kernel PCA for illumination invariant face recognition *International Advance Computing Conference* pp 1177–82

[29] Huang S C, Cheng F C and Chiu Y S 2013 Efficient contrast enhancement using adaptive gamma correction with weighting distribution *IEEE Trans. Image Process.* **22** 1032–41

[30] Zhang X 2009 A Novel Color Image Fusion Method Based on the Multi-scale Retinex and DWT *International Conference on Information Management and Engineering* pp 395–8

[31] Fan C N, Wang H bin and Zhang F Y 2009 Improved Wavelet-based Illumination Normalization Algorithm for Face Recognition *International Conference on Information Science and Engineering (ICISE2009)* vol 1 pp 583–6

[32] Emadi M, Khalid M, Yusof R and Navabifar F 2012 Illumination normalization using 2D wavelet *Procedia Engineering* vol 41 pp 854–9

[33] Matsuyama E, Tsai D-Y, Lee Y, Tsurumaki M, Takahashi N, Watanabe H and Chen H-M 2013 A modified undecimated discrete wavelet transform based approach to mammographic image denoising. *J. Digit. Imaging* **26** 748–58

[34] Tsai D-Y, Matsuyama E and Chen H-M 2013 Improving image quality in medical images using a combined method of undecimated wavelet transform and wavelet coefficient mapping. *Int. J. Biomed. Imaging* **2013** 797–924

[35] Lee Y, Tsai D Y and Suzuki T 2008 Contrast enhancement of medical images using sigmoid-type transfer curves for wavelet coefficient weighting adjustment *Med. Imaging Inf. Sci.* **25** 48–53

[36] Gunaseelan K and Seethalakshmi E 2013 Image Resolution and Contrast Enhancement Using Singular value and Discrete wavelet Decomposition *J. Sci. Ind. Res.* **72** 31–5

[37] G.Tirupati and Rao B S 2014 A Novel Technique for Fundus Image Contrast Enhancement *Int. J. Bio-Medical Informatics e-Health,* **2** 1–6

[38] Iqbal M Z, Ghafoor A, Siddiqui A M, Riaz M M and Khalid U 2014 Dual-tree complex wavelet transform and SVD based medical image resolution enhancement *Signal Processing* **105** 430–7

[39] Iqbal M Z, Ghafoor A and Siddiqui A M 2013 Satellite image resolution enhancement using dual-tree complex wavelet transform and nonlocal means *IEEE Geosci. Remote Sens. Lett.* **10** 451–5

[40] Lee W L, Yang C C, Wu H T and Chen M J 2009 Wavelet-based interpolation scheme for resolution enhancement of medical images *J. Signal Process. Syst.* **55** 251–65

[41] Demirel H, Anbarjafari G and Jahromi M N S 2008 Image equalization based on singular value decomposition *International Symposium on Computer and Information Sciences* pp 1–5

[42] Hu H 2011 Multiscale illumination normalization for face recognition using dual-tree complex wavelet transform in logarithm domain *Comput. Vis. Image Underst.* **115** 1384–94

[43] Bhandari A K, Kumar A and Singh G K 2015 Improved knee transfer function and gamma

correction based method for contrast and brightness enhancement of satellite image *AEU - Int. J. Electron. Commun.* **69** 579–89

[44] Zhoa X, Lin Y, Ou B and Yang J 2015 A Wavelet-Based Image Preprocessing Method for Illumination Insensitive Face Recognition *J. Inf. Sci. Eng.* 1711–31

[45] Chen T, Yin W, Zhou X S, Comaniciu D and Huang T S 2006 Total variation models for variable lighting face recognition *IEEE Trans. Pattern Anal. Mach. Intell.* **28** 1519–24

[46] Gross R and Brajovic V 2003 An image preprocessing algorithm for illumination invariant face recognition *Audio Video Based Biometric Pers. Authentication* 10–8

IOP Conf. Series: Materials Science and Engineering **557** (2019) 012028 doi:10.1088/1757-899X/557/1/012028

# Performance analysis of Software Defined Network (SDN) in link failure scenario

**M A I M Sakari[1], N Yaakob[1], A Amir[1], R B Ahmad[1,2], M N M Warip[1] and Z Ibrahim[3]**

[1]School of Computer and Communication Engineering (SCCE), Universiti Malaysia Perlis (UniMAP), 01000 Arau Perlis
[2]Faculty of Informatics and Computing, Universiti Sultan Zainal Abidin (UniSZA), Besut Campus, 22200 Besut Terengganu
[3]Technopreneur at UniMAP Sdn Bhd, Blok D, Bangunan PPPIT UniMAP, Jalan 3 Pengkalan Asam, 01000 Kangar Perlis

**Abstract.** The proliferation and advancement in network technologies has brought unprecedented opportunities towards the development of Software Defined Network (SDN) which presents an alternative network architecture that separates data forwarding and control functions to simplify global network management. As SDN is developed to handle large-scale network, addressing network dynamics such as link failure remains challenging. Link failure in SDN could still cause major performance disruption which can lead to severe performance degradation for underlying applications. Therefore, this paper analyses several link failure circumstances in SDN and show how it reacts to such situations. As SDN decouples data and control planes, routing algorithm is determined by its controller. In this paper, routing algorithm is used to control the behaviour of the network traffic when one or more links are disconnected. Ryu controller which is one of the open controllers in SDN, acts as a brain to adapt how traffic is controlled for the whole network. Performance of SDN in different network topologies is compared in terms of throughput and Round-Trip-Time (RTT) which shows desirable SDN performance.

## 1. Introduction

Ensuring uninterrupted Internet services and providing high Quality-of-Services (QoS) is becoming burgeoning demand in fulfilling industry-specific regulations. This alternative is towards improving future Internet and service mobility to enhance overall user experience. As such, networking protocols have evolved significantly over the last few decades. However, traditional network configuration is time-consuming and error-prone due to manual configuration by network administrator. Moreover, the traditional network configurations also complicates network segmentation where all devices are placed in the same 'zone'. Such configuration faces high risk of route diversion during link failure scenarios. Figure 1 shows a traditional network that is made up of packets switches and several clients connected through Ethernet connection. This design complicates administrative task for consistent policies deployment. As a result, organizations are more likely to encounter some failures in network links. To resolve this issue, a new paradigm on network management which is called Software Defined Networking (SDN) has been introduced as shown in Figure 2 [1].

SDN is a new networking architecture which is proposed to overcome the limitations in the traditional network. The aim of SDN is to ease network management and allow network administrators to quickly react on dynamic topology change such as link failure [2]. In addition, SDN introduces new networking architecture which decouples network controller and data transmission functions which make it directly programmable. Due to its promising features, SDN offers a cost-effective, dynamic, manageable and adaptable networking, making it well suited to handle dynamic nature of today's applications [3]. Moreover, the adaptable nature in SDN architecture makes it very convenient for network administrators to handle large and unprecedented program, automation and fully control of the network.

IOP Conf. Series: Materials Science and Engineering **557** (2019) 012028 doi:10.1088/1757-899X/557/1/012028

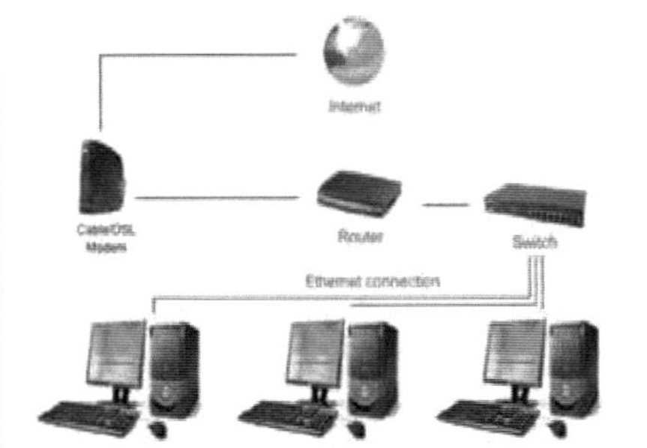


**Figure 1.** Traditional Network

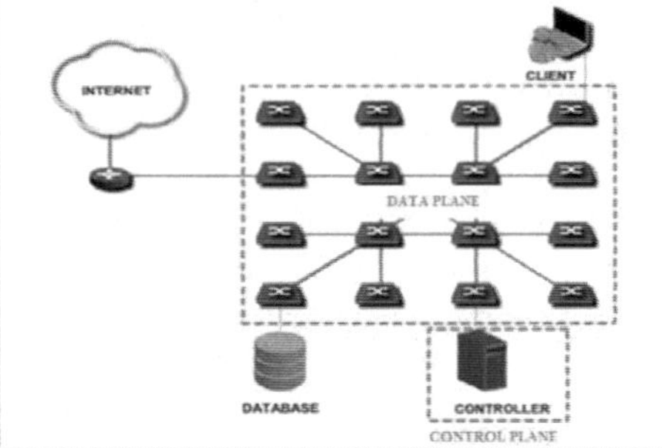


**Figure 2.** Architecture of SDN

This kind of network will also allow better network management and optimization of network resources. Figure 2 shows the connection between client and database server in SDN which consists of control plane (controller) and data plane. Unlike the traditional networking, SDN can simulate link failure scenarios without affecting the network architecture.

Link failure is one of the common problems in both the traditional network as well as SDN [4]. This problem is sometimes inevitable especially in network with large data centre. In SDN, a link failure in network architecture happens when switches are disconnected with controller or during switch replacement. Link failure scenario could generate large amount of unexpected traffic in the network due to the rerouting and retransmission of packets [5]. In link failure scenario, a network administrator in SDN handles traffics from a centralized controller without having to look at real network switches. However, link failure in SDN will still cause major performance disruption which can severely affect the underlying applications. Nevertheless, the performance of traditional network in link failure in terms of packet loss, throughput, end-to-end delay, and packet delivery ratio will be much affected compared to SDN. This is due to difficulties in identifying and analysing the problem in link failure scenarios due to large number of routers over the Internet and variations in network topology. This paper investigates SDN's performance and reaction in several link failure scenarios and topologies. Performance analysis of several link failure scenarios in SDN have been simulated and analysed using Mininet environment and Ryu controller.

## 2. Simulation Design and Testing Environment

To simulate SDN performance in link failure scenarios, simulation environment have been designed in several topologies; basic and custom. The basic topology includes the single and linear topology, while custom topology consists of triangle and fat-tree design. Each topology is simulated with OpenFlow-enabled network architecture using Mininet simulator. The default topology is minimal topology which is predefined with one OpenFlow kernel switch connected to two hosts and OpenFlow reference controller, whereas number of switches and hosts can be changed for other topologies using the command-line interface (CLI).

### *2.1. Single Topology*

Figure 4 shows a single topology which consists of a single OpenFlow-enabled switch connected with multiple hosts. The switch is connected to the OpenFlow controller via a secure channel. A linear topology with 8 hosts connected with its own switch in linear fashion is clearly shown in Figure 3. All switches are interconnected with each other and are in turn connected with OpenFlow controller.

### *2.2. Linear Topology*

Linear topology requires n number of switches for n number of hosts. In other words, each host will be connected with their respective switch as shown in Figure 4. For example, host H1 will connect with switch S1, host H2 with switch S2. All the switches are connected with one another which in turn connected with a common controller. A linear topology in this paper is designed to have 8 hosts and thus 8 switches.

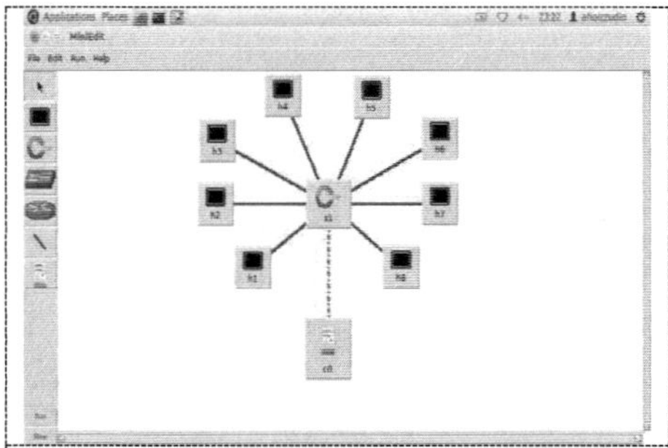

**Figure 3.** OpenFlow-based Single Topology having 8 Hosts

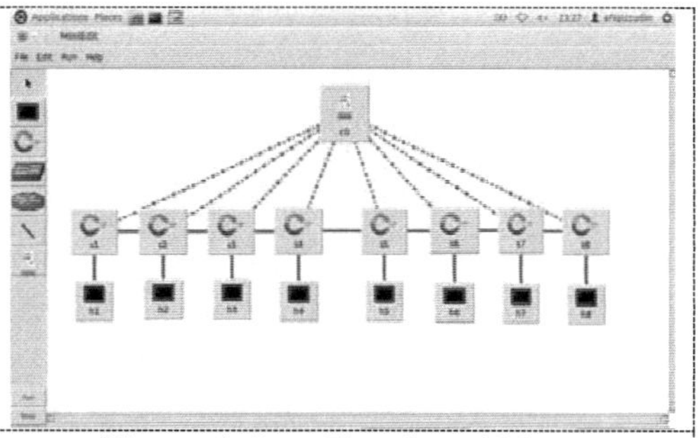

**Figure 4.** OpenFlow-based Linear Topology having 8 Hosts

### 2.3. *Triangle Topology*

Triangle topology consists of three switches which are interconnected with each other. As shown in the Figure 5, only switch S1 has two hosts, the other 2 switches (*S2* and *S3*) has 3 hosts. All IP addresses is changed from the default set by Mininet in order to resemble real-world networks. Each switch will have its own default and static route.

### 2.4. *Fat Tree Topology*

Fat tree topology is one of data center topology that provides efficient communication between hosts. This custom topology is using multipath routing with 4-pod fat tree as shown in Figure 6.

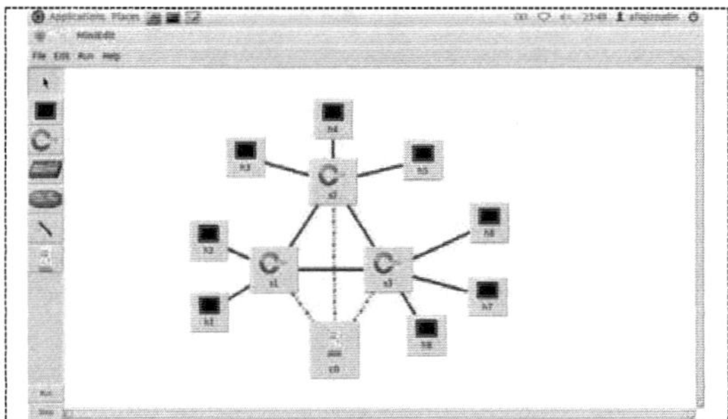

**Figure 5.** OpenFlow-custom Triangle Topology having 8 Hosts

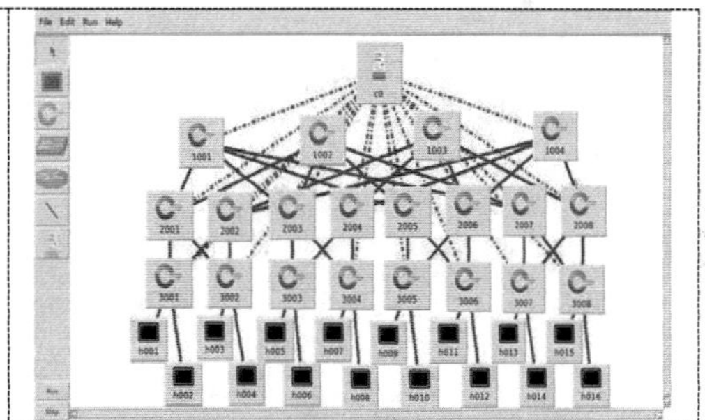


**Figure 6.** OpenFlow-custom Fat Tree Topology with 4 pods having 16 Hosts

The hierarchical layers of the network includes the core, aggregation, edge switch and hosts. Each switch has been renamed to differentiate switch for each layer. Assuming that the controller is at the top of the layer, then the 4 core switches are numbered from 1001~1004. Both second layer (aggregation switch) and third layer (edge switch) has 8 switches starting from 2001~2008 and 3001~3008 respectively. Hosts are numbered from h001~h016 and IP addresses are ranging from 10.1.0.1~10.8.0.2.

## 3. Link Failure Scenario Testing

Two methods are used to simulate link failures in this paper. The first method is by disabling a specific link interconnected between switches or hosts and the second method is to completely shut down all switches which disable communication between hosts in the network. *Ping* tests were run inside Mininet CLI to test the connection in each scenario. The connection was tested between the first host and last host in each of topology. The performance are measured in terms of Round Trip Time (RTT), number of packet transmit and receive, percentage of packet loss and transmission time. Figure 7 and Figure 8 show examples of link failure scenario in triangle and combination scenarios respectively.

IOP Conf. Series: Materials Science and Engineering **557** (2019) 012028 doi:10.1088/1757-899X/557/1/012028

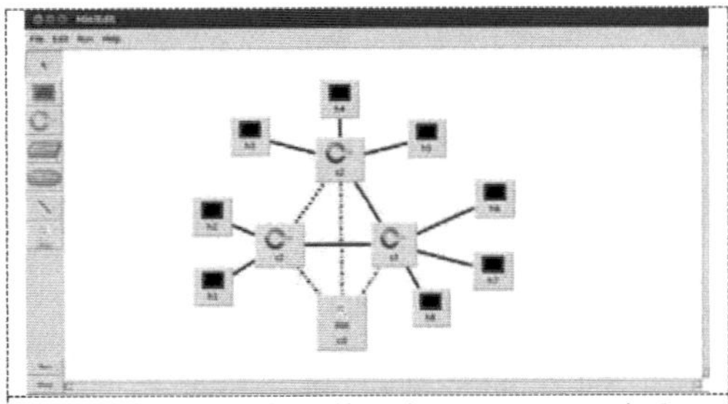

**Figure 7.** Link Failure between s1 and s2

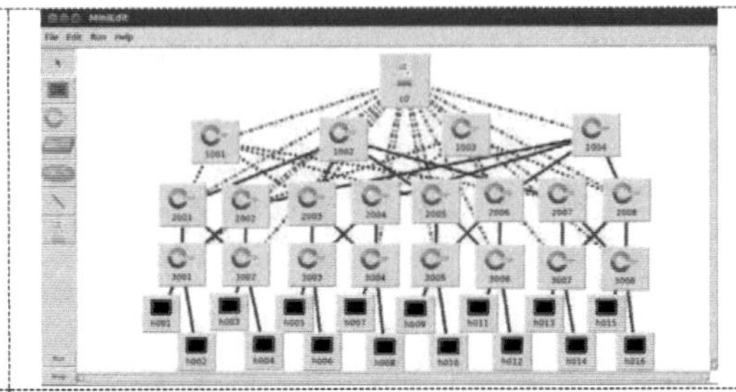

**Figure 8.** Example of Switch Failure in fat-tree

## 4. Performance Analysis

A comparative performance analysis of a single, linear, triangle and fat-tree topology were conducted in Mininet simulator. The basic topologies are simulated with fully working link since a link failure in such network will cause a major breakdown. Hence, performance analysis for customs networks is done by comparing all network topologies on the basis of packet transmission rate, Round-Trip-Time (time required to transmit packet from source node to destination node) and throughput.

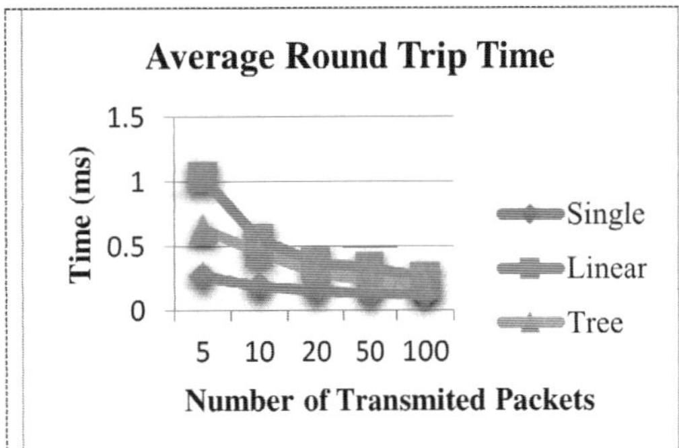


**Figure 9.** Average RTT for Single, Linear and Tree Topology

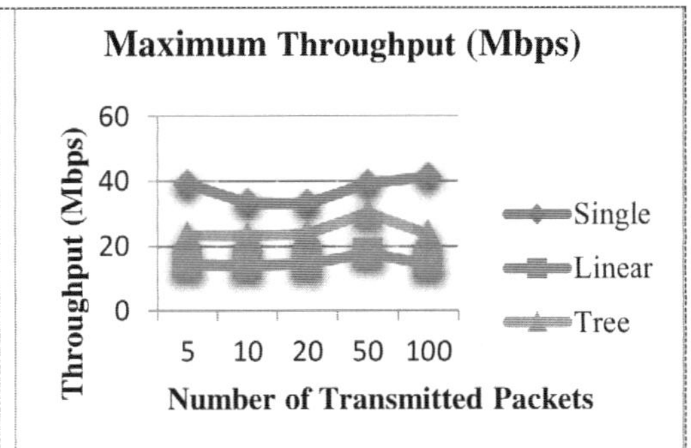


**Figure 10.** Maximum Throughput Chart for Single, Linear and Tree Topology

Figure 9 shows the average RTT for basic network topologies of single, linear and tree topologies while Figure 10 shows the resulted throughput for the same topologies. From the figure, it is clearly shown that a linear topology is taking more time for transmission of packet as compared to single and tree topology. This is due to the increase in the number of hops between end nodes thus resulting in longer propagation time for intermediate nodes to deliver packet to destination.

However, single topology is taking minimum amount of time to deliver packets to its destination since all nodes are connected with a single OpenFlow-enabled switch. As such, packet transmission is faster in single topology. Figure 10 shows the maximum throughput with respect to packet transmission rate. The results shows that linear topology obtained the lowest throughput compared to the other two topologies because the bandwidth utilization is less and the overall round-trip propagation delay between nodes is higher. Whereas, single topology network is having maximum throughput since the single topology only have one switch and between any two nodes in a network consists of only two hops. Thus, single topology incur less delay and higher throughput whereas linear topology incur more delay and less throughput.

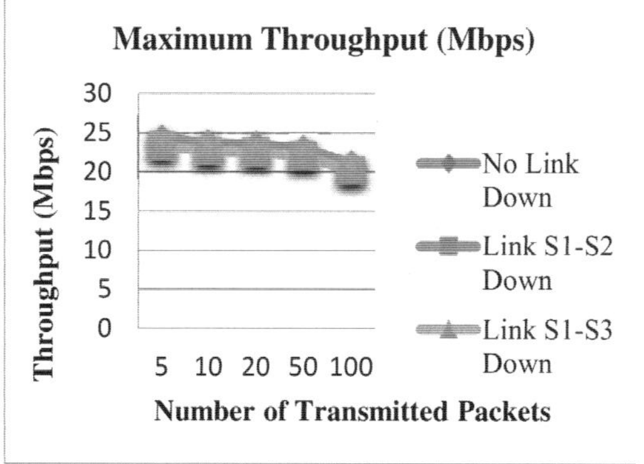


**Figure 11.** Maximum Throughput for Triangle Topology

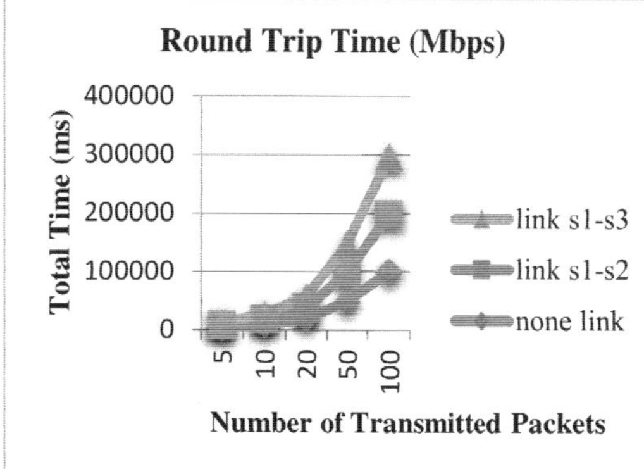


**Figure 12.** RTT for Triangle Topology

Maximum throughput and packet transmission rate is as shown in Figure 11 and Figure 12 respectively. Figure 11 shows the throughput comparison in 3 scenarios. As shown, the throughput when *s1-s3* link is down in triangle topology is very less as compared to other two topologies. This is because the bandwidth utilization is less and thus the overall round-trip propagation delay between nodes is higher. Obviously, when no link is down, maximum throughput is achieved since the network has no failure so all packets can be transmitted. Referring to Figure 12, total time taken for *h1* to transmit packets to *h8* when link *s1-s3* is down is more than when there is no link failure. When link *s1-s3* fail, the controller will provide a new routing table for *h1* to reach *h8*. As the rerouting will incur some delay, it is proven a link down will affect the total time taken for transmitting packets. In link failure scenarios, switch in triangle topology will follow routing table provide from Ryu Controller.

## 5. Conclusion

This paper investigates performance of SDN in different link failure scenarios and topologies. Performance comparison in term of throughput, and Round-Trip-Time (RTT) have been presented and discussed. Based on the results obtained, it is observed that SDN is capable of handling link failure scenarios by rerouting traffics to alternative links. In such cases, packets can still be transmitted to their destinations with the increase in RTT and lower throughput.

**Acknowledgments**
This paper is financially funded by Technopreneur at UniMAP Sdn. Bhd. (TUSB) [Grant number: 2017/08/0006].

## References

[1] "Traditional VS Software Defined Networking." [Online]. Available: http://www.mavenspire.com/blog/traditional-vs.-software-defined-whats-the-differences.

[2] Huang Liaoruo, Shen Qingguo and Shao Wenjuan, "A source routing based link protection method for link failure in SDN," *2016 2nd IEEE International Conference on Computer and Communications (ICCC)*, Chengdu, 2016, pp. 2588-2594.

[3] T. Theodorou and L. Mamatas, "CORAL-SDN: A software-defined networking solution for the Internet of Things," *2017 IEEE Conference on Network Function Virtualization and Software Defined Networks (NFV-SDN)*, Berlin, 2017, pp. 1-2.

[4] Huang Liaoruo, Shen Qingguo and Shao Wenjuan, "A source routing based link protection method for link failure in SDN," *2016 2nd IEEE International Conference on Computer and Communications (ICCC)*, Chengdu, 2016, pp. 2588-2594.

[5] S. Waleed, M. Faizan, M. Iqbal and M. I. Anis, "Demonstration of single link failure recovery using Bellman Ford and Dijikstra algorithm in SDN," *2017 International Conference on Innovations in Electrical Engineering and Computational Technologies (ICIEECT)*, Karachi, 2017, pp. 1-4.

# Prediction of Color Level and Chlorophyll Content of Corn (Zea mays L.) Leaves by Using Mobile Phone Cameras

**Astika I W[1*], Khayati N N[1]**
[1]Department of Mechanical and Biosystem Engineering, Bogor Agricultural University, Bogor, Indonesia
[2]Alumnus of Department of Mechanical and Biosystem Engineering, Bogor Agricultural University, Bogor, Indonesia

E-mail: [*]wayanastikaipb@yahoo.co.id

**Abstract**. This study aimed at designing a method to determine color level of corn leave and to estimate chlorophyll content using mobile phone cameras. The image of the leafs are captured with hand phone camera, and then the R, G, B color components are extracted. Four levels of color levels were used as those of leaf color card issued by IRRI. As for the natural light compensation, the colors of the Indonesian national inhabitant ID card were used as color references. The leafs were placed on the ID card, captured together, and then the leaf color level is determined by the variation of the grey levels of the leaves and the color references. The chlorophyll content were measured with SPAD chlorophyll meter. The average accuracy of leaf color prediction were 81.48% for sweet corn and 76.82% for hybrid corn under various field luminance, while the accuracy were 94.45% under a fixed 1500 lux luminance. The accuracy of chlorophyll prediction was very low with and R2= 0.4762 for sweet corn and R2= 0.5284 with national ID card references for hybrid corn. This low results were relatively contrast with the prediction from manual measurement of leaf color level, where the correlation coefficients were R2 = 0.8466 for sweet corn and R2 = 0.8506 for hybrid corn.

## 1. Introduction

Early stage of precision farming is characterized by variable rate fertilizer application referring to the plant appearance. Leaf color is an easy way to predict of nutrient sufficiency status of plant, even with human eyes. Leaf color chart (LCC) is a simple technology created for determining the leaf color level (LCL) and then used as a reference for determining the N fertilizing dose. Variable N fertilizing rate based on LCC reference has been applied in India, it was found that the N fertilizer and pesticide requirement decreased [1]. For maize, the use of LCC for fertilizing dose give better yield as compared to blanket dose [2].

With current wide spread of mobile phone technology, and the possibility of using mobile phone camera to capture leaf images, the LCC tool can be replaced by the phone camera. At the same time, related applications can be added into the mobile phone. The merit of using mobile phone camera is its practicability, and it is available at all time as it is always brought by farmers.

Some researches have been done to recognize the color level of leaf, as well to determine the fertilizing dose based on the predicted leaf color level. An Android application to predict leaf color level and determine N fertilizing dose has been developed [3]. The leaf color is determined based color value (CV) which is derived from RGB channels. A considerable high accuracy was obtained. Another researcher extracted the RGB, HSV, and L*, a*, b* channels to predict leaf color level [4]. It was found that the G, b*, and V single channel give good accuracy, while combination of two channels give better accuracy. In another research [5] found that L*,a*,b* channels give a low RMSE and high determination coefficient in predicting SPAD values.

Along with using automatic digital camera of mobile phone, environment light condition much affect the captured images. Therefore the data of leaf color is difficult to obtain [6]. In order to recognize the leaf color, reference colors are required. [7] used a white paper and palm skin as the reference colors. White paper is not practical since it is not available at all time for the farmers, while the palm skin is reliable since its color may vary over different farmers. This research try to overcome those problems by introduction new tools for reference colors. The objectives of the research was to develop methods to predict leaf color level of corn leaves and to predict chlorophyll content (SPAD index) of the leaves by using hand phone cameras.

## 2. Research Method

### 2.1. *Materials and Instruments*

Materials used in this research were corn leaves with various color levels. The leaves were taken from corn plants which were intentionally cultivated with various fertilizer treatments making various levels of color levels from yellowish to dark green. Fives brands of Android mobile phones were used to capture the images of the leaves. As for the standard color levels, the 4-levels leaf color chart issued by IRRI (International Rice Research Institute) were used. The chart contains 4 levels of color, from yellowish, light green, green, and dark green named levels '2' , '3' , '4' , and '5' respectively. The hand phones and the leaf color chart are shown in figure 1.

As reference colors, the Indonesian electronic ID card were used. The card has 3 colors in its drawing: blue (ocean), yellow (eagle), and green (island). The card were used because of its practicability, all farmers bring it all the time. For chlorophyll content measurement, the Minolta SPAD were used.

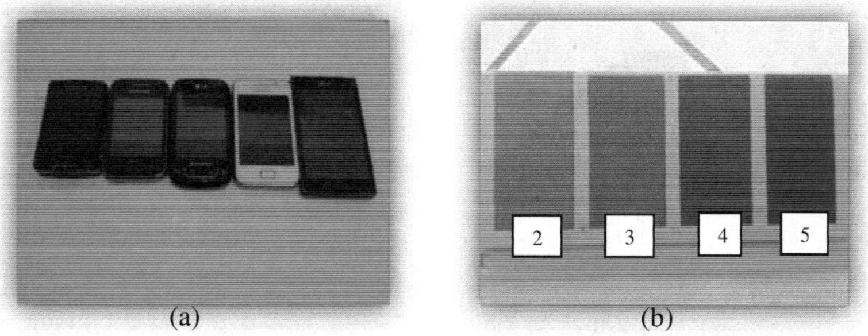


**Figure 1.** The five brands of Android hand phones (a) and IRRI 4-level leaf color charts (b)

### 2.2. *Image Capturing and Color Level Formulation*

Images of corn leaves were captured as the following way. A part of corn leaf was placed on the electronic ID card as shown in figure 2. The leaf and the card are captured together, the color of four points are taken: leaf, ocean, eagle, and island colors. The ocean, eagle, and island colors will become the reference color for the leaf color. As the natural light vary, the change in leaf color will be followed by the changes of other colors.

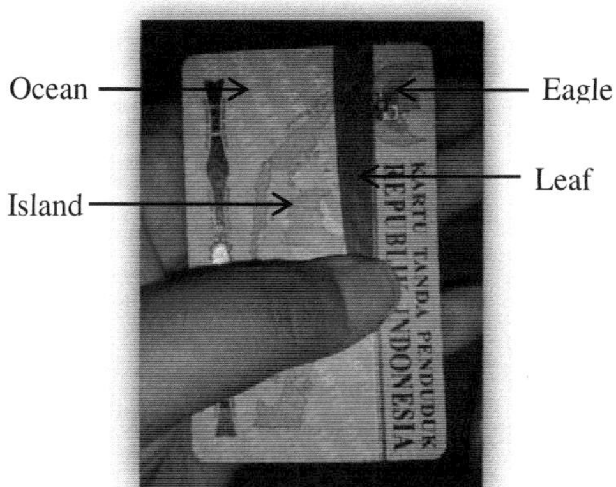


**Figure 2.** Position of corn leaf on the ID card for image capturing

Over the changes of the four color, the trend of changes are predicted to be in regular order. When one color tends to become lighter, other colors will become lighter. On the opposite, when one color tends to become darker, other colors will become darker too. Therefore, the difference of the darkness (indicated by grey scale) of two colors are seemly constant. The formulation are as follows:

IF GO - GL = [ $a_i$ , $b_i$ ]; GE - GL = [ $c_i$ , $d_i$ ]; GE - GL = [ $e_i$ , $f_i$ ] THEN CL = i ............... (1)

GL, GO, GE, GI = (R + G + B) / 3 ............................................................... (2)

Where GL : grey scale of leaf image
GO : greasy scale of ocean image
GE : grey scale of eagle image
GI : grey scale of island image
$a_i,b_i,\ldots,f_i$ : constants for color level i
i = predicted color level : 2,3,4,5

The true leaf color levels were measured manually using the IRRI 4-level leaf color chart. Thousands of corn leaves were collected for calibration process until a consistent $a_i$, $b_i$, $c_i$, $d_i$, $e_i$, $f_i$ constants were obtained, and thousands other were collected for validation process.

Image capturing were done in the morning 6.00-10.00 AM and afternoon 3.00-5.00 PM under various environment brightness ranging from 200 to 12000 lux. Three crop ages were selected: before the first fertilizing (9-12 days after sowing), before the second fertilizing (21-25 days after sowing), and before the third fertilizing (32-36 days after sowing).

### 2.3. *Chlorophyll Content Formulation*

With the same samples used for calibration of equation 1 and 2, the leaves were also measured with SPAD instrument in order to measure the chlorophyll content. The relationship of leaf color level and chlorophyll content were formulated with linear regression, both the manual measurements and the image-based predictions.

## 3. Results and Discussions

### 3.1. *Calibration and Validation of Color Level Prediction*

Calibration process is intended to obtain the constants $a_i$, $b_i$, $c_i$, $d_i$, $e_i$, $f_i$ in equation 1. The best combinations of the constants are shown in table 1.

**Table 1.** Constants obtained from calibration process

| Leaf color level - LCL | Hybrid corn | | | | | | Sweet corn | | | | | |
|---|---|---|---|---|---|---|---|---|---|---|---|---|
| | GE-GL | | GI-GL | | GO-GL | | GE-GL | | GI-GL | | GO-GL | |
| (i) | $a_i$ | $b_i$ | $c_i$ | $d_i$ | $e_i$ | $f_i$ | $a_i$ | $b_i$ | $c_i$ | $d_i$ | $e_i$ | $f_i$ |
| 2 | 24 | 55 | 49 | 79 | 79 | 99 | 44 | 65 | 64 | 84 | 84 | 104 |
| 3 | 56 | 74 | 80 | 89 | 100 | 115 | 66 | 75 | 85 | 99 | 105 | 115 |
| 4 | 75 | 85 | 90 | 99 | 116 | 129 | 76 | 85 | 100 | 112 | 116 | 126 |
| 5 | 86 | 110 | 100 | 126 | 130 | 150 | 86 | 100 | 113 | 125 | 127 | 152 |

After the calibration process, images of new leaves were captured for validation process. As shown in table 2, for example, in predicting 376 leaves with LCL 2, 315 leaves of hybrid corn were predicted correctly, while 60 leaves were recognized as LCL 3 making the accuracy was 83.78%. Low accuracy was shown by LCL 5, many of the leaves were recognized as LCL 4. This is because the fact that the colors of LCL 4 and LCL 5 are similar (Figure 1b). Here the natural light affect the result of prediction as the true color data is difficult to be obtained.

A special experiment was carried out in order to eliminate the various environment effect. The image capturing were done in a constant 1500 lux light condition. The accuracy is shown in table 3. Higher accuracies were obtained, but still the accuracy of LCL 5 was still the worst.

**Table 2**. Validation of the prediction model

| | Hybrid corn | | | | | | | Sweet corn | | | | |
|---|---|---|---|---|---|---|---|---|---|---|---|---|
| Measured LCL | Sample number | Number of Sample Predicted as LCL | | | | Accuracy | Sample number | Number of Sample Predicted as LCL | | | | |
| | | 2 | 3 | 4 | 5 | | | 2 | 3 | 4 | 5 | |
| 2 | 376 | 315 | 60 | 0 | 0 | 83.78 | 756 | 613 | 122 | 2 | 0 | 81.08 |
| 3 | 4064 | 234 | 3505 | 320 | 3 | 86.25 | 2171 | 149 | 1833 | 167 | 4 | 84.43 |
| 4 | 1926 | 10 | 176 | 1615 | 121 | 83.85 | 2714 | 42 | 288 | 2219 | 146 | 81.76 |
| 5 | 136 | 2 | 11 | 25 | 98 | 72.06 | 105 | 2 | 11 | 22 | 63 | 60.00 |
| Average | | | | | | 81.48 | Average | | | | | 76.82 |

**Table 3.** Accuracy of LCL prediction at 1500 lux luminance

| Measured LCL | Sample number | Number of Sample Predicted as LCL | | | | Accuracy |
|---|---|---|---|---|---|---|
| | | 2 | 3 | 4 | 5 | |
| 2 | 232 | 222 | 10 | 0 | 0 | 95.69 |
| 3 | 292 | 2 | 11 | 271 | 7 | 92.81 |
| 4 | 254 | 0 | 10 | 240 | 4 | 94.49 |
| 5 | 131 | 0 | 0 | 15 | 115 | 87.79 |
| Average | | | | | | 92.69 |

Accuracies among telephone brands were similar. The different were probably because of the different in automatic color compensation, in which different brand may have different self light adjustments. As shown in table 4, capturing in random brightness showed lower accuracies as compared to those in constant brightness.

**Table 4.** Accuracy of LCL prediction among telephone brands

| Telephone Brand | Accuracy at various condition (%) | | |
|---|---|---|---|
| | Hybrid corn, Random brightness | Sweet corn, Random Brightness | Random corn, Fixed brightness 1500 lux |
| 1 | 78.00 | 76.00 | 94.50 |
| 2 | 78.25 | 72.50 | 93.75 |
| 3 | 76.00 | 73.50 | 93.50 |
| 4 | 84.00 | 70.50 | 92.50 |
| 5 | 75.50 | 72.50 | 89.75 |

### *3.2. Prediction of Chlorphyl Content*

Relation of LCL and SPAD Index are formulated in linear regression models. Figure 3 shows the relation between manually measured LCL and SPAD Index. The determination coefficient are relatively high: 0.8506 for hybrid corn, and 0.8466 for sweet corn. These determination coefficients are considerably higher those of telephone-measured LCL as shown in figure 4 figuring 0.4762 for hybrid corn and 0.5284 for sweet corn. The lower values those of telephone-measured LCL are clearly caused by errors in the prediction of the LCL (table 3 and table 4)

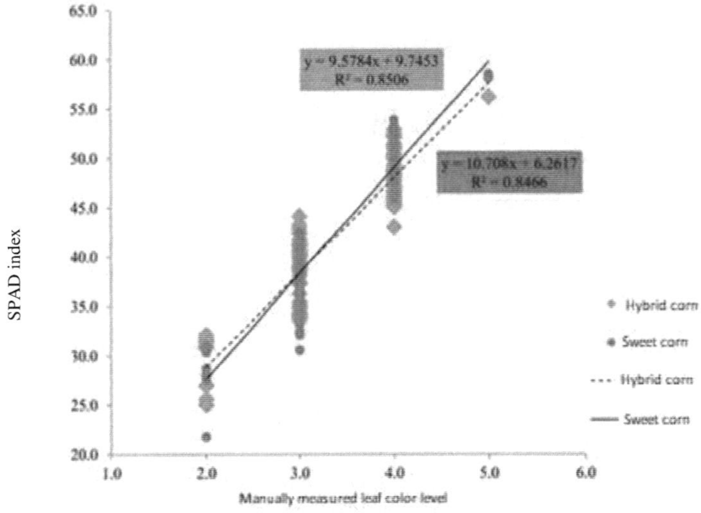


**Figure 3.** Relationship between manually-measured LCL and SPAD index

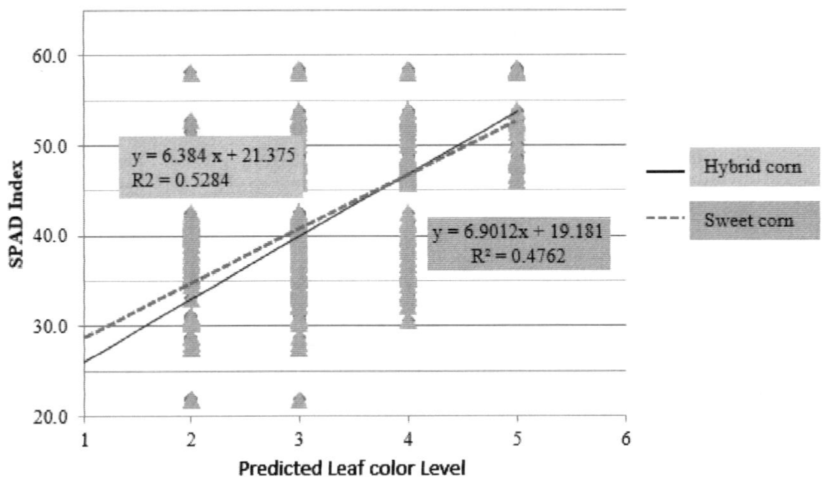


**Figure 4.** Relationship between telephone-measured LCL and SPAD Index

## 4. Conclusions

It can be concluded that:

1. Different brand of mobile phone give different color detection as expressed with R,G,B components making different results of leaf color level prediction. A unique calibration process is required for each hand phone brand.
2. Reference colors using national ID Card gave a moderate agreement in predicting leaf color level as compared with manual measurement. Natural light variation still affects the color prediction.
3. Prediction of SPAD index by using leaf color chart has a relatively good linear agreement with $R^2$ = 0.8506 for hybrid corn and $R^2$=0.8466 for sweet corn, but with mobile phone camera it has low agreements with $R^2$= 0.5284 for hybrid corn and $R^2$= 0.4762 for sweet corn.

**References**

[1] Islam, Z., Bhanudeb Bagchi, Mahabub Hossain. 2007. Adoption of leaf color chart for nitrogen use efficiency in rice: Impact of assessment of a farmer-participatory experiment in West Bengal, India. Field Crops Research 103(2007) 70-75

[2] Singh, V, Yadvinder Singh, Bijay Singh, H.S.Thind, Ajay Kumar, Monika Vasistha. 2011. Calibrating the leaf color chart for need based fertilizer nitrogent management in different maize (Zea mays L.) genotypes. Field Crops Research 120 (2011) 276-282.

[3] Intaravanne Y., Sarun Sumriddetchkajorn. 2015. Android based rice leaf color nalyzer for estimating the need amount of nitrogent fertilizer. Compuers and Electronic in Agriculture 116 (228-233)

[4] Reyes, J. F., Christioan Correa, Javier Zuniga. 2017. Reliability of different color spaces to estimate nitrogen SPAD values in maize. Computers and Electronics in Agriculture 143(2017)14-22.

[5] Agarwal, A., Snehasis Dutta Gupta. 2018. Assessment of spinach seedling health status and chlorophyl content by multivariate data analysis and multiple linear regression of leaf image features. Computers and Electronics in Agriculture 143(2017)14-22.

[6] Zhang, Y., Tang Liang, Liu Xiao-jun, Liu, Lei-lei, Cao Wei-xing, Zhu Yan. 2014. Modelling dynamic of leaf color based on RGB value in rice. Journal of Integrative Agriculture. 13(4) 749-759.

[7] Astika I. W. 2010. The use of hand phone camera to determine paddy leaf color level as a reference for fertilizing dosage. Procceding of Asian Federation of Information Technology in Agriculture (AFITA) Conference, Bogor October 2010: 105-108.

# Analysis of EEG Spectrum Bands Using Power Spectral Density for Pleasure and Displeasure State

**Anis Ameera[1], A.Saidatul[1] and Z Ibrahim[2]**

[1] Biosignal Processing Research Group (BioSIM), School of Mechatronic Engineering, Universiti Malaysia Perlis, Arau, Perlis, Malaysia
[2]Faculty of Technology, University of Sunderland, St Peter's Campus, Sunderland, SR6 0DD, United Kingdom

**Abstract.** The technology of reading human mental states is a leading innovation in the biomedical engineering field. EEG signal processing is going to help us to explore the uniqueness of brain signal that carries thousands of information in human being. The aim of this study is to analyze brain signal features between pleasure and displeasure mental state. Brainwaves is divided into 5 sub frequency bands namely alpha (8 – 13 Hz), beta (13 – 30 Hz), gamma (30 – 100 Hz), theta (4 – 8 Hz) and delta (1 – 4 Hz). However, in this study, alpha and beta waves were analyzed to investigate the mental states. Twenty subjects were recruited from undergraduate engineering student's education background in UniMAP with age ranging between 19 to 23 years old. The subject must be healthy and right-handed. The subject was required to view a series of pleasure and displeasure images for 10 minutes and take rest for 30 seconds between pleasure and displeasure view. Truscan EEG device (Deymed Diagnostic, Alien Technic, Czech Republic) with 19 channels were used to acquire EEG data with frequency sampling of 1024 Hz and impedance is kept below 5 kΩ. A bandpass filter was used to extract alpha and beta waves. The signal was segmented and PSD value using Welch and Burg method was calculated for both mental states. 7 statistical features (mean, mode, median, variance, standard deviation, minimum and maximum) were obtained from PSD value and used as an input for the classifier. K-Nearest Neighbour (KNN) and Linear Discriminant Analysis (LDA) were used to classify into two mental states. As a result, Welch method gives the highest classification accuracy which is 99.3 % for alpha waves followed by 97.5 % for beta waves from channel F4. It can be concluded that alpha waves are the most potential waves to be used in order to differentiate pleasure and displeasure features.

## 1.0 Introduction

A human brain is a complex nature and understanding the nature of interconnected neurons needs information signals from the brain. Electroencephalography (EEG) is non-invasive neurological diagnostic modalities which record brain signal. EEG data provide brain activity information. Different brain state depends on daily activity [1].

BCI is a technology advancement which enhances the communication between the human brain and external devices. The BCI reads different brain signals produced by the various parts of the human brain, interpret these signals into commands and actions for various applications [2]. The existence of BCI provides an alternative communication for anyone that suffers from any kind of neurological disorder that affects major muscle activity such as talking and swallowing [3] and allows them to communicate by distinguishing their brain signal.

The process of methodologies for EEG signal processing is divided into 4 stages which are data acquisition, preprocessing, feature extraction and classification. Data acquisition involves experimental protocol. Previous researcher has proposed different protocol in reading pleasure state [4].

During preprocessing stage, butterworth bandpass filter was implemented to filter into different sub-band frequency which in this research focus on alpha and beta waves for discriminating pleasure and displeasure mental state. As previous research [5] [6],suggest that the involvement of frontal cortex for beta and gamma waveform of brain is stable in reading emotion. Subsequently, in feature extraction, a linear feature extractor which are Power Spectral Density (PSD) was implemented and compared between two methods which are parametric (Burg method) and non-parametric (Welch method). Both of this method assumes the EEG signal is stationary random process in nature.

Lastly, in classification stage, The system learns to classify a set of features for a certain time duration to differentiate each mental states. Apparently, in this research, k-Nearest Neighbour (k-NN) and Linear Discriminant Analysis (LDA) were applied to compare the classification accuracy in discriminating two mental states.

## 2.0 Experiment

### *2.1. Subject*

Twenty subjects aged between 19 to 24 years old have participated in this study. All subjects are right handed, good health condition and no mental history. This criterion of subject selection is compulsory to standardized subject's state.

### *2.2. Experimental Set Up*

EEG signal were recorded using Truscan EEG device by Deymed Diagnostic. 19 electrode cap channel was used in recording raw EEG signal. The EEG signals were recorded by Truscan software and display brain signals from 19 electrodes (F3, F4, T3, T4, C3, C4, P3, P4, FZ, CZ, PZ, Fp1, Fp2, F7, F8, T5, T6, O1 and O2). Electrode A1 and A2 act as a reference point. The electrode placement is according to the international 10- 20 system as shown in Figure 1. The signals were recorded in a quiet and calm environment where the subjects were asked to sit comfortably on a chair. The subjects were given 5 minutes to relax for pre- experimental procedures. The sampling frequency was set to 1024 Hz and the impedances were kept below 5 kΩ. The cutoff range signal was set as -80 to 80.

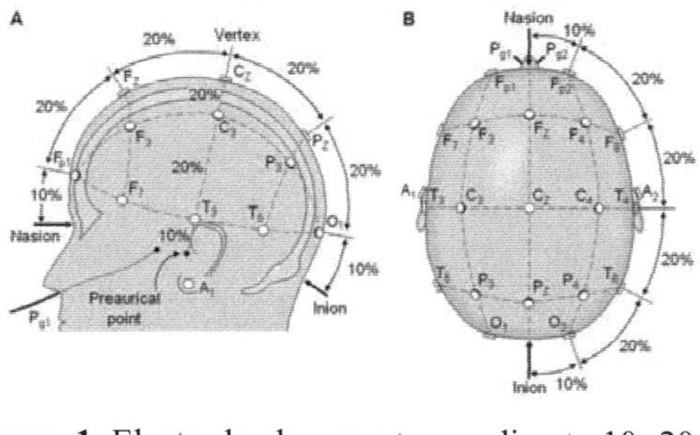


**Figure 1.** Electrode placement according to 10- 20 international system [7]

### *2.3. Experimental Protocol*

First, subject was required to fill in consent form and demographic form which contains questions related to the protocol of study. The process takes 10 minutes. Afterwards, subject was asked to sit comfortably on a chair in front of computer screen. The task is divided into three. Each task consists of a different set of picture. Task 1 consists a set of a picture to simulate pleasure, task 2 consist of the pictures to stimulate displeasure and task 3 consists of both pleasure and displeasure picture to compare which picture actually stimulate pleasure towards the subject. Subject requires to view each of the picture shown.

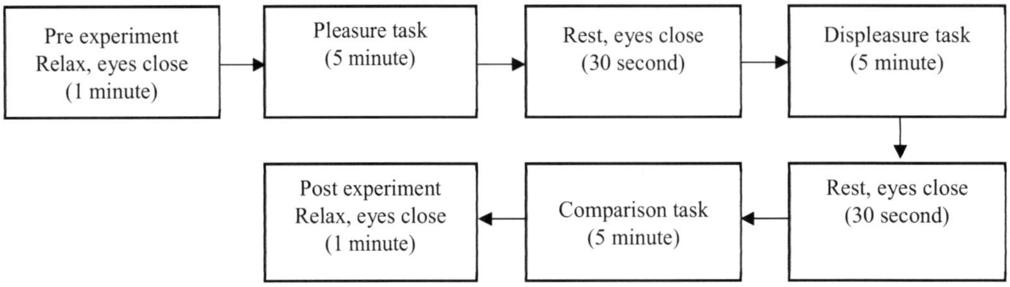


**Figure 2.** Experimental protocol

### 3.0 EEG Signal Processing

The recorded signal undergoes signal processing which comprises of 3 stages. Starting with preprocessing, feature extraction and lastly is classification. Based on Figure 2 shows the process for data analysis.

*3.1. Preprocessing Method*

Each raw signals have amplitudes of microvolts. Hence, preprocessing is required to denoise input raw data to maintain the information recorded. Butterworth bandpass filter with order 2 have been use to filter according to 2 sub-band frequency components which are alpha and beta.

*3.2. Welch method*

Welch method is applied to estimate power spectrum of determined time sequence. Data window is used in each sequence and enable it to overlap [8]. Let {xd(n)} be the sequence, where d=1,2,3...L (signal intervals), M is interval length. Thus, power spectral density using Welch method is defined by,

$$\hat{p}d(f) = \frac{1}{MU} \left|\sum_{n=0}^{M-1} xd\,(n)w\,(n)\, e^{-j2\pi f}\right|^2 \quad (1)$$

which U stands for normalization factor for power in window function

$$U = \frac{1}{M}\sum_{n=0}^{M-1}|w(n)|^2 \quad (2)$$

Where w(n) stands for windowed data. Welch power spectrum is the average over modified periodograms defined by,

$$\hat{P}Welch(f) = \frac{1}{L}\sum_{i=0}^{L-1}\hat{P}d(f) \quad (3)$$

*3.3. Burg Method*

This method creates an estimation of reflection coefficients by minimizing the average of forward and backward linear prediction error fulfilling the Levinson Durbin recursion. The advantages of this method are estimating short data records and resolve closely space sinusoids in signal with low level [8]. PSD using the Burg method is defined by,

$$P_{xx}^{BU}(f) = \frac{\hat{e}p}{\left|1+\sum_{l=1}^{p}\hat{a}p\,(l)e^{-2jfl}\right|^2} \quad (4)$$

$\hat{e}p$ indicates the total least square error and it is the sum of forward and backward prediction errors.

$$\hat{e}f, p(n) = x(n) + \sum_{i=1}^{p}\hat{a}p, ix(n-i), \quad n = p+1, \ldots\ldots, U \quad (5)$$

$$\hat{e}b, p(n) = x(n-p) + \sum_{i=1}^{p}\hat{a}p, ix(n-p+i), \quad n = p+1, \ldots. U \quad (6)$$

*3.4. k-Nearest Neighbour (k-NN) Classifier*
*k*-NN act as a linear classifier that within specified data point x, detects a hypersphere around it that comprises of K points and assign x to the class that has the largest number of representative inside the hypersphere [9]. The process occurs as k-NN finds k neighbourhood in training data and assign recurrent class in the neighbourhood of k. The value of k is quantified from 1 to 10. Subsequently, the value of k will be wide-ranging as per sequence number until 10. The highest value of classification rate with the value of k will be chosen without contemplating the average value for 10. However, in this study, the distance used is correlation and the rule set is random as it provides the highest accuracy to the given data.

*3.5. Linear Discriminant Analysis (LDA)*
Consist of a linear combination of variable [9]. The concept uses hyperplanes to differentiate training feature vectors shows for dissimilar classes. The orientation and location of this hyperplane are decided by training data. Additionally, for a two class problem, the class of unseen feature vector is based on which side of hyperplane the feature vector situated.

**4.0 Results and Discussion**
The results show characterization of alpha and beta waves using channel F4 during pleasure and displeasure state.

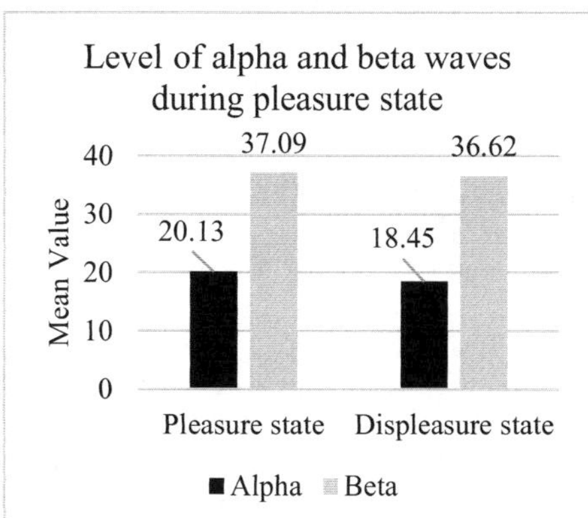


**Figure 3.** Level of alpha and beta waves during pleasure and displeasure state

Based on Figure 3, it indicates that alpha waves in pleasure state is higher compare to displeasure state due to alpha waves are more dominant during relax mind as in pleasure state while the value of alpha wave decreases during displeasure state as the mind is not in relax phase anymore [10]. As for beta waves, in this study the value is not significant but significant value maybe obtain using advance feature extractor to extract better features.

Comparison of classification accuracy are tabulated in Table 1. K-NN and LDA classifiers were applied to classify two mental states.

**Table 1.** Comparison between k-NN and LDA classifier between alpha and beta waves

| Feature extractor | Signal wave | Channel | Classifier | |
|---|---|---|---|---|
| | | | ***k*-NN (%)** | **LDA (%)** |
| Welch | ∝ | **F3** | 12.09 | 37.14 |
| | | **F4** | **99.29** | 57.14 |
| | | **F3& F4** | 93.92 | 56.42 |
| | $\beta$ | **F3** | 13.21 | 35.36 |
| | | **F4** | 97.50 | 61.79 |
| | | **F3&F4** | 91.43 | 57.14 |
| Burg | ∝ | **F3** | 96.07 | 60.00 |
| | | **F4** | 95.36 | 54.29 |

| | | | |
|---|---|---|---|
| | **F3 &F4** | 97.50 | 58.21 |
| $\beta$ | **F3** | 92.50 | 61.07 |
| | **F4** | 95.36 | 63.57 |
| | **F3 & F4** | 96.43 | 65.71 |

According to Table 1, it can be shown that highest classification accuracy is 99.3% by using Welch method for alpha waves followed by 97.5 % for beta waves from channel F4.

## 5.0 Conclusion

The method proposed in this study was based on feature extraction using PSD with Welch and Burg method for classifying pleasure and displeasure mental state. Two classifiers were compared which are *k*-NN and LDA to test the efficacy of proposed features. The advantage of using PSD is easy to applied. The simulation results indicate that alpha waves are most potential waves in discriminating pleasure and displeasure states. Also, the highest classification accuracy (99.3%) was achieved by using Welch method in channel F4.

## 6.0 Acknowledgement

The authors would like to acknowledge the support and encouragement given by the Vice Chancellor of University of Malaysia Perlis (UniMAP), Dato' Profesor Dr. Zul Azhar Zahid Jamal. This work is assisted by Biosignal Processing Research Group (BioSIM).

## References

[1] R. F. Ahmad, A. S. Malik, H. U. Amin, N. Kamel, and F. Reza, "Classification of cognitive and resting states of the brain using EEG features," in *2016 IEEE International Symposium on Medical Measurements and Applications, MeMeA 2016 - Proceedings*, 2016, pp. 1–5.

[2] D. R. Edla, K. Mangalorekar, G. Dhavalikar, and S. Dodia, "Classification of EEG data for human mental state analysis using Random Forest Classifier," *Procedia Comput. Sci.*, vol. 132, pp. 1523–1532, Jan. 2018.

[3] F. Lotte *et al.*, "Electroencephalography ( EEG ) -based Brain-Computer Interfaces." 2015.

[4] M. Islam, T. Ahmed, M. S. U. Yusuf, and M. Ahmad, "Channel selection and feature extraction for cognitive state estimation with the variation of brain signal," in *2013 International Conference on Electrical Information and Communication Technology (EICT)*, 2014, pp. 1–6.

[5] S. N. Light, J. A. Coan, C. Frye, H. H. Goldsmith, and R. J. Davidson, "Dynamic variation in pleasure in children predicts nonlinear change in lateral frontal brain electrical activity.," *Dev. Psychol.*, vol. 45, no. 2, pp. 525–533, 2009.

[6] S. K. Hadjidimitriou and L. J. Hadjileontiadis, "Toward an EEG-based recognition of music liking using time-frequency analysis," in *IEEE Transactions on Biomedical Engineering*, 2012, vol. 59, no. 12, pp. 3498–3510.

[7] J. S. Kumar and P. Bhuvaneswari, "Analysis of electroencephalography (EEG) signals and its categorization - A study," in *Procedia Engineering*, 2012, vol. 38, pp. 2525–2536.

[8] S. Noshadi, V. Abootalebi, M. T. Sadeghi, and M. S. Shahvazian, "Selection of an efficient feature space for EEG-based mental task discrimination," *Biocybern. Biomed. Eng.*, vol. 34, no. 3, pp. 159–68, 2014.

[9] A. Zammouri, A. Ait Moussa, and Y. Mebrouk, "Brain-computer interface for workload estimation: Assessment of mental efforts in learning processes," *Expert Syst. Appl.*, vol. 112, pp. 138–147, 2018.

[10] Lal, S. K., & Craig, A. (2001). A critical review of the psychophysiology of driver fatigue. Biological psychology, 55(3), 173-194.

IOP Conf. Series: Materials Science and Engineering **557** (2019) 012031 doi:10.1088/1757-899X/557/1/012031

# Exploring EEG based Authentication for Imaginary and Non-imaginary tasks using Power Spectral Density Method

**Tze Zhi Chin**[1], **A Saidatul**[1] **and Z Ibrahim**[2]

[1] Biosignal Processing Research Group (BioSIM), School of Mechatronic Engineering, Universiti Malaysia Perlis, Arau, Perlis, Malaysia

[2]Faculty of Technology, University of Sunderland, St Peter's Campus, Sunderland, SR6 0DD, United Kingdom

**Abstract.** Biometric technology has swiftly emerged as a go-to solution for improving cyber security especially in financial fraud and security threats. EEG based-authentication is best of security in cyber security application as it is unique and cannot be replicated. The aim of this study is to investigate the possibility of adopting imaginary or non-imaginary task for human authentication. In this study, twenty subjects were recruited from undergraduate students with age ranging from 19 to 30 years old. The subject must be healthy and right-handed. The subjects were required to perform non-imaginary task (left hand or right hand movement) and imaginary task (just need to imagine the movement of left hand or right hand). Duration for each task is 1 minute and take rest for 1 minute between the tasks. Truscan EEG device (Deymed Diagnostic, Alien Technic, Czech Republic) with 19 channels were used to collect EEG data with 1024 Hz frequency sampling and the impedance is kept below 5 kOhm. Bandpass filter was used in pre-processing to extract alpha (8-13Hz) and beta (14-30Hz) waves. The signal was segmented and the power spectral density were calculated by Welch's method and Burg's method. The statistical features (mean, median, mode, variance, standard deviation, minimum and maximum) were obtained from PSD were used as input of classifier. K-nearest neighbour classifier (KNN) and Linear Discriminant Analysis (LDA) were applied for classification. In conclusion, Welch method gives the highest classification accuracy which is 98% for beta waves from channel C4 with the K-nearest neighbour classifier. Imaginary task shows the higher classification accuracy which is 98.03% instead of non-imaginary task which is 94.95%. Thus, imaginary task is more suitable for authentication.

## 1. Introduction

Biometric technology has swiftly emerged as a go-to solution for improving cyber security especially in financial fraud and security threats. Biometric approaches have obtained energetically rising interest for individual authentication and identification as they are associated with physiological as well as behavioural of individual [1]. Biometric authentication systems presently in use have some vulnerabilities. For instance, iris recognition can be cheated by high resolution images while palm-prints and fingerprints can be replicated. Thus, it is greatly desired to find robust biometric approaches that can probably overcome aforementioned weaknesses [2]. In the drive to improve highest level of security, EEG signal is one of the alternative method to replace conventional biometric system.

Electroencephalogram (EEG) is a non-invasive method that records the electrical activity of brain by measuring the small voltage fluctuations with placement of the electrodes on the scalp surface [3]. The communication of brain cells will generate electrical impulse and active the whole time [4]. Brain signals can be influenced by individual's mental state which makes them very hard to be collected under coercion and force. If a mental state is chosen for authentication, it can be inhibited by a user in threatening situation [5]. Moreover, brain signals are related to the genetic information of individuals so it is unique for each individual and stable over time [6].

EEG based authentication is measuring and statistical analysis of bio signal characteristics of individual. It is important to proof individual's unique identity especially in security purpose [7]. In most EEG-related studies, features are extracted from the frequency bands of the signal. Five main frequency bands have been recognised. They are arranged in increasing order which are called delta (δ), theta (θ), alpha (α), beta (β) and gamma (γ). The amplitudes and frequencies of signals change from one state to another [2].

The goal of this research is to investigate the possibility of adopting imaginary or non-imaginary task for human authentication. Motor and motor imaginary activities will generate electrical signals on the motor cortex on the brain [8]. Due to the artifact and complexity of signals, advanced signal processing is needed to improve the EEG signals. Pre-processing, feature extraction and classification are the three major stages of signal processing [9].

## 2. Data Acquisition

A total number of twenty subjects were recruited from undergraduate students with the range of age from 19 to 30 years old. The subject must be healthy and right-handed. All subjects are free from mental injury or mental disorder. Before conducting the experiment, the subject was requested to fill in the Consent Form to let subject understands what it means to be in the study and get approval from subject to take part in it. Next, Demographic Information Form was filled by subject to know the general background of subjects.

After that, the EEG cap is placed correctly on the subject's scalp. This device captures EEG signals across 19 channels with 1024Hz sampling rate and 256 data are collected per second. The voltage is set as 70uV and frequency range from 1 to 80 Hz to get the EEG signals frequency bands for analysis. The electro gel is injected into every electrodes to reduce impedance below 5 kΩ. Then the subject adjusts to a comfortable position and close the eyes for whole process.

There are four tasks in the experimental protocol which are left hand movement, imaginary left hand movement, right hand movement and imaginary right hand movement. Duration for each task is 1 minute and take rest for 1 minute between the tasks. Subject should open and close the hand according to the metronome track at 60 BPM as well as imagine open and close the hand to ensure that subject moves and imagines at constant speed. The duration of each task is 2 minutes while for the whole process is 10 minutes.

## 3. EEG Signal Processing

EEG signal is complex, random and not stationary in nature, it is unable to interpret signal by visual inspection. Thus, advanced signal processing is required to investigate the EEG signals by extracting hidden information from the brain signal [10]. The EEG signal processing aid separated into three stages including pre-processing, feature extraction and classification.

### *3.1 Pre-processing*

The raw signals generated from non-cerebral system are filtered and removed unwanted frequencies to enhance the embedded information in the brain signals [11]. Butterworth bandpass filter was used as the pass band is maximally flat with no ripple [9][12]. To obtained the two desired frequency which are alpha wave (8 -13Hz) and beta wave (13-30Hz), raw signals required to filter through Butterworth band pass filter.

*3.2 Feature Extraction*
Feature extraction is a vital part of signal processing as accuracy of classification is determined by quality selection of the features [12].

*3.2.1. Welch Method.* Welch's method is a non-parametric method. It is usually used to estimate the power spectrum of a given time sequence [13][14]. The sequences are allowed to overlap and windowed each data segment. Thus, the SNR for Welch method is high.

*3.2.2. Burg Method.* Burg method is depended on reducing forward and backward prediction errors whilst fulfilling the Levinson-Durbin recursion. By using this method, reflection coefficients are estimated straight away instead of calculating auto correlation. The benefits of this method are resolving nearly spaced sinusoids in low noise levels signals and estimating short data records as AR PSD estimates are proximate to the true values. Moreover, the Burg method confirms that a stable AR model and is computational efficiently. The accuracy of this method is reduced for high-order models, long data records and high SNR as it can cause line splitting or generate irrelevant peaks in the spectrum estimate. The spectral density estimate computed by the Burg method is also influenced by frequency shifts which relative to the true frequency resulting from the initial phase of noisy sinusoidal signals. This effect is enhanced when analysing short data sequences [13][15].

*3.3 Classification*
Classification is the final stages of EEG signal processing. The statistical features (mean, median, mode, variance, standard deviation, minimum and maximum) were obtained from output of feature extraction were used as input of classifier.

*3.3.1. k-Nearest-Neighbors (k-NN).* K-Nearest-Neighbors (k-NN) is a non-parametric, non-linear and supervised learning algorithms that used for classification. Among several types of supervised statistical learning algorithms, k-NN algorithms obtains typically high performance without any assumptions about the distributions from which the training examples are drawn [16].

The minimum distance from the signal to the training set is decided to find the k-NN category of the training dataset. Distance metrics are an approach to investigate the distance between a new data point and existing training dataset. The distance based on correlation is a measure of statistical dependence between two vectors [17]. The correlation distance is calculated as follow:

$$d_{st} = \left(1 - \frac{(x_s - \bar{x}_s)(y_t - \bar{y}_t)}{\sqrt{(x_s - \bar{x}_s)(x_s - \bar{x}_s)'}\sqrt{(y_t - \bar{y}_t)(y_t - \bar{y}_t)'}}\right) \quad (1)$$

$$\bar{x}_s = \frac{1}{n}\sum_j \bar{x}_{s\,j}$$

$$\bar{y}_t = \frac{1}{n}\sum_j \bar{y}_{t\,j}$$

*3.3.2. Linear Discriminant Analysis (LDA).* Linear Discriminant Analysis (LDA) is one type of statistical classification that used in machine learning and pattern recognition to get the features that linearly combined more than two classes of objects [18]. It is a linear transformation techniques that are commonly used for dimensionality reduction. LDA is supervised learning algorithm and computes the directions that will represent the axes that maximize the separation between multiple classes [19].

## 4. Result and Discussion
*4.1 Determination of Most Informative Electrode for EEG based Authentication*
There are two electrodes were selected for comparison, the electrodes are located on motor cortex area which related to our imaginary and non-imaginary activities[8].

**Table 1.** Classification Result based on Channel

| Frequency band | C3 | C4 | Combination C3 and C4 |
|---|---|---|---|
| Alpha | 89.5% | **96.98%** | 95.33% |

From Table 1, by comparing the alpha frequency band, it can be shown that C4 is the most informative electrode with highest accuracy of 96.98% and followed by combination C3 and C4 with 95.33% and the last is C3 electrode with accuracy of 89.5%.

*4.2 Determination of Best Feature Extraction for EEG based Authentication*

There are two types of feature extraction that used for analysis. The feature extracted are based on one second segmentation time of filtered signal.

**Table 2.** Classification Result based on Feature Extraction

| Feature | Frequency band | Result |
|---|---|---|
| Welch | Alpha | 96.4% |
| | **Beta** | **97.94%** |
| | Combination of alpha and beta | 87% |
| Burg | Alpha | 96.98% |
| | Beta | 95.15% |
| | Combination of alpha and beta | 94.15% |

As shown in Table 2, Welch method with beta features show the highest accuracy which is 97.94% compared to Burg method with beta frequency is 95.15%.

*4.3 Determination of Best Task for EEG based Authentication*

**Table 3.** Classification Based on Task

| Channel | Non-imaginary task | Imaginary task |
|---|---|---|
| C4 | 95.22% | 97.63% |
| | 96.4% | 98.17% |
| | 94.91% | 98.95% |
| | 92.45% | 97.36% |
| Average | 94.75% | **98.03%** |

From Table 3, imaginary task shows the highest accuracy with 98.03% while non-imaginary tasks shows only 94.75%.

## 5. Conclusion

In conclusion, Welch method gives the highest classification accuracy that almost 98% for beta waves from channel C4 with the K-nearest neighbour classifier. Imaginary task shows the higher classification accuracy which is 98.03% instead of non-imaginary task which is 94.75%. Thus, imaginary task is more suitable for biometric authentication.

**Acknowledgments**

We would like to thank Universiti Malaysia Perlis for supporting our research.

**References**

[1] K. P. Thomas and A. P. Vinod, "Utilizing individual alpha frequency and delta band power in EEG based biometric recognition," in *2016 IEEE International Conference on Systems, Man, and Cybernetics (SMC)*, 2016, pp. 004787–004791.

[2] I. Jayarathne and M. Cohen, "Survey of EEG-Based Biometric Authentication," no. January, 2018.

[3] J. Galbally, C. McCool, J. Fierrez, S. Marcel, and J. Ortega-Garcia, "On the vulnerability of face verification systems to hill-climbing attacks," *Pattern Recognit.*, 2010.

[4] R. A. Ramadan and A. V. Vasilakos, "Brain computer interface: control signals review," *Neurocomputing*, vol. 223, no. October 2016, pp. 26–44, 2017.

[5] Y. Chen *et al.*, "A High-Security EEG-Based Login System with RSVP Stimuli and Dry Electrodes," *IEEE Trans. Inf. Forensics Secur.*, 2016.

[6] Q. Gui, Z. Jin, and W. Xu, "Exploring EEG-based biometrics for user identification and authentication," *2014 IEEE Signal Process. Med. Biol. Symp. IEEE SPMB 2014 - Proc.*, 2015.

[7] O. Nieves and V. Manian, "Automatic person authentication using fewer channel EEG motor imagery," in *2016 World Automation Congress (WAC)*, 2016, pp. 1–6.

[8] V. Asanza, E. Pelaez, and F. Loayza, "EEG signal clustering for motor and imaginary motor tasks on hands and feet," *2017 IEEE 2nd Ecuador Tech. Chapters Meet. ETCM 2017*, vol. 2017–Janua, pp. 1–5, 2018.

[9] Z. Y. Ong, A. Saidatul, and Z. Ibrahim, "Power Spectral Density Analysis for Human EEG-based Biometric Identification," in *2018 International Conference on Computational Approach in Smart Systems Design and Applications (ICASSDA)*, 2018, pp. 1–6.

[10] S. A. Awang, M. P. Paulraj, and S. Yaacob, "Implementing eigen features methods/neural network for EEG signal analysis," in *2013 7th International Conference on Intelligent Systems and Control (ISCO)*, 2013, pp. 201–204.

[11] Siuly, "Analysis and Classification of EEG Signals," 2012.

[12] P. Kumari and A. Vaish, "Brainwave Based Authentication System: Research Issue and Challenge," in *International Journal of Computer Engineering and Applications*, 2014, vol. IV.

[13] D. P. Subha, P. K. Joseph, R. Acharya U, and C. M. Lim, "EEG Signal Analysis: A Survey," *J. Med. Syst.*, vol. 34, no. 2, pp. 195–212, Apr. 2010.

[14] A. Saidatul, M. P. Paulraj, S. Yaacob, and M. A. Yusnita, "Analysis of EEG signals during relaxation and mental stress condition using AR modeling techniques," in *2011 IEEE International Conference on Control System, Computing and Engineering*, 2011, pp. 477–481.

[15] S. Bhattacharya, R. J. Haddad, and M. Ahad, "A multiuser EEG based imaginary motion classification using neural networks," *Conf. Proc. - IEEE SOUTHEASTCON*, vol. 2016–July, 2016.

[16] S. K. Bashar and M. I. H. Bhuiyan, "Motor Imagery Movements Detection of EEG Signals using Statistical Features in the Dual Tree Complex Wavelet Transform Domain," *2015 IEEE Int. Conf. Telecommun. Photonics*, no. May, pp. 1–5, 2015.

[17] K. Chomboon, P. Chujai, P. Teerarassammee, K. Kerdprasop, and N. Kerdprasop, "An Empirical Study of Distance Metrics for k-Nearest Neighbor Algorithm," in *The Proceedings of the 2nd International Conference on Industrial Application Engineering 2015*, 2015, pp. 280–285.

[18] H. N. Oon, A. Saidatul, and Z. Ibrahim, "Analysis on Non-Linear Features of Electroencephalogram (EEG) Signal for Neuromarketing Application," in *2018 International Conference on Computational Approach in Smart Systems Design and Applications (ICASSDA)*, 2018, pp. 1–8.

[19] R. A. FISHER, "The Use of Multiple Measurements in Taxonomic Problem," *Ann. Eugen.*, vol. 7, no. 2, pp. 179–188, Sep. 2014.

# PSD-Based Features Extraction For EEG Signal During Typing Task

**Wei Bin Ng[1], A Saidatul[1], Chong Y.F[1] and Z Ibrahim[2]**

[1] Biosignal Processing Research Group (BioSIM), School of Mechatronic Engineering, Universiti Malaysia Perlis, Arau, Perlis, Malaysia

[2] Faculty of Technology, University of Sunderland, St Peter's Campus, Sunderland, SR6 0DD, United Kingdom

**Abstract.** Electroencephalograph (EEG) is an electrical field that generated by our brain incessantly. The EEG signal released by the brain is different when a people is performing different activities in their daily life. And such EEG signals consist complicated information that can be interpreted. The aims of this study is to analyse the specific EEG channels of a user when they are performing a typing task with laptop. Meanwhile, this research also aimed to verify the performance of the different sub frequency band which is alpha and beta to recognize the specified tasks. The frequency sampling was set at 1024 Hz and the impedance was kept below 5k ohm of each channels. The Truscan EEG (Deymed, Diagnostic, Czech Republic) device consists of 19 channels and only selected channels which is F3 and F4 is filtered through butterworth bandpass filter (1Hz-80Hz) in the pre-processing stage. Power Spectra Density was calculated by using Welch and Burg Method to extract the features from filtered data. K-Nearest Neighbour (KNN) classifier and Linear Discriminant Analysis (LDA) were used in classification. It is found that the combination of channel F3 and F4 for Alpha frequency using Welch method gives the highest accuracy which is 98.45%.

## 1. Introduction

Brain is an unique structure that is very essential to human being. The brain cells communicate to each other through electrical impulses and it can be measured by the electrode place on the human scalp. The EEG signals of a people is different when they are perform different kind of activity in life, and it is vary from one person to other person. In [1], the EEG is showing the different level of electrical activity in the different stage of sleep. Besides that, the anatomy of brain is also divided into 4 main lobes which is Frontal lobe, Temporal lobe, Parietal lobe and Occipital lobe and all these lobe are respective to different functions [2], [3].

In this paper, the EEG signals of subjects are recorded during they are performing the typing tasks. Then, it is evaluated based on the use of different feature extraction method and classification. The structure of this paper is organized as few sections which are experiment protocol and methodology, result and discussion, and lastly the conclusion.

## 2. Experiment Protocol and Methodology

### *2.1. Data Set and Protocol*

Twenty subjects with right-handed and without mentally disorder or brain injury were enrolled from the undergraduates of UniMAP in this research. The Truscan EEG device that consists of 19 channels were

IOP Conf. Series: Materials Science and Engineering **557** (2019) 012032 doi:10.1088/1757-899X/557/1/012032

used to collect the EEG signal of the subjects. The sampling rate during the data collection was set at 1024 Hz and the sensitivity of the voltage was set at 70μV. The subjects were asked to perform the 3 times of the typing task in this experiment. Figure 1 and Figure 2 show the photo during the data collection and the flow of the experiment. Then, EEG signal processing will be applied to extract most informative features from the raw data.

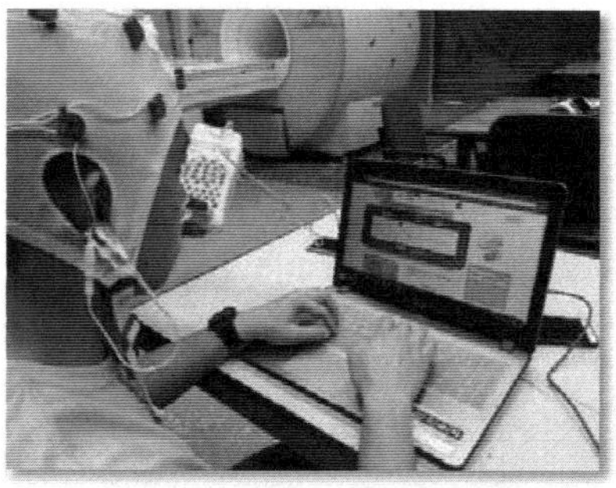

**Figure 1.** Subject performing typing task.

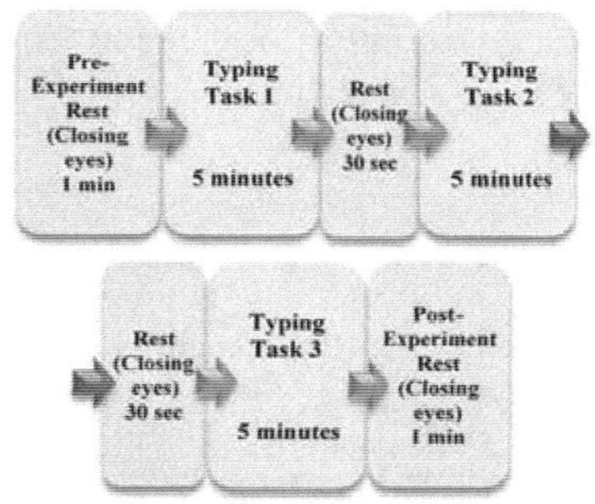


**Figure 2.** The flowchart of experiment protocol.

*2.2. EEG Signal Processing*
The collected EEG signals were converted to the ASCII form which can be readable easily by the MATLAB software. Basically, the EEG signal processing consists of 3 stages which is pre-processing, feature extraction and classifier. In this research, the F3 and F4 channel is selected.

*2.2.1. Pre-processing*
The Butterworth bandpass filter was implemented during the pre-processing stage. The butterworth filter provides a response that is no ripple and maximally flat in the passband and stopband [4]. It provide good average transient characteristics and expense wide transition region from band pass to band stop thus enable a good compromise between amplitude response selectivity [5]. In fact, the collected EEG raw data consist of all frequency band such as Delta, theta, alpha, beta, and gamma. It is then pre-processing with the Butterworth bandpass filter to separate the signal accordingly. The frequency band chosen is the alpha which is 8 Hz to 13 Hz, beta that are from 13 Hz to 30 Hz. In this paper, the raw EEG signals from frontal channel F3 and F4 for typing task 1, typing task 2 and typing task 3 were extracted for EEG signal processing. The frontal channels F3 and F4 were selected because the frontal lobe of the brain is related with the voluntary action [2].

*2.2.2. Feature Extraction*
The power spectral density (PSD) was used as the feature extraction method in this study. PSD is a good tool for stationary signal processing and suitable for narrowband signal. It is a common signal processing technique that distributes the signal power over frequency and show the strength of the energy as function of frequency [6]. Under the PSD, the Welch method and Burg method were adopted in this study.

A. Welch Method
Welch method is a modified segmentation scheme and used to evaluate the average periodogram. Generally the Welch method of the PSD can be describe by the equations below, the power spectra density equation is defined first. Then Welch Power Spectrum that mean average of the periodogram for each interval is expressed.

$$P(f) = \frac{1}{MU} \left| \sum_{n=0}^{M-1} x_i(n)\, w(n) e^{-j2\pi f} \right|^2 \quad (1)$$

$$P_{welch}(f) = \frac{1}{L}\sum_{i=0}^{L-1} P(f) \tag{2}$$

B. Burg Method

Burg method is a method that diminishing the forward and backward prediction errors so it satisfy the Levinson-Durbin Recursion [7]. With higher order of Burg Model, the accuracy become lower, and false peaks will be inferred in the spectra. Which is that, the Burg method is highly suitable for short data records as it can generate accurate prediction and always produces a stable model [7]. Overall, the Burg method of PSD can be computed through following equation:

$$P_{burg}(f) = \frac{\widehat{E_p}}{\left|1+\sum_{k=1}^{p}\hat{a}_p(k)e^{-2j\pi f}\right|^2} \tag{3}$$

### *2.2.3. Classifier*

a. K-Nearest Neighbour (K-NN)

K-Nearest Neighbour is known as non-parametric method and instant-learner because it does not make any assumptions on the input data and directly take the available data as training [8]. The principle of K-NN is classification through recognizing the nearby neighbour (data) by the mean of similarity or the distant between the distributed data [9]. The word "K" mean the number of the referred neighbour. K-NN is used by many researcher as its simplicity, fast, and able to achieve high recognition accuracy [9].

b. Linear Discriminant Analysis (LDA)

It is a method to explain the distinctive nature of more than two classed of group by finding a linear combination of the extracted features [10], [11]. In the Linear Discriminant Analysis, it assumes that the conditional probability density functions of the classes are normally distributed with same covariance matrix [12]. The criteria of separability for class is done through the within-class and between-class scatter [13]. After that, the transformation of data is done to provide the definite boundaries to classify the data. The LDA draws the linear decision region between the given classes [14].

## 3. Result and Discussion

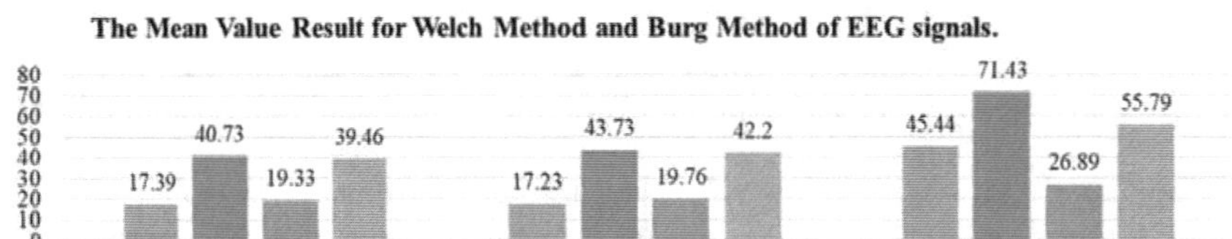


| | Typing Task 1 | Typing Task 2 | Typing Task 3 |
|---|---|---|---|
| Alpha (Welch) | 17.39 | 17.23 | 45.44 |
| Beta (Welch) | 40.73 | 43.73 | 71.43 |
| Alpha (Burg) | 19.33 | 19.76 | 26.89 |
| Beta (Burg) | 39.46 | 42.2 | 55.79 |

**Figure 3.**The mean value result for Welch method and Burg Method of EEG signals.

The mean value obtain from after the feature extraction is displayed in Figure 3. After the feature extraction, the EEG signals exhibits a similar level in typing task 1 and typing task 2. However, a significant increase is observed in mean value of typing task 3.The value of beta is greater as the frequency band is domain to the human active state and motor movement. The Alpha frequency band exhibit the best accuracy 98.45% for classification in the combination of F3 and F4 channels. The graph of performance for Welch Method EEG data is shown in Figure 4. Whereas for the Burg Method EEG data, the combination of the Alpha and Beta frequency band exhibit the best accuracy 94.12% for classification in the combination of F3 and F4 channels as shown in Figure 5 below.

In an overview, the classification of the signal is better when combination of channels F3 and F4 is used. The Welch method feature extraction is suitable use for classification because it give higher in percentage in general. Meanwhile, the LDA classification has been carried out. However, the obtained result isn't as good as the K-NN classification. It can be said that the classifier LDA is generally not good with few category variables though it performs very well with linear data [15]. A summary of the result is shown in the Table 1. Based on the Table 1, it can be shown that highest result is 98.45% by the combination of F3 and F4 channels in the Welch Method Alpha frequency Band in K-NN classifier. While the highest result of Burg Method EEG data in K-NN classifier is 94.12% by the combination of F3 and F4 channels for combination of Alpha and Beta frequency Band.

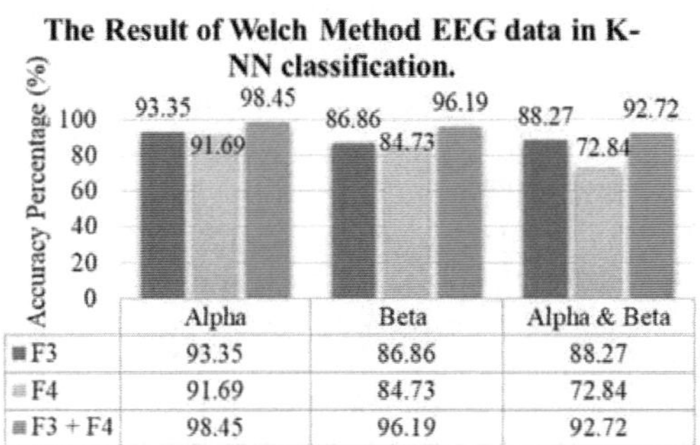


Figure 4. Result of Welch Method EEG data.

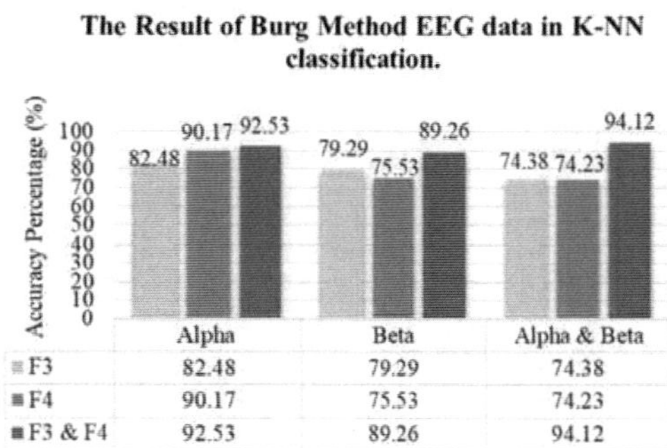


Figure 5. Result of Burg Method EEG data.

**Table 1.** Classification result for Welch method and Burg method EEG signals.

| Frequency Band | Channel | KNN | LDA |
|---|---|---|---|
| **Alpha (Welch)** | F3 | 93.35 | 52.16 |
| | F4 | 91.69 | 45.58 |
| | **F3 & F4** | **98.45** | 37.26 |
| Beta (Welch) | F3 | 86.86 | 53.37 |
| | F4 | 84.73 | 50.01 |
| | F3 & F4 | 96.19 | 34.18 |
| **Alpha & Beta (Welch)** | **F3** | 88.27 | **57.81** |
| | F4 | 72.84 | 36.83 |
| | F3 & F4 | 92.72 | 31.07 |
| Alpha (Burg) | F3 | 82.48 | 44.77 |
| | F4 | 90.17 | 49.06 |
| | F3 & F4 | 92.53 | 51.49 |
| **Beta** (Burg) | F3 | 79.29 | 44.59 |
| | F4 | 75.53 | 50.67 |
| | **F3 & F4** | 89.26 | **53.21** |
| **Alpha & Beta** (Burg) | F3 | 74.38 | 47.81 |
| | F4 | 74.23 | 46.35 |
| | **F3 & F4** | **94.12** | 30.08 |

## 4. Conclusion

In this study, the PSD is used as feature extraction and the Welch method and Burg method is used. The average value in the first time user and second time user is low and almost similar. However, it is observed that the average mean value has been lifted significantly in third time repeated user. In general, the combination of both frontal channels F3 and F4 give better result in KNN classification. The highest accuracy can go up to 98.45% for the combination of F3 and F4 channels in the alpha frequency band. The Welch method feature extraction is more suitable as it can give better classification result than Burg method in this case. For the performance of sub frequency band, the combination of Alpha and Beta frequency band exhibit better result in the Burg Method feature extraction in K-NN classifier than single frequency band is used.

**Acknowledgments**

We would like to thank Universiti Malaysia Perlis (UniMAP) for supporting our research works.

**References**

[1] Larry R.Squire, Floyd E. Bloom, Nicholas C. Spitzer, Sascha du Lac, Anirvan Ghosh, and Darwin Berg, *Fundamental Neuroscience Third Edition*, 3rd ed. ELSEVIER, 2008.

[2] Maris A Patestas and Leslie P. Gartner, *A Textbook of Neuroanatomy*. Blackwell Publising Ltd, 2006.

[3] L.-E. Notevarp Bjørge and T. Hübertz Emaus, "Identification of EEG-based signature produced by visual exposure to the primary colors RGB," no. July, 2017.

[4] K. A. I. Aboalayon, W. S. Almuhammadi, and M. Faezipour, "A comparison of different machine learning algorithms using single channel EEG signal for classifying human sleep stages," in *2015 Long Island Systems, Applications and Technology*, 2015, pp. 1–6.

[5] M. Lewandowski and J. Walczak, "Comparison of single-level and multi-level filtering systems of EEG signal," in *2015 Signal Processing: Algorithms, Architectures, Arrangements, and Applications (SPA)*, 2015, pp. 34–37.

[6] Z. Y. Ong and Z. Ibrahim, "Power Spectral Density Analysis for Human EEG- based Biometric Identification," in *2018 International Conference on Computational Approach in Smart Systems Design and Applications (ICASSDA)*, 2018, pp. 1–6.

[7] A. Saidatul, M. P. Paulraj, S. Yaacob, and M. A. Yusnita, "Analysis of EEG signals during relaxation and mental stress condition using AR modeling techniques," *Proc. - 2011 IEEE Int. Conf. Control Syst. Comput. Eng. ICCSCE 2011*, pp. 477–481, 2011.

[8] M. Manjusha and R. Harikumar, "Performance analysis of KNN classifier and K-means clustering for robust classification of epilepsy from EEG signals," in *2016 International Conference on Wireless Communications, Signal Processing and Networking (WiSPNET)*, 2016, pp. 2412–2416.

[9] R. M. Isa *et al.*, "EEG brainwave behaviour due to RF Exposure using kNN classification," in *2013 IEEE 3rd International Conference on System Engineering and Technology*, 2013, pp. 385–388.

[10] J. Ye and Q. Li, "A two-stage linear discriminant analysis via QR-decomposition," *IEEE Trans. Pattern Anal. Mach. Intell.*, 2005.

[11] R. Masoomi and A. Khadem, "Enhancing LDA-based discrimination of left and right hand motor imagery: Outperforming the winner of BCI Competition II," in *2015 2nd International Conference on Knowledge-Based Engineering and Innovation (KBEI)*, 2015, pp. 392–398.

[12] S. Yang and F. Deravi, "On the Usability of Electroencephalographic Signals for Biometric Recognition: A Survey," *IEEE Trans. Human-Machine Syst.*, vol. 47, no. 6, pp. 958–969, 2017.

[13] S. Balakrishnama and a Ganapathiraju, "Linear Discriminant Analysis - a Brief Tutorial," *Compute*, 1998.

[14] A. Ishfaque, A. J. Awan, N. Rashid, and J. Iqbal, "Evaluation of ANN, LDA and Decision trees for EEG based Brain Computer Interface," in *ICET 2013 - 2013 IEEE 9th International Conference on Emerging Technologies*, 2013.

[15] A. Bhardwaj, A. Gupta, P. Jain, A. Rani, and J. Yadav, "Classification of human emotions from EEG signals using SVM and LDA Classifiers," in *2015 2nd International Conference on Signal Processing and Integrated Networks (SPIN)*, 2015, pp. 180–185.

# Non Linear Features Analysis between Imaginary and Non-imaginary Tasks for Human EEG-based Biometric Identification

**Zhi Ying Ong[1], A Saidatul[1], V Vijean[1] and Z Ibrahim[2]**

[1] Biosignal Processing Research Group (BioSIM), School of Mechatronic Engineering, Universiti Malaysia Perlis, Arau, Perlis, Malaysia

[2] Faculty of Technology, University of Sunderland, St Peter's Campus, Sunderland, SR6 0DD, United Kingdom

**Abstract.** Electroencephalogram (EEG) is a signal contains information of brain activities. Nowadays, many types of research regarding EEG have been done such as neuromarketing. The human brain is very complicated but it contains a lot of information. EEG signal is a non-stationary signal, it changes over time and it also depends on human's emotion, thinking and activities. Due to the uniqueness of the EEG signal, the EEG signal has the potential to be used in the human authentication system. In this paper, an imaginary task and a non-imaginary task were studied to find out which type of task is possible to be used in authentication system. In preliminary study, five subjects were volunteered and performed the motor imagery and motor execution tasks. EEGO$^{TM}$ sports device (ANT Neuro, Enschede, Netherlands) with 32 channels was used to record the EEG signal and the sampling frequency is set to 512 Hz. The EEG signal was analysed by using EEG signal processing namely pre-processing, feature extraction and classification. Power line interference was removed by using a notch filter. Daubechies 8 wavelet family with 5$^{th}$ level decomposition has been applied to remove baseline wander noise. The performance of non-linear features such as Empirical Mode Decomposition (EMD), Hurst Exponent and Entropy were examined. Random forest gives good classification accuracy for imaginary task and non-imaginary task which are 83.53% and 87.06% respectively, thus, it shows non-linear features is possible to be employed in biometric identification.

## 1. Introduction

A signal that contains information of brain activities is known as electroencephalogram. It is defined as electrical activity of an alternating type recorded from scalp surface [1]. Electroencephalography (EEG) is a way to measure the electroencephalogram. Recently, research on EEG is getting more and more. Many types of research regarding EEG have been done such as neuromarketing [2]. The human brain is very complicated but it contains a lot of information. EEG signals are electrical activities generated from activities in the neurons. Local current flows when neurons activated. During synaptic excitations of dendrites of many pyramidal neurons in the cerebral cortex, the currents flow mostly measured by EEG [3].

Human brain contains of 4 lobes namely frontal lobe, temporal lobe, occipital lobe and parietal lobe. Each lobe has its specific function. Frontal lobe is used for movement, planning and problem solving. Temporal lobe is used for auditory, memory and speech signal. Occipital lobe is used for visual processing. Parietal lobe is used for sensing, perception, spelling and arithmetic [4]. Figure 1 shows basic structural of a neuron and figure 2 shows the picture of brain.

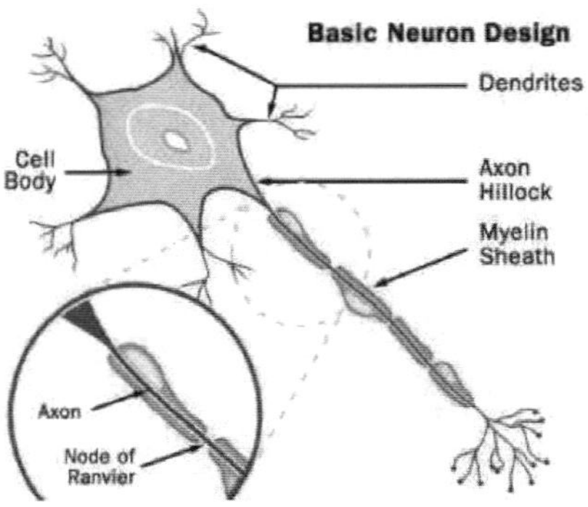


**Figure 1.** Basic structural features of a neuron [3].

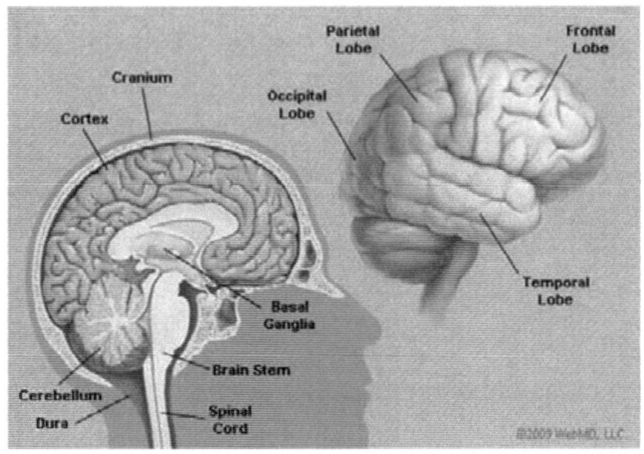


**Figure 2.** Picture of brain [5].

EEG signal is a non-stationary signal, it changes over time and it also depends on human's emotion, thinking and activities. Due to uniqueness of the EEG signal, the EEG signal has the potential to be used in the human authentication system. In this paper, an imaginary task and a non-imaginary task were studied in order to find out which type of task is possible to be used in authentication system.

## 2. Data Acquisition

Five healthy and right-handed subjects were volunteered in this preliminary study. They are students from Universiti Malaysia Perlis (UniMAP) with age 20 to 30 years old. Before the experiment start, the subjects required to sign the informed consent form, answer the questionnaire and take part in Standardized Mini-Mental State Examination (SMMSE). The informed consent form is to prove that the subjects willing to volunteer in this study and give confidence to the subjects about the safety issues. The questionnaire and the Standardized Mini-Mental State Examination (SMMSE) is to obtain the background of the subjects and make sure the subjects are in healthy mental condition with no neurological disorder respectively [6].

The subjects were asked to sit comfortably on the chair. The environment of the room was quiet. EEGO$^{TM}$ sports device (ANT Neuro, Enschede, Netherlands) with 32 channels was used to acquire EEG signal and the sampling frequency was set to 512 Hz. The EEG cap was placed on the scalp and the gel was inserted in all channels to reduce the impedance. The impedance kept below 5 kΩ to minimise the noise. The EEG cap was connected to amplifier and tablet with EEGO64 software.

After experimental set up, the subjects were asked to relax, close eyes and avoid body movement. Then, they were performed motor imagery and motor execution tasks according to the sound of 60 Beats Per Minute (BPM). During motor imagery task, the subjects required to imagine opening and closing left fist for 1 minute, right fist for 1 minute and followed by both fist for 1 minute. After performing the imaginary task, the subjects were asked to rest for 30 seconds and took part in non-imaginary task. For non-imaginary task which is motor execution, the subjects were asked to open and close their fists instead of imagine. The procedures similar with motor imagery task. Figure 3 and figure 4 shows the photo during data collection.

## 3. EEG Signal Processing

There are 3 key stages namely pre-processing, feature extraction and classification to analyse he raw EEG signals [7].

### *3.1. Pre-processing*
Pre-processing is the stage of remove unwanted noises. Notch filter was used to remove power line interference. The power line interference in Malaysia is at 50 Hz [8]. Therefore, the notch filter was set to filter the frequency at 50 Hz [1], [9], [10], [11]. This filter is to minimize the possibility of power line interference. Baseline wander noise was de-noise by using Daubechies 8 wavelet family with 5$^{th}$ level decomposition. Daubeches of db8 is the most effective Daubechies among others. It performed de-noise effectively [12], [13].

### *3.2. Feature Extraction*
Non-linear feature was used in this study due to EEG signal is a non-stationary signal. EEG signal changes over time. In this preliminary study, 4 different methods of non-linear feature were compared. First, Empirical Mode Decomposition (EMD) was applied to decompose the signal and Hilbert transform forms the Intrinsic Mode Functions (IMF) components from EMD. Then Hurst Exponent and Approximate Entropy were applied to extract the features of IMF component respectively. Hurst Exponent and Approximate Entropy also applied to extract the features of original signal after filtered directly.

*3.2.1. Empirical Mode Decomposition (EMD).* Empirical Mode Decomposition (EMD) decomposes the signal into small function which is known as Intrinsic Mode Functions (IMF). IMF is a significant function. Hilbert transform is useful in calculating instantaneous frequency and amplitude. In this preliminary study, it has been used to form IMF components which are instantaneous frequency and instantaneous amplitude respectively [14]. The IMF components were used to extract features of Hurst Exponent and Approximate Entropy.

*3.2.2. Hurst Exponent.* Hurst Exponent is a measure of long range correlation in a time-series. It is popular due to its capability of as a measure of the smoothness of a fractal time-series based on asymptotic behavior [14].

*3.2.3. Entropy.* Approximate Entropy is a measure that quantifies the complexity or irregularity of a time-series data [15], [16]. Larger value of Approximate Entropy will have more complexity and irregularity of the signal [17].

### *3.3. Classification*
Random Forest is also known as random decision tree. It is capable to work with large data sets. It does not required normalization. It only required simple and little features parameter [18]. The accuracy based on class is calculated using equation (1) and the overall accuracy is calculated using equation (2).

$$A = \frac{T_P + T_N}{T_P + T_N + F_P + F_N} \tag{1}$$

$$TA = \frac{Total\ correct}{Total\ of\ predicted\ values} \tag{2}$$

A denotes the accuracy based on class, TA denotes the overall accuracy, $T_P$ denotes the number of true positives, $T_N$ denotes the number of true negatives, $F_P$ denotes the number of false positives and $T_N$ denotes the number of true negatives.

## 4. Results and Discussions
The classification accuracy for different method of non-linear features were examined. The non-linear features that used in this preliminary study were Empirical Mode decomposition (EMD), Hurst

Exponent and Approximate Entropy. The features of 32 channels were extracted and the EEG signals had divided into 170 segments.

The classification accuracy using different method of non-linear features were analysed. EMD followed by HE is extracted the filtered EEG signal using EMD then extracted the features of EMD using Hurst Exponent. EMD followed by ApEn is extracted the filtered EEG signal using EMD then extracted the features of EMD using Approximate Entropy. HE is used only Hurst Exponent to extract the features of filtered EEG signal. ApEn is used only Approximate Entropy to extract the features of filtered EEG signal. Table 1 and table 2 show the classification result for motor imagery task and motor execution task using different method of non-linear features. Random Forest is the method of classification that had been used in this preliminary study. The subject is referred to class.

**Table 1.** Classification result for motor imagery task using different method of non-linear features.

| | Subject 1 | Subject 2 | Subject 3 | Subject 4 | Subject 5 | Overall |
|---|---|---|---|---|---|---|
| **EMD followed by HE** | 60.00 | 60.38 | 65.31 | 74.42 | 62.75 | 37.65 |
| **EMD followed by ApEn** | 71.64 | 75.41 | 75.81 | 96.00 | 84.48 | 57.65 |
| **HE** | 89.71 | 93.94 | 89.04 | 95.39 | 97.06 | 77.65 |
| **ApEn** | 92.21 | 93.33 | 91.03 | 100.00 | 100.00 | 83.53 |

**Table 2.** Classification result for motor execution task using different method of non-linear features.

| | Subject 1 | Subject 2 | Subject 3 | Subject 4 | Subject 5 | Overall |
|---|---|---|---|---|---|---|
| **EMD followed by HE** | 58.49 | 58.33 | 60.47 | 61.22 | 60.47 | 36.47 |
| **EMD followed by ApEn** | 78.38 | 84.06 | 84.06 | 90.48 | 93.55 | 68.24 |
| **HE** | 88.89 | 96.97 | 88.00 | 95.89 | 95.89 | 82.35 |
| **ApEn** | 89.16 | 97.37 | 88.46 | 100.00 | 100.00 | 87.06 |

As shown in table 1 and table 2, the classification accuracy for both tasks using Approximate Entropy is the highest among others. On the other hand, the classification accuracy is low when using Empirical Mode Decomposition. The accuracy based on class only calculated the predicted values within the class, it does not consider predicted values of other classes. Overall accuracy consider all predicted values. Therefore, the accuracy based on class is higher than overall accuracy. From table 1 and table 2, the classification accuracy based on class and overall between motor imagery task and motor execution task not much different.

## 5. Conclusion

From this study, Approximate Entropy is most suitable to be used in human authentication system among others. Empirical Mode Decomposition (EMD) is not suitable in this case because it limited the features of EEG signal although the extracted features is significant. The classification accuracy for both tasks not much different and it is above 80%. Therefore, this can concluded that motor imagery and motor execution are suitable for human authentication system. In future, fusion features between time frequency domain analysis and non-linear method are expected can perform in a more effective way for human authentication system. The recognition accuracy process need to further enhance by developing new algorithm.

**Acknowledgments**

The author would like to acknowledge the support from the Fundamental Research Grant Scheme (FRGS) under a grant number of **FRGS/1/2018/TK04/UNIMAP/02/11** from the **Ministry of Education Malaysia**.

**References**

[1] M. Teplan, "Fundamentals of EEG measurement," *Meas. Sci. Rev.*, vol. 2, no. 2, pp. 1–11, 2002.

[2] A. Stasi, G. Songa, M. Mauri, A. Ciceri, F. Diotallevi, G. Nardone, and V. Russo, "Neuromarketing empirical approaches and food choice: A systematic review," *Food Res. Int.*, vol. 108, no. May 2017, pp. 650–664, 2018.

[3] W. Khalifa, A.-B. M. Salem, M. I. Roushdy, and K. Revett, "A survey of EEG based user authentication schemes," in *The 8th International Conference on INFOrmatics and Systems (INFOS2012)*, 2012, pp. 55–60.

[4] L. S. Hooi, H. Nisar, Y. V. Voon, and S. M. Ieee, "Comparison of Motion Field of EEG Topo-maps for Tracking Brain Activation," *2016 IEEE EMBS Conf. Biomed. Eng. Sci.*, pp. 251–256, 2016.

[5] M. Matthew Hoffman, "Piture of the Brain," *WebMD*, 2014. [Online]. Available: https://www.webmd.com/brain/picture-of-the-brain#1. [Accessed: 10-Nov-2017].

[6] A. Vertesi, J. A. Lever, D. William Molloy, I. Tuttle, L. Pokoradi, E. Principi, and B. Sanderson, "Standardized Mini-Mental State Examination," *Can. Fam. Physician*, vol. 47, pp. 2018–2023, 2001.

[7] S. Munian, S. Sivalingam, V. Jayaraman, and S. Nordebo, "Analysis of Real Time EEG Signals," 2014.

[8] R. Sudirman, N. A. Zakaria, M. N. Jamaluddin, M. R. Mohamed, and K. N. Khalid, "Study of electromagnetic interference to ECG using faraday shield," *Proc. - 2009 3rd Asia Int. Conf. Model. Simulation, AMS 2009*, pp. 745–750, 2009.

[9] K. S. Bayram, M. A. Kizrak, and B. Bolat, "Classification of EEG signals by using support vector machines," in *2013 IEEE International Symposium on Innovations in Intelligent Systems and Applications, IEEE INISTA 2013*, 2013, no. June, pp. 48–51.

[10] S. Vaid, P. Singh, and C. Kaur, "EEG signal analysis for BCI interface: A review," *Int. Conf. Adv. Comput. Commun. Technol. ACCT*, vol. 2015–April, no. August, pp. 143–147, 2015.

[11] M. Ferdjallah and R. E. Barr, "Adaptive Digital Notch Filter Design on the Unit Circle for the Removal of Powerline Noise from Biomedical Signals," *Eng. Educ.*, vol. 41, no. 6, pp. 529–536, 1994.

[12] K. Asaduzzaman, M. B. I. Reaz, K. S. Sim, and M. S. Hussain, "A Study on Discrete Wavelet-Based Noise Removal from EEG Signals," in *Advances in Computational Biology*, vol. 680, 2010, pp. 593–599.

[13] H. A. Akkar and F. Ali Jasim, "Optimal Mother Wavelet Function for EEG Signal Analyze Based on Packet Wavelet Transform," *Int. J. Sci. Eng. Res.*, vol. 8, no. 2, pp. 1222–1227, 2017.

[14] U. R. Acharya, S. Vinitha Sree, G. Swapna, R. J. Martis, and J. S. Suri, "Automated EEG analysis of epilepsy: A review," *Knowledge-Based Syst.*, vol. 45, pp. 147–165, 2013.

[15] D. Abásolo, R. Hornero, P. Espino, J. Poza, C. I. Sánchez, and R. De La Rosa, "Analysis of regularity in the EEG background activity of Alzheimer's disease patients with Approximate Entropy," *Clin. Neurophysiol.*, vol. 116, no. 8, pp. 1826–1834, 2005.

[16] H. Bo, Y. Fusheng, T. Qingyu, and C. Tin-cheung, "Approximate Entropy and It's Preliminary Application in the Field of EEG and Cognition," in *Proceedings of the 20th Annual International Conference of the ZEEE Engineering in Medicine and Biology Society*, 1998, vol. 20, no. 4, pp. 2091–2094.

[17] Y. Kumar, M. L. Dewal, and R. S. Anand, "Features extraction of EEG signals using approximate and sample entropy," *2012 IEEE Students' Conf. Electr. Electron. Comput. Sci. Innov. Humanit. SCEECS 2012*, pp. 1–5, 2012.

[18] C. Donos, M. Dümpelmann, and A. Schulze-Bonhage, "Early Seizure Detection Algorithm Based on Intracranial EEG and Random Forest Classification," *Int. J. Neural Syst.*, vol. 25, no. 05, p. 1550023, 2015.

IOP Conf. Series: Materials Science and Engineering **557** (2019) 012034 doi:10.1088/1757-899X/557/1/012034

# Estimating Greenhouse Gas Emissions from Irrigated Paddy Fields in Indonesia under Various Water Managements

**Chusnul Arif[1*], Budi Indra Setiawan[1], Nur Aini Iswati Hasanah[2], Masaru Mizoguchi[3]**

[1]Department of Civil and Environmental Engineering, IPB University, Indonesia.
[2] Department of Environmental Engineering, Islamic University of Indonesia (UII), Yogyakarta 55584, Indonesia.
[3] Department of Global Agricultural Sciences, The University of Tokyo, Tokyo 113-8657, Japan.

*Email: chusnul_arif@apps.ipb.ac.id

**Abstract**. Methane ($CH_4$) and nitrous oxide ($N_2O$) are two main greenhouse gasses emitted from paddy irrigated paddy fields. Their fluxes are commonly affected by water managements in the fields. However, the main problem in the study of greenhouse gas emissions in paddy fields is the instrumentation for measuring emissions. Measurements of greenhouse gas emissions are costly and complicated. The current study proposes estimating method to quantify greenhouse gas emissions by an artificial neural network (ANN) model. They are estimated based on easily measurable parameters such as soil moisture, soil temperature, soil electrical conductivity (EC), soil redox potential (Eh) and soil pH. The model was verified based on field experiments that were conducted in Bogor, West Java, Indonesia during 26 March – 24 June 2015. Here, three regimes of water management, i.e. continuous flooded (FL), moderate (MR) and dry (DR) regimes, were performed in the field. The DR regime released the lowest total greenhouse gas emissions; however, it reduced grain yield by 58% and 12% compared to the FL and MR regimes respectively. The developed model showed high accuracies for both greenhouse gasses estimation where the coefficients of determination ($R^2$) values were 0.84 and 0.76 for $CH_4$ and $N_2O$ prediction respectively.

## 1. Introduction

Water management in rice production affects greenhouse gas emissions, including methane ($CH_4$) and nitrous oxide ($N_2O$). Many research findings have reported $CH_4$ emissions increasing under flooded water when soil is saturated [1]. On the other hand, $N_2O$ emissions are released at a higher rate when there is limited water under aerobic soil conditions [2]. $CH_4$ is commonly formed during decomposition of organic materials when there is limited oxygen and sulfate in the field [3]. Organisms that play a major role in the process of forming this gas are called methanogenic bacteria. Meanwhile, $N_2O$ gas is primarily formed by nitrification and denitrification processes in the soil [4].

Research on greenhouse gas emissions in paddy fields were first conducted in 1998 in Indonesia. The two gasses in discussion are commonly measured using a closed chamber box with specific dimensions placed over a single/multiple paddies. Then, the gas sample is collected from the chamber and it is quantified using a gas chromatograph in the lab [5]. The method is time consuming and complicated however, and uses more expensive equipment. In addition, gas sampling is usually conducted in the afternoon, which does not represent all conditions in a day.

The quantifying method for the two greenhouse gases from paddy fields comes with limitations, which need to be recognized. Greenhouse gas emissions released to the atmosphere are affected by environmental parameters including soil moisture, soil temperature and soil pH [6]. In addition, those two gasses are produced under soil reduction-oxidation (redox) processes and intermediated by the soil microbes as and fertilizer application. The fertilizer application is an important factor in influencing greenhouse gas emissions from the soils, particularly $N_2O$ emission [7]. The environmental parameters that showed those activities are represented by soil redox potential and soil electrical conductivity.

The relationship between greenhouse gasses and the environment are complex and very difficult to develop mathematical models. Because of this limitation, artificial neural networks (ANN) would be a more appropriate method to use. It can learn and recognize the patterns of empirical data without explicit physical consideration [8]. It has been used in many agricultural aspects such as irrigation planning and application [9], especially to estimate subsurface wetting by drip irrigation [10]. The objective of this study is to propose an ANN model to estimate greenhouse gas emissions, i.e., $CH_4$ and $N_2O$ based on easily measurable parameters such as soil moisture, soil temperature, soil electrical conductivity (EC), soil redox potential (Eh) and soil pH. The developed model is then validated by the empirical data.

## 2. Materials and Methods

### *2.1 Field experiment and design*

The model to estimate $CH_4$ and $N_2O$ emissions was developed based on field experiments that were conducted in paddy fields located in Bogor, West Java, Indonesia from 26 March to 24 June 2015. Here, three plots were supplied with different irrigation regimes with two replications. The total area for the first replication was 237 $m^2$, and the second was 185.2 $m^2$ (Figure 1), the first being under the continuous flooding regime (FL) for conventional rice farming. The water level was kept at 2 and 5 cm in depth above the soil surface from the beginning of the cultivation period until approximately 7 days before harvesting, and then the field was conditioned dry in the harvesting. For the second plot the moderate irrigation regime (MR) was used, in which the soil moisture was kept at a saturated level or water level in the soil surface at the beginning until approximately 7 days before harvesting, and then the field was conditioned dry in the harvesting. The last plot was dry irrigation regime (DR) in which the water level was kept at soil surface from the beginning to 20 days after transplanting, then the water level was drained at -5 cm water depth until approximately 7 days before harvesting, and then the field was conditioned dry in the harvesting.

The soil texture was identical in all plots and there was no significant difference between them, it was classified as clayey, having a clay content of 46.1 ± 3.0%, silt of 29.7 ± 2.3% and sand of 24.2 ± 4.8%. One day before planting, compost (manure) was sown to the field with doses of 5 ton/ha, then Phonska and Urea fertilizers in identical doses of 75 kg/ha were sown 20 days after transplanting to the fields. Silica fertilizer was then sown 30 days after transplanting with a dose of 100 kg/ha. Finally, Phonska and Urea fertilizers were sown again 40 days after transplanting with each dose of 75 kg/ha.

IOP Conf. Series: Materials Science and Engineering **557** (2019) 012034 doi:10.1088/1757-899X/557/1/012034

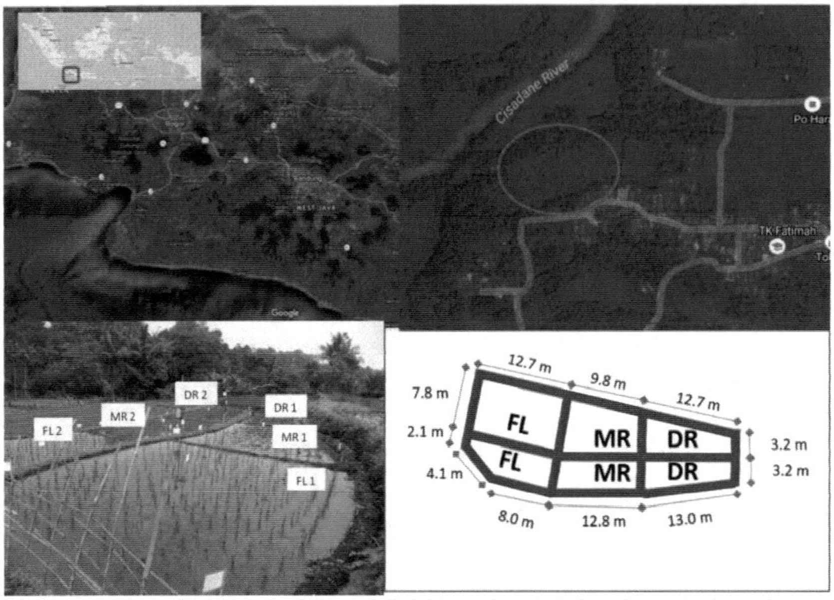


**Figure 1.** Experimental design and field location in West Java, Indonesia.

### *2.2 Environmental parameters measurements*

Soil parameters consisted of soil moisture, soil temperature, soil EC, soil redox potential (Eh) and soil pH. Soil moisture, soil temperature and soil EC were measured by a 5-TE sensor from Decagon Device Corp at 5 cm soil depth, while soil pH and Eh were measured by pHmeter (pH 3310 SET 2 incl. SenTix® 41), and ORP meter (WTW Sentix), respectively. In particular for soil Eh measurement, the output of ORP meter is first converted by normalizing the redox value according to a SHE (*Standard Hydrogen Electrode*) reference electrode on the redox observed by the instrument. Soil Eh values range from -300 mV to 700 mV, in which 300 mV is border of aerobic and anaerobic conditions [11].

Meanwhile, greenhouse gas emissions were measured manually using a closed chamber box. The size of the chamber was 30 x 30 x 120 cm and equipped with a fan to circulate air inside the box. There are four samples were taken on 0, 10, 20, 30 minutes after the chamber is placed in the fields. Then, the sampling gas was quantified using a GC (Micro GC CP 4900) with flame ionization detector (FID) in the lab.

### *2.3 Artificial Neural Network (ANN) Model*

$CH_4$ and $N_2O$ emissions were estimated by using the ANN, a computational mode that learns the dataset given to produce a model that represents the nature of the data. It behaves like a black box that receives input, and produces an output. The ANN model only suggested system modeling where high complexity or uncertainty occurs. It will provide a good prediction, even if the nonlinearity of the system being predicted is high [12]. Since ANN is a learning data model, the more data provided for training, the more accurate the result of prediction is. However, nearly a hundred-percent-accurate modeling dataset should also be avoided, as it is over-fitting phenomena that damage the generalization [13].

The current model consisted of three layers, i.e. input, hidden and output layer. As input, there were five nodes consisting of soil moisture, soil temperature, soil EC, soil Eh and soil pH respectively. Those parameters were selected because of their effects on greenhouse gas emissions from paddy fields. $CH_4$

and $N_2O$ gasses were set as output nodes (Figure 2). Each layer was connected to a hidden layer that was represented by weight values, which are the output of the model that will be used to estimate greenhouse gas emission. Finding optimal weight values, the learning method is back propagation that has two main steps. The first step is propagation (forward and backward propagation), and the second is weight updated by minimizing the errors. Here, for the activation, a sigmoid function was used with the following equations:

$$f(y) = \frac{1}{1+e^{-gy}}, \tag{1}$$

$$y = \sum_{i=0}^{n} x_i w_i \tag{2}$$

where f(y), xi, wi, n, y, g are activation values, the inputs, weights, number of inputs, total signal input and gain parameter, respectively.

The algorithm was developed by Visual Basic Application in MS. Excel 2007. The performance was evaluated by comparing observed and estimated data that was represented by a coefficient of determination ($R^2$). The $R^2$ value ranged 0 to 1. Higher $R^2$ and its score closer to 1 indicated that the model was accurate.

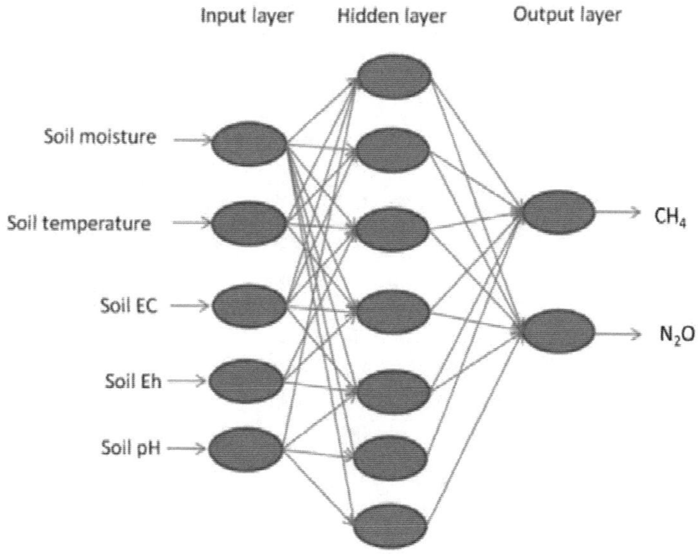


**Figure 2.** The ANN model to estimate greenhouse gas emission from paddy fields.

## 3. Results and Discussion

### *3.1 Greenhouse Gas Emissions and Environmental Parameters*

Figure 3 shows seasonal variations of greenhouse gas emissions under different water levels, soil moisture and soil Eh. The FL regime released more $CH_4$ emission, particularly during 0-21 days after transplanting and reached its peak on 35 days after transplanting. It occurred when standing water materialized in the field, with an average of 2.1 cm above soil surface, the soil moisture was 0.631 $m^3/m^3$ caused by an anaerobic condition that was presented by soil Eh value (<300 mV) with its interval being 129.9 – 171.9 mV. This condition promotes more methanogen activities during organic matter decomposition, resulting in more $CH_4$ emissions when oxygen and sulfate were limited [14]. On the other hand, when average water level reached 0.3 cm below soil surface and the average soil moisture was 0.542 $m^3/m^3$ during that period in the DR regime, the flux of $CH_4$ was negative, which indicated an oxidative process in aerobic condition (soil Eh > 300 mV) had occurred in the fields.

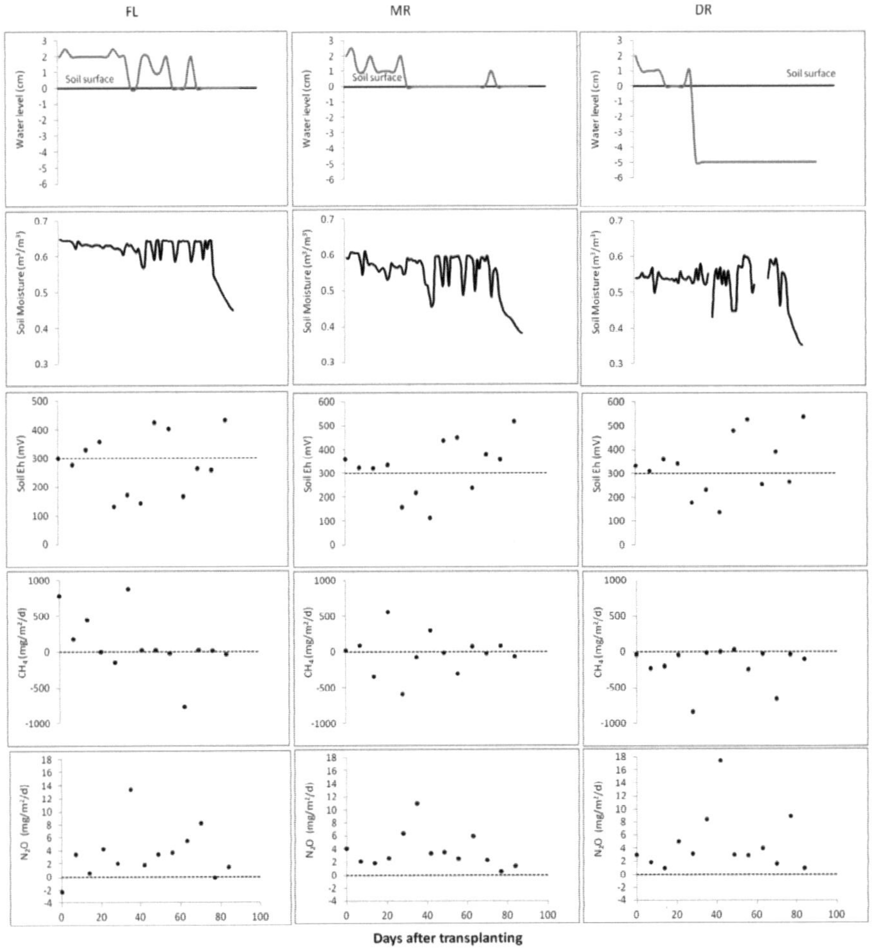


**Figure 3**. Seasonal variations of greenhouse gas emission, water level and soil moisture

For $N_2O$ emission, the flux was comparable and there was no significant difference among the regimes, particularly during the first 28 days after transplanting. The average of $N_2O$ emission in this period was 2.0, 2.9 and 2.9 $mg/m^2/d$ for FL, MR and DR, respectively. $N_2O$ emission had different characteristics with $CH_4$ emission; it mainly being affected by fertilizer application [14, 28]. 28 days after transplanting, fertilizer was sown at regular doses for all plots, soil EC had no significant difference among regimes. $N_2O$ is a byproduct of nitrification and denitrification from biological processes. In aerobic soil where $O_2$ level is relatively high, $N_2O$ is mainly formed by nitrification process [15]. This probably occurred during the DR regime, when the water level was dropped after 31 days after transplanting became 5 cm below soil surface and soil Eh was 342.1 mV, the $N_2O$ emission increased significantly. It was revealed that aerobic conditions under DR treatment promotes more microbial processes, i.e. nitrification and denitrification in soil, producing $N_2O$. This result was similar with previous finding that stated its flux is limited in the flooded, on the other hand the flux reaches maximum at the beginning of the disappearance of flooding water in the soils. Also, when N fertilizer is supplied to the field, $N_2O$ flux rises dramatically [16].

**Table 1.** Total greenhouse gas emissions and rice productivity in each water regime.

| Parameters | Irrigation Regimes FL | MR | DR |
|---|---|---|---|
| $CH_4$ (kg/ha) | 65.3 | -16.7 | -163.4 |
| $N_2O$ (kg/ha) | 3.2 | 3.1 | 4.2 |
| GWP* (kg $CO_{2\text{-eqv}}$) | 2591.7 | 516.2 | -2821.8 |
| Grain yield (ton/ha) | 4.16 + 0.68 | 2.96 + 1.02 | 2.64 + 1.02 |
| Biomass yield (ton/ha) | 12.96 + 1.81 | 10.4 + 0.91 | 13.36 + 0.57 |
| Irrigation water (mm) | 510.4 | 447.3 | 434.6 |

* GWP: Global Warming Potential at the 100-year time horizon of 25 and 298 for $CH_4$ and $N_2O$, respectively (IPCC 2007).

In total, greenhouse gases were emitted at different levels under three water management regimes (Table 1). The FL regime with maintaining higher water levels released more $CH_4$ emissions, particularly in the early growth stage to mid-season stage. $CH_4$ emissions were released to the atmosphere during flooding conditions when soil moisture was at a saturated level. Meanwhile, the DR regime released the highest $N_2O$ emission when drought conditions occurred by maintaining water levels at -5 cm below the soil surface, starting from 20 days after transplanting. This result indicated that $CH_4$ and $N_2O$ emissions have an opposite trend. global warming potential (GWP) is therefore used as an indicator to represent total greenhouse gas emissions. Table 1 shows that DR regime has the lowest GWP, but it significantly decreases the rice yield. The grain yield of DR regime was 58% and 12% lower than that FL and MR regimes, respectively. It was indicated that under limited water, more spikelet sterility occurred, particularly around flowering time. FL regime produced the highest yield of 29% and 37% higher than that of the MR and DR regimes. It was revealed that to obtain optimal grain yield, water stress should be avoided. Meanwhile, the MR regime released greenhouse gas emission at a lower rate of 80% than that of the FL. In addition, it saved more water irrigation, by being 12.4% lower than the FL regime, also decreasing rice yield by 29%. Therefore, the MR regime may be the most effective option for water management in Indonesia for mitigating greenhouse gas emissions.

### *3.2 Model Validation*

The ANN model has the ability to capture the non-linearity of a system, including as a solution to making a model of $CH_4$ and $N_2O$ flux emissions form paddy fields that have a complex relationship with the soil parameters. Figure 4 shows validation results of the developed model, with high accuracies for both greenhouse gas predictions where the coefficients of determination ($R^2$) values for $CH_4$ and $N_2O$ were 0.84 and 0.76. The values, nearly 1 ($R^2 \approx 1$), indicated a good relationship between the model and the observed data, thus the developed model is acceptable [17]. $N_2O$ emissions were estimated less accurate than $CH_4$ emission, most likely caused by overestimation since insignificant differences of $N_2O$ emission were observed in a previous study [18].

The model was accurate in predicting greenhouse gas emissions from a paddy field. Since the field experiments were conducted under certain soil and climatic conditions, we recommend conducting more field measurements that represented any soil and climatic conditions to enrich the observed data, so the model can be trained under a wider interval of soil and climatic conditions.

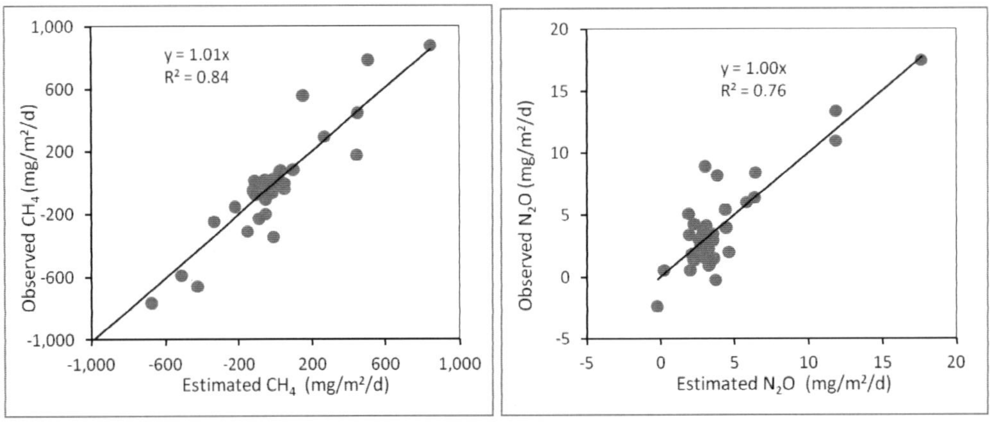


**Figure 4.** Validation results of the developed ANN model.

## 4. Conclusion

Based on our experiments, water management was correlated to rice productivity and greenhouse gas emissions. Maximum grain yield can be obtained under sufficient water supply as represented by the FL. However, this regime released more greenhouse gas emissions as indicated by the highest amount of GWP. MR may be an effective option for water management in Indonesia for mitigating greenhouse gas emissions without losing significant yield. The developed ANN model was effective to predict greenhouse gas emissions, i.e. $CH_4$ and $N_2O$, using easily measurable inputs such as soil moisture, temperature, electrical conductivity, Eh and pH. The developed model showed excellent accuracies for both greenhouse gas predictions where the $R^2$ values for $CH_4$ and $N_2O$ were 0.84 and 0.76. To conclude, this model can be applied to other rice cultivation systems with similar soil and climate characteristics. For wider application, it is recommended that more field experiments be conducted to better train the model under a wider interval of soil and climatic conditions.

## References

[1] Nishimura S, Sawamoto T, Akiyama H, Sudo S and Yagi K, 2004 Methane and nitrous oxide emissions from a paddy field with Japanese conventional water management and fertilizer application, *Global Biogeochemical Cycles* Vol 18, GB2017.

[2] Akiyama H, and Yagi K, 2005 Direct $N_2O$ Emissions from Rice Paddy Fields: Summary of Available Data. *Global Biogeochemical Cycles* Vol 19, GB1005

[3] Epule E.T, Peng C and Mafany N.M, 2011 Methane Emissions from Paddy Rice Fields: Strategies Towards Achieving a Win-Win Sustainability Scenario between Rice Production and Methane Emission Reduction. *Journal of Sustainable Development* 4(6), 188-196.

[4] Mosier A R, Duxbury J M, Freney J R, Heinemeyer O, and Minami K, 1996 Nitrous Oxide Emissions from Agricultural Fields: Assessment, Measurement and Mitigation. *Plant Soil.* 181(1): 95-108.

[5] Sander B.O and Reiner Wassmann R, 2014 Common practices for manual greenhouse gas sampling in rice production: a literature study on sampling modalities of the closed chamber

method, *Greenhouse Gas Measurement and Management*, 4(1):1-13.

[6] Setiawan B.I, Imansyah A, Arif C, Watanabe T, Mizoguchi M and Kato H, 2014 SRI Paddy Growth and GHG Emissions at Various Groundwater Levels. *Irrigation and Drainage*, 63(5), 612-620.

[7] Cai Z.C, Xing G.X, Yan X.Y, Xu H, Tsuruta H, Yagi K and Minami K, 1997 Methane and Nitrous Oxide Emissions from Rice Paddy Fields as Affected by Nitrogen Fertilisers and Water Management. *Plant Soil*, 196(1), 7-14.

[8] Basheer I.A and Harmeer M, 2000 Artificial Neural Networks: fundamentals, computing, design, and application. *Journal of Microbiological Methods* 43:3-31

[9] Raju S K, Kumar D N and Duck L, 2006 Artificial neural networks and multicriterion analysis for sustainable irrigation planning. *Computers and Operations Research*, 33:1138-1153

[10] Hinnell A.C, Lazarovitch N, Furman A, Poulton M and Warrick A.W, 2010 Neuro-Drip: estimation of subsurface wetting patterns for drip irrigation using neural networks. *Irrigation Science,* 28 (6):535-544

[11] Inglett P.W, Reddy K.R and Corstanje R, 2005 *Anaerobic soils*. In: *Encyclopedia of Soils in The Environment;* Hillel D, Eds, Amsterdam, Netherland, Elsevier

[12] Silva I.N, Spatti D.H, Flauzino R.A, Liboni L.H.B and Alves S.F.R, 2017 *Artificial Neural Networks: A Practical Course.* Berlin, Germany, Springer

[13] Srivastava N, Hinton G, Krizhevsky A, Sutskever I and Salakhutdinov R, 2014 Dropout: A Simple Way to Prevent Neural Networks from Overfitting. *Journal of Machine Learning Research*, 15(1), 1929–1958

[14] Bouwman A.F, 1990 *Introduction. In Soil and the Greenhouse Effect;* Bouwman, A.F., Eds.; John Wiley & Sons, New York, United States

[15] Hou A.X, Chen G.X, Wang Z.P, Cleemput O.V and Patrick W.H, 2000 Methane and Nitrous Oxide Emissions from A Rice Field in Relation to Soil Redox and Microbiological Processes. *Soil Science Society of America Journal*, 64, 2180–2186

[16] Snyder C.S, Bruulsema T.W and Jensen T.L, 2007 Best Management Practices to Minimize Greenhouse Gas Emissions Associated with Fertilizer Use. *Better crops*, 19(4), 16-18

[17] Kim H.K, Jang T.I, Im, S.J and Park, S.W, 2009 Estimation of Irrigation Return Flow From Paddy Fields Considering The Soil Moisture. *Agricultral Water Management*, 96, 875–882.

[18] Setiawan B.I, Imansyah A, Arif C, Watanabe T, Mizoguchi M and Kato H, 2013 Effects of Groundwater Level on $CH_4$ and $N_2O$ Emissions under SRI Paddy Management in Indonesia. *Taiwan Water Conservancy*, 61(4), 135-146.

# Preliminary study on EEG based typing biometrics for user authentication using nonlinear features

**Intan Amalina[1], A Saidatul[1], CY Fook[1] and Z Ibrahim[2]**

[1] Biosignal Processing Research Group (BioSIM), School of Mechatronic Engineering, Universiti Malaysia Perlis, Arau, Perlis, Malaysia

[2] Faculty of Technology, University of Sunderland, St Peter's Campus, Sunderland, SR6 0DD, United Kingdom

**Abstract.** The application of password usage in databased user authentication method is unsafe, due to higher chance of off-line guessing attacks. Hackers are able to save our personal information and can detect them until the perfect match is acquired. Thus, the aim of this research is to examine EEG signals during typing task from three engineering students (right handed) with age of between 19 to 23 years old to identify the regular human typing pattern to differentiate between one user to another. All the subjects were asked to perform three times typing tasks for 5 minutes with rest in between for 30 seconds. The subjects were required to rest before and after performing the typing tasks for one minute. Truscan EEG device (Deymed Diagnostic, Alien Technic, Czech Republic) with frequency sampling of 1024 Hz and 19 channels was used. This study applied Infinite Impulse Response (IIR) notch filter to remove 50Hz powerline artefacts and proposed the implementation of nonlinear features such as Distribution Entropy (DistEn) and Approximate Entropy (ApEn). The features were extracted from channels F3 and F4. The extracted entropies features vector are used as an input to k-Nearest Neighbour (kNN) classifier. As a conclusion, the kNN and LDA classifier giving promising accuracy which are 82.22% (F3), 88.52% (F4), 94.81% (F3 and F4) and 80.37% (F3), 81.48% (F4), 89.63% (F3 and F4) respectively.

## 1. Introduction

Biometric is used to identify the identity of an input sample when compare to a template to identify specific user by certain characteristics[1]. Those individual unique characteristics of the users can be in a form of DNA and fingerprint. Authentication is a procedure to confirm the identity of user and validate the reachability of the service to that user.

Authentication based on passwords involves username or ID that is rare to the user[2]. The passwords must be unrepeated, long, complex, consist of both letters and numerals[3]. A secure password is necessary to have lowercase and uppercase letters, special characters, as well as symbols. The full length of the password must be up to 8 characters or more.

Nowadays, passwords are unlimited to databased user authentication. It requires a great measure of performance security. The password is not secured because when users enter their password in public area and observed by spy cameras, this can cause password theft. In order to overcome this problem, the brain wave patterns are recorded during typing task to identify the regular human typing pattern to differentiate between one user and another.

IOP Conf. Series: Materials Science and Engineering **557** (2019) 012035 doi:10.1088/1757-899X/557/1/012035

## 2. Previous Study

Research was done by C. Lin and C. Wu to chance the possibility of using multi-channel electroencephalography (EEG) being collected before movements in order to predict human errors. The task done was hear-and-type task which imitated everyday numerical data entry tasks performed by phone operators. A computer program read out 30 nine-digit, randomly generated numbers without decimals in each trial. Those subjects typed out the numbers. For signal processing, the raw data at 1000 Hz sampling rate were filtered by to 30 Hz bandpass filter. Then 3 temporal features were extracted in a 150-ms time window. For classification, LDA classifier detected 74.34% of numerical typing errors [4].

Research done in [5] was to investigate human-fatigue using smartphone. Each of the subjects involved would have eight writing parts to be done at separate time. In each parts, 1193 taps would take 11.6 minutes. They are asked to type pre-specified writing using their smartphones. The session is separated into 10 segments. 4 features were extracted from the EEG signals. Support vector machine (SVM) and artificial neural network (ANN), are used detect the tiredness level of users. The accuracy obtained were 88.8% and 77.5% respectively.

## 3. Data Acquisition

Three male participants aged of between 19 and 23 years old have involved in this experiment. Right handed with no health issues are chosen. Furthermore, left-handed subjects, those with high blood pressure problem, those with exposure to general anaesthesia and smokers are excluded.

EEG signals were taken by using Truscan EEG device (Deymed Diagnostic, Alien Technic, Czech Republic) with 19 channels to track the brain signals. EEG signals were measured at 19 locations which are Fp1, Fp2, F7, F3, Fz, F4, F8, T3, C3, Cz, C4, T4, T5, P3, Pz, P4, T6, O1 and O2. The position of electrodes follows the international 10-20 system as in Figure 1. EEG signals were taken by EEG machine consists of 16 bits A/D resolution. The participants took 5-10 minutes to remain calm before starting the trials. The sampling frequency was 1024 Hz. The impedances were remained lower than 5 kΩ. The cutoff range signal was 1Hz to 80Hz.

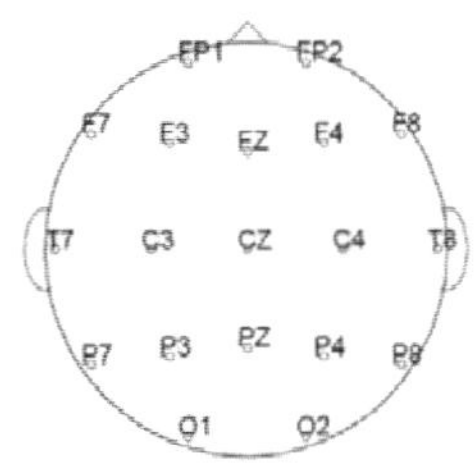


**Figure 1.** The location of electrodes based on 10-20 International System[6]

At first, the subjects filled in the demographic form and questionnaires. Then, they were given 5 minutes to rest. After that, the subjects were told to take a seat in front of LCD computer screen. All subjects were required to perform three times typing tasks for 5 minutes per each trial with rest in between for 30 seconds. The subjects were given a text for them to copy. The subjects were asked to rest before and after performing the typing tasks for a minute.

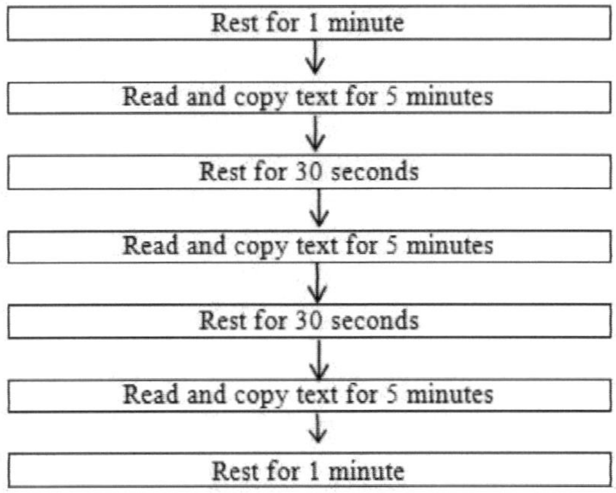


**Figure 2.** Experimental procedure

## 4. Signal Processing

In EEG signal processing, the raw signals were processed to identify the human regular typing patterns. The brain signals went through a few stages in order to be classified. To filter out the noise in the signal, the pre-processing method is applied. Then the informations in filtered signal are removed to describe their behaviours.

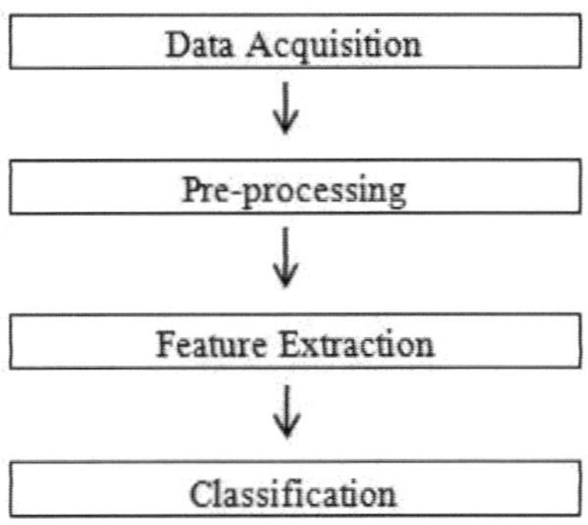


**Figure 3.** Flowchart of signal processing

### *4.1. Pre-processing*

Raw EEG signals that have been taken have amplitudes in microvolts and consists of frequency up to 80 Hz. Thus, pre-processing is needed which involves filtering, and removal of noise. Notch filter of 50 Hz cutoff frequency were applied to eliminate power line interference[7]. The artefacts are particularly caused by wires, light fluorescence and electrode that involves in the data acquisition[8].

### *4.2. Feature Extraction*

In this study, 2 types of nonlinear features are used due to the nature of EEG signal that is non-stationary signal. EEG signals change over time. The nonlinear features are tested separately and by combination to achieve a high accuracy.

4.2.1. *Approximate Entropy (ApEn).* It was applied to measure nonlinear difference contained in filtered signal. It identifies any alterations in the behaviour and differentiate the sample resemblances by pattern length and similarity coefficient[9]. Time series with high irregular pattern results in higher ApEn value whereas time series with highly same patterns results in less ApEn value.

*4.2.2. Distribution Entropy (DistEn).* DistEn evaluates the complexity of time-series by approximating empirical probability density function (ePDF) in the state space[10]. It has been revealed, that time-series with disorganized regime contains dispersive ePDF. This will make ePDF become concentrative for both random and periodic information. Furthermore, ePDF periodic data can concentrate in some separate distances. Therefore, it can be concluded that chaotic series will have higher values of DistEn whereas in periodic series, lower values of DistEn will be obtained.

### *4.3. Classification*

*4.3.1 k Nearest Neighbour (kNN).*The kNN classifier applies a feature distance in a data set toassign the group of the data belongings[11]. A group is formed when the distance within the data is near to each other whereas many groups are created when the distance within the data is far. kNN differentiate a testing data with a training data. The objects are arranged by greatest vote of its neighbours. k values vary to detect the match class between training and testing data.

*4.3.2 Linear Discriminant Analysis (LDA).* LDA is easy in computational procedure and produces best results. Training feature vectors are divided by linear hyper planes to have a smaller variance inside the clusters. The data supposed to have normal distribution. The classes also need to have same covariance matrices. Furthermore, maximization in Fisher's criterion provides the greatest value of the distance among the classes means and the lowest value of variance[12].

## 5. Results and Discussions

Both extracted entropies are used as an input to classifiers such as k Nearest Neighbour (kNN) and Linear Discriminant Analysis (LDA). The entropies are extracted from F3 channel, F4 channel and both F3 and F4 channels.

**Table 1.** kNN and LDA classifier results based on nonlinear features and EEG channels

| **Classifiers** | **Nonlinear Features** | **Channels** | **Percentage (%)** |
|---|---|---|---|
| kNN | ApEn & DistEn | F3 | 82.22 |
| | | F4 | 88.52 |
| | | F3 and F4 | 94.81 |
| LDA | ApEn & DistEn | F3 | 80.37 |
| | | F4 | 81.48 |
| | | F3 and F4 | 89.63 |

Based on the Table 1, the combination of approximate entropy (ApEn) and distribution entropy (DistEn) shows a high accuracy. This means that the nonlinear features need to be combined in order to reach a high accuracy. By comparing both classifiers, it can be said that the kNN classifier shows the higher accuracy than LDA classifier. This is due to its nature which is effective to be used when the training data is large enough. Lastly, the results of this study display that the frontal channels F3 and F4 produce unique brainwave signals in individuals while performing the typing tasks.

## 6. Conclusion

In this research, it can be concluded that channels F3 and F4 give significant and unique brainwave signals in individuals. In order to ensure each individual's identity can be attained, repeating typing tasks are performed. Overall, by comparing both classifiers, the kNN and LDA classifier, the highest performance accuracy can be achieved by kNN which is 94.81%.

## Acknowledgments

We would like to acknowledge the support and encouragement given by the PPK Mechatronic School, University of Malaysia Perlis (UniMAP). We also thank those colleagues from UniMAP who provided guidance in assisting the research.

## References

[1] R. Malathi and R. Jeberson Retna Raj, "An Integrated Approach of Physical Biometric Authentication System," *Procedia Comput. Sci.*, vol. 85, no. Cms, pp. 820–826, 2016.

[2] Y. Zhao, S. Li, and L. Jiang, "Secure and Efficient User Authentication Scheme Based on Password and Smart Card for Multiserver Environment," *Secur. Commun. Networks*, vol. 2018, 2018.

[3] R. Wash *et al.*, "Understanding Password Choices : How Frequently Entered Passwords Are Re-used across Websites This paper is included in the Proceedings of the Understanding Password Choices : How Frequently Entered Passwords are Re-used Across Websites," *Symp. Usable Priv. Secur.*, no. Soups, pp. 175–188, 2016.

[4] C. J. Lin and C. Wu, "Predicting numerical data entry errors by classifying EEG signals with linear discriminant analysis," *Behav. Inf. Technol.*, vol. 34, no. 8, pp. 787–798, 2015.

[5] H. Al-Libawy, A. Al-Ataby, W. Al-Nuaimy, M. A. Al-Taee, and Q. Al-Jubouri, "Fatigue Detection Method Based on Smartphone Text Entry Performance Metrics," *Proc. - 2016 9th Int. Conf. Dev. eSystems Eng. DeSE 2016*, pp. 40–44, 2017.

[6] "EEGcomplexityAD_Kliman."

[7] B. Harender and R. K. Sharma, "EEG signal denoising based on wavelet transform," *Proc. Int. Conf. Electron. Commun. Aerosp. Technol. ICECA 2017*, vol. 2017–Janua, pp. 758–761, 2017.

[8] Hallmarq, "MRI artifacts - and how to avoid them," pp. 1–10, 2017.

[9] U. R. Acharya, H. Fujita, V. K. Sudarshan, S. Bhat, and J. E. W. Koh, "Application of entropies for automated diagnosis of epilepsy using EEG signals: A review," *Knowledge-Based Syst.*, vol. 88, pp. 85–96, 2015.

[10] P. Li, C. Yan, C. Karmakar, and C. Liu, "Distribution entropy analysis of epileptic EEG signals," *Proc. Annu. Int. Conf. IEEE Eng. Med. Biol. Soc. EMBS*, vol. 2015–Novem, pp. 4170–4173, 2015.

[11] M. Pozzi and A. Der Kiureghian, "Response Spectrum Compatible PSD for High-Frequency Range," *22nd Conf. Struct. Mech. React. Technol.*, no. 1, pp. 5–8, 2013.

[12] K. E. C. Vidyasagar, M. Moghavvemi, and T. S. S. T. Prabhat, "Performance evaluation of contemporary classifiers for automatic detection of epileptic EEG," *2015 Int. Conf. Ind. Instrum. Control. ICIC 2015*, no. Icic, pp. 372–377, 2015.

IOP Conf. Series: Materials Science and Engineering **557** (2019) 012036 doi:10.1088/1757-899X/557/1/012036

# Spectral Efficiency Improving Technique in OCDMA-VLC system for IoT Application using Catenated-OFDM Modulation Scheme

**N. M. Nawawi[1] , Anuar M. S.[1] Junita M. N.[1] and M. F. Talib[1]**

[1] Optical Group Communication, Advanced Computing Engineering (ACE) Centre, School of Computer and Communication Engineering, Universiti Malaysia Perlis (UniMAP), Pauh Putra, 02600, Perlis.

E-mail: norizan@unimap.edu.my; anuarms@unimap.edu.my; junita@unimap.edu.my; fauzan.talib@gmail.com

**Abstract.** Visible light communication (VLC) technology offers unregulated, unlicensed and huge bandwidth to meet the growing demand in future indoor communication system. One of the major challenge in VLC system is in improving transmission speed with optimize the bandwidth capacity. The method to improve spectral usage is using advanced modulation formats such as orthogonal frequency division multiplexing (OFDM). In this work, a new approach for modulation technique based on multiband OFDM has been developed which capable to improve the spectral efficiency and optimize the data rate. A proposed technique is then adapted in OCDMA-VLC environment to analyse the reliability of the proposed scheme in high capacity network. Different selection of parameter values such as bit rates, number of bands and effective received power will have different effect on the system performance. Numerical result shown that the catenated-OFDM outperforms conventional OFDM by five times spectral efficiency improvement and eight times spectral efficiency improvement at their five and eight number of bands respectively. It is also shown that the proposed catenated-OFDM design is working well even at very high bit rate of 15 Gbps. Moreover, there are enhancement in receiver sensitivity with 7.3 dB power penalty compared to previous work.

## 1. Introduction

The growing demand for internet-connected of smart devices and high speed networks shapes the internet of things (IoT) environment. The IoT interconnected different devices with ubiquitous accessibility and built-in intelligence. It has gained wide acceptance, growth popularity and predicted to reach seven billion IoT transmission devices in the forthcoming fifth generation (5G) networks by 2025 [1]. According to CISCO, there will be sevenfold increase in mobile data traffic in 2021 compared to 2016 as shown in Figure 1. The growth in the number of devices accessing the mobile networks is the main reason for the drastic increase in mobile data traffic. Along with this, the development of online social medias (such as facebook, instagram and twitter) has further increased the mobile data traffic. The proliferation of these devices has created a huge pressure on the currently established third generation (3G) and fourth generation (4G) mobile communication networks. These technologies become unable to cope with the tremendous growth of the mobile service demands. Therefore, the research and industrial communities are targeting that the new standard of mobile technology in 5G environments will improved the security, enhanced the capacity and transmission performance by 1000 times compared to the existing technologies [2]. One of the option to overcome this issue is by using higher ranges of electromagnetic spectrum in order to replace the radio wave spectrum in 5G networks.

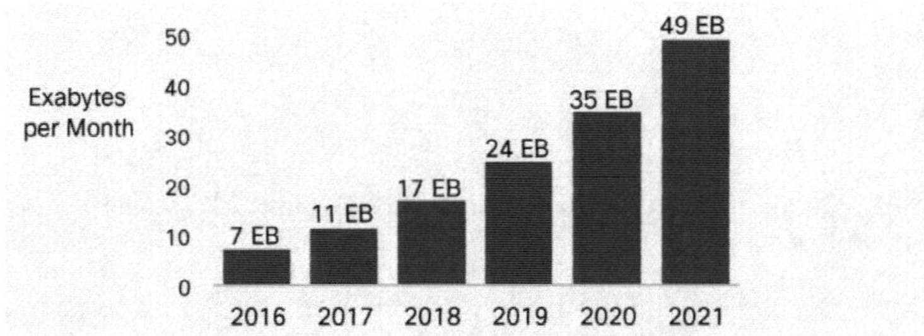


**Figure 1.** Global mobile data traffic [3]

Recently, optical wireless communication (OWC) came along taking the attention because of several importance characteristics such as large bandwidth, high security, low cost, and license free operation. OWC especially visible light communication (VLC) can be a potential candidate solution for 5G networks. The VLC technology utilizes a visible light spectrum in the electromagnetic spectrum which has large bandwidth, safe for eyes, more flexible and integrity than other communication systems in many regards. VLC system has been investigated for about one and half decade and received a lot of interest [4]. Since the medium for transmission in VLC system is visible light and not radio waves that can penetrate walls, the issue of security is inherently solved. It is because light cannot leave the room, containing data and information in one location. There is no way to retrieve and access the information unless a user is in a direct path of the light being used to transmit the data.

Although OFDM is an attractive modulation technique that can achieved the requirement for 5G standard performance, another issue to enable VLC system suited for future 5G networks is access control in this multiuser environment [2]. Therefore, the efficient backbone design network to support higher capacity distribution of connectivity between VLC access points are needed. Optical code division multiple access (OCDMA) is one of the promising technique that offers multiple access control based on optical encoding technique that can enable the OFDM access in the VLC network. OCDMA is a networking technique that provides multiple access and offers various advantages such as large effective bandwidth utilization, asynchronous transmission with low latency access ability to provide privacy and security [5] and soft capacity in demand. The unprecedented increasing of traffic volumes in wireless network, the data privacy and confidentiality is becoming more important. Therefore, in this paper proposed a combination design of VLC and OCDMA technologies which is useful to envision the most promising candidate for future mobile communication. This paper is divided into five sections, after introduction in Section 1; Section 2 discusses the system architecture, Section 3 gives theoretical development and simulation analysis, Section 4 discusses the analysis results followed by conclusion in Section 5.

IOP Conf. Series: Materials Science and Engineering **557** (2019) 012036 doi:10.1088/1757-899X/557/1/012036

## 2. System Architecture

Future telecommunication systems must be spectrally efficient to support a number of high data rate users. For all of the above reasons, OFDM has already been accepted by many of the future generation systems. Another issue to enable VLC suits for 5G requirement is an efficient backbone design of an access control in this multiuser environment. This paper discussed the catenated-OFDM as a new modulation technique for OCDMA-VLC system. A ZCC code has been used as code generation technique in optical code encoder and decoder. ZCC code is proposed to reduce the impact of system impairment and multiple access interference (MAI) in spectral amplitude coding OCDMA [6]. In cat-OFDM modulation scheme, M-QAM mapping is chosen for modulation formats taken on the data stream and shifted into parallel format. The parallel data is then mapped and transmitted by assigning one symbol for each carrier. An inverse Fourier transform is performed into the mapped spectrum to find the corresponding time waveform. To make sure any multipath interference not affected the orthogonality of the data channels, the cyclic prefix (guard period) can be added at the beginning of each symbol. Finally, the parallel data is shifted back into the serial symbols which is called OFDM signals.

**Table 1.** Parameter used in numerical calculation.

| Parameter | Value |
|---|---|
| Number of subcarrier | $N_{sc} = 512$ |
| Number of FFT | $N_{FFT} = 1024$ |
| Position array | 256 |
| Subcarrier spacing | $\Delta f_c = 9.76$ MHz |
| FFT sampling frequency | $f_{FFT} = 40$ GHz |
| Responsivity | $R = 0.32$ |
| Visible laser wavelength | $\lambda_o = 480$nm |
| Receiver noise temperature | $T = 300$K |
| Receiver load resistor | $R_L = 1030\Omega$ |
| Boltzmann's constant | $K = 1.8 \times 10^{-23}$ W/K/Hz |
| Electron charge | $e = 1.60217646 \times 10^{-19}$ C |
| Ambient current | $I_{r_ambient} = 5100\ \mu$A |
| Noise bandwidth factor | $I_2 = 0.562$ |

The system architecture of the proposed catenated-OFDM scheme for OCDMA-VLC system based on ZCC code is given in Figure 2. At the transmitter, an electrical adder is used to concatenate several OFDM bands before entering the optical domain. The most advantage of this scheme is it is fully utilized the available electrical bandwidth so more data can be sent through the medium compared to a traditional OFDM scheme. An electrical catenated-OFDM signal is then being modulated by Mach-Zehnder external modulator with respective unique code from ZCC encoder. As a ZCC code encoder has been adapted into this proposed design, $n$-band of catenated-OFDM are spread by ZCC codes simultaneously over visible light medium based on several optical channels. Each channel has specific spreading code refer to different wavelength which means the generated catenated-OFDM signal band is modulated and carried out through the communication medium in code domain. After passing through the visible spectrum medium, the signal is detected by ZCC decoder and demultiplexed to individual optical channel by applying specific code. The PIN photodetector at receiver part is used to detect the optical signal of each optical channel and pass through to the splitter to separate and extract back each catenated-OFDM signal band by filtering at their respective frequency band. At a very high data rate, the required electrical bandwidth also would be wider and is not cost-effective to implement even with the best commercial digital-to-analog converters (DACs) and analog-to-digital converters (ADCs) in silicon integrated circuit (IC) [7]. Instead of having high spectral efficiency, this technique also has a great capacity to accommodate exponential growth in communications mainly for internet users [8]. This process may eventually lead to the realization of commercial transmission products based on optical OFDM in the future, with the potential benefits of high spectral efficiency and flexible network design. An Optisystem Version 11.0 was used as a simulation tool to simulate the proposed catenated-OFDM modulation technique in OCDMA-VLC system shown in Figure 2 based on parameter on Table 1. The optical channel is set to two using four number of weight for each encoder.

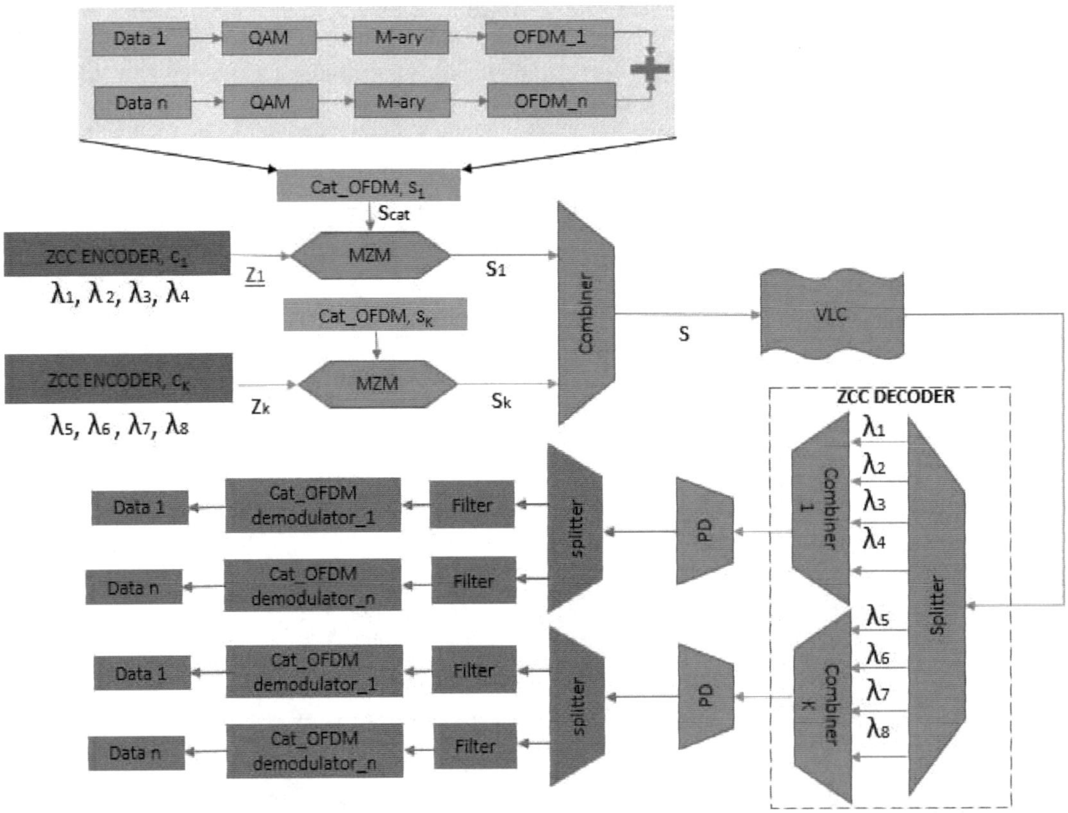


**Figure 2.** System architecture of Catenated-OFDM modulation technique in OCDMA-VLC system based on ZCC code.

## 3. Theoretical Development

For the noise variances, there are two types of noises considered dominant in this analysis which are quantum shot noise and thermal noise. The quantum shot noise is essentially represented by the variance of the photocurrent that contains a distortion to the desired signal. The signal and ambient light induced quantum shot noise for catenated-OFDM VLC system and catenated-OFDM OCDMA-VLC system can be written as below

| Catenated-OFDM VLC system | Catenated-OFDM OCDMA-VLC system |
| --- | --- |
| $\sigma_{shot}^2 = 2eRP_r c_k m_n B + 2eI_{r\_ambient} I_2 B$ (1) | $\sigma_{shot}^2 = 2eR\frac{P_r w}{L} c_k m_n B + 2eI_{r\_ambient} I_2 B$ (2) |

where e is the electron charge, $R$ is the responsivity of the photodetector, $P_r$ is an effective received power, $w$ and $L$ are the ZCC code weight and code length respectively, $c_k$ is an OFDM information symbol, $m_n$ is the modulation index for respective sub-band, $I_{r\_ambient}$ is an ambient currents in the photodetector, and $I_2$ is the noise bandwidth factor which is set to 0.562 [9]. When the generated signal current is much smaller than the ambient noise current, the induced shot noise in the photodetector can be usually treated as additive white Gaussian noise (AWGN), to which category the electrical pre-amplifier noise also belongs [10].

The mean square of the random current is given by

$$\sigma_{thermal}^2 = \frac{4kTB}{R_L} \quad (3)$$

where $T$ is an absolute temperature, $k$ is the Boltzmann's constant, $B$ is the receiver bandwidth and $R_L$ is load resistance. Note that, the thermal noise can be reduced by using large value or load resistor but still consistence with the receiver bandwidth requirement.
The signal to noise ratio (SNR) is given by

$$SNR = \frac{Total\ signal\ power}{Total\ noise} = \frac{I_r^2}{\sigma_{Total}^2} \quad (4)$$

The SNR in Equation (5) for catenated-OFDM VLC system by substituting Equation (1) and Equation (3) in Equation (4) and the SNR in Equation (6) for catenated-OFDM OCDMA-VLC system by substituting Equation (2) and Equation (3) in Equation (4).

| Catenated-OFDM VLC system | Catenated-OFDM OCDMA-VLC system |
| --- | --- |
| $SNR = \frac{R^2 P_r^2 c_k^2 m_n^2}{2eRP_r c_k m_n B + 2eI_{r\_ambient} I_2 B + \frac{4kTB}{R_L}}$ (5) | $SNR = \frac{R^2 \frac{P_r^2 w^2}{L^2} c_k^2 m_n^2}{2eR\frac{P_r w}{L} c_k m_n B + 2eI_{r\_ambient} I_2 B + \frac{4kTB}{R_L}}$ (6) |

As the proposed system consists of catenated-OFDM OCDMA-ZCC model with M-ary QAM in each sub-channel, thus, the bit error rate (BER) can be obtained from SNR as below:

$$BER = \frac{\sqrt{M}-1}{\sqrt{M}\log_2\sqrt{M}} erfc\sqrt{\frac{3\log_2 M}{2(M-1)}SNR} \tag{7}$$

## 4. Results and Discussion

The effect of spectral efficiency due to the changes in catenated-OFDM modulation technique is compared to the conventional OFDM. Figure 3 shows the comparison of signal spectrum as a function of frequency for conventional OFDM, 5-bands catenated-OFDM and 8-bands catenated-OFDM. The link parameters used for numerical calculation is in Table 1. Bandwidth of each OFDM band can be calculated as $BW_{SC} = \Delta f\ x\ N_{sc\_used} = \Delta f\ x\ N_{position\_array}$ = 2.5 GHz. The maximum OFDM bands can be carried are eight with no guard band. For a system with 1024 of FFT size at FFT sampling frequency of 40 GHz, the available electrical bandwidth is 20 GHz (half of the sampling rate) as shown in Figure 3. Catenated-OFDM modulation technique aims to fully utilized the available electrical bandwidth by concatenating several OFDM bands into the whole spectrum.

The spectrum of catenated-OFDM consist of several bands that contain of OFDM data at their representative center or carrier frequencies. Each band brings 10 Gbps data rate. Figure 3(a) shows the OFDM band at carrier frequency of 3.75 GHz. The carrier frequency for second, third, fourth and fifth band are then set accordingly as 7.5 GHz, 11.25 GHz, 15.0 GHz and 18.75 GHz, respectively as shown in Figure 3(b). Each band occupied 2.5 GHz bandwidth thus accommodated five different bands of data. When we set the band spacing equal to subcarrier spacing (9.76 MHz), catenated-OFDM can accommodate eight number of bands and fully utilized the available electrical bandwidth as shown in Figure 3(c).

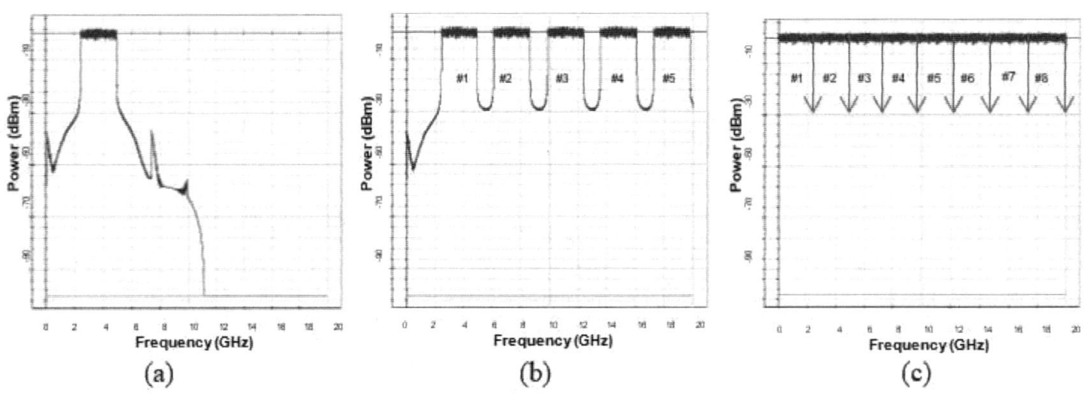


**Figure 3.** Spectrum of (a) conventional OFDM signal and (b) five bands of catenated-OFDM signal and (c) eight bands of catenated-OFDM signal. IFFT size is 1024 is applied.

The catenated-OFDM outperformed conventional OFDM by five times spectral efficiency improvement and eight times spectral efficiency improvement at their five and eight bands catenated-OFDM respectively. Higher number of bands in a spectrum means that more data can be transferred with benefits of high bandwidth efficiency to the system.

**Table 2.** Conventional OFDM and catenated-OFDM spectral efficiency improvement

| | Spectral Efficiency (SE) (b/s/Hz) | Improvement |
|---|---|---|
| Conventional OFDM | 0.5 | - |
| 2 bands catenated-OFDM | 1 | 2 times |
| 5 bands catenated-OFDM | 2.5 | 5 times |
| 8 bands catenated-OFDM | 4 | 8 times |

The results in Figure 4 and Figure 5 are based on SNR in Equation (5). Figure 4 shows the effect of effective received power on the system performance at data rate of 10 Gbps. According to typical acceptable BER of $10^{-9}$ error floor [11], the effective power for conventional OFDM is approximately -16.8dBm. The effective power of catenated-OFDM technique for two, five and eight bands are approximately to -17 dBm, -15 dBm and -13.4 dBm respectively. Compared to conventional OFDM performance, two bands catenated-OFDM outperformed the conventional OFDM by 0.2 dB with double improvement in spectral efficiency. While the power penalty is only ~1.8 dB with five times improvement in spectral efficiency for five bands catenated-OFDM performance and eight times spectral efficiency improvement in eight bands catenated-OFDM performance with only ~3.4 dB power penalty (the spectral efficiency improvement is shown in Table 2). This indicates that the catenated-OFDM technique is not only able to achieve high spectral efficiency but still capable to detect weak power (less than -10 dBm) at the same time. Thus, it proves that the proposed catenated-OFDM modulation technique can be used in visible light communication system with low power penalty.

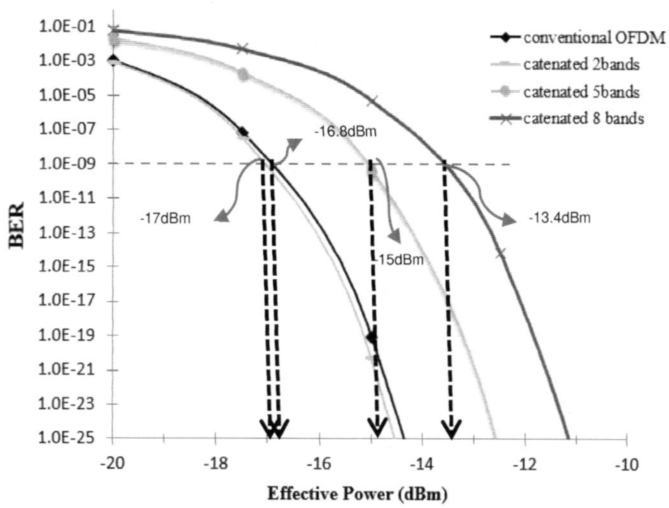


**Figure 4.** Performance of BER against the effective power for catenated OFDM and conventional OFDM

Figure 5 depicts the corresponding BER with respect to the effective power level as the bit rate varies from 5 Gbps to 15 Gbps. In this analysis, the number of band is fixed to five. At BER of $10^{-9}$, the photodetector sensitivity for 5 Gbps, 10 Gbps and 15 Gbps are at -16.5 dBm, -15 dBm and -14 dBm respectively. As can be seen the lower the bit rate, less bit errors have occurred. The power penalty loss is about 1 dB when the bit rate increases from 10 Gbps to 15 Gbps in the same environments. Between 5 Gbps to 10 Gbps the penalty loss is approximately 1.5 dB. This graph proved that the proposed catenated-OFDM design is working well even at a very high bit rate of 15 Gbps.

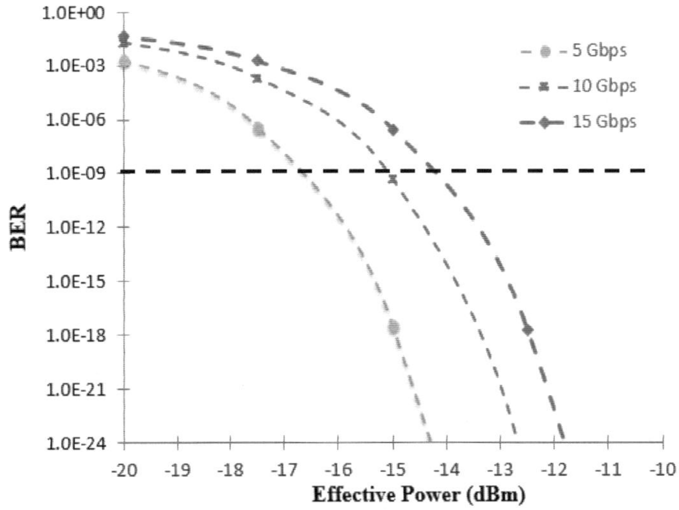


**Figure 5.** Performance of BER against the effective power with different bit rate

Figure 6 gives the BER performance from the SNR Equation (6). This result compared the performance of catenated-ZCC-VLC (Band = 8), OZCZ-VLC [12] and ZCC-VLC systems with respect to the effective power received for the same number of optical channels. In order to quantify the performance in VLC system, the BER is evaluated using similar parameters setup to give an indication of minimum effective power for three different systems. For this result, the bit rate is fixed to 622 Mbps with number of weight is 3 and the number of channels is 4. It can be observed that, the proposed catenated-OFDM in ZCC-VLC system performs better than OZCZ-VLC and conventional ZCC-VLC systems. At BER = $10^{-9}$, the received effective power is approximately -10.2 dBm for conventional ZCC-VLC system, -13.2 dBm for OZCZ-VLC system and -20.5 dBm for catenated-ZCC-VLC system. This result reveals improvement in term of receiver sensitivity of catenated-ZCC-VLC, which the power penalty of 7.3 dB compared to OZCZ-VLC and 10.3 dB compared to ZCC-VLC system. It shows that the catenated-ZCC can advanced the performance of conventional ZCC-VLC and OCZC-VLC. This condition is due to the proposed catenated-ZCC-VLC system has a large number of subcarriers in which increase the signal power and the SNR.

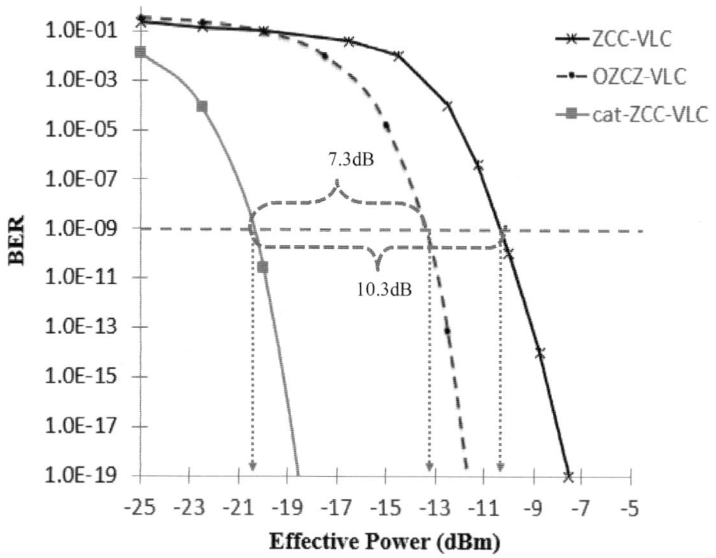


**Figure 6.** BER performance comparison versus effective power for W = 3 and

## 5. Conclusion

In summary, the performance of catenated-OFDM has been analyzed and presented for VLC system and also its integration with OCDMA system based on ZCC code. The proposed catenated-OFDM modulation technique indicates performance improvement in term of spectral efficiency, bit rate and effective power. Different selection of parameter values such as bit rates, number of bands, effective power, weights and number of user will have different effect on the system performance. The combination of OCDMA and VLC can be regarded as a new attractive research area as OCDMA is an effective multiple access network to support higher capacity distribution of connectivity between VLC access points This result reveals improvement in term of receiver sensitivity of catenated-OFDM OCDMA-VLC, which approximately -20.5 dBm for 8 bands. Using catenated-OFDM modulation and zero cross correlation between users, the proposed system totally eliminated the PIIN noise thus significantly improved the overall system performance.

## References

[1] I. Farris, A. Orsino, L. Militano, A. Iera, and G. Araniti 2018 Federated IoT services leveraging 5G technologies at the edge *Ad Hoc Networks* vol 68 pp 58–69

[2] W. Abdallah and N. Boudriga 2016 Enabling 5G Wireless Access Using Li-Fi Technology : An OFDM Based Approach *18th Int. Conf. Transparent Opt. Networks* pp 1–6

[3] Cisco 2017 Cisco Visual Networking Index: Global Mobile Data Traffic Forecast Update 2016-2021 *Cisco* pp 2016–2021

[4] D. C. O'Brien 2011 Visible Light Communications: Challenges and potential *IEEE Photonic Soc. 24th Annu. Meet. PHO 2011* vol 3 pp 365–366

[5] G. Kaur and G. S. Bal 2017 Performance analysis of SAC-OCDMA in free space optical medium using DDW code *Optik (Stuttg).* vol 133 pp 36–42

[6] M. S. Anuar, S. A. Aljunid, N. M. Saad, and S. M. Hamzah 2009 New design of spectral amplitude coding in OCDMA with zero cross-correlation *Opt. Commun.* vol 282 no 14 pp 2659–2664

[7] W. Shieh 2009 Experimental Demonstration and Numerical Simulation of 107-Gb/s High

Spectral Efficiency Coherent Optical OFDM *J. Light. Technol.* vol 27 no 3 pp 168–176

[8] N. M. Nawawi, M. S. Anuar, and M. N. Junita 2018 Cardinality improvement of Zero Cross Correlation (ZCC) code for OCDMA visible light communication system utilizing catenated-OFDM modulation scheme *Optik (Stuttg).* vol 170 pp 220–225

[9] T. Komine, J. H. Lee, and S. Haruyama 2009 Adaptive Equalization System for Visible Light Wireless Communication Utilizing Multiple White LED Lighting Equipment *IEEE Trans. Wirel. Commun.*, vol 8 no 6 pp 2892–2900

[10] K. Cui, G. Chen, Z. Xu, and R. D. Roberts 2010 Line-of-sight visible light communication system design and demonstration *Commun. Syst. Networks Digit. Signal Process (CSNDSP) 2010* pp 621–625

[11] G. Traven and I. Dreven 2010 Limits of data transfer rate in optical communications.

[12] L. Feng, J. Wang, R. Q. Hu, and L. Liu 2015 New design of optical zero correlation zone codes in quasi-synchronous VLC CDMA systems *EURASIP J. Wirel. Commun. Netw.* vol 2015 no 1 pp 120

**Acknowledgments**

This work has been funded and supported by the Department of Higher Education, Ministry of Higher Education Malaysia under Fundamenal Research Grant Scheme, FRGS (9003-00609) leads by Prof. Madya Ir. Dr. Anuar Mat Safar, Universiti Malaysia Perlis.

# Technologies required to improve agricultural issues of Indonesia

**D Miyamoto[1,2], Desrial[2] and IM Edris[2]**

[1]Yanmar CO.,LTD, Osaka, Japan, 5730-8311

[2]Yanmar Agricultural Research Institute-IPB (YARI-IPB), IPB Darmaga Campus, Bogor, West Java, Indonesia, 16680

[1] Corresponding author: daisuke_miyamoto@yanmar.com

**Abstract.** Acceleration of mechanization is one of the effective methods to increase farming productivity. Mechanization status in Indonesian agriculture is not sufficient yet and the current objectives of mechanization is in the stage of intensification of production and relocation of manual labor from farming activities. However, mechanization should also support farm management to become industrialized to increase farm household welfare. By use of ICT technologies equipped on the advanced machinery can also contribute to optimize the effect of using machinery and maximize productivity in their activities as well as proper maintenance of the machine.

**Keywords:** agriculture, machinery, Indonesia, SMARTASSIST

## 1. Introduction

Indonesia is known as the most populated country in ASEAN countries and it is one of the most important policy to increase productivity of agricultural production especially of staple food products. Mechanization would improve the production, however, mechanization status in Indonesia is still not sufficient. This paper would like to suggest some of the technologies which will be required to improve farm management by utilizing machineries.

## 2. Method

This paper was written based on authors' knowledge about the condition in the field of agricultural machineries application gathered in our research survey in 2015 and 2016.

## 3. Result

### *3.1. The existing status of agricultural mechanization development in Indonesia*

The first stage of farm mechanization is called as subsistence farming where no machineries are involved in this stage. Then through industry revolution marked with the invention of machines and development of the world's first compact diesel engines by Yamaoka (1933), agriculture entered to the next stage of developing diesel-powered, labor-saving, reduced-manpower machinery and to opening new markets around the world for agricultural, marine, construction and industrial uses [1].

In Indonesia, small farm mechanization system development is started with material transfer in the year of 1950, followed by design transfer in 1970 and the capacity transfer beginning in 1980 [2]. The objectives of mechanization in Indonesia are to increase productivity through intensification, reduce post harvest losses, increase added value and maintain the quality of farm product [3]. In addition to those multiple objectives, the ultimate goals are to increase farm household welfare and create employment opportunity in the rural area [3].

The evidence of agriculture machineries contributes to increasing of production is by measuring the production of rice as staple food for Indonesia. With assumption of two-wheels tractor can increase cropping index by 34% to 41%, it contributes to rice production of about 19.47% to 20.81% during year of 2011 to 2015 [4]. Realizing this opportunity and responsibility to feed 319 million people in 2045 [5] MoA started in 2014 sets optimist goal to bring Indonesia as a world food barn in 2045. Thus, government distributed more budget for agricultural machineries grant as shown in Table 1.

**Table 1.** Selected agricultural machineries grant from MoA in 2015-2017 [6].

| Selected agricultural machineries grant | Unit | | |
|---|---|---|---|
| | 2015 | 2016 | 2017 |
| Two-wheels tractor | 27,812 | 31,734 | 14,615 |
| Four-wheels tractor | 1,472 | 2,250 | 1,572 |
| Transplanter | 5,879 | 5,854 | 1,730 |
| Harvester | 3,246 | 12,893 * | 4,631 * |

Note:
*: including power thresher; small, medium and large combine harvester; vertical dryer and building

### *3.2. Agricultural machineries utilization level in Indonesia*

In order to understand the application of agricultural machineries by farmers in Indonesia, Yanmar CO.,LTD collaborated with Bogor Agricultural University (IPB) through Yanmar Agricultural Research Institute-IPB (YARI-IPB) to conduct research survey in strategic food crops i.e. rice, maize and soybean. This research was conducted in production center of the crops in 2015 for rice and 2016 for maize and soybean. The result is shown in Table 2 and 3.

In term of rice production, it can be seen that land preparation is already fully mechanized with two-wheels tractor but farmers haven't applied four-wheels tractor yet. Rice transplanter is still not commonly used by almost areas of survey, which only 6% of farmers in Cilacap, Central Java Province) uses rice transplanter. Special condition is for rice harvester where this machines are already used by all farmers in Pinrang, South Sulawesi Province where only 3% and 32% of farmers used rice combine harvester in other survey areas.

**Table 2.** Level of agricultural mechanization in rice production activity (Unit in %) [9].

| Activity and machines | Location of survey | | |
|---|---|---|---|
| | Cilacap, Central Java Province | East Lampung, Lampung Province | Pinrang, South Sulawesi Province |
| Land preparation by: | | | |
| • Two-wheels tractor | 100 | 100 | 100 |
| • Four-wheels tractor | 0 | 0 | 0 |
| Rice transplanting by four-rows transplanter | 6 | 0 | 0 |
| Rice harvesting by combine harvester | 3 | 32 | 100 |

Slightly similar result of figures is shown in maize and soybean production. All farmers respondent in Nganjuk of East Java Province used two-wheels tractor for land preparation. However, only 32% and

27% of farmers uses two-wheels tractor for land preparation in Pohuwato of Gorontalo Province but as much as 26% and 40% of farmers uses four-wheels tractor.

**Table 3.** Level of agricultural mechanization in maize and soybean production activity (Unit in %) [10].

| Activity and machines | Location of survey | | | |
|---|---|---|---|---|
| | Nganjuk, East Java Province | | Pohuwato, Gorontalo Province | |
| | Maize | Soybean | Maize | Soybean |
| Land preparation by: | | | | |
| • Two-wheels tractor | 100 | 100 | 32 | 27 |
| • Four-wheels tractor | 0 | 0 | 26 | 40 |
| Harvester | 0 | 0 | 3 | 0 |

Based on our study on the application of rice transplanter, it is found that it is difficult to run rice transplanter machine on the wet paddy field area especially in Bogor, West Java Province due to alluvial soil type, soft and deep hard pan [11].

### *3.3. Future trend of agricultural machineries technology*

As mentioned in the previous chapter, agriculture is started in a form of subsistence farming. Affected by industrialization technology, economic growth of the country and lifestyle change, agriculture machineries gradually steps forward to industry 3.0 and 4.0 where the application of sensor and automation, robot technology as well as ICT embedded on the machine.

*3.3.1. Needs of IoT technologies to improve productivity.* Utilization of agricultural machinery can relocate manual labor force from farming activities dramatically, however, to improve farm management by use of machinery, machines should be maintained properly and various information from daily work should be recorded and utilized in the convenient way to improve the operation. This requirements leads to the necessity of Smart Machinery system using ICT technologies.

*3.3.2. Introduction of Smart Machinery system.* To improve utilization and expect effective work performance of agricultural machine, data collection during the operation besides proper use of machine is required. To maintain agricultural business by using machine, machine should be under proper maintenance and the usage of machine should also be well managed to improve productivity of their business. To perform these objectives, several services based on IoT technology has already been introduce in this business field. One of those service which can support advanced farming management will be Yanmar SMARTASSIST system.

Yanmar SMARTASSIST system consists of various functions which can improve daily farm work efficiency such as machine information monitoring equipped with GPS, Field information system, Work log management, and analysing functions to suggest improvement as well as antitheft monitoring function.

*3.3.2.1. Proper maintenance of machine.* To maintain advanced farming business and prevent unexpected interval by machine service, it is very important for machine owners to monitor their machine condition every day and hour to realize advance warning from the machine in case of trouble before machine down. To monitor and make warning to the machine owners, various information should be collected from their machine.

*3.3.2.2. Daily work management.* Managing daily machine usage under control of machine owner is one of the important process to improve their business. By recording operational information (Figure 1) from the machine, owner can evaluate their machine and their worker's performance and result for daily tasks. Not only to evaluate the performance, they can use their record to continuously improve process which causing loss in their process by visualization of each tasks.

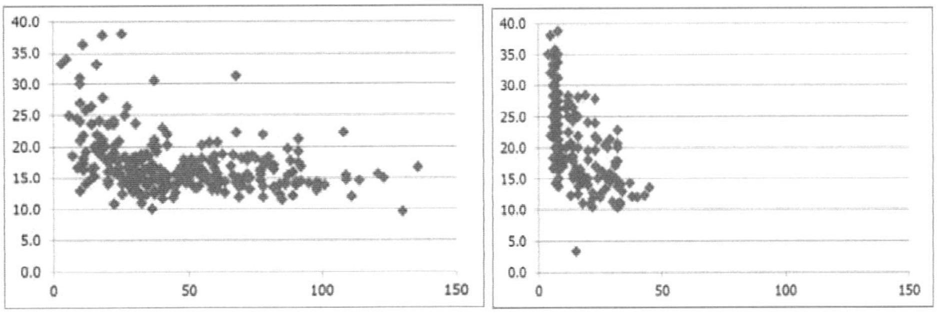


**Figure 1.** Work efficiency depends on the operator's skill and condition of the field.

*3.3.2.3. Field and work log information.* Information about the field (Figure 2) where machines work largely affects their work efficiency. Information such as the field location, actual area, crop variety, and condition of the field need to be stored and reviewed before the next season's business.

**Figure 2** . Field information can be visualized for easy review.

Work log and plants growing status in the field also need to be recorded to determine the schedule of next work. Daily recording can be done through smartphone application which can be analysed by using computer. This can also support to fulfil GAP requirement.

*3.3.2.4. Importance of support from machine suppliers.* Besides monitoring machine condition all the time by machine owners, Yanmar Remote Support Center also support monitoring the machine condition every hour and inform the owners in case of any machine trouble. This can decrease machine down time and unexpected theft cases.

*3.4. Do Indonesia can apply this technology?*

In 2010, Indonesia enters bonus demography where the dependancy ratio getting smaller in the history of a country. When bonus demography ended in Japan in 1970, Japan became the third world largest economy after USA and Uni Soviet. This is the starting and critical point for Indonesia to be stronger in economy.

In this era, Indonesia is a country where internet utilization is very huge. In 2017, the internet users grow by 8% to 143.26 million people or 55.68% of total population in Indonesia [12]. From this number, 49.52% is those who ages 19-34 yo, 29.55% is 35-54 yo, 16.68% is 13-18 yo and 4.24% is over than 54 yo. This figures shows that ICT is all-well accepted by young generation means that we are sure that agricultural machineries with ICT will be accepted by young generation.

## 4. Conclusion

To improve productivity of farming activities is needed to bring Indonesia as a world food barn in 2045. In this stand point, more mechanization accelerated by technological improvement of suppliers to adapt farm condition in Indonesia also supported by government subsidy would be the most important factor to achieve improvement. However, to improve farming productivity by use of machinery, farm management should also be industrialized to maximize the effect of using machinery in their business.

By use of ICT technologies in the future machines should also support their management by suggesting maximum productivity in their activities as well as proper maintenance of the machine.

## 5. Reference

[1] Yanmar Co.,LTD 2015 *100 Years of Yanmar 1912-2012* (Japan: Yanmar CO.LTD)

[2] Handaka 2003 *Sustainable farm mechanization development: an alternative solution for technology development* (Serpong: Indonesian Centre for Agricultural Engineering Research and Development)

[3] Handaka 2005 *Agricultural engineering R&D in Indonesia: challenge and prospect toward sustainable agriculture and APCAEM programme* (Country paper for APCAEM TC/GC Meeting in New Delhi, 21-24 November 2005)

[4] Ministry of Agriculture 2016 *Laporan Kinerja Direktorat Jenderal Prasarana dan Sarana Pertanian Tahun Anggaran 2016* (*Performance Report of Directorate General of Facilities and Infrastructure of Agriculture of Fiscal Year 2016,* in bahasa) (Jakarta: Direktorat Jenderal Prasarana dan Sarana Pertanian (*Directorate General of Facilities and Infrastructure of Agriculture*))

[5] Statistics Indonesia. Statistical Yearbook of Indonesia 2018. Jakarta: Statistics Indonesia

[6] Ministry of Agriculture 2016 Laporan Kinerja Kementerian Pertanian Tahun 2015 (*Annual Report of Ministry of Agriculture Year 2015*) (Jakarta: Ministry of Agriculture)

[7] Ministry of Agriculture 2017 Laporan Kinerja Kementerian Pertanian Tahun 2016 (*Annual Report of Ministry of Agriculture Year 2015*) (Jakarta: Ministry of Agriculture)

[8] Ministry of Agriculture 2018 Laporan Kinerja Kementerian Pertanian Tahun 2017 (*Annual Report of Ministry of Agriculture Year 2015*). (Jakarta: Ministry of Agriculture)

[9] Yanmar Agricultural Research Institute-IPB 2015 *Investigation of Indonesian Farmers and Agricultural Villages Status: Trend and Future* (*Case Study in Central Java, Lampung and South Sulawesi Province*) (Bogor: YARI-IPB) (*unpublished*)

[10] Yanmar Agricultural Research Institute-IPB 2016 *Investigation of Indonesian Farmers and Agricultural Villages Status: Trend and Future on Maize and Soybean Farming (Case Study in East Java and Gorontalo Province).* (Bogor: YARI-IPB) (*unpublished*)

[11] Desrial, Miyamoto D and Edris IM 2018 Modifikasi dan uji kinerja mesin penanam padi delapan baris (Yanmar VP8DN) (*Modification and performance test of eight rows rice transplanter* Yanmar VP8DN) (*Nat. Conf.* of Indonesian Society of Agricultural Engineers in Yogyakarta, 29-31 August 2018)

[12] Indonesia Internet Service Provider Association 2017 Penetrasi & Perilaku Pengguna Internet Indonesia (https://apjii.or.id/content/read/39/342/Hasil-Survei-Penetrasi-dan-Perilaku-Pengguna-Internet-Indonesia-2017, accessed on Dec 13, 2018)

IOP Conf. Series: Materials Science and Engineering **557** (2019) 012038 doi:10.1088/1757-899X/557/1/012038

# Integration of Asset Tracking System through Trilateration Method as Detection Mechanism

**M A Fadzilla[1, 3], Shahriman, A.B[1, 5], A. Harun[1, 3], M.S.M Hashim[1, 2], Z.M Razlan[1,2], W.K Wan[1, 2], Z. Ibrahim[5], Abdullah, A. H[2],. J.S.C Turner[1, 3] , K.A.A Kassim[4], M.S.A Khalid[4], Z. Jawi[4] and M.H.M Isa[4], R.C Ismail[3]**

[1]Motorsports Technology Research Unit (MoTECH), Universiti Malaysia Perlis, Pauh Putra Campus, 02600 Arau, Perlis, Malaysia

[2]School of Mechatronic Engineering, Universiti Malaysia Perlis, Pauh Putra Campus, 02600 Arau, Perlis, Malaysia

[3]School of Microelectronic Engineering, Universiti Malaysia Perlis, Pauh Putra Campus, 02600 Arau, Perlis, Malaysia

[4]Malaysian Institute of Road Safety Research (MIROS), Lot 125-135, Jalan TKS 1, Taman Kajang Sentral, 43000 Kajang, Selangor, Malaysia

[5] Faculty of Technology, University of Sunderland, St Peter's Campus, Sunderland, SR6 0DD,United Kingdom

*Corresponding author: shahriman@unimap.edu.my

**Abstract.** Demands for localization system has been growing rapidly in the last several years both for an outdoor and indoor area. In conjunction with this, the capability and reliability of this system to precisely locate and track objects of interest for the indoor area has catered researchers and study on how to do so. One of the major ideas on making it more advance is by incorporating the use of wireless devices into the system. There are numbers of issues that could interrupt the efficiency and success of the system. One of the main problems is the signal loss mainly caused by the attenuation of the signal as they propagate through from the transmitter to the receiver. These attenuations are mostly due to the surface types the signal are traveling on and the objects that are in the Line of Sight in between the transmitter and receiver. In order to ensure the most reliable and efficient wireless connection between transmitter and receiver, a propagation study on the signal is needed for us to analyze and find the best way to trade off the signal attenuation based on the environment surrounding the system. By doing so, a thorough system that has models that can work efficiently even if we are to consider the attenuation factors. The system consists of nodes installed inside the research institute that acts as both transmitter and receivers. The transmitter and receiver will then process the signal that will then determine their location. The receiver is connected to the laptop in order to get a real-time reading so that we will be able to locate the transmitter. A networked of nodes are installed inside the research institute for experiment and the layout of the research is conferred for future references. Data from the experiment are then analyzed and a model for the signal propagation alongside the research institute is created. This model will be able to apprehend the signal attenuation despite the surrounding environment such as furniture and walls. A completed asset tracking system with models of signal attenuation will be built in the future for a more efficient signal transmission.

## 1. Introduction

Demands for localization system has been growing rapidly in the last several years for both indoor and outdoor area. Hand in hand with this, the capability and preciseness of the system to locate the objects of interest has been an important research topic and study.

Asset tracking has been a major solution for multiple localization problems in the industry and daily lives of people. The development of asset tracking system, however, has been slow for the past years due to less usage in consumers' daily lives[1]. Nevertheless, the application of asset tracking in the industry is growing rapidly especially in warehouses, logistics management, plantation or even manufacturing lines. Large companies in the industry might benefit from the system as they need to manage a huge number of company's assets and other equipment.

As an example, a shipping company requires an efficient asset tracking system to packages and handles shipments. Some other companies tracks certain types of asset such as construction vehicles, the packaging itself, scanner for identification of packages and even the smallest things such as files and documents. Management of these asset is crucial as the assets are movable and will be brought out by workers as they transfer the packages from one location to another. Considering that human errors such as misplacing of scanners and misplacing of packages this will lead to a huge localization task as they need to track and find out the missing asset by themselves without knowing where they are. Hence to reduce the time-consuming method of finding the items manually, asset tracking system meant to reduce tracking time and also increase the efficiency of the company. This paper focuses on designing an asset tracking system to locate indoor assets or items.

Trilateration is a method that is alternative to that of a triangulation that relies on the distance measurements only. Electronic devices distance measurement makes trilateration a cost-efficient localization technique for control surveys[2]. It can also be used to determine the location coordinate and location by using both the transmitter and receiver. It has the same working concept as that of triangulation[3].

Multiple nodes will be placed inside a location of interest to create a network that will then be analyzed to locate a moving node that is inside or within the reach of the nodes. The same concept was used in Global Positioning System (GPS) localization considering that the nodes placed acts as a satellite while the moving nodes are the object of interest that is to be located[4].

For indoor areas, developments and advancements have emerged from researchers by building a system that can precisely locate and track an object of interest on a real-time basis. Indoor area has been more variegated in terms of application due to the fact that the typical GPS would not be able to be used in an indoor area due to its limitation[5]. This has brought upon the idea of implementation of radio technologies and also the application of IEEE 802.11 (Wi-Fi)[6]. However, rather than being a hassle, this is a huge advantage for developers and also researchers due to the fact that the technology is common and widespread making it easier and more flexible to be adjusted and improvised[7].

## 2. Materials and Method

### *2.1. Equipment*

The system consists of multiple nodes placed inside our research institute for monitoring of either moving or stationary objects. The system was carried out by using a pair of transmitter and receiver of XBEE S2 nodes from Digi which will be represented as nodes. These nodes operate at 2.4 GHz and antennas with gain of 2 dBi omnidirectional [8]. Both transmitter and receiver were mounted at 0.8 m height above ground. The real reason for allocation of height is such that on average the height of furniture inside the research institute is at 0.8 m. By doing so, we will be able to perform a functional asset tracking system with consideration of furniture and walls as obstacles and eliminates the possibility of the signal attenuation due to the floor surface by placing it at 0.8 m above ground.

IOP Conf. Series: Materials Science and Engineering **557** (2019) 012038 doi:10.1088/1757-899X/557/1/012038

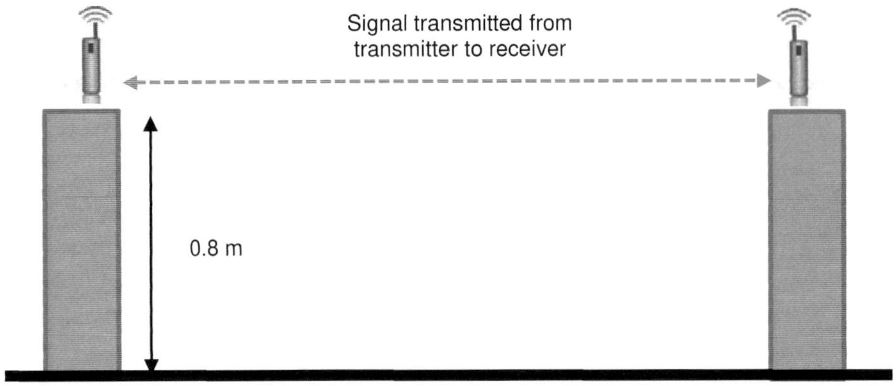


**Figure 1.** Nodes placement and height.

For the receiver, it will be connected to a laptop that will display the current location of the nodes placed inside the research institute and also the location of the moving node whenever the user requested it. As shown in Figure 1, the transmitter stated as Tx was mounted on a mobile object that will be moving around the research institute. Meanwhile, another set of nodes consisting of 4 XBEEs will then be attached to a stand that represented as receiver, Rx which will be stationary. The receivers will be connected to a transmitter and keep on receiving signals from the transmitter as long as it is powered up.

*2.2. System Setup*

For this system, signal strength is implemented to indicate the received power when carrying out the pilot test. Accordingly, a Received Signal Strength Indicator (RSSI) can be predicted and measured based on the average signal strength over the distance of radius centered on the receiver[1]. During the whole experiment, the furniture allocation and placement are kept fixed as shown in Figure 2. Meanwhile, the number of people inside the research institute during the experiment were kept fixed and minimum as only the person in charge of reading the signal strength and another one as the person allocated to moving the transmitter inside the research institute

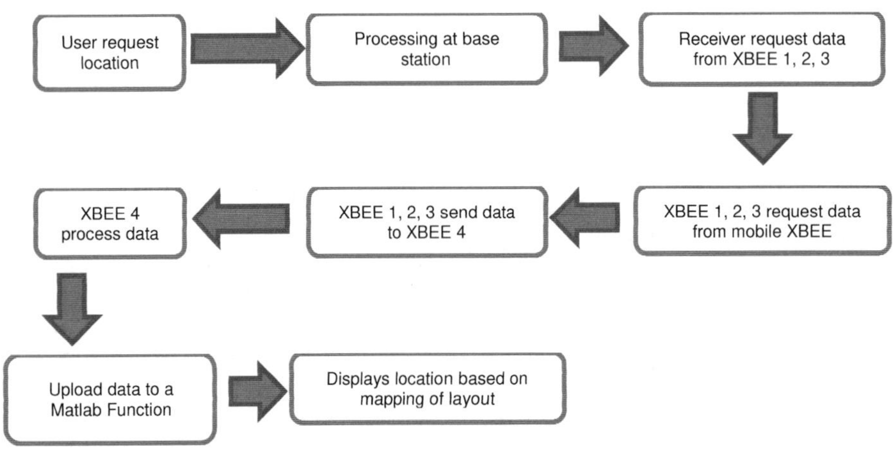


**Figure 2.** Flow chart showing the process designed for the system to collect and transmit data.

IOP Conf. Series: Materials Science and Engineering **557** (2019) 012038 doi:10.1088/1757-899X/557/1/012038

For the indoor localization system, a mesh network was designed in order to perform a localization system application for the experiment. XBEE 1, 2, 3 are receivers that are connected to both the receiver and base station, while XBEE4 is a transmitter which only connects to the receiver at the base stations. Mobile XBEE only connects to XBEE1, 2 and 3. Whenever the user request for the location of the tagged item, XBEE 4 will send a request to all receivers to feed XBEE4 the current RSSI of the mobile XBEE in order to determine its current position. The mesh network design can be seen in Figure 3.

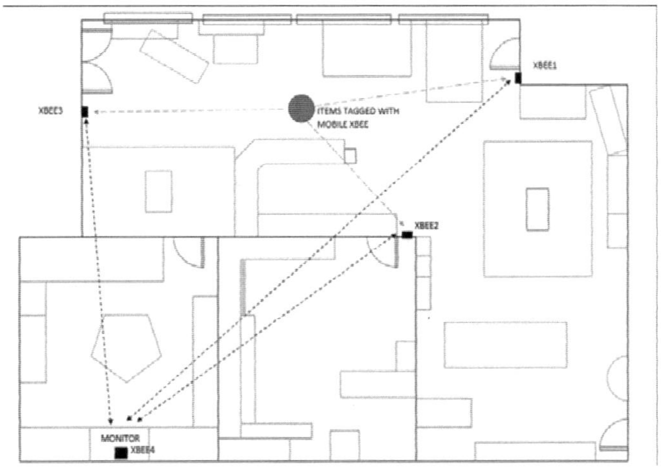


**Figure 3.** Mesh Network Design.

From this design, we can expect that whenever the user request, the current location of mobile XBee can be shown and located on the monitor that is connected to the XBEE 4. For the localization system experiment, the mobile XBee will be carried along by a human to stroll out along the research institute and then set to appear on a real-time basis

## 3. Results and Discussion

For the indoor localization system, a signal propagation was studied and analyzed based on a previous study [9], [10]. The properties of signal that propagates inside the research institute were analyzed and studied to create a model that will then ensure an effective and efficient signal transmission inside the research institute. Theoretically, the complexity of the environment causes the fluctuations in the signal power while the objects surrounding its path influence the signal strength [11]. However in real time application, this might not apply as every room or indoor area are different according to its purposes such as warehouse, offices, pantries and leisure room. Each of these rooms possesses its own purpose with its own furniture and equipment. Hence a proper signal propagation study is needed beforehand[12].

From the propagation study intended before, we have come out with a layout of the research institute with specific measurements as the intended distances needed for trilateration method can be clarified either the distance measured and predicted by the system is the same as that of the real measurements. This is compulsory as we will be able to identify the major factors affecting the signal propagation as the signal is transmitted from a transmitter to a receiver.

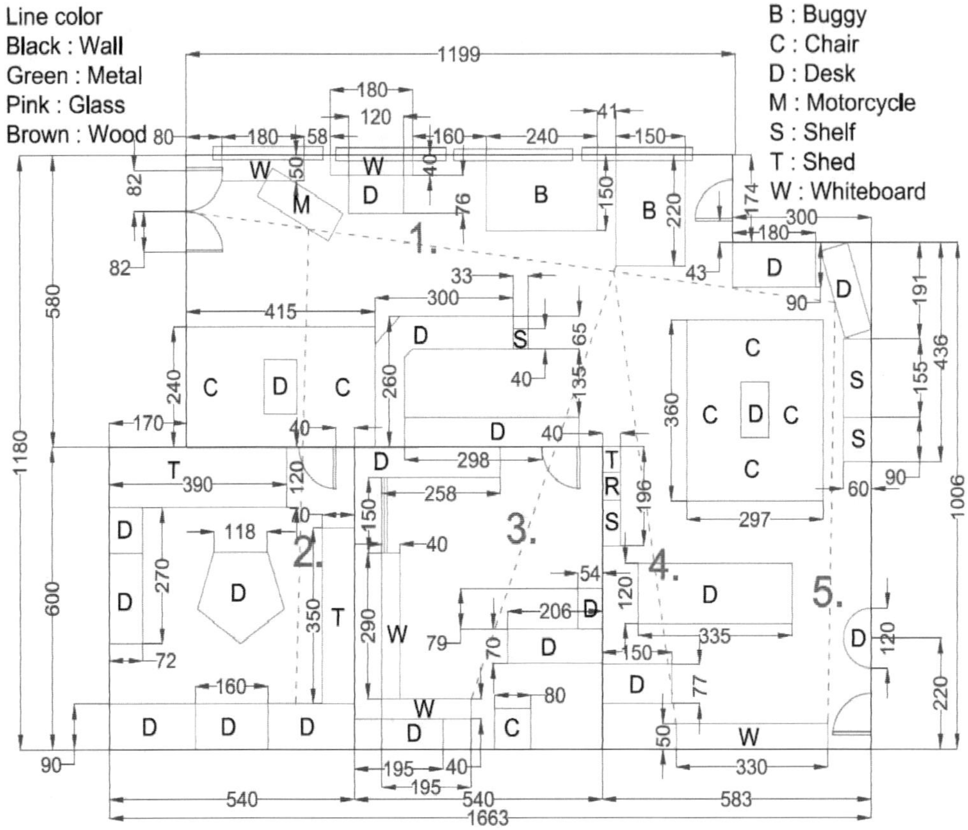


**Figure 4.** Layout of the Research Institute.

For the localization system, however, we need to be able to identify the major signal attenuating factor that will need to be considered in determining the proper propagation study on the signal. These factors are mainly known as a propagation mechanism. Propagation mechanism of the signal transmitted refers to how the signal is being propagated. Scattering, reflection, and diffraction are the most common mechanism that encountered in a typical signal transmission[13][12]. Scattering occurs when the transmitted signal is faced with large numbers of small dimension objects like that of metal cabinets, lampposts. The reflected energy during the scattering process is spread out in multiple directions before reaching the receiver[6]. This shows that the signal power transmission does not efficiently happen even though the signal is received in the end. For reflection, it happens whenever the transmitted signal encounters an object with large dimensions as compared to its wavelengths such as walls, ceilings, and sofas[10]. Some of the signals that have been transmitted will be absorbed through these medium and the remaining will be reflected off of the medium's surface [14]. While diffraction occurs whenever there are obstacles or objects that have sharp edges resulting in a secondary signal that in effect bend around the obstruction [6]. In normal situations, these aspects need to be taken into consideration whenever analyzation of signal propagation is performed.

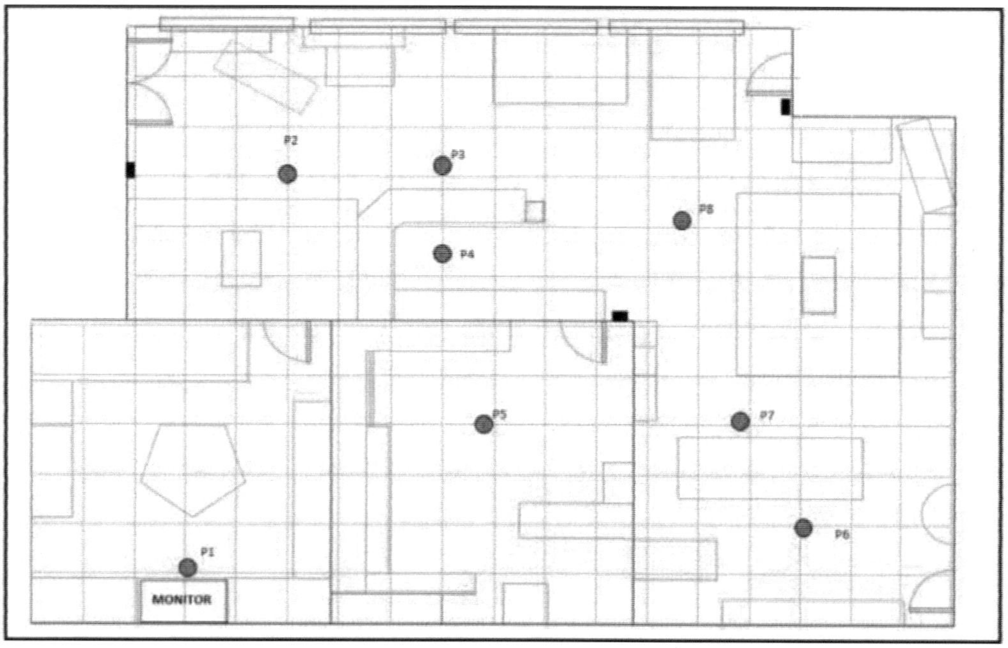


**Figure 5.** Simulation by using Matlab for 8 XBEEs for indoor localization system.

The indoor localization system was executed by using 8 different receivers with the same settings while being at different positions. The user requested for the monitoring module to display all 8 location of tagged items at the same time and display it at the monitor. The result can be seen as shown in Figure 5. It is observed that the localization of the tagged items was almost accurate with an error of up to 2 meters. This shows that the indoor localization system can be performed by using XBEE. This, however, does not mean that XBEE is the best option in implementing the most efficient indoor localization system. However, in terms of performance, the signal propagation does not perform well based on the signal profiling done in my previous paper. This is mainly due to the interference of furniture, walls and the surface of the floor.

For each of the localization intended for each points starting from point P1 to point P8, there is some attenuation in the measurements of the distance they are away from the XBEE 1, 2 and 3. This may be due to the positioning of the furniture. For example for point P6, the distance predicted by the system was that they are to be 14.2 m away from XBEE 1 meanwhile the actual measurement is that they are 12.7 m away from XBEE 1. This has confirmed us that the system still has an error in the prediction of location for up to 2 m. This shows that the system still has flaws in them in terms of prediction of distance. This has allocated us some major improvisation on the system accuracy and also prediction ability. This can be handled more efficiently by making a more subjugated study and more specific modeling in future system development.

A more subjugated and variegated system design is still needed in order for us to perform the best and most efficient asset tracking system. This is a need in order to ensure the highest accuracy in the system and also the best in terms of robustness, and complexity of the design itself. Other than that, we may need to consider other wireless devices such as LoRa, MEMSIC, and WLAN.

## 4. Conclusion

This paper presented an integrated wireless asset tracking system designed to locate and track desired asset or objects around area of interest in this case the indoor area of our research institute. The design was done in consideration of the furniture, walls and also the layout of the research institute in order to minimize the signal attenuation caused by the obstacles and also minimize the signal loss making it more reliable and accurate. This paper however looked into only the application of wireless asset tracking for indoor area only.

Based on the preliminary data collection on controlled surrounding, it is observed that the localization of the tagged items was almost accurate with an error of up to 2 meters. This shows that the indoor localization system can be performed by using XBEE. This, however, does not mean that XBEE is the best option in implementing the most efficient indoor localization system. However, in terms of performance, the signal propagation does not perform well based on the signal profiling done in my previous paper. This is mainly due to the interference of furniture, walls and the surface of the floor. Further improvements and studies are needed for us to design a better and more reliable wireless asset tracking system in the future.

**Acknowledgments**
The authors would like to acknowledge Eastern Pacific Industrial Corporation Berhad (EPIC) for financial and administrative support on this project through the academic-industry collaboration programme.

**References**

[1] S. Shukri *et al.*, "Analysis of RSSI-based DFL for human detection in indoor environment using IRIS mote," *2016 3rd Int. Conf. Electron. Des. ICED 2016*, no. August 2016, pp. 216–221, 2017.

[2] "11. Trilateration | The Nature of Geographic Information." [Online]. Available: https://www.e-education.psu.edu/natureofgeoinfo/c5_p12.html. [Accessed: 08-Jan-2019].

[3] S. S. LATHIKUMARI, "WALAN INDOOR BASED LOCALIZATION," UNIVERSITY OF TEXAS, 2011.

[4] P. Sedlacek, M. Slanina, and D. Kovac, "An Overview of Indoor and Outdoor Positioning Technologies with Focus on their Precision - An Overview of Indoor and Outdoor Positioning Technologies with Focus on their Precision," 2016.

[5] A. A. Khudhair, S. Q. Jabbar, M. Q. Sulttan, and D. Wang, "Wireless indoor localization systems and techniques: Survey and comparative study," *Indones. J. Electr. Eng. Comput. Sci.*, vol. 3, no. 2, pp. 392–409, 2016.

[6] A. H. Ali, M. R. A. Razak, M. Hidayab, S. A. Azman, M. Z. M. Jasmin, and M. A. Zainol, "Investigation of indoor WIFI radio signal propagation," *ISIEA 2010 - 2010 IEEE Symp. Ind. Electron. Appl.*, no. Isiea, pp. 117–119, 2010.

[7] J. S. C. Turner *et al.*, "Modelling indoor propagation for WSN deployment in smart building," *2014 2nd Int. Conf. Electron. Des. ICED 2014*, pp. 398–402, 2011.

[8] MaxStream, "XBee ™ / XBee-PRO ™ OEM RF Modules," *MaxStream*, no. 801, pp. 1–33, 2005.

[9] I. Ibrahim, M. K. Faizi, M. A. M. Saad, Z. Ibrahim, and N. A. M. A. Hambali, "Wireless signal propagation study in an experiment building for optimized wireless asset tracking system development .," pp. 1–5.

[10] I. Ibrahim, M. K. Faizi, M. A. M. Saad, Z. Ibrahim, and N. A. M. A. Hambali, "Localization assessment for asset tracking deployment by comparing an indoor localization system with a possible outdoor localization system," pp. 1–6.

[11] W. Rong-Hou, L. Yang-Han, T. Hsien-Wei, J. Yih-Guang, and C. Ming-Hsueh, "Study of characteristics of RSSI signal," *Proc. IEEE Int. Conf. Ind. Technol.*, pp. 3–5, 2008.

[12] D. L. Ndzi *et al.*, "Signal propagation analysis for low data rate wireless sensor network applications in sport grounds and on roads," *Prog. Electromagn. Res.*, vol. 125, no. November 2011, pp. 1–19, 2012.

[13] D. L. Ndzi *et al.*, "Wireless sensor network coverage measurement and planning in mixed crop farming," *Comput. Electron. Agric.*, vol. 105, pp. 83–94, 2014.

[14] A. Asmussen, R. Rodriguez, and L. Roshak, "MC1319x Range Performance," *Free. Appl. note AN2902*, pp. 1–14, 2004.

# Software Defined Networks *Security: Link Failure Analysis* in SDN

**M A H. Al-Junaid[1], Mohd Nazri Bin Mohd Warip, R Badlishah Ahmad[2],Anuar Mat Safar[1], I Zunaidi[3], Wan Khairunizam[1], Z M Razlan[1], A B Shahriman[1].**

[1]Universiti Malaysia Perlis (UniMAP), Kampus Pauh Putra, 02600 Arau, Perlis, Malaysia.
[2]Universiti Sultan Zainal Abidin.
[3]Faculty of Technology, University of Sunderland, St Peter's Campus, Sunderland, SR60DD,United Kingdom

**Abstract.** SDN is the new network technology. Although SDN still new, it is inseparable from the issue of a link failure. By using SDN, link failure scenario can be simulated and collection of data can be analysed. Before creating link failure scenario, several topologies were needed to show similarities with traditional network. There are many topologies that can be created by using SDN because SDN already provide virtual network devices. Each network device will be build inside virtual network topology. With use of desired topology then link failure scenario can be implemented. SDN also need a controller that acts a "brain" to the network. The controller will be provide a routing algorithm to control the behaviour of the network when one or more links become fail to connect with host or controller. One of suitable controller was Ryu which can adapt to any traffic handler and already provide application resemble to network behavior. All software components will be simulated inside Mininet emulator, one of realistic virtual network application.

## 1. Introduction

In order to comply with all of these criteria, networking protocols have evolved significantly over the last few decades. However, the way traditional networks set up the deployment of one protocol to realize these needs in the worldwide organization is quite challenging. Traditional network configuration is indeed a time-consuming and error-prone because many steps are needed for the network administrator to add or remove a single device. A traditional network that consists of a modem, router, switch and several clients which have an Ethernet connection. This configuration approach makes it much more complex for a network administrator to deploy a consistent set of policies. As a result, organizations are more likely to encounter some failure in network links. To resolve this issue on a traditional network, it is now the time to introduce a new viewpoint on network management which is called Software Defined Networking (SDN) [1].In recent years, the growth of SDN has gathers the attention of the research community. Mostly because of the flexibility it presents in its model of centralized management, which facilitates the development of solutions according to demand of the network by client [2]. Currently, traditional network does not have a good solution to solve the link failure scenario as soon as possible. It takes time for network administrator to identify the problem before handling the situation in traditional network.The challenge for large data center are getting complex because of the public demand for internet services. In fact, large data center network are prone to have a serious network problem such as data link failure. Whenever the link failure scenario happens, network administrator has a hard time to identify or analyze the problem. The reason is because there are thousands of routers over the Internet and it also depends to the network topology. Thus, it is difficult to overcome the problem during network failure in many arrangements of the various elements of a computer network. the objective, to study the performance of SDN in the case on link failure in terms of packet loss, throughput, end-to-end delay, and packet delivery ratio and how the link failure affects the overall performance and, to understand and analyse several related methods used to solve the link failure scenario in SDN and, to evaluate the overall performance of the selected methods in custom network topologies. The network topology will be design in the simulation and the data will be analysed.

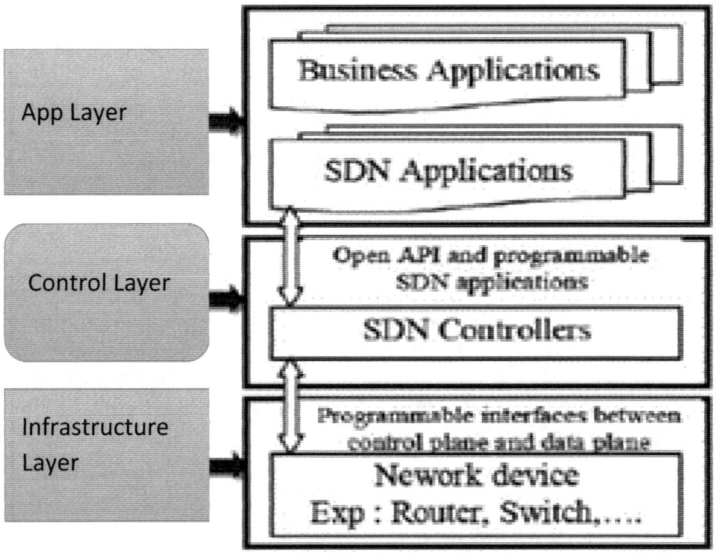


**Figure 1.** SDN architecture [11]

Network company's endeavoured to adapt to distributed situations by growing new specialized arrangements of routing protocols. Be that as it may, there is just restricted data on the execution of each, and reasonable execution correlations are not generally accessible. Recently, the designs of utilizing Mininet was not frequently exist in execution assessments of OpenFlow. The simulation apparatus show the reasonable undertaking of planning, building and testing for clients with down to word systems when growing true frameworks. This will enable framework fashioners to decide the rightness and productivity of a plan before the framework is really conveyed. Another favourable position of test systems are they permit the assessment among different system measurements and approval instruments to acquire comes about that are not tentatively quantifiable on bigger geologically appropriated designs.

*1.1. Mininet is an emulator platform*
using OF protocol, runs a collection of end-hosts, switches, routers and links on a single Linux kernel by using lightweight virtualization. Mininet components function as real network components [3].

*1.2. Floodlight Controller*
Floodlight Controller In [4] experiment, authors choose Floodlight to compare and analyse the routing convergence time of legacy network. Floodlight also been used in [5] for multipath routing. The GA and Dijkstra algorithm for solving the BCMPO problem has been load inside Floodlight controller. With the algorithm, Floodlight can choose routing paths automatically according to the whole network status scenario. Figure 1 show the structure of core module, which is responsible for transforming received OF packets into events. [6].

## 2. Motivations

SDN components proceeding with Mininet, the Mininet creating Software Defined Networks (SDN) components, alter, share SDN and Mininet with different networks and make cooperations. Selection.

Controller Ryu is a component-based of SDN framework. Ryu needs to be installed so Mininet has a controller component to allow communication between devices. To begin ensure system has python installed, by executing python on the terminal, which would turn into the python shell, besides displaying the version.

### *2.1 Wireshark*

Wireshark is a network analysis tools that will captures packets in real time and display the readable data. Other than that, Wireshark also includes filters, color-coding and other features that show network traffic and inspect individual packets. This section will discuss about the system flowchart, simulation parameter, topology design, and the routing algorithm used in the system.

### *2.2 OpenFlow*

In this section, the OpenFlow-enabled network architecture for basic OpenFlow network topologies is implemented using Mininet. Mininet supports 5 built-in network topologies, namely, minimal, single, linear, tree and reversed topology [7]. The default topology is minimal topology which is predefined with one OpenFlow kernel switch connected to two hosts and OpenFlow reference controller, whereas number of switches and hosts can be changed for other topologies using the command-line interface (CLI).In this section will be focus only to three basic which are single, linear and tree topology. Table 1 shows the value of parameter and explanation inside CLI.

### *2.3 Network topology*

Network topology is one of data centre topology that provides the efficient communication between the hosts. This custom topology has a capability for multipath routing. The topology used will be 4-pod topology as shown in Figure 2 The hierarchical layers of the network include core switch, aggregation switch, edge switch and host. A topology having 16 hosts is designed in Mininet using CLI command as below:

**$ sudo mn –topo=custom, 4.py –mac –switch=ovsk –controller=remote**

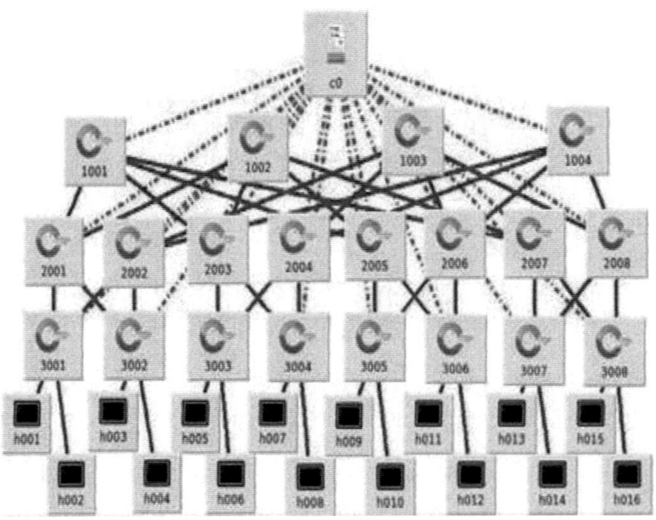


**Figure 2.** OpenFlow-custom Topology, 4 pods having 16 H

IOP Conf. Series: Materials Science and Engineering **557** (2019) 012039 doi:10.1088/1757-899X/557/1/012039

Each switch been renamed to show different between switch in each layer. Assume the controller at the top of the layer, core switch will become first layer contain of 4 switch start from 1001~1004. Both second layer (aggregation switch) and third layer (edge switch) each has 8 switch and will start from 2001~2008 and 3001~3008 respectively. Host also will be rename from default h1-h16 to h001~h016. For the host IP address also will be change from default 10.0.0.1~10.0.0.16 to 10.1.0.1~10.8.0.2. The links connection of all switches and hosts are shown in Figure 2.

*2.4 Ryu Application Routing Algorithm*

Ryu Application Routing Algorithm the most important thing in simulating SDN network is the controller. As been discussed in previous section, Ryu will be main controller used to run the simulation. This section will cover about routing algorithm in the Ryu controller. The selected routing algorithm will determine the behavior of host or switches when a link failure scenario happens. Ryu controller has already provides many applications to be run inside controller. All Ryu applications have their own function and behavior for switches and also hosts so that it can meet user requirements. In the next section, each application will be introduced to be used for the desired topologies.

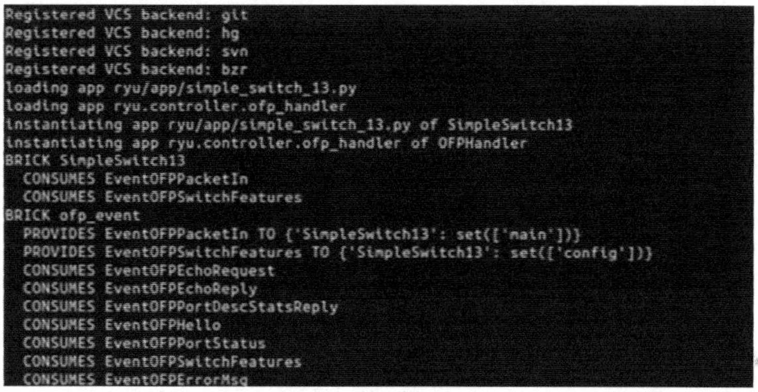


**Figure 3.** Ryu Application

## 3. Simulating SDN on mininet and floodlight controller.

A standout amongst the most important communication protocols is the Internet Protocol Suite, TCP/IP, utilized for correspondence over the web and comparative systems. This systems administration shows an availability between two hubs in the system, by determining how information ought to be transmitted, designed, tended to, directed and got at the goal. The TCP/IP framework architecture was produced after the OSI reasonable model on the grounds that the internet was executed utilizing the favored TCP/IP suite amid the spread of the World Wide Web [8]. In addition, most sellers and large programming consumers favored using this model instead of OSI because of the way that TCP/IP can meet worldwide benchmarks in which the OSI model couldn't give. Table 1 demonstrates a next to each other examination of the OSI, TCP/IP, and SDN models.

IOP Conf. Series: Materials Science and Engineering **557** (2019) 012039 doi:10.1088/1757-899X/557/1/012039

```
info( '*** Add hosts\n')
h1 = net.addHost('h1', cls=Host,ip='172.16.20.10/24', defaultRoute = 'via 172.16.20.1')
h2 = net.addHost('h2', cls=Host,ip='172.16.20.11/24', defaultRoute = 'via 172.16.20.1')
h3 = net.addHost('h3', cls=Host,ip='172.16.10.10/24', defaultRoute = 'via 172.16.10.1')
h4 = net.addHost('h4', cls=Host,ip='172.16.10.11/24', defaultRoute = 'via 172.16.10.1')
h5 = net.addHost('h5', cls=Host,ip='172.16.10.12/24', defaultRoute = 'via 172.16.10.1')
h6 = net.addHost('h6', cls=Host,ip='192.168.30.10/24', defaultRoute = 'via 192.168.30.1')
h7 = net.addHost('h7', cls=Host,ip='192.168.30.11/24', defaultRoute = 'via 192.168.30.1')
h8 = net.addHost('h8', cls=Host,ip='192.168.30.12/24', defaultRoute = 'via 192.168.30.1')
```

**Figure 4.** IP Host Address for Topology

The floodlight [9] is controller of an OpenFlow for big company networks in light of programming of Java language and dispersed under the Apache permit. The main undertaking began as of the Beacon controller [10] and now it is bolstered by an engineers' group and furthermore by Big-Switch Networks, a jump up that products business equipment controller.

In this section, a comparative performance analysis of a topology is done based on results obtain after execution of networks. No link failure simulation will be done in the basic topologies because result give almost guarantee failure. Thus, the basic topology will be simulated with fully working link. The performance analysis is done by comparing all network topologies on the basis of bandwidth utilization, packet transmission rate, end-to-end delay (time required to transmit packet from source node to destination node) and throughput. Here, performance analysis for topologies discussed for a topological designed network of eight hosts. Bandwidth utilization for topologies is tabulated in Table 1 and is shown in Figure 5.

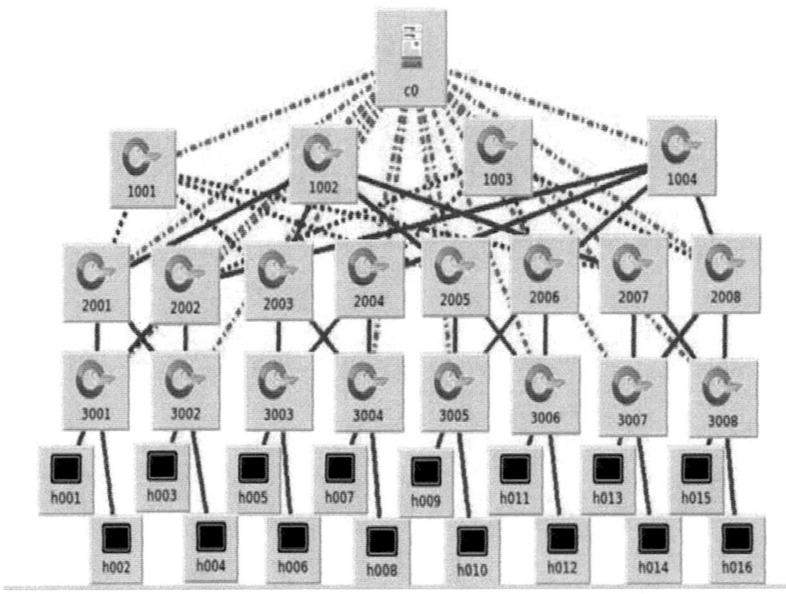


**Figure 5.** Switch Failure Combination Scenario

## 4. Results and discussions.

This section will discuss the analysis of link failure scenario in topology. Topology with 4 pods will be simulated in Mininet environment. Several scenarios have been created to satisfy the link failure scenario so that the performance can be clearly measured. the performance analysis is done by comparing the analysis with and without link failure based on throughput, end-to-end delay, packet loss rate, and packets delivery ratio. For topology, several switches will be created to be stopped thus make all interconnected links become fail so it can match with the scenario that already been discuss in the previous chapter. Topology, 10Mbps of bandwidth will be set in the network. There are 3 selected aggregation switches from the total number of 8 switches in layer 2 will be turned off and also for the 2 core switches from the total number of 4 core switches. Table 1 shows the collected bandwidth using Iperf tool for each switch scenario with and without failure.

**Table 1.** Comparison of Bandwidth in Switch Failure Scenario

| Scenario | Average Utilized Bandwidth (Mbps) | Overall Performance (%) |
|---|---|---|
| All switch start | 8.95 | 89.5% |
| 1 aggregation switch down | 8.31 | 83.1% |
| 2 aggregation switch down | 7.82 | 78.2% |
| 3 aggregation switch down | 7.66 | 76.6% |
| 1 core switch down | 7.58 | 75.8% |
| 2 core switch down | 7.99 | 79.9% |

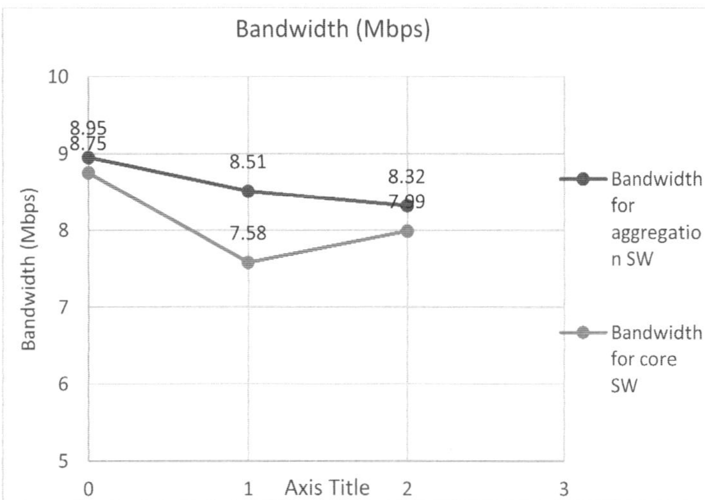


**Figure 6.** Bandwidth Chart for Topology

IOP Conf. Series: Materials Science and Engineering 557 (2019) 012039 doi:10.1088/1757-899X/557/1/012039

Based on the result in Table 1, core switches will give more influence for the decrease in bandwidth compared to aggregation switches. It is because of the core switches placed at the first of layer and has more influence links connected with other switches during new routing data path algorithm. In Figure 6 clearly, show that more switches are turned off will incur the decrease of bandwidth utilization in the network topology. The comparison of the number of switches down with performance can be seen in Figure 6. It is because aggregation switch connects closer to the host so the connection possibility between two hosts will reduce. The more aggregation switches down; the connection possibility reduces more. So it affects the connection bandwidth in the entire network. The same can be applied with the core switch that also has been also down. Thus, the more core switch is down, the bandwidth will also decrease.

**Table 2.** Throughput Data Calculation for Topology

| | | Throughput (kbps) | |
|---|---|---|---|
| No of Transmitted Packet | All switch start | 1 aggregation switch down | 2 aggregation switch down |
| 100 | 90 | 83.1 | 78.2 |
| | | Throughput (kbps) | |
| | 3 aggregation switch do | 1core switch down | 2core switch down |
| | 77 | 76 | 80 |

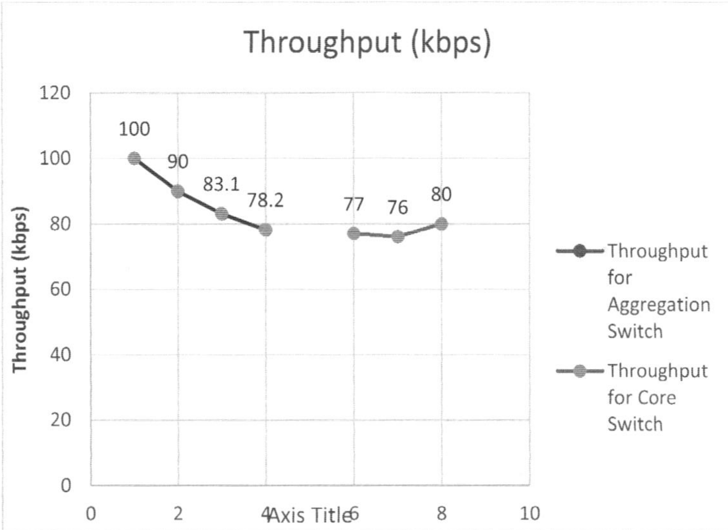


**Figure 7**. Throughput Chart for Topology

Throughput for topology will be calculated based on bandwidth and RTT. The result can be referred to in Table 2. Total throughput when there are none switch failures are 90 kbps. By referring to Table 2, the more switch been turned off the less outcome of throughput value. The value of throughput will affect accordingly to the value of bandwidth simulation result. The value of 1 core switche down versus 1 aggregation switch down clearly shown that core switch gives more effect to the network when been turned off. Figure 7 shows the throughput values in graphically.

**Table 3.** Packet Delivery Ratio Chart for Topology

| | | Packet Delivery Ratio (%) | | |
|---|---|---|---|---|
| No of Transmitted Packet | All switch start | 1 aggregation switch down | 2 aggregation switch down | 3 aggregation switch down |
| 100 | 100% | 31% | 18% | 11% |
| | | | | |
| | core switch dov | core switch down | | |
| | 60% | 15% | | |

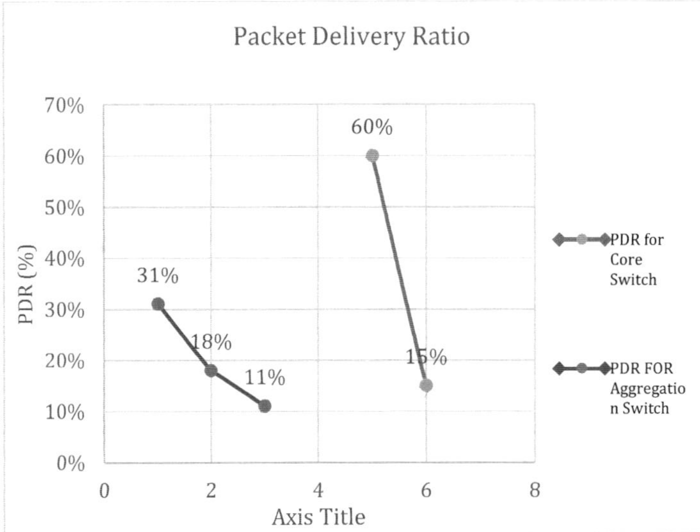


**Figure 8**. Packet Delivery Ratio Chart for Topology

The packet delivery ratio is the ratio of packets successfully received by the host to the total transmit the packet. the total number of success packet received and the total number of the packet transmitted can be referred in ping statistics from the ping test in Mininet environment. Table 3 shows the packet delivery ratio in topology based on the simulation scenario. By referring to Figure 8, the packets delivery ratio will decrease when there are more switches been turned off. The highest PDR is for 1 core switch in layer 1 of topology which is 60%. The lowest PLR is for 3 aggregation switch down which deliver only 8% from 100 packets been transmitted to the host 016.

## 5. Conclusions

The purpose of the performance analysis of SDN in link-failure scenario study has been achieved with the help of research from journals and papers by previous researchers. As the advanced technology approach, SDN is the trend that will replace existing traditional networks, especially in data center. Study of SDN with link failure scenario will greatly improve aspects in computer networking. Although SDN still new to the computer networking world, with applying this SDN proved it can offer many data analysis in long terms for the future. There are so many methods to implement link failure scenario in SDN. Each method has already been compared and the suitable method has been successfully used to simulate link failure scenario. By using the chosen method, the study of performance analysis in case of link failure in term of throughput, end-to-end delay, packet loss rate and packet delivery has been simulate and the data has been analyze and discuss throughout this research project. There are five topology that has been discuss in this research analysis which are single, linear, tree, triangle and fat tree topology. Each of topology can be used to simulated link failure scenario but only several topologies will give a detail result of network performance. All topology has been designed and data from each of it has been discussed in previous chapter. Hence, the research performance analysis objective has been fully achieved.

## 6. REFERENCES

[1] "Traditional VS Software Defined Networking." [Online]. Available:

http://www.mavenspire.com/blog/traditional-vs.-software-defined-whats-the-differences.

[2] Open Networking Foundation. (2013). SDN Architecture Overview. Onf, 1–5.

[3] Kaur, K., Singh, J., & Ghumman, N. S. (2014). Mininet as Software Defined Networking Testing Platform. International Conference on Communication, Computing & Systems (ICCCS–2014), 3–6.

[4] Zhang, H., & Yan, J. (2015). Performance of SDN Routing in Comparison with Legacy Routing Protocols. Proceedings - 2015 International Conference on Cyber-Enabled Distributed Computing and Knowledge Discovery, CyberC 2015, 491–494.

[5] Liu, Y., Pan, Y., Yang, M., Wang, W., Fang, C., & Jiang, R. (2015). The Multi-Path Routing Problem in the Software Defined Network, 250–254.

[6] Isaia, P., & Guan, L. (2016). Performance benchmarking of SDN experimental platforms. IEEE NETSOFT 2016 - 2016 IEEE NetSoft Conference and Workshops: Software-Defined Infrastructure for Networks, Clouds, IoT and Services, 116–120. https://doi.org/10.1109/NETSOFT.2016.7502456

[7] Kaur, K., Singh, J., & Ghumman, N. S. (2014). Mininet as Software Defined Networking Testing Platform. International Conference on Communication, Computing & Systems (ICCCS–2014), 3–6.

[8] Henderson, T. R., & Riley, G. F. (2006). Network Simulations with the ns-3 Simulator. Proc. Sigcomm, 527. https://doi.org/10.1.1.415.6550

[9] Pal, C., Veena, S., Rustagi, R. P., & Murthy, K. N. B. (2014). Implementation of simplified custom topology framework in Mininet. 2014 Asia-Pacific Conference on Computer Aided System Engineering, APCASE 2014, 48–53. https://doi.org/10.1109/APCASE.2014.6924470

[10] S. Liao, X. Hong, C. Wu, B. Wang and M. Jiang, "Prototype for customized multicast services in software defined networks," 2014 22nd International Conference on Software, Telecommunications and Computer Networks (SoftCOM), Split, 2014, pp. 315-320.

IOP Conf. Series: Materials Science and Engineering **557** (2019) 012040 doi:10.1088/1757-899X/557/1/012040

# FBMC as 5G candidate for high speed mobility

**Montadar Abas Taher, Khalid Hussein Kutheir**

Department of Communications Engineering, University of Diyala, Ba'aqubah, 32001, Diyala, Iraq

montadar@ieee.org

**Abstract**. Beyond 4G, communication systems will be highly flexible and more congested. This is because of the increased demand for wireless applications, such as machine type communications (MTC) aside from the cellular mobile communications. Hence the need for the next generation, 5G, becomes essential. In the 4G systems, the cyclic-prefix orthogonal frequency division multiplexing (CP-OFDM) has been adopted as a modulation technique. However, CP-OFDM suffers from the spectrum waste problem, because of the high out of band (OOB) emission. Thus, not all frequency spectrums will be useful. On the other hand, the required, increased demand, for high data rate motivated the industry to find another candidate as a modulation technique to be adopted in the next generation. Different candidates have been introduced; the Filter Bank Multi-Carrier (FBMC) is one of the promising candidates for that job. In this paper, we will show that the FBMC can be employed in 5G communication systems and can exploit almost, all, the advantages of the previous, CP-OFDM, system with little increment in the complexity, such as the implementation of multiple input-multiple output (MIMO) antenna diversity, and the high throughput with noteworthy mobility speeds.

## 1. Introduction

Next wireless systems, which is represented by 5G generation, have to be able to support services to individuals continually moving, driving amongst home and work, getting together with family and companions or visiting malls and relaxation offices. During achieving these human tasks, individuals utilize cellular phones to associate with the Internet, using venture time for business-related assignments, diversion, or associating in online networks. In other words, mobility, and sometimes high speed should be one of the important capabilities of the 5G communication systems [1], where the last generation, 4G, cannot provide it efficiently.

Moreover, machine type communication (MTC), wireless sensors, internet of thing (IoT) devices, and vehicle to vehicle communication etc. [2] are all additional requirements to be connected to the 5G networks. Thus, the need for another capable system better than the 4G is on demand now. Furthermore, 5G must be compatible with all previous generations, fourth, third, second (4G, 3G, 2G), WiFi, to provide to the client massive data rate and services [3].

In other words, and as quantities, the 5G systems must fulfil the following features: the network should provide, approximately, ten times data rate higher than 4G, 1~10 Gbps, where 4G theoretical peak value is 150 Mbps [3], round-trip latency in 4G is around 10 ms, while in 5G it should be around 1 ms, in other words, approximately, ten times reduction in the latency [4]. It is expected to empower a huge number of associated gadgets with higher transmission capacities for longer spans in a particular zone, thus, higher bandwidth in the unit area [4]. Keeping in mind the end goal to understand the vision of IoT, developing 5G systems need to give availability to a large number of devices [5], furthermore, 5G remote systems need to guarantee finish scope independent of clients' areas. 5G imagines that system ought to for all intents and purposes be constantly accessible, availability of 99.999% [4]. Moreover, the energy reduction, better battery life, and the mobility speed increment. However, in 4G, the mobility can reach up to 350 kmph [6], while in 5G it is required to reach higher than 500 kmph [7].

From the aforementioned requirements, the vital remark to design 5G networks is the modulation and access scheme, which must support huge data rates and access capacity. Hence, the spectrum must be almost perfectly utilized. The dominant signaling scheme for most of applications nowadays is the CP-OFDM [3] as long as it is robust to fading and frequency selective channels, can be implemented efficiently using inverse/fast Fourier transform (I/FFT), which leads to high reduction in the computational complexity, and effortless integration with the multiple antenna techniques (MIMO). Despite the aforementioned strong points of CP-OFDM, it does not have the capacity to cope with the 5G requirements.

The key point in 5G systems, the time and frequency must be well localized. Because of the rectangular pulse at the transmit and receiving devices of the CP-OFDM, the frequency will not be localized, i.e., high out of band (OOB) emission will appear significantly. To overcome this problem, numerous techniques have been introduced in the literature [8-11], right now 4G harnesses windowing and filtering, such a scheme is called weighted overlap and add OFDM (WOLA) [12], in this approach, the adjacent symbols are overlapped in the time-domain. Thus, and increased cyclic prefix (CP) duration must be available to consider the windowing at both the transmitter and the receiver, that is, the spectrum efficiency is reduced.

On the other hand, different modulation and access schemes have been proposed, such as Universal Filtered Multicarrier (UFMC) of universal filtered OFDM (UF-OFDM) [13], where sub-band filtering applied, the second proposal is the filtered OFDM (F-OFDM) [14], where the same technique of UFMC is used but the number of sub-bands is higher. The third suggestion is the Generalized Frequency Division Multiple Access (GFDM) [15], in which modulation and filtering of independent blocks of subcarriers achieved, where the filter is circularly shifted in time and frequency, results in reduced OOB emission. However, this operation leads to non-orthogonal subcarriers which might produce inter-symbol interference (ISI) and inter-carrier interference (ICI). Filter Bank Multi-Carrier (FBMC) is another recommendation as a 5G physical layer modulation and access scheme [16]. This last candidate, FBMC, requires the orthogonality to be for the adjacent subcarriers only, unlike CP-OFDM, where orthogonality must be achieved for all subcarriers.

In this paper, an introduction to FBMC schemes will be introduced. The compatibility of 4G systems and vehicular speed capabilities with applicability to MIMO systems will be shown. It will be shown that FBMC can be efficiently implemented using I/FFT operations and the spectral usage is significantly improved compared to CP-OFDM, which is already employed in 4G systems. The rest of this paper is organized as follows: section 2 introduces the mathematical model of the CP-OFDM and FBMC. In section 3 an engineering phenomenon for the design procedures for the FBMC will be suggested. Based on the procedures in section III, simulation results and discussions will be given in section 4. Concluding remarks will be shown in section 5.

## 2. Mathematical model

Presently, OFDM is the dominant scheme in current communication systems. The third generation partnership project (3GPP) suggested continuing to use OFDM with little modification in some essential parts for the next generation of communication systems, 5G. This suggestion will make the 5G network to be more compatible with 4G systems [17]. However, we still need to investigate another candidate other CP-OFDM. Here the FBMC will be introduced mathematically compared to the traditional CP-OFDM. Different alternates were given to FBMC [18] such as FBMC-QAM and FBMC-OQAM. Furthermore, there is more than one name for the FBMC-OQAM [19] but all refer to the same technique.

Nevertheless, the configurations of CP-OFDM and FBMC can be shown both in one structure as shown in Figure 1, where both systems use a transmit and receive prototype filters, $\psi T(t)$ and $\psi R(t)$, respectively. Thus, the system of Figure 1 will be fed with symbols s(t) [19]. That is, the signal after transmission can be expressed mathematically as [20]

$$s(t) = \sum_N \sum_K p_{T_{k,n}}(t) x_{k,n} \tag{1}$$

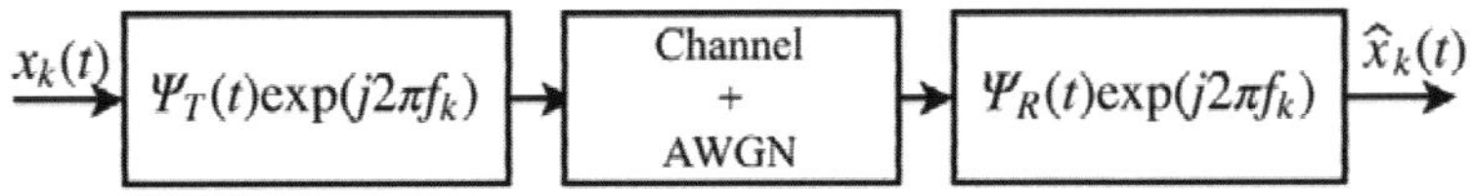


**Figure 1.** OFDM and FBMC unified configuration, showing the transmitter, receiver, and the channel, where AWGN stands for additive white Gaussian noise.

In (1), $n$=0,1..., $N$-1 stands for the subcarrier index, $k$=0,1...$K$-1 is the time index. The modulating symbols $x_{k,n}$ are randomly drawn from $M$-QAM, and $M$ represents the modulation (mapping) order, while the transmitted pulse $p_{Tk,n}(t)$ can be expressed as

$$p_{T_{k,n}}(t) = \psi_T(t - kT) e^{j2\pi n F(t-kT)} \tag{2}$$

where $\Psi_T(t)$ is the transmit filter as shown in Figure 1. The transmitted pulse $p(t)$ is shifted in time and frequency, where $T$ and $F$ are time and frequency shifts, respectively. After passing the signal through the channel, the received symbols are

$$z_{n,k}(t) = \int_{-\infty}^{\infty} \hat{s}(t) p_{n,k}^{*}(t) dt \tag{3}$$

where $\hat{s}(t)$ is the received signal, the superscript * indicates complex-conjugate operation. Before going to show the usage of the last expression with respect to traditional CP-OFDM and FBMC, it is imperative to show that it is not possible practically to achieve maximum symbol density, $T \times F = 1$, time localization $\xi_t < \infty$, $\xi_F < \infty$, and perfect orthogonality at the same time [19, 21]. Hence, it is by nature, we have to lacerate one of the above conditions during design time.

As stated above, CP-OFDM has got different applications, which employed a rectangular pulse in both transmitter and receiver ends. Thus, the transmit/receive prototype filters in CP-OFDM can be written as [21]

$$\psi_T(t) = \begin{cases} \frac{1}{T_o} & -\frac{T_o}{2} - T_{CP} \le t < \frac{T_o}{2} \\ 0 & otherwise \end{cases} \tag{4}$$

and

$$\psi_R(t) = \begin{cases} \frac{1}{T_o} & -\frac{T_o}{2} \le t < \frac{T_o}{2} \\ 0 & otherwise \end{cases} \tag{5}$$

where $T_o$ is the time scaling constant which is depends on the subcarrier spacing. Hence, to achieve orthogonality, $T$ must be equals to $T_o + T_{CP}$, where $F = 1/T_o$, while the time localization $\xi_t = (T_o + T_{CP})/2\sqrt{3}$, but unfortunately, the frequency localization will be distorted $\xi_F = \infty$ [19, 21, 22].

On the other hand, FBMC has different definitions [23-25]. Nam et.al, [23] scratches the most important parameter, which is the localization in the frequency-domain, where the OOB emission has been distorted significantly. While the design of Yun et.al. [24] and Kim et.al. [25] lost its orthogonality. However, different prototype filter can be used to construct FBMC such as PHYDYAS [16] and Hermite pulse [26, 27]. The Hermite pulse can be the best fit as will be shown in the discussion, thus, Hermite pulse can be written as

$$\psi(t)=\frac{1}{\sqrt{T_o}}e^{-2\pi\left(\frac{t}{T_o}\right)^2}\sum_c u_c H_c\left(2\sqrt{\pi}\frac{t}{T_o}\right) \tag{6}$$

where $c$ = 0, 4…20. In [26] the coefficients have been determined as $u_0$=1.41269, $u_4$=-0.0030145, $u_8$= -0.0000088041, $u_{12}$=-2.2611×10$^{-9}$, $u_{16}$=-4.457× 10$^{-15}$, and $u_{20}$=1.8633× 10$^{-16}$. If $TF$=2, then the orthogonality can be achieved where $T=T_o$, $F=2/T_o$. Accordingly, time and frequency localization $\xi_F\times\xi_t$=1.02× ¼π. While the PHYDYAS prototype filter can achieve time and frequency localization of 1.13× ¼π, which is degraded by a factor of 1.107, this is because the Hermite filter is based on the Gaussian pulse,

$$\psi(t)=e^{-\pi t^2} \tag{7}$$

It is worth mention that the Gaussian pulse has crucial property, which is

$$\Psi(f)=\psi(t) \tag{8}$$

That is meant that the frequency domain properties of the filter are the same properties in the time domain of the same filter. Moreover, it is shown in [28] that the best time and frequency localization condition is

$$\zeta_t\times\zeta_F\geq\frac{1}{4\pi} \tag{9}$$

Only when the pulse is Gaussian, the equality condition holds. Thus, the Hermite filter (6) has better performance than the PHYDYAS filter. In the next section, a design procedure of an efficient behaviour FBMC-OQAM will be given based on the abovementioned requirements.

## 3. Design criteria

To design an efficient FBMC-OQAM, the first step is to design the appropriate filter. Various filters have been introduced in the literature, as stated in the last section. But all designing engineers must keep in mind that the best design is; when the localization of the filter both in time and frequency must match the channel dispersion [29]. Nevertheless, the time localization, accordingly, can be defined as follows [29]

$$\zeta_t=\left[\int t^2\left|\psi(t)\right|^2dt\right]^{\frac{1}{2}} \tag{10}$$

while the frequency localization can be defined as [29]

$$\zeta_F=\left[\int f^2\left|\Psi(f)\right|^2df\right]^{\frac{1}{2}} \tag{11}$$

with (10) and (11) in mind, one should decide the better filter which should be $\psi(t)=\psi(-t)$, and according to (9), the Hermite filter (6) is the best choice. Here the prerequisite for orthogonality will be achieved for real orthogonality not complex. Thus, we have to exploit the reduced size of time and frequency spacing, hence,

$$T=\frac{T_o}{2}$$
$$F=\frac{1}{T_o} \tag{12}$$

From (12) the symbol density, *TF* is 0.5, this is because we will harness only real parts of the transmitted symbol. In other words, real part of the symbol will occupy, say, the first part of the original symbol period and the imaginary part will occupy the rest of the period, hence, the name of Offset Quadrature Amplitude Modulation (OQAM), at the end, the resultant density, *TF*, will be unity.

Based on previous introductions, an efficient implementation will be through the utilization of the IFFT in the transmitter and FFT in the receiver sides. This type of implementation enables the 4G system to be compatible with 5G devices after non-essential modifications to the 4G equipment. It is shown in [16, 27, 30] that the FBMC-OQAM can be implemented using the IFFT, whereby factorizing the prototype filter in (1), then, (1) will represent an IFFT of size $N$, the result is repeated $\Omega$ times, where $\Omega$ is the overlap factor [16]. A large vector is created; this vector is element-wise multiplied by the factorized prototype filter. The same sequence has to be repeated but in a reverse manner at the receiver.

Conventional CP-OFDM has no problems with MIMO applicability, because of the availability of the complex orthogonality,

$$\langle p_{k1,n1}(t), p_{k2,n2}(t)\rangle = \delta_{(k2-k1),(n2-n1)} \tag{13}$$

where $\delta$ is the Kronecker delta function. Thus, different channel estimation techniques can be employed to recover the transmitted signal from the received one. Comparatively to FBMC-OQAM, we have only real orthogonality,

$$\mathrm{Re}(\langle p_{k1,n1}(t), p_{k2,n2}(t)\rangle) = \delta_{(k2-k1),(n2-n1)} \tag{14}$$

Thus, it is not conceivable to implement the MIMO techniques directly in FBMC-OQAM, unless the complex orthogonality (13) taken back. It is conceivable to repossess complex orthogonality as in [31-34]. Thus, different methods for the spreading can be implemented either in time or in frequency. [31, 34] spreading can be employed efficiently because it uses Hadamard sequences, while [33] used the fast Fourier Transform matrix for spreading, which increases the computational complexity dramatically.

## 4. Simulation results and discussion

In this section, we will set up the simulation parameters for FBMC-OQAM, which will be comparable to the conventional CP-OFDM, with respect to the number of subcarriers, a number of transmitting and receiving antennas, the carrier frequency, modulation order, $M$, etc. At the beginning, the carrier frequency will be fixed for both systems, which is 2.5GHz. Figure2 shows three types of prototype filters, PHYDYAS, Hermite, and rectangular. The frequency response of the rectangular prototype filter shows the worst performance since it is not localized in frequency as shown in Figure 2, the second prototype filter is the PHYDYAS, also behaves worse but outperforms the rectangular prototype filter, while the last prototype filter, Hermite prototype, produces better performance than the other two prototype filters.

The frequency localization, as can be seen in Figure2, is better than the PHYDYAS and rectangular prototype. The symbol density, i.e., *TF* will be 0.5, where $F$=1/$T_o$, and $T$= $T_o$ /2, this is for the FBMC-OQAM configuration, where $F$=15K kHz. According to the results of Figure2, the following results will be depicted using the Hermite prototype filter, in other words, the rest of the simulations will be done using Hermite prototype filter since it has the best frequency localization than the other two prototype filters, as shown in Figure2.

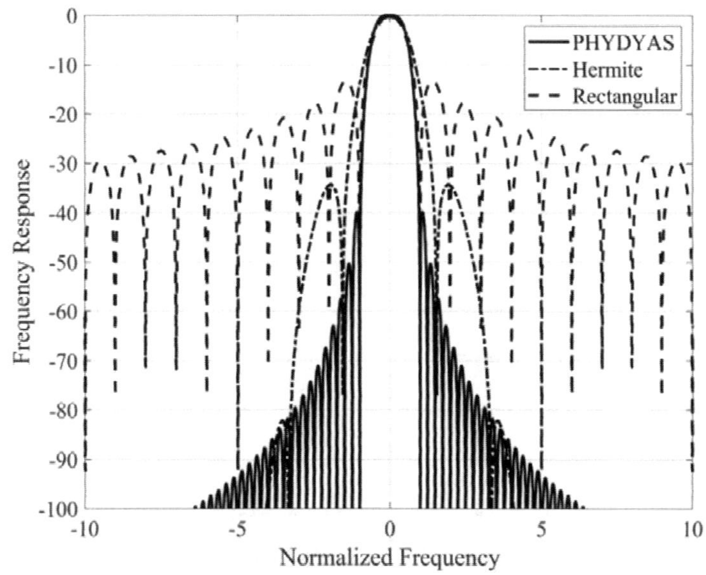


**Figure 2.** The frequency response of PHYDYAS, Hermite, and rectangular prototype filters.

In Figure 3, the power spectral density of the FBMC-OQAM outperforms that of the CP-OFDM. It is shown that the out of band emission of the CP-OFDM starts at -9 dB and extends evidently, while the power spectral density of the FBMC-OQAM starts, evidently, at -65 dB, thus, a difference of about -56 dB between the two schemes. This gives evidence to the FBMC-OQAM to be a strong candidate for the 5G applications, where it is obvious that the frequency localization of the FBMC-OQAM is more efficient than the CP-OFDM.

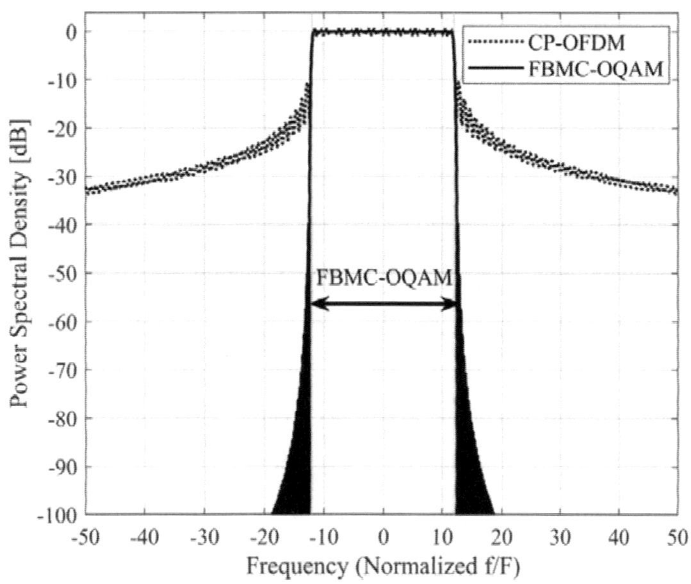


**Figure 3.** Power spectral density comparison between CP-OFDM and FBMC-OQAM.

The capability to improve the throughput with respect to the velocity can be seen in Figure 4. It can be seen that at the signal to noise ratio (SNR) level of 15 dB and velocity of 120 km per hour (kmph), the throughput is 3.4 Mbit/s for the FBMC-OQAM and 2.75 Mbit/s for the conventional 4G scheme, which is represented by the CP-OFDM. At 400 kmph and SNR level of 10 dB, the throughputs were 0.35 Mbit/s and 0.2 Mbit/s for the FBMC-OQAM and CP-OFDM, respectively.

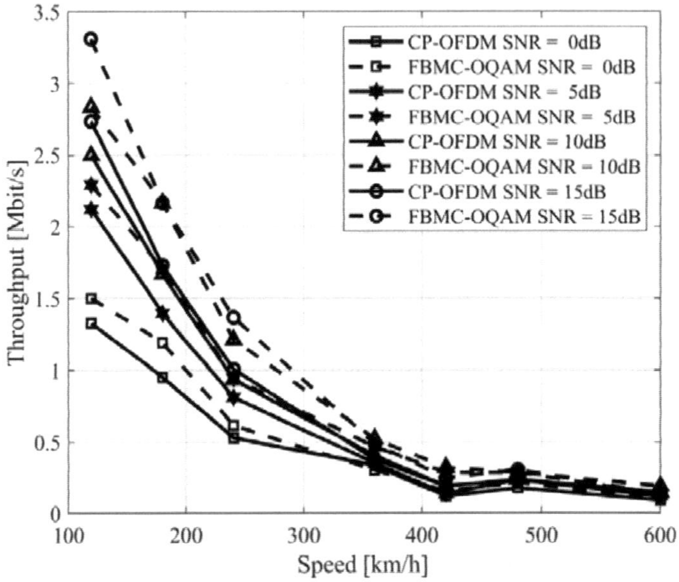


**Figure 4.** Throughput comparison between CP-OFDM and FBMC-OQAM, with respect to velocity.

It is obviously shown that the upper limit of the 4G scheme has been exceeded and still the 5G candidate, which is represented here by FBMC-OQAM, outperforms the 4G scheme. However, at 600 kmph, the throughput of both systems is almost the same, in fact, the difference is 0.05 Mbit/s, but this is at the expense of the quality of the service of the CP-OFDM scheme.

One more averment is shown in Figure 5. In Figure 5 the throughput is clearly shown even at higher velocities. For instance, at 20 dB SNR and 540 kmph speed, the throughputs of the two systems were 0.65 Mbit/s and 0.9 Mbit/s for the CP-OFDM and FBMC-OQAM, respectively. Another case is when the SNR level equals to 15 dB with a velocity of 420 kmph, the throughput reached 0.15 Mbit/s and 0.25 Mbit/s for the CP-OFDM and FBMC-OQAM, respectively. In both figures, Figure 4 and 5, the power delay profile is Vehicular-A with Jakes Doppler model. The root mean square (*rms*) Doppler spread, $\tau_{rms}$, is 370 ns [35].

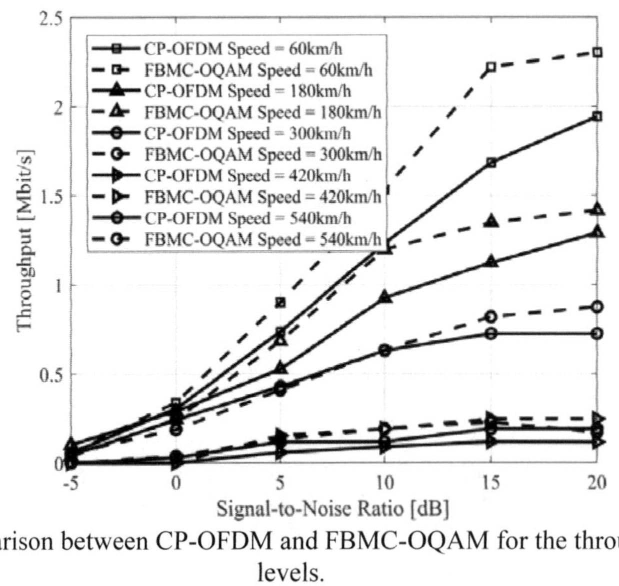


**Figure 5.** A comparison between CP-OFDM and FBMC-OQAM for the throughput versus SNR levels.

It was shown in the last section that the applicability of MIMO techniques is not as direct as possible. Thus, a simple modification must be achieved before the employment of the MIMO systems. In this paper, spreading in frequency has been accomplished, that is, the complex orthogonality has been recovered. Therefore, MIMO systems can be used in FBMC-OQAM again, as can be used in CP-OFDM directly. However, the conduction of spreading will put more computational complexity to the system, which can be considered as a drawback in this system.

The bit error rate (BER) performance will be shown for three different scenarios, when the modulation order $M = 4$, $M = 16$, and $M = 64$. Two transmit antennas, Tx, and two receive antennas, Rx, MIMO system.

Figure 6 presents Alamouti's space-time block coding approach. It can be seen that both the traditional CP-OFDM and FBMC-OQAM have the same BER performance behaviours, where the last is first coded (spreaded in frequency) then 2-Tx and 2-Rx antennas are employed, while the former, CP-OFDM, did not need modification. A higher modulation order, say, $M = 16$, the same behaviours of Figure 6 can be seen in Figure 7. It is shown that since the mapping order became higher, more SNR is required for both systems, but both systems, CP-OFDM and FBMC-OQAM, showed the same BER performance. The same scenario in Figure 6 has been used in Figure 7, which includes 2-Tx and 2-Rx MIMO Alamouti's space-time block coding technique employment. Furthermore, Figure 8 implements the same configurations of Figure 6 and Figure 7 but the modulation order $M$ increased to 64. It is shown in Figure 8 that both systems have the same BER performance as depicted in previous two scenarios when $M = 4$ and $M = 16$.

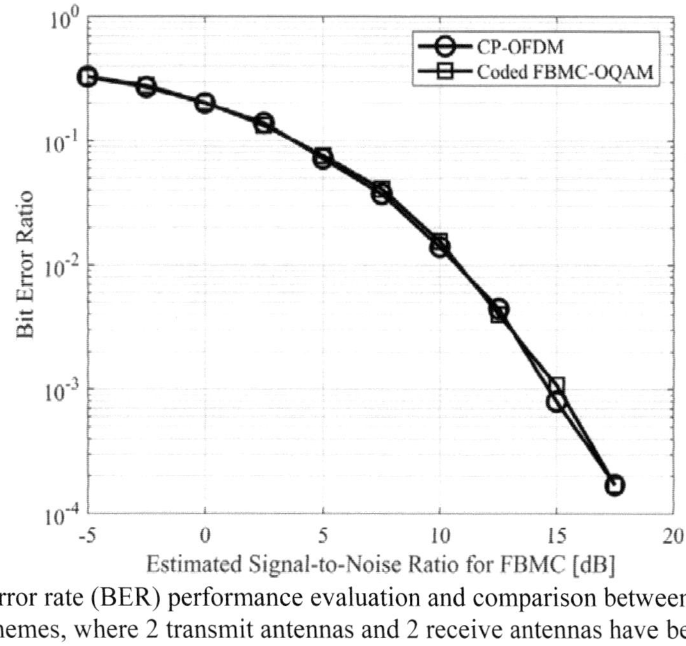


**Figure 6.** Bit error rate (BER) performance evaluation and comparison between CP-OFDM and FBMC-OQAM schemes, where 2 transmit antennas and 2 receive antennas have been employed for both systems with modulation order $M = 4$.

IOP Conf. Series: Materials Science and Engineering **557** (2019) 012040 doi:10.1088/1757-899X/557/1/012040

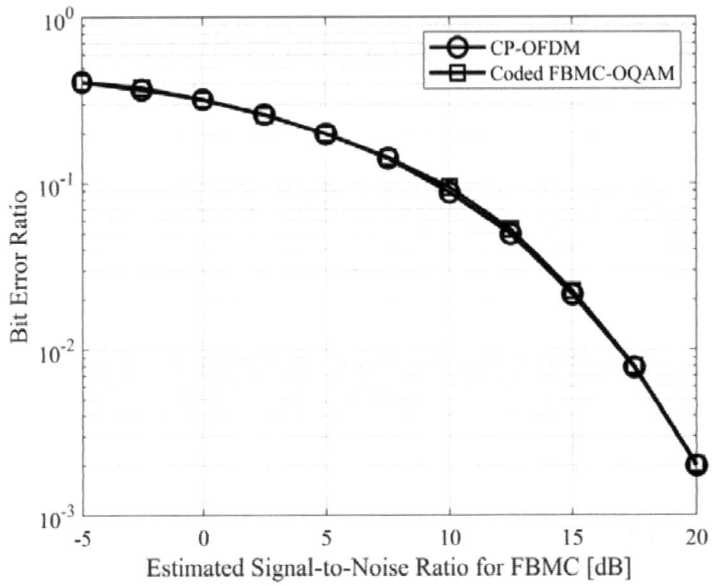


**Figure 7.** Bit error rate (BER) performance evaluation and comparison between CP-OFDM and FBMC-OQAM schemes, where 2 transmit antennas and 2 receive antennas have been employed for both systems with modulation order $M = 16$.

All of the previous depictions were compared in a fair manner, where a number of subcarriers, guard period, power, frequency spacing, carrier frequency etc. are assigned values such that no one of the systems can outperform the other due to the simulation parameter settings. These parameters were explained in the section of design criteria, Section 3. Hence, with evidence, it can be said that the FBMC-OQAM outperforms the CP-OFDM with respect to mobility and throughput and can be recommended to be the optimal candidate for 5G systems.

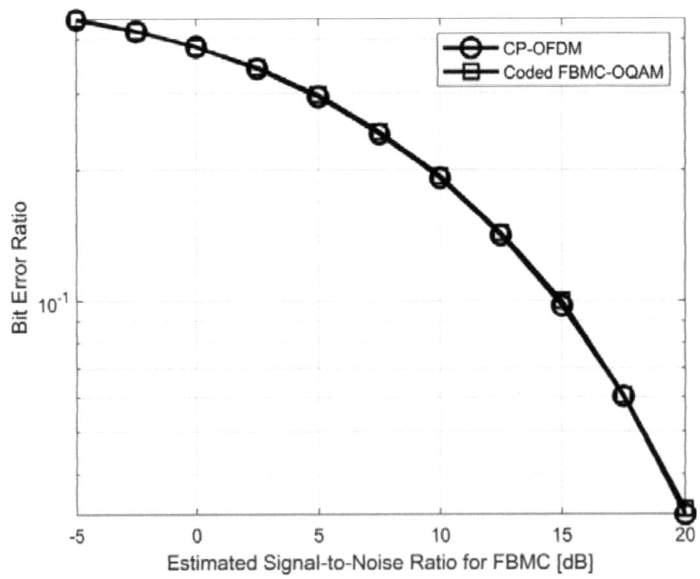


**Figure 8.** Bit error rate (BER) performance evaluation and comparison between CP-OFDM and FBMC-OQAM schemes, where 2 transmit antennas and 2 receive antennas have been employed for both systems with modulation order $M = 64$.

## 5. Conclusions

This paper introduced one of the 5G physical layer modulation techniques, FBMC-OQAM. It is shown that FBMC-OQAM outperforms the previous generation modulation technique, CP-OFDM, in different aspects such as; frequency localization, where the power spectral density of the FBMC-OQAM was well localized compared with the CP-OFDM, mobility with high speeds, and the capacity to implement MIMO systems, after simple modifications. Thus, FBMC-OQAM can be recommended as the next generation candidate to replace the CP-OFDM, since the throughput, with higher speeds than 4G limits, and the BER performance shows respectable behaviours.

**Acknowledgment**

We would like to express our special thanks of gratitude to all members of the Space Navigation and Control Laboratory (SNCL) who gave us the golden opportunity to do this work, which also helped us in doing a lot of research and we came to know about so many new things we are really thankful to them.

## References

[1] Schwarz S, Philosof T and Rupp M, 2017, Signal processing challenges in cellular-assisted vehicular communications: efforts and developments within 3gpp LTE and beyond *IEEE Signal Process Mag* **34** pp 47–59.

[2] Wunder G, Jung P, Kasparick M, Wild T, Schaich F, Chen Y, Brink S T, Gaspar I, Michailow N, Festag A, Mendes L, Cassiau N, Ktenas D, Dryjanski M, Pietrzyk S, Eged B, Vago P and Wiedmann F, 2014, 5GNOW: non-orthogonal, asynchronous waveforms for future mobile applications *IEEE Commun Mag* **52** pp 97–105.

[3] Andrews J G, Buzzi S, Choi W, Hanly S V, Lozano A, Soong A C K and Zhang J C, 2014, What will 5G be? *IEEE J Sel Areas Commun* **32** pp 1065–82.

[4] Agiwal M, Roy A and Saxena N, 2016, Next generation 5G wireless networks: A comprehensive survey *IEEE Commun Surv Tutor* **18** pp. 1617–55.

[5] Intelligence GSMA, 2014, Understanding 5G: Perspectives on future technological advancements in mobile *White paper* pp 1–26.

[6] Sesia S, Baker M and Toufik I, 2011, LTE-the UMTS long term evolution: from theory to practice, 2nd ed. Chichester, U.K.: John Wiley & Sons.

[7] Xiang W, Zheng K and Shen X S, 2017, 5G mobile communications, Switzerland: Springer International Publishing.

[8] Taher M A, Mandeep J, Ismail M, Samad S A and Islam M, 2014, Reducing the power envelope fluctuation of OFDM systems using side information supported amplitude clipping approach *Int J Circ Theor App* **42** pp. 425–35.

[9] [8]Taher M A, Mandeep J, Ismail M, Samad S A and Islam M, 2013, Reducing the PAPR of OFDM systems by random variable transformation *ETRI J* **35** pp. 714–17.

[10] Taher M A, Mandeep J, Ismail M, Samad S A, Islam M and Mahdi H F, 2015 Post-IFFT-modified selected mapping to reduce the PAPR of an OFDM system *Circ Syst Signal Pr* **34** pp 535–55.

[11] Taher M A, Mandeep J, Ismail M, Samad S A and Islam M, 2013, Sliding the SLM-technique to Reduce the non-linear distortion in OFDM systems *Elektron Elektrotech* **19** pp 103–11.

[12] Qualcomm Incorporated, 2016, Waveform candidates, *in 3GPP TSG-RAN WG1 #84b*, Busan, Korea.

[13] Schaich F, Wild T, and Chen Y, 2014, Waveform contenders for 5G - suitability for short packet and low latency transmissions, *in Conf. 2014 IEEE 79th IEEE VTS VEH TECHNOL*, Seoul, South Korea, pp 1–5.

[14] Zhang X, Jia M, Chen L, Ma J, and Qiu J, 2015, Filtered-OFDM - enabler for flexible waveform in the 5th generation cellular networks, *in Conf. 2015 IEEE GLOB TELECOMM CONF*, San Diego, CA, USA, pp 1–6.

[15] Michailow N, Matthé M, Gaspar I S, Caldevilla A N, Mendes L L, A. Festag, G. Fettweis, 2014, Generalized frequency division multiplexing for 5th generation cellular networks *IEEE T*

*Commun* **62** pp 3045–61.

[16] Bellanger M, D Ruyet Le, Roviras D, Terré M, Nossek J, Baltar L, Bai Q, Waldhauser D, Renfors M, Ihalainen T, Viholainen A, Stitz T H, Louveaux J, Ikhlef A, Ringset V, Rustad H, Najar M, Bader C, Payaro M, Katselis D, Kofidis E, Merakos L, Merentitis A, Passas N, Rontogiannis A, Theodoridis S, Triantafyllopoulou D, Tsolkas D, Xenakis D, Tanda M, Fusco T, Huchard M, Vandermot J, Kuzminskiy A, Schaich F, Leclair P and Zhao A, 2010, FBMC physical layer: a primer *PHYDYAS Journal* pp 1–31.

[17] 3GPP, 2017, Study on new radio access technology; Physical layer aspects, *TR 38.802*.

[18] Bölcskei H, 2003, Orthogonal frequency division multiplexing based on offset QAM *in Advances in Gabor Analysis*: Birkhäuser Boston, pp. 321–52.

[19] Farhang-Boroujeny B, 2011, OFDM versus filter bank multicarrier *IEEE Signal Proc MAG* **28** pp. 92–112.

[20] Matz G, Schafhuber D, Grochenig K, Hartmann M, and Hlawatsch F, 2007, Analysis, optimization, and implementation of low-interference wireless multicarrier systems *IEEE T Wirel Commun* **6** pp 1921–31.

[21] Nissel R, Schwarz S, and Rupp M, 2017, Filter bank multicarrier modulation schemes for future mobile communications *IEEE J Sel Area Comm* **35** pp 1768–82.

[22] Sahin A, Guvenc I, and Arslan H, 2014, A survey on multicarrier communications: prototype filters, lattice structures, and implementation aspects *IEEE Commun Surv Tutor* **16** pp 1312–38.

[23] Nam H, Choi M, Han S, Kim C, Choi S, and Hong D, 2016, A new filter-bank multicarrier system with two prototype filters for QAM symbols transmission and reception *IEEE T Wirel Commun* **15**, pp 5998–09.

[24] Yun Y H, Kim C, Kim K, Ho Z, Lee B, and Seol J, 2015, A new waveform enabling enhanced QAM-FBMC systems *in Conf. 2015 IEEE 16th IEEE INT WORK SIGN P*, Stockholm, Sweden, pp 116–20.

[25] Kim C, Yun Y H, Kim K, and Seol J, 2016, Introduction to QAM-FBMC: from waveform optimization to system design *IEEE Commun Mag* **54** pp 66–73.

[26] Haas R and Belfiore J-C, 1997, A time-frequency well-localized pulse for multiple carrier transmission *Wireless Pers Commun* **5** pp 1–18.

[27] Nissel R and Rupp M, 2018, Filter bank multicarrier modulation *in multiple access techniques for 5G wireless networks and beyond*: Springer International Publishing pp 63–92.

[28] Boashash B, 2015, Time-frequency signal analysis and processing: a comprehensive reference: *Academic Press*.

[29] Floch B L, Alard M, and Berrou C, 1995, Coded orthogonal frequency division multiplex [TV broadcasting] *in Proc. of the IEEE* **83** pp 982–96.

[30] Qu D, Wang F, Wang Y, Jiang T, and Farhang-Boroujeny B, 2017, Improving spectral efficiency of FBMC-OQAM through virtual symbols *IEEE T Wirel Commun* **16** pp 4204–15.

[31] Nissel R and Rupp M, 2017, Enabling low-complexity MIMO in FBMC-OQAM *in Conf. 2016 IEEE Globecom Workshops* pp 1–6.

[32] Lélé C, Siohan P, and Legouable R, 2010, The alamouti scheme with CDMA-OFDM/OQAM *EURASIP J Adv Sig Pr* **2010**.

[33] Zakaria R and Ruyet D L, 2012, A novel filter-bank multicarrier scheme to mitigate the intrinsic interference: application to MIMO systems *IEEE T Wirel Commun* **11** pp 1112–23.

[34] Lélé C, Siohan P, Legouable R, and Bellanger M, 2007, CDMA transmission with complex OFDM/OQAM *EURASIP J Wirel Comm* **2008**.

[35] ITU, 1997, Recommendation ITU-R M.1225: guidelines for evaluation of radio transmission technologies for IMT-2000 *ITU, Technical Report*.

IOP Conf. Series: Materials Science and Engineering **557** (2019) 012041 doi:10.1088/1757-899X/557/1/012041

# Tensile Properties and Impact Strength of RHDPE/BF Composites: The Effects of Chemical Treatment

**Aini Asifa Ahmad Kamal[1], N Z Noriman[1], S T Sam[2], Awad A. Al-rashdi[3], I. Johari[4], Z M Razlan[5], Shahriman A B[5], I Zunaidi[6] and Wan Khairunizam[5]**

[1]Center of Excellence Geopolymer and Green Technology (CEGeoGTech), Faculty of Engineering Technology (FETech), Universiti Malaysia Perlis (UniMAP), UniCITI Alam Campus, 02100, Perlis, Malaysia.
[2]School of Bioprocess Engineering, Universiti Malaysia Perlis, Kompleks Pengajian Jejawi 3, 02600 Arau, Perlis, Malaysia
[3]Chemistry Department, Umm Al-Qura University, Al-qunfudah University College, Al-qunfudah Center for Scientific Research (QCSR), Saudi Arabia
[4]School of Civil Engineering, Engineering Campus, Universiti Sains Malaysia, Penang, Malaysia
[5]School of Mechatronic Engineering, Universiti Malaysia Perlis (UniMAP), Pauh Putra Campus, 02600 Arau, Perlis, Malaysia
[6]Faculty of Technology, University of Sunderland, St Peter's Campus, Sunderland, SR6 0DD, United Kingdom

**Abstract.** Recycled high-density polyethylene reinforced with bamboo filler were compounded with twin-screw extruder and injection molding process. The main objective of this study is to investigate the effect of different chemical treatment with different concentration for bamboo filler reinforced with recycled high-density polyethylene composites via injection molding. The fillers were reinforced with plastic for different concentration of sodium hydroxide and acetic acid at 2.5 %, 5.0 % and 7.5 %. Mechanical measurements will show the presence of different chemical treatment with different concentration insignificantly effect in the composites tensile properties and also impact properties. The result of the mechanical analysis carried out showed that the presence of natural filler in composites will improve the properties of the material.

## 1. Introduction

Modern technologies required materials with unusual combination properties that cannot be met by conventional materials [1]. The composite material is invented by combining two different materials in new materials which may lead for a be better suited particular application either of the original materials alone and also to new desirable properties [1].

Actually, development of polymer composites starts in the 19th century that introduced new age research with the new option by using natural fibers as a part of more differentiated fields [2].

The increase in community interest and environmental consciousness, the unsustainable consumption of petroleum and new environmental regulations, led to thinking of the use of environmentally friendly materials. Natural fiber is considered one of the environmentally friendly materials which have good properties compared to synthetic fiber (Islam et al., 2015). Then, the use of natural fibers leads to cost reduction and lightweight composites, through the mechanical properties of natural fiber composites are much lower compared than the synthetic fiber composites [4].

Natural filler has a lot of advantages compared with synthetic filler. However, there is one major challenge in using natural filler include the bamboo filler which is the inherent flaws within fibres which is weak interfacial bonding to the thermoplastic bonding and high moisture absorption that reduce their compatibility with polymer matrices resulting in poor mechanical properties of the composites [5][6][7][8].

To avoid this problem, surface modification is required prior to composite fabrication. The properties of filler reinforced polymer composites may improve by using chemical treatment (sodium hydroxide, peroxides, isocyanides and maleic anhydride organosilanes) and physical (corona treatment, cold plasma treatment) [1][9].

Chemical modification on natural fiber will give a great approach for the establishment of covalent bonding between the filler matrix [10]. Here, it is used to alter the structure of the fiber, at the same time the interfacial adhesion between the polymer matrix and filler will improve the physical and mechanical properties of the composite material. Based on the previous research, surface modification by using a chemical treatment on filler can stop the moisture absorption, increase the surface roughness and clean the bamboo fiber surface [11][12][13].

## 2. Experimental Procedures

### *2.1. Materials*

The recycled high-density polyethylene (RHDPE) will be supplied by Zarm Scientific and Supplies Sdn. Bhd. in solid pellet form with the melt flow rate of 0.236 g / 10 min at 200 ° C, 5 kg load. RHDPE is the main matrix used for this study. Then, bamboo that was used as filler will be obtained around Perlis, Malaysia area. The bamboo flour will be sieved based on the required size using a vibratory sieve shaker AS 200 controls before undergoing the particle size analyzer (Malvern Sciro 2000 Mastersizer) for the particle size confirmation. Others chemical like sodium hydroxide (NaOH) and acetic acid ($CH_3COOH$) will be supplied by ADV System Technology.

### *2.2. Chemical Treatment*

Bamboo filler (BF) of 150-300 μm range size was used in this investigation. Two different chemical treatments by using NaOH and $CH_3COOH$ solution with concentrations ranging from 2.5 %, 5.0 % and 7.5 % [14][15] was prepared for alkaline and acid treatment. The BF was immersed in chemical solutions for 2 hours at the room temperature. After immersion, BF was washed several times with running distilled water to remove any excess of surface modification solution on the filler's surface. The pH of the filler was checked until it turns neutral. Finally, BF was dried in an oven at 45 °C for 12 hours period before proceeding to the fabrication stage of composites [16].

### *2.3. Composites Preparation*

Different treatment of BF will be used in this study according to the formulation in Table 1. The composites were prepared by compounding using the twin screw extruder with a barrel temperature of 170 ° C. The RHDPE mixed with BF will be loaded into the feeder and run through the twin screw extruder and produce composites pallet. Then, composites pallet will undergo an injection molding process for the final product and ready for further testing.

**Table 1** The filler loading of the composites

| Parameters | Materials | Formulation |
|---|---|---|
| Treated with NaOH | RHDPE / BF untreated | RHDPE / BF untreated |
| | RHDPE / BF treated 2.5 % | RHDPE / BF treated 2.5 % |
| | RHDPE / BF treated 5.0 % | RHDPE / BF treated 5.0 % |
| | RHDPE / BF treated 7.5 % | RHDPE / BF treated 7.5 % |
| Treated with $CH_3COOH$ | RHDPE / BF untreated | RHDPE / BF untreated |
| | RHDPE / BF treated 2.5 % | RHDPE / BF treated 2.5 % |
| | RHDPE / BF treated 5.0 % | RHDPE / BF treated 5.0 % |
| | RHDPE / BF treated 7.5 % | RHDPE / BF treated 7.5 % |

### *2.4. Tensile Test*

The tensile test specimens were prepared according to ASTM D638 with the common specimen for ASTM D638 that has a constant rectangular cross-section. The specimen size is 170 mm × 1 mm × 0.5 mm. The tensile testing was conducted using Axial-Torsion Universal Testing Machine (INSTRON-5982) at room temperature and all tensile specimens will be pulled under a constant crosshead speed of 50 mm/min. Each test was performed until tensile failure occurred. Before the test, the thickness of specimens will be measured using Vernier caliper. Five specimens will be conducted and average values were tabulated. The tensile test is the most common mechanical test to determine the tensile properties of natural fillers like tensile strength, Young's modulus and % of elongation.

### *2.5. Impact Test*

Charpy impact test was conducted according to ASTM 790 at room temperature by using Compact Charpy Impact Tester (Model MAT 23). The specimen size is 80 mm × 1 mm × 0.5 mm. From this test, the fracture energy (Joule) will be determined. Five specimens will be conducted and the impact energy values will be averaged to obtain a mean value.

## 3. Results and Discussion

### *3.1. Tensile Properties*

The effect of surface treatment on the mechanical properties by tensile analysis on tested material can be referred in Figure 1, Figure 2 and Figure 3 for tensile strength, elongation at break and Young's modulus result. Two different chemicals were used for chemical treatment which is sodium hydroxide (NaOH) and acetic acid (CH3COOH) with three different concentrations.

Figure 1 illustrates the tensile strength RHDPE / BF composites of different chemical treatment with different concentration. The trend for NaOH show, increasing concentration of NaOH will decrease the tensile strength, while for CH3COOH vice versa, which is increasing the CH3COOH concentration will increase the tensile strength. Based on the result, it shows 2.5 % of NaOH displayed the highest tensile strength (25.195 MPa) compared to CH3COOH at 7.5 % with (23.781 MPa). The treatment for alkaline improves the BF with small concentration used compared acid treatment.

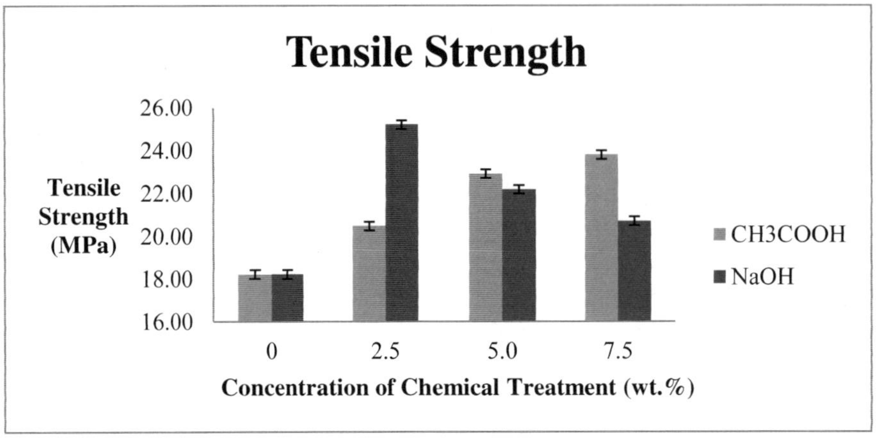


**Figure 1:** Effect of treatment on tensile strength.

The treatment gives better tensile strength compared to untreated BF due to the bonding of filler with matrix thereby improving the fiber-matrix interaction [17][18][19]. The significant of alkaline treatment is the disruption of hydrogen bonding in the filler surface, thus increasing surface roughness [19][20]. So, it improved filler-matrix interaction by the removal of hemicellulose and lignin that lead to the better incorporation of filler with matrix [21][22][23]. Unfortunately, tensile strength drop at 5 % treated with NaOH due to the excess removal of cellulose content at the surface of the filler by the high concentration of alkali. So, it would damage the filler's structure [7][21]. Apart from that different type of natural filler and chemical have different optimum concentration during the treatment process which later affects the performance of the composites [21][24].

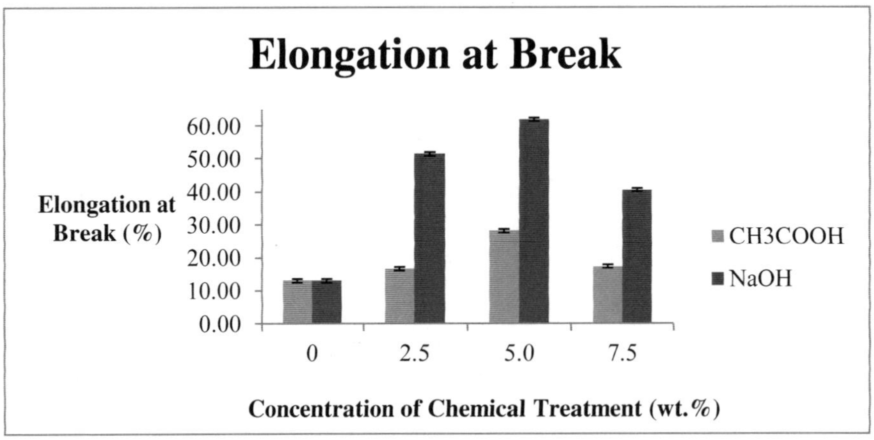


**Figure 2:** Effect of treatment on elongation at break.

Figure 2 shows the elongation at break for different chemical treatment with different chemical concentrations. The optimum value of elongation at break for RHDPE / BF composites for both chemical treatments was a value of 28.16 %, $CH^3COOH$ and 61.92 %, NaOH with 5 % of chemical concentration. The elongation at break of composites depends on the filler and matrix interaction which is composite with higher elongation, indicates higher ductility of materials and lower elongation indicates lower ductility of materials [25][26].

Based on Figure 3 below, it is clear that chemical treatment improves Young's modulus of the RHDPE / BF composites. The trend was increasing from 2.5 % until 7.5 % concentration of both chemicals. The 7.5 % concentration NaOH and $CH_3COOH$ of treated composites exhibit the highest modulus elasticity of 627.58 MPa and 806.08 MPa [7][27][28].

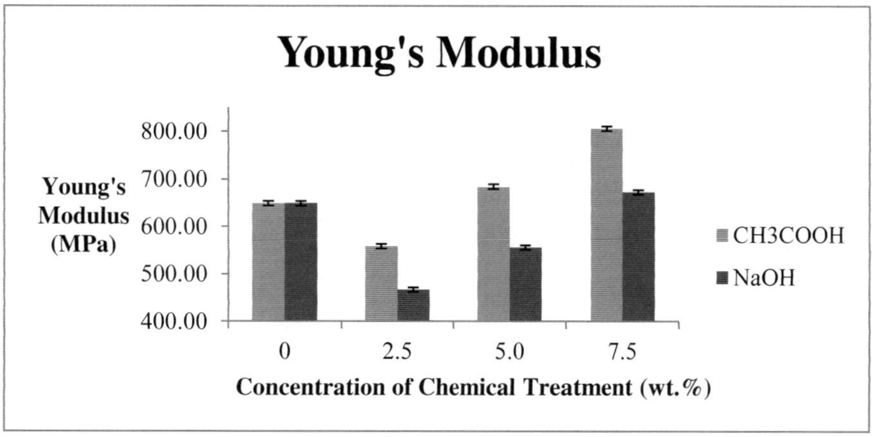


**Figure 3:** Effect of treatment on Young's modulus.

### *3.2. Impact Properties*

The detailed analysis of the Charpy impact test for all composites specimens is presented in Figure 4. The impact strength of alkaline treatment by NaOH give better impact strength compared with acid treatment by CH3COOH. The highest impact strength is for NaOH treatment at 2.5 % with 2.662 J/m. Then, impact strength decrease with increasing of concentration for both treatments by NaOH and CH3COOH. At 7.5 % concentration of NaOH and CH3COOH, the value of impact strength was the lowest which is NaOH, 1.659 and CH3COOH, 0.63. The excess of removal cellulose content with a high concentration of chemical treatment will damage the filler and increase the brittleness of composite hence decrease the impact strength due to restricted mobility of polymer chains [29].

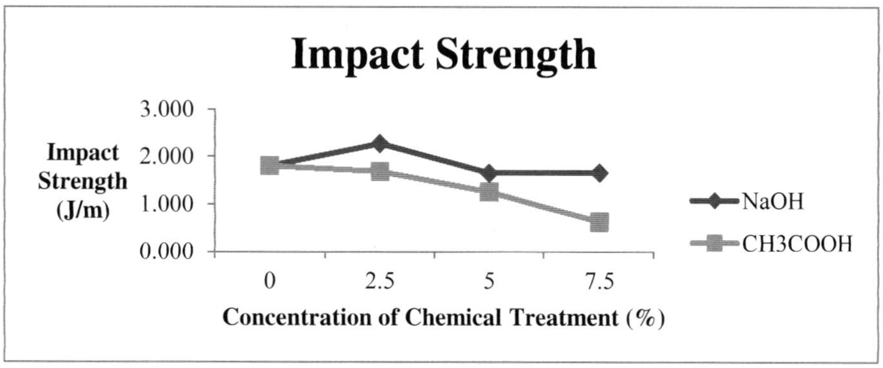


**Figure 4:** Effect of treatment on impact energy.

## 4. Conclusions

In this research, the effect of chemical treatment on tensile and impact properties of bamboo filler reinforced recycled high-density polyethylene composites were investigated. It is clearly observed that both treatment either with alkaline (NaOH) or acidic (CH3COOH) treatment improves the tensile properties. The 2.5 % NaOH treated BF / RHDPE composite has optimum tensile strength while CH3COOH treated BF/ RHDPE composite at 7.5%. As for impact strength, the optimum impact strength at 2.5 % of NaOH treated.

**References**

[1] Luna, I. Z., Dam, K. C., Chowdhury, A. M. S., Gafur, A., Khan, N., & Khan, R. A. (2015). Physical and Thermal Characterization of Alkali Treated Rice Husk Reinforced Polypropylene Composites. *Advances in Materials Science and Engineering, 2015,* 1–7.

[2] Swain, P. T. R., & Biswas, S. (2017). Influence of Fiber Surface Treatments on Physico-Mechanical Behaviour of Jute / Epoxy Composites Impregnated with Aluminium Oxide Filler. *Journal of Composite Materials, 51*(28), 1–14.

[3] Islam, S., Kovalcik, A., Hasan, M., & Thakur, V. K. (2015). Natural Fiber Reinforced Polymer Composites. *International Journal of Polymer Science,* 2–4.

[4] Mohammed, L., Ansari, M. N. M., Pua, G., Jawaid, M., & Islam, M. S. (2015). A Review on Natural Fiber Reinforced Polymer Composite and Its Applications. International Journal of Polymer Science.

[5] Kabir, M. M., Wang, H., Lau, K. T., & Cardona, F. (2012). Chemical Treatments on Plant-Based Natural Fibre Reinforced Polymer Composites *: An Overview. Composites: Part B, 43,* 2883–2892.

[6] Ma, H., & Joo, C. W. (2011). Influence of Surface Treatments on Structural and Mechanical Properties of Bamboo Fiber-Reinforced Poly (Lactic Acid) Biocomposites. *Journal of Composite Materials, 45*(23), 2455–2463.

[7] Manalo, A. C., Wani, E., Zukarnain, N. A., Karunasena, W., & Lau, K.-T. (2015). Effects of Alkali Treatment and Elevated Temperature on The Mechanical Properties of Bamboo Fibre-Polyester Composites. *Composites: Part B, 80,* 73–83.

[8] Zhang, X., Wang, F., & Keer, L. M. (2015). Influence of Surface Modification on the Microstructure and Thermo-Mechanical Properties of Bamboo Fibers. Materials, *8*, 6597–6608.

[9] Luo, S., & Netravali, A. N. (1999). Mechanical and Thermal Properties of Environment-Friendly "Green" Composites Made From Pineapple Leaf Fibers and Poly(hydroxybutyrate-co-valerate) Resin. *Polymer Composites, 20*(3), 367–378.

[10] Cao, X. V., Ismail, H., Rashid, A. A., Takeichi, T., & Vo-huu, T. (2014). Effect of Filler Surface Treatment on the Properties of Recycled High-Density Polyethylene /(Natural Rubber)/(Kenaf Powder) Biocomposites. *Journal of Vinyl & Additive Technology,* 1–7.

[11] Gwon, J. G., Lee, S. Y., Chun, S. J., Doh, G. H., & Kim, J. H. (2010). Effect of Chemical Treatments of Wood Fibers on The Physical Strength of Polypropylene Based Composites. *Korean J. Chem. Eng, 27*(2), 651–657.

[12] Kallakas, H., Shamim, M. A., Olutubo, T., Poltimae, T., Suld, T. M., Krumme, A., & Kers, J. (2015). Effect of Chemical Modification of Wood Flour on The Mechanical Properties of Wood-Plastic Composites. *Agronomy Research, 13*(3), 639–653.

[13] Tan, M. Y., Kuan, H. T. N., & Lee, M. C. (2017). Characterization of Alkaline Treatment and Fiber Content on the Physical, Thermal, and Mechanical Properties of Ground Coffee Waste / Oxobiodegradable HDPE Biocomposites. *International Journal of Polymer Science, 2017,* 1–12.

[14] Benyahia, A., Merrouche, A., Rokbi, M., & Kouadri, Z. (2013). Study the Effect of Alkali Treatment of Natural Fibers on the Mechanical Behavior of the Composite Unsaturated Polyester-Fiber Alfa. *21eme Congres Francais de Mecanique, 26(*30), 1–6.

[15] Oushabi, A., Sair, S., Hassani, F. O., Abboud, Y., Tanane, O., & Bouari, A. El. (2017). The Effect of Alkali Treatment on Mechanical, Morphological and Thermal Properties of Date Palm Fibers (DPFs): Study of the Interface of DPE-Polyurethane Composite. *South African Journal of Chemical Engineering.*

[16] Ma, L., He, L., & Zhang, L. (2017). Effect of surface treatments on tensile properties of hemp fiber reinforced polypropylene composites, *AIP Conference Proceedings 1829*, 020016.

[17] Kannan, R., & Megat, M. M. H., (2013). A Review on Mechanical Properties Of Bamboo Fiber Reinforced Polymer Composite, *Australian Journal of Basic and Applied Sciences, 7*(8), 247-253.

[18] Rokbi M., Osmani H., Imad A., Benseddiq N., Effect of Chemical treatment on Flexure Properties of Natural Fiber-reinforced Polyester Composite, *Procedia Engineering, 10,* 2092-2097.

[19] Benyahia, A., Merrouche, A., Rahmouni, Z. E., Rokbi, M., Serge, W., & Kouadri, Z. (2014). Study of the alkali treatment effect on the mechanical behavior of the composite unsaturated polyester-Alfa fibers. *Mechanics & Industry, 15*(1), 69-73.

[20] Yan L., Chouw N., Yuan X., Improving the mechanical properties of natural fiber fabric reinforced epoxy composites by alkali treatment, *Journal of Reinforced Plastics and Composites, 31,* 425-437.

[21] Ain U. Md Shah., H, M. T., Sultan, M. J., F. C., & Abu, A. R., Talib. (2016). A Review on the Tensile Properties of Bamboo Fiber Reinforced Polymer Composites. *BioResources, 11*(4), 1-23.

[22] Zhang, X., Wang, F., & Keer, L. (2015). Influence of Surface Modification on the Microstructure and Thermo-Mechanical Properties of Bamboo Fibers. *Materials,8*(10), 6597-6608.

[23] Nor Azowa I., Kamarul Arifin H., Khalina A., (2010). Effect of fiber treatment on mechanical properties of Kenaf fiber-Ecoflex composites, *Journal of Reinforced Plastics and Composites, 29,* 2192-2197.

[24] Boopathi, L., Sampath, P. S., & Mylsamy, K. (2012). Investigation of Physical, Chemical and Mechanical Properties of Raw and Alkali Treated Borassus Fruit Fiber, *Composite Part B: Engineering, 43*,(8), 3044-3052.

[25] Chandra Mohan, H.K, Dr. G.V.Gnanendra Reddy, Dr. M. Chowde Gowda. (2015). Mechanical Properties of Untreated and Alkali Treated Sida Acuta Stem Fibre, *International Journal of Scientific & Engineering Research, 6*(2), 1352-1359.

[26] Senthil Kumar, P., Prakash, S., & Prakash, J. (2018). The Effect of Chemical Treatment On The Tensile Properties Of Sisal Fiber Reinforced Epoxy Composite, *International Research Journal of Engineering and Technology, 5*(8), 1431-1435.

[27] Ariffin, A., & Yusof, Y. (2017). Effect of Extraction Process and Surface Treatment on the mechanical properties in Pineapple Leaf Fibre. *MATEC Web of Conferences,135*, 00042.

[28] A. Khalina, K. Shaharuddin, M. S. Wahab, M. P. Saiman, H. A. Aisyah. (2017). The Effect of Alkaline Treatment on Tensile Strength and Morphological Properties of Kenaf Fibres for Yarn Production, *International Journal of Agricultural and Bio systems Engineering, 11*(7), 525-529.

[29] Mubarak, Y. A. (2018). Tensile and Impact Properties of Microcrystalline Cellulose Nanoclay Polypropylene Composites. *International Journal of Polymer Science,2018*, 1-13.

# Path Loss Analysis Considering Doppler Shift Effect on Cellular Communication for Connected Car Application at Rural Area

**A F A Hamid, M T A Rahman*, A Rahman and M M M Zabidi**

School of Mechatronic Engineering, Universiti Malaysia Perlis (UniMAP), Kampus Pauh Putra, 02600 Arau, Perlis, Malaysia

*Corresponding author: tasyrif@unimap.edu.my

**Abstract.** In this paper, the study on path loss of cellular communication is conducted by experiment and simulation on three different communication modules at rural area. This three-communication module will have different carrier frequency to determine the effect of doppler shift to communication path loss of moving vehicle. To enhance the accuracy of these studies, both simulation and experiment output will be compared. Simulation studies will be based on proposed modified path loss model. For the experimental portion, the Path Loss will be determined based on two things: the measured received power on communication modules antenna and transmit power from base station. From this experiment, the best selection of communication module by considering the fading effect from doppler shift is selected to be used in connected car.

## 1. Introduction

Connected car technology ought to be a keystone of revolutionizing in the automotive area. The use of wireless communication technologies on vehicle is the key to connecting vehicle to become one important part in Intelligent Transportation System (ITS) [1]. From basic information such as location of the car and now the vehicle engine performance itself can be monitored. Due to rapid advancement of electronic system for vehicle engine, previously every vehicle using only one Engine Control Unit (ECU) but now several control unit systems have been developed to control every part in vehicle by sharing the information each other's [2]. In the 1990s, with the introduction of 802.11 wireless communication standard, ECU not only communicates within the internal vehicle system, it can also share information between different vehicles [3]. This technology made the connected car system can be achieve to support the ITS.

According to Kleberger et al. [2011], "the connected car consists of three domains: the vehicle itself, consisting of the in-vehicle network and the ECUs, the portal at the automotive company, and the communication link between them" [4]. The studies in this paper is focusing on communication link between vehicle and database to ensure the reliability of the information sharing through cellular network system. As suggested by Junyeong et al. [2014], it is important to provide better communication link quality due to increasing of vehicle speed and data packet size [5]. To overcomes the poor communication link during vehicle moving in urban area road, experiment on path loss and handover delay of cellular communication is made in this paper.

## 2. Existing Studies

There are few proposed solutions to provide better quality of communication. As suggested by Jyothi et al. [2014], it's better to use Global System for Mobile communications (GSM) SIM900A rather than Bluetooth and Zigbee communication system for advance mobile communication [6]. Before the communication module is implemented on connected car, two main parameters need to be analysed; as path loss and doppler shift effect [7]. Yet, despite there are many existed models for analysing the path loss model but not all of it can be compatible to be used in certain unique environment [8]. For instance, Shooting and Bouncing Rays (SBR) method in Ray tracing path loss model was used by Amornthipparat et al. [2008] to study non-Line of Sight (NLOS) path loss in street cell environment [9]. Similarly, there have been studies on path loss between transmitter and receiver at urban and suburban environment [10]. However, so far there are only few researches that consider the influence of doppler effect in the path loss.

## 3. Methodology

In this paper, we are comparing path loss that happen on three different communication modules that have different frequency carrier by analysing the measured path loss data. Additionally, the tuned free space path loss model is used for the simulation. As for the experiment, we need to complete three phases of development and analyses in order to find suitable communication module that can be implemented in connected car system.

### 3.1 Communication Node Development.

Communication node hardware has been developed with three devices that are Arduino Uno, GSM modem and GPS. For GSM, three different models (with separate band and coverage) have been selected to test their communication performance; SIM900, SIM800L and SIM5360E. Note that SIM900 and SIM800L are using 2G signal for their communication coverage. In this experiment, SIM900 was using 1800MHz as the system band. Alternatively, for SIM800L and 900MHz, they were set as network band by using the AT+CBAND AT command. Lastly, SIM5360E was set to using UMTS 2100MHz as network band due to the modem 3G coverage capability. The experiment setup for these three GSM modems can be referred in Figure 1. Each communication node will perform the AT+CSQ AT command to obtain the communication modem Received Signal Strength Indicator (RSSI) which determines the received power in the GSM Antenna and perform ping request to test the latency of communication link. In this communication node GPS GY-NEO6MV2 has been used to get a latitude and longitude of the vehicle in deciding the distance between vehicle location and base station. At the end, these three GSM modems have sent all the data to webserver at www.000webhost.com.

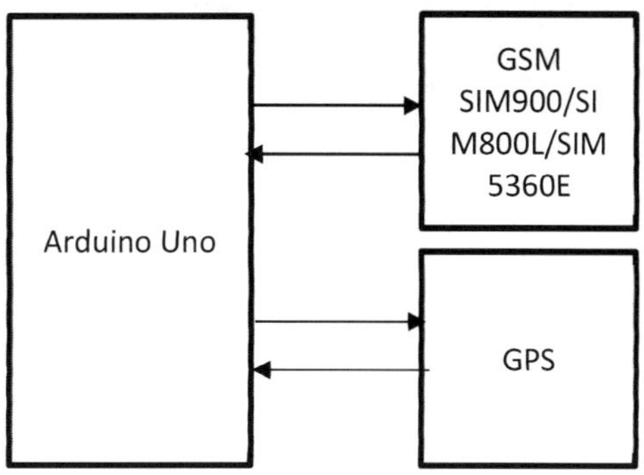


**Figure 1.** SIM900 Communication node

In this communication node GPS GY-NEO6MV2 was used to get a latitude and longitude of the vehicle in deciding the distance between vehicle location and base station.

3.2 Webserver deployment

For recording, the data collected from the communication node. Webserver was deployed at www.000webhost.com by using Hypertext Preprocessor (PHP) and My Structured Query Langue (MySQL) algorithm. This webserver used to store longitude, latitude, RSSI and vehicle speed data as displayed in Figure 2.

+ Options

| TIME | LATITUDE | LONGITUDE | RSSI | LATENCY | SPEED |
|---|---|---|---|---|---|
| 0000-00-00 00:00:00.000000 | 20.21 | 30.57 | -31 | 37 | 100 |
| 2018-07-19 20:35:27.235288 | 20.21 | 30.57 | -31 | 37 | 100 |
| 2018-07-19 20:35:31.781555 | 20.21 | 30.57 | -31 | 37 | 100 |
| 2018-07-19 20:37:32.035466 | 20.21 | 30.57 | -31 | 37 | 100 |
| 2018-07-19 20:42:17.046072 | 20.22 | 30.56 | -46 | 16 | 89 |
| 2018-07-19 20:42:35.053954 | 20.22 | 30.56 | -60 | 16 | 110 |
| 2018-07-19 20:42:56.104383 | 20.22 | 30.56 | -46 | 16 | 172 |
| 2018-07-19 20:43:14.256229 | 20.22 | 30.56 | -43 | 18 | 140 |
| 2018-07-19 20:43:32.302869 | 20.22 | 30.56 | -41 | 16 | 86 |
| 2018-07-19 20:43:50.579899 | 20.22 | 30.56 | -41 | 16 | 133 |
| 2018-07-19 20:44:08.843778 | 20.22 | 30.56 | -39 | 19 | 98 |
| 2018-07-19 20:44:26.895334 | 20.22 | 30.56 | -44 | 16 | 82 |
| 2018-07-19 20:44:45.117941 | 20.22 | 30.56 | -52 | 16 | 165 |
| 2018-07-19 20:45:03.274390 | 20.22 | 30.56 | -45 | 18 | 107 |
| 2018-07-19 20:45:21.607261 | 20.22 | 30.56 | -48 | 17 | 159 |
| 2018-07-19 20:45:40.082625 | 20.22 | 30.56 | -50 | 16 | 171 |
| 2018-07-19 20:45:58.030297 | 20.22 | 30.56 | -49 | 17 | 131 |
| 2018-07-19 20:46:16.057617 | 20.22 | 30.56 | -42 | 16 | 153 |
| 2018-07-19 20:46:34.205018 | 20.22 | 30.56 | -54 | 16 | 102 |
| 2018-07-19 20:47:22.440602 | 42.00 | 30.00 | -52 | 16 | 98 |
| Console 20:47:40.487426 | 48.00 | 37.00 | -49 | 16 | 106 |

**Figure 2.** Data collection table for communication node.

3.3 Analysis on measure and simulation path loss of communication that correlate to distance and vehicle speed.

For the analysis on communication link performance, the experiment only using one method ; measuring path loss based on RSSI data in each types of communication module. This communication module tested by moving the vehicle along the rural area as shown in Figure 3. Three different velocity has been used in this test; 60km/h, 90km/h and 120km/h.

**Figure 3.** Jalan Arau-Changlun rural area beside paddy field road.

The distance between vehicle and base station along the road was determined by referring to the Open Signal App as illustrated in Figure 4 that contains base station location information.

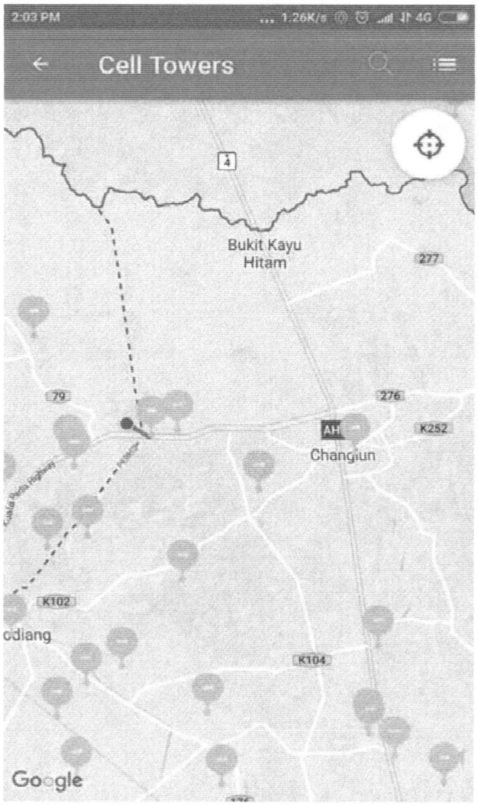


**Figure 4.** OpenSignal Apps to determine base station location.

As for the simulation , the Tuned Free Space Path Loss model proposed as modified path loss model calculation. This model was suggested by Hong et al. as modified path loss model to simulate the experiment for high velocity moving communication node and considering doppler shift effect [11]. Doppler shift is a fading path loss effect that happens due to moving communication node. Based on this experiment, the received signal may have a Doppler shift of,

$$f_D = \frac{v \cos \theta}{\lambda} \quad (1)$$

associated with it, where θ is the arrival angle of the received signal relative to the direction of motion, v is the receiver velocity, and

$$\lambda = \frac{c}{f_c} \quad (2)$$

is the signal wavelength (c =299 792 458 m/s is the speed of light) and ($f_c$ is a frequency band of signal). The geometry associated with the Doppler shift is shown in Figure 5.

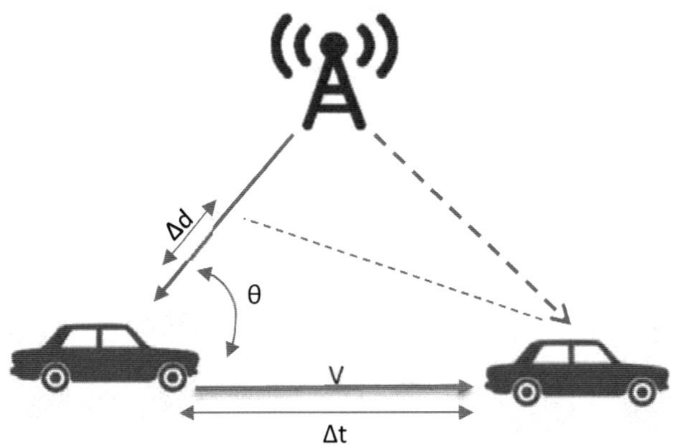


**Figure 5.** Doppler shift effect model.

The Doppler shift results from the fact that transmitter or receiver movement over a short time interval Δt causes a slight change in distance

$$\Delta d = v\Delta t cos\theta \quad (3)$$

that the transmitted signal needs to travel to the receiver. The phase change due to this path length difference is

$$\Delta\varphi = \frac{2\pi v\Delta t cos\theta}{\lambda} \quad (4)$$

The Doppler frequency is then obtained from the relationship between signal frequency and phase:

$$f_D = \frac{1}{2\pi} \times \frac{\Delta\emptyset}{\Delta t} = v\cos\theta/\lambda \quad (5)$$

This doppler shift $f_d$ parameter was calculated based on three different frequency carrier from communication module. Arrived angle, $\theta$ and distance, $\Delta d$ was measured based on the GPS data and OpenSignal Apps information.

Next, this doppler shift parameter was employed in calculating path loss in Tuned Free-Space path model using the following formula:

$$L = K_1 \times log_{10}(f_d) + K_2 \times log_{10}(d) + K_3 + 32.45 + 20 \times log_{10}(d) + 20 \times log_{10}(f) \quad (6)$$

Where L is a path loss result that obtain from Tuned Free-Space model by using $K_1$, $K_2$ and $K_3$ that is median of optimum values minimum error between tuned path loss model and measured data obtained from Hong et al. analysis result on signal measurement at rural area. The value of $K_1 = 5.0$, $K_2 = 19.2$ and $K_3 = 9.0$.

## 4. Result

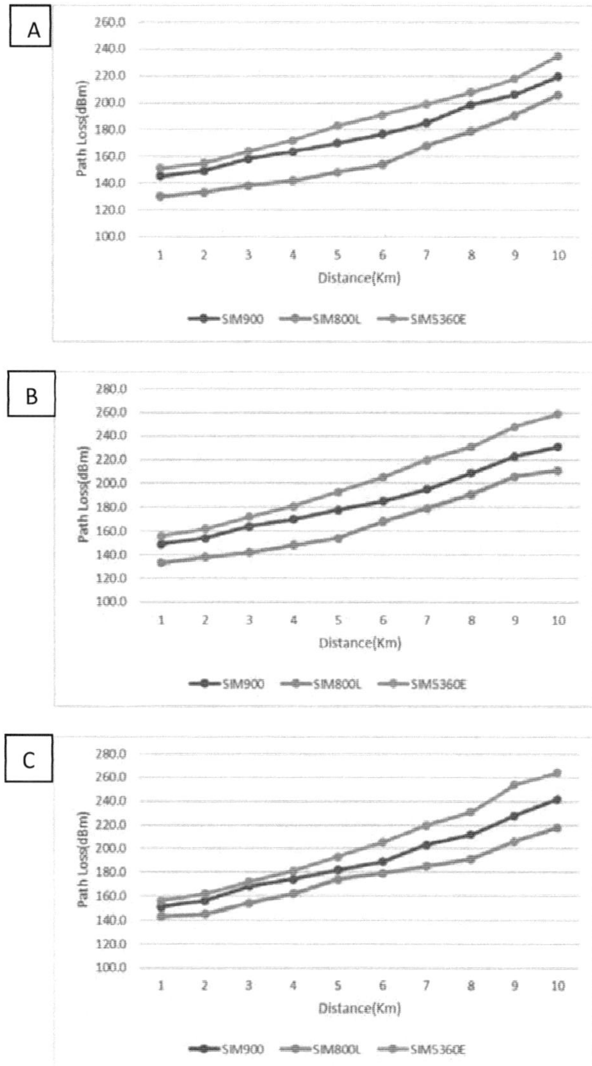


**Figure 6.** Path Loss vs Distance measurement data for three different communication modules with vehicle speed at A=60KmH, B=90KmH and C=120KmH

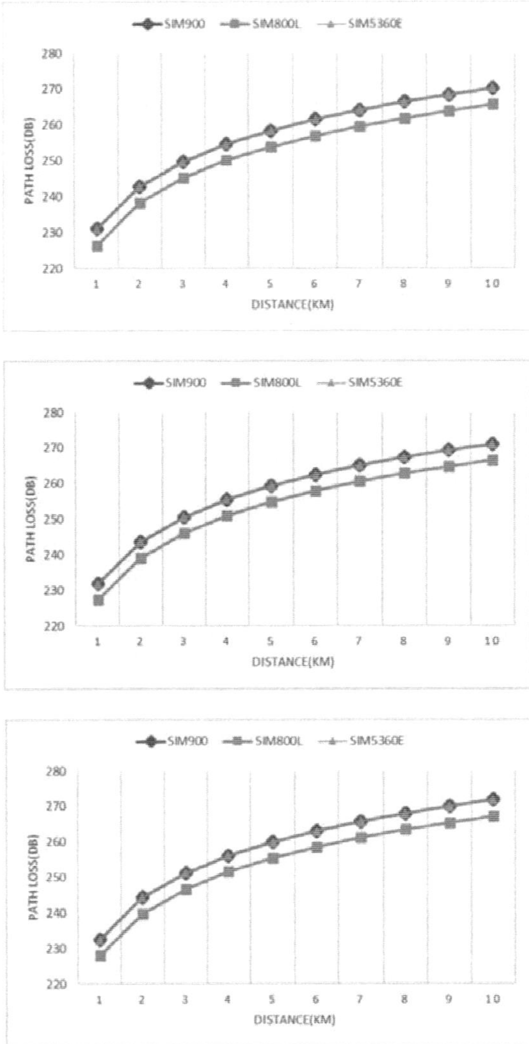


**Figure 7.** Path Loss vs Distance Tuned Free-Space path loss model data for three different communication modules with vehicle speed at A=60KmH, B=90KmH and C=120KmH

The results above are referring to the 3rd Generation Partnership Project (3GPP) in Technical Specification Group Radio Access Network, Evolved Universal Terrestrial Radio Access (E-UTRA) and Base Station (BS) radio transmission and reception (Release 15) that stated, the standard transmitting power for macro cellular communication base station is 46dBm. The result on Figure 6, where the path loss data are generated based on measuring RSSI from communication module and transmitting power from base station. It is showed the differences in path loss that happen in communication link will increasing if the frequency is increase. Whereas, in Figure 7 showed that the calculated data chart from the trend of path loss is just same with measuring data.

## 5. Conclusion

In this paper, the propagation path loss effect on communication link of connected car was investigated by considering doppler effect. From the experiment conducted, result shown that SIM800 is the most suitable, because of the low frequency from the SIM 800 communication module. This showed that the lower the frequency band, the better communication link in the signal propagation aspect. However, further study are needed to enhance the understanding of the communication link in urban or suburban area, before any conclusions can be made. Therefore, data from this study can be applied to network technologies specializing in connected car communication systems with another part, such as signal encoding algorithm and network technology.

## 6. Acknowledgement

We would like to extend our sincere gratitudes for the funding provided by the Universiti Malaysia Perlis (UniMAP) (Geran Penyelidikan Khas - No.9004-00063) and UniMAP Automotive Racing Team, which enable this research to be carried out.

## References

[1] Lu N, Cheng N, Zhang N, Shen X, Mark JW. Connected vehicles: Solutions and challenges. IEEE Internet Things J. 2014;1(4):289–99.
[2] Prashant A. Shinde, Y.B. Mane. Real Time Vehicle Monitoring and Tracking System based on Embedded Linux Board and Android Application. International Conference on Circuit, Power and Computing Technologies [ICCPCT]. 2015.
[3] Coppola R, Morisio M. Connected Car: Technologies, Issues, Future Trends. ACM Comput Surv [Internet]. 2016;49(3):1.
[4] Kleberger P, Olovsson T, Jonsson E. Security aspects of the in-vehicle network in the connected car. IEEE Intell Veh Symp Proc. 2011;(Iv):528–33.
[5] Bok J, Ryu HG. Path loss model considering Doppler shift for high speed railroad communication. Int Conf Adv Commun Technol ICACT. 2014;1–4.
[6] MtechG.Harish PJ. Controlling System. Des Implement Real Time Veh Monit Track Control Syst. 2014;
[7] Phillips C, Sicker D, Grunwald D. A survey of wireless path loss prediction and coverage mapping methods. IEEE Commun Surv Tutorials. 2013;15(1):255–70.
[8] Alam N, Balaie AT, Dempster AG. Dynamic path loss exponent and distance estimation in a vehicular network using Doppler effect and received signal strength. IEEE Veh Technol Conf. 2010;
[9] A. Amornthipparat, H. Shirai, K. Yonezawa, T. Inoue, and Y. Nakamura, "Estimation of High Frequency NLOS Path loss in StreetCell Environment," ICCE 2008. Second International Conf., pp. 336340, Jun. 2008.
[10] Rappaport TS, Blankenship K, Xu H. Propagation and radio system design issues in mobile radio systems for the GloMo project. Mob portable radio Res … [Internet]. 1997;1–26.
[11] Wei H, Zhong Z, Guan K, Ai B. Path loss models in viaduct and plain scenarios of t he High-speed Railway. 5th Int Commun Netw China ICST Conf. 2010;1–5.

IOP Conf. Series: Materials Science and Engineering **557** (2019) 012043 doi:10.1088/1757-899X/557/1/012043

# Path Loss with Doppler Shift Effect and Handover Delay Analysis on Cellular Communication for Connected Car Application in Urban Area

**A F A Hamid, M T A Rahman*, A Rahman and M M M Zabidi**

School of Mechatronic Engineering, Universiti Malaysia Perlis (UniMAP), Kampus Pauh Putra, 02600 Arau, Perlis, Malaysia

*Corresponding author: tasyrif@unimap.edu.my

**Abstract.** In this paper, the study on path loss and delay of cellular communication is conducted by considering doppler shift fading effect, shadowing effect and handover session in urban area. The experiment will consist observation on three different wireless communication module. This three-wireless module will have different carrier frequency and radio interface technologies to determine the effect of doppler shift to communication path loss and handover delay of a moving vehicle. The path loss of a moving vehicle will be determined based on measured received power on communication modules antenna and transmit power from base station. Handover delay that experienced by a moving vehicle will be measured based on data transmission latency during communication session transferring from previous cell network area that pass by the vehicle to a new cell network area. From this experiment, the best selection of wireless communication module by considering doppler shift effect and handover delay in urban area is determined as suited module to be use on connected car system.

## 1. Introduction

Connected car technology become one important part in Intelligent Transportation System (ITS) [1]. From basic information such as location and now the vehicle engine performance itself can be monitored. Due to fast advancement of electronic system for vehicle engine, previously every vehicle using one Engine Control Unit (ECU) and now multiple control unit system has been developed to control every part in vehicle by sharing information each other's [2]. In 1990s with 802.11 wireless communication standard introduced, ECU not only able to communicate within internal vehicle system but it's can share information between different vehicle [3]. This technology made the connected car system can be achieve to support the ITS.

According to Kleberger et al. [2011], "the connected car consists of three domains: the vehicle itself, consisting of the in-vehicle network and the ECUs, the portal at the automotive company, and the communication link between them" [4]. The studies in this paper is focusing on communication link between vehicle and database to ensure the reliability of the information sharing through cellular network system. As suggested by Junyeong et al. [2014], it is important to provide better communication link quality due to increasing of vehicle speed and data packet size [5]. To overcome the poor communication link during vehicle moving in urban area road, experiment on path loss and handover delay of cellular communication is made in this paper.

## 2. Experiment Setup

To test cellular network reliability in communication link, communication node hardware was developed with three devices that are Arduino Uno, GSM modem and GPS to perform communication between vehicle and database. For GSM, three different types of modem model (with distinct band and coverage) were selected to be tested on their communication performance; SIM900, SIM800L and SIM5360E. Note that SIM900 and SIM800L use 2G signal for their communication coverage. In this experiment, SIM900 was using 1800MHz as the system band. Alternatively, for SIM800L and 900MHz, they were set as network band by using the AT+CBAND AT command. SIM5360E was set to using UMTS 2100MHz as network band due to the modem 3G coverage capability. The experiment setup for these three GSM modems can be referred to in Figure 1. Each of the communication node will perform AT+CSQ AT command to get the Received Signal Strength Indicator (RSSI) of communication modem that determine the received power in GSM antenna and perform ping request to test latency of communication link. In this communication node GPS GY-NEO6MV2 was used to get a latitude and longitude of the vehicle in deciding the distance between vehicle location and base station. At the end, these three GSM modems will send all the data to webserver was deployed at www.000webhost.com.

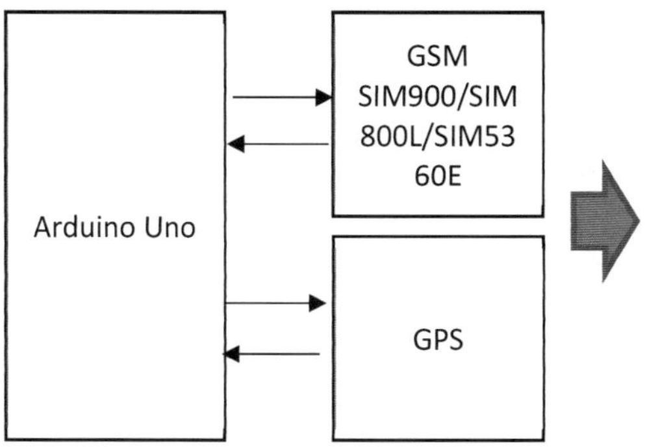


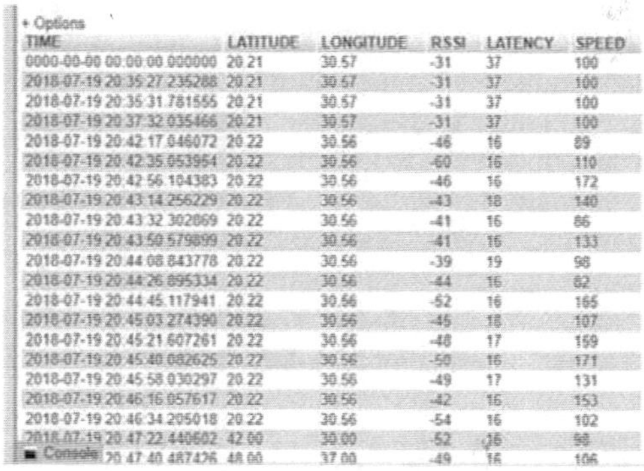
+ Options

| TIME | LATITUDE | LONGITUDE | RSSI | LATENCY | SPEED |
|---|---|---|---|---|---|
| 0000-00-00 00:00:00.000000 | 20.21 | 30.57 | -31 | 37 | 100 |
| 2018-07-19 20:35:27.236288 | 20.21 | 30.57 | -31 | 37 | 100 |
| 2018-07-19 20:35:31.781555 | 20.21 | 30.57 | -31 | 37 | 100 |
| 2018-07-19 20:37:32.035466 | 20.21 | 30.57 | -31 | 37 | 100 |
| 2018-07-19 20:42:17.046072 | 20.22 | 30.56 | -46 | 16 | 89 |
| 2018-07-19 20:42:35.053954 | 20.22 | 30.56 | -60 | 16 | 110 |
| 2018-07-19 20:42:56.104383 | 20.22 | 30.56 | -46 | 16 | 172 |
| 2018-07-19 20:43:14.256229 | 20.22 | 30.56 | -43 | 18 | 140 |
| 2018-07-19 20:43:32.302869 | 20.22 | 30.56 | -41 | 16 | 86 |
| 2018-07-19 20:43:50.579899 | 20.22 | 30.56 | -41 | 16 | 133 |
| 2018-07-19 20:44:08.843778 | 20.22 | 30.56 | -39 | 19 | 98 |
| 2018-07-19 20:44:26.895334 | 20.22 | 30.56 | -44 | 16 | 82 |
| 2018-07-19 20:44:45.117941 | 20.22 | 30.56 | -52 | 16 | 165 |
| 2018-07-19 20:45:03.274390 | 20.22 | 30.56 | -45 | 18 | 107 |
| 2018-07-19 20:45:21.607261 | 20.22 | 30.56 | -48 | 17 | 159 |
| 2018-07-19 20:45:40.082625 | 20.22 | 30.56 | -50 | 16 | 171 |
| 2018-07-19 20:45:58.030297 | 20.22 | 30.56 | -49 | 17 | 131 |
| 2018-07-19 20:46:16.057617 | 20.22 | 30.56 | -42 | 16 | 153 |
| 2018-07-19 20:46:34.205018 | 20.22 | 30.56 | -54 | 16 | 102 |
| 2018-07-19 20:47:22.440602 | 42.00 | 30.00 | -52 | 16 | 98 |
| ■ Console 20:47:40.487426 | 48.00 | 37.00 | -49 | 16 | 106 |

**Figure 1.** Communication node hardware setup and My Structured Query Langue (MySQL) table that will receive longitude, longitude, RSSI, latency and vehicle speed data from communication node on moving vehicle.

## 3. Experiment Environment

The communication link is tested on urban area at Jalan Kangar, Perlis as in figure 2. This area is selected based on the dense of buildings and it's tall that more than three floors. Additionally, from open signal apps as in figure 3 the area also has a large amount of base station. This will give advantages in minimizing propagation problem based on the distance between vehicle and base station.

**Figure 2.** Kangar road.

IOP Conf. Series: Materials Science and Engineering **557** (2019) 012043 doi:10.1088/1757-899X/557/1/012043

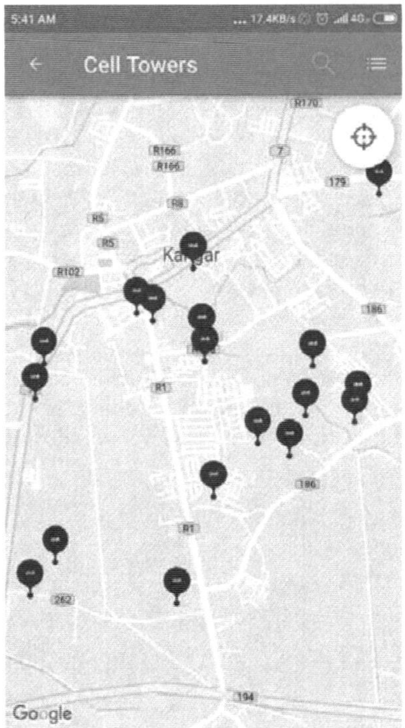


**Figure 3.** Base station location by referring on Open Signal apps.

## 4. Path Loss with Considering Doppler Shift Effect

In mobile communication propagation studies for high velocity communication node such as vehicle there two main parameters need to be analysed that is path loss and doppler shift effect [6]. Yet, despite there are many existed models for analysing the path loss model but not all of it can be compatible to be used in certain unique environment [7]. For example, Amornthipparat et al. [2008] used Shooting and Bouncing Rays (SBR) method in Ray tracing path loss model to study non-Line of Sight (NLOS) path loss in street cell environment [8]. Similarly, there have been studies on path loss that happen between transmitter and receiver in urban and suburban environment [9]. However, so far there are only few researches that consider the influence of doppler effect in the path loss. Doppler shift is a frequency distortion effect that influence carrier frequency by transmitter. Doppler shift is related with velocity of vehicle, the angle of Line-of-Sight (LOS) and direction of vehicle moving as in figure 4 [10]. This effect gives signal uncertainties to signal modulation encoding phase and that will create a data loss in communication link [7]. The path loss experiment is conduct by applying AT+CSQ command on communication module to get RSSI data. RSSI data is calculated with Okumura propagation model as in equation 1 that was state the best model for path loss calculation in urban area [11].

$$L_p(dB) = L_f + A_{mu}(f, d) - G(h_{te}) - G(h_{re}) - G(AREA) \quad (1)$$

Where $L_p$, is the median value of the propagation Path loss.

$L_f$ is the free space propagation loss.
$A_{mu}$ is the median attenuation relative to the free space.
$G(h_{te})$is the base station antenna height gain factor = 46 dBm [12].
$G(h_{re})$is the mobile antenna height gain factor that was get from RSSI.
G (AREA) is the gain due to the type of the environment.

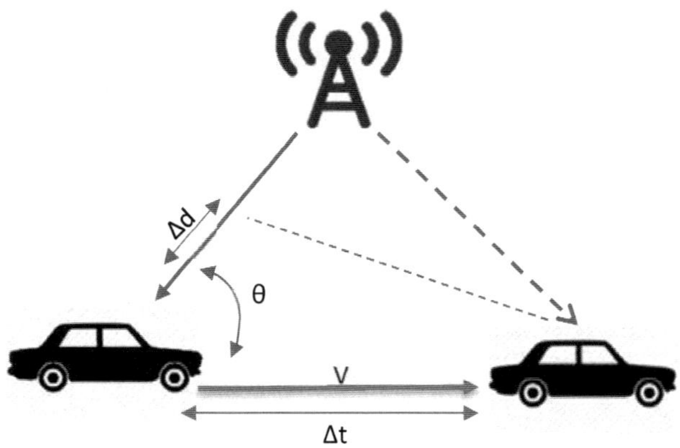


**Figure 4.** Doppler shift effect model. Where V = velocity, Δd = distance and θ = LOS angle.

## 5. Handover Delay

Beside propagation issues in mobile communication, handover delay will happen when the communication node is moving between two or more cell network that cover by base station [13]. Communication handover is a process when the base station cell network transferring ongoing communication link form current cell network area into a new cell network area that pass by the vehicle like in figure 5 [14]. During this handover process, the ongoing communication link will have a delay in data transferring sequence. The experiment on three communication modules is to determine the differences of average delay that effect on communication latency between them. The experiment is conducted by ping on server www.google.com with 32-bit of data, the latency is a time for the Hypertext Transfer Protocol (HTTP) request to be successful. Handover delay is measured when the handover process between two base station is occurs.

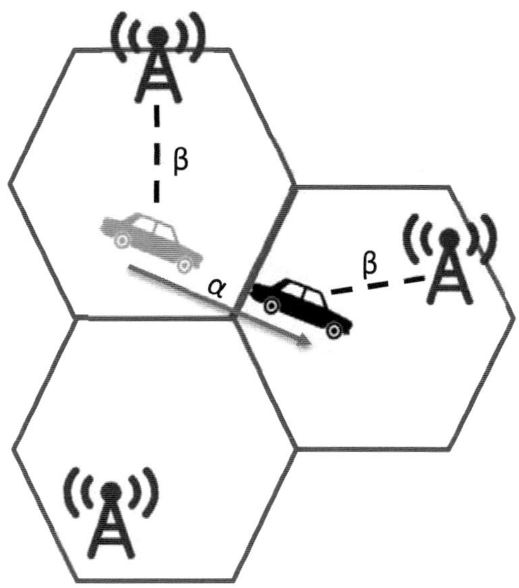


**Figure 5.** The vehicle that move from one cell network area to another cell network area where the α is direction of vehicle movement and β is communication link between the vehicle and base station.

IOP Conf. Series: Materials Science and Engineering **557** (2019) 012043 doi:10.1088/1757-899X/557/1/012043

## 6. Experiment Result and Discussion

The path loss result from three communication module experiment is separate to three different vehicle velocity 40Km/H, 60Km/H and 80Km/H as in figure 6. The result shows increasing trend of path loss corresponding to the increase of distance between vehicle and base station.

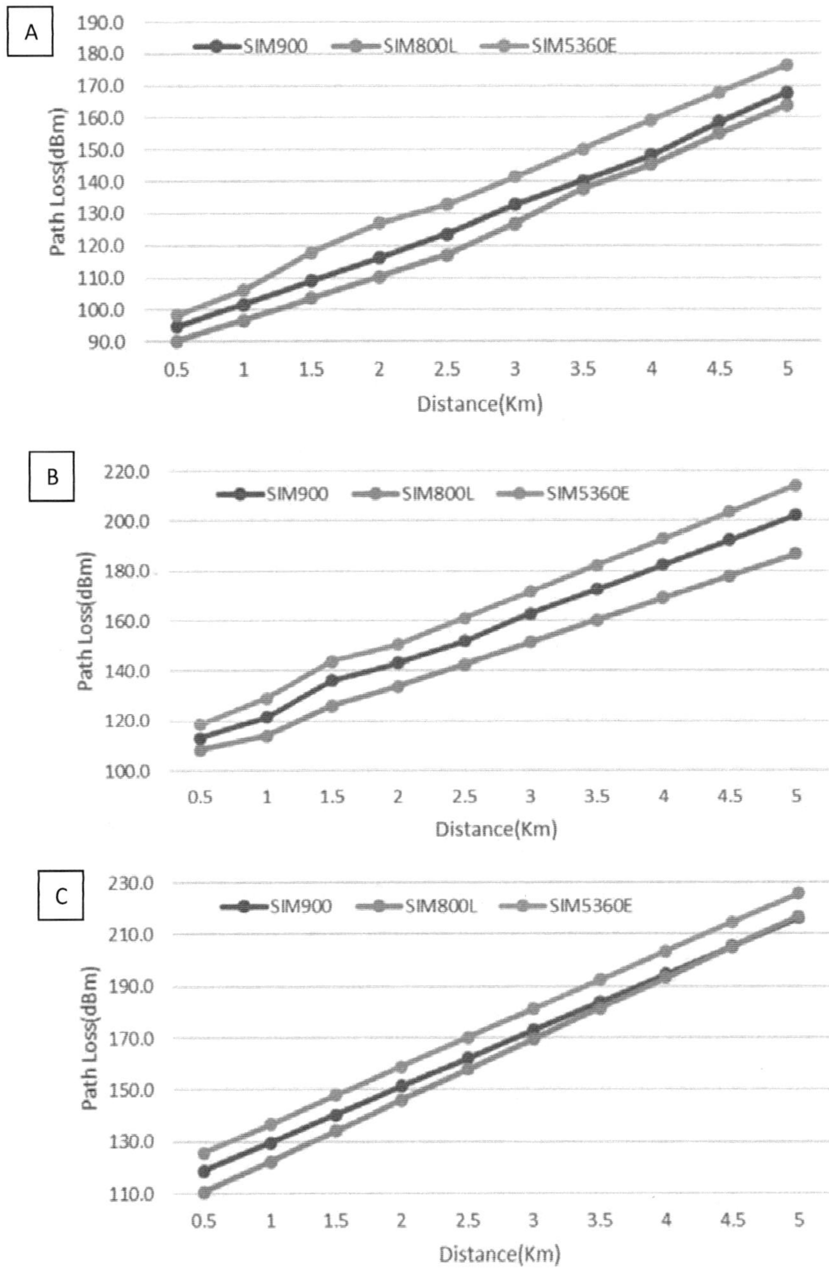


**Figure 6.** Path Loss vs Distance measurement data for three different communication modules with vehicle speed at A=40Km/H, B=60Km/H and C=80Km/H

The latency result from three communication module experiment is separate to three different vehicle velocity 40Km/H, 60Km/H and 80Km/H as in figure 7. The result shows handover process that occur when $\log(L_p)$ become zero and delay of the communication link in this situation is $80mS > delay > 20mS$. However, the SIM5360E receive lowest affect of handover delay in this experiment.

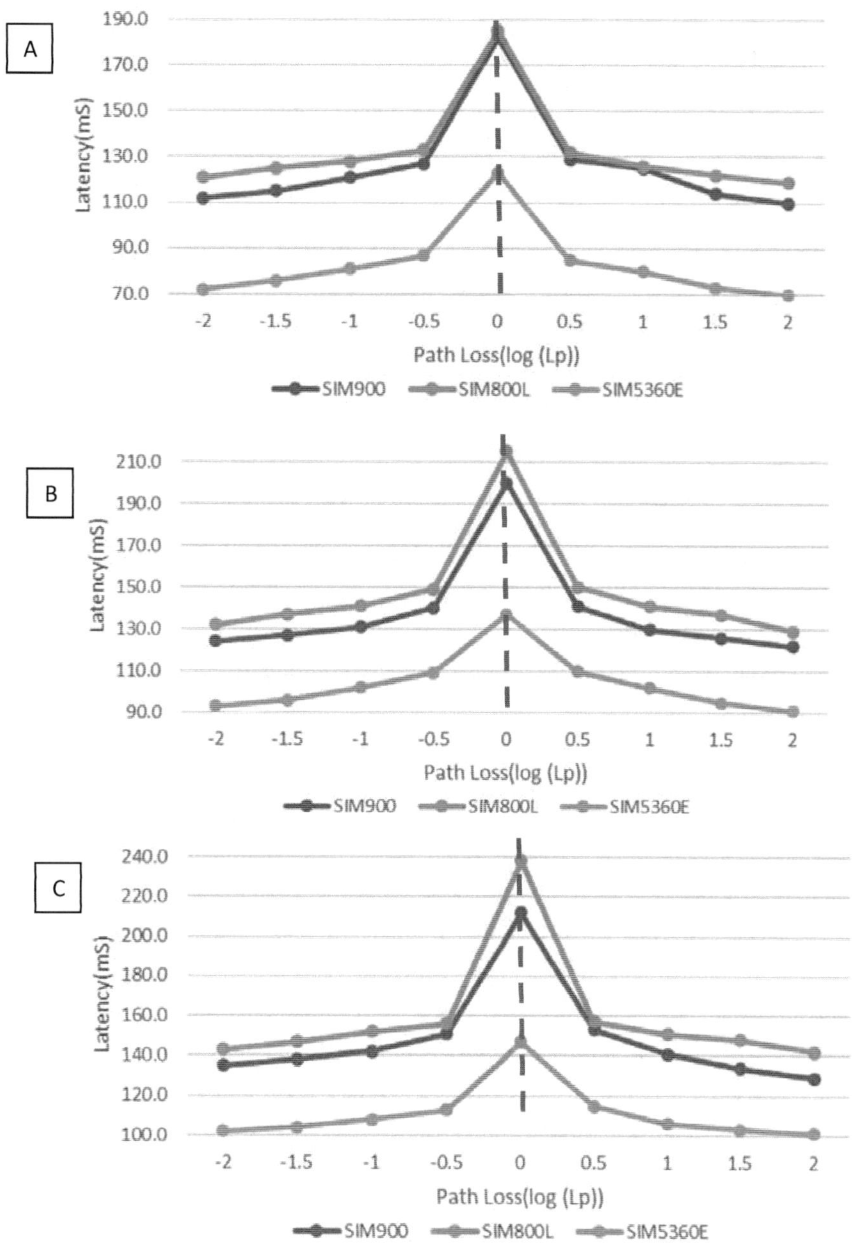


**Figure 7.** Path Loss vs Latency data for three different communication modules with vehicle speed at A=40Km/H, B=60Km/H and C=80Km/H. The red dash line is where handover process happen.

## 7. Conclusion

This study has explored full potential of each communication module. Communication module SIM800, SIM900 and SIM5350E is analysed based on the path loss data and handover delay that take place on urban environment. It was found out that SIM800 with 900Mhz carrier frequency had a lowest path loss despite path loss in SIM800 is increasing when velocity of vehicle is increase due to doppler shift effect. However, the studies for handover delay shows that SIM5360E significantly had the lowest latency of communication link and better handover delay process. In the meanwhile, 5G technology will be introduced in next few years. With this implementation of communication technology, the amount base station will increase too and it's made path loss effect become less important. With huge amount cell network of base station, the performance of handover process become significantly important. From this observation, it can be concluded that the best communication module to be placed in connected car system is SIM5360E.

## References

[1] Lu N, Cheng N, Zhang N, Shen X, Mark JW. Connected vehicles: Solutions and challenges. IEEE Internet Things J. 2014;1(4):289–99.
[2] Prashant A. Shinde, Y.B. Mane. Real Time Vehicle Monitoring and Tracking System based on Embedded Linux Board and Android Application. International Conference on Circuit, Power and Computing Technologies [ICCPCT]. 2015.
[3] Coppola R, Morisio M. Connected Car: Technologies, Issues, Future Trends. ACM Comput Surv [Internet]. 2016;49(3):1.
[4] Kleberger P, Olovsson T, Jonsson E. Security aspects of the in-vehicle network in the connected car. IEEE Intell Veh Symp Proc. 2011;(Iv):528–33.
[5] Bok J, Ryu HG. Path loss model considering Doppler shift for high speed railroad communication. Int Conf Adv Commun Technol ICACT. 2014;1–4.
[6] Phillips C, Sicker D, Grunwald D. A survey of wireless path loss prediction and coverage mapping methods. IEEE Commun Surv Tutorials. 2013;15(1):255–70.
[7] Alam N, Balaie AT, Dempster AG. Dynamic path loss exponent and distance estimation in a vehicular network using Doppler effect and received signal strength. IEEE Veh Technol Conf. 2010;
[8] A. Amornthipparat, H. Shirai, K. Yonezawa, T. Inoue, and Y. Nakamura, "Estimation of High Frequency NLOS Path loss in StreetCell Environment," ICCE 2008. Second International Conf., pp. 336340, Jun. 2008.
[9] Rappaport TS, Blankenship K, Xu H. Propagation and radio system design issues in mobile radio systems for the GloMo project. Mob portable radio Res … [Internet]. 1997;1–26.
[10] Wei H, Zhong Z, Guan K, Ai B. Path loss models in viaduct and plain scenarios of the High-speed Railway. 5th Int Commun Netw China ICST Conf. 2010;1–5.
[11] Alumona TL NKN. Path Loss Prediction of Wireless Mobile Communication for Urban Areas of Imo State, South-East Region of Nigeria at 910 MHz. Int J Sens Networks Data Commun [Internet]. 2015;04(01):1–4.
[12] 3rd Generation Partnership Project (3GPP), 3rd Generation Partnership Project; Technical Specification Group Radio Access Network; Evolved Universal Terrestrial Radio Access (E-UTRA); Base Station (BS) radio transmission and reception (Release 15), p. 61-73, 2018.
[13] Li M, Fei Y, and Victor CML, A N EW M ETHOD TO S UPPORT UMTS / WLAN V ERTICAL H ANDOVER U SING SCTP. 2004;(August):44–51.
[14] Doppler K, Rinne M, Wijting C, Ribeiro CB, Hug K. Device-to-device communication as an underlay to LTE-advanced networks. IEEE Commun Mag. 2009;47(12):42–9.

## Acknowledgement

The authors would like to extend their sincere gratitude for the funding provided by the Universiti Malaysia Perlis (UniMAP) (*Geran Penyelidikan Khas* - No.9004-00063) which enable this research to be carried out.

# Root-Zone Cooling in Tropical Greenhouse: a Review

**A G Niam[1] and H Suhardiyanto[1]**

[1] Department of Mechanical and Biosystem Engineering, Bogor Agricultural University.
* Email: ghautsun@apps.ipb.ac.id

**Abstract**. To prevent the decrease of vegetable production during the rainy season in a tropical region, it is important that cultivation is carried out in greenhouses. However, air temperature inside the greenhouse tends to be very high during the daytime in clear days. To overcome these problems, root zone cooling system has been developed as an energy efficient cooling system for tropical greenhouses. In this paper, performances of the root zone cooling system for tropical greenhouse are described. The application of the root zone cooling system for plant production in tropical greenhouses is still very potential to be developed. This is because the most significant temperature influencing plant growth of some plants was the root-zone temperature. It is important to evaluate the application of root zone cooling in tropical greenhouse for protected plant production in the tropics. It has been shown that CFD simulations analysis are able to visualize the temperature distribution in a tropical greenhouse equipped during root zone cooling. This has a positive potential for improving performance analysis, design, construction with more careful attention to environmental climate factors, type of screen, dimensions, arrangement of plant layout, and types of plants produced.

## 1. Introduction

Air temperature in the greenhouse during sunny days in the tropics tends to be too high for the growth of plants inside the greenhouse. Cooling the greenhouse as a whole requires very large amount of energy. It has been proven that cooling load to reduce the air temperature inside the greenhouse as a whole to 6 °C below the outside air temperature can reach 0.32 MJ $m^{-2}$ $h^{-1}$ in the summer at night [1]. Therefore, root zone cooling system has been developed as energy efficient cooling system for tropical greenhouse.

## 2. Root zone temperature and plant response

Research on plant response due to differences in temperature in roots began in the late 70s, where the roots of tomato plants warmed around 23-26 °C in winter (night air temperature around 9 °C) had the best growth compared to roots without heat [2]. The same thing has also been tested on potatoes in vitro, where at 28/13 °C day/night the potato plants can produce a significantly higher zeatin riboside hormone as compared to potato plants which are given a temperature treatment of 30/28 °C day/night [3]. In addition, effects of root-zone and shoot-zone temperature have also been tested on lettuce plants that are cultivated in floating hydroponic. The most significant temperature influencing the biomass growth of lettuce was the root-zone temperature, where from a temperature variation between 12 °C to 31 °C, the best temperature was around 24 °C [4]. All of the above evidences occur in the subtropical region, and have a positive impact on the quality and productivity of plants. The development of root zone cooling system for plant production in tropical greenhouse still remain a big challenge.

## 3. Tropical greenhouse

Tropical greenhouse has its characteristic to allow natural ventilation at maximum level. Natural ventilation occurs due to the existence of two factors that trigger the mechanism of fluid movement. The first trigger factor is caused by thermal buoyancy, which is often referred to as the stack effect, where the temperature difference that occurs in the fluid inside the greenhouse comes from the heat convection process, solar radiation flux and metabolism of organisms in the greenhouse [3]. Heated air will reduce its density so that the air mass gets lighter and with the influence of gravity can cause air parcels that are getting lighter. The air then tends to move up or float. The second trigger factor is the presence of wind which causes a difference in pressure on the walls and cover of the greenhouse building because of the pressure drop that forces the air inside the greenhouse to move through the gap of the vent openings [4]. Therefore, thermal factors play a dominant role when the air velocity is low, resulting in air movement due to differences in temperature and air density inside and outside the greenhouse.

The wind speed limit where thermal factors can still play a dominant role is 1 ms-1 [5], whereas when the air velocity is more than 1.8 ms-1 the thermal effect on the ventilation rate can be ignored [5,6]. If the wind speed outside the greenhouse is quite high and the difference in air temperature inside and outside the greenhouse is small, then the dominant wind factor and the influence of thermal factors can be ignored. The minimum ventilation openings area is 20% of the greenhouse floor area so that the temperature in the greenhouse can approach ambient temperatures outside the greenhouse [5]. However, if the area of ventilation openings is more than 40% of the area of the greenhouse, then the rate of natural ventilation is quite good and can avoid extreme temperature increases in tropical greenhouses [3]. Therefore, the use of screens as a cover for ventilation openings helps to reduce the number of disturbing pest attacks into the greenhouse, but their use will reduce the rate of ventilation and increase the air temperature in the greenhouse.

## 4. Simulation of temperature distribution in the root zone

It is important to predict temperature distribution in the root zone. One method for describing the temperature distribution pattern in the plant root zone is a numerical simulation approach. Today many numerical simulations are approached based on the finite volume method. The numerical calculation is solved by a computational approach that has considered three main aspects of physics, namely in the form of mass, momentum and energy equilibrium. Then, this is known as Computational Fluid Dynamics (CFD) [7]. CFD is a field of science that studies ways to predict fluid flow, heat transfer, chemical reactions, and other phenomena by solving mathematical and physical equations, so that this can be a design tool as well as research tools [8].

Temperature distribution in tropical greenhouses equipped with refrigerated NFT hydroponic systems has been simulated using CFD-based software. Root zone cooling was applied through nutrient solution which was channeled into NFT gutters in the greenhouse and used for tomato cultivation (Figure 1.a). Through an experimental approach and reliability test, the simulation results of the distribution of temperature in the greenhouse have a maximum error of 8.06% with a temperature distribution of around 31.3 °C to 32.0 °C (Figure 1.b). In addition, the value of the uniformity coefficient obtained was around 98.2%. The simulation results show that root zone cooling through nutrient solution also gives rise to the cold effect on the air around nutrient gutters in Figure 1.b. Therefore, the concept of zone cooling, in addition to reducing energy consumption, can also have a positive effect on the leaf zone around the cooled area.

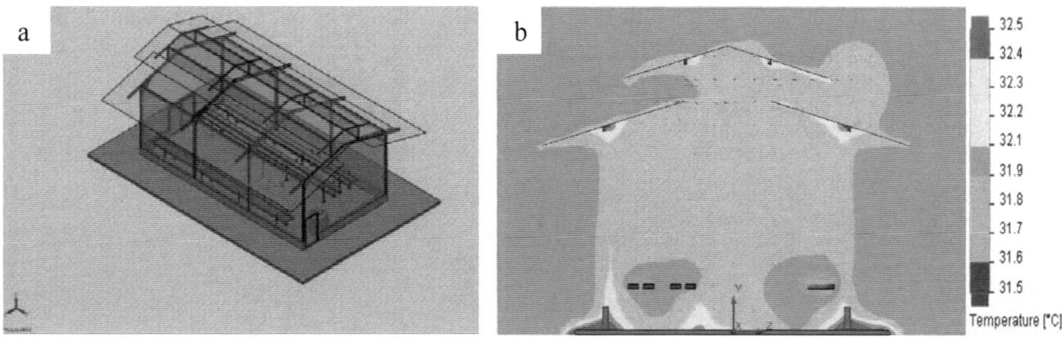


**Figure 1**. a) greenhouse geometry, b) cut plot temperature distribution at the front view

## 5. Root zone cooling on leaf vegetable cultivation

Lettuce and capsicum have been successfully grown in the tropical greenhouse in Singapore, by applying the root zone cooling system between 15-25 °C while their shoots were maintained at hot ambient temperatures ranging from 26-42 °C [9]. The results of this study prove that the treatment of root zone cooling at 25 °C yields a yield of 37.7% higher than the treatment temperature of 30 °C, and 61.4% higher than without root zone cooling treatment. The interaction of root zone cooling application and P-induction can increase the average growth rate by about 31% [9]. Similar results have also been proven in the hydroponic nutrient film technique (NFT) in tropical climate regions of Brazil, for lettuce plants. The plants were tested at root zone temperature around 24-30 ° C and interact with the addition of nutrient solution concentrations from 1 to 3 dS $m^{-1}$. The test results concluded that cooling the root zone around 26 °C was the optimum temperature for the growth of lettuce in tropical climate regions. However, the addition of EC to nutrient from 1 to 3 $dSm^{-1}$, cause reduction in growth of lettuce [10]. In addition, the butterhead lettuce plant also shows photosynthetic $CO_2$ assimilation, stomatal conductance, relatively mid-content midday leaf higher at constant rooting temperatures of around 20 °C, compared to rooting temperatures which are allowed to follow fluctuations in ambient air temperature of 23-40 °C [11].

## 6. Root Zone Cooling on Tuber Plants Cultivation

Root zone cooling for the cultivation of tuber plants has been carried out also for shallots. With a floating hydroponic system, the temperature distribution around plant holes is simulated using CFD. Root zone cooling in shallots is intended to accelerate flowering initiation as occurred in the highlands. The application of root zone cooling on shallots was adopted from the vernalization process, where the bulb of shallots is stored first at a temperature of around 10 °C for 40 days before being planted on the ground. This can accelerate flower induction. The cooling process is carried out in line with the tubers planted. A study has reviewed the root zone cooling effect on hydroponic floating for the production of shallot seeds in tropical lowland greenhouses. The study was carried out at three root temperature set points, namely low (8-10 °C), medium (13-15 °C) and ambient (23-26 °C), where the ambient air temperature fluctuated around 32-38 °C [12]. The results of the study showed that the onion plant had not been able to produce flower initiation at the temperature setting variation. However, another response appears in the form of multiplication of tubers (figure 2). Root zone cooling on hydroponic floating was able to increase the number of tubers almost 2 times more than ambient conditions, namely 12 tubers per clump at low temperatures, 9 tubers per clump at medium temperatures, and 5 tubers per clump at ambient temperatures. The same ass true for aeroponic systems in tropical lowlands, the results obtained for each set of low, medium and ambient temperature points are 5.00, 4.50, and 1.50 cloves per clump, respectively [13].

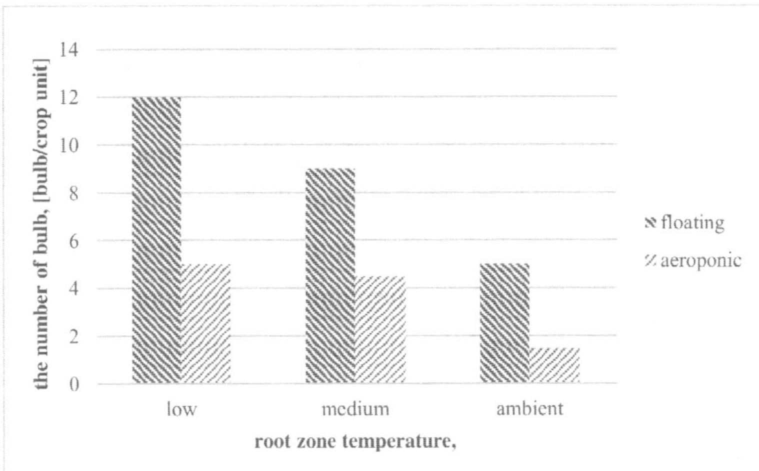


**Figure 2.** The number of shallot bulbs produced on floating and aeroponic systems with root zone cooling.

Root zone cooling has also been applied to aeroponic systems for mini potato seed production [14]. The set temperature points in the chamber for root cooling were 10 °C, 15 ° C, 20 °C, and ambient fluctuation (30.3-32.6 °C). The best tuber production was obtained in the aeroponic system with 10 °C zone cooling which included average productivity (579 tubers $m^{-2}$), average tuber number (14.85 tubers per plant), and average tuber weight (409.15 mg per tuber). Zone cooling with a temperature of 15 °C produced 55 tubers $m^{-2}$ with an average number of tubers 1.67 tubers per plant, tuber weight 205.44 mg per tuber. The number of tubers per plants at zone cooling 20 °C was not significantly different from 15 °C zone cooling. Potato plants at non-cooling temperatures (30.3-32.6 °C) did not produce tubers.

## 7. Root zone cooling on fruiting plants cultivation

Amaliah (2018) conducted a study on cooling the root zone for hydroponic curly chili cultivation using cooling elements in the form of water flowed through galvanized pipes with a set point temperature of 14 °C and the temperature of the planting medium reached below 26.9 °C. Chili plants grown in hydroponic systems with cooling zones produce higher fruit weights per plant than plants without zone cooling treatment. Chili plants with zone cooling have a fruit weight per plant of 873.6 g and chili plants without zone cooling have a fruit weight per plant of 546 g.

## References

[1] Kozai T, Quan Z, Hayashi M dan Watanabe I 1985 Cooling Load of a Greenhouse: Characteristics of Cooling Load in Summer Nighttime *Journal of Agricultural Meteorology* **41** 121–30

[2] Moorby J dan Graves C J 1979 Root and air temperature effects on growth and yield of tomatoes and lettuce *Symposium on Research on Recirculating Water Culture 98* hal 29–44

[3] Mousa W A Y, Lang W, Auer T dan Yousef W A 2017 A pattern recognition approach for modeling the air change rates in naturally ventilated buildings from limited steady-state CFD simulations *Energy and Buildings* **155** 54–65

[4] Molina-Aiz F D, Valera D L, Peña A A, Gil J A dan López A 2009 A study of natural ventilation in an Almería-type greenhouse with insect screens by means of tri-sonic anemometry *Biosystems Engineering* **104** 224–42

[5] Kumar K S, Tiwari K N dan Jha M K 2009 Design and technology for greenhouse cooling in tropical and subtropical regions: A review *Energy and Buildings* **41** 1269–75

[6] Willits D H 2003 Cooling fan-ventilated greenhouses: A modelling study *Biosystems Engineering* **84** 315–29

[7] Malalasekera W dan Versteeg H K 2006 *An Introduction to Computational Fluid Dynamics - The Finite Volume Method* vol 44

[8] Molina-Aiz F D, Fatnassi H, Boulard T, Roy J C dan Valera D L 2010 Comparison of finite element and finite volume methods for simulation of natural ventilation in greenhouses *Computers and Electronics in Agriculture* **72** 69–86

[9] Luo H Y, Lee S K dan He J 2009 Integrated Effects of Root-Zone Temperatures and Phosphorus Levels on Aeroponically-Grown Lettuce (Lactuca sativa L.) in the Tropics *The Open Horticulture Journal* **2** 6–12

[10] Cometti N, Bremenkamp D, Galon K, Hell L dan Zanotelli M 2013 Cooling and concentration of nutrient solution in hydroponic lettuce crop *Horticultura Brasileira* **31** 287–92

[11] He J, Lee S K dan Dodd I C 2001 Limitations to photosynthesis of lettuce grown under tropical conditions : alleviation by root-zone cooling *Journal of experimental botany* **52** 1323–30

[12] Niam A, Suhardiyanto H, Seminar K dan Maddu A 2017 Simulation of Temperature Distribution on the Planting Hole of Floating Hydroponic for Shallot Production in Tropical Lowland *JTEP* **5** 235–44

[13] Nurwahyuningsih, Suhardiyanto H dan Sobir 2017 Application of Root Zone Cooling system to Improve Bulbs Formation of Shallot (Allium cepa var. aggregatum) *JTEP, Jurnal Keteknikan Pertanian* **5** 107–12

[14] Sumarni E 2013 *Development of Zone Cooling System for Aeroponically Grown Potato Seed Production in Humid Tropical Lowland* (Bogor Agricultural University)

[15] Amaliah W 2018 *Root Zone Cooling System for Hydroponic Curly Chili Cultivation in Tropical Greenhouse* (Bogor Agricultural University)

# Spouted Bed Drying of White Pepper (*Piper nigrum L.*) with Microwave Preheating Treatment

**E. Hartulistiyoso[1], Y.A. Purwanto[1] and A.M.A. Mukhlis[1]**
[1] Department of Mechanical and Bio-system Engineering, Bogor Agricultural University

**Abstract.** Drying is an important process in the production of high quality of white pepper commodities that have a high economic value. The objective of this study was to investigate the drying process and the quality of white pepper after spouted bed drying with microwave preheating treatment based on quality parameters of total oil content and microbial contamination. The results of the study indicated that the drying process of spouted bed was able to reduce the moisture content of pepper on average by 59.94% d.b. for 31 minutes on treatment of non-preheating, 60.38% d.b. for 37 minutes on treatment of preheating of 320 watts, and 59.35% d.b. for 32 minutes at treatment of preheating of 640 watts. The initial white pepper seeds that was boiling and soaking for 3 days had a total microbial contents of $2.5\times10^7$ CFU/g. Total microbial of white pepper after spouted bed drying process without preheating was $1.54 \times 10^5$ CFU/g, while the total microbes of white pepper with preheating treatment of 320 and 640 watts were $3.0\times10^4$ and $6.0\times10^3$ CFU/g, respectively. Oil content in all treatments was not significantly different. Oil content obtained in the treatment of non-preheating, preheating of 320 watts and preheating of 640 watts were 2.88%, 3.21% and 2.86%, respectively. It can be concluded that, in all treatments, the amount of moisture reduction and the drying time were relatively similar. However, preheating treatment can reduce microbial content to the IPC quality standard.

## 1. Introduction

Indonesia is the second largest pepper producing country after Vietnam with the total production of 75,000 Mt, comprises of 44,5000 Mt black pepper and 30,000 Mt white pepper. Export of white pepper of Indonesia amounted 21,000 Mt in valued of USD 19,5000 was considered to be the largest one (IPC, 2016). As a high economic value commodity, the quality of white pepper should remain high and hence drying has an important role. Drying of white pepper is practically done after soaking 3-10 days, therefore the initial moisture content of pepper is quite high. High water content conditions are very susceptible to fungal growth if the drying process is slow, so that it can reduce the quality and even damage the pepper. The drying process at the farm level is carried out by sun drying, where it depends on the weather conditions, it can last 2-3 days until the moisture content reaches 15% w.b. Research conducted by Usmiati and Nurdjannah (2007) shows that sun drying of white pepper requires an average time of 13 hours for 2 days and mechanical drying with temperatures ranging from $40^{o}$-$65^{o}$C lasts 4 - 4.5 hour. One method that can be developed in the mechanical drying process is the use of microwaves.

Microwaving in white pepper can be faster, better essential oils content and lower microbial contamination than hot air ovens (Hartulistiyoso and Sudarmaji 2005). In addition, microwaves were also developed in the pepper sanitation process (Emam et al. 1995) and evaluation results showed that the method still maintained the main aroma compounds of pepper (Plessi et al. 2002). However, the application of microwaves singly can cause uneven heating and inhomogeneous field distribution on certain products, depending on the dielectric and thermo-physical properties. In addition, drying with a microwave oven singly can be very expensive, both in terms of equipment and operational costs. To overcome this problem, the application of microwaves can be combined with drying hot air. This will usually increase the efficiency and economic value of the drying process (Schiffmann 2006). Evaporating the free water content on the surface of the material is relatively efficient by using hot air, while to eliminate the free internal water content and the water content bound, microwaves provide an efficient way. Drying of hot air can be done through a spouted bed process, so that the drying process can be carried out in a relatively shorter time because of the high rate of loss of water content and ensure uniform heating of the material due to good particle movement. In this study, microwave treatment was carried out at the beginning before the drying process of spouted bed. By applying microwave treatment at the beginning of the drying process (preheating), it is expected to shorten the drying process and provide better quality results of drying white pepper.

## 2. Material and Method

### *2.1. Sample Preparation and experimental procedure*

Pepper was obtained from community plantations in the Enrekang district of South Sulawesi province with an age of approximately 8-9 months after flowering. The drying process was carried out on white pepper seeds after the immersion process for 3 days. After the process of soaking and washing the white pepper seeds, the pepper was left on top of the filter so that the water that attaches to the surface of the seeds that can be released by gravity decreases. The wet white pepper was then weighed to get 400 grams of white pepper and put into a spouted bed drying room for non-preheating treatment, while for preheating treatment pepper was stored in a container. The preheating process was done using a microwave oven which has a working frequency of around 2450 Hz. The power used was 320 watts for 2 minutes and 640 watts for 2 minutes. The temperature before and after the preheating was measured, then the preheating pepper was transferred to the spouted bed drying room and the convection drying process was carried out. During the drying process, several parameters for drying analysis are measured and the quality of white pepper is measured after the drying process was complete.

### *2.2. Measurement and Data Analysis*

The final moisture content of the material is determined using the distillation method. This method is a water content test method for white pepper based on SNI 0004: 2013. This method was used because the ingredients (pepper seeds) contain volatile compounds.

Total microbes were calculated using the scatter count method (Hadioetomo 1993). This method was based on the assumption that every cell that can live will develop into a single colony. The number of living organisms contained in the sample was based on the total colonies that appear on the cup. Total microbes were determined by taking a sample of 1 gram of pepper.

The desired drying air flow was generated from the blower then passes through the electric heating element which was placed on the air flow path before entering the drying chamber. An electric heater heated the air that passes through it and the resulting air temperature was maintained at temperatures ranging from 40-65$^{o}$C with a thermostat.

Measurement of material mass includes the initial mass, mass during the drying period and the final mass. The mass measurement of the material was carried out every two minutes during the drying process and the drying process is stopped when it reaches the desired moisture content (about 10% bb). When retrieving mass data, the blow dryer air was turned off for a moment, then the pepper and drying chamber are weighed with digital scales (accuracy of 0.01 gram). Pepper mass was obtained by subtracting the data from the mass of the drying chamber.

Temperature measurements were carried out on the air entering and exiting the drying room, the ambient air temperature, and the temperature of the material during the drying process every 2 minutes. In the flow of air in and out, the temperature measured was the temperature of the wet ball and the temperature of the dry ball. The measurement of the temperature of the material was done by stopping the air flow for a moment Temperature measurement using a T-type thermocouple connected to a hybrid recorder to display temperature data.

Drying time was the total time used to dry the material from the initial water content to the desired final moisture content. The time used for measuring the mass of the material does not count as drying time.

Determination of essential oil content was based on the SNI 0004: 2013 test method carried out by distillation using water as its solvent.

**Figure 1.** Experimental setup of spouted bed drying

## 3. Results and Discussion

### 3.1. *Temperature in the Drying Process*

As shown in Figure 2, the temperature of the material with non-preheating and preheating treatment has increased gradually to reach a relatively constant temperature of around 50 °C in the range of the 20$^{th}$ minute drying process. It shows that the temperature of the drying air at 40 to 65°C did not significantly affect the temperature of the material. The initial temperature of pepper seeds on non-preheating drying was around 28.1°C. In the treatment of preheating with 320 watts of power for 2 minutes, the pepper temperature was increased to 41.1 °C, while at 640 watts for 2 minutes, the pepper temperature increased to 63.8 °C. The rapid increase of the temperature was caused by the principle of microwave heating which vibrated polar compounds (water) found in pepper seeds.

After the preheating process in a microwave oven, the pepper was transferred into a spouted bed drying chamber and convective drying was carried out. Since the water content of the material was still high, the temperature tend to decrease at the beginning of the process to the wet bulb temperature. It was seen that in the second minute drying with a preheating of 320 and 640 watts, the pepper temperature was only in the range of 33.1 and 31.2 °C respectively (Figures 2b and 2c).

The outlet air temperature has a value that was not much different from the temperature of pepper seeds as shown in Figure 16. This can indicate that the interaction between the inlet hot air and the material enabled even heat transfer from the air to the pepper seeds.

### 3.2. *Drying Process*

The drying process of spouted bed reduced pepper water content by an average of 59.94% d.b. for 31 minutes in the non-preheating treatment, 60.38% d.b. for 37 minutes on the 320 watt preheating treatment, and 59.35% d.b. for 32 minutes on the 640 watt preheating treatment. The decrease in water content occured quickly because the high air flow rate, which was around 15 ms$^{-1}$. The circulation of pepper seeds in the drying room enabled good contact between the air and pepper seeds. A good circulation of pepper seeds increased the surface area of the material exposed to hot air, so that the heat transfer from the air to the material take place well and at the end it increased the drying rate of the material. A fast air flow bring water vapor to the environment air quickly, thus eliminating air humidity faster, increasing the gradient of water vapor pressure and ultimately increasing the rate of drying (Fellows 2009). Mujumdar (2006) explained that external factors that influence the release of water as steam from the surface of the material are temperature, humidity and air flow, surface area, and pressure.

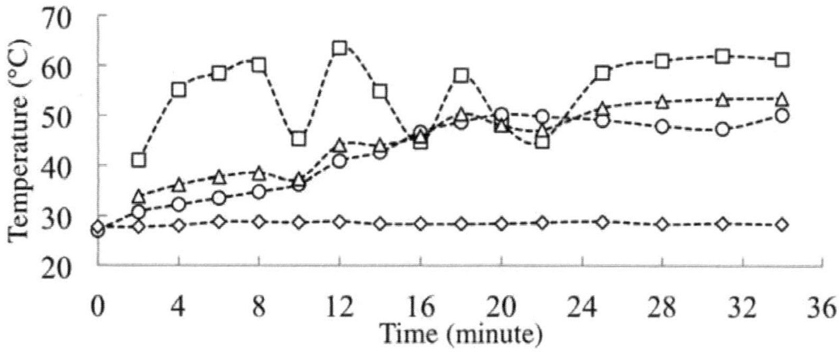


(a)

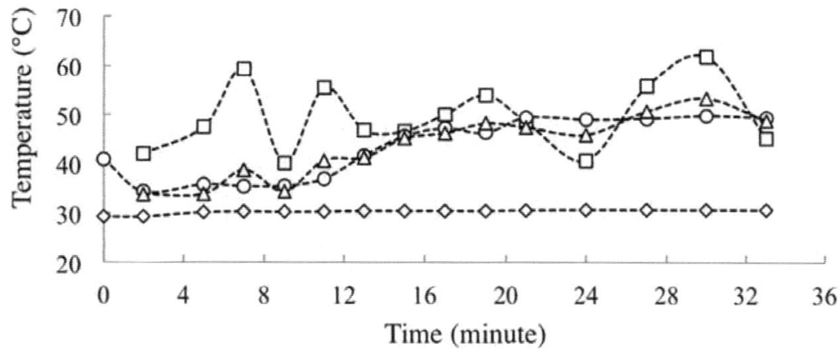


**(a)**

(b)

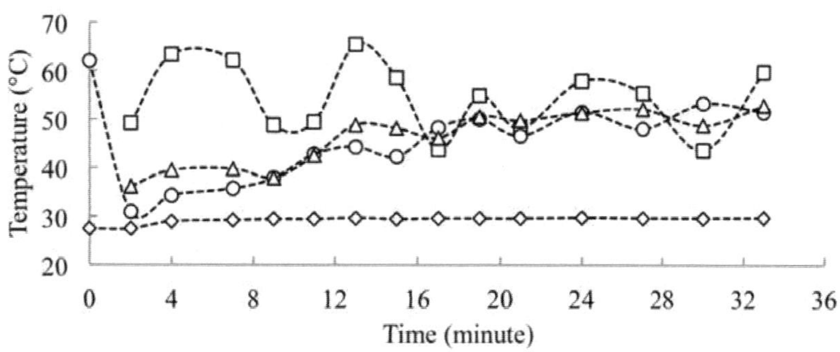


(c)

**Figure 2.** Temperature of air and material during drying process: (a) non-*preheating*; (b) *preheating* 320 watt; (c) *preheating* 640 watt. --□—air inlet temperature, --Δ—air outlet temperature, --○—material temperature , --◊-- ambient temperature.

To compare the drying characteristics of various treatments, it can be done by reviewing the graph of the ratio of water content ($M_t/M_o$) to the drying time (Figure 3). The moisture content ratio is the ratio between the seed moisture content at the specific drying time and the initial moisture content of the seed. The graph shows that, the preheating treatment does not adequately affect changes in the moisture content of the material during the drying process. These results are different from those described by Schiffmann (2006) that preheating will cause the drying curve to be steeper. This difference can be occurred due to its short preheating process, which was applied to avoid material damage caused by high temperature. It indicated that the drying characteristic of the seeds was relatively the same in all treatments, which formed an exponential pattern as the case for drying of cork stoppers (Magalhães and Pinho 2007).

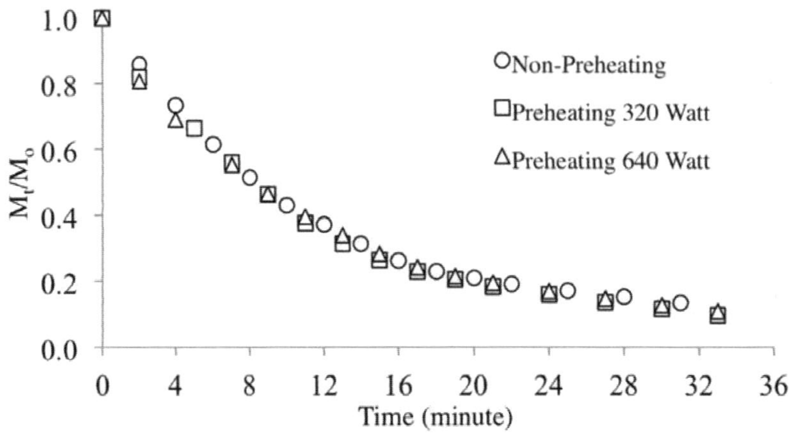


**Figure 3.** Moisture ratio of white pepper after various drying treatments

### *3.3. Microbial Content*

The microbiological test was important to determine the effect of the preheating and drying process on total microbes on white pepper seeds. White pepper seeds from 3 days of soaking have total microbes (TPC) of $2.5 \times 10^7$ CFU/g. This result was as explained by Usmiati and Nurdjannah (2007) that the value of TPC pepper at the end of immersion was very high, ranging from $10^7$ to $10^8$ CFU/g. They further explained that, high microbes could be the result of enzymatic processes that allow decay by microbes that may be present in soaking water or fresh pepper seeds. The results showed that the preheating treatment was able to provide a positive influence on total microbes after the drying process, namely by reducing the total microbial content. The total microbes with the preheating treatment were significantly different from those in the treatment without preheating, while in the preheating of 320 and 640 watts, the total value of microbes was not significantly different. Total white pepper microbes after spouted beds drying without preheating was averaged $1.54 \times 10^5$ CFU/g, whereas total microbes in white pepper with the treatment of preheating of 320 and 640 watts were averaged $3.00 \times 10^4$ and $6.00 \times 10^3$ CFU/g respectively. The results showed that white pepper through a preheating process and then continued by drying spouted beds at an average temperature of around 55°C had met the International Pepper Community (IPC) total microbial quality standards, which was $5 \times 10^4$ CFU/g.

### *3.4. Essential Oil Content*

Essential oils have an important role in the distinctive aroma of white pepper. The essential oil content in all treatments was not significantly different at the test level of 5%. Essential oil content obtained from the treatment of non-preheating, 320 watt preheating, and 640 watt preheating was in average of 2.88%, 3.21%, and 2.86% respectively. In all of these treatments, the essential oil content was above the minimum limit of ISO quality requirements, which was 1%. In addition, the essential oil content obtained was not much different from the level of essential oils produced by traditional methods of 2.49% and the results of mechanical processing of 3.11% (Nurdjannah 2005). Thus, the drying process with a spouted bed both non-preheating and preheating still maintains essential oils on white pepper. Essential oils can evaporate together with water when the drying process takes place. The essential content can be maintained at the drying temperature of below 70°C (Hidayat et al. 2009).

## 4. Conclusion

The two minutes treatment of 320 and 640 W microwave in preheating of spouted beds drying of white pepper seeds did not affect its drying characteristics. The spouted bed drying reduced the water content considerably in the range of 57-64% d.b. in a short time of about 30 minutes. Preheating treatment with microwave was able to reduce microbes' content of white pepper better than drying without preheating and meet the IPC standard of below 5.00×10 CFU/g. The essential oil content of the dried pepper was in all treatments was not significantly different and higher than the ISO standard.

## 5. References

[1] Emam OA, Farag SA, Aziz NH. 1995. Comparative effects of gamma and microwave irradiation on the quality of black pepper. *Z Lebensm Unters Forsch*. 201: 557-561
[2] Fellows PJ. 2009. *Food Processing Technology*. 3rd ed. Cambridge (UK) : Woodhead Publishing
[3] Hadioetomo RS. 1993. *Mikrobiologi Dasar dalam Praktek: Teknik dan Prosedur Dasar Laboratorium*. Jakarta (ID): Gramedia Pustaka Utama
[4] Hartulistiyoso E dan Sudarmaji K. 2005. Pengeringan lada putih (*Piper nigrum* L.) menggunakan gelombang mikro (*microwave*). *Prosiding Seminar Nasional Teknologi Inovatif Parcapanen untuk Pengembangan Industri Berbasis Pertanian*.
[5] Hidayat T, Nurdjannah N, Usmiati S. 2009. Analisis teknis dan finansial paket teknologi pengolahan lada putih (*white pepper*) semi mekanis. *Bul Littro*. 20(1):77 - 91
[6] [IPC] Internasional Pepper Community. 2016. *Pepper Statistical Yearbook 2016*.
[7] Magalhães A dan Pinho C. 2007. Spouted bed drying of cork stoppers. *Chemical Engineering and Processing* 47 (2008): 2395–2401. doi: 10.1016/j.cep. 2007.11.009
[8] Mujumdar AS. 2006. Principles, classification, and selection of dryers. Di dalam: Mujumdar AS, editor. *Handbook of Industrial Drying*. Ed ke-3. Florida (US) : CRC Pr
[9] Nurdjannah N. 2005. Use of antioxidant to inhibit browning on white pepper decorticating process. *Jurnal Littri* 11(2): 78 - 84
[10] Plessi M, Bertelli D, Miglietta F. 2002. Effect of microwaves on volatile compounds in white and black pepper. *Lebensm.-Wiss. u.-Technol*. 35(2002): 260–264
[11] Schiffmann RF. 2006. Microwave and Dielectric Drying. Di dalam: Mujumdar AS, editor. *Handbook of Industrial Drying*. Ed ke-3. Florida (US) : CRC Pr
[12] Usmiati S dan Nurdjannah N. 2007. Pengaruh lama perendaman dan cara pengeringan terhadap mutu lada putih. *J Tek Ind Pert*. 16(3):91-98

# Effect of Voltage and Frequency in Pasteurization Pulsed Electric Field (PEF) Continous System of Pineapple (Ananas comosus [L.] Merr) Juice

**D W Indriani[1], S Amalia[1], S H Sumarlan[1] and N Barunawati[2]**

[1]Department of Agricultural Engineering, Faculty of Agricultural Technology, Universitas Brawijaya Malang

[2]Department of Agricultural Cultivation, Faculty of Agriculture, Universitas Brawijaya Malang

**Abstract.** Pineapple which is one of the perishable commodities with high water content about 90.73% in 100 gr of fresh material. Thermal pasteurization has a weakness that can degrade the quality of fruit juice products. In contrast, non thermal processes have expectations which is don't make decreasing quality of the product. One of the non-thermal pasteurization type called Pulsed Electric Field (PEF). Nowadays, PEF system have two type batch and continous, which respectively used in this research is PEF continous system. This study uses research methods group random design (GRD) with two factorial that voltage (20, 30 and 40 kV) and frequency (10, 20, 30 and 40 kHz). Based on these studies it is known that the voltage and frequency is significantly different to the value of Total Plate Count (TPC), and absorbance. However there aren't significantly different with the pH, vitamin C, viscosity, total dissolved solid (TDS), total sugar and color. The treatment based on the microbial test are at a voltage of 40 kV and a frequency of 40 kHz which can reduce bacteria significantly $1.01 \times 10^3$ cfu / mL or 0.94 by log cycle with the effectiveness of microbial death by 88.5%. The treatment based on the microbial test are at a voltage of 40 kV and a frequency of 30 kHz which The result indicates that physical and chemical characteristic of pineapple juice become 3.98 of pH, total dissolved solid (TDS) of 6.25 °brix, total sugar of 14.75 % viscosity of 3 cp, vitamin C of 7.6 mg/100ml, absorbance of 0.67, color on the brightness (L*) value of 24.05, redness (a*) of 6.35, and yellowish (b*) of 7.5. This result shows that PEF continuous respectively work system to pasteurize the pineapple juice.

**Keywords:** Frequency, Fruit Juicy, Pulsed Electric Field (PEF), Pineapple, Voltage

## 1. Introduction

Horticultural commodities continue to developed, because the commodity has a high economic value howeverless realized variety, quantity, quality. Pineapple (Ananas comosus [L.] Merr) as one of the mainstay horticultural commodities in tropical fruit trade especially in Indonesia, the average consumption of pineapple in Indonesia is 6.132.695 tons per year [15]. Pineapple fruit is a perishable commodity with a high water content of 90.73% in 100 grams of material.

One effort that can be done is processing ie in the form of juice, as for the procedure of the process of making fruit juice is process of fresh fruit into fruit juice product by destroying the fruit into a essence and add additional ingredients to support the taste of the product and then do some treatment for extend shelf life [9]. The extended shelf life of pineapple juice did pasteurization can be done two methods which are thermal and non-thermal. According to [10], pasteurization a process that has been done in the food processing industry because able to inactivate the enzyme and suppress the amount of microbes. Thermal pasteurization there are weaknesses, it can damage the texture, taste, color and other physical properties that can reduce the quality of fruit juice products. When compared with the thermal pasteurization process, the non thermal pasteurization process is able to maintain the quality of food products but in the reduction of the number of microbes causing damage to food needs a further assessment.

In addition to the voltage factor on the Pulsed Electric Field (PEF) pasteurization process there are other factors that influence the final product result that is the distribution of frequency. Frequency contributes to processing time this is because the distribution of the frequency causes a reduction in cell membrane resistance resulting in cell damage to the food product [11]. Therefore the the need for further research on the effect of differences in the frequency of Pulsed Electric Field (PEF) pasteurization process, in addition to being able to get the best product results is also able to minimize processing time. With result that the research titled is Effect of Voltage and Frequency on Pasteurization of Pulsed Electric Field (PEF) Continuous System on Quality of Pineapple (Ananas comosus [L.] Merr) Juice.

## 2. Material and Methods

The tool used in the research is circuit of Pulsed Electric Field (PEF) continous system tools, bottles, knives, gloves, aluminum foil, cotton, autoclave, basin, scissors, tube measuring, and paper labels. Tools used for the analysis of pipettes, micropipettes, petri dishes, cotton, brown paper, measuring tubes, measuring cylinders, Laminar Air Flow (LAF), scales, gloves, aluminum foil, test tube, cotton, blue tip, electric stove, And bionex brand of incubators. The materials used in this research are pineapple fruit was obtained in Blimbing Market, Malang, East Java, 96% alcohol, and aquades. The material used for the analysis is the plate count agar (PCA) merk brand.

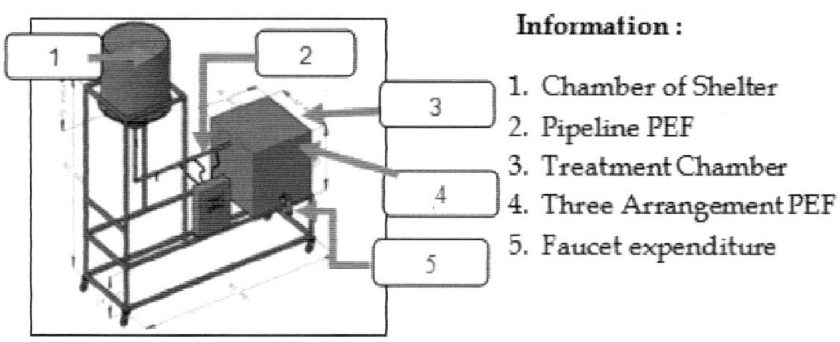


**Figure** 1. PEF Continous System Pasteurization Tool Circuit

### *2.1. Experimental Design*

This study uses research methods Group Random Design (GRD) with two factorial that voltage (20 kV, 30 kV and 40 kV) and frequency (10 kHz, 20 kHz, 30 kHz and 40 kHz). Each treatment was repeated 2 times.

### *2.2. Implementation Reasearch*

IOP Conf. Series: Materials Science and Engineering **557** (2019) 012046 doi:10.1088/1757-899X/557/1/012046

The process of making pineapple fruit ; first pineapple fruit washed with water then peeled the skin, then pineapple fruit split into four parts. Furthermore, pineapple fruit that has been cleaved then done blanching process. Pineapple fruit is done by blender then filtering and immersion for 1 hour Fruit juice is then added sugar and citric acid, then heated for 10-15 minutes. After the cold the pineapple juice is put into a sterile bottle

Non-thermal process using PEF, place the pineapple juice on the chamber, connect it to the electric current. Set the voltage at the first treatment of 20 kV and frequency of 10 kHz, then press the ON button. Open the output faucet and place it in a sterilized bottle and cover with a sterilized bottle cap to avoid contamination with the environment. Clean the treatment room with 96% alcohol after each research. Repeat for any combination of voltage and frequency of PEF.

### *2.3. Observation*

The test parameters performed are Total Plate Count, absorbance, pH, vitamin C, viscosity, total dissolved solid (TDS), total sugar and colour

### *2.4. Data Analysis*

The data obtained were analyzed using Two-Way ANOVA. If there is a real difference then BNT advanced test and if the interaction of both factors show real difference, then continued with LSD test.*Formatting the title*

## 3. Result and Discusss

### *3.1. Material Characteristics of Treatment*

Based on the results of pineapple juice analysis before treatment with Pulsed Electric Field (PEF) shown on **Table 1**.

**Table 1. Characteristic Value of Pineapple Fruit**

| Characteristics | Control | Thermal Pasteurization | Literatur | unit |
|---|---|---|---|---|
| ***Total Plate Count* (TPC)** | $8{,}8x10^3$ | $9{,}35x10^2$ | $1.\ 10^4$ | Cfu/ml |
| **pH** | 3,99 | 4,09 | Max 4* | - |
| **TDS** | 6,75 | 5 | 10-11* | $^0$brix |
| **Total Sugar** | 16,19 | 17,91 | Max 5 | % |
| **Viskosity** | 4,5 | 5,5 | - | Cp |
| **Vitamin C** | 7,66 | 3,83 | 34,77** | Mg/100ml |
| **Absorbance** | 0,79 | 0,52 | - | abs |
| **Brightness(L*)** | 23,45 | 23,20 | - | - |
| **Redness (a*)** | 6,65 | 6,25 | - | - |
| **Yellowish(b*)** | 7,70 | 5,95 | - | - |

Information * SNI 01-3719-1995 **Daniela (2015)

Based on Table 1. At TPC value, research of pineapple juice has lower value than TPC literature, According to [7] the difference of TPC value is caused by the difference of treatment to material and environmental conditions. TPC, pH, color, viscosity, vitamin c, total sugar, TDS and absorbance on control factor and thermal pasteurization have difference caused by thermal pasteurization using heat treatment at the treatment so as to cause undesired reactions. In accordance with the statement of [5], during the process of thermal pasteurisation large amounts of energy are transferred to food. This energy can cause unwanted reactions, such as loss of essential nutrients, discoloration, odor and taste.

IOP Conf. Series: Materials Science and Engineering **557** (2019) 012046 doi:10.1088/1757-899X/557/1/012046

*3.2 Total Plate Count (TPC)*

Total Plate Count (TPC) is the total amount of microbes contained in pineapple juice. The calculation of the amount of contamination will be related to the effectiveness of decreasing the number of microorganisms. The result showed that giving 40 kV (voltage) and 40 kHz (frequency) can reduce total microbial in pineapple juice equal to 1.01 x $10^3$ while giving 20 kV and 10 kHz frequency can only decrease total of pineapple juice nutrient equal to 6,35 x $10^3$. Based on these results then obtained the amount of microbial contaminants presented in Table 2.

**Table 2. Number of Microbial Contaminants**

| Treatment | Total Plate Count cfu/mL) | | Averange (cfu/mL) |
|---|---|---|---|
| | I | II | |
| **Control** | 9,3 x$10^3$ | 8,4 x $10^3$ | 8,8 x $10^3$ |
| **Thermal Pasteurization** | 6,7 x $10^2$ | 1,2 x $10^2$ | 9,35 x $10^2$ |
| **$V_1F_1$** | 5,2 x $10^3$ | 7,4 x $10^3$ | 6,3 x $10^3$ |
| **$V_1$ $F_2$** | 3,8 x $10^3$ | 2,9 x $10^3$ | 3,35 x $10^3$ |
| **$V_1$ $F_3$** | 2,2 x $10^3$ | 2,2 x $10^3$ | 2,2 x $10^3$ |
| **$V_1$ $F_4$** | 1,5 x $10^3$ | 1,1 x $10^3$ | 1,3 x $10^3$ |
| **$V_2$ $F_1$** | 2,2 x $10^3$ | 6,2 x $10^3$ | 4,2 x $10^3$ |
| **$V_2$ $F_2$** | 2,9 x $10^3$ | 2,5 x $10^3$ | 2,7 x $10^3$ |
| **$V_2$ $F_3$** | 1,6 x $10^3$ | 1,1 x $10^3$ | 1,35 x $10^3$ |
| **$V_2$ $F_4$** | 1,3 x $10^3$ | 9,1 x $10^2$ | 1,1 x $10^3$ |
| **$V_3$ $F_1$** | 2,7 x $10^3$ | 2,1 x $10^3$ | 2,4 x $10^3$ |
| **$V_3$ $F_2$** | 2,2 x $10^3$ | 1,7 x $10^3$ | 1,95 x $10^3$ |
| **$V_3$ $F_3$** | 2,4 x $10^3$ | 1,4 x $10^3$ | 1,9 x $10^3$ |
| **$V_3$ $F_4$** | 1,4 x $10^3$ | 6,2 x $10^2$ | 1,01 x $10^3$ |

*3.3 The Effectiveness of Microbial Deaths*

The result of Two Way ANOVA test shows that the voltage and frequency give significant difference, ie sig V (0.032) and sig F (0.001). Then the next test is LSD test indicates that the voltage affecting and frequency, while the interaction between the two factors does not give Influence "no real difference" ($P > 0.05$). When viewed on the value of ANOVA, it can be concluded that the voltage and frequency can reduce the number microbial in pineapple fruit

The smallest microbial death effectiveness is owned by 20 kV and 10 kHz voltage treatment with a percentage of 28.8%. While the greatest microbial effectiveness of 40 kV and 40 kHz voltage treatment with a percentage of 88.5%. An increase in the percentage effectiveness of microbial mortality is shown in **Figure 2.**

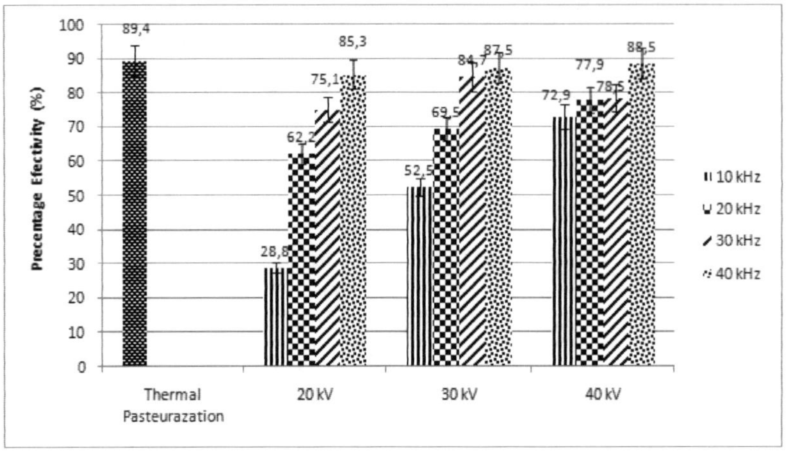


**Figure 2.** Graph of Percentage Value of Effectiveness of Decreasing the Number of Microorganisms

Giving high voltage shocks causes damage to cell structures such as the destruction of cytoplasmic cell membranes where cell mobilization does not occur resulting in microbial inactivation [8]. According to [12], Frequency on Pulsed Electric Field (PEF) continuous system contributes to the processing timebecause the giving of the frequency causes the reduction of cell membrane resistance so that the arising of cell damage in food products.

*3.4 Decrease in Number of Microbes and Log Cycle*

The calculation of the decrease in the number of microbes will be related to Log Cycle microorganisms. The lowest decrease was obtained at 20 kV voltage treatment with frequency of 10 kHz that is equal to 0.14 log cycle. While the highest decrease is in the treatment voltage of 40 kV with a frequency of 40 kHz is equal to 0.94 log cycle. When viewed from the acquisition of the largest D value, pineapple juice with PEF treatment get a D value of less than 90% or 1 log cycle then no need to calculate D value. This is because in the PEF process the lack of treatment time in which the material only passes through the process of giving voltage and high frequency. The value of microbial log cycles or microbial decline is shown in **Figure 3**.

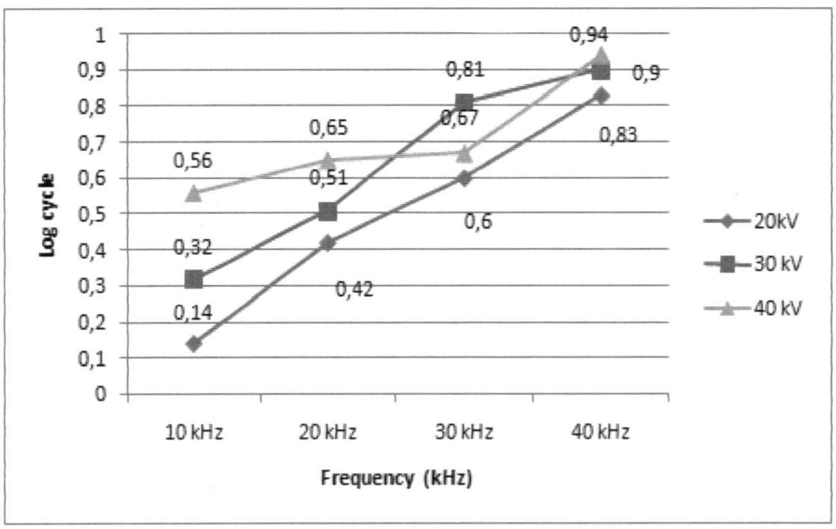


**Figure 3.** Graph of Relation Frequency and Voltage to Logarithmic Reduction of Microbes

Level of acidity (pH) in pineapple juice that has been subjected to voltage and frequency treatment in the study ranged from 3.96-3.99. The graph of voltage and frequency relationship to pH of pineapple juice is shown in **Figure 4**.

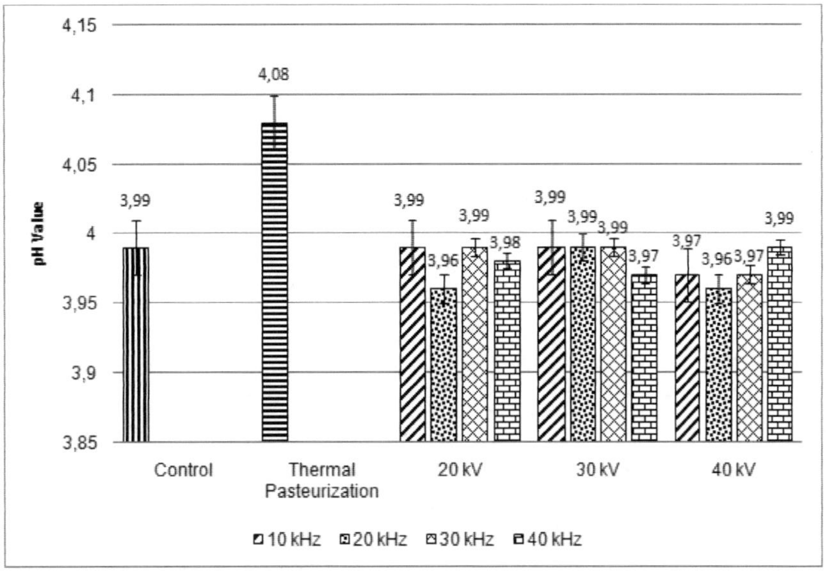


**Figure 4.** Graph Relation of Voltage (kV) and Frequency (kHz) of Average Value of pH Pineapple Juicy

Based on the result of analysis of Two Way ANOVA fingerprint showed that the voltage, frequency and interaction are not significantly different (P> 0.05) to pH in pineapple juice so it is concluded that PEF pasteurization has no effect in pH value of pineapple juicy

In **Figureure 4**, the data obtained only range from less than pH 4 which is known that juice usually has a low pH because it is rich in organic acids [10]. Factors of pH damage occur because of high temperatures where high temperatures bring great energy as a result of chemical damage in food [4]

*3.5 Total Dissolve Solid (TDS)*

Total soluble solids in pineapple research were between brix 6-6.25 °brix the relationship of voltage and frequency to TDS value is shown in **Figure 5.**

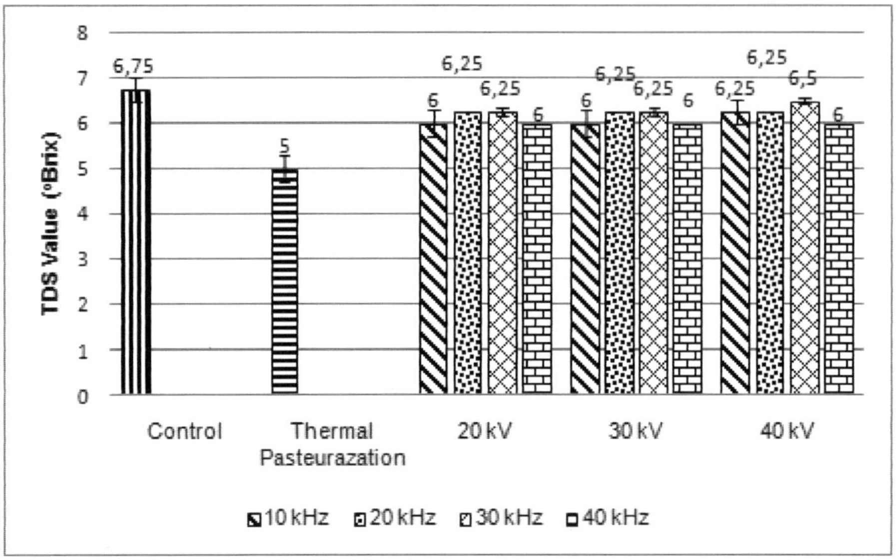


**Figure 5**. Graph of Voltage Relation (kV) and Frequency (kHz) to mean of TPT (°Brix) Pineapple juice

From Two Way Test ANOVA showed that the voltage, frequency and interaction were not significantly different (P> 0.05) so it can be concluded that PEF pasteurization has no effect on the total Dissolve solids (TDS) of pineapple juice. In **Figure 5**, according to the data, it can be concluded that the voltage and frequency of PEF does not have a big effect on the TDS value. According to SNI 01-3719-1995, the value of TPT on the juice is 10-11 °Brix which is very high value when compared to the value obtained that is about 6-6.25 °Brix. According to [13], pineapple juice if left for a day will form a sediment. This precipitate when no homogenization will affect in the total test of dissolved solids.

### *3.6 Total Sugar*

The largest total sugar value was PEF treatment with 20 kV and 10 kHz of 15.49% and the smallest sugar value was PEF treatment with 40 kV and 20 kHz frequency of 14.67. The average graph of total sugar can be seen in **Figure** 6.

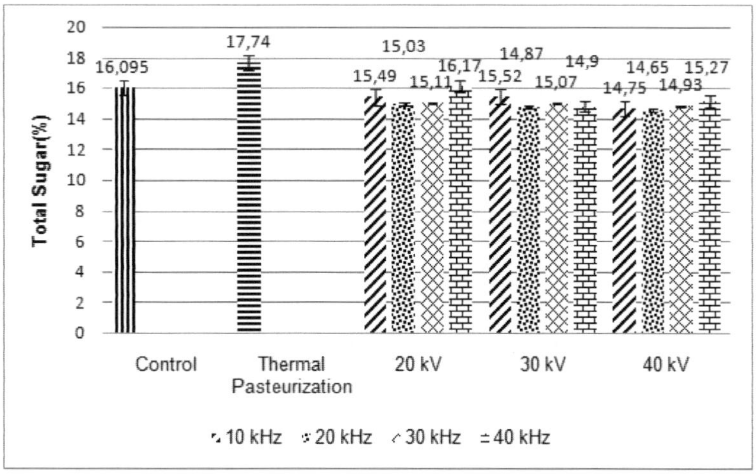


**Figure 6.** Graph of Voltage Relation (kV) and Frequency (kHz) to mean Total Sugar (%) Pineapple Juicy

Two way ANOVA test showed that the voltage, frequency and interaction are not significantly different (P> 0.05) so that PEF pasteurization cannot be concluded that the total value of pineapple juice. In Figure 6 obtained data exceeds the value of SNI 01-3719-1995 is max 5% this is because the total test of sugar made 3 days after the process of making pineapple juice resulting in a change in chemical content during the storage process. According to [6], the longer storage of fruit products will affect the sugar content when the storage is done 5 days and 10 days increased sugar levels when compared with control values, this is due to the polysaccharide split into sugar (sucrose, glucose , Fructose).

**Viscosity**

The highest viscosity value at 20 kV and 30 kHz is 5 cp and the lowest viscosity value is 2.5 cp of 30 kV and 30 kHz with control value 4.5 cp and thermal pasteurization 5.5 cp. The result of stress and frequency test of viscosity can be seen in **Figure 7**.

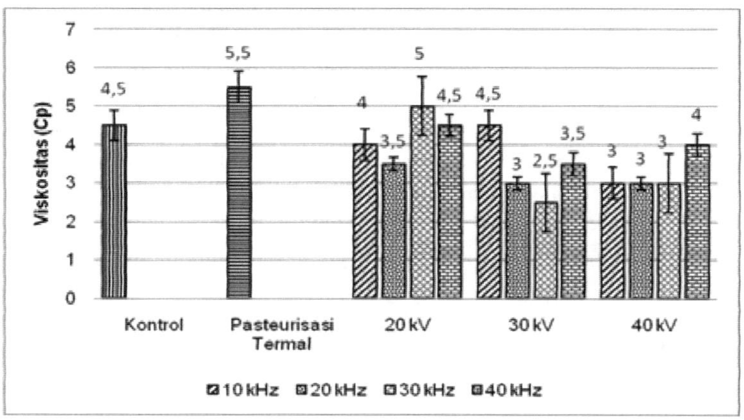


**Figure 7.** Graph of Voltage Relation (kV) and Frequency (kHz) to Average Viscosity (Cp) of Pineapple Fruit

Two way ANOVA test showed that the voltage, frequency and interaction were not significantly different (P> 0.05) so that PEF (non-thermal pasteurization) PEF (non-thermal pasteurization) was not affected in the viscosity value of pineapple juice. Viscosity of pineapple juice after PEF treatment when seen in **Figure 7**. to get a fluctuating value that is a decrease and increase in graphics this can happen because the testing process is too long so that cause the deposition of the pineapple juice. According to [1], the greater the viscosity of the juice make the more concentrated ingredients . This change occurs because the more the amount of water added to the extract the viscosity of the product will be smaller and vice versa, the less water added to the extract the viscosity will be higher.

*3.7 Vitamin C*

The content of vitamin C in fruit juice experienced significant value, the highest vitamin C value occurred at 20 kV and 10 kHz frequency of 7.73 mg / 100ml and the lowest vitamin C value occurred at 40 kV at 40 kHz frequency of 7.63 mg / 100ml. The relationship of voltage and frequency to the value of vitamin C can be seen in **Figure 8.**

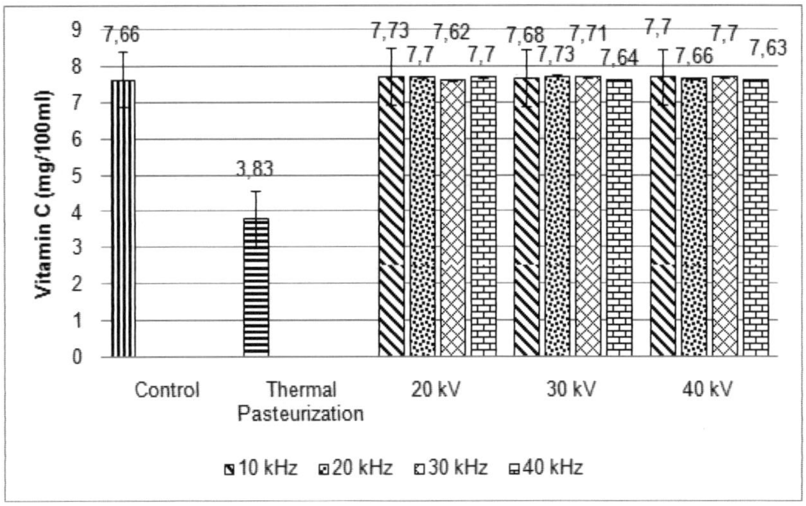


**Figure 8.** Graph of Voltage Relation (kV) and Frequency (kHz) to the average of Vitamin C of Pineapple Fruit

The Two Way ANOVA test showed that the stress, frequency and interaction were not significantly different (P> 0.05) to the value of vitamin C in pineapple juice, so it could mean that non-thermal Pasteurization of PEF (Pulsed Electric Field) does not affect the value of Vitamin C pineapple juice. In Figure 8, vitamin C was obtained with ± 7 mg / 100ml, when compared with the literature [3], stated that the vitamin C content of pineapple fruit was 34.766 mg / 100ml which in vitamin C yield decreased. Value occurs Vitamin C is an easily damaged vitamin, vitamin damage factor is heating, material destruction with blender, cutting material, the presence of metal, pH and light that makes oxidized of vitamin C [8].

*3.8 Absorbance*

The greatest absorbance value is 0.69 at 20 kV and 40 kHz and the smallest absorbance value is 0.62 at 40 kV and 20 kHz where control value is 0.78 and thermal pasteurization value is 0.51. Following is a graph of the absorbance relationship shown in **Figure 9**.

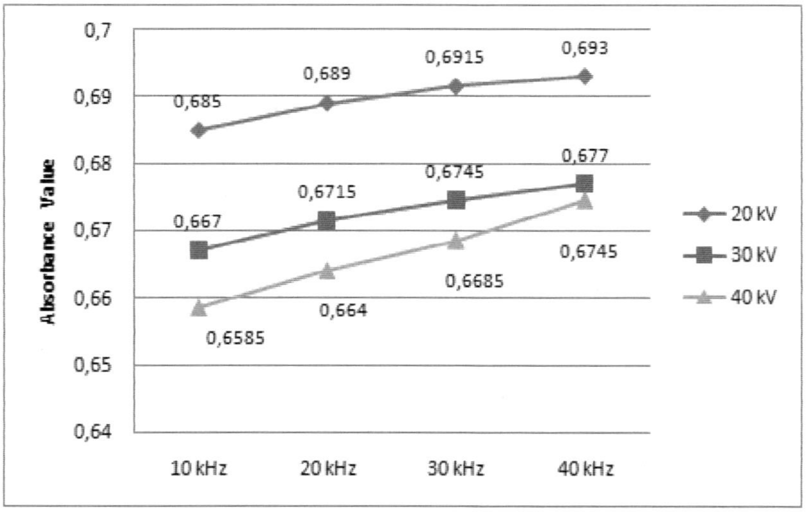


**Figure 9.** Graph of Voltage Relation (kV) and Frequency (kHz) to the average absorbance of Pineapple Juicy

Two Way ANOVA test result showed that voltage and frequency gave significant difference between sig V (0,000) and sig F (0,002), while the interaction between the two factors did not give "no real difference" (P> 0,05) with Sig . 0.346 to the absorbance of pineapple juice so that the voltage and frequency can increase the quantity of substances in the pineapple juice.

When viewed graphic pattern it can be concluded the higher the given frequency then the absorbance value obtained greater and vice versa giving voltage treatment can reduce the absorbance value. According to [11], the lower absorbance range in absorbance values indicates higher absorption of particles and higher absorbance values of lower particle absorption.

*3.9 Colour*

The measurement of color values is known using a color reader, ie the measurement uses three brightness parameters (L), redness (a *), and yellowish (b *). From the results of color testing obtained the color test in the category has a brightness level of bright, faded reddish and yellowish

Two way ANOVA test showed that the voltage, frequency and interaction were not significantly different (P> 0.05) so that PEF (non-thermal pasteurization) PEF (non-thermal pasteurization) was not affected in the colour value of pineapple juice.

*3.10 Brightness Value (L*)*

The smallest brightness value that is 21.85 is at 20 kV and 20 kHz voltage treatment and the greatest brightness value is applied voltage of 20 kV and 40 kHz equal to 24,35. The relationship of PEF voltage and frequency to the brightness value can be seen in **Figure 10.**

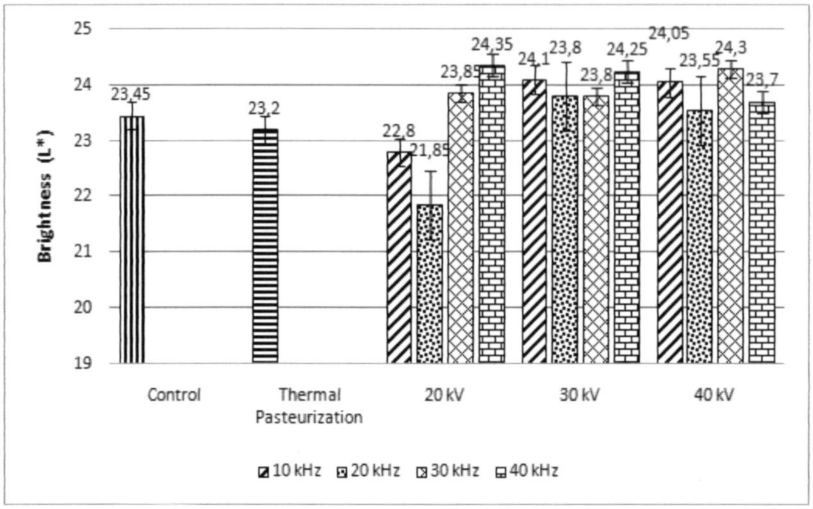


**Figure 10.** Graph of Voltage Relation (kV) and Frequency (kHz) to Average Brightness (L *) Pineapple Juicy

In Figure 10, there is a fluctuating value based on [2] statement. The increase in brightness occurs due to the variation of the voltage treatment where the particles contained in the product decreases slightly, causing the increase in color and the brightness color change can also be affected by the non-enzymatic browning counter PEF treatment of some enzymes develops thus affecting the brightness level

*3.11 Redness Value (a*)*

After the treatment with PEF obtained the largest reddish value at 20 kV at 30 kHz frequency has an average increase of 6.6 and the smallest reddish value at 20 kV at 40 kHz frequency has the smallest average value of 5.8. From the above results it is found that the reddish value is included in the reddish fade category PEF voltage relationship and frequency to the redness value can be seen in **Figure 11.**

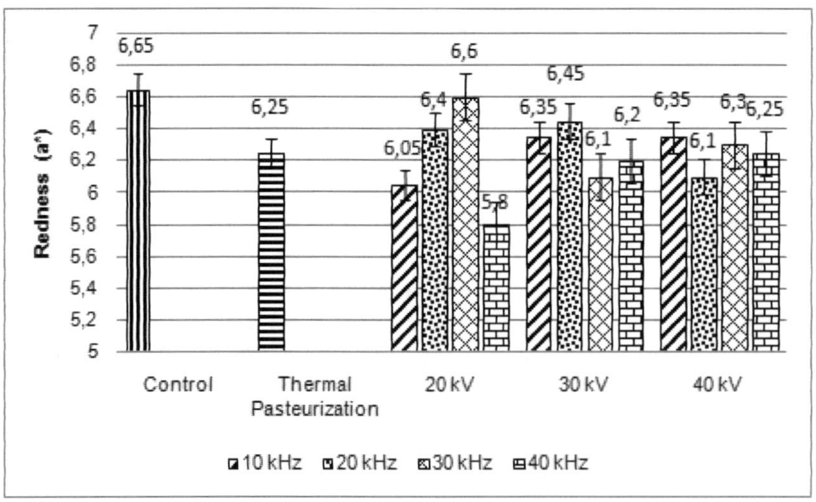


**Figure 11.** Graph of Voltage Relation (kV) and Frequency (kHz) to mean of Redness (a *) Pineapple Juicy

**Figure 11** shows that the reddish values with different PEF stress treatments have fluctuating values. The existence of fluctuating value in pineapple juice that has been done PEF due to high voltage causes hydrolysis of simple sugars so that the sugar content will go down which in the presence of sugar changes will occur fructose and glucose solubility and form browning process/ browning [4]

*3.12 Yellowish Value (b*)*

The value of b* in pineapple juice treated with PEF has a value of 5.85-7.55. The result of stress and frequency test to yellowish value can be seen in **Figure 12.**

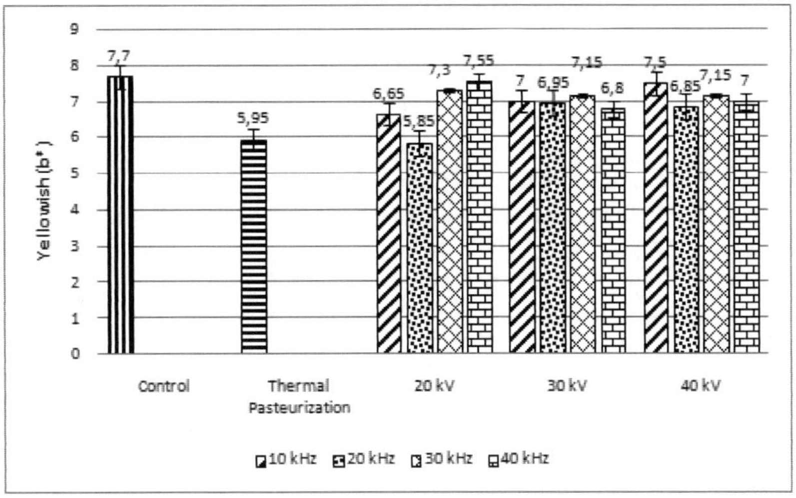


**Figure 12.** Graph of Voltage Relation (kV) and Frequency (kHz) to mean of Yellowish (a *) Pineapple Juicy

In **Figure 12**. Shows that According to [2], yellowish values tend to fluctuate. This decline is due to differences in test time from the process of making pineapple juice. There is a decrease in yellowish s due to precipitation so that the yellowish color decreases

## CONCLUSION

In the research of pineapple juice with pulsed electric field (PEF) voltage and pulsed electric field (PEF) effect significantly to Total Plate Count (TPC) and absorbance value but no significant effect on total soluble solid, total sugar, viscosity, and color of pineapple juice. The best treatment is a 40 kV (voltage) and 40 kHz (frequency). The best treatment results get TPCis 1,01 x$10^3$ cfu/mL or 0,94 *log cycle* with the effectiveness of microbial mortality is 88,5%. The treatment based on the microbial test are at a voltage of 40 kV and a frequency of 30 kHz which Physical and chemical become pH of 3.98, total dissolved solid (TDS) of 6.25 °brix, total sugar of 14,75 % viscosity of 3 Cp, vitamin C of 7.6 mg/100ml, absorbance of 0.67, color on the brightness (L*) value of 24,05, redness (a*) of 6.35, and yellowish (b*) of 7.5.

## REFRENCES

[1] Afrianti, L.H, Y. Taufik dan H. Gustianova. **Karakteristik Fisiko-Kimia Dan Sensorik Jus Ekstrak Buah Salak (*Salacca edulis Reinw*) Varietas Bongkok**. Jurnal Teknologi Pangan, Vol 1, No 1.

[2] Cahyanti, R.N. 2015. **Aplikasi *Pulsed Electric Field* (Pef) Pada Sari Tebu Hijau (*Saccharum Officinarum* L.) (Kajian Tegangan dan Frekuensi PEF).** Thesis. Faculty of Agricultural Technology. Universitas Brawijaya

[3] Daniella, C., L. M. Lubis dan R.J. Nainggolan.2015. **Pengaruh Perbandingan Sari Buah Nenas Dengan Melon Serta Konsentrasi Gula Terhadap Mutu Permen Jahe (Hard Candy).** Jurnal Rekayasa Pangan dan Pert., Vol.3 No.3

[4] Estiasih, T dan Ahmadi K. 2009. **Teknologi Pengolahan Pangan**. Jakarta : Bumi Aksara

[5] Hawa, L.C. dan R.I. Putri. 2011. **Penerapan Pulsed Electric Field Pada Pasteurisasi Sari Buah Apel Varietas Ana: Kajian Karakteristik Nilai Gizi, Sifat Fisik, Sifat Kimiawi dan Mikrobia Total**. AGRITECH, Vol. 31, No. 4.

[6] Helmiyesi, Rini B.H dan Erma P. 2008. **Pengaruh Lama Penyimpanan Terhadap Kadar Gula dan Vitamin C pada Buah Jeruk Siam (Citrus nobilis var. microcarpa).** Buletin Anatomi dan Fisilogi. Vol XVI (2).

[7] Hidayani, N.D. 2015. **Aplikasi *Pulsed Electric Field* (Pef) Pada Sari Tebu Hijau (*Saccharum officinarum* L.) (Kajian Tegangan PEF dan Kosentrasi Natrium Benzoat).** Skripsi. Fakultas Teknologi Pertanian. Universitas Brawijaya

[8] Lehninger, A.L. 1996. **Principles of Biochemistry**. Worth Publisher.Inc, New York.

[9] Lembaga ilmu pengetahuan indonesia. 2009**. Balai Informasi Teknologi LIPI (Pangan&Kesehatan)**. www.bit.lipi.go.id

[10] Liani, Riska D.A Rini. 2014**. Pengaruh Tegangan Dan Frekuensi Terhadap Karakteristik Dan Penurunan Jumlah Mikroorganisme Sari Buah Belimbing (*Averrhoa Carambola* L) Menggunakan Pulsed Electric Field (PEF).** Skripsi. Fakultas Teknologi Pertanian Universitas Brawijaya. Malang.

[11] Neldawati, Ratnawulan, dan Gusnedi. 2013. **Analisis Nilai Absorbansi dalam Penentuan Kadar Flavonoid untuk Berbagai Jenis Daun Tanaman Obat**. *Pillar Physics* Vol 2

[12] Nizar, M.N. 2014. **Efek *Pulsed Electric Field* (PEF) Pada Rendemen dan Kualitas Minyak Bunga Melati (*Jasminum sambac*) (Kajian Rasio Bahan Berbanding Pelarut dan Frekuensi).** Skripsi. Fakultas Teknologi Pertanian, Universitas Brawijaya.

[13] Santoso, H. B. 1998. **Sari Buah Nanas.** Yongyakarta : KANISIUS

[14] SNI 01-3719-1995. **Minuman Sari Buah**. Badan Standarisasi Nasional Peraturan BPOM No. 36 Tahun 2013.

[15] Zohratun, E., Yuli N., dan Rusdianjah. 2013. **Pengaruh Suhu dan Waktu Terhadap Hasil Ektraksi Pektin dari Kulit Buah Nanas.** Simposium Nasional RAPI XII. Surakarta : Universitas Sebelas Maret.

IOP Conf. Series: Materials Science and Engineering **557** (2019) 012047 doi:10.1088/1757-899X/557/1/012047

# Simulation On Natural Convection Semi-Glazed Flat Plate Solar Collector For Green Bell Pepper Drying Application

**Leopold O. Nelwan[1], Yulfi N. Maulia[1], Delymi Oktariski[1]**

[1]Departemen Teknik Mesin dan Biosistem Fateta IPB, Gedung Fateta IPB, PO Box 220, Bogor.
*email: lonelwan@ipb.apps.ac.id*

**Abstract.** A simulation on semi-glazed flat plate solar collectors under natural convection mode used for green bell pepper drying has been conducted. The solar collector has a 60% glazed area which was installed to the drying chamber with a 45o slope. The simulation consisted of the heating and drying process. For the heating process, a semi-empirical model of the buoyancy effect and heat balance was proposed. To determine the constants of the model, the non-linear least square method based on the experimental data of irradiation, ambient air temperature and the drying chamber temperature for no-load operation was used. Furthermore, for the drying process simulation was carried out based on the mass and energy balance of the product and the drying air as well as the drying characteristics of green bell pepper. The simulation results showed that in clear weather conditions, the temperature and air flow rate produced by the collector can be used to dry a stack of ten trays within effective time less than 20.6 hours or equal to less than three days sunshine.

## 1. Introduction

Drying with flat plate solar collector is one of the solar drying technologies that is often used for drying agricultural products. In this drying method, air is heated and then passed through to the product for the drying process to take place. In general, air flow through the product is performed by using a fan that requires a mechanical energy source. Certainly, this increases energy requirements while also causing the dependence of the dryer's operation on mechanical / electrical energy sources.

To eliminate the use of blower, it is often to utilize free convection (passive mode) phenomena of the solar flat plate collector. This can be done by arranging the flat plate solar collector and drying container in such a way that a sufficient amount of hot air flow can occur. The arrangement is generally carried out by installing a flat plate solar collector mounted on the bottom of the dryer at a certain angle. The obstacle of this type of dryer is its low capacity. In addition, the drying rate is very dependent on weather conditions so that the quality may be poor [1].

Flat plate solar collectors generally use transparent cover in reducing radiative and convective heat loss. However, the use of cover actually reduces solar energy entering into the collector especially at large solar angles. Some results of the unglazed type collector (without transparent cover) testing indicate that the performance of this type of collector is quite good [2-3], [4], [5].

Testing of semi-glazed collectors with some levels of transparent-coverage area and various slope angles of the collector have been carried out by Nelwan et al. [6]. The result showed that the 60% transparent-coveage with $45^{o}$ slope has the useful energy which was almost the same as the collector with 100% coverage, even better at a steeper collector slope.

The potential of the drying system for drying agricultural products such as green bell pepper drying is interesting to be investigated. The purpose of this study was to simulate the potential of the semi-glazed flat plate solar collector for green bell pepper drying. The simulation was based on buoyancy models of the system with heat input only from solar radiation as well as the energy and mass balance of the drying process.

## 2. Method

### *2.1 Drying system design*

The flat plate solar collector designed in this study was the back-pass type in which the heated air flows below the absorber plate. The size of the solar collector was 1 m long, 0.3 m wide with 0.05 m thickness. The insulation used was flexible elastomeric foam with a thickness of 0.01 m. The channel cross section size was 0.02 x 0.3 m. The transparent cover was made of glass with a thickness of 5 mm. The inlet and outlet of the collector was located in the shorter side. The absorber used in the collector is a zinc-plated iron plate and painted black. The insulation applied at the bottom of the collector had a thickness of 25.4 mm. The solar collector designed in this drying system was a semi glazed solar collector where part of the collector closer to the inlet was left without transparent cover while the other part was equipped with transparent cover where the coverage area was 60% of the collector absorber area.

The drying chamber was designed with a length and width of 0.3 x 0.3 m and a height of 0.3 m. There was one stack of 4 trays where the distance between trays was 0.05 m. The drying chamber was also insulated with flexible elastomeric foam. The solar collector outlet was mounted to the drying container inlet at an angle of $45^{o}$ in accordance with the results of experiment in the previous study to utilize the buoyancy force of air flow into the collector [6].

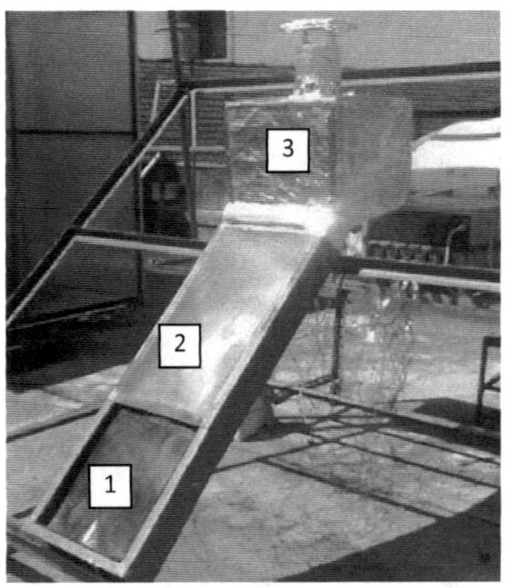


**Figure 1.** The semi-glazed solar collector assembled with drying chamber (1: unglazed part of the semi-glazed solar collector, 2: the glazed part, 3: the drying chamber)

### *2.2 Testing of the drying system*

The collector that had been assembled with drying chamber were tested without load from morning to evening. During testing, measurements of the ambient air temperature and air temperature inside the drying chamber were carried out using T-type thermocouples while solar irradiation was measured using a pyranometer. Both temperature and irradiation were recorded at 15-minute intervals.

*2.3 Drying system simulation*

*2.3.1 Air heating*

The model used for the simulation included the energy balance at the collector and the mass flow rate due to the buoyancy effect. With the ambient air temperature ($T_{amb}$ in K) used as the incoming air temperature into the collector, the quasi-steady energy balance in solar collectors can be expressed as:

$$\dot{m}c_p\left(T_o - T_{amb}\right) = IAR\,\tau\alpha - UA_T\left(T_{avg} - T_{amb}\right) \quad (1)$$

where m is mass flow rate, cp is specific heat of air, $T_o$ is collector outlet temperature (K), I is solar irradiation ($W/m^2$), A is the collector area ($m^2$), R is the ratio of the incident solar irradiation on the tilted collector to that on a horizontal surface (dimensionless), is the effective transmittance-absorptance product of the collector (dimensionless), U is the overall heat loss coefficient of the collector ($W/m^2$-$^oC$), $A_T$ is the total area of the collector ($m^2$). The air temperature in the collector ($T_{avg}$ in K) is defined as the arithmetic mean of the air temperature (environment) with the outlet air temperature.

The mass flow rate due to buoyancy effects in the collector was determined by:

$$\dot{m} = CA_D \frac{P}{R_s}\frac{1}{T_{avg}}\sqrt{2gH\frac{T_{avg} - T_{amb}}{T_{amb}}} \quad (2)$$

where C is the shrinkage coefficient ranged between 0 and 1, H is the total height of the drying system (m), P denotes the atmospheric pressure (Pa), $R_s$ is the ideal gas constant for air (kJ/kg$^o$C). The coefficients C and U are very dependent on the conditions of the flow and geometry of the system while are different between covered and opened areas. Because the cross-sectional areas ($A_D$) are different in the drying chamber and the collector, the product of C and $A_D$ is defined as one coefficient $K_1$. For simplification $UA_T$ was combined into one coefficient, $K_2$ and the coefficient R was also combined into $K_3$. Therefore equations (1) and (2) are expressed as:

$$\dot{m}c_p\left(T_o - T_{amb}\right) = K_1 I - K_2\left(T_{avg} - T_{avg}\right) \quad (3)$$

and

$$\dot{m} = K_3 \frac{P}{R_s}\frac{1}{T_{avg}}\sqrt{2gH\frac{T_{avg} - T_{amb}}{T_{amb}}} \quad (4)$$

To determine the three coefficients ($K_1$, $K_2$ and $K_3$), the non-linear least square method of with equations (3) and (4) was used by using the data of temperature of the drying chamber, solar irradiation and ambient air temperature. The method was performed by using a Microsoft Excel solver based on Generalized Reduced Gradient Algorithm.

*2.3.2 Drying*

The drying model is developed based on heat balance and energy and the drying characteristics of paprika with 2.5 cm in diameter and 3–4 mm in thickness. Since the direction of air flow is assumed to be uniform upward and it is assumed that no heat leakage from the chamber's wall, then the rate of change of water vapor and the heat balance in the drying air and the rate of energy change in the product are approached from the deep bed equation by Bala (1999) which is expressed in the following equations:

$$0 = G_a\frac{\partial H}{\partial x} + \rho_d\frac{\partial M}{\partial \theta} \quad (5)$$

$$0 = G_a\left(c_{p,a} + c_{p,s}H\right)\frac{\partial t_a}{\partial x} + \left(-\rho_d c_{p,s}\frac{\partial M}{\partial \theta} + h_{cv}\right)\left(t_a - t_g\right) \tag{6}$$

$$\rho_d\left(c_{pg} + c_{pw}M\right)\frac{\partial t_g}{\partial \theta} + \rho_d t_g c_{pw}\frac{\partial M}{\partial \theta} = h_{cv}\left(t_a - t_g\right) + \rho_d\left(h_{fg} + c_{ps}t_g\right)\frac{\partial M}{\partial \theta} \tag{7}$$

where G is the air mass flux (kg/s-m$^2$) generated from free convection, M is the moisture content (%d.b.), Cp is the specific heat (kJ/kg-$^o$C), H is the absolute humidity (kg/kg dry air), h is the heat transfer coefficient (kW/m$^2$-$^o$C), hfg is the latent heat of evaporation (kJ/kg), A is the contact area (m$^2$), t is the temperature ($^o$C), whereas subscript a denotes air, amb is ambient air, g is for the product.

A thin layer drying model was applied to predict the moisture content of the product in each tray during the drying process, which is expressed as:

$$\frac{dM}{d\theta} = -k\left(M - Me\right) \tag{8}$$

where k is the drying constant (1 / hour) and Me is the equilibrium moisture content (% d.b.). The drying constant of green bell pepper used was obtained from the results of Faustino et al. [7], while the equilibrium moisture content as the functions of temperature and humidity was obtained from Kiranoudis et al. [8].

The heat transfer coefficient (h) between air and green-bell pepper slice was approximated from the equation:

$$\frac{hD}{\lambda} = 0.228\left(\frac{c_p\mu}{\lambda}\right)^{1/3}\left(\frac{\rho v D}{\mu}\right)^{0.731} \tag{9}$$

where cp, $\mu$ and $\lambda$ are the specific heat, viscosity (kg/m-s) and thermal conductivity of air (W/m-oC), respecitvely, while v denotes the air velocity and D denotes the diameter of the green-bell pepper slice (m).

## 3. Results and discussion

*3.1Performance test and model validation*

Figure 2 shows the data of irradiation, ambient air temperature and drying air inside the chamber during the experiment used to obtain the K1-K3 values in equations (3) and (4). The range of irradiation values was at 300 to 900 W/m2, while the ambient air temperature was between 33 and 40oC. By using the non-linear least squares method, the obtained values of K1, K2 and K3 were 0.2417, 0.0053 and 0.0115 respectively. Figure 2 also shows the calculated drying chamber temperature and air flow rate, which ranges from 32-47oC and 0.0039-0.0111 kg/ s, respectively. It can be seen that the fluctuating temperature of the drying chamber and the fluctuating air flow rate are affected by the solar irradiation. The plot between the data and the calculated of air temperature drying in Figure 3 shows that the model could predict the air temperature quite well.

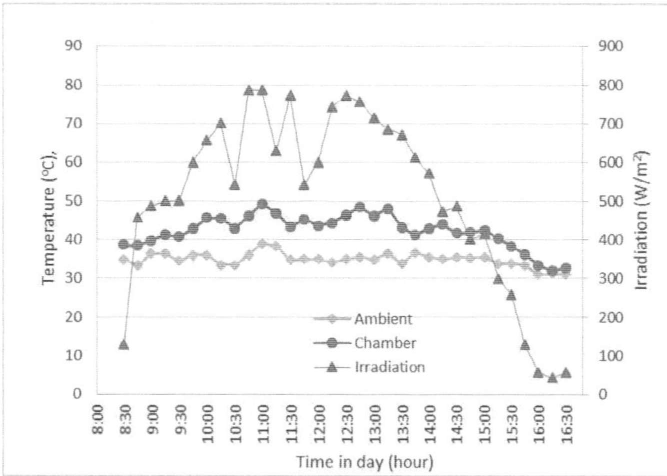


**Figure 2.** Irradiation, ambient temperature and drying chamber (data) with respect to time in day

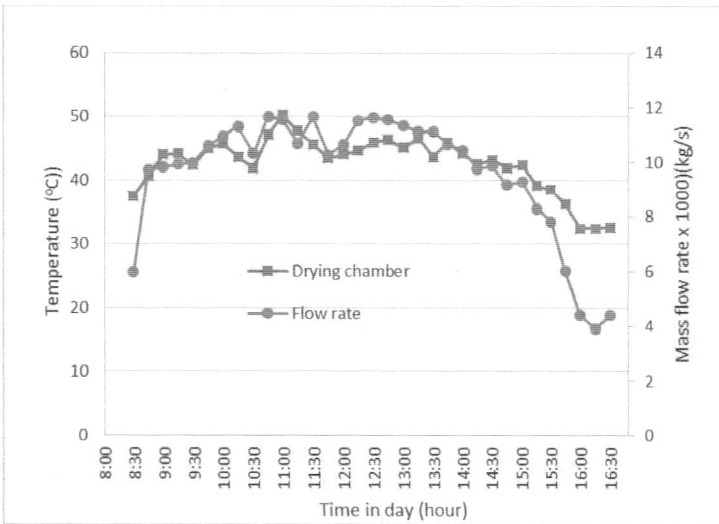


**Figure 3.** Calculated drying chamber temperature and air mass flow rate with respect to time in day

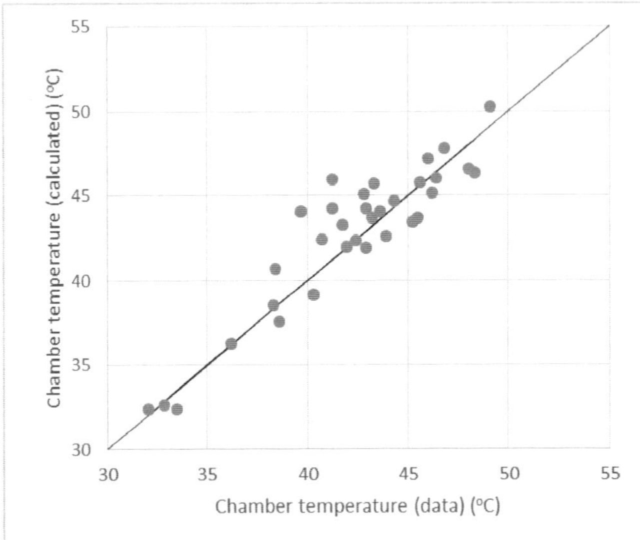


**Figure 4.** Chamber temperature (calculated and data)

*3.2 Drying*

The conditions of solar irradiation used for simulation during the drying process are assumed to be identical every day with a total reaching 5.8 kWh per day with an average ambient air temperature of 32.3°C and RH of 50-70%. The sliced green-bell pepper was spread on ten trays arranged in a vertical direction. With the dimensions of the drying rack and the thickness of 1 layer for each tray, the total of fresh product slices was 0.9765 kg. In such conditions, the simulated temperature of the drying air from the collector was in the range of 37-48°C or 7-9°C above the ambient air temperature.

The evolution of moisture content of the product during drying is shown in Figure 4. Products located at the bottom of the tray experience the fastest decrease in moisture content, i.e. within about 5 hours the lowest moisture content of the product has reached 60% w.b. while the top one was still above 80% w.b. However, the difference in moisture content decreases with respect to the drying time . After approximately after 14 hours of effective drying, the moisture content of the bottom green-bell pepper slightly increases due to the occurrence of readsorption because the equilibrium moisture content of the condition was quite high. Furthermore, the moisture content in all trays was almost the same and the average reached 14% w.b. after 20.6 hours.

The simulation results show that the total solar energy consumed reached 19 MJ / kg of water evaporated. Such the energy consumption is quite high compared to the drying method in general. The drying potential is still quite high, because the difference in moisture content between the top / bottom shelf was not too large. An increase of the amount of the load can still be applied, but the increase in pressure drop must be taken into account.

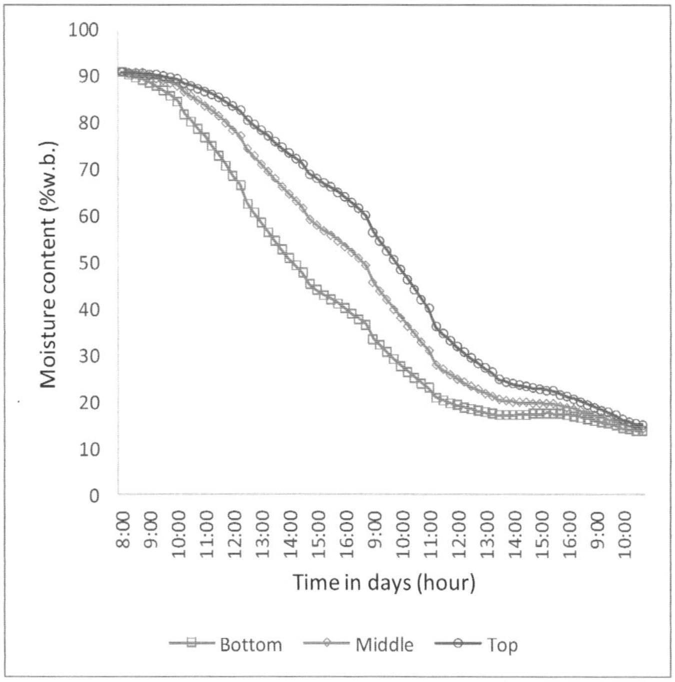


**Figure. 5.** Evolution of the moisture content (%w.b.) during the drying process

## Conclusion

A drying simulation using a flat collector with natural convection mode for cabinet green bell pepper slices drying has been developed. Simulation on the air heating section was based on the buoyancy model and the energy balance on the flat collector. Assuming solar irradiation of 5.8 kWh per day and the average ambient air temperature of 32°C the temperature of the drying air from the collector produced is in the range 37-48°C or 7-9°C above the ambient air temperature. With a load of 0.9765 kg of green bell pepper slices, drying can be accomplished within effective time of 20.6 hours.

## REFERENCES

[1] Mustayen A.G.M.B., S. Mekhilef and R. Saidur. 2014. Performance study of different solar dryers: A review. Renewable and Sustainable Energy Reviews 34 (2014) 463–470

[2] Burch, J. ,Salasovich, J. and Hillman T. 2005. An Assessment of Unglazed Solar Domestic Water Heaters. ISES Solar World Congress, Florida August 6–12, 2005

[3] Chan H.Y., S. Riffat, and J. Zhu. 2011. Experimental performance of unglazed transpired solar collector for air heating. World Renewable Energy Congress, Sweden, 8-11 May 2011.

[4] Leon, M.A., and S. Kumar. 2007. Mathematical modeling and thermal performance analysis of unglazed transpired solar collectors Solar Energy 81 (2007) 62–75.

[5] Van Decker, G.W.E., K.G.T. Hollands, and A.P. Brunger, 2001. Heat exchange relations for unglazed transpired solar collectors with circular holes on a square or triangular pitch. Solar Energy 71 (1), 33–45.
[6] Nelwan, L. O., Y. N. Maulia, D. Oktariski. 2012. Thermal study of a semi-glazed solar collector for various tilt angle. (in Indonesian). Prosiding Seminar Nasional Perteta, Denpasar, 13-14 Juli 2012.
[7] Faustino, J. M. F., M. J. Barroca and R. P. F. Guine. 2007. Study of the drying kinetics of green bell pepper and chemical characterization. Food and Bioproducts Processing, Trans IChemE, Part C, September 2007. Vol 85 (C3) 163–170
[8] Kiranoudis, C. T., Z. B. Maroulis, E. Tsami & D. Marinos-Kouris. 1993. Equilibrium Moisture Content and Heat of Desorption of Some Vegetables. Journal of Food Engineering 20 (1993) 55-74 circular holes on a square or triangular pitch. Solar Energy 71 (1), 33–45.

# Experimental Study of Compact Multi-Pass Evaporator Two Phase Flow

**R Heng[1], Z M Razlan[1,*], I Zunaidi[2], Shahriman A B[1], W K Wan[1], A Harun[3], M S M Hashim[1], M K Faizi[1], I Ibrahim[1] and N S Kamarrudin[1]**

[1]School of Mechatronic Engineering, Universiti Malaysia Perlis, Pauh Putra Campus, 02600, Arau, Perlis, Malaysia
[2]Faculty of Technology, University of Sunderland, St Peter's Campus, Sunderland, SR6 0DD, United Kingdom
[3]School of Microelectronic Engineering, Universiti Malaysia Perlis, Pauh Putra Campus, 02600, Arau, Perlis, Malaysia

*Corresponding author: zuradzman@unimap.edu.my

**Abstract.** The distribution of the liquid and gas flows in multi-pass channel that illustrate a compact evaporator for air-conditioning system was experimentally studies. The dividing header are design horizontal with a square cross section of 20mm x 20mm and length is 290mm, also ten curve multi-pass channel with length of 300mm each connected between the circle cross section 50mm combination tank. The distance between the dividing header and combing tank were 120mm. The test was conduct under uniform backpressure conditions; additionally, the working fluid used was water and air. This experiment conducted to examine the influence of inlet-flow at the header entrance (stratified flow and annular-mist flow) under uniform backpressure condition to improve the uniform of the water distribution inside every curve channels. It was found that the inlet-flow condition at the entrance header has relatively influence the water distribution. By comparison between two flow conditions, great value of uniformity for water distribution in each channel is improved under annular-mist flow condition.

## 1. Introduction

Major progress in the development of compact evaporator has been made by aerospace, automotive and cryogenic industries. Nowadays the global warming issue is the most significant aspect toward greater heat transfer rate per unit volume is needed to increase thermal duty and energy efficiency during this period. Macro compact multi-pass evaporator is one example of compact evaporator that base on multiple branching conduits to distribute refrigerant. At present, most of the problems in two phase flow are based on unsteadiness between the parallel multi-pass branches

This research is aimed to improve the Coefficient of Performance (COP) of the evaporator, which is, experiment on distribution characteristic of liquid and gas flow, the effect of the inlet-flow distribution (annular mist flow and stratified flow) and pressure lost condition.

A multi-pass evaporator usually creates a mal-distribution of gas-liquid flow among channel. This problem causes non-uniform evaporator heat transfer and mass transfer absorption. The problem may effect on evaporator Coefficient of Performance of the evaporator, which can deliver a high amount of energy usage.

The objective of this study are mainly to design experimental apparatus to simulate macro compact multi-pass evaporator with liquid-gas phase flow in uniform backpressure condition, by using air and water as working fluid. Also, to study experimentally base on the two phase liquid-gas flow distribution characteristic of (liquid phase and gas phase) in dividing header.

This study is conduct by limitation scope of experimental conditions as follows. The multi pass channel of branches from the dividing header is fixed with vinyl tube sized 2/8" or 6.35 mm of diameter. To generate the desired flow pattern, i.e., stratified and annular mist flow, some flow generator devices shall be used. Also, in this experimental study, a uniformed backpressure condition at the outlet of multi pass channel shall be applied.

## 2. Literature Review

Two-phase flow happens when distribution of liquid and gases in flow channel that have some commonly observe flow structures. This flow structure is known as two phase flow pattern, which have particular identifying characteristic. Prediction on pressure drop and heat transfer coefficient is nearly related to two-phase flow structure of fluid [1].

Mal-distribution occur when fluid and gas flow through header and divide into branches in refrigerant tube, it is frequently occurring in multi-flow evaporator and in worst case scenario it can cause no liquid provided into some of the branch [2].

Mader et al. [3] systematically investigate on mal-distribution in different path can cause decrease of COP and heat pump capacity as well as increased annual operating cost. Uneven distribution can likely reduce the effectiveness of heat exchanger and little impact on device mechanical failure also hydraulic instabilities. For example, Non-Uniform flow of heat exchanger-reactor among individual or differential channels may result substantial temperature deviations, resulting in thermal runaway, which could arise from an exothermic reaction studies by Rebrov et al [4].

Two phase flow of liquid and gas characteristic in one conduit was review significantly by Barnea D and Y. Taitel, main focus was on flow pattern identification which is to classify of various characteristic of flow pattern to sort into several groups also explanation on the method used to develop and conducted on flow pattern of two phase flow detection in conduit [5].

Two-phase flow pattern in different position of the tube cases, some researchers shows the definition which are nearly close to the resent pattern because of the flow pattern designation is not yet reach accurately standardize. The large amount of study and development have been done, deal with two-phase flow in vertical or horizontal tube, yet there is a small amount of studies with the whole range of inclination. The most acceptable flow patterns for vertical tubes with upward flow are defined by Bennett A.W [6] and from G. F. Hewitt D and N. Roberts [7]. Mandhane et al, is the most frequently use for liquid-gas flow pattern maps in horizontal tube [8].

The first step to predict or to understanding the characteristic of two-phase flow is to know the type of the flow, which is the pattern of the flow. Each flow has common different in physical phenomena such as, shear for annular flow and acceleration of gravity for stratified flow. The different correlation and model only can be applied to specific flow pattern. Thank to development of flow pattern map, thus can be used to determine the configuration of two phase flow depending on the flow condition [9]. The flow pattern prediction can be made by mapping the experiment data on flow pattern and transition boundaries using two dimensional plots. Two common type of coordinate that most-use are dimensionless coordinate's map and dimension coordinate's map [2, 5, 7, 8].

Dimension coordinate's map is the basic form of coordinate condition that uses mass superficial, superficial velocity and momentum flux, but this mapping method are limited to the range of data collected as well as flow condition beneath which experiment were conducted.

This mapping method is straightforward dimension analysis approach to select or predict the physical variable that affected the flow pattern. The two phase flow transition boundaries must be a function of all variable. The reason why is that these set of variable already been reduced as it is assumed that geometric entrance has no effect on the steady state flow pattern. Baker (1954), use to account for the different physical properties and introduced non-dimensional parameters, which called Baker's coordinates map [10]. Baker is the one proposed top lot the flow map with two parameter set $\dot{m}_G/\lambda$ and $\dot{m}_L/\psi$ where $\dot{m}_G$ and $\dot{m}_L$ are gas and liquid mass velocities flow rates, respectively and $\lambda$ and $\psi$ are defined as [11].

Jesus Moreno Quiben (2005), superficial velocities can be defined as the volumetric flow rate of the phase across the entire cross-sectional area of the liquid-gas flow. It is possibly being expressed as the phase velocity if it would flow alone in the total cross section. He describes the formulation of superficial gas velocity and superficial; liquid velocity [12]. Vapour quality ( $\chi$ ) is the ratio between of vapour mass flow rate ($\dot{m}_G$) to the sum of mass flow rate ($\dot{m}_G + \dot{m}_L$) [12, 13, 17]. Mass fluxes (G) is the ratio between mass flow rate ($\dot{m}$) divided with cross section (A) of the tube [12, 17]. The cross-sectional area of the channel or pipe can be subdivided into an area subject to the vapour phase (V) and the area led by the liquid phase (L) [17]. The mass flow rate ($\dot{m}$) of the liquid-gas flow mixture is the sum of the mass flow rates of vapour (V) and liquid (L)[17]. Void fraction is the one of most important parameter used to characterize two phase flow properties. It used as the key to unlock the physical value for determining various other significant parameters [14]. Jesús Moreno Quibén, void fraction is defined by the ratio between cross section divide with total of cross section [12].

The important hydrodynamic aspects of two-phase pressure drop in horizontal mini- and micro-channels have been studied by different researchers. Similar with the ordinarily sized channel, frictional pressure drop in the micro-channel still plays the dominant role when compared with other pressure drop components such as acceleration pressure drop and pressure drop due to an abrupt flow area [15].

Void fraction is one of the key parameters needed to evaluate the gravitational and acceleration terms in the total pressure drop of two-phase gas–liquid flow in different channels. Different methods available for void fraction estimation have been used by previous researchers [15, 16]. It is representing the ratio of the domain occupied by the vapour to the total domain that is occupied by the two-phase flow mixture. Void fraction can be defined as the ratio of the cross-sectional area occupied by vapor to the total cross sectional area of the conduit [17].

## 3. Methodology

This experiment was conducted to investigate the influence of inlet-flow header on two phase flow mal-distribution. In this experiment, water and compress air are used as a working fluid to simulate actual meal-distribution of two phase flow that occur in multi-pass branches in the evaporator. The investigations are based on several parameters which is, inlet-flow pattern in the header and the superficial velocities of both gas-liquid flow at the entrance of the header and uniform back-pressure at the dividing channel outlet.

### *3.1. Experimental study methodology flow*

This investigation starts by viewing other researcher paper to investigate the importance factor that can give impact to mal-distribution in the dividing channels. Than the experiment apparatus need to be design base on uniform back-pressure condition at the outlet of the branches before contracting the apparatus. For the next step, the installation of annular mist flow generator and stratified flow generator device as the inlet-flow header, to ensure the entrance of two phase flow in the test channel in a desire flow condition; annular-mist flow and stratified flow.

Two phase flow is based on gas-liquid flow, since this test using room temperature water and compressed air as a fluid running in the test channels. The generating flow that generate at the inside of test channel can be examined using Baker's flow pattern maps to ensure the mal-distribution that need. The difference inlet flow at the header may give effect to the mal-distribution of two phase flow in the multi-pass branches. From the data receive on two phase flow under particular aspect of inlet flow that is presumed that this data can develop more understanding about two phase flow distribution behavior in multi-pass parallel channel.

### *3.2. Experimental apparatus and setup*

Figure 1 shown experimental apparatus schematic diagram during this research. The experimental study will simulate actual gas-liquid flow in multi-pass channel on evaporator using isothermal air-water two phase flow test channel. Water supply and gas volumetric flow rate is controlled by the air flow meter and water flow meter.

To simulate the actual situation of evaporation, the test channel has been developed using transparent acrylic plate with 20mm wide, 20mm height also with 290mm of length that is set horizontally and equip with ten curve 5mm branches with 300mm length also distance between dividing header and combining tank is 120mm shown in figure 2.

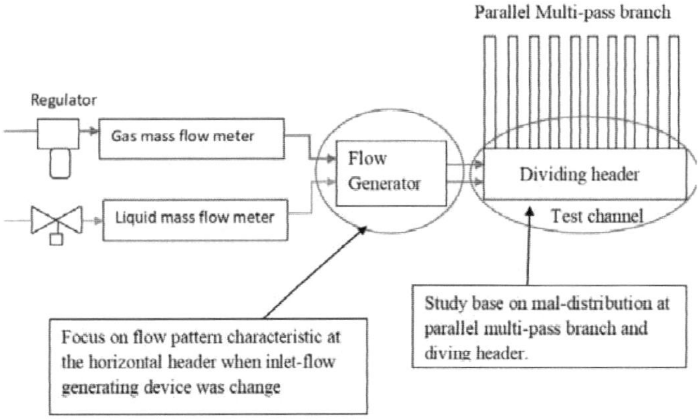


**Figure 1.** Experimental apparatus schematic diagram

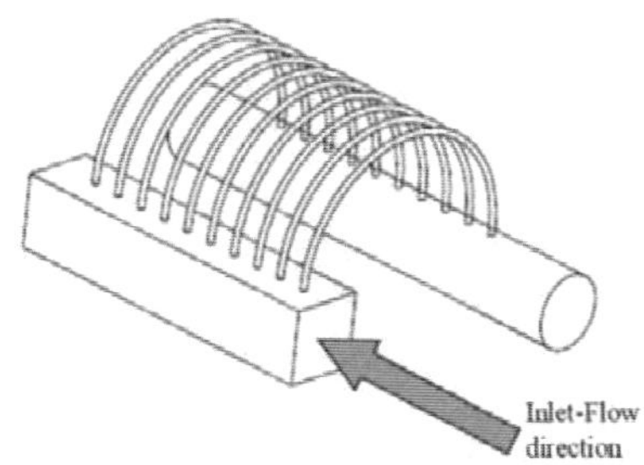


**Figure 2.** Display the details of test channel

The investigation will present the stratified flow and annular mist flow pattern. In this case, the inlet flow pattern at the entrance of the header is considered to give an effect on two phase mal-distribution among the channel. In this experimental study, two inlet flow condition at the entrance of the header are implemented which is an annular mist flow and stratified flow. The differences on flow pattern mal-distribution of both inlet flow conditions can be observed visually trough test channel. The flow generator device is attached at the header entrance of the dividing channels to create desire flow pattern. The developing region happen in the test channel that has the same cross section as the flow generator.

The actual evaporator refrigerant can be determined using vapour quality (quality x) and mass flow rate, due to the value of superficial velocity of gas (compress air), $J_G$ and superficial velocity of liquid (water), $J_L$ at the entrance of the header. Refrigerant R-134a has the value of 1.0m/s to 0.5m/s in superficial velocity of liquid and 0.015m/s to 0.045m/s in superficial velocity of gas, further with mass flow rate of M = 40kg/h to 160kg/h as well as a vapour quality range x = 0.2 up to 0.7 [2]. Addition, this experimental investigation is done in a room temperature.

**Table 1.** The experiment condition summary

| | |
|---|---|
| Running fluid | Air and water |
| Header superficial air velocity | 1.0m/s, 3.0m/s, 5.0m/s |
| Header superficial water velocity | 0.015m/s, 0.03m/s, 0.045m/s |
| Back of the branches pressure condition | Uniform outlet pressure |
| Inlet-flow pattern at the header | Annular mist flow and Stratified flow |
| Condition of the header | Horizontal header |
| Condition of the channel | Parallel channel |
| Channel characteristic | Circle tube and Multi-channel tube |

To claim the impact of the pressure at the back of the multiple branches on flow distribution, the experiment is done based on uniform pressure at the back of the multiple branches were conducted. Figure 3 shown the experimental apparatus for determine pressure at the back of the multiple branches. The experiment conduct by using either one of the multiple branches is connected to the separator and other branch is connected to combining tank, which have a great capacity or volume. The pressure in the combining tank and at the separator must have zero in pressure differences getting from u-tube or manometer. To ensure the pressure difference is at zero, the valve at the combining tank and separator outlet is used to regulate the pressure to get uniform pressure. Combining tank at the outlet of the divided branch is shown in Figure 4.

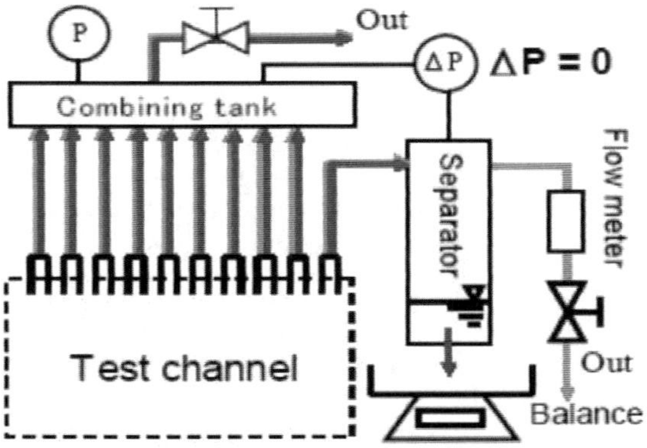


**Figure 3.** Uniform condition of pressure at the back of multi-pass channel test apparatus [2]

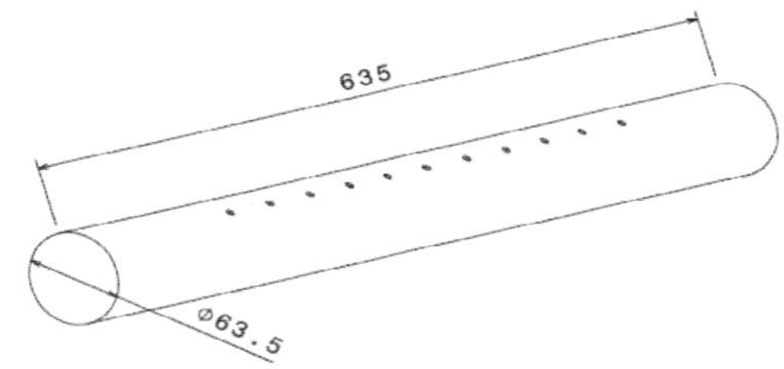


**Figure 4.** Combining tank

## 4. Results

In this section, the complete contraction of the experiment apparatus and experimental result are revealed. On the other hand, these study deliver two main result components that base of (i) design experiment apparatus for macro compact multi-pass channel with two phase flow and (ii) evaluation result depend on the impact of the entrance flow pattern before entering the dividing channel in water distribution ratio.

### *4.1. Completion of experimental apparatus for compact multi-pass evaporator*

In this section, the author will give clear vision for the reader about the experimental apparatus arrangement to do the test. Figure 5 shows the arrangement of experimental apparatus.

1. Separator tank.
2. Air flow speed (l/min).
3. Water flow speed (l/min).
4. Flow generator.
5. Experimental apparatus table 2” x 1’ 1/2’ x 3”.
6. Flexible rubber tube 1/2".
7. Dividing header (in this figure stratified flow generator is use).
8. Flexible transparent tubing 5mm.
9. Combining Tank 2”.
10. Pressure differential manometer.

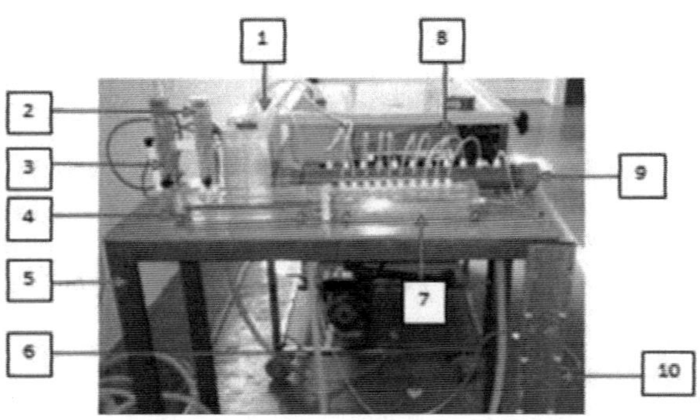


**Figure 5.** Experiment apparatus arrangement

*4.2. Comparison base on water distribution uniformity improvement occur between stratified flow and annular mist flow condition*

From the result obtain under stratified flow and annular mist flow condition, as the result show in figure 6. Figure was pick from the highest result that possibly obtain under stratified and under annular mist flow condition, in other word by comparing the best from the best result to get obtain the greater improvement on uniform flow of the water distribution in every channel. Moreover, these two results can be compared on difference uniformity of the water distribution through every channel.

Firstly, the comparison can be made on the difference of the superficial air and water velocity of the experiment. The uniformly distribute was occurred when the superficial air and water velocity, $J_G$, $J_L$ is increase to the highest value which is at $J_G$ = 5.0 at $J_L$=0.045m/s for Case B (Stratified flow condition). The difference of result show in Case A (Annular-mist flow condition), when the superficial air velocity, $J_G$ = 3.0m/s in all different value of superficial water velocity, the water distribution ratio is uniformly distribute in every channel was improve.

The improvement of uniformity water distribution occurs in Case A can easily be observe by the staged or horizontal line occur over the graph compare to the result obtain in Case B. As a result, shown in Case B, the curving line on the graph happened in every conditions, it is shown the sudden increment occur on the water distribution ratio. On the other hand, the smooth horizontal line shows the uniformity of the water distribution through every channel is represent in Case A.

Base on the result and the observation made on the influence of the superficial air velocity, $J_G$ and superficial water velocity, $J_L$. Under annular-mist flow condition the uniformity of the water distribution inside the channel is improve especially when high superficial air velocity is implement [10].

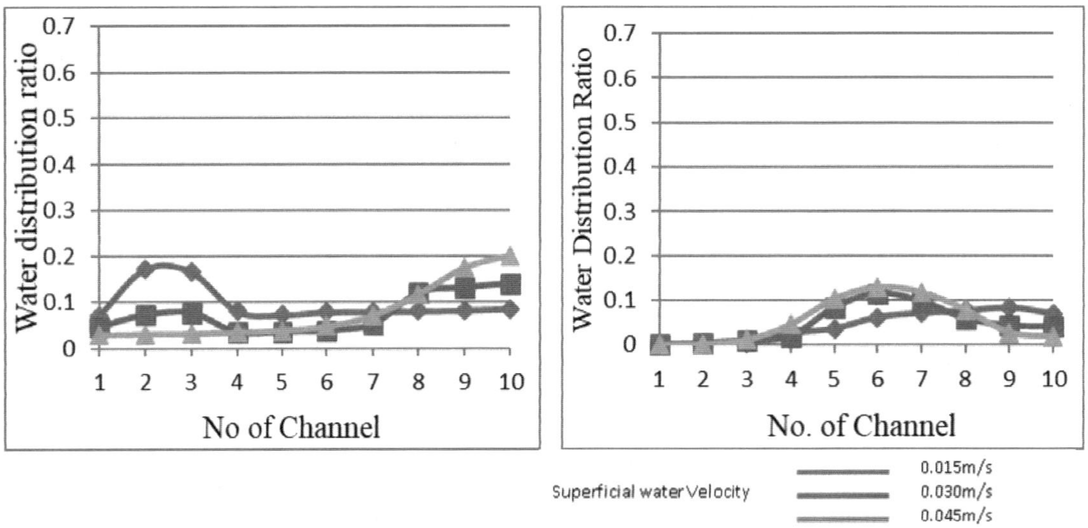


(a) Case A (Annular-mist Flow condition (b) Case B (Stratified Flow Condition

**Figure 6.** Water distribution ratio under (a) Annular-mist flow (curve tube $J_G$ = 0.3 m/s) and (b) Stratified flow (curve tube $J_G$ = 5.0m/s)

## 5. Conclusion and recommendation

First of all, the experimental study was start by design the experiment apparatus for multi-pass curve channel under uniform backpressure condition to simulate the actual compact multi-pass evaporator. The design process start by design selection on the experimental apparatus, it involves experiment apparatus schematic diagram, header-entrance condition and the uniform backpressure condition at the outlet of the channels. Next, the material selection and fabrication process for each equipment, also the specific measuring device used in this experimental apparatus.

Stratified-flow condition is the first two-phase flow condition that examine. The water distribution at the multi-pass channel under stratified-flow condition is approve to have effect went the adjustment of superficial gas velocity is increase

Furthermore, for the annular-mist flow header entrance under uniform backpressure condition, this flow was observed to have an effect to the adjustment made on superficial air and water velocity is show that superficial air and water velocity have a certain effect to the distribution of water in the dividing channel under annular mist flow conditions.

On the other hand, the comparison can be made between these two different flows to obtain the greater uniformity improvement on water distribution of multi-pass channel. Generally, under annular mist flow the effect of superficial water and air velocity was give greater improvement to uniformly distribute the water through the multi-pass channel compared to stratified flow especially at high superficial air velocity.

These experimental studies have a lot of potential for future work. The uniform backpressure can be compare to un-uniform back pressure condition, to know the influence of the outlet channel pressure condition to the distribution liquid and gas in the dividing header. On only that, the adjustment in curve channel high can be made, to know the influence of the gravitational force when the two-phase flows start to distribute. In addition, the experiment can be test on velocity of the air inside the dividing channel by adding digital air flow meter and use of digital pressure differential meter get the exact measurement of uniform pressure between combining tank and separator tank. Besides that, to have better observation result for air flow, high speed camera and particle image (PIV) is suggest to be used in future. Not only that, to have greater accuracy of measurement and adjustment on the inlet-flow condition, the digital mass flow controller is suggested.

**Acknowledgments**
TUSB Research UniMAP supports this research under the research grant 2017/08/0006.

**References**

[1] John R. Thome., (2007), Engineering Data Book III, 1st Edition, “Chapter 12: Two Phase Flow Pattern”, *Wolverine Tube, Inc., Switzerland.*
[2] Zuradzman bin Mohamad Razlan, Hiroaki Goshima, Masafumi Hirota, Ryota Isobe,Yasuhiro Mizuno, Naoki Maruyama and Akira Nishimura,(2011).Gas-Liquid Flow Distributions in Multipass Channels with Vertical Upward Branches, *The Open Transport Phenomena Journal,* 3, pp. 17-30.
[3] Mader, G., Palm, B., Elmegaard, B., (2014). Maldistribution in Air-Water Heat Pump Evaporators. Part 1: Effects on Evaporator, Heat Pump and System Level.
[4] E.V. Rebrov, J.C. Schouten, M.H.J.M. de Croon (2011), “Single-phase fluid flow distribution and heat transfer in microstructured reactors”,. *Chemical Engineering Science* 66, pp. 1374-1393.
[5] D. Barnea and Y. Taitel (1986), "Flow Pattern Transition in Two-Phase Gas Liquid Flows", Chapter 16, the Encyclopedia of Fluid Mechanics, 3, "Gas Liquid Flows", Gulf Publishing, N.P. Cheremisinoff, Editor, pp. 403-474 (1986)
[6] Bennett, A.W., Hewitt, G.F., Kearsey, H.A., Keeys, R.K.F. and Lacey, P.M.C. (1965). "Flow visualisation studies of boiling at high pressure". *Proc.Irut.Mech.Eng.,* 180, (Part 3c), 260.
[7] Hewitt, G.F., Roberts, D.N., (1969).” Studies of two-phase flow patterns by simultaneous X-ray and flash photography. UKAEA, Harwell: Report, AERE-M 2159.
[8] Mandhane, J.M., Gregory, G.A., Aziz, K., (1974).” A flow pattern map for gas–liquid flow in horizontal pipes*”. Int. J. Multiphase Flow* 1, pp. 537–553.
[9] Stéphane Lips, Josua P. Meyer. (2011), “Two-phase flow in inclined tubes with specific reference to condensation: A review”, *International Journal of Multiphase Flow*, 37, pp. 845–859.
[10] C. Foletti, S. Farise, B.Grassi, D.Strazza, M.Lancini, P.Poesio (2011). “Experimental investigation on two-phase air/high-viscosity-oil flow in a horizontal pipe”. *Chemical Engineering Science*, 66, pp.5968–5975.

[11] Jesús Moreno Quibén (2005). "Experimental and analytical study of two-phase pressure drops during evaporation in horizontal tubes". PhD thesis, University Pierre ana Marie Curic, France and Spanish.

[12] Bingdong Zhang, Xingkai Zhang, Dong Wang, Shanfang Huang (2013) "Equal quality distribution of gas–liquid two-phase flow by partial separation method". International *Journal of Multiphase Flow*, 57. pp.66–77.

[13] John R. Thome., (2007), Engineering Data Book III, 1st Edition, "Chapter 17: Void Fraction in Two-Phase Flow", *Wolverine Tube*, Inc., Switzerland, pp.1-33.

[14] Sira Saisorn and Somchai Wongwises. (2009). An experimental investigation of two-phase air–water flows through a horizontal circular micro-channel". *Experimental Thermal and Fluid Science* 33, pp.306–315.

[15] Kawahara, P.M.-Y. Chung, M. Kawaji. (2002), Investigation of two-phase flow pattern, void fraction and pressure drop in a microchannel". *International Journal of Multiphase Flow*, 28, pp.1411–1435.

[16] Julian Winkler, Jesse Killion, Srinivas Garimella, Brian M. Fronk (2012)." Void fractions for condensing refrigerant flow in small channels: Part I literature review". *International Journal of Refrigeration*, 35, pp.219 -245.

[17] F. Poggi, H. Macchi-Tejeda, A. Maréchal, D. Leducq, A. Bontemps (2007). Experimental study of single and two-phase adiabatic flow distribution in compact heat exchangers. 18th France Congress of Mechanics Engineering, Grenoble, France, pp.27-31.

IOP Conf. Series: Materials Science and Engineering **557** (2019) 012049 doi:10.1088/1757-899X/557/1/012049

# Analysis of Relation Between Propulsion Angle and Thrust Generation of SEA-EYE

**I Eswanto[1], Z M Razlan[1,*], I Zunaidi[2], Shahriman A B[1], W K Wan[1], A Harun[3], H Desa[1], M K Faizi[1], I H W Nordin[1] and M S Muhamad Azmi[1]**

[1]School of Mechatronic Engineering, Universiti Malaysia Perlis, Pauh Putra Campus, 02600, Arau, Perlis, Malaysia
[2]Faculty of Technology, University of Sunderland, St Peter's Campus, Sunderland, SR6 0DD, United Kingdom
[3]School of Microelectronic Engineering, Universiti Malaysia Perlis, Pauh Putra Campus, 02600, Arau, Perlis, Malaysia

*Corresponding author: zuradzman@unimap.edu.my

**Abstract.** Analysis of the relation between propulsion angle and thrust generation of SEA-EYE was presented and discussed with three different angles of the boat propeller. The results of this analysis has been decided to solve the problem of the study. Features of SEA-EYE boat and the boat has been analysed in the Ansys Workbench (Fluid Flow CFX). The resulting decision was presented in the form of diagrams, tables and graphs. The results for the three boat propeller angle is almost close to perfect, but the final decision was made after carefully researched and as far as possible.

## 1. Introduction

SEA-EYE is an Unmanned Surface Vehicle (USV), for sea observation and checking applications. Unmanned Surface Vehicles (USV) or Autonomous Surface Vehicles (ASV) is a vehicle without a driver operating in the water. Harbour protection technologies are assessed through a survey of use Unmanned Surface Vehicles (USV) especially SEA-EYE is used for spying. The SEA-EYE can be controlled in two ways. It can either be remotely controlled physically, or it can work self-sufficiently utilizing GPS. The most exceptional peculiarity of its outline is the way it does away with the requirement for adjustment frameworks that are regularly coordinated into USV to keep them from overturning.

The speciality here is that it can move upside down. There are two sets of propellers, top and bottom, so when it flips over, the top propellers will switch off and the bottom ones will switch on automatically. It's powered by an electric motor, it also produces very little noise, which gives it an added advantage when it needs to approach another boat covertly, according to its developers. Since it doesn't use a fuel engine, it's also claimed to be inexpensive to maintain and operate.

The battery-powered USV can be recharged using the onboard solar panels, helping ensure that it always has enough power to turn back. It can be modified to explore to a specific set of coordinates through GPS, with the administrators mindful of its area at all times as it consistently transmits its coordinates over to the base station.

The advantages: -

- ✓ Anti-capsize system
- ✓ Day and night vision system
- ✓ Ground controller with user friendly GUI

✓ Solar powered
✓ Autonomous path planning and navigation system
✓ Can overcome the problem of using manpower to do the tedious task of spying with alternating same area

## 2. Unmanned Surface Vehicle (USV)

Unmanned Surface Vehicle (USV) use the electrochemical power sources have classified in four types, such as a seawater battery, inside pressure ship body is standard batteries and working at ordinary pressure, fuel cell, for electrically insulated from seawater is compensated battery working pressure at ambient pressure [1].

There are several techniques to boost the powers and energy conversion system for unmanned surface vehicle is tether method, battery method, environmental energy conversion and docking station establishment [3]. The propeller behind marine hull (astern autonomous underwater vehicles) will cause various effects on the performance because multiple angles. Three separate conditions, including straight movement two steady approaches equivalent to $\alpha = 5^{\circ}$ and $15^{\circ}$. The general extent for the approach somewhere around $0^{\circ}$ and $15^{\circ}$. Comparable hydrodynamic proficiency for propeller at diverse approaches, the created push and subsequently the required propeller torque were altogether higher than those in straight [9].

Propellers there is the utilization of electric power to drive the torque and power control. The parts of the study were the sort or state of the propellers, including peculiarities, element stream and push misfortune impacts. Thruster controller for low level there are two various of typical and great conditions. In typical conditions, when encountering low to direct push misfortunes and in amazing conditions, when encountering substantial and sudden push misfortunes because of ventilation and done and finished with water impacts. In the affectability to push misfortunes, when the loss of propeller stacking may, though changes in propeller push and torque, lead to changes in shaft speed and power [5].

## 3. Thrust

According to the second and third laws of Newton's, thrust is the mechanical force resulting from the reaction accelerating a mass of gas. System that accelerates mass in one direction causes power equal in magnitude but opposite in direction. Thrust in mechanical engineering is of force orthogonal.

In the direction perpendicular to the surface normal either be used to power and called thrust. Force or thrust, measured in units of the International System of Units (SI) as the Newton (N), and thrust is also used to accelerate 1 kilogram of mass at the rate of 1 meter per second squared.

From Newton's second law of motion: -

$$F = \frac{[(m * V)2 - (m * V)1]}{(t2 - t2)} \tag{1}$$

The simple force equation: -

$$F = m * a \tag{2}$$

A boat produce thrust when the propeller is turned on by pushing the water backwards and overall force contained in the water moving through the propeller.

$$\mathbf{T} = \mathbf{v}\frac{dm}{dt} \tag{3}$$

where: -

T = thrust (force)

$\frac{dm}{dt}$ = the rate of change of mass with respect to time (mass flow rate of exhaust)

v = the speed of the exhaust gases

## 4. Momentum Flux Correction Factor

The velocity across most inlets and outlets is not uniform. Dimensionless correction factor β, called the momentum flux correction factor.

Fixed Control Volume: -

$$\Sigma \vec{F} = \frac{d}{dt} \int_{CV} \rho\vec{V}\, dv + \int_{CS} \rho\vec{V}\, (\vec{V}.\vec{n})dA$$

$$\Downarrow$$

$$\sum \vec{F} = \frac{d}{dt} \int_{CV} \rho\vec{V}\, dv + \sum_{out} \beta\dot{m}\, \vec{V}\, avg + \sum_{in} \beta\dot{m}\, \vec{V}\, avg \tag{4}$$

$\beta$ produced either into or out of control surfaces for the inlet or outlet of the cross sectional area $A_c$ is then shown to the mass flow rate m and an average $\vec{V}_{avg}$ through the inlet or outlet.

Momentum flux across an inlet or outlet: -

$$\int_{Ac} \rho\vec{V}\, (\vec{V}.\vec{n})dAc = \beta\dot{m}\, \vec{V}\, avg \tag{5}$$

* Along the inlet or outlet density is uniform while $\vec{V}$ is in the same direction as $\vec{V}avg$

$$\beta = \frac{\int_{Ac} \rho\vec{V}\, (\vec{V}.\vec{n})dAc}{\dot{m}\, Vavg} = \frac{\int_{Ac} \rho\vec{V}\, (\vec{V}.\vec{n})dAc}{\rho\, Vavg\, Ac\, Vavg} \tag{6}$$

where: -

$V_{avg}$ = constant
substituted $\boldsymbol{\rho}\, V_{avg} Ac = \dot{m}$

At the inlet or outlet area can happen the control surface slices normal: -

$$(\vec{V}.\vec{n})dAc = V\, dAc \tag{7}$$

Momentum flux correction factor: -

$$\beta = \frac{1}{Ac} \int_{Ac} (\frac{V}{Vavg})^2\, dAc \tag{8}$$

* $\beta$ greater than or equal to unit

## 5. Calculation

The value of Sea-Eye boat such as: -

Dimension of propeller,
$D$=2.5 in → 0.0635 m
Mass of overall boat, $m$ = 7 kg
Density of water, $\boldsymbol{\rho}$ = 998 kg/m$^3$
Gravity, $g$ = 9.81 m/s$^2$
Reynold number for water, $Re$ = 2300

Area

$A = \frac{\pi D^2}{4}$ ➡ for 4 propellers, A' = A x 4

Velocity

Velocity inlet, $V_{inlet}$ ⟹ $Re = \frac{\rho \, V_{inlet} \, D}{\mu}$

Velocity outlet, $V_{outlet} = \sqrt{\frac{m\,g}{\rho\,A'}}$

Volume flow rate

$V_{outlet} = A' \, V_{outlet}$

Mass flow rate

$\dot{m} = \rho \, V_{outlet}$

## 6. Results and discussions

Table 6.1 shows the value of pressure and velocity for every angle of propeller. From the value can find the value of thrust and the results will get a difference results. The results will be able to know which angle the propeller most suitable for a boat moving smoothly. Thus, the pressure and velocity of water distributions analysis are shown in figure 6.1 and 6.2.

**Table 6.1.** Results of pressure and velocity

| Angle of SEA-EYE propeller | Pressure (Pa) | Velocity (m/s) | Thrust (N) |
|---|---|---|---|
| 0° | 127900 | 26.04 | 1478.81 |
| 30° | 261800 | 25.98 | 1475.40 |
| 50° | 318900 | 24.46 | 1389.08 |

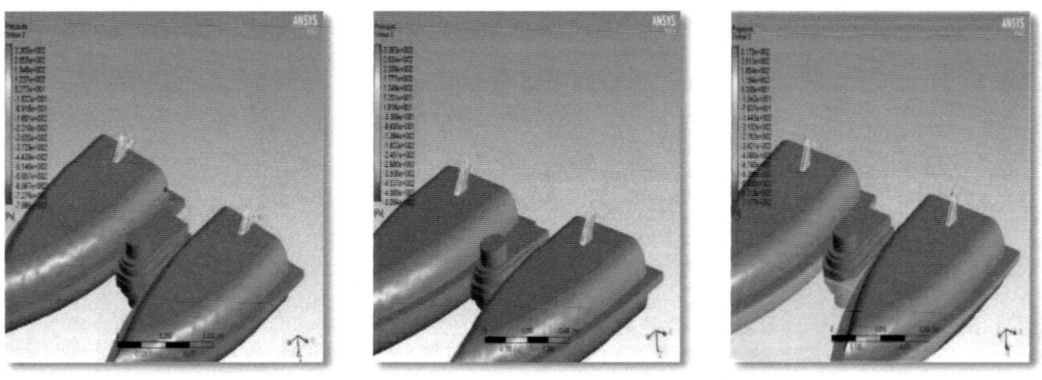

(a) angle of propeller 0° (b) angle of propeller 30° (c) angle of propeller 50°

**Figure 6.1.** The pressure at propeller for the water (angle of propeller 0°, 30°, 50°)

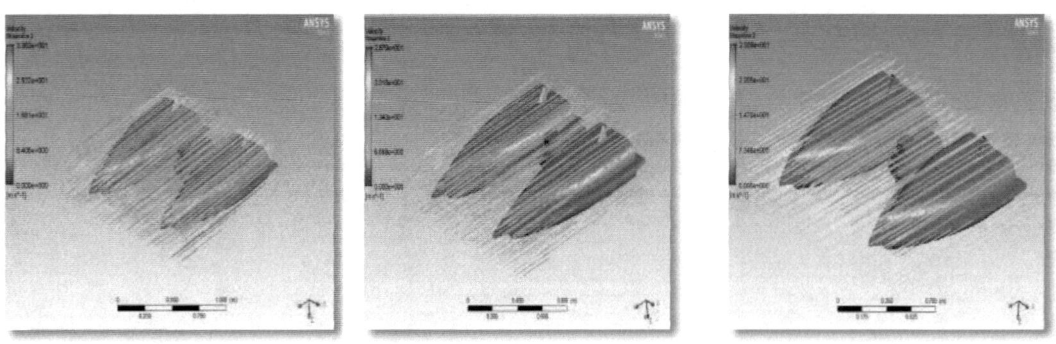

(a) angle of propeller 0° (b) angle of propeller 30° (c) angle of propeller 50°

**Figure 6.2.** The velocity distribution of the water (angle of propeller 0°, 30°, 50°)

IOP Conf. Series: Materials Science and Engineering **557** (2019) 012049 doi:10.1088/1757-899X/557/1/012049

Figure 6.3 describe the value of velocity vs. angle of SEA-EYE propeller. The velocity of each angle is not much difference but there are variations. For angle 0° the velocity is 25.22 m/s, for angle 30° the velocity decrease to 22.62 m/s and for angle 50° the velocity also decreases to 22.05 m/s. Thus, will be able to solve this problem. From the value of velocity can know the pressure at SEA-EYE boat high or low.

Figure 6.4 describe the value of pressure vs. angle of SEA-EYE propeller. The pressure of each angle is not much difference but there are variations. For angle 0° the pressure is 123.7 Pa, for angle 30° the pressure increase to 230.6 Pa and for angle 50° the pressure also increases to 251.3 Pa. Thus, will be able to solve this problem. From the value can know the pressure at SEA-EYE boat is high or low.

Figure 6.5 describe the value of thrust vs. angle of SEA-EYE propeller. The velocity of each angle is not much difference but there are variations. To get the value of thrust from the value of velocity multiply the value of mass flow rate. For angle 0° the thrust is 742.5 N, for angle 30° the thrust decrease to 665.96 N and for angle 50° the thrust also decreases to 649.17 N. From the results will know which angle is the best.

From the analysis, it has been decided to choose the angle of propeller is 0°. This is because the pressure at this angle is low and the thrust is high. Relation between this two value is also have connection with the high value of velocity. For the other two angles 30° and 50° can also be used but are not suitable and not achieve the objectives of this project to get the speed of boat is smoothly.

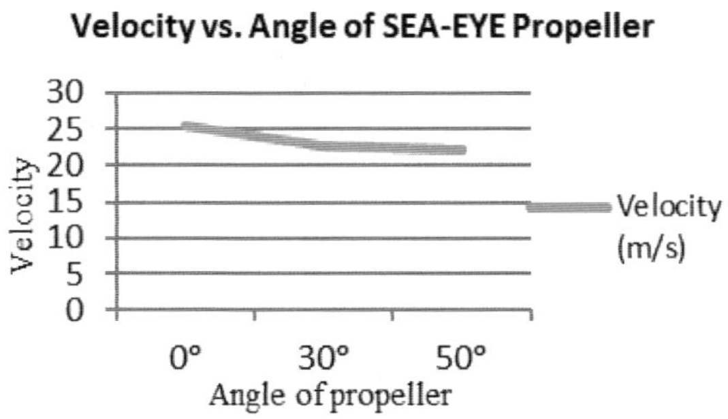


**Figure 6.3.** Velocity vs. Angle of SEA-EYE Propeller

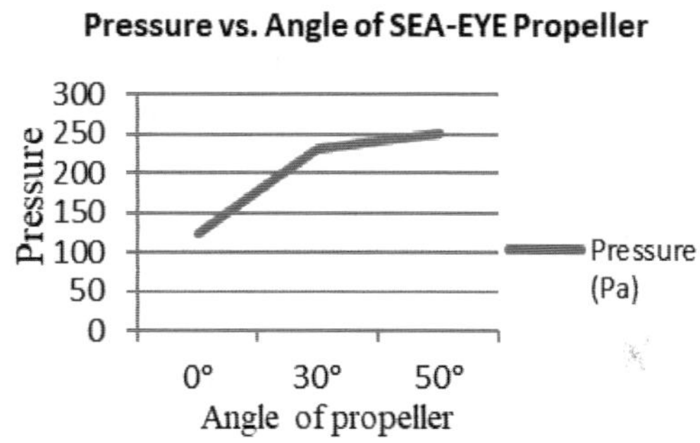


**Figure 6.4.** Pressure vs. Angle of SEA-EYE Propeller

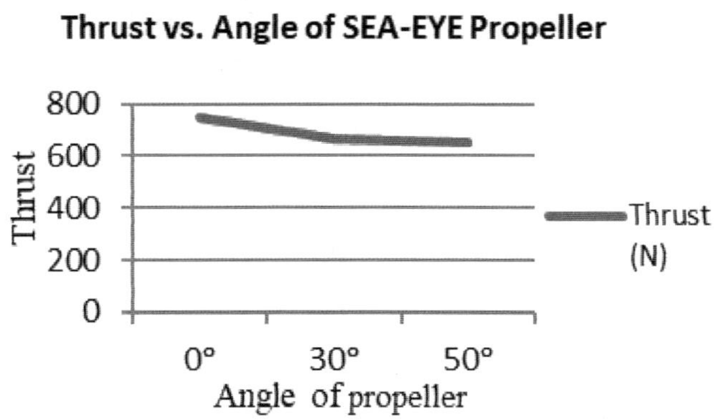


**Figure 6.5.** Thrust vs. Angle of SEA-EYE Propeller

## 7. Conclusion and recommendation

This study has analysed the relation between propulsion angle and thrust generation of SEA-EYE. The analysis has assisted by Ansys Workbench (Fluid Flow CFX) simulation and the drawing with Catia. For the objective one has achieved. In Catia software all parts of SEA-EYE were drawn using it. Then, the boat drawing transfer to Ansys software. In Ansys Workbench software have many systems to analysis any problems. For example, the systems are Fluid Flow CFX, Fluid Flow Fluent, etc. For this project only use system Fluid Flow CFX. The objective number four has achieved with use this software.

This analysis, describe about relation between pressure, velocity and thrust of SEA-EYE boat. The results show details information. The differences of high and low values of parameters are very important to make the right decision. By doing this analysis, we can know more about the boat, pressure, velocity and more. Besides, the causes and why it is so it can be identified.

As a conclusion of this study, the selection of the angle of propeller is $0^{\circ}$ is good because before making any decision expires research has been made as far as possible. For other two angle $30^{\circ}$ and $50^{\circ}$ also good to use but both is not much suitable for this problem.

**Acknowledgments**

TUSB Research UniMAP supports this research under the research grant 2017/08/0006.

**References**

[1] Hasvold, O., and Storkersen, N. (2001). Electrochmical power sources for unmanned underwater vehicle used in deep sea survey operation. Journal of Power Sources, 96, 252-258.

[2] Mantia, M.L., and Dabnichki, P. (2011). Influence of the wake model on the thrust of oscillating foil. Engineering Analysis with Boundary Elements, 35, 404-414.

[3] Wang, X., Shang, J., Luo, Z., Tang, L., Zhang, X., and Li, J. (2012). Reviews of power systems and environmental energy conversion for unmanned underwater vehicles. Renewable and Sustainable Energy Reviews. 16, 1958–1970

[4] Robert, Y., Leroyer, A., Barreb, S., Rongere, F., Queutey, P., and Visonneau, M. (2014). Fluid Mechanics in rowing: the case of the flow around the blades. Procedia Engineering, 72, 744 – 749.

[5] Sorensen, A. J., and Smogeli, O. N. (2009). Torque and power control of electrically driven marine propellers. Control Engineering Practice, 17, 1053–1064.

[6] Carreno, J. E., Mora, J. D., and Perez, F. L. (2013). Mathematical model for manoeuvrability of a riverine support patrol vessel with a pump-jet propulsion system. Ocean Engineering, 63, 96–104.

[7] Ashraf, M. A., Young, J., and Lai, J. C. S. (2011). Reynolds number, thickness and camber effects on flapping airfoil propulsion. Journal of Fluids and Structures, 27, 145–160.

[8] Horisawa, H., Sumida, S., Yonamine, S., and Funaki, I. (2013). Thrust generation through low-power laser-metal interaction for space propulsion applications. Vacuum, 88, 75-78.

[9] Hayati, A.N., Hashemi, S.M., and Shams, M. (2013). A study on the behind-hull performance of marine propellers astern autonomous underwater vehicles at diverse angles of attack. Ocean Engineering, 59, 152–163.

[10] Gierusz, W., and Tomera, M. (2006). Logic thrust allocation applied to multivariable control of the training ship. Control Engineering Practice, 14, 511–524.

[11] Xiao, Q., and Liao, W. (2010). Numerical investigation of angle of attack profile on propulsion performance of an oscillating foil. Computers & Fluids, 39, 1366–1380.

[12] Harmon, F. G., Frank, A. A., and Joshi, S. S. (2005). The control of a parallel hybrid-electric propulsion system for a small unmanned aerial vehicle using a CMAC neural network. Neural Networks, 18, 772–780.

[13] Coletti, M. (2012). A thrust formula for an MPD thruster with applied-magnetic field. Acta Astronautica, 81, 667–674.

[14] Grecenko, A. (2010). Thrust and slip of a low-pressure tire on compressible ground by the compression-sliding approach. Journal of Terramechanics, 47, 249–259.

[15] Liao, L., and Pasternak, I. (2009). A review of airship structural research and development. Progress in Aerospaceace Sciences, 45, 83–96.

# High Enhancement of Unidirectional UWB Array Antenna's Gain Using Flat Reflector Structure for Microwave Applications

**M A Jamlos[1], W A Mustafa[1], Syed Zulkarnain Syed Idrus[2], Wan Khairunizam[3], Z M Razlan[3], Shahriman A B[3] and I Zunaidi[4]**

[1]Faculty of Engineering Technology, Kampus Sg. Chuchuh, Universiti Malaysia Perlis, 02100 Padang Besar, Perlis, Malaysia.
[2]Center of Excellence Geopolymer and Green Technology, Universiti Malaysia Perlis, 01000 Kangar, Perlis, Malaysia.
[3]Advanced Intelligent Computing and Sustainability (AICoS) Research Group, School of Mechatronics Engineering, Universiti Malaysia Perlis, 02600 Arau, Malaysia.
[4]Faculty of Technology, University of Sunderland, St Peter's Campus, Sunderland, SR6 0DD, United Kingdom

mohdaminudin@unimap.edu.my

**Abstract.** UWB array antenna with the high enhancement of gain is proposed. New additional flat reflector with a distance of 20 mm enhances the proposed antenna gain. UWB antenna performance with reflector is better than UWB antenna without reflector structure. Antenna with reflector recorded wider operating bandwidth; 1.9GHz-10.6GHz with higher gain ranged from 3.5dB until 14.2dB compared with without reflector antenna operated from 2.2GHz to 10.6GHz with lower gain ranges of 1.9dB to 9.8dB. The reflector structure enhanced the antenna gain by 50% where the average gain throughout operating frequency for an antenna with reflector and antenna without a reflector is 10.2dB and 6.8dB respectively.

**Keywords:** Flat reflector, Ultra-wide band, Array antenna, High gain, Unidirectional antenna

## 1. Introduction

Microwave imaging could be defined as seeing the internal structure of an object by exposing electromagnetic fields at microwave frequencies of 300MHz- 30GHz [1]. It is done by producing images consist of distributed electrical property in the human body called human microwave imaging. Techniques for microwave imaging generally divided by three major parts involving active, passive and hybrid techniques. As for active microwave imaging, several antennas transmit microwave signals into tissues and reflected signals known as backscattered signals are used to generate microwave images. Active microwave imaging can be further separated into ultra-wideband (UWB) radar imaging and tomography imaging. At first, UWB radar imaging technique is introduced purposely only for breast tumor detection [2] and then followed by emerging modification technique used comparable principles technique for stroke and tumor detection in the brain later [3]. Radar-based techniques are preferable since they only needs to identify strong scattering point caused by inclusions with high frequent occasion [4]. This kind of technique is less complicated which involve less sophisticated algorithm of delay and sum confocal microwave imaging algorithm. This algorithm is the most common algorithm used in ultra-wideband radar-based microwave imaging technique.

Currently, microstrip UWB antennas have to gain great interested from researches due to their advantages such as simple structure, low profile, high data rate, easy integration with monolithic microwave integrated circuits (MMICs) and simple to fabricate. Thus, UWB antenna has turned to be the ideal candidate for future short-range (10 m) high-data wireless communication applications, peer-to-peer ultra-fast communications and a lot of other applications. This has inspired researchers to thoroughly investigate the design of UWB antennas [5]. Due to great benefits provided by UWB technology, various fields and domains with potential applications are deployed using this kind of technology such as in communications, imaging, radar, landmine detection, localization and biomedical systems [6, 7]. Application of microwave energy in the medical field is currently gaining interest by the research community where microwave-based systems have the high potential of being simple, safe, portable and cost-effective.

Due to high gain and better return loss offered by UWB antenna that well suitable for medical applications, the utilization of UWB antenna promising significant function for detection purpose compared with most of the conventional compact antennas which have low gain and poor return loss. In microwave imaging, high gain and good directional beam-width for wide frequency band antennas are really necessary for target detection, localization systems and cancer screening applications [8]. High gain and wide operated bandwidth antenna that boosts the focus of the electromagnetic energy are crucial in microwave imaging for interacting and penetrating the multilayer structures with different characteristics [9] contributes in providing images with high details. Clear and sharp images lead to accurate diagnostic and avoid false treatment given to the patient.

Currently, most of the UWB antenna structures explained in literature are consist of single patch element exhibited nearly omnidirectional radiation patterns and considerably low gain. [10]. Furthermore, a single UWB antenna has been explored comprehensively before and already established worldwide. According to [11, 12], most of the compact UWB antennas demonstrated omnidirectional radiation patterns with moderately low gain and noticeable distortion within the impulse response. UWB arrays antenna can be considered to be applied once the directional radiation pattern and higher value of antenna gain are required [10]. Apart from the directional and high gain antenna, UWB array antenna also required to overcome the disadvantages of microstrip single antenna which is narrow bandwidth with low efficiency [13]. Partial ground technique and present of parasitic element applied in the design lead to the realization of UWB array characteristic. The arrays structures are very effective in capture the scattered signal that can be analyzed to solve the inverse problem and obtain the constructed images of the preferred tissues [14].

Apart from array structure and coaxially fed, the introduction of additional copper reflector contribute significantly in increasing the proposed antenna gain in order to penetrate multi-structure human head so that the signal could reach the tumor inside the brain. It is done by reflecting the backward radiation towards frontwards radiation and hence increases the gain by summing up the original frontwards radiation with reflected one. Reflector in the antenna is a structure that reflects electromagnetic waves towards the desired direction. Reflectors also work to stop undesirable signals from interfering either the side or back of the antenna [15]. The reflector is extremely necessary for single direction transmission and reception or maximum gain possible required. Reflectors could be either a standalone device or be integrated with the antenna as a whole structure. Standalone reflector function is to redirect electromagnetic (EM) energy where corner reflector is one of the most common standalone reflector types. This kind of reflector generally used for radar application as a passive repeater.

IOP Conf. Series: Materials Science and Engineering **557** (2019) 012050 doi:10.1088/1757-899X/557/1/012050

## 2. Antenna Design

### *2.1 Antenna Design Simulation*

Fig. 1 demonstrated the simulated design of UWB array antenna using taconic (TLY-5) with a dielectric constant of $\varepsilon r = 2.2$, a thickness of $t = 1.5748+/-0.02$ and tangent loss of $\tan \delta = 0.0009$ as the substrate. The antenna is printed with 4×1 copper radiating patch array properly connected with quarter wave transformer transmission line associated with a copper parasitic element for the front side as shown in Fig. 1(a). The patches comprise of four identical circulars with a diameter of 15 mm. Parasitic element is placed on very close to feeding line with the gap only 0.2 mm. As shown in Fig. 1 (a), each quarter-wave transmission line has its own specific wide dimension for 50Ω, 70.71Ω and 100Ω to ensure equal current distribution towards all four patches could be realized. Quarter-wave transformers of 70.71 Ω are used to have an ideal match between the 100 Ω lines and the 50 Ω lines [13].

On the other hand, Fig. 1(b) shows the copper partial ground plane is printed at the back side with 50Ω SMA connector coaxially fed in the middle of the lower part of the antenna back where the signal is fed directly to the radiating patch. Meanwhile Fig. 1 (c) and (d) show the copper reflector with 20 mm gap functioned to reduce the side lobe and realized the directional antenna which has higher gain property by reflecting the backward radiation towards frontwards radiation. The size of the reflector is similar to the size of the sensor.

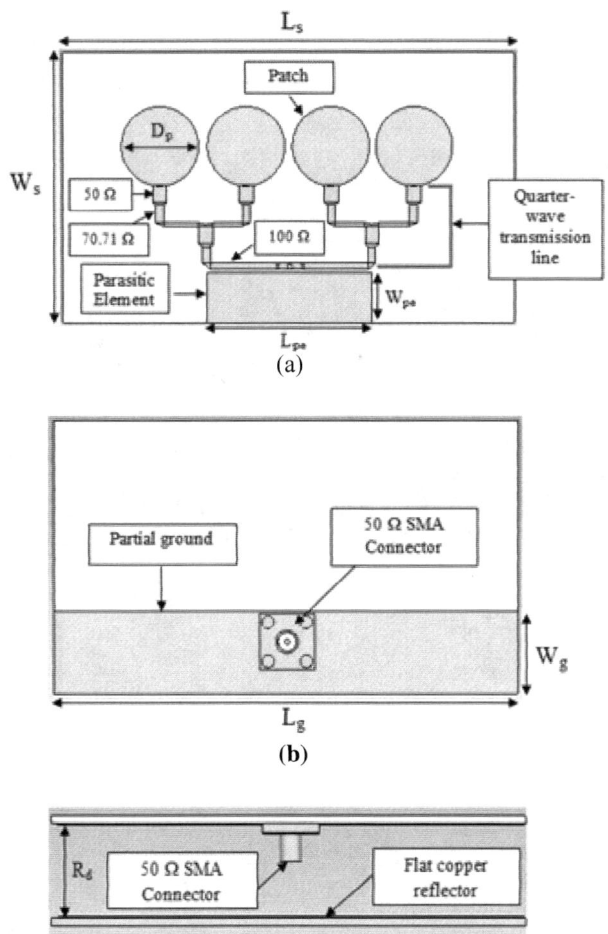


(c)

**Figure 1.** Quarter wave transformer The simulated geometry of the proposed UWB array antenna, a) front view, b) transparent back view (without reflector) c) top view

IOP Conf. Series: Materials Science and Engineering **557** (2019) 012050 doi:10.1088/1757-899X/557/1/012050

All important parameters of the designed antenna are optimized to obtain the best result in term of compact size, high gain and wide bandwidth. The optimized dimensions for the antenna are listed in Table 1. Measurements of gain, patterns and s-parameter have been performed using the setup consisting of Agilent ENA 8051C and anechoic chamber. The horn antenna is used as the transmitting antenna while antenna under test (AUT), UWB array as the receiver with a distance of 0.84 m between them.

**Table 1.** Optimized Antenna Parameter

| Symbol | Quantity |
|---|---|
| Ls | 90.0 mm |
| Ws | 45.0 mm |
| Dp | 7.50 mm |
| Wpe | 8.00 mm |
| Lpe | 32.0 mm |
| Rd | 20.0 mm |
| Wg | 18.0 mm |
| Lg | 90.0mm |
| Wr | 90.0 mm |
| Lr | 45.0 mm |
| Ls | 90.0 mm |
| Ws | 45.0 mm |

*2.2 Antenna Fabrication*

Fig. 2 shows the comparison image between the fabricated UWB antenna with reflector and UWB antenna without reflector as shown in Figure 2(a) and (b) respectively.

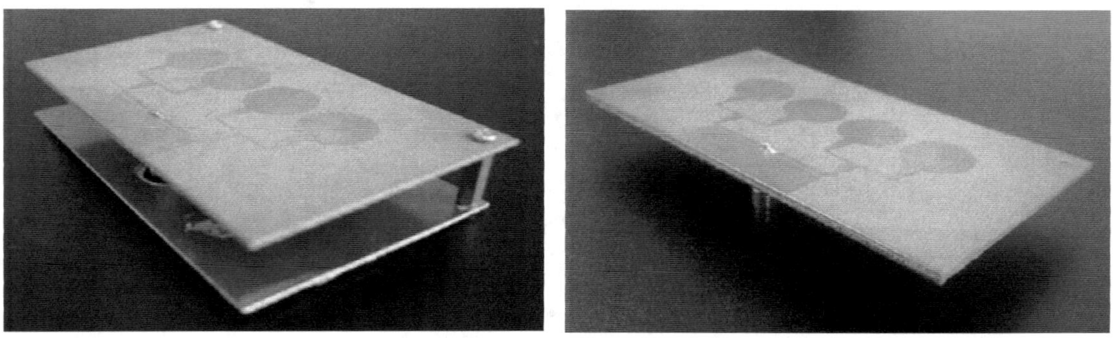

(a) (b)

**Figure 2.** Fabricated UWB-Array antenna, a) with reflector, b) without reflector

On the other hand, Figure 3 demonstrated the geometry of the fabricated UWB array antenna with reflector in detail.

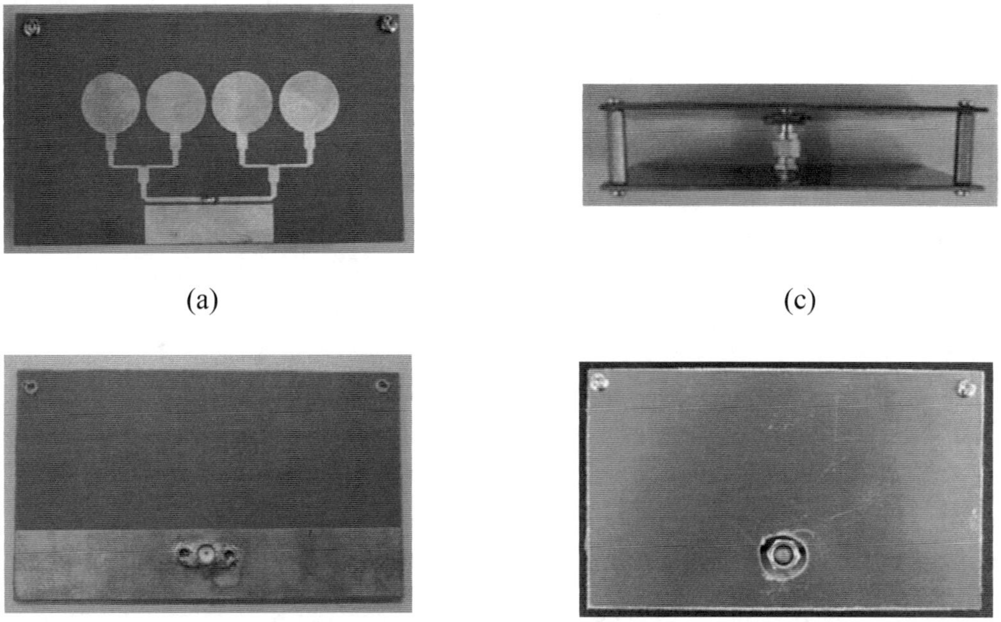

(a) (c)

(b) (d)

**Figure 3.** The fabricated UWB array antenna with reflector. a) front view, b) transparent back view (without reflector), c) top view, d) back view

## 3. Results and Discussions

Simulation and measurement results for both antennas with reflector and without reflector in term of reflection coefficient are shown in Fig. 4. Both simulation and measurement results for each antenna fulfilling the requirement for UWB characteristic; 3.1 GHz until 10.6 GHz. The antenna with reflector performance is better than the antenna without reflector performance. From the figure, it shows the antenna with reflector recorded a wider range bandwidth of operating frequency compared with an antenna without reflector for both simulated and measured results. For simulated, an antenna with reflector recorded UWB operated frequency started from 1.8 GHz until 10.8 GHz while antenna without reflector started from 2.0 GHz until 10.8 GHz. On the other hand, an antenna with reflector and without reflector recorded 1.9 GHz until 10.6 GHz and 2.2 GHz until 10.6 GHz respectively for measured results. Reflection coefficient less than -10 dB is selected due to the condition where 90% of the signals are successfully transmitted while only the left 10% is reflected back [17]. Partial ground technique and additional of parasitic element assure lower reflection coefficient achieved.

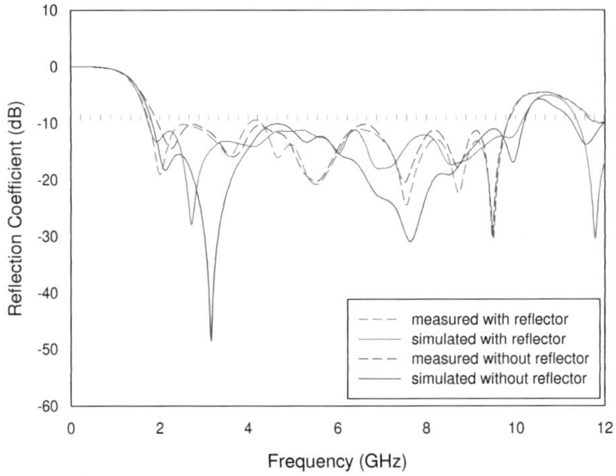


**Figure 4.** Simulated and measured reflection coefficient for UWB array antenna with reflector and without reflector

The antenna with reflector exhibits higher gain than antenna without reflector over the whole operated frequency both for simulated and measured as depicted in Fig. 5. Simulated and measured results recorded almost similar readings and considered as an adequate agreement between them. The measured gain for an antenna with reflector and without reflector ranged from 3.5 dB until 14.2 dB and 1.9 dB until 9.8 dB respectively. The reflector structure enhanced the antenna gain by 50% where the average gain throughout operating frequency for an antenna with reflector and antenna without a reflector is 10.2 dB and 6.8 respectively. The significant gain enhancement is done with the function of reflector in reflecting the backward radiation towards frontwards radiation and executed as a unidirectional beam associated with high gain.

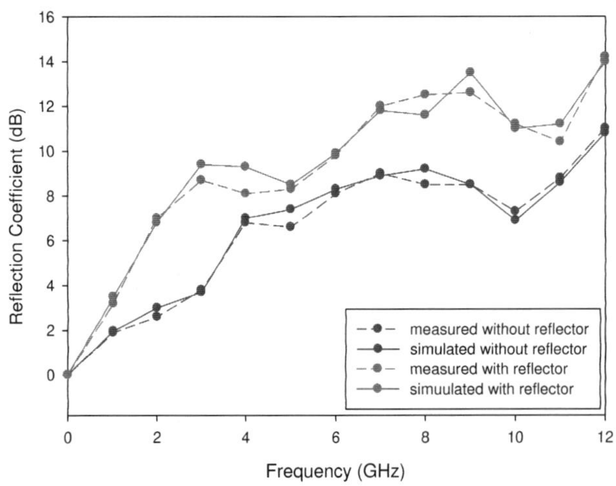


**Figure 5.** The simulated and measured gain for UWB array antenna with reflector and without reflector

In this design, one of the efficient method to obtain the directional sensor and optimize the gain while maintaining the UWB frequency is by adjusting the reflector distance (Rd) as shown in Fig. 6.

Meanwhile, Fig. 7 illustrates that the proposed UWB array's reflection coefficient differ significantly among the various Rd. Compared to other Rd, which is 5 mm, 10 mm, 15 mm, 25 mm and 30 mm, only Rd = 20 mm successfully achieves the reflection coefficient within the targeted operating frequencies. Hence, the reflector with a specific distance of Rd = 20 mm has been chosen in the UWB sensor design with a dimension of 90 mm × 45 mm.

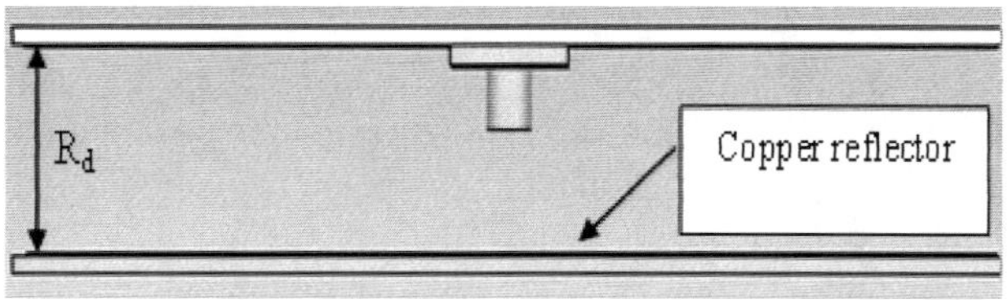


**Figure 6.** Reflector distance for UWB array antenna (top view)

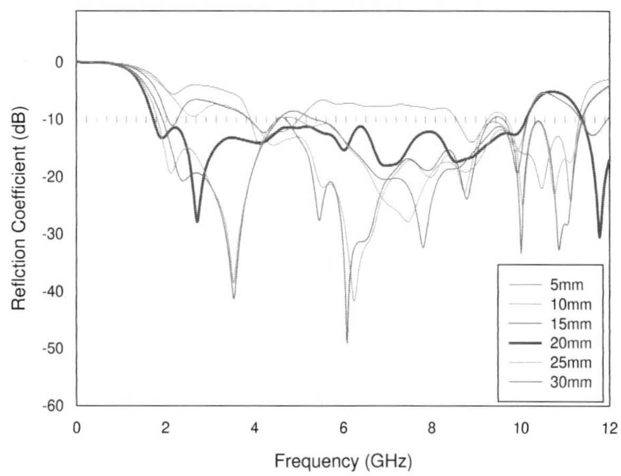


**Figure 7.** The reflection coefficient of various Rd for UWB array antenna

Instead of bandwidth and gain, the radiation pattern is the other essential parameter to evaluate the proposed antenna. The measurement and simulated radiation pattern results of Azimuth-Plane for both antennas are shown in Fig. 8. The figure demonstrates the polar radiation pattern for the proposed antenna at the frequency of 2 GHz and 3 GHz that are essential frequencies for brain microwave imaging application [1]. The radiation pattern indicates both antenna with reflector and without reflector does radiate over a wide frequency band [12] where antenna with reflector recorded better radiation pattern indicated by the wider area covered especially the main lobe. Antenna with reflector also could be categorized under unidirectional antenna since the radiated wave radiates more on the main lobe compared to the side and back lobe. In addition, simulation and measurement results indicate acceptable agreement between them. The sensor exhibits the averaged total radiation efficiency of 95% and 87% for an antenna with reflector and antenna without reflector respectively. The total radiation efficiency of the antenna promising the good probe or sensor for human brain microwave imaging. The pattern, gain and frequency ranges of the antenna are the parameters as a good probe or sensor for human brain microwave imaging.

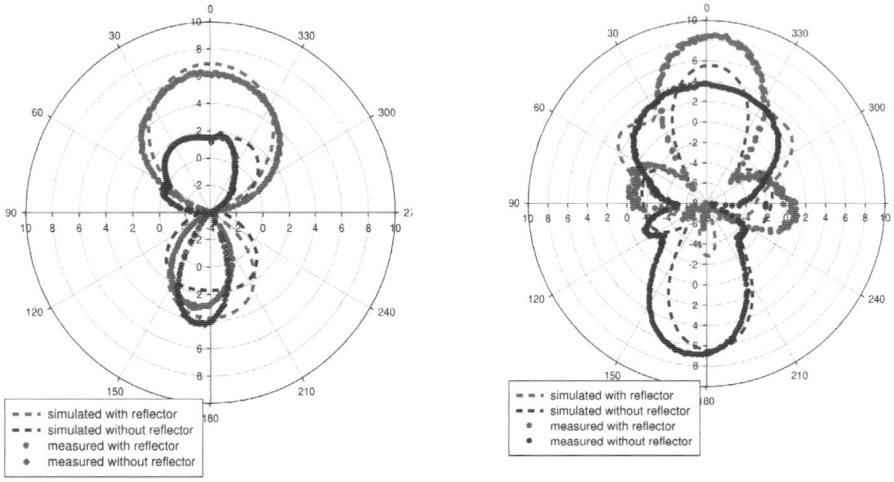


(a) (b)

**Figure 8.** Measured and simulated polar radiation pattern in Azimuth plane for an antenna with reflector and without reflector; (a) 2 GHz and (b) 3 GH

## 4. Conclusion

A UWB array antenna with the high enhancement of gain for brain microwave imaging is proposed. Additional flat reflector structure contributes significantly to improving the overall antenna performance especially the gain and radiation pattern. The antenna with reflector demonstrated wider reflection coefficient of less than -10dB started from 1.8 GHz until 10.6 GHz with the ranges of gain between 3.5 dB to 14.2 dB as compared with an antenna without reflector that recorded 2.2 GHz to 10.6 GHz and 1.9 dB to 9.8 dB for the reflection coefficient and gains respectively. Moreover, measurement results have identical behavior compared to simulations. Thus, in a nutshell, the proposed antenna is very suitable to be implemented in brain microwave imaging applications due to high gain and wide bandwidth.

## References

[1] K. M. Chew, R. Sudirman, N. H. Mahmood, N. Seman, and C. Y. Yong, "Human Brain Microwave Imaging Signal Processing: Frequency Domain (S-parameters) to Time Domain Conversion", Engineering, vol. 5, p. 31. 2013

[2] S. C. Hagness, A. Taflove, and J. E. Bridges, "Two-dimensional FDTD analysis of a pulsed microwave confocal system for breast cancer detection: fixed-focus and antenna-array sensors", Biomedical Engineering, IEEE Transactions, 45(12), 1470-1479.1998

[3] D. Ireland, and M. E. Bialkowski, "Microwave head imaging for stroke detection", Progress In Electromagnetics Research M, 21, 163-175. 2011

[4] L. M. Camacho and S. Tjuatja, "FDTD simulation of microwave scattering from a lung tumor," Antennas and Propagation Society International Symposium, IEEE, pp. 815-818. 2005

[5] A. K. Gautam, S. Yadav, and Binod Kr Kanaujia, "A CPW-Fed Compact UWB Microstrip Antenna", IEEE Antennas and Wireless Propagation Letters, Vol. 12. 2013

[6] X. N. Low, Z. Chen, and S. P. Terence, "A UWB Dipole Antenna with Enhanced Impedance and Gain Performance", IEEE Transactions on Antenna and Propagations, Vol. 57, No. 10,

October. 2009

[7] A. Eusuola, Y. Chen, and G. Y. Tian, "Novel Ultra-Wideband Directional Antenna for Microwave Breast Cancer Detection", AP-S/ URSI. 2011

[8] A. J. Jasmine and J. M. T. Anita, "Design of Vivaldi Antenna for Brain Cancer Detection", International Conference on Electronics and Communication Systems (ICECS -2014), Coimbatore, INDIA, Feb.13 -14. 2014

[9] F. S. G. B. Barnes, "Handbook of biological effects of electromagnetic fields", Boca Raton: CRC/Taylor & Francis. 2007

[10] M. Garbaruk, "Analysis of ultra-wideband linear antenna arrays", (Electrical Review), ISSN 0033-2097, R. 88 NR 8. 2012

[11] A. M. Abbosh, "Directive Antenna for Ultrawideband Medical Imaging Systems", International Journal of Antennas and Propagation. 2008

[12] N. Ojaroudi, M. Ojaroudi, F. Geran, and Sh. Amiri, "Omnidirectional Antenna with Multi-Resonance Performance for Microwave Imaging Systems", Telfor Journal, Vol. 5, No. 2. 2013

[13] N. A. Wahab, Z. Maslan, W. Norsyafizan W. Muhamad, N. Hamzah, "Microstrip Rectangular 4x1 Patch Array Antenna at 2.5GHz for WiMax Application", in Second International Conference on Computational Intelligence, Communication Systems and Networks, Liverpool, pp. 164-168. 2010

[14] A. F. Sheta, I. Elshafiey, A. Mohra, Z. Siddiqui, and A. R. Sebak, "A Compact Antenna for Microwave Imaging and Hyperthermia Treatment of Brain Tumor", IEEE, 978-1-4673-0292-0112. 2012

[15] A. Imani, Soleimani, M. Soleimani, S. Amiri, "Analysis of Dual-Reflector Antenna for Radar Applications", Applied Computational Electromagnetics Society Journal, Vol. 30, Issue 12, p1294-1300. 2015

[16] L. J. Zhang, S. G. Choi, B. C. Ahn, J. H. Bang, D. H. Kim, Y. T. Choi, "Simplified Feed Prime Focus Reflector Antenna", Microwave Journal, Vol. 59, Issue 2, p108-114. 2016

[17] M. I. Jais, M. F. Jamlos, M. Jusoh, T. Sabapathy, M. K A. Nayan, A. Shahadah, H. A. Ramli, I. Ismail, F. A. A. Fuad, "1.575 GHz Dual-Polarization Textile Antenna (DPTA) for GPS Application", Wireless Technology and Applications (ISWTA), pp. 376-379. 2013

# Comparison Between Performance of De-ionized Water and $Al_2O_3$/De-ionized Water on Enhancing the Transportation of Heat in Small Space

**Nazih A Bin-Abdun[1,2], Z M Razlan[1, *], S A Bakar[1], C H Voon[3], W K Wan[1], I Zunaidi[4], M K Albzeirat[5] and N Z Noriman[6]**

[1]School of Mechatronic Engineering, Universiti Malaysia Perlis, Pauh Putra Campus, 02600 Arau, Perlis, Malaysia

[2]Seiyun Community College, Department Air-Conditioning, Hadhramoot, Yemen

[3]Institute of Nano Electronic Engineering, University Malaysia Perlis, Seriab, 01000, Kangar, Perlis, Malaysia

[4]Faculty of Technology, University of Sunderland, St Peter's Campus, Sunderland, SR6 0DD, United Kingdom

[5]School of Manufacturing Engineering, University Malaysia Perlis, Pauh Putra Campus, 02600 Arau, Perlis, Malaysia

[6]Faculty of Engineering Technology, Universiti Malaysia Perlis, UniCITI Alam Campus, Sungai Chuchuh, 02100 Padang Besar, Perlis, Malaysia

[a] Corresponding author: zuradzman@unimap.edu.my

**Abstract.** In this article, the influence of deionized water and $Al_2O_3$/deionized water nanofluid to cooling battery ambient temperature is shown in figure 1. The battery temperature is observed to be decrease as the $Al_2O_3$/deionized water nanofluids volume concentration and high flow rate is incremented. The $Al_2O_3$/deionized water nanofluid exhibits enhancement as compared to deionized water under laminar flow conditions. The 0.60 vol. % concentration of $Al_2O_3$ with 10g surfactant and 1 L/min flow rate gives the highest heat transfer rate value among all with 65 % higher as compared to deionized water at laminar flow was observed. It has been observed that 24 hr of ultra-sonication was the best duration in the presence of a surfactant, where it gives the best stability and improved thermal conductivity, this improvement is due to decrease of aggregates within nanoparticle.

## 1. Introduction

Transportation of heat is one of the significant manufacturing processes [1]. Long-established heat transfer fluids such as oil engine, ethylene glycol, pure water, etc, have not given enough enhancement for cooling systems because of their comparatively low thermal conductivity [2], [3]. Thence, the development of highly efficient working fluids is disbanding the disadvantages of conventional fluids has become one of the most significant priorities in the cooling systems. It was determined in the literature review that using nanofluids instead conventional fluids enhances the transportation of heat by incrementing the thermal conductivity of the nanofluid compared to the conventional fluids [4].

## 2. Experimental Apparatus

Figure 1 shows the experimental preparation diagram for measuring the heat transfer performance. The experimental devices employed involves fluid flow pipes, DC pump (Max. Flow: 9 L/Min), a flowmeter (1.2 L/Min), a stainless steel tank used of fluids ($Al_2O_3$/ deionized water or deionized water) storage (1 litre), two-Temperature Bath (one to control the fluid temperature (cooler, 20 °C), and another is for controlling the water temperature (heater, 40 °C), which represents heat load of battery, heat exchanger made from copper coils (two coils, length every one equal 2.2 m), ten-thermocouples (K-type, accuracy equal ±0.1 °C ) are installed on experiment parts (represented by red dots). The fluid was pumped to the heat exchanger in the test compartment (battery compartment) from a stainless-steel tank that the temperature was controlled by the temperature bath (cooler). A water that the temperature was being controlled from the temperature bath (heater) was pumped toward battery model in the battery compartment. Heat exchange occurs in the battery compartment. Differences in temperature among inlet and exit of fluid flow (deionized water or $Al_2O_3$/deionized water) were measured. Details of the conditions examined in Table 1 were explained.

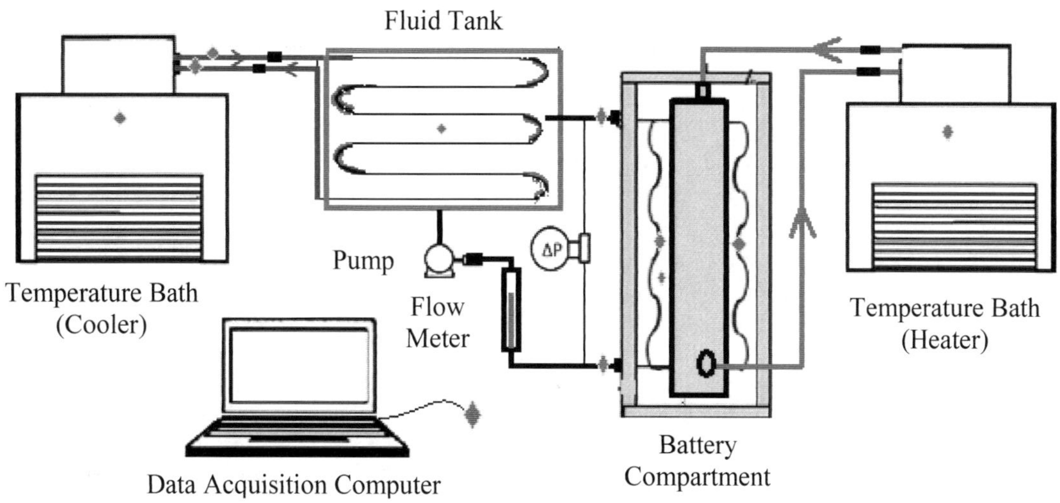


**Figure 1**. Schematic diagram for experimental setup system

A two-step method was utilized in the nanofluid preparation. Where $Al_2O_3$ nanoparticle were dispersed in deionized water by stirring the mixture (nanofluid) for 3 hours into bottle as shown in figure 2, and after that put the bottle in ultrasonic bath used to prevent aggregation of $Al_2O_3$ nanoparticle into deionized water for. And SDS surfactant was added to raise dispersion.

**Figure 2**. Magnetic stirring

IOP Conf. Series: Materials Science and Engineering **557** (2019) 012051 doi:10.1088/1757-899X/557/1/012051

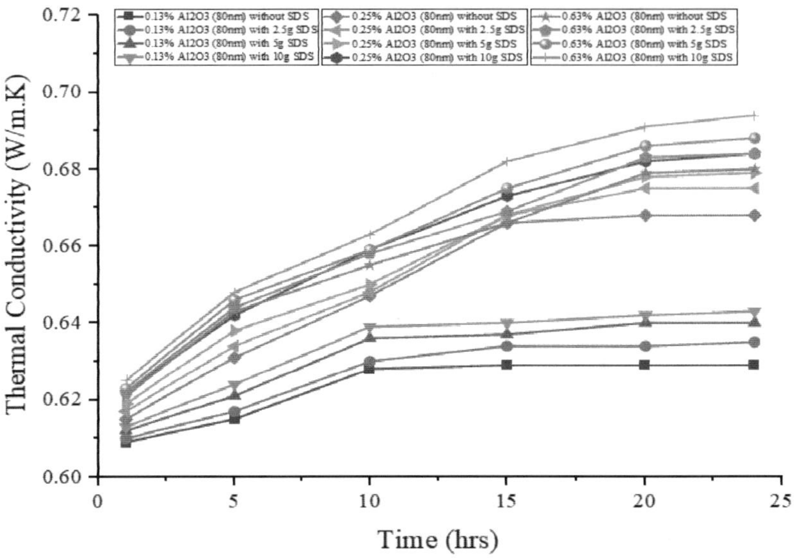


**Figure 3**. Thermal conductivity of $Al_2O_3$ /deionized water nanofluids without and with SDS as a function of sonication time and various volume concentration

Figure 3 represents the sonication time impacts of nanofluid was studied at different 6 times intervals, carried out to investigate the thermal conductivity of $Al_2O_3$/deionized water nanofluids with and without surfactants (sodium dodecyl sulphate) at three various volume concentrations (0.13%, 0.25% and 0.63%) with constant temperature equal 30 °C. The values obtained were presented that the thermal conductivity of nanofluids increases with increase of sonication time and to some extent with the quantity of surfactant. Where the thermal conductivity improvement was 9% with 0.63 % volume concentration and 10 g from surfactant for $Al_2O_3$ (80 nm)/deionized water nanofluid. But the difference of thermal conductivity values of $Al_2O_3$/deionized water nanofluid prepared with 1 to 10 hours of sonication time were high, while with 10 to 24 hours was low relative. It can be concluded from figure 3 that 24 hr of ultra-sonication is the best duration in the presence of a surfactant, this improvement is due to decrease of cluster sizes inside nanoparticle.

**Table 1.** Specifications and test conditions

| Elements | Specifications |
|---|---|
| Volume flow rate of the pump (Q) | 0.5, 0.8 and 1 L/min. |
| The surface area of copper tube (one coil) ($\mathbf{A_{Si}}$) | $0.0346\ m^2$ |
| Cross-sectional area of copper tube (one coil) ($\mathbf{A_{ci}}$) | $1.9635 \times 10^{-5} m^2$ |
| The ambient temperature in the compartment | 40 °C |
| Heat load of battery model | 91 W |
| Battery compartment size | $( 0.4 \times 0.31 \times 0.08 )\ m^3$ |

## 3. Shape and Size Nanoparticles

The morphology of the $Al_2O_3$ nanoparticle was evaluated through Scanning Electron Microscope (SEM) imaging [5]. The SEM analysis is conducted to characterize the morphology as shown in figure 4, the shape of nanoparticles was semi spherical and the mean size equal 85 ±5 nm. The nanoparticles ($Al_2O_3$) was purchased from US-Research Nanomaterial.

**Figure 4.** SEM image of dry $Al_2O_3$ nanoparticles

## 4. Experimental Apparatus

The values for thermophysical properties of the nanofluid (Al2O3/deionized water) at various volume concentrations are taken at the average temperature of fluid in heat exchanger.
The density of Al2O3/deionized water is the determined values by using Pak and Cho (1998) equation, while thermal conductivity, viscosity, and specific heat have been measured by using a KD2 Pro, viscometer (LVDV- Pro, Brookfield) and differential scanning calorimetry (DSC) respectively.

### *4.1. Influence of surfactant and particle concentrations with various temperature on thermal conductivity*

After checking the accuracy of the apparatus, the fluids were tested and thermal conductivity measured. Since nanofluid was prepared from deionized water and ($Al_2O_3$) nanoparticles. The thermal conductivity of fluids was measured for the same temperature range (20 to 50 °C) at atmospheric pressure.

Figure 5 shows the change in thermal conductivity of deionized water for different temperature. This graph indicates that as the temperature increases, there is an increase in thermal conductivity. The measurement carried out two time to check from relative error for the reading, where results showed that test 2 value error less.

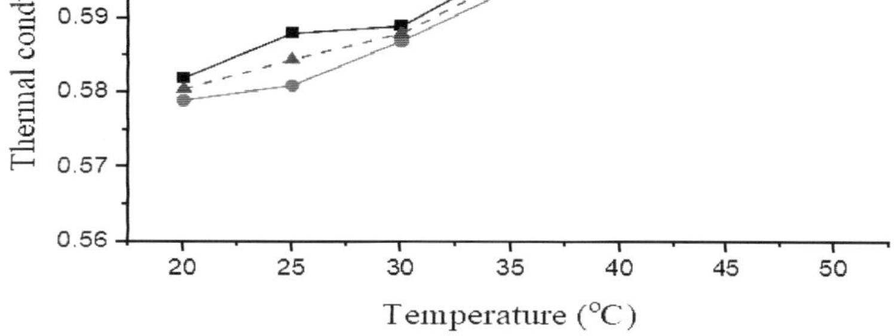


**Figure 5.** Thermal conductivity measurements for deionized water with a various temperature at atmospheric pressure

IOP Conf. Series: Materials Science and Engineering **557** (2019) 012051 doi:10.1088/1757-899X/557/1/012051

Figures from 6, 7 and 8 presents the experimental measured data change of thermal conductivity with the differences in the SDS surfactant (0g, 2.5g, 5g and 10g) and volume concentration of the $Al_2O_3$ nanoparticles. The experiment was repeated 3 times for each sample to obtain more accurate values and the average values were taken for analysis study. The theoretical data was calculated by (Hamilton and Crosser) equation. Also show that thermal conductivity of $Al_2O_3$/de-ionized water increase with temperature increment.

Figures 6 (a), (b), (c) and (d) show that the thermal conductivity of $Al_2O_3$/ de-ionized water, when volume concentration 0.13%, increases with the SDS surface a slight increase from 0 g to 10 g reach to 1%.

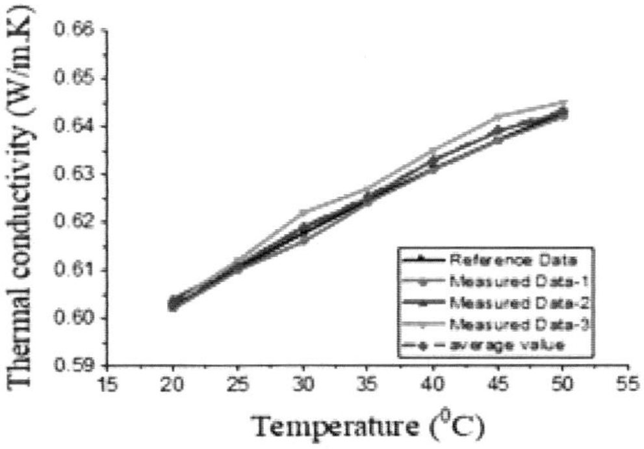


**(a) 0.13% $Al_2O_3$/deionized water without SDS**

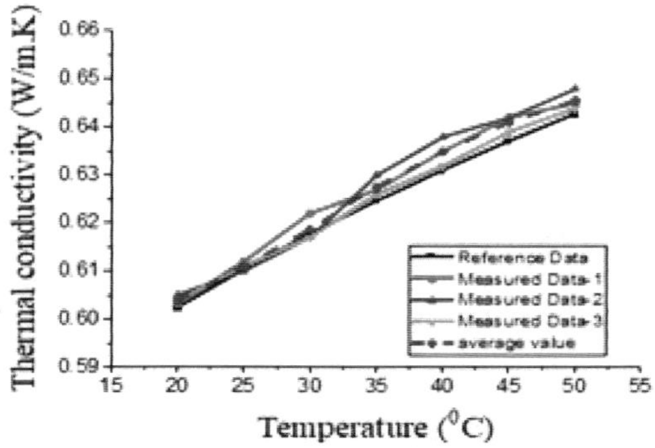


**(b) 0.13% $Al_2O_3$/deionized water with 2.5 g SDS**

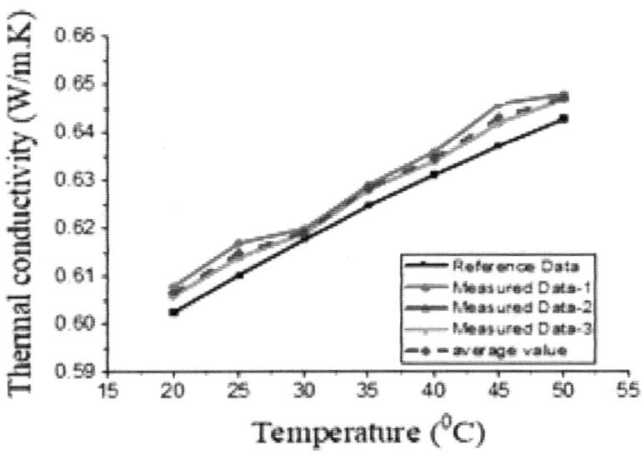


**(c) 0.13% $Al_2O_3$/deionized water with 5 g SDS**

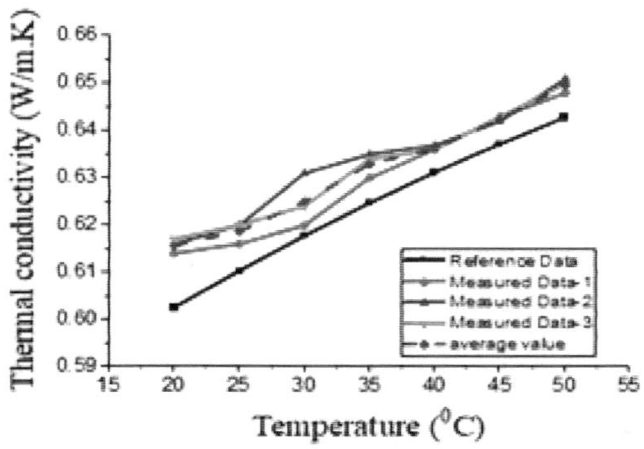


**(d) 0.13% $Al_2O_3$/deionized water with 10 g SDS**

**Figure 6.** Thermal conductivity measurements for $Al_2O_3$ (5g)/deionized water nanofluid with a various temperature at atmospheric pressure

Figures 7 (a), (b), (c) and (d) show that the thermal conductivity of $Al_2O_3$/ de-ionized water, when volume concentration 0.25%, increases with the SDS surface an increase from 0 g to 10 g reach to 5.6 %.

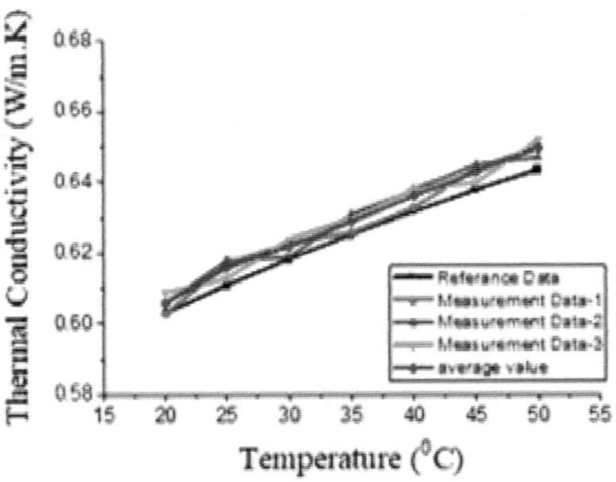


**(a) 0.25% $Al_2O_3$/deionized water without SDS**

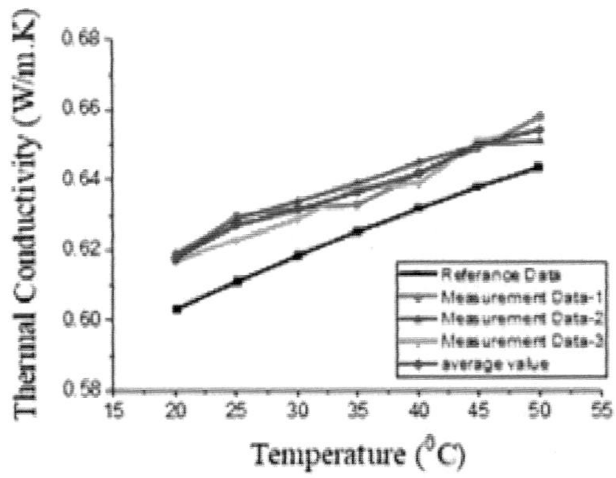


**(b) 0.25% $Al_2O_3$/deionized water with 2.5 g SDS**

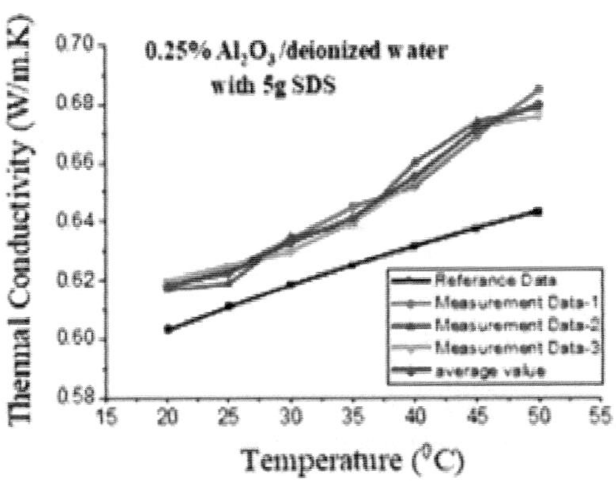


**(c) 0.25% $Al_2O_3$/deionized water with 5 g SDS**

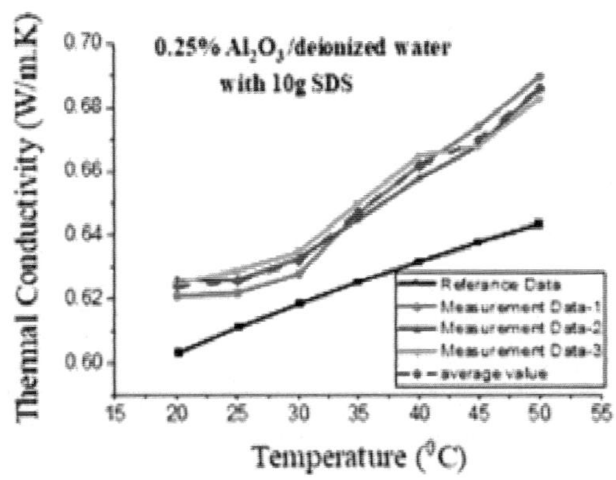


**(d) 0.25% $Al_2O_3$/deionized water with 10 g SDS**

**Figure 7.** Thermal conductivity measurements for $Al_2O_3$ (10g)/deionized water nanofluid with a various temperature at atmospheric pressure

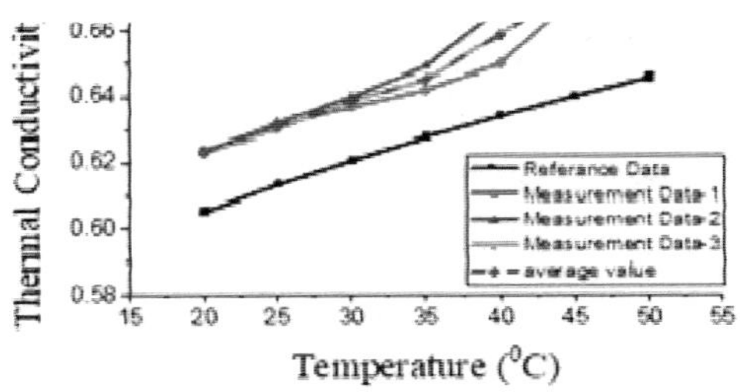


**(c) 0.60% $Al_2O_3$/deionized water with 5 g SDS**

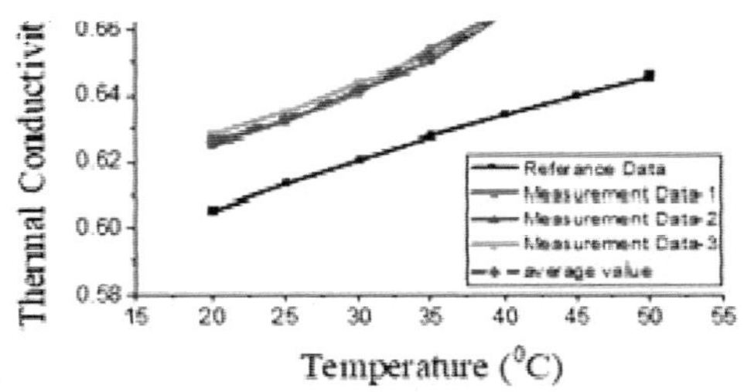


**(d) 0.60% $Al_2O_3$/deionized water with 10 g SDS**

**Figure 8.** Thermal conductivity measurements for $Al_2O_3$ (25g)/deionized water nanofluid with a various temperature at atmospheric pressure

### *4.2. Influence of surfactant and particle concentrations with various temperature on viscosity*

After checking the accuracy of the apparatus, the fluids were tested and dynamic viscosity measured. The dynamic viscosity of fluids was measured for the same temperature range (20 to 50 °C) with volume concentrations of 0.13, 0.25 and 0.60 vol. %. at atmospheric pressure. The experiments were conducted once per sample with a change in SDS.

Experiments displayed in figure 9 that nanofluid viscosity decreases with an increment in temperature and increase with an increasing volume concentration and SDS surfactant, and this increasing of nanofluid viscosity lead to incrementing the pressure drop, up made to consumed a larger pumping power.

Figure 9 shows that the maximum of viscosity of $Al_2O_3$/ deionized water was decreased when increase the temperature from 20 °C to 50 °C, reach to 41.9% in case 0.13 vol. % concentration. However, there are increasing of nanofluid viscosity due to the increase the nanoparticle concentration and surfactant, reach to 41.5% in 0.60 vol. % concentration and 10g SDS surfactant.

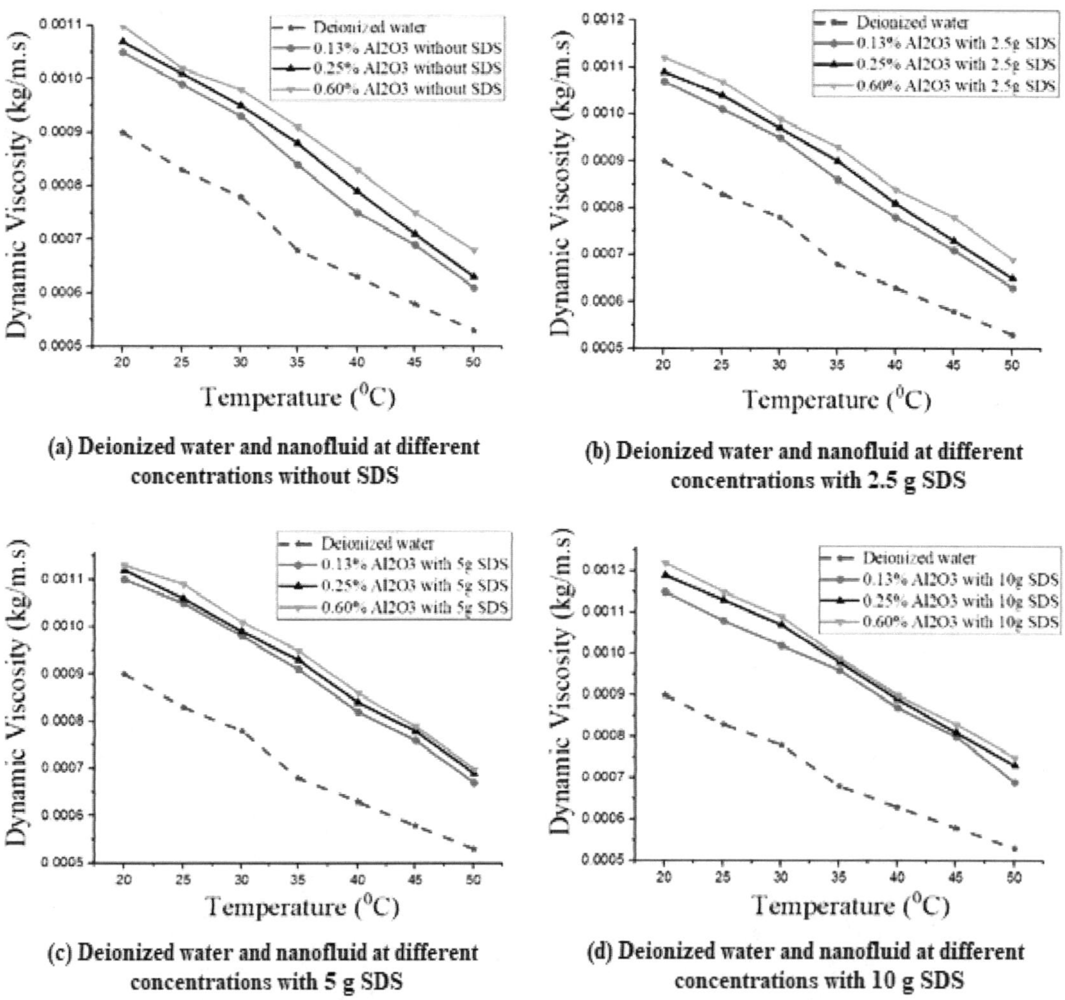


**(a) Deionized water and nanofluid at different concentrations without SDS**

**(b) Deionized water and nanofluid at different concentrations with 2.5 g SDS**

**(c) Deionized water and nanofluid at different concentrations with 5 g SDS**

**(d) Deionized water and nanofluid at different concentrations with 10 g SDS**

**Figure 9.** Dynamic viscosity measured for an $Al_2O_3$/de-ionized water nanofluid at different concentrations and temperatures with change of SDS

*4.3. Influence of surfactant and particle concentrations with various temperature on specific heat*

After checking the accuracy of the apparatus, the fluids were tested and specific heat measured. The specific heat of fluids was measured for the same temperature range (20 to 50 °C) at atmospheric pressure.
Figure 10, it is noted that the specific heat capacity of $Al_2O_3$/ de-ionized water decreases moderately with increasing temperature to 35 °C, then there happens a slight increase to 50 °C, and decreases substantially as high nanoparticle volume concentration, reach to 1.6% in case 0.60 vol. % concentration.

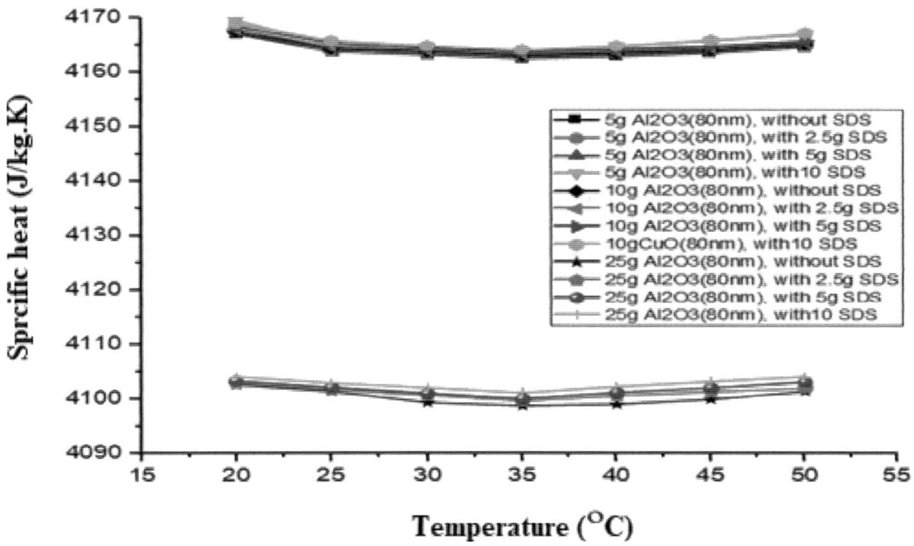


**Figure 10.** Specific heat measured for an $Al_2O_3$/deionized water nanofluid at different concentrations and temperatures with change of SDS

## 5. Calculation of Heat Transfer Procedure

In this paper employ a forced convective heat transfer experimental to study the transportation of heat effectiveness and it influence on pressure drop of $Al_2O_3$/deionized water nanofluid. To determine the Reynolds number, Nusselt number, heat transfer coefficient and power consumption, the following procedure has been performed:

The dimensionless Reynolds number (Re) and Prandtl number are:

$$Re = \frac{\rho_f V_m d_i}{\mu_f} \qquad (1)$$

$$Pr = \frac{Cp_f \times \mu_f}{K_f} \tag{2}$$

The following relation can be used to calculate the experimental Nusselt number:

$$Nu = \frac{h_{ex.}\, d_i}{k_f} \tag{3}$$

The heat transfer coefficient is determined by the relation [6], [7]:

$$h_{ex.} = \frac{\dot{m}\, Cp_f\, (T_{in} - T_{out})}{A_S\, \Delta T_{lm}} \tag{4}$$

The logarithmic mean temperature difference is calculated by the relation:

$$\Delta T_{lm} = \frac{\Delta T_h - \Delta T_c}{\ln\left(\frac{\Delta T_h}{\Delta T_c}\right)} \tag{5}$$

The formula of the peripheral area is given by

$$A_S = 2\,\pi\, d_i\, L \tag{6}$$

The heat transfer rate is calculated from the following formula [8], [9]:

$$Q = \dot{m}\, Cp_f\, \Delta T = \dot{m}\, Cp_f\, (T_{out} - T_{in}) \tag{7}$$

Pressure drop throughout the heat exchanger is specified by $\Delta p$, which is determined by pressure differential gauge as shown in Figure 1. Now, the pumping power is calculated from the following formula [10]:

$$\text{Pumping power} = \frac{\dot{m} \times \Delta p}{\rho_f}$$

Where, $\rho_f$ is the density of fluids, $V_m$ is the mean velocity of fluids, $A_c$ is cross section area of the tube, $d_i$ is inside diameter of the tube, $\mu_f$ is the viscosity of fluids, , k is the thermal conductivity of fluids, $T_{in}$ and $T_{out}$ are inlet and outlet temperatures of heat exchanger, $Cp_f$ is the specific heat of fluids, $A_s$ is the surface area of the tubes and $\Delta T$ is the temperature variation of the cooling fluids.

## 6. Results and Discussions

Firstly, before conducting methodical experiments on the application of Al2O3/deionized water nanofluid as fluid in the heat exchanger, some examinations were done on deionized water for check the heat exchanger and reliability of the experimental apparatus, and it comparison with $Al_2O_3$/de-ionized water nanofluid. The results of the experiments were obtained at constant inlet temperature of $20 \pm 0.5$°C.

Table 2, show the thermophysical properties measurement results after the preparation of the nanofluid directly. And noted that 0.13 vol. %, 0.25 vol. % and 0.60 vol. % with 10 g SDS surfactant were more stable, so will be done examinations on these concentrations only with 10 g SDS surfactant.

IOP Conf. Series: Materials Science and Engineering **557** (2019) 012051 doi:10.1088/1757-899X/557/1/012051

**Table 2.** Summary of thermophysical properties of fluids at bulk mean temperature with change concentrations and surfactant

| Thermophysical properties | Volume concentration (%) | SDS Surfactant (g) | Results | Notes |
|---|---|---|---|---|
| Thermal conductivity (W/m.K) | 0 | 0 | 0.579 | It is low of de-ionized water |
| | 0.13 | 0 | 0.603 | Increases with the raise of nanoparticle volume concentration, SDS surfactant and temperature. |
| | | 2.5 | 0.604 | |
| | | 5 | 0.60667 | |
| | | 10 | 0.61567 | |
| | 0.25 | 0 | 0.606 | |
| | | 2.5 | 0.61767 | |
| | | 5 | 0.61833 | |
| | | 10 | 0.624 | |
| | 0.60 | 0 | 0.61167 | |
| | | 2.5 | 0.62 | |
| | | 5 | 0.6233 | |
| | | 10 | 0.62633 | |
| Viscosity ( kg/m.s ) | 0 | 0 | 0.0009 | It is low of de-ionized water |
| | 0.13 | 0 | 0.00105 | Increases with the raise of nanoparticle volume concentration, SDS surfactant and decreasing with raise of temperature. |
| | | 2.5 | 0.00107 | |
| | | 5 | 0.0011 | |
| | | 10 | 0.00115 | |
| | 0.25 | 0 | 0.00107 | |
| | | 2.5 | 0.00109 | |
| | | 5 | 0.00112 | |
| | | 10 | 0.00119 | |
| | 0.60 | 0 | 0.0011 | |
| | | 2.5 | 0.00112 | |
| | | 5 | 0.00113 | |
| | | 10 | 0.00122 | |
| Specific heat (J/kg K) | 0 | 0 | 4183 | It is high of de-ionized water |
| | 0.13 | 0 | 4167.07 | There is a slight increase with the raise of SDS surfactant and significant decrease with high concentrations of nanofluid in deionized water. |
| | | 2.5 | 4167.44 | |
| | | 5 | 4168.05 | |
| | | 10 | 4169.61 | |
| | 0.25 | 0 | 4167.48 | |
| | | 2.5 | 4167.98 | |
| | | 5 | 4168.51 | |
| | | 10 | 4168.99 | |
| | 0.60 | 0 | 4102.68 | |
| | | 2.5 | 4102.98 | |
| | | 5 | 4103.38 | |
| | | 10 | 4104.18 | |

**Table 3.** Summary of results ($Al_2O_3$/ de-ionized water) at bulk mean temperature with change concentrations and surfactant

| Mass flow rate L/min | Volume Concentrations | SDS Surfactant | Reynolds number | Prandtl number | Heat transfer rate ($Q^{\cdot}$) W |
|---|---|---|---|---|---|
| 0.5 | 0 % | | 1178.65 | 7.33 | 53.5 |
| | 0.13% | 5 | 966.882 | 7.57 | 72.4 |
| | | 10 | 924.844 | 7.85 | 73.1 |
| | 0.25% | 5 | 953.144 | 7.65 | 77.2 |
| | | 10 | 897.077 | 8.1 | 77.6 |
| | 0.60% | 5 | 955.146 | 7.53 | 87.8 |
| | | 10 | 884.685 | 8.1 | 88.3 |
| 0.8 | 0% | | 1885.85 | 7.33 | 85.5 |
| | 0.13% | 5 | 1547.01 | 7.57 | 115.8 |
| | | 10 | 1479.75 | 7.85 | 117 |
| | 0.25% | 5 | 1525.03 | 7.65 | 123.5 |
| | | 10 | 1435.323 | 8.1 | 124.1 |
| | 0.60% | 5 | 1528.234 | 7.53 | 140.5 |
| | | 10 | 1415.496 | 8.1 | 141.3 |
| 1 | 0% | | 2357.31 | 7.33 | 106.9 |
| | 0.13% | 5 | 1933.76 | 7.57 | 144.8 |
| | | 10 | 1849.69 | 7.85 | 146.2 |
| | 0.25% | 5 | 1906.288 | 7.65 | 154.4 |
| | | 10 | 1794.154 | 8.1 | 155.1 |
| | 0.60% | 5 | 1910.293 | 7.53 | 175.6 |
| | | 10 | 1769.369 | 8.1 | 176.6 |

Experimental results showed in table 3 that the Reynolds number decremented with increasing volume concentration of nanoparticles and the SDS surfactant, and it was observed that Prandtl number increased with both increase of volume concentration and surfactant and not change with rising flow rate. And dispersion nanoparticles, Reynolds number and SDS surfactant on thermal small areas were investigated. Furthermore, the influence of flow laminar on the heat transfer rate was studied. Found that the heat transfer rate was incremented with rising flow rate, volume concentration of nanoparticle, SDS surfactant, and Reynolds number.

## 7. Conclusions

In this article, the heat transfer rate of fluids in the simulated battery cooling system (heat exchanger) has been measured experimentally. The fluids are base fluid (de-ionized water) and nanofluid ($Al_2O_3$/de-ionized water) at different volume concentrations without and with surfactants. The results are concluded as follows:

- The influence of ultra-sonication time where it was improved slowly for thermal conductivity and stability, due to used indirect ultra-sonic.
- It was observed that the nanofluid at the concentrations of 0.13 vol.% and 0.25 vol.%, had stabilized for two months after preparation when adding the 10g from SDS surfactant. While 0.60 vol.% had stabilized for 5 weeks with same SDS surfactant.
- The thermal conductivity values of nanofluids were found to be increased with increasing temperature and increasing with rising of volume concentration nanoparticles and surfactant.

- The viscosity values of nanofluids were found to be decreased with increasing temperature and increasing with rising of volume concentration nanoparticles.
- The specific heat values of nanofluids were found to be decreased with increasing temperature to certain point, and then rises slowly. And also, specific heat values of nanofluids slightly decreasing with small concentration nanoparticles and surfactant, while clearly decreasing with high concentration.
- Experimentally, it was observed that heat source (battery model) equal 91 W, when using the deionized water at 1 l/min flow rate gives improvement by 17.5%, while using $Al_2O_3$/de-ionized water nanofluid at same flow rate and 0.60% vol.% concentration with 10 g SDS surfactant gives improvement by 94%. This means that $Al_2O_3$/de-ionized water nanofluid gives better improvement than deionized water by 65%.

## References

[1] I. Manna, "Synthesis, characterization and application of nanofluid - An overview," *Journal of the Indian Institute of Science*, vol. 89, no. 1, pp. 21–33, 2009.

[2] N. A. Bin-Abdun, Z. M. Razlan, A. B. Shahriman, K. Wan, D. Hazry, S. Faiz Ahmed, N. H. Adnan, R. Heng, H. Kamarudin, and I. Zunaidi, "Base fluid in improving heat transfer for EV car battery," in *AIP Conference Proceedings*, 2015, vol. 1660.

[3] X. Wang, X. Xu, and S. U. S. Choi, "Thermal Conductivity of Nanoparticle - Fluid Mixture," *Journal of Thermophysics and Heat Transfer*, vol. 13, no. 4. pp. 474–480, 1999.

[4] I. M. Shahrul, I. M. Mahbubul, R. Saidur, and M. F. M. Sabri, "Experimental investigation on Al2O3-W, SiO2-W and ZnO-W nanofluids and their application in a shell and tube heat exchanger," *International Journal of Heat and Mass Transfer*, vol. 97, pp. 547–558, 2016.

[5] Babita, S. K. Sharma, and G. S. Mital, "Preparation and evaluation of stable nanofluids for heat transfer application: A review," *Experimental Thermal and Fluid Science*, vol. 79, pp. 202–212, 2016.

[6] K. Y. Leong, R. Saidur, M. Khairulmaini, Z. Michael, and A. Kamyar, "Heat transfer and entropy analysis of three different types of heat exchangers operated with nano fl uids ☆," vol. 39, pp. 838–843, 2012.

[7] C. J. Ho and W. C. Chen, "An experimental study on thermal performance of Al 2 O 3 / water nano fl uid in a minichannel heat sink," vol. 50, pp. 516–522, 2013.

[8] T. L. Bergman, F. P. Incropera, D. P. DeWitt, and A. S. Lavine, *Fundamentals of heat and mass transfer*. John Wiley & Sons, 2011.

[9] R. S. Vajjha, D. K. Das, and D. P. Kulkarni, "International Journal of Heat and Mass Transfer Development of new correlations for convective heat transfer and friction factor in turbulent regime for nanofluids," *International Journal of Heat and Mass Transfer*, vol. 53, no. 21–22, pp. 4607–4618, 2010.

[10] P. Taylor, M. Kahani, S. Z. Heris, and S. M. Mousavi, "Journal of Dispersion Science and Technology Effects of Curvature Ratio and Coil Pitch Spacing on Heat Transfer Performance of Al 2 O 3 / Water Nanofluid Laminar Flow through Helical Coils Effects of Curvature Ratio and Coil Pitch Spacing on Heat Transfer," no. November 2014, pp. 37–41, 2013.

## Acknowledgments

The authors fully acknowledged Universiti Perlis Malaysia for the support which makes this important research viable and effective.

# Experimental study of UAV propeller Thrust generation for various Propeller Blade-Span and the relation with mouth-ring

**M N K Othman[1]*, Z M Razlan[1], Z A Wahid[1], Shahriman A B[1], I Zunaidi[2], Wan Khairunizam[1] and N Z Noriman[3]**

[1]School of Mechatronic Engineering, Universiti Malaysia Perlis, Pauh Putra Campus, 02600 Arau, Perlis, Malaysia.
[2]Faculty of Technology, University of Sunderland, St Peter's Campus, Sunderland, SR6 0DD, United Kingdom
[3]Faculty of Engineering Technology, Kampus Sg. Chuchuh, Universiti Malaysia Perlis, 02100 Padang Besar, Perlis, Malaysia.

**Abstract.** In this study, the main focus is to investigate the thrust generation of various propeller blade width-spans with the relation of mouth-ring application. By using the parameter, the thrust was calculated based on the value of velocity, voltage and rpm. The testing was performed in two condition that is with mouth-ring and without mouth-ring and 2 size of propeller been tested in this experiment. The application of mouth-ring is based on the gap between the wall of mouth-ring and the blade tip. There are 3 size of gap; 10mm, 8mm and 5mm. The result shows that 10mm mouth-ring have better thrust generation other than two gap. In addition, if compare to with and without mouth-ring, the application of mouth-ring also has proved it has a better thrust. Furthermore, the application of mouth-ring is evaluating to be good base on this study.

## 1. INTRODUCTION

Electric UAVs continue to be popular applications in roles ranging from traffic monitoring to filming and others. Propeller is use to generate thrust for a comfortably lift and to generate thrust under most flight conditions is well understood due to this long history of use and development.

The study requires evaluating the thrust generation of the propeller as the various width-span of the propeller blade being tested. If the thrust increase, it can help to lift the UAV comfortably. We can withstand the problem by reducing the power usage. In these applications, UAV needed large amount of required thrust force to perform a good lifting and maintain a good stability. It is important that enough force need to be produced in order to stabilized the UAV.

With various width-span of the propeller, we can get a suitable blade based on the result. Thrust can be changed by either changed the width-span or changing the rotational rate of the motor drive. As for the relation with mouth-ring, the study included the materials of the mouth-ring made of. Heavy material will likely affect the power usage and the thrust required in order to lift up the UAV. In this case plastic and carbon fibre been used to study the relation between materials and various width-span of the propeller blade.

## 2. METHODOLOGY

### *2.1. Project development Flow*

At the first stage of the testing, the experimental have been setup. The apparatus has been connected to each device. Two size of propeller use in this experiment; 10 inch and 9 inch. The exact voltage is used to run the propeller: 4V, 7V, 9V and 11V.

IOP Conf. Series: Materials Science and Engineering **557** (2019) 012052 doi:10.1088/1757-899X/557/1/012052

Anemometer and tachometer will be used to take the velocity and rpm value on the rotating propeller. The reading was taken to make the calculation to know the value of the thrust generate.
Next, the experiment repeated using 10-inch propeller with the application of 3 mouth-rings. Each of the mouth-ring have the gap between 5mm, 8mm and 10mm. Same step of taking the data used in this mouth-ring application test.

Since all the tested have been done, the thrust value is being calculated and compares the data based on the thrust value. A few graphical explanations needed to show the data comparison between the application of using mouth-ring and without using mouth-ring.

### *2.2. Design Modelling*
In this study, the mouth-ring is made of using a tube that is made to ring. Each of the mouth-ring inner diameters is 27.4cm, 27cm and 26.4 cm.

The size of the mouth-ring is based on the gap size between mouth-ring and blade tip.

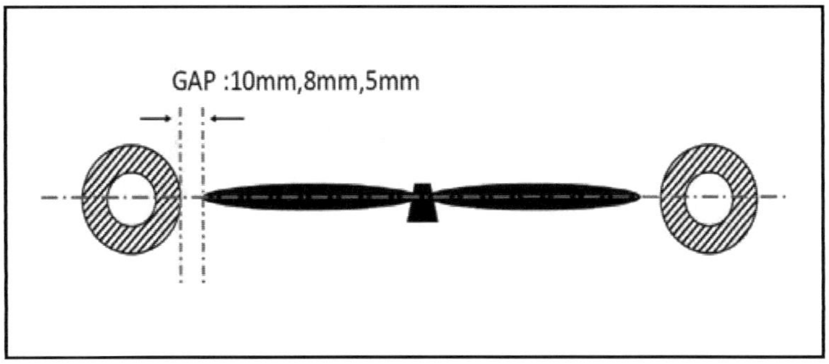


**Figure 1.** The gap of the mouth-ring and the blade tip

## 3. SELECTION OF PART
For the part, several devices were used in order to make the testing. The device is brushless motor 1000kV, electric speed controller (ESC), remote control, receiver, anemometer, and tachometer and dc power generator.

There are two size of propeller used that is made of plastic used in this test.

**Table 1.** Propeller diameter.

| Material | Diameter |
|---|---|
| Plastic | 9 inches (22.86 cm) |
| | 10 inches (25.4 cm) |

### *3.1. Equations*
The equation used is based on the static thrust equation

$$T = \dot{m}V$$

Where $\dot{m} = \rho AV$

$$\therefore \; T = \rho AV^2$$

Where;

$$A = \frac{\pi d^2}{4}$$

$\rho$, density of air $= 1.225\ kg/m^3$

$V = velocity\ of\ air\ accelerated\ by\ the\ propeller\ (m/s)$

$d = propeller\ diameter$

Here, the thrust produced by the propeller depends on the air density, RPM, diameter, shape and area of the blade and lastly the propeller pitch that also related to the angle of attack.

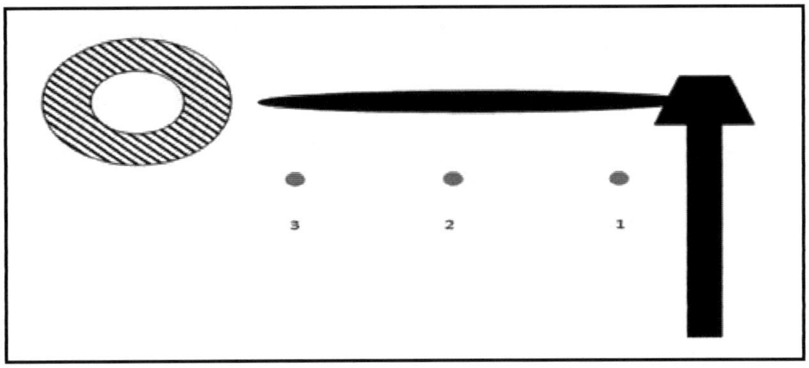


**Figure 2.** Point of velocity reading

## 4. Result
Based on the testing, data below will explain about all the findings

### *4.1. Case Scenario 1: Free propellers*

**Table 2.** Thrust value for free propellers

| Volt (V) | Thrust (N) | |
|---|---|---|
| | 10 inch (25.4cm) | 9 inch (22.86 cm) |
| 4 | 0.1701 | 0.1008 |
| 7 | 0.394 | 0.305 |
| 9 | 0.5211 | 0.3882 |
| 11 | 0.7066 | 0.5022 |

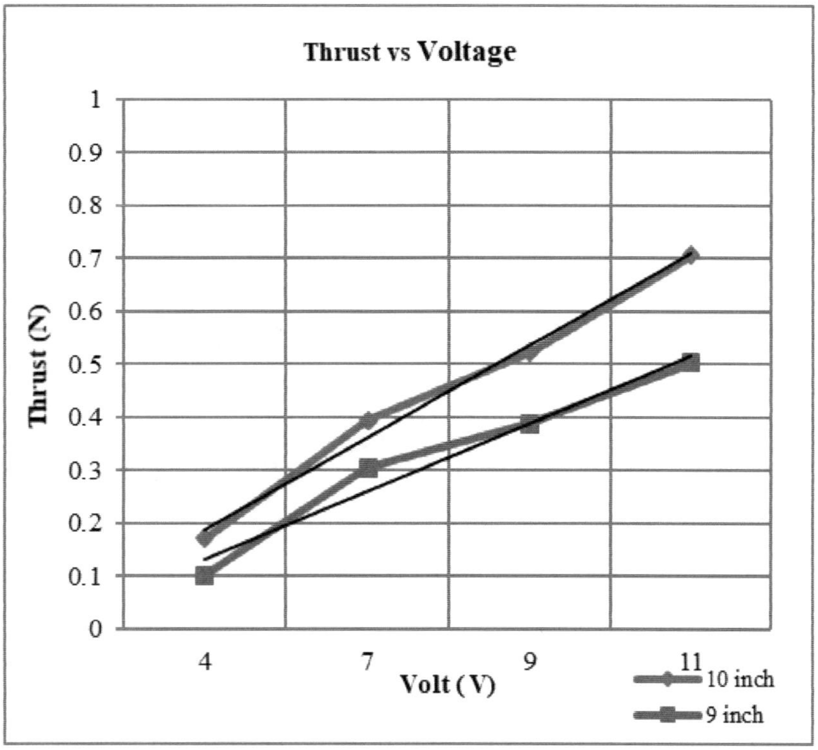


**Figure 3.** Thrust value of free propeller without mouth-ring

*4.2. Case Scenario 2: With mouth-ring application; 10mm,8mm and 5mm gap.*

**Table 3.** Thrust value with mouth-ring application.

| Volt | Thrust (N) | | |
|---|---|---|---|
| | 10mm | 8mm | 5mm |
| 4 | 0.1824 | 0.1636 | 0.1999 |
| 7 | 0.3889 | 0.3619 | 0.3649 |
| 9 | 0.6936 | 0.5600 | 0.5135 |
| 11 | 0.9193 | 0.6986 | 0.5833 |

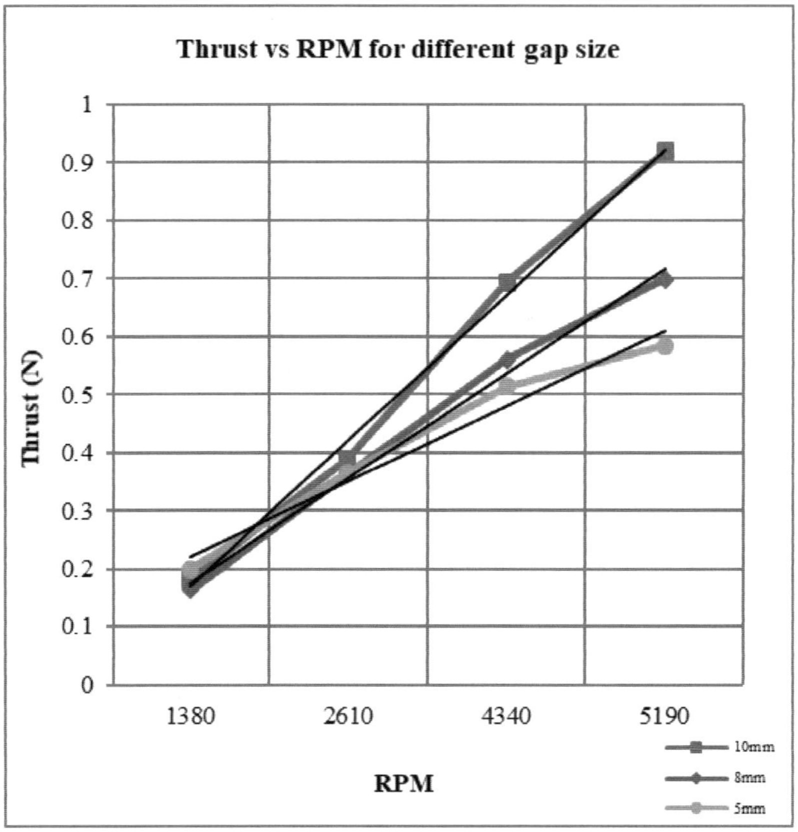


**Figure 4.** Graph for Thrust value with mouth-ring application

### *4.3. Case Scenario 3: Comparison between without mouth-ring and with-mouthring.*

**Table 4.** Thrust value of with mouth-ring and without mouth-ring

| Volt (V) | Thrust (N) | |
|---|---|---|
| | without mouth-ring | with mouth-ring |
| 4 | 0.1701 | 0.1824 |
| 7 | 0.394 | 0.3889 |
| 9 | 0.5211 | 0.6936 |
| 11 | 0.7066 | 0.9193 |

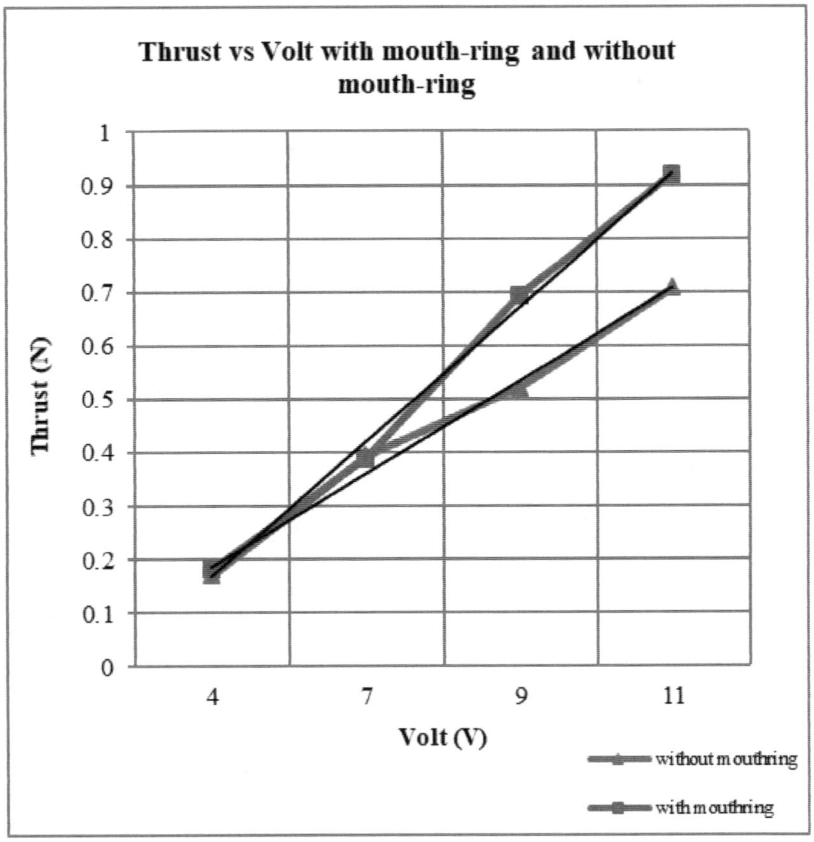


**Figure 5.** Graph Thrust value of with10 mm mouth-ring and without mouth-ring

## 5. Discussion

The thrust generation in Case 1 shows that bigger propeller can generate more thrust than small propeller because of the large surface area of the propeller

While in case 2, it shows that 10mm gap has the highest thrust generate than the other two gaps. Why is bigger gap was generated more thrust is because of the application of the mouth-ring prevent the turbulence of air occurs at the blade tip.

Case 3 shows the comparison of using mouth-ring and with-out using mouth-ring. The graph shows that application of mouth-ring helps a lot in generates more thrust. It shows that, at this particular point the air flow rate is increase with the application of mouth-ring.

## 6. CONCLUSION

The thrust generation of various propeller blade width-spans with the application of mouth-ring was examined through this research study. Both of tests with mouth-ring or without mouth-ring were tested also with the applied of gap between mouth-ring and the propeller. The conclusions which can be drawn from this research are:

a) The thrust can be large by using a bigger diameter of propeller. The bigger the area of the propeller, the higher the velocity which lead to thrust increase. Bigger diameter propeller also gave better air flow across the surface of the blade.

b) Application of mouth-ring can increase the thrust produced with the right design and size. It also can act as a protection in surroundings. Use of mouth-ring (duct) help in velocity increase with the good streamline of along the air flow through the duct.

c) Size of the blade tip clearance would also effect the thrust generation. If the gap is too large, it rejects all the advantages of using the duct.

## 7. RECOMMENDATION

There is recommendation that can be improved in this project in order to get better results:

a) Use various diameter of mouth-ring for testing. This study only focus on 1inch diameter of mouth-ring, maybe for the next study will be test on various diameters.

b) Increasing the range of blade tip clearance (gap size). Result show that using 10mm has higher thrust than using 8mm and 5mm. The gap can be test using more than 10mm to see if there any increasing of thrust id the gap become bigger.

## ACKNOWLEDGMENT

The authors gratefully acknowledge the support from project supervisor Dr. Zuradzman B. Mohd Razlan whose help, simulating suggestions and encouragement in all time of process and for useful discussions.

## References

[1] M. Cutler and J. P. How, "Analysis and Control of a Variable-Pitch Quadrotor for Agile Flight," vol. 137, no. October, pp. 1–14, 2015.
[2] J. P. How, T. Supervisor, and E. H. Modiano, "Design and Control of an Autonomous Variable-Pitch Quadrotor Helicopter," 2012.
[3] A. Smedresman, D. Yeo, and W. Shyy, "Design, Fabrication, Analysis, and Testing of a Micro Air Vehicle Propeller," no. June, 2011.
[4] B. Michini, J. Redding, N. K. Ure, M. Cutler, and J. P. How, "Design and Flight Testing of an Autonomous Variable-Pitch Quadrotor."
[5] M. Cutler, B. Michini, and J. P. How, "Comparison of Fixed and Variable Pitch Actuators for Agile Quadrotors," pp. 1–17.
[6] D. Witt, P. Y. L. J. Young, N. Architecture, and M. Engineering, "Comparative Study of FPP vs . CPP for an All-Electric Naval Combatant."
[7] A. Balachandran, A. Shah, and J. Challa, "Propulsion Selection and Analysis for Unmanned Aerial Vehicles for SAE Aero Design Series," pp. 29–35, 2014.
[8] M. Cutler and J. P. How, "Actuator Constrained Trajectory Generation and Control for Variable-Pitch Quadrotors," pp. 1–15.
[9] Y. Naidoo and G. Bright, "Rotor Aerodynamic Analysis of a Quadrotor for Thrust Critical Applications," no. November, pp. 23–25, 2011.
[10] B. D. Rutkay, "A Process for the Design and Manufacture of Propellers for Small Unmanned Aerial Vehicles by Affairs in partial fulfillment of the requirements for the degree of Master of Applied Science," 2014.
[11] S. Yilmaz, D. Erdem, and M. S. Kavsaoglu, "Performance of a ducted propeller designed for UAV applications at zero angle of attack flight : An experimental study," Aerosp. Sci. Technol., vol. 45, pp. 376–386, 2015.
[12] D. Yongle, S. Baowei, and W. Peng, "Numerical investigation of tip clearance effects on the performance of ducted propeller," Int. J. Nav. Archit. Ocean Eng., vol. 7, no. 5, pp. 795–804, 2015.
[13] G. M. Hoffmann and C. J. Tomlin, "Quadrotor Helicopter Flight Dynamics and Control : Theory and Experiment ∗," no. August, pp. 1–20, 2007.
[14] A. I. Abrego, A. Engineer, M. Field, R. W. Bulaga, and C. Engineer, Performance Study of a Ducted Fan System," pp. 1–6, 2002.
[15] R. J. Weir, "Ducted Propeller Design and Analysis," vol. 29000, no. October, 1987.
[16] K. Regmi, "Investigation of Perforated Ducted Propellers to use with a UAV," 2013.
[17] S. Yılmaz and D. Erdem, "Effects of Duct Shape on a Ducted Propeller Performance," pp. 1–9.
[18] Y. Tamura, Y. Nanke, and M. Matsuura, "Development of a High Performance Ducted Propeller," no. 7, pp. 1–8.
[19] "Unmanned Air Vehicle ( UAV ) Ducted Fan Propulsion System Design and Manufacture Submitted by Wah Keng Tian Department of Mechanical Engineering," 2010.
[20] M. Rutkowski and W. Krusz, "Design and analysis of ducted fan for," no. Icas 2002.
[21] J. M. Stubblefield and R. W. Kimball, "Numerically-based Ducted Propeller Design Using Vortex LatticeLiftig Line Theory," no. 1993, 2008.

# Dynamic Analysis of UAV's Motor Support Bar Length Control System

**M F Zulkifli[1]*, Z M Razlan[1], Shahriman A B[1], I Zunaidi[2], Wan Khairunizam[1] and N Z Noriman[3]**

[1]School of Mechatronic Engineering, Universiti Malaysia Perlis (UniMAP), Pauh Putra Campus, 02600 Arau, Perlis, Malaysia
[2]Faculty of Technology, University of Sunderland, St Peter's Campus, Sunderland, SR6 0DD, United Kingdom
[3]Faculty of Engineering Technology, Kampus Sg. Chuchuh, Universiti Malaysia Perlis, 02100 Padang Besar, Perlis, Malaysia.

**Abstract.** UAV (Unmanned Aerial Vehicle) can be described as aircraft that do not need any presence of pilots inside it. Basically, UAV is come out in a small aircraft so that the aircraft can be easily controlled by the people from afar[1]. The UAV's motor support bar length control systems are the UAV's control systems that move according to the variable arm length movement and also a constant revolution of the propeller speeds. The purpose of the study is to run the dynamic analysis at the UAV's motor support bar length control systems and also to enhance the UAV's mathematical modelling by using the SOLIDWORKS® software which involved in using both CAD and CAE systems[2]. The detailed design is used SOLIDWORKS® software to conduct the static and dynamic analysis of UAV's motor support bar length control systems. The design is restricted to the arm due to the critical part that has the highest vibration at the UAV's motor support bar length control systems. The results that obtain from the study from the static and dynamic analysis are the displacement of the motor, Von Misses stress of the arm, and also the resonance frequency that will give the modes shape to the systems.

## 1. Introduction

The quadrotor is widely used of UAV's in the industry or worldwide. Quadrotor's main components are motor (brushless motor) and also the propeller. These two elements will give the quadrotor to hover and manoeuvring around of the area. The propeller from the quadrotor will create a thrust force and in aid the quadrotor to lift. In the meantime, the control system to stabilize and manoeuvre the quadrotor consumes too much battery and also the vibration that occur when the changeable position of the brushless motor. This research is to study the dynamic analysis of UAV's motor support bar length control systems.

## 2. Background of UAV's motor support bar length control system

UAV is an aircraft that does not need any presence of pilots inside of it due to it is a small aircraft that can be controlled from afar to manoeuvre at the open air. UAV is also involving the environment such as in a significant competition that challenges the people around the world to make the best aircraft in the world. Also, UAV is also helping in the military to spy the outsiders that can be a threat to the country [3]. Most of the UAV are made in quadrotor due to its easy to design, small rotors, and excellent manoeuvrability [4].

*2.1. Control Movement of Variable Arm Length (VAL) quadrotor*

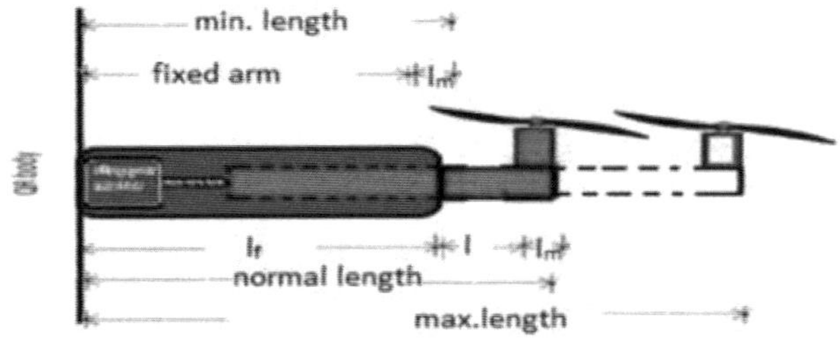


**Figure 1.** The arm design with respect to the length [3]

The idea of the motor support bar length control systems is come from the idea of the variable arm length system which the system can counteract the energy depletion during the flight time[3]. The problem is because of the motor speed control where the rate of the two motor are increasing, and the other two must be decreased to obtain the yaw movement. In order to cope this situation, the recent research by [3] has shown that when changing the position of the arm's length will counter the energy consumption of quadrotor when implied the theory of increasing and decreasing the motor to generate the required torque in yaw, pitch, and roll attitude as shown in Figure 1 [3].

Based on the above figure [3], we can show that the arm involves of two parts which are fixed and non-fixed part. Furthermore, based on the fixed part contains quadrotor frame at the outer part and stepper motor at the inner parts. In addition, the stepper motor revolves in persistent speed. Moreover, the fixed arm of the quadrotor frame is built in a hollow shape. On the other hand, the threaded shaft is using to connect to the stepper motor and move the sliding arm of the non-fixed part to slide along the fixed hollow arm. The increasing and decreasing of the arm's length from the design cause the formula as follow:

$$\max length = (l_f + l_m + L) + \Delta d \cdot (l_f + l_m) \quad (1)$$
$$\min length = (l_f + l_m + L) - \Delta d \cdot (l_f + l_m) \quad (2)$$

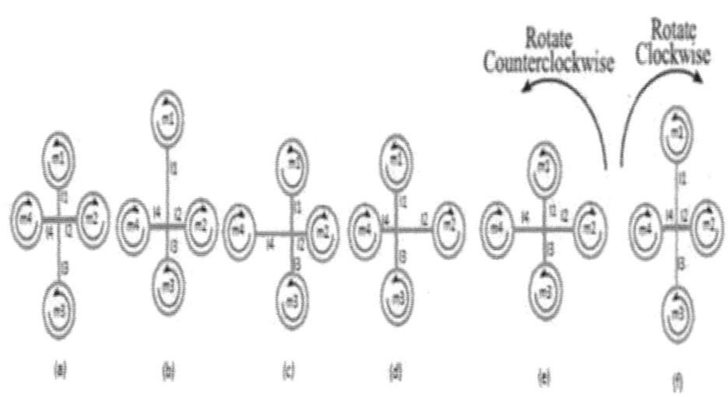


**Figure 2.** Pitch, Roll, and Yaw attitude with respect arm length [3]

From the above figure [3], the attitude of the quadrotor movement which is determined by pitch angle, θ, roll angle, ∅, and also the yaw angle, φ. Moreover, the quadrotor attitude based on Figure 2.4(a) and Figure 2.4(b) will cause the quadrotor to pitch backward and pitch forward when there is increase in position of the arm's length of one of the motor with respect of y-axis, respectively. However, the quadrotor attitude that based on Figure 2(c) and Figure 2(d) shows that the roll movement of the quadrotor body when there is increase in position of the arm's length of one of the motor with respect of x-axis position which is in roll leftward and roll rightward, respectively. Meanwhile, the yaw attitude of quadrotor that rotating counter-clockwise when there is increase in position of each respected motor that is in x-axis in Figure 2(e), while the quadrotor will be rotating clockwise when there is increase in position of each respected motor that is in y-axis in Figure 2(f).

*2.2. Vibration during changeable position of the motor*

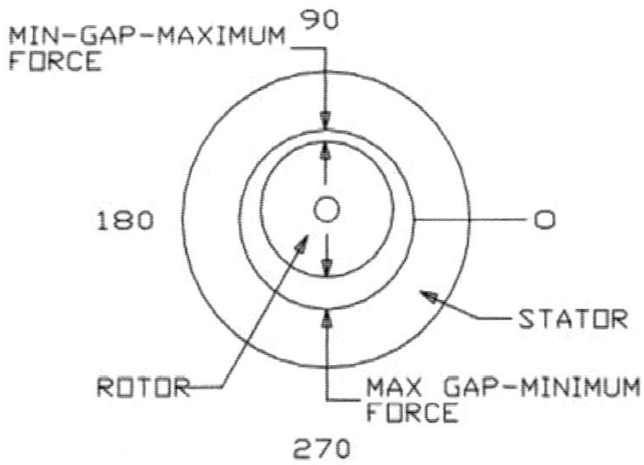


**Figure 3.** Non-symmetrical air gap around the rotor[5]

Based on the figure above, it shows the source of vibration that occurs from the rotor. The non-symmetrical air gap around the rotor causes some vibration as in our case most likely occur at the brushless motor and also the stepper motor [5]. In addition, the changeable position of the arm of the quadrotor effect on the COG of the quadrotor itself which is the point of support on which the body would stable and balance.

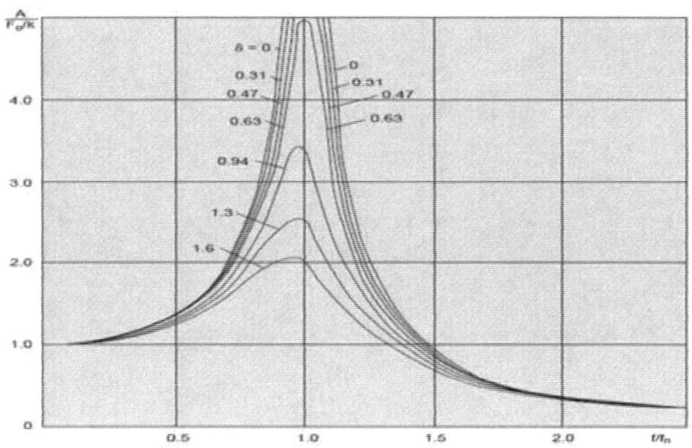


**Figure 4.** Graph of amplitude-frequency of enormous block motion

The resonance that occurs based on unbalanced mass due to the changeable position of the centre of gravity it is shown based on the above graph above. The forced vibrations that occur that occur to the rotor are due to the external force that applied at the weight of the rotor and shows the harmonic motion of force with respect to the time. On the other hand, the overload COG of an aircraft gives serious vibration for the aircraft and thus occurs fatigue failure to the aircraft. The research has been done by the method of regression analysis and the rain flow tallying strategy was embraced to record the heap cycles on the amplitude and mean value [6]. Besides that, they are also estimated an aircraft vibration using static and dynamic measurements to determine COG position of the main rotor. It shows the main rotor will have the highest uncertainty when compared to other position of COG.

## 3. Methodology

In the methodology of this study, we have to find the free-body diagram (FBD) of the quadrotor body system. Generally, the FBD is needed when wanting to do the schematic diagram or overview of the whole quadrotor body. Furthermore, the quadrotor body system is necessary to find its dynamic algorithm where the force and torque moment will be the priority of the quadrotor body system.

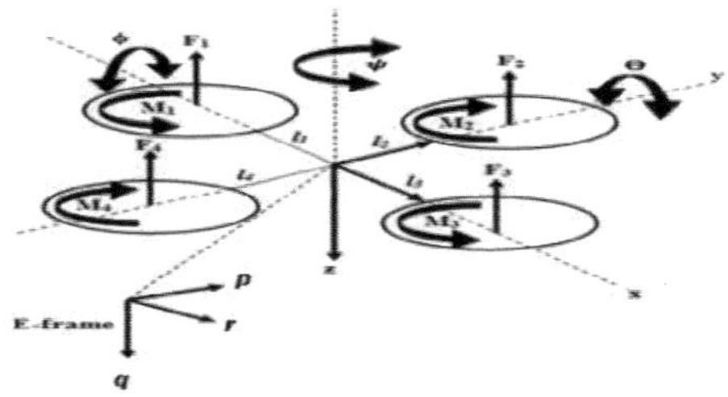

**Figure 5.** Quadrotor FBD with respect to Center of Gravity (COG)[3]

In addition, the COG of the quadrotor body system will be at the center of the quadrotor body mass system which the COG is needed to be in this state for a steady movement. Next, we will sketch the detail model of the quadrotor in the SOLIDWORKS® since the SOLIDWORKS® provided the CAD system. To determine the COG location of the quadrotor, we will be using the SOLIDWORKS® software, which the SOLIDWORKS® software provided every detail mass properties of the quadrotor by using the evaluate function. Afterward, the static analysis and the vibration analysis is also determined by the SOLIDWORKS® [7]. SOLIDWORKS® has developed its software to the CAE system in which the system can run the various analysis of the design based on the real-time simulation[2]. The design is limited to the arm itself which the vibration mostly occurs on that part. Lastly, we compared the result of both static analysis and vibration analysis of the quadrotor by the variable of the arm length.

### *3.1. Design development*

The development to find the mass properties and also to run the static and vibration analysis on the system, we have to design the model of the structure before proceeding with the analysis. The design is made by using the SOLIDWORKS® and consists of few parts for model the quadrotor which is the body of the quadrotor with two stands, four fixed arms, four moving arms, four stepper motor, four brushless motors.

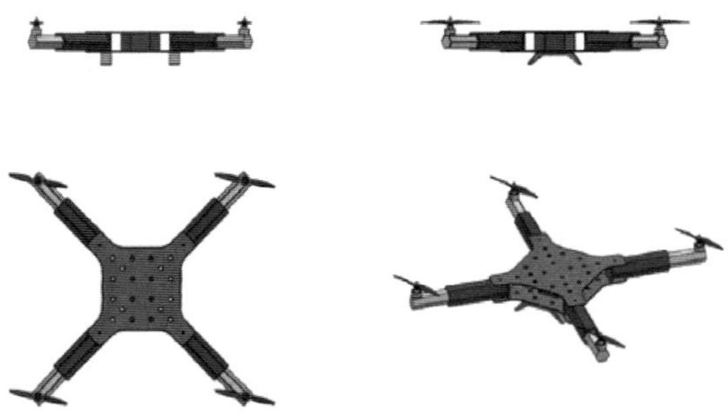

**Figure 6.** Full model VAL-quadrotor using SOLIDWORKS®.

The assembly of the design is shown based on the above figure where the stepper motor will act as a mechanism to move the moving arm to certain position to give the attitude of quadrotor movement without any change of the rotation of propeller at the brushless motor.

IOP Conf. Series: Materials Science and Engineering **557** (2019) 012053 doi:10.1088/1757-899X/557/1/012053

### *3.2. Mass properties of the quadrotor*

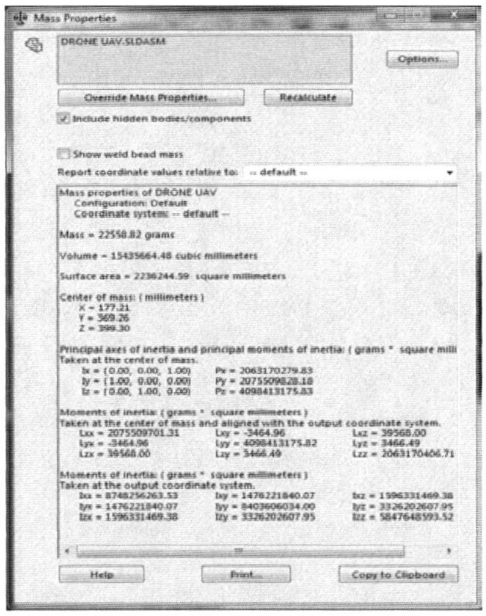


**Figure 7.** Mass properties of quadrotor using SOLIDWORKS®

The mass properties of the quadrotor can be evaluated inside the SOLIDWORKS® software. This function can be done by went to the "Tools" tab then click the "Evaluate" and after that select the mass properties icon of the model. Based on the mass properties, we can find the total mass of the quadrotor, center of mass, principle axes of inertia and also the moment of inertia.

### *3.3. Static and dynamic analysis using SOLIDWORKS®*

The SOLIDWORKS® Simulation will be using to find the static and dynamic analysis of the quadrotor. First, we will run the static analysis due to we suspect there will be a dynamic analysis based on multiple static responses at the model. The static analysis will be run at the most critical part of the design of the model which at the arm of the quadrotor where it is most likely occur the vibration.

**Figure 8.** Configuration of the load, fixed geometry and the gravity applied to the arm of the quadrotor using SOLIDWORKS®

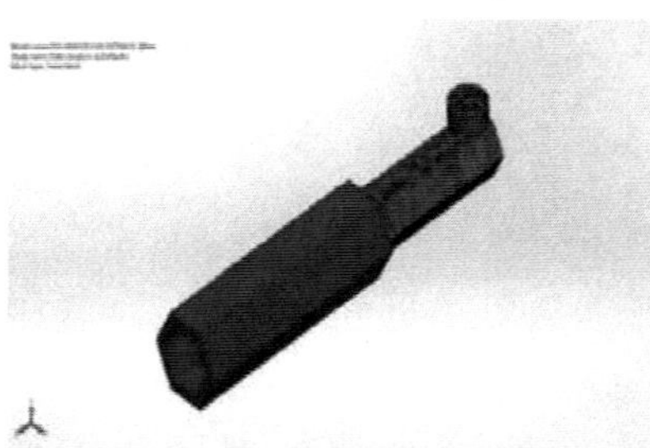

**Figure 9.** Mesh model of the arm of the quadrotor using SOLIDWORKS®.

The finite element method (FEM) of static analysis must determine the parameter to the arm since we make it as the simplified design of the quadrotor to show the dynamic analysis of the system. The end of the fixed shaft was given a fixed geometry since it is attached to the frame of the quadrotor. Meanwhile, the remote mass was determined at the moving arm which the remote mass that was given is actually the mass of the brushless motor. The gravity acceleration is set at the top plane of the design. This parameter is used only for the static analysis only with respect to the changeable position of the arm. The position of the arm will start from the initial point of the moving arm until 60 mm with the increment of 20mm. After that, we run the mesh analysis before run the simulation for the static analysis. From the static analysis, we can figure out the Von Misses stress and maximum displacement in our cases of study. Meanwhile, the dynamic analysis for SOLIDWORKS® does not need the parameter for gravity acceleration. The steps for setting before run the simulation is almost the same with the static analysis. The result that we get based on the dynamic analysis is the mode shapes along with the resonance frequency.

## 4. Result and discussion

For the simulation of the study on the static and dynamic analysis, the results more focus on the comparing the effect of changeable position to maximum displacement, Von Misses stress and also the mode shapes with respect of the resonance frequency at the quadrotor arm.

### *4.1. Static Analysis of quadrotor*

The resulting force based results from the simulation as shown from below table:

**Table 1.** Resultant Force at the quadrotor arm calculated by SOLIDWORKS®

| Selection set | Units | Sum X | Sum Y | Sum Z | Resultant |
|---|---|---|---|---|---|
| Entire Model | N | 1.66893e-006 | 61.4167 | 1.14441e-005 | 61.4167 |

The resultant force at the quadrotor arm is 61.4167N based on calculation made by SOLIDWORKS® software with applying the parameter to the arm of the quadrotor. The maximum displacement results show that the peak displacement happens at the end of the moving arm which is the remote mass is apply.

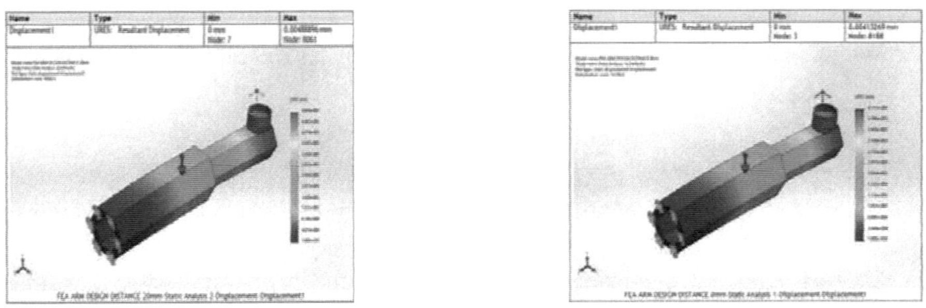

**Figure 10.** Displacement of 0mm of the arm of the quadrotor (left) and 20mm of the arm of the quadrotor (right) using SOLIDWORKS®.

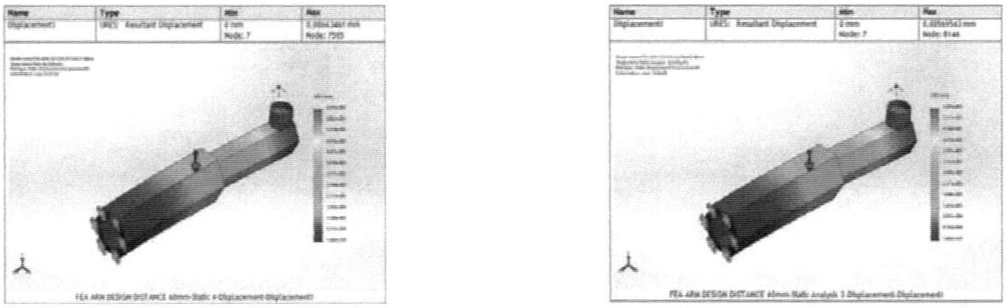

**Figure 11.** Displacement of 40mm of the arm of the quadrotor (left) and 60mm of the arm of the quadrotor (right) using SOLIDWORKS®.

The results of the changeable position of the arm with respect of maximum displacement at the arm of the quadrotor can be seen from the graph below:

**Figure 12.** Graph displacement against distance arm.

Based on the above graph, it shows the displacement of the quadrotor movement along Y-axis against the distance arm along X-axis movement. It shows the distance of arm influenced the displacement of arm movement along Y-axis. The displacement of the arm along Y-axis increase steadily against the distance arm. In addition, based on the static analysis of the quadrotor also shows the Von Misses stress at the quadrotor arm design.

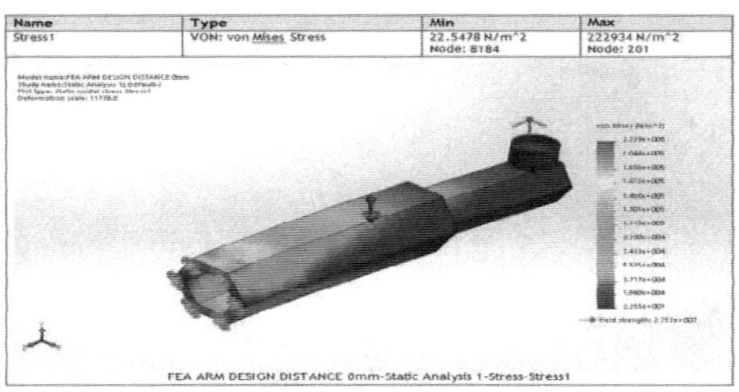


**Figure 13.** Von Misses stress at 0mm distance of arm using SOLIDWORKS®.

IOP Conf. Series: Materials Science and Engineering **557** (2019) 012053 doi:10.1088/1757-899X/557/1/012053

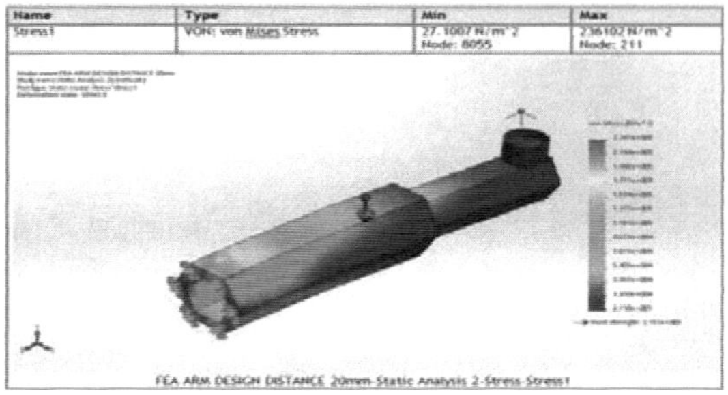

**Figure 14.** Von Misses stress at 20mm distance of arm using SOLIDWORKS®.

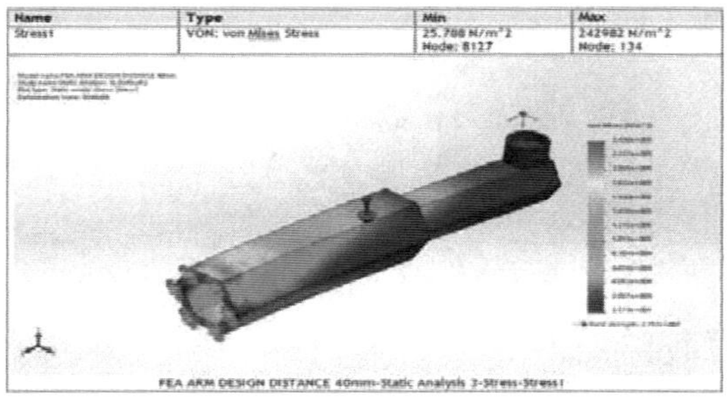

**Figure 15.** Von Misses stress at 40mm distance of arm using SOLIDWORKS®.

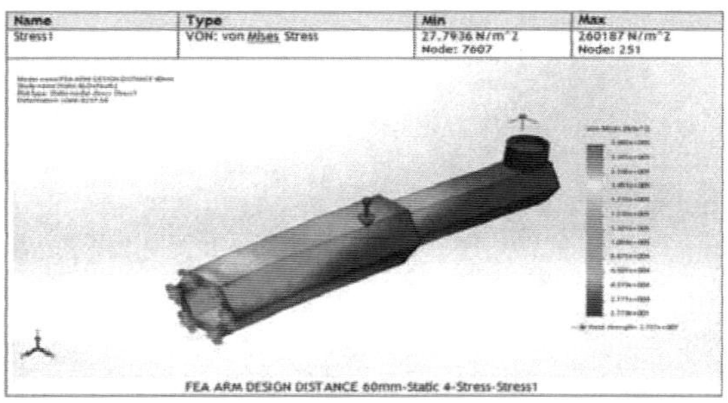

**Figure 16.** Von Misses stress at 60mm distance of arm using SOLIDWORKS®.

The Von Misses stress results shows that the stress most likely occur at the fixed arm. In addition, the Von Misses stress also increase when there is an increase of the arm length. The red arrow in the figures indicated that the yield stress of the arm is 27.57 MPa due to the material properties we have selected at the arm is 1060 Aluminum alloy. Besides that, the highest amount of stress happen at the arm are 22.29 KPa, 23.61Kpa, 24.30 KPa, and 26.20 KPa at the distance of arm 0mm, 20mm, 40mm, and 60mm respectively. Based on the highest amount of the stress, it is shows that it is still below of the yield strength of the arm of the quadrotor.

### *4.2. Dynamic Analysis of quadrotor*

The dynamic analysis of the quadrotor that we want to find in this study is the resonance frequency of the quadrotor with its mode of shapes.

**Table 2:** Table of resonance frequency and distance of arm length according to its mode shape.

| Modes 1 | | Modes 2 | | Modes 3 | |
|---|---|---|---|---|---|
| Frequency(Hertz) | Arm Length, mm | Frequency(Hertz) | Arm Length, mm | Frequency(Hertz) | Arm Length, mm |
| 318.64 | 0 | 318.8 | 0 | 1275.3 | 0 |
| 291.77 | 20 | 292.24 | 20 | 1152.7 | 20 |
| 268.9 | 40 | 269.39 | 40 | 1055.2 | 40 |
| 248.84 | 60 | 249.07 | 60 | 975.12 | 60 |

According to the table, it shows the modes of resonance frequency against with the distance of moving arm. In the modes 1, it shows the resonance frequency is decreasing when increasing the arm length from 318.64Hz to 248.84Hz. In the meantime, the modes 2 show the same decreasing when increasing the arm length to 60mm which are from 318.8Hz to 249.07Hz. On the other hand, the resonance frequency also decreasing in modes 3 where the resonance frequency drops from 1275.3Hz to 975.12Hz.

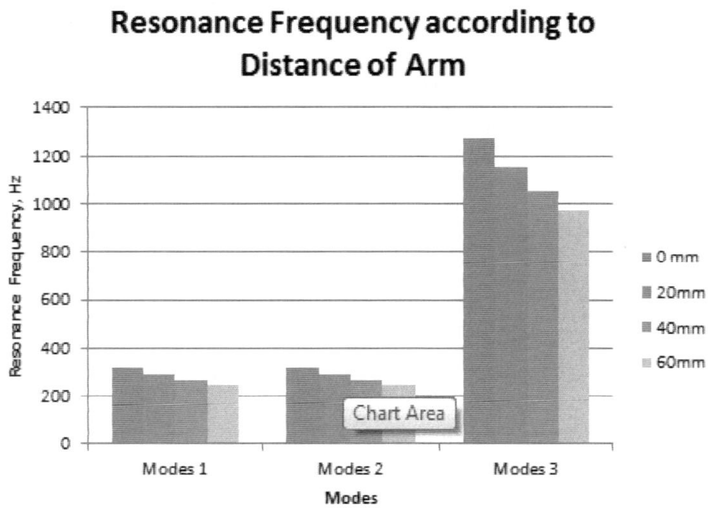


**Figure 17.** Graph resonance frequency according to distance of arm.

According to the graph shown, it shows the graph of resonance according to its modes shape. From the observation, it shows that modes 1 and modes 2 almost give the identical resonance frequency and shows increasing in the modes 3 for resonance frequency by according to the distance of arm length. In addition, based on its modes shape, it shows that each modes showing a decreased in value of resonance frequency against with the arm length of quadrotor and obviously the modes shape 3 shows the decreasing with drastic.

## 5. Conclusion

### *5.1. Summary*

Throughout the experiment that we have conduct from the analysis of SOLIDWORKS® Simulation, we have achieved the results the results of the static and dynamic analysis are maximum displacement, Von Misses stress and the modes shape of the resonance frequency. The maximum displacement and the Von Misses stress show increasing trends towards the increment of the position of the arm. Meanwhile, the mode shapes of the resonance frequencies decrease with respect of increasing of the arm position. The analysis shows that the Von Misses stress still in the range of the yield strength which the quadrotor can be controlled in the sky but must be taken care with force frequency which can meet the natural frequency of the quadrotor when the position of the arm is increasing.

## References

[1] S. Norouzi Ghazbi, Y. Aghli, M. Alimohammadi, and A. A. Akbari, "Quadrotors unmanned aerial vehicles: A review," *Int. J. Smart Sens. Intell. Syst.*, vol. 9, no. 1, pp. 309–333, 2016.

[2] L. M. Gómez-López, V. Miguel, A. Martínez, J. Coello, and A. Calatayud, "Simulation and modeling of single point incremental forming processes within a solidworks environment," *Procedia Eng.*, vol. 63, no. 2005, pp. 632–641, 2013.

[3] N. Yasameen Kamil, D. Hazry, K. Wan, and Z. M. Razlan, "A novel VAL: Quadrotor control technique for trajectory tracking based on varying the Arm's Length," *ARPN J. Eng. Appl. Sci.*, vol. 11, no. 15, pp. 9195–9204, 2016.

[4] S. A. Raza, "Design and control of a quadrotor unmanned aerial vehicle," *ProQuest Diss. Theses*, vol. MR61280, p. 122, 2010.

[5] W. R. Finley, M. M. Hodowanec, and W. G. Holter, "An analytical approach to solving motor vibration problems," *IEEE Trans. Ind. Appl.*, vol. 36, no. 5, pp. 217–232, 2000.

[6] Y. Li, Z. Wang, Y. Chen, Y. Zhang, and W. Sun, "Research on compiling fatigue load spectrum of individual aircraft and analysis of fatigue life based on flight data," *Proc. IEEE 2012 Progn. Syst. Heal. Manag. Conf. PHM-2012*, 2012.

[7] S. Elhibir, "Dynamic Analysis of a Rotor-Bearing System," *Pr. magisterska*, p. 155, 2009.

IOP Conf. Series: Materials Science and Engineering **557** (2019) 012054 doi:10.1088/1757-899X/557/1/012054

# Design Modification of Greenhouse Effect Solar-Hybrid Rack Type Dryer

**D Wulandani[1], S E Agustina[1], M Yulianto[1], L O Nelwan[1], and L M Rahman[1]**

[1] Department of Mechanical and Biosystem Engineering, Faculty of Agricultural Technology, Bogor Agricultural University, Bogor, Indonesia

**Abstract.** Hybrid solar-type rack dryers are modified to improve efficiency and to save the drying energy. Modifications are made to the main components, namely; furnace and heat exchanger. The dimension of dryer was 1.8 m length, 0.7 m width and 1.8 m high. The dryer was equipped by 22 racks with dimension of 0.5 m x 0.65m. The furnace was modified to a cylindrical shape instead of rectangular shape and was designed with optimization of heat utilization and comfortable to be operated. To reduce heat loss, the wall of furnace was designed contact directly the drying room. The furnace material was iron galvanized plate with a thickness of 2 mm. The foregoing design of heat exchanger was tube bank type with horizontal fins attached on tubes. In this research, the adoption of vertical fins has enlarged the heat transfer area and increase the effectiveness of heat exchanger. The smoke flow from the combustion chamber of furnace was directed at the opposite side of the ambient air inlet for drying, so that smoke does not enter the drying chamber.

## 1. Introduction

Drying of agricultural product have an important role for products preservation and make it easier for further processing, as applied to the process of making sweet potato flour. The hygienic drying process is a necessity in the sweet potato processing industry. Mechanical dryers can overcome these problems, due to the dryers can be used all the time, both day and night, when the sun is clear or cloudy, even during the rainy season, so that the continuity of production can be maintained. Hybrid solar dryer become the alternatif dryer for the need.

This research proceed of a double rack type greenhouse effect (GHE) solar dryer that has been investigated previously [1]. In order to reduce the consumption levels of drying energy, several modifications were conducted appropriate to the drying needs of sweet potatoes in small household industries. The purpose of this study is to modify the GHE solar dryer, especially in the size of the drying capacity, solar energy conversion system, heat exchanger, a small portion of the furnace and rack placement.

## 2. Method

### *2.1. Material and apparatus*

The apparatus used in the study consisted of workshop tools, such as electric welding, iron cutting tools,ect. While the materials needed were iron plates, iron pipes, holow iron, aluminum perforate, wood, fans, electrical equipment, tool kits etc. The product dried was slice sweet potato.

IOP Conf. Series: Materials Science and Engineering **557** (2019) 012054 doi:10.1088/1757-899X/557/1/012054

*2.2. Procedure*
Research follows the stages as shown in Figure 1.

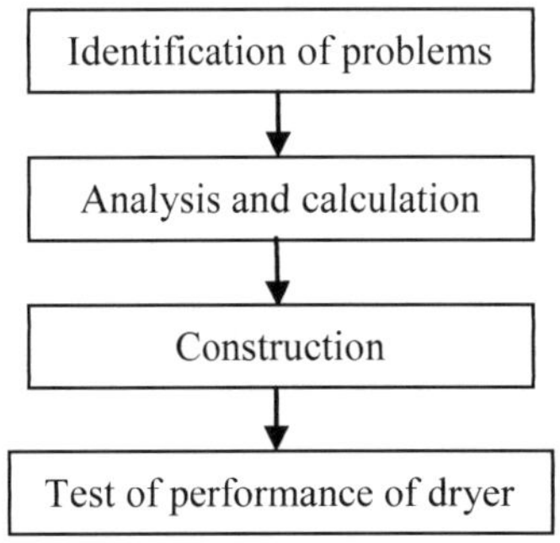


**Figure 1.** Research steps

## 3. Results

*3.1. Hybrid solar dryer before the modification*
Figure 2 showed the hybrid solar dryer designed in previous research [1]. The dryer utilized solar energy and biomass energy for drying of herbal product, such as roselle [1], "temu lawak" or wild ginger [2], slice of modified cassava [3]. Dimension of dryer is 4.4 m length, 1.8 m width and 3 m high. The dryer consist of:

1. Racks as product holder: There are 144 trays with dimension of 0.5 m x 0.6 m. The distance between trays is 0.12 m. The capacity of dryer is 400 kg of roselle pod or 200 kg of ginger slice.
2. Inlet: Ambient air is intake through the inlet. There are three inlets with diameter of 0.2 m which located on the center of dryer
3. Outlet: The outlet is use to stream out the vapor evaporated from the product. There are two rectangular outlets of 0.2 m x 0.3 m on the right and left side of dryer, respectively.
4. Blower: Three blowers are implemented in inlet position with a power of 80 W each.
5. Furnace: It have trapezoidal configuration and made of 2 mm thick iron plate. The furnace is equipped with the iron rod grille for placing the firewood.
6. Heat exchanger (HE): HE is consist of several vertical pipes (0.25 m diameter, 2 m length). There are two heat exchanger system; the first is air-air HE system and the second is water-air HE system. The first used combustion air, whilst the second utilized hot water from solar concentrator.
7. Chimney: It is the prolongation of heat exchanger that have a function for releasing the smoke produced from the combustion chamber of furnace
8. Solar concentrator: Pair of solar concentrator were placed on the top of dryer to collect solar radiation in the black absorber pipe to heat the water. Solar tracking is powered by 1 Hp motor. Solar tracking is controlled by light dependent resistor. The water is circulated entire the pipe (from absorber pipe to heat exchanger pipe) by two pumps motor (1/4 Hp).
9. Water tank: It used as water container made of cyllindrical iron (2 mm thick).

The problems have been identified from the several experiments of the dryer with agricultural products could be summarized below.

1) The dryer need the high power for solar tracking. The solar tracking is absolute requirement for solar concentrator to capture the maximum solar radiation.
2) The energy consumption to evaporate the water vapor from the drying product is high of 29 MJ/kg of vapor until 57 MJ/kg of vapor [2], due to the high of heat loss through the furnace wall and the heat exchanger didn't yet transfered heat optimally.

**Figure 2.** The dryer before modification [2]

*3.2. The dryer after modification appropriated with the need of slice of sweet potato drying*

The dryer is modified to dried wet slice of sweet potato. The dimension of dryer is L x W x H = 1.8 m x 0.7 m x 1.8 m, in order to obtain the dryer capacity of 100 kg of sweet potato slice. The structure of dryer modified is illustrated in Figure 3.

Below is the components and material of dryer modified:

1. Rack: The racks consist of 24 trays with dimension of 0.7 m x 0.5 m. The interval among the trays is 0.1 m. The drying room and racks is designed adjacent to the furnace wall, in order to make use of heat loss from the furnace wall,.
2. Inlet: One rectangular inlet is designed which placed on the opposite position with the inlet of furnace, so the smoke flow entered to the inlet could be avoided. Dimension of inlet is 0.6 m and 0.15 m. This inlet position make it possible to intake ambient air and flow it through the HE pipe longer than previous design.
3. Outlet: There are no change of outlet design.
4. Blower: There are three blower (12 W) implemented on the left and right of HE wall respectively. Temperature of drying room more uniform than previous design.
5. Furnace: Cillindrical furnace is designed consist of iron plate 2 mm and 0.80 m of diameter. The grilles is mounted for placing the firewood. The cillindrical shape could flowing combustion air easily in to HE pipe. Furnace is equipped with cover and handle for safety and to easily operated.
6. Heat exchanger: There is one system of air-air HE. The HE is crossflow type, where combustion air enter the HE pipe, while the ambient air flow across the outer wall of the pipes. The vertical fins is attached along the outer pipe to increase heat transfer area.
7. Chimney: No significant change of chimney design.
8. Drying room: Drying room is function as solar collector. Solar irradiation is transmitted by tansparent wall of dryer. The dryer floor and HE wall performe the function of absorber.

**Figure 3.** Dryer modified

### *3.3. Heat transfer analysis of furnace*

The previous design of furnace [1] have a trapezoidal shape without cover. Rahari [4] have modified a cyllidrical furnace from an iron plate (3.2 mm thick). The furnace system efficiency was 27 %.

In this research, design modification of furnace applied Rahari's cyllindrical furnace design. The material of furnace is iron plate (2 mm thick) which is used to reduce weigh of furnace. The position of furnace in dryer system is unity with dryer floor and adjacent to drying room to take advantage of heat loss from furnace wall. The efficiency of furnace system modified is 45 %.

### *3.4. Design modification of heat exchanger*

Heat exchanger is the important component of dryer to facilitate the exchange of heat between combustion air which contain smoke at high temperatures while keeping it from mixing with drying air at lower temperature, so the dried product is free from smoke.

The modification of HE in this study was carried out on the shape, dimensions and addition of fins. The number of fins, the position and the dimension in the HE pipe were simulated and compared to get the best results. HE performance with the addition of fins is expressed by the value of fin efficiency and the effectiveness of fins and the amount of heat that HE can transfer. This analysis has been submitted in a separate paper [3]. Wulandani and Arnanda [3] have added horizontal fins on the pipe wall of HE. The design have decreased energy consumption until 43 % compared to the previous design [1], however HE effectieness still low of 4.0.

Modification have been carried out at current research was by added vertical fins along the vertical pipe to enhance heat transfer area. Based on the 10 design scenarios of HE heat transfer simulation , it was found that the best design for vertical fin shape was with 26 pipes, 3.5 cm pipe diameter, 1 m pipe length. The number of vertical fins is 64 pieces, with dimensions of 2.5 cm wide, 1 m long and 1.2 mm thick. The heat trasfered by HE is 6.8 kW, with fin effectiveness of 4.9. This result is greater than that of the horizontal fin shape.

### *3.5. Drying performance*

The dryer have been tested for drying slice of sweet potato of 50 kg for 16 hours. The average fresh sweet potato water content was 69.9% bb and dry sweet potato sawut water content was 10.2% bb. Table 1 shows performance data from the results of drying of sweet potatoes slice. Dryers have an average thermal efficiency of 40%, and a drying system efficiency of 15% and energy consumption of 17 MJ / kg of vapor. The value shows drying performance improvement compared to the previous dryer [1]; [2]; [3].

**Table 1.** Drying performance of GHE hybrid solar dryer

| NO | Parameters | 1 | 2 | 3 | 4 |
|---|---|---|---|---|---|
| 1 | Product | Roselle | Wild ginger | Modified cassava chip | Sweet potato |
| 2 | Capasity (kg) | 32 | 60 | 18 | 50 |
| 3 | Initial moisture content (% wb) | 89 | 81.3 | 70 | 69.9 |
| 4 | Final moisture content (% wb) | 12 | 8.5 | 12.6 | 10.2 |
| 5 | Drying time (hour) | 28 | 30.5 | 11.5 | 16 |
| 6 | Drying temperature (°C) | 43 | 41.8 | 43.5 | 53.7 |
| 7 | Thermal efficiency (%) | 14 | 14.5 | 24.8 | 40.1 |
| 8 | Drying efficiency (%) | 3.5 | 8.5 | 9.4 | 15.2 |
| 9 | Energy consumption (MJ/kg vapor) | 55 | 28.6 | 28 | 17.0 |

Note:
1. Wulandani and Nelwan, 2009 [1]
2. Elsamila and Wulandani, 2014 [2]
3. Wulandani and Arnanda, 2016 [3]
4. This research

## 4. Conclusion

Design modification of hybrid solar dryer have resulted better performance of drying compared to previous design. System efficiency of furnace is 45 %. This value increase 1.6 times comparing to that of previous design. Modification of heat exchanger have enlarge the heat transfer area by adding the vertical fins. Fin effectiveness is obtained of 4.9, greater than the horizontal fins.

The slice of sweet potato with a capacity of 50 kg with initial moisture content of 70% wb was dried to the final moisture content of 10% wb for 16 hour. Drying energy is supplied by solar energy and biomass during drying process simultaneously. The energy consumption is 17 MJ/kg vapor. The dryer modified have reduced energy consumed significantly about 69 % comparing to previous dryer. The drying system efficiency is 15.2 % encreased about four times comparing to that of previous dryer design.

## References

[1] Wulandani D and Nelwan L O 2009 Rancang bangun kolektor surya tipe plat datar dan konsentrator surya untuk penghasil panas pada pengering produk-produk pertanian (Design of plat plate solar collector and concentrator as heating of dryer for agricultural products) *Prosiding Seminar Hasil-hasil Penelitian IPB* chapter 5 pp 568-578

[2] Aritesty E and Wulandani D 2014 Performance of the rack type-Greenhouse effect solar dryer for wild ginger (*Curcuma xanthorizza* Roxb.) drying *Energy Procedia* (Elsevier), vol 47 pp. 94-100

[3] Wulandani D and Arnanda C V 2016 Heat Transfer Analysis of Heat Exchanger for Additional Heating of the Rack Type Hybrid Solar Dryer *Proceeding of The 2nd International Symposium on Agricultural and Biosystem Engineering*

[4] Rahari W 2016 Desain tungku biomassa pada alat pengering efek rumah kaca hybrid tipe rak *Thesis of undergraduate- IPB*

## Acknowledgments

Authors wishing to acknowledge assistance from technician and student of Renewable Energy Engineering Division, Department of Mechanical and Biosystem, IPB and financial support from Department of Mechanical and Biosystem, IPB and Kemenristekdikti Republik Indonesia.

# The Effect of Modification Vertical Partition On Kiln Drum Performance For Coconut Shell Carbonization

**L Sucahyo[1] and M Mustaqimah[1]**

[1]Department of Mechanical and Biosystem Engineering IPB University Indonesia

email: lilissucahyo@apps.ipb.ac.id

**Abstract.** Coconut shell as an organic residue that could be re-utilized to charcoal raw material through carbonization or pyrolysis process. Kiln is a thermally insulated chamber that produces enough temperatures to complete some process such as carbonization. This study aimed to conduct a performance test of modification drum kiln with vertical partition for coconut shell carbonization. These partitions were added in order to limit the direct contact between coconut shells and husk, thus the temperature inside each chamber can be easier to maintain. The data obtained with placing the thermocouple in five places, while also measure wind velocity, relative humidity and environment. The drum kiln has 6 main components, which are pyrolysis main chamber, initial combustion chamber, air inlet, chimney, unloading and vertical partition. Optimum performance of kiln with vertical partition was in average temperature of 295.67 °C, maximum temperature 531.33 °C, charcoal yield 21.67 - 26.80%, carbonization time 78.33 minute and calorific value of 7,505.33 kcal/kg. Vertical partition able to limit and control kiln temperature also increased pyrolysis rate as well as charcoal yield.

**Keywords**. Kiln, Pyrolysis, Carbonization, Coconut Shell.

## 1. Introduction

Coconut (*Cocos nucifera*) is a famous tropical fruit originated from tropical region. It is belonging to the palm family. According to the statistical data by the Asian Pacific Coconut Community (APCC), the global coconut area in 2015 was 11.98 million ha, while 75% of the coconut world production comes from Indonesia, Philippines, and India [1]. The major product of coconut is coconut flesh, coconut water and others biomass residues i.e. coconut shell, husk and leaves. The utilization of coconut residue especially coconut shell still limited and many of them became wasted. Coconut shell has a strong structure and harder to be decomposed by microorganisms. It is slightly similar with wood in terms of composition that contains lignin (29.4%), hemicellulose (27.7%), cellulose (26.6%), extractive component (4.2%), anhydrate (3.5%), water (8%), ash (0.6%), and nitrogen (1%) [2].

Carbonization is one of the pyrolysis processes that decomposed material using heat, with more than 150°C in temperature and limited or without oxygen. The material will crack out in to carbon (C), carbon monoxide (CO), methane, water vapor, hydrocarbon gases, and the other volatile substances. Charcoal is a pure carbon formed in pyrolysis process. Processing the coconut shell into a charcoal using pyrolysis process will certainly add its calorific value, compared with the original coconut shell. According to [3], coconut shell charcoal can have a calorific value up to 7.332 kcal/kg, while the original coconut shell can only produce 4.950 kcal/kg.

Drum kiln is one of the simple pyrolysis tools that commonly used by society to produce charcoal. Pyrolysis process occur inside the closed drum with a limited oxygen supply from air inlets in the side and below of the drum. Heating energy comes from inside the system by combustion of the raw material. Some research about coconut shell kiln already done before. The designed of venturi-drum kiln with optimum capacity of 12.45 kg coconut shell and succeeded to reach pyrolysis temperature up to 908.56 °C [3]. Other design made by metal drum kiln shown the optimum capacity of 11 kg and pyrolysis temperature of 401.80 °C [4]. The result of pyrolysis process highly depends on the pyrolysis temperature occurred in the kiln. This study aimed to design and conduct a performance test of drum kiln with vertical partition for coconut shell carbonization. The vertical partition was added inside the as a modification kiln to limit the direct combustion and high thermal heating and air circulation inside the bulk of coconut shell. Thus, using this vertical partition, the temperature inside each chamber can be easier to maintain.

## 2. Material and Methods

### *2.1 Design Approach*
As the vertical partition limit the direct contact between the coconut shells, it was expected that the temperature and air circulation in every chamber would also be limited and able to be controlled.

### *2.2 Tools and Materials*
The tools required to make the kiln available on workshop i.e electric welding, grinding, metal cutting, etc. For the performance test were type K thermocouples; autonics recorder; KANOMAX climomaster and scale weight. Materials used to make this kiln were iron plate with 2 mm thickness for the kiln wall and 3 mm thickness for kiln base. The coconut shell and husk were used as the raw material for pyrolysis process.

### *2.3 Research Procedure*
This research started by collecting information regarding coconut shell and pyrolysis process characteristics. Design formulation and analysis were done by using SolidWorks 2016 software. The desired temperature for this slow pyrolysis were 400 °C. The kiln was fabricated in the workshop, and tested for the pyrolysis time, temperature, charcoal yield and calorific value., The performances of kiln with vertical partition and without vertical partition were evaluated. The research procedure can be seen as in Fig. 1.

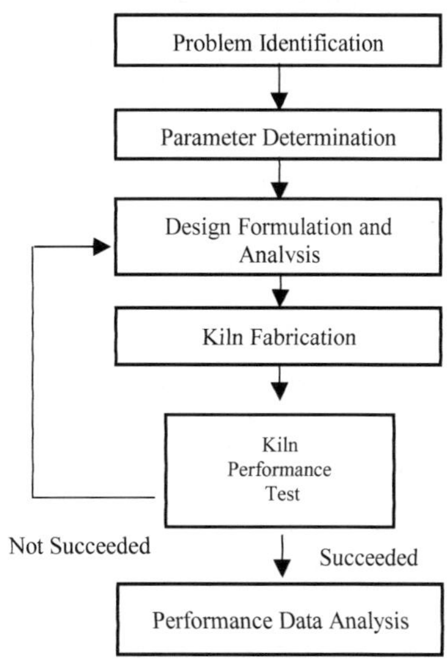


**Figure. 1** Research procedure

*2.4 Performance Test Parameters*

The kiln parameters were determined in order to represent the real functional and structural performance.

(a) Pyrolysis Rate

Pyrolysis time were measured from the first time the loading path closed until the smoke diminished and the charcoal yield were taken from the kiln. Pyrolysis rate is the ratio between the charcoal yield and the pyrolysis time.

(b) Pyrolysis Temperature

Temperature is the most important factor determining the entire pyrolysis process and the charcoal yield result. There were five measurement points; the left, right, and back pyrolysis chambers; kiln wall, and kiln chimney.

(c) Charcoal Yield

The charcoal yield was determined from the percentage of charcoal formed from the coconut shell raw material. It also shown the efficiency of pyrolysis process.

(d) Charcoal calorific value

As the charcoal result is a solid fuel, the calorific value should be tested in order to know the ability of the charcoal to produce energy. Calorific value was tested in Research and Development Center of Forest Products (P3HH) Laboratory, Ministry of Environment and Forestry, Bogor, Indonesia.

*2.5 Kiln Performance Test*

The performance test was done in several times. The data obtained by placing the thermocouple in several places; each of pyrolysis chamber, kiln wall, and kiln chimney. The wind velocity, relative humidity and environment temperature were also measured by using KANOMAX Climomaster around 1 m in front of the main oxygen inlet. The temperature was recorded every 5 minutes. After the smoke diminished, the charcoal was taken through the unloading door and measured for the further analysis.

## 3. Result and Discussion

*3.1 Functional Design*

The main function of kiln is where the pyrolysis process taking places. There should be a uniform temperature distribution inside the kiln to produce a high yield of charcoal. Thus, the initial ignition chamber and some air inlets on the wall provided to support an adequate air flow inside the kiln. The function explanation for the kiln parts is shown in Table 1.

**Table 1.** Functional design of the kiln

| No | Part | Function |
|---|---|---|
| 1 | Initial combustion chamber | A place to ignite the first combustion process inside the kiln. |
| 2 | Main chamber | The coconut shell and husk placed inside this chamber. This is also where the pyrolysis process take place. |
| 3 | Air inlets | Provide limited oxygen so that pyrolysis process can be occurred. |
| 4 | Chimney | This part acted as pyrolysis smoke outlet. |
| 5 | Unloading | To take the charcoal yield out from the main chamber. |
| 6 | Vertical partition | This partition was added in order to limit the direct contact between coconut shells and husks. |

### *3.2 Structural Design*
The detail part of the kiln is shown in the Figure 2 below.

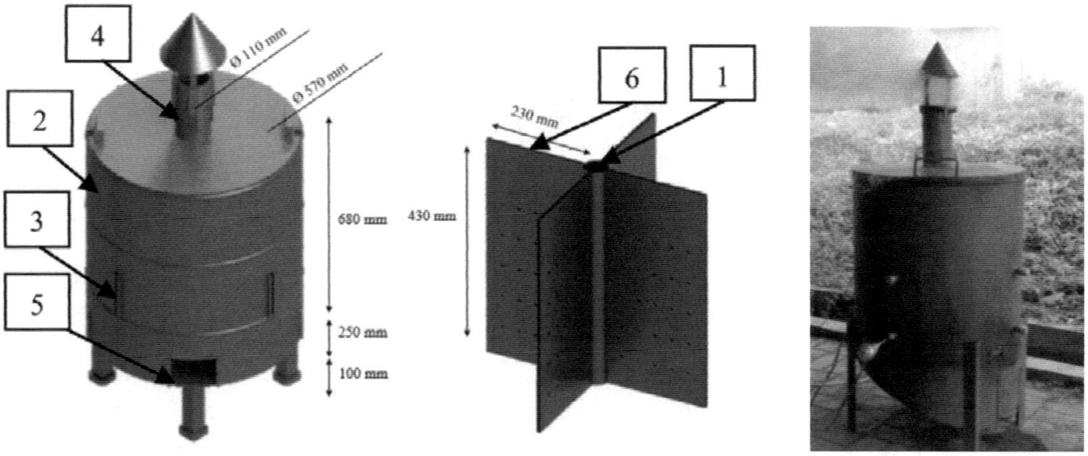


**Figure 2.** Kiln part and measurement design and actual fabrication

(1) Initial combustion chamber
This initial combustion chamber used to ignite the first combustion. There were around 5 cm of coconut shell and husk placed inside this chamber as the first part of the raw material to be burnt. This initial chamber shaped like a hollow tube, with diameter of 100 mm and height of 480 mm.

(2) Main chamber
The pyrolysis process was conducted inside this main chamber. In order to made the proper dimension of the kiln, there were some calculation ensued. Bulk density of coconut shell is 510 kg/m$^3$ [5]. The mass for one batch of pyrolysis is 15 kg and volume of the kiln main chamber was 174 liter.

(3) Air inlets
The base part of the kiln was made of iron plate with 2 mm holes which were distributed uniformly. This main air inlet was not directly in contact with the environment, but it was bordered by the unloading part so that the user can easily control the amount of the air going to the kiln. There were another eight air inlets in the kiln wall, with diameter 0.5 inch. This side air inlet helps the circulation of oxygen right into the main chamber in hope that the temperature would be evenly distributed.

(4) Chimney
This chimney had 100 mm in diameter. The main function was creating a turbulence effect for the high temperature smoke inside the kiln and it would distribute the temperature even more uniform.

(5) Unloading
In order to simplify the process for getting the charcoal out, an unloading part provided downside the main chamber. There was horizontal partition between the kiln main chamber and unloading chamber, which also acted as the base plate with the main air inlet distributed in it. After the pyrolysis process were done, as the horizontal partition removed, the charcoal would fall into unloading chamber and could be taken.

(6) Vertical partition
There were total of four vertical partition in this kiln. It was made from iron plate with 3 mm thickness. There were also some holes created in order to let the heat keep exchanged from one chamber to another as the air flowed. The dimension of each partition was 230 mm x 430 mm.

### *3.3 Kiln Performance Test*
There were two kind of test conducted in this research; performance test for kiln with vertical partition, and performance test for kiln without partition. Each of it has three times different performance tests. The results are shown in Table 2 and 3 below:

**Table 2.** Performance test results of kiln with vertical partition

| Test | Initial Mass (kg) | Average Temperature (°C) | Maximum Temperature (°C) | Charcoal Mass (kg) | Yield (%) | Time (min) | Wind speed (m/s) | Environment Temp (°C) | RH (%) | Calorific Value (kcal/kg) |
|---|---|---|---|---|---|---|---|---|---|---|
| 1st | 15 | 263.67 | 512 | 3.25 | 21.67 | 80 | 0.34 | 31.03 | 67.57 | 7,603 |
| 2nd | 15 | 324.64 | 548 | 3.4 | 22.67 | 85 | 0.17 | 29.55 | 74.35 | 7,518 |
| 3rd | 15 | 298.71 | 534 | 4.02 | 26.8 | 70 | 0.35 | 28.2 | 80.6 | 7,395 |
| Average | 15 | 295.67 | 531.33 | 3.56 | 23.71 | 78.33 | 0.29 | 29.59 | 74.17 | 7,505.33 |

**Table 3.** Performance test results of kiln without vertical partition (control)

| Test | Initial Mass (kg) | Average Temperature (°C) | Maximum Temperature (°C) | Charcoal Mass (kg) | Yield (%) | Time (min) | Wind speed (m/s) | Environment Temp (°C) | RH (%) | Calorific Value (kcal/kg) |
|---|---|---|---|---|---|---|---|---|---|---|
| 1st | 15 | 254.92 | 684 | 2.49 | 17.58 | 85 | 0.3 | 28.33 | 75.83 | 7,629.00 |
| 2nd | 15 | 363.37 | 626 | 2.95 | 20.67 | 85 | 0.54 | 28.2 | 79.15 | 7,318.00 |
| 3rd | 15 | 345.27 | 642 | 2.72 | 18.15 | 80 | 0.73 | 31.45 | 64.73 | 7,428.00 |
| Average | 15 | 321.19 | 650.67 | 2.72 | 18.80 | 83.33 | 0.52 | 29.33 | 73.24 | 7,458.33 |

Pyrolysis process depends on some factors, such as the raw material composition, properties (density, thermal conductivity, thermal capacity, material permeability), size of particle, temperature and heating rate. Temperature and heating were very important in term of material conversion rate and the result composition of pyrolysis process [6]. The coconut shell was already been crushed into smaller pieces. The vertical partitions were placed in the center of two side air inlets, in order to let all the chamber filled with the same amount of oxygen, as shown in Figure 3.

**Figure 3.** Coconut shell and husk before combustion in the kiln with vertical partition (left) and without vertical partition (right)

Kiln with vertical partition had average pyrolysis rate of 0.191 kg/min, little bit higher than the kiln without vertical partition which was 0.182 kg/min. As the vertical partition limit the contact between four chambers of coconut shell, the pyrolysis process would concentrate into each chamber. With lower amount of raw material that should be turned into charcoal, the heat would easily spread inside each particular chamber and the pyrolysis rate and time would be shorter, thus the pyrolysis rate can be increased.

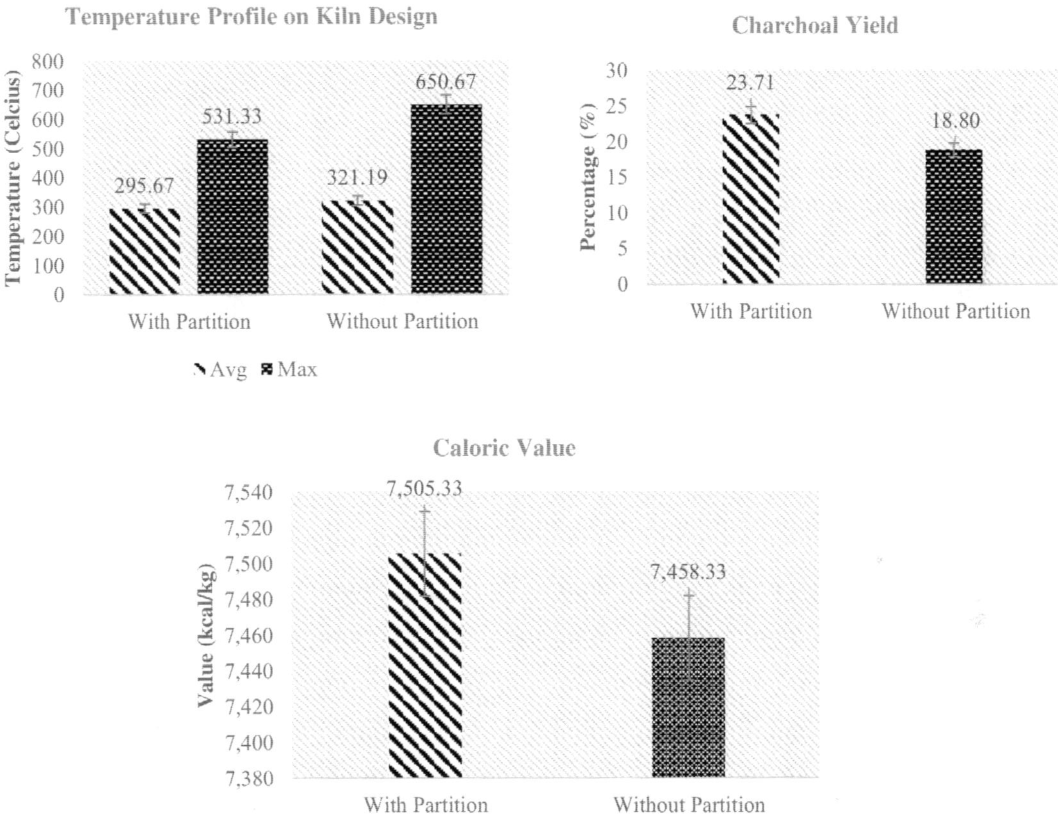


**Figure 4.** Kiln performance data result (a) Temperature; (b) Charcoal Yield; (c) Caloric Value.

As the vertical partition limit the direct contact between coconut shells, can be seen in Figure 4(a) that the partition could maintain the average temperature in 295.67°C while kiln without vertical partition went up to 321.19°C. For maximum temperature, the vertical partition was able to hold the heat in 531.33°C when the temperature inside kiln without vertical partition increased up to 650.67°C. Pyrolysis occurred in 240-350°C is the decomposition of cellulose; in 200-260 °C is the decomposition of hemicellulose and in 280-500 °C is the decomposition of lignin. This lignin decomposition resulted in higher carbon composition, up to 45.7% out of all carbon contained in a material. The heat above 500 °C will decompose the material into the other forms like synthetic gas and bio oil [7]. As the temperature went higher, there would be more material decomposed. With this vertical partition, the charcoal yield result was 23.71 % more higer than the kiln without vertical partition 18.80; as can be seen in Figure 4(b). The average calorific value of kiln with vertical partition is 7,505.33 kcal/kg, which also higher than the control, 7,458.33 kcal/kg.

## 4. Conclusion

The drum kiln has 6 main components, which are pyrolysis main chamber, initial combustion chamber, air inlet, chimney, unloading part and vertical partition. Kiln with vertical partition best performance was in average temperature of 295.67 °C, average maximum temperature 531.33 °C, charcoal yield 21.67 - 26.80%, average carbonization time 78.33 minute and calorific value of 7,505.33 kcal/kg, while kiln without vertical partition best performance was in average temperature of 321.19 °C, average maximum temperature 650.67 °C, charcoal yield 17.58 - 20.67%, average carbonization time 83.33 minute and calorific value of 7,458.33 kcal/kg. Vertical partition able to limit and control kiln average and maximum temperature also increase charcoal yield.

## References

[1] Pirmansah, Alit. (2015). Coconut Statistical Year Book 2015: Asian and Pacific Coconut Community (APPC), Jakarta Indonesia. pp 10-13

[2] Suhardiyono, H. (1998). Coconut: Cultivation and its Benefit. Jakarta: Kanisius. pp 23-26

[3] Hasanah, N. (2013). Design and Performance Test of Venturi Drum Type Metal Kiln for Coconut Shell Carbonization. Indonesia: Faculty of Agricultural and Engineering Technology, Bogor Agricultural University.

[4] Isriyanto. (1993). Modification and Performance Test of Metal Kiln as Carbonization Tool. Indonesia: Graduate School, Bogor Agricultural University.

[5] Kumar, N. and D. Kumar (2014). Utilization of coconut shell in different forms in concrete. International Journal for Scientific Research and Development, vol 2, no. 7, pp. 158-160

[6] Rahimah, F. (2014). Kiln Design for Palm Oil Shell Carbonization. Indonesia: Faculty of Agricultural Engineering and Technology, Bogor Agricultural University.

[7] Antal M.A. and G. Varhegyi (1995). Cellulose pyrolysis kinetics: the current state of knowledge. Ind. Eng. Chem. Res., vol 34, pp. 703-717

# Effect of Roadways Plantation on Signal Propagation Analysis in Connected Autonomous Vehicle Communication

**J S C Turner[1,3], Shahriman A B[1, 2], A Harun[1, 3,5], M S M Hashim[1, 2], Z M Razlan[1, 2], W K Wan[1, 2], Z Ibrahim[1, 2], Abdullah A H[2], M A Fadzilla[1, 2], K A A Kassim[4], M S A Khalid[4], Z Jawi[4] and M H M Isa[4], S A Z Murad[3] and R C Ismail[3]**

[1]Motorsports Technology Research Unit (MoTECH), Universiti Malaysia Perlis, Pauh Putra Campus, 02600 Arau, Perlis, Malaysia
[2]School of Mechatronic Engineering, Universiti Malaysia Perlis, Pauh Putra Campus, 02600 Arau, Perlis, Malaysia
[3]School of Microelectronic Engineering, Universiti Malaysia Perlis, Pauh Putra Campus, 02600 Arau, Perlis, Malaysia
[4]Malaysian Institute of Road Safety Research (MIROS), Lot 125-135, Jalan TKS 1, Taman Kajang Sentral, 43000 Kajang, Selangor, Malaysia
[5] Corresponding author: aziziharun@unimap.edu.my

**Abstract.** At present, the development of autonomous vehicle has altered the outlook of modern transportation worldwide. The state-of-the-art vehicular communication for transportation system is advancing, especially in vehicle to infrastructure (V2I) communication. An effective communication between vehicle and infrastructure has become a significant part of autonomous transportation criteria. The necessity for high quality of service communication inspire for good planning and preparation in communication process. Per se, this paper proposes vegetation attenuation models for advance planning of communication process between vehicle to infrastructure, defined mainly by plants, trees and vegetation along the roadways in Malaysia. The channel measurement performed in Universiti Malaysia Perlis test-bed having large tall trees and low shrubs along the routes resulted in several interesting results which would shape the planning of CAV communication. It is observed that communication close to low plantation or shrub requires high power consumption as the range is significantly reduced. It is also learned that certain types of plantations allows for different level of signal attenuation depending on the antenna heights. The research also found out that the attenuation profile follows strictly the log normal distribution and as such certain planning could be made to reshape the communication process to cater for this.

## 1. Introduction

A reliable, sustainable and robust communication network is a fundamental aspect of connected transportation system especially in connected autonomous vehicle program. Such fundamental is very critical in vehicle-to-vehicle as well as vehicle-to-infrastructure communication and data transmission. In vehicle-to-infrastructure communication, which is typically wireless and bi-directional, robust and reliable communication could be the deciding factor for accident avoidance and smooth transportation. With the advent of autonomous vehicle, roadways would soon become more sophisticated transportation alleys and as such communication with lane markings, road signs, and traffic lights would have to be properly managed.

To manage such complexity, strong communication foundation is required. Although there are several communication technologies available, such as WIMAX, Wi-Fi, LTE and DSRC, these technologies may not be able to support low latency, high accuracy and reliable data transmission required by Connected Vehicle Technology (CVT) applications. [1]. While Dedication Short Range Communication (DSRC) provides low latency, fast network connectivity, various application of CVT may not be supportable [2]. In order to ascertain the communication and the reliable network connectivity, a thorough study has to be done to understand the factors affecting the V2I communication. One of the critical area of communication that require special attention is channel communication in V2i [1-2].

The concept of V2I communication and V2V communication is the latest advanced topic in autonomous connected vehicle field. The interaction of V2I uses wireless communication technologies which can provide wireless exchange of operational data between vehicles and highway infrastructure. These are intended to to make driving safe, improve mobility, to reduce greenhouse gas and maintenance cost. V2I communication can be applied to all types of vehicle and road thus transforming them into smart-infrastructure. Each car can act as a moving sensor, this car can provide information to other car and infrastructure wirelessly, exchange information and many more. For example, cars can communicate with each other and alert drivers to roadside hazards ahead, wirelessly. It comprises a wireless network where automobiles send messages to each other with information about what they're doing. This data would include speed, location, and direction of travel, braking, and loss of stability. This interaction must require research in creations of algorithms for managing traffic, avoiding collisions and other vehicular tasks that use data exchanged between vehicles and infrastructure elements to perform calculations that can recognize high risk situations [3]. This result in warning and driver alerts through specific actions. For example, the ability for traffic signal systems to communicate information such as signal phase and timing (SPAT) to the vehicle in order to signal safety and warning to drivers. Most of V2I system are closely related and rely on V2V system which uses Intelligent Transportation System (ITS). These applications are usually based on roadside sensors and require observation on toll control or speed measurement and many more.

Several wireless technologies such as DSRC, LTE and VANET supports vehicular communication and prior studies suggested that most measurements in V2I communication use Wi-Fi such as WAVE or IEEE802.11p protocol, IEEE 802.11b, etc. as it can support both V2V and V2I communication. WiMAX is another recent technology that some researchers use to compare with Wi-Fi for evaluating the feasibility of vehicular network that focuses in V2I communication [4]. Apart from that, various research works related to protocol and security issues has been highlighted as one of challenges in vehicular communication due to large volume of information being access and exchanged in V2I communication [5]. However, it is also important to consider other sensing challenges such as source of blockage in communication path which include millimeter wave link, vehicles, pedestrians, static objects, trees and buildings. There are prior studies that have been disclosed regarding significant impact of trees, vegetation and antenna orientation as factors influencing the coverage areas [6]. As such, preliminary studies in areas where source of blockage in communication path is necessary. A good number of studies have been carried out on wireless sensor network architecture of vegetation and agriculture [6][7][8][9]. There many models being adapted and used for agricultural and vegetation and up to present, limited studies and lack of appropriate models is found in the studies of transportation environment with vegetation for CAV applications. This research looks into understanding the attenuation factors affecting such communication via channel measurement. Therefore, this paper evaluated a generic form of vegetation attenuation models, path loss and proposes appropriate parameters in the implementations of models for CAV communication.

This paper is organized as follows; in section 2 describes vegetation models and parameters used in path loss and received power calculations. Section 3 presents measurement location, setup and device that were used. Section 4 presents experimental designs and followed by results and discussions in section 5. Conclusions are drawn in Section 6.

IOP Conf. Series: Materials Science and Engineering **557** (2019) 012056 doi:10.1088/1757-899X/557/1/012056

## 2. Background

Channel measurement has been used extensively to provide comprehensive understanding of the surrounding area of interest for communication planning purpose. Many works has been performed in the infrastructure deployment planning as well as in the agricultural field. Authors in [7-11] have been performing such studies in agricultural field specifically for sensor deployment and monitoring system execution. Nevertheless, there has not been much work done in the area of road transport planning especially concerning the deployment of autonomous vehicle. In this study, a good understanding of signal propagation in roadways is essential for establishing reliable communication standard.

For the radio-wave propagation is free space, the path loss (L) can be predicted using the free space loss (FSL) equation,

$$L_{FSL}(\text{dB}) = -27.56 + 20\log_{10}(f) + 20\log_{10}(d) \tag{1}$$

Where $f$ is the frequency in MHz, $d$ is the distance between the isotropic transmitting and receiving antennas in meters.

The 1st Fresnel Zone [12-14] is known to be the most to affect the performance of wireless network as obstacle in this zone can create out of phase signal from 0 to 90 degrees, thus it is important to estimate the maximum radius of 1st Fresnel Zone before conducting channel measurements on the test-bed. The radius, $r$ of Fresnel Zone at the greatest distance is at the center (B) of line-of-sight (LOS) between the transmitting and receiving antennas as depicted in Figure 1. According to authors in [13] and [14], the maximum allowable obstruction height must be within 60% of clear Fresnel Zone for optimum radio performance. The 1st Fresnel Zone can be calculated using the general equation as in (2). In equation (2), $n$ is the $n^{th}$ Fresnel Zone radius in meter, $d_1$ is the distance of point B (center) from the transmitter ($T_x$) in meter, $d_2$ is the distance of B from the receiver ($R_x$) in meter and $\lambda$ is the wavelength of the signal in meters as in (3).

$$F_n = \sqrt{\frac{n\lambda d_1 d_2}{d_1+d_2}} \tag{2}$$

$$\lambda = \frac{c}{f} d_1 = d_2 \tag{3}$$

where, $c$ is the speed of light in vacuum ($3\text{x}10^8 \text{ms}^{-1}$), ($d_1 + d_2$) is the total distance ($D$) in meter and $f$ is the signal frequency in Hertz. Based on Figure 1, the ($r$) at its maximum distance and the allowable height for obstruction can be calculated using (4) and (5), respectively.

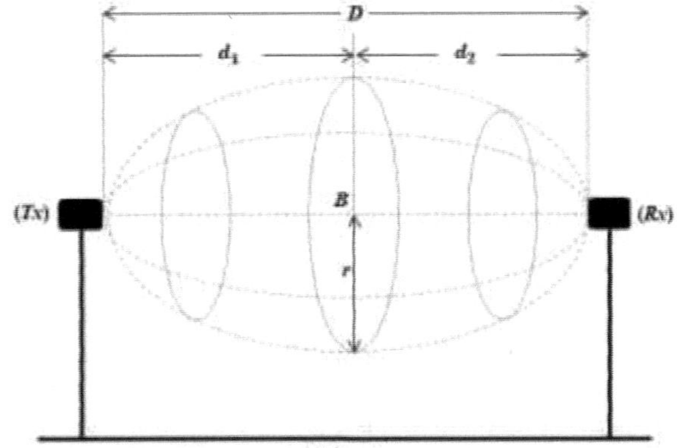


**Figure 1.** The 3D Ellipsoid of the 1st Fresnel Zone

$$R = 8.657\sqrt{\frac{D(km)}{F(GHz)}} \quad (4)$$

$$R = 8.657\sqrt{\frac{0.6D}{F}} \quad (5)$$

where, $R$ is the Fresnel Zone Clearance Radius in meter, $D$ is the total distance in kilometers and $F$ is the frequency transmitted in gigahertz.

Signal reflections, diffractions and scattering from medium and surfaces impact propagation in mobile communication system [12]. As such, the reflection from the ground and surface from surrounding objects must be taken into account by considering the ground-bounce reflection and the surface roughness [13]. The ground-bounce reflection generates a mirror-like or specular signal reflection with an angle of incidence, $\theta_{inc}$ equals to the angle of departure, $\theta_{dep}$ or ($\phi_1 = \phi_2$) as in (6), where the signal is reflected on a smooth and flat surface as in Figure 2 [12]. Hence, knowing both $T_x$ (transmitter) and $R_x$ (receiver) height, distance $d_1$ and $d_2$ can be determined. In Figure 2, $H_t$ and $H_r$ is the transmitting and receiving antennas height, respectively, $d$ is the total distance between the transmitting and receiving antennas and $d_{slant}$ is the total distance of direct wave (slant range) as in (7) and (8), respectively. It is also understood that the received signal can be determined by summing the direct reflected wave at the receiver whereas the magnitude and phase of reflected signal can be determined by ground's reflection coefficient. The reflection coefficient is denoted by rho (ρ = -1) with smooth surface and small angle incidence.

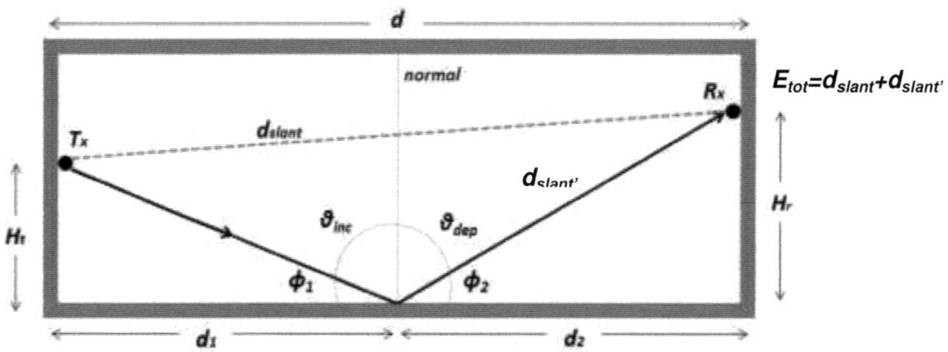

**Figure 2.** Two-ray ground reflection model [10][12]

$$\phi_1 = tan^{-1}\left(\frac{H_t}{d_1}\right), \phi_2 = tan^{-1}\left(\frac{H_r}{d_2}\right) \quad (6)$$

$$d = d_1 + d_2 \quad (7)$$

$$d_{slant} = \sqrt{d_2 + (H_t - H_r)^2} \quad (8)$$

Figure 2 shows the signal reflection on smooth and flat surface and this model is commonly known for predicting large-scale signal strength over distance in kilometres for mobile radio system and line of sight in urban environment. This model considers both direct path and ground reflected propagation path between the transmitter and receiver. It is also a widely used model in wireless sensor network when vegetation is present with antenna measurement at significant height and outside the vegetation. In addition, based on Figure 2, the received power, $P_r$ at a distance $d$ from the transmitter for Two-ray ground bounce model can be calculated using equation (9). Where $P_t$ is the transmitted power in dB,

$G_t$ is the transmitter antenna gain, $G_r$ is the receiver antenna gain, $d$ is ($T_x$-$R_x$) separation distance in meters . Antenna gain, $G$ can be calculated using equation (10) and the (11).

$$P_r = P_t G_t G_r \frac{H_t^2 H_r^2}{d^4} \quad (9)$$

$$G = \frac{4\pi A_e}{\lambda^2} \quad (10)$$

$$\lambda = \frac{c}{f} = \frac{2\pi c}{\omega_c} \quad (11)$$

Where $A_e$ is the effective aperture which is related to physical size of antenna, $f$ is the carrier frequency in Hertz, $\omega_c$ is carrier frequency in radius per sec and $c$ is speed of light in meters per second. At large values of $d$ the received power and path loss is independent of frequency. Path loss for Two-ray model with antenna gain is in equation (12).

$$PL(dB) = 40logd - (10logG_t + 10logG_r + 20logH_t + 20logH_r) \quad (12)$$

At small $d$, an addition of equations must be used to compute the total E-field using equation (13). $E_{tot}(d, t)$ is the total electric field, $E_0$ is the free space E-field in units of V/m, at a reference distance $d_0$ from the transmitter and $d_{slant}$ and $d_{slant'}$ represents the two propagating waves arriving at receiver which is the direct wave and the reflected wave that travel at distance $d_{slant}$ and $d_{slant'}$, respectively as shown in Figure 2.

$$E_{tot}(d,t) = \frac{E_0 d_0}{d_{slant}} cos\left(\omega_c\left(t - \frac{d_{slant}}{c}\right)\right) + (-1)\frac{E_0 d_0}{d_{slant'}} cos\left(\omega_c\left(t - \frac{d_{slant'}}{c}\right)\right) \quad (13)$$

Another method of determining the ground reflection is by taking the surface roughness into consideration. The Rayleigh criterion [13], $H_R$ can be used to quantify the flatness of the surface depending on the terrain variations ($\Delta_h$) as in (14), where $d$ is the distance between transmitting and receiving antennas, $\lambda$ is the wavelength (in meters), $H_t$ is the transmitting antenna height and $H_r$ is the receiving antenna height. The terrain variations can cause signal to experience 90 degree phase shift at the reflection point [13]. That is, if the surface is having roughness which is almost negligible ($\Delta_h < H_R$), then the surface can be considered as smooth and reflection may occur. With severe roughness ($\Delta_h > H_R$) the surface will experience scattering instead.

$$H_R = \frac{\lambda d}{8(H_r + H_t)} \quad (14)$$

In addition, there a number of models that provides prediction of excess attenuation due to existence of vegetation. One of them is Non-zero Gradient (NZG) model. Proposed by Seville and Craig, this model is used for frequencies above 5GHz and currently under ITU-R recommendation [18-19][24]. The NZG model is given by equation (15) where $R_0$ is the initial gradient,$R_\infty$ is final gradient of attenuation curve, $k$ is the offset of the final gradient and $d$ is the vegetation depth in meters [18]. The extension of this model that takes into account the antenna beam width and frequency is given by equation (16), where $W$ is the maximum effective coupling width between the transmitting and receiving antennas that lies between vegetation medium (largest measured vegetation depth), $f$ is the signal frequency in GHz and $a$, $b$, $c$, $k$, $R_0$ and $R_\infty$ are constants given in [19].

$$Atten = R_\infty d + k\left(1 - exp\left(-\frac{R_0 - R_\infty}{k} d\right)\right) \quad (15)$$

$$A = \frac{R_\infty}{f^a W^b} d + \frac{k}{W^c}\left(1 - exp\left(-\frac{(R_0 - R_\infty)W^c}{k} d\right)\right) \tag{16}$$

Modified Exponential Decay (MED) model is another vegetation model that was proposed by Weissberger and the modified version is in CCIR recommendations [20]. This model is known to be very simple but does not include the measurement of propagation mechanisms but can be used for radio-wave propagation through the foliage medium. A more simplified and generalized form of MED vegetation model is given by equation (17) [6].

$$P_r(dB) = X f^Y d^Z \tag{17}$$

Where *f* is the frequency in megahertz (MHz), *d* is the vegetation depth in meters and X, Y and Z are the model parameters.

Weissberger's model [7] represented by equation (18) and it is applicable where a ray path is blocked by dense, dry, in-leaf trees found in temperate climates. It is applicable in the situations where the propagation is likely to occur through a grove of trees rather than by diffraction over the canopy top, and is given by

$$L_W(\mathrm{dB}) = \begin{cases} 1.33 \times f^{0.284}\, d^{0.588}\; 14m < x \leq 400m \\ 0.45 \times f^{0.284} d_f,\; 0m \leq d_f < 14m \end{cases} \tag{18}$$

Where *f* is the frequency in GHz, and $d_f$ is the foliage depth in meters.

Other modified vegetation models under MED models are such as ITU-R model [24][25], COST235 model [21][25] and FITU-R model [23][25] as in equation (19), (20) and (21), respectively. For these models, *f* represents its frequency in megahertz (MHz) and *d* represents vegetation depth in meter (m) based on Annex 1 in ITU-R recommendation [24]. ITU-R model is applicable for distance below 400 meters and its *X* and *Y* parameters can be changed for improvement in the model to find the best fit [19]. In other models such as COST235 and FITU-R model, they consider more specific geometry measurements with given parameters for in-leaf and out-of-leaf conditions.

$$L_{ITU-R}(dB) = 0.2 \times f^{0.3} d^{0.6}\;\; (d < 400) \tag{19}$$

$$L_{COST}(dB) = \begin{cases} 26.6 \times f^{-0.2} d^{0.5}\;\; out - of - leaf \\ 15.6 \times f^{-0.009} d^{0.26}\;\; in - leaf \end{cases} \tag{20}$$

$$L_{FITU-R}(dB) = \begin{cases} 0.37 \times f^{0.18} d^{0.59}\;\; out - of - leaf \\ 0.39 \times f^{0.39} d^{0.25}\;\; in - leaf \end{cases} \tag{21}$$

Moreover, Harun et al. have also shown that WSN channel can be modeled using a log-normal model. The log-normal model is described by (22) [9]

$$P_r(d) = P_{r_0} - 10\alpha log_{10}(d) + X_\sigma \tag{22}$$

Where $P_r(d)$ is the received power (in dBm) at a distance *d* (in meters) from the transmitter, $P_{r_0}$ is the signal strength at 1 m antenna separation, α is the path loss exponent and $X_\sigma$ represents a Gaussian random variable with zero mean and standard deviation of $\sigma\ dB$.

IOP Conf. Series: Materials Science and Engineering **557** (2019) 012056 doi:10.1088/1757-899X/557/1/012056

## 3. Measurement Location and Equipment

The measurement has been conducted on Universiti Malaysia Perlis campus route highlighted in yellow line as depicted in Figure 3. This route is chosen because there are two types plantations planted on each roadside along the route highlighted in yellow. Tall trees with thick leaves and trunk are located on the left lane while low plantation or shrubs are located on the right lane.

It is very crucial to simulate Malaysia roadways with existing plantations in conducting this research especially as Malaysia roadways in the Northern state mostly have heavy plantations along the roadside, highways, streets, etc. This particular route has been identified for data collection to represent similar existence of plantations around Malaysia roadways. Shrub in Figure 4 has been chosen to represent low plantations normally found along Malaysian road.

**Figure 3.** Universiti Malaysia Perlis campus route. The route highlighted in yellow has tall trees and shrub along the roadside.

**Figure 4.** Shrub or low plantations in a row along the roadside.

**Figure 5.** Tall trees with thick leaves and trunks in a row along the roadside.

Another important field to be measured is trees which is very prevalent around the roadways in Malaysia. Figure 5 shows tall trees with thick leaves and trunk. The trees represent similar plantations that can be found along roadsides, highways and streets of urban areas in Malaysia. Thus, this type of plantation is used in this research for data collection.

Data collection was performed using XBEE S2C wireless sensor nodes as the receiver and transmitter. The illustration of the methodology is shown in Figure 6. The transmitter and receiver are placed on top of metal pole that can be adjusted for multiple antenna heights. In this experiment, the data collection is focused on trees and shrub plantation areas along the roadside with antenna height of 0.3 meter and 1.3 meter, respectively. The software used for programming the wireless nodes is XCTU Software [16]. The receiver was programmed as remote while the transmitter was programmed as local or as base station which is connected to the laptop running with XCTU Software. The nodes operated at 2.4 GHz frequency band with 3 dB antenna gain [15]. The receiver sensitivity is at -95 dBm and the rest of node's specification is as given in Table 1.

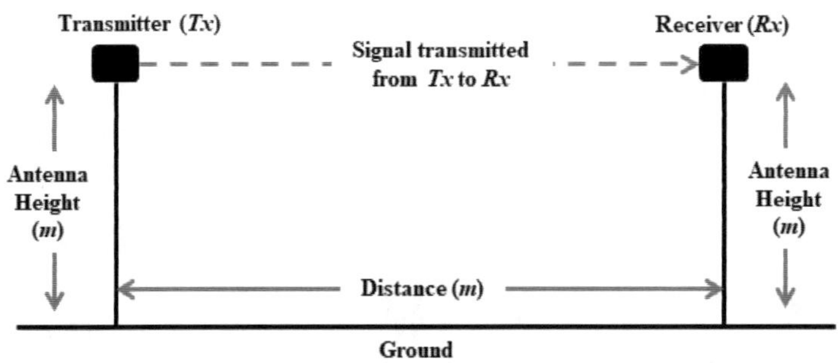


**Figure 6.** Transmitter and receiver measurement setup

**Table 1.** XBEE S2C specification used in data collection.

| Specification | Xbee Series 2 |
|---|---|
| Performance | |
| **Indoor/urban Range** | up to 133 ft. (40 m) |
| **Transmit Power Output** | 2mW (+3dBm) |
| **RF data Rate** | 250,000 bps |
| **Serial interface data rate** | 1200 - 230400 bps (non-standard baud rates also supported) |
| **Receiver sensitivity** | -95 dBm (1% packet error rate) |
| Power Requirement | |
| **Supply Voltage** | 2.8 – 3.4 V |
| **Operating current** | 40mA (@ 3.3 V) |
| **Power down current** | < 1 uA @ 25oC ISM |
| General | |
| **Operating Frequency Band** | ISM 2.4 GHz |
| **Dimensions** | 0.960" x 1.087" (2.438cm x 2.761cm) |
| **Operating temperature** | -40 to 85° C |
| **Antenna options** | Integrated whip, Chip, RPSMA, or U. FL Connector |
| Networking and security | |
| **Supported network topologies** | Point-to-Point, Point-to-Multipoint, Peer-to-Peer & Mesh |
| **Number of channels** | 16 Direct sequence channels |
| **Addressing options** | Pan ID and Addresses, Cluster IDs and Endpoints |

## 4. Experimental Design

The measurement for both plantations was performed by placing the transmitter (*Tx*) at fixed location. The receiver (*Rx*) was placed at every 1 meter distance (*d*) increment from the transmitter along the plantation as depicted in Figure 7 and Figure 8, respectively. The measurement geometry for tree is in line of trees while shrub is into vegetation. Received Signal Strength Indicator (RSSI) values are collected and recorded. Since data collection is focused on plantations along the roadside, the measurements were collected during no traffic condition. In other words, there were no vehicles passing by the road during data collection to ensure stable measurements. The measurement was taken with increment 1 meter until the value recorded has reached its receiver's sensitivity for both shrub and trees. At each plantation, the measurement was taken with 2 different antenna heights, 0.3 meter and 1.3 meter respectively. The measurement was taken for two minutes at each point and is repeated for two times and averaged. The recorded values were concluded and presented. In this experiment, the maximum distance for data collection that can be achieved is 45 meters and 63 meters, respectively.

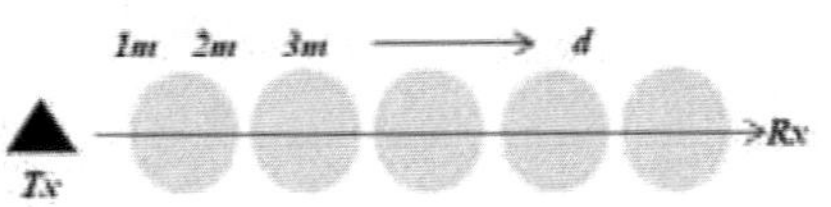


**Figure 7.** Measurement geometry for trees (line of tree)

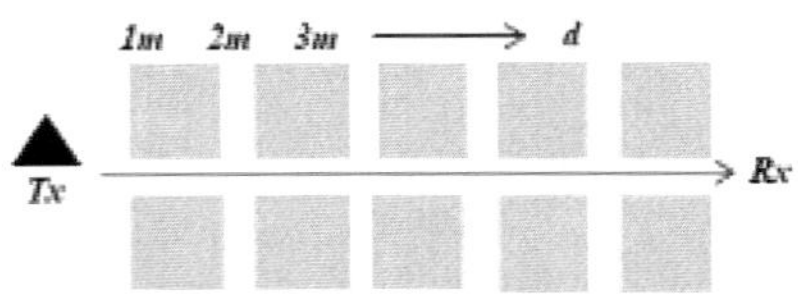


**Figure 8.** Measurement geometry for shrub (into vegetation)

## 5. Results and Discussions

First measurement was performed along the roadside area with shrub plantations. The lane with shrub plantation consisted of thick leaves, was planted and grew in a row along the roadside about 45 meters length. Measurement was also carried out over the path length of 0 meter to 45 meters with 1 meter increment at a time. Figure 9 and Figure 10 shows the result of the measurement.

As signal propagates along shrub path, it is observed that significant drop happened after 6 m distance for 0.3 m antenna height which brought the signal power from -51 dBm to -91 dBm. Low antenna height (almost on the ground) limited the propagation path thus ground reflection and scattering took effect earlier. The different in power loss for the first 3 m distance is 33 dB compared to antenna height at 1.3 m which is 28 dB. The range of the propagation for 1.3 m antenna height is extended further compared to 0.3 m antenna height which shows almost -91 dBm reading at 28 m distance while for 1.3 m antenna height, the signal power is at -77 dBm for the same range.

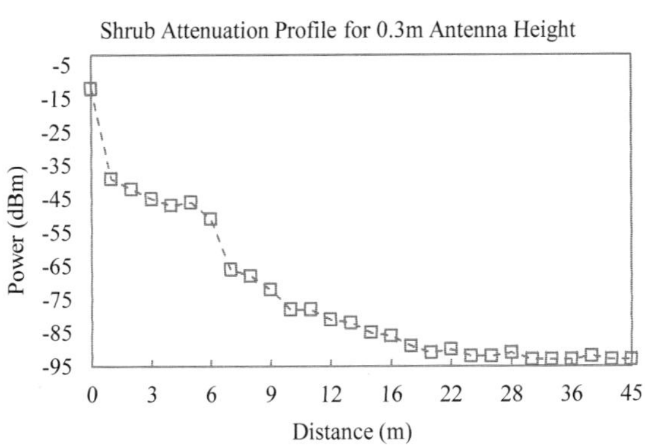


**Figure 9.** Shrub attenuation profile for antenna height 0.3 m

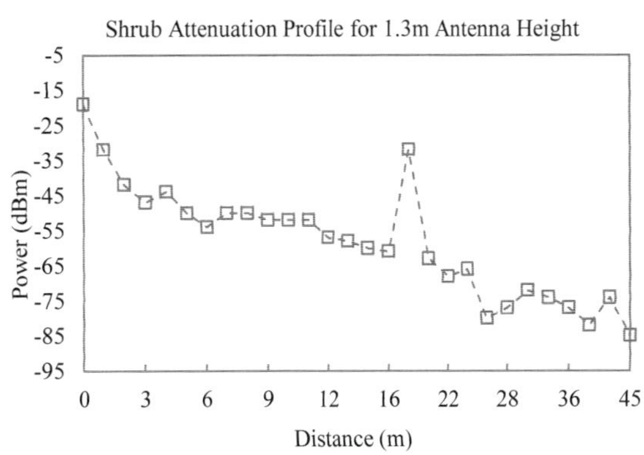


**Figure 10.** Shrub attenuation profile for antenna height 1.3 m

Second experiment was performed at tall trees path which has a range of about 63 m in length. Figure 11 and Figure 12 shows the result from the measurement for roadside with tall trees. As signal propagates along tall trees, power drops significantly for both antenna heights although 0.3 m antenna height recorded extensive power drop compared to that of 1.3 m. This is mainly due to the existence of trees with trunks and leaves that initiated high level of scattering. This phenomena cause signals to diverted in various ways and many find the way back to receiver's end at multiple times due to delays variation as the signal propagates in the Fresnel Zone. Nevertheless, as the distance increases, the difference in signal drops for both antenna heights become more incoherent. It is observed that the drop for 0.3 m antenna height is significant which is at 19 dB from 10 m to 25 m distance while the drop for 1.3 m antenna height grows significantly for the same range which is at 29 dB. Moreover, significant fluctuations in signal can be observed starting at 8 m up to 60 m for 0.3 m antenna height and at 12 m up to 60 m for 1.3 m antenna heights. The presence the trees in between have affected the Fresnel Zones that cause multipath fading in signal propagation. The trees having larger dimension in wavelength compared to the signal has caused the signal to bend around the trees, absorbed or reflected towards each other. Hence, signal combines both constructively and destructively which results in random fluctuations.

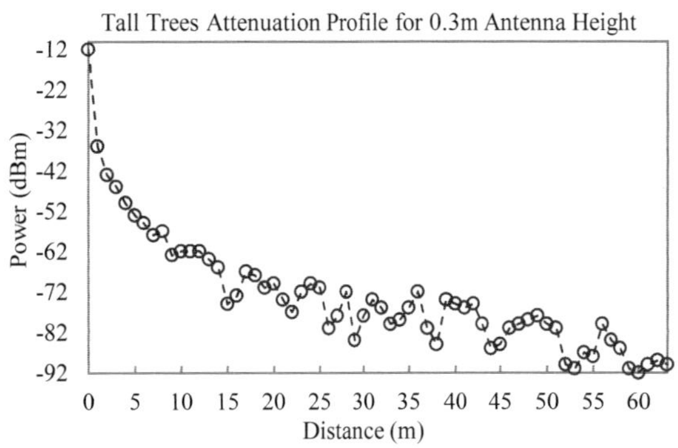


**Figure 11.** Tall trees attenuation profile for antenna height 0.3 m

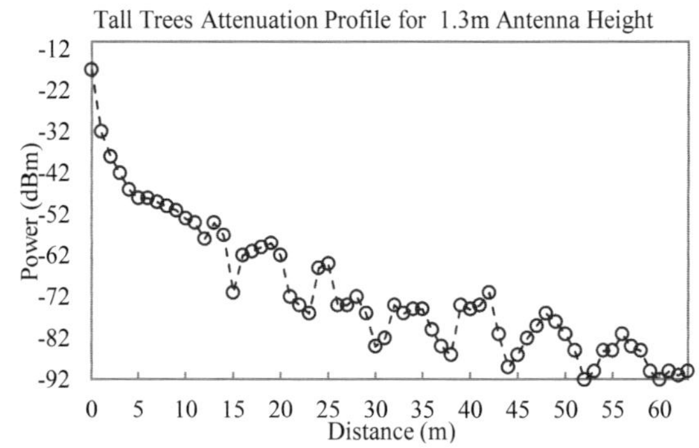


**Figure 12.** Tall trees attenuation profile for antenna height 1.3 m

A comparison between the two surface medium of propagation has been done and the result is as shown in Figure 13 and Figure 14. Based on Figure 9, it is observed that at 0.3 m antenna height the difference between two plantations is very obvious and shrub seemed to be the one causing higher attenuation compared to trees. For 1.3 m antenna height however, the separation is less obvious and it seemed as though the effect is almost similar for both plantations.

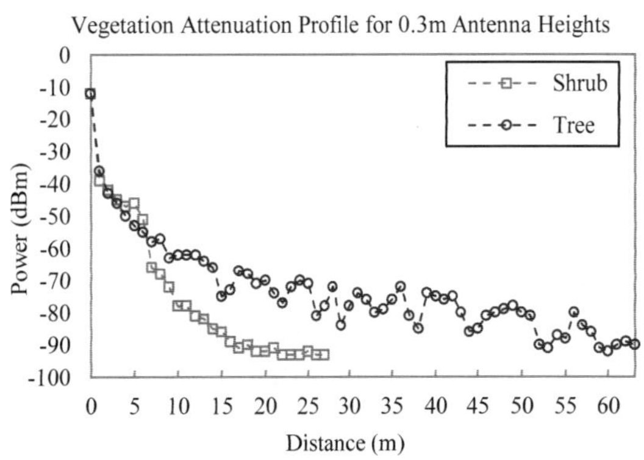


**Figure 13.** Vegetation attenuation profile for antenna height 0.3 m

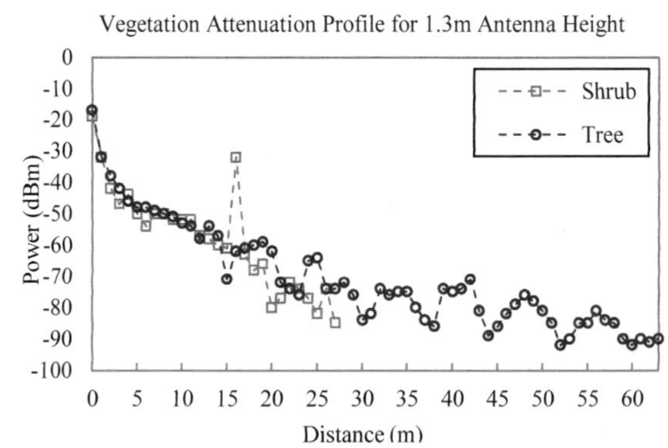


**Figure 14.** Vegetation attenuation profile for antenna height 1.3 m

In order to understand further the attenuation factor and effect of these plantations on signal propagation, a modelling was done for each of the profiles and the results are in the following figures.

Graphs in Figure 15, Figure 16, Figure 17 and Figure 18 describe the distribution of data for both shrub and trees attenuation profiles using log normal distribution, NZG [18],Weissberger's Model [20], COST235 [21] (Out-of-Leaf), ITU-R [22] and Fitted ITU-R (FITU-R) [23] models. The RMSE [17] values for vegetation attenuation models are as shown in Table 2. Figure 15 shows the modelling for shrub at 0.3 m antenna height while Figure 16 shows the modelling for 1.3 m antenna height. Based on the models, shrub signal profile at 0.3 m antenna height is following log normal distribution with RMSE of 8.40. Modelling against COST235 (out-of-leaf) attenuation model gives RMSE of 36.87 while ITU-R model, NZG model and Weissberger's model give RMSE of 39.59, 33.23 and 40.76, respectively. Shrub signal modelling at 1.3 m antenna height follows log normal distribution as well with RMSE 9.05. Modelling shrub signal at 1.3 m antenna height against COST235 (out-of-leaf) attenuation model gives RMSE of 20.88 while ITU-R model, NZG model and Weissberger's model give RMSE of 23.29, 17.88 and 24.43, respectively. Line of tree was also modelled with log normal, COST235, ITU-R, NZG, and Weissberger's model. Based on the attenuation profile, both 0.3 m and 1.3 m antenna heights show the error close to log normal distribution at 6.43 and 4.87 respectively as shown in Figure 17 and Figure 18. Tree modelling for 0.3 m and 1.3 m antenna height yield similar result where the profile follows log normal distribution. RMSE for other attenuation models such as COST235 in-leaf and both FITU-R for both in-leaf and out-of-leaf were also calculated as stated in Table 2. Based on Table 2, COST235 model shows smallest RMSE value for both shrub and tree at 0.3 m and 1.3 m antenna height, respectively. Shrub attenuation profile at 1.3 m antenna height gives smallest RMSE value as compared to other vegetation attenuation models. NZG model gives the smallest RMSE value which is 17.88, followed by COST235 (In-Leaf) model with RMSE value of 20.26, COST235 (Out-of-Leaf) model with RMSE value of 20.88, ITU-R model with RMSE value of 23.29, Weissberger's model with RMSE value of 24.43 and FITU-R model with RMSE value of 24.55.

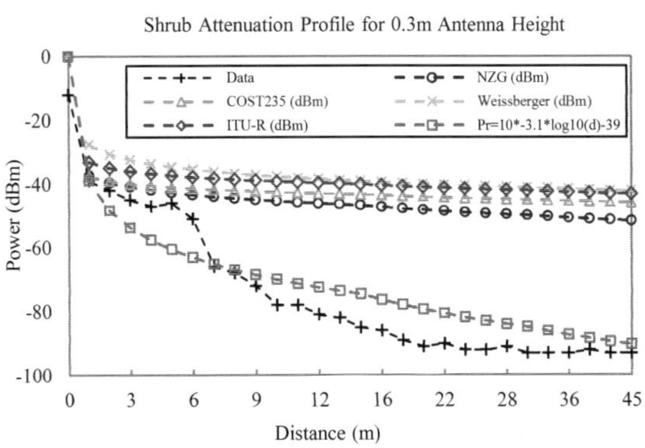


**Figure 15.** Shrub attenuation profile modeling for antenna height 0.3 m

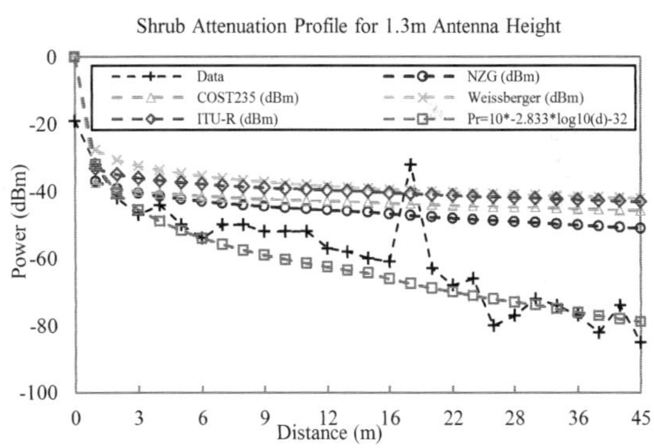


**Figure 16.** Shrub attenuation profile modeling for antenna height 1.3 m

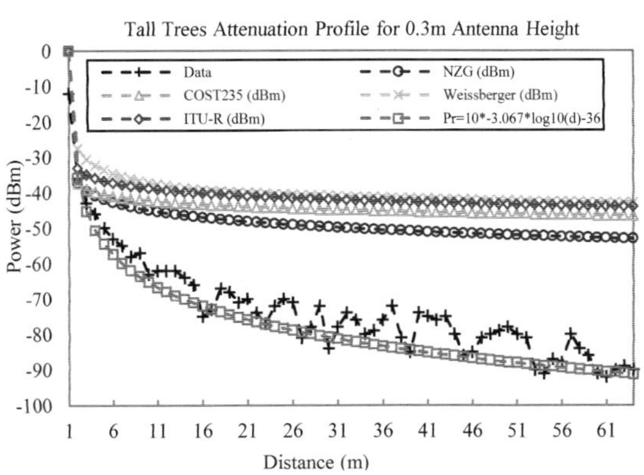


**Figure 17.** Trees attenuation profile modeling for

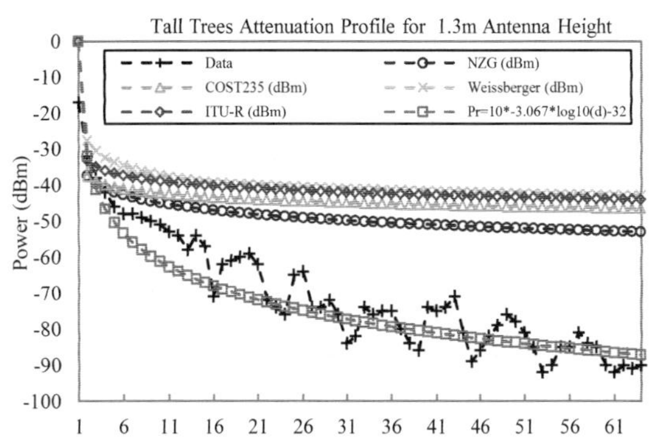


**Figure 18.** Trees attenuation profile modeling for

antenna height 0.3 m antenna height 1.3 m

**Table 2.** RMSE of Vegetation Attenuation Models at Roadside Area

| | Antenna Height (m) | RMSE (dB) | | | | | | |
|---|---|---|---|---|---|---|---|---|
| | | | | | In-Leaf | Out-of-Leaf | In-Leaf | Out-of-Leaf |
| | | $VL_{Weissb.}$ | $VL_{ITU-R}$ | $VL_{NZG}$ | $VL_{COST235}$ | | $VL_{FITU-R}$ | |
| Shrub | 1.3 | 24.43 | 23.29 | 17.88 | 20.26 | 20.88 | 45.52 | 24.55 |
| | 0.3 | 40.76 | 39.59 | 33.23 | 35.99 | 36.87 | 62.49 | 40.95 |
| Tree | 1.3 | 41.57 | 40.54 | 32.01 | 37.85 | 37.92 | 66.73 | 42.08 |
| | 0.3 | 41.19 | 40.16 | 31.62 | 37.48 | 37.54 | 66.36 | 41.70 |

## 6. Conclusion

Vehicle-to-Infrastructure communication is very crucial in implementing a robust and reliable CAV program. In order for such communication to be established, proper study and planning need to be performed taking into consideration the environment in and around roadways in Malaysia. By looking into the test bed for potential CAV deployment in Universiti Malaysia Perlis, a few types of roadside plantations were identified as key contributor to wireless signal attenuation affecting the communication. Shrub and tall trees with trunk and thick leaves are always available along the urban and rural road transport environment in Malaysia. As such identifying the signal attenuation factor for these plantations would certainly be an important task in this work. Based on the studies performed, this research identified that for low plantation such as shrub, the difference in power loss is almost 33 dB compared to antenna height at 1.3 m which is 28 dB throughout the studied path. As for trees a more significant difference found for both 0.3 m and 1.3 m antenna heights as signal tends to fluctuates throughout the path. Comparing both shrub and trees signal profile, it is observed that at 0.3 m antenna height the separation between both plantations. It is very obvious shrub seemed to be the one causing highest attenuation at 0.3 m antenna height compared to shrub at 1.3 m and trees for similar antenna heights. For 1.3 m antenna height however, the separation is less obvious and it seemed that the effect is almost similar for both plantations. Thus, for any communication exchange the effect of these plantations along the roadside could be quantified and the profile could be sufficient to provide understanding of the effect. Signal profile modeling shows that both shrub and trees profile at both antenna heights follows log normal distribution and none follows Weissberger's distribution. For a reliable communication between vehicle-to- infrastructure to be established, all these models and results have to be properly incorporated into the planning and execution.

## 7. References

[1] Dey K C, A. Rayamajhi, Mashrur C, Bhavsar P, Martin J 2016 *Vehicle to vehicle (V2V) and vehicle to infrastructure (V2I) communication in a heterogenous wireless network – Performance evaluation* Elsevier Transportation Research Part C Vol 68 pp 168-194

[2] Dar K, Bakhouya M, Gaber J, Wack M, Lorenz P 2010 *Wireless communication technologies for ITS applications [topics in automotive networking]* IEEE Communication Magazine 48 (5), 156-162

[3] Goel S, Yuan Y 2015 *Emerging Research in Connected Vehicles* IEEE Intelligent Transportation Systems Magazine Guest Editorial

[4] Chou C M, Li C Y, Chien W M, and Lan K 2009 *A Feasibility Study on Vehicle-to-Infrastructure Communication: WiFi vs. WiMAX* pp 397–398

[5] Ghori M R, Zamli K Z, Quosthoni N, Hisyam M, Montaser M 2018 *Vehicular Ad-hoc Network (VANET): Review* IEEE International Conference on Innovative Research and Development (ICIRD)

[6] Ndzi D L, Harun A, Ramli M F, Kamarudin L M, Zakaria A, Shakaff A Y M, Jaafar M N, Zhou S, Farook R S 2014 *Wireless Sensor Network Coverage Measurement and Planning in Mixed Crop Farming* Computers and Electronics in Agriculture Elsevier

[7] Harun A, Shakaff A Y M, Zakaria A, Jaafar M N, Kamarudin L M, Ndzi D L 2011 *Wireless sensor networks mapping and deployment in tropical precision farming* Third International Conference on Computational Intelligence, Modeling and Simulation (CIMSiM), pp. 346-350

[8] Harun A, Ramli M F, Ndzi D L, Kamarudin L M, Shakaff A Y M, Jaafar M N, Zakaria A 2013 *Antenna Positioning Impact on Wireless Sensor Networks Deployment in Agriculture* Australian Journal of Basic and Applied Sciences, 7(5): 55-60

[9] Harun A, Ramli M F, Kamarudin L M, Ndzi D L, Shakaff A Y M, Zakaria A 2012 *Comparative performance analysis of wireless RSSI in wireless sensor networks motes in tropical mixed crop precision farm* Intelligent System Modelling and Simulation (ISMS)

[10] Turner J S C, Kamarudin L M, Ndzi D L, Harun A, Zakaria A, Shakaff A Y M, Mamduh S M 2014 *Modelling indoor propagation for WSN deployment in smart building* 2nd International Conference on Electronic Design 398–402

[11] Ndzi D L, Arif M, Shakaff A Y M, Ahmad M N, Harun A, Kamarudin L M, Razalli M S 2012 *Signal Propagation Analysis for Low Data Rate Wireless Sensor Network Applications in Sport Grounds and On Roads* Progress in Electromagnetic Research **125** 1-19

[12] Rappaport T S 2002 *Wireless Communications: Principles and Practice* 2nd Edition Prentice-Hall PTR

[13] Seybold J S 2005 *Introduction to RF Propagation* John Wiley & Sons Ltd

[14] Fresnel Zone *4Gon.Htm* http://www.4gon.co.uk/solutions/technical_fresnel_zones.php

[15] Network M, Requirement H, Modules Z, Board X P C B, Board U, Requirement S, Double I Y *Configure the Zigbee XBee S2C Modules as – Coordinator Router End Device Configure the Zigbee as Coordinator*

[16] XCTU User Guide Configuration and Test Utility Software *https://www.digi.com/resources/documentation/digidocs/pdfs/90001458-13.pdf* Digi

[17] Chai T, Draxler R R 2014 *Root Mean Square Error (RMSE) or Mean Absolute Error (MAE)?* Geoscientific Model Development Discussions Vol 7 pp 1525-1534

[18] Seville A, Craig K H 1995 *Semi-empirical model for millimeter-wave vegetation attenuation rates* Electron. Lett. 31(17) 1507–1508

[19] Seville A 1997 *Vegetation attenuation: Modeling and measurements at millimetric frequencies* 10th International Conference of Antennas Propagation Conf. Publ. 407

[20] Weissberger M A 1982 *A model for estimating electromagnetic wave attenuation in a forest (EWAF) environment* Annual Revision Proceedings Computer Electromagnetic 801-808

[21] COST235 1996 *Radio propagation effects on next-generation fixed service terrestrial telecommunication systems* Final Report Luxembourg

[22] CCIR 1986 *Influences of terrain irregularities and vegetation on troposphere propagation* CCIR Report 235–236 Geneva

[23] Al-Nuaimi M O, Stephens R B L 1998 *Measurements and prediction model optimization for signal attenuation in vegetation media at centimetre wave frequencies* IEEE Proceedings in Microwave of Antennas and Propagation Vol 145 No 3 201–206

[24] ITU-R P.833-3 2001 *Attenuation in vegetation* International Telecommunication Union Geneva

[25] Meng Y S, Lee Y H 2010 *Investigations of foliage effect on modern wireless communication systems: a review* Proceedings Electromagentic Research 105 313-332

IOP Conf. Series: Materials Science and Engineering **557** (2019) 012057 doi:10.1088/1757-899X/557/1/012057

# Pellet Mill Fixed Dies Type for Production of Solid Fuel Pellets from *Acacia mangium* Bark

**Rengga Arnalis Renjani[1,2] and Dyah Wulandani[2]**

[1]Department of Agricultural Engineering, Stiper Agricultural University (INSTIPER), Nangka 2 Street, Maguwoharjo, Depok, Sleman, Indonesia
[2]Department of Mechanical and Biosystem Engineering, Bogor Agricultural University (IPB), Bogor, Indonesia
E-mail: rengga_tepins@instiperjogja.ac.id; dwulandani@yahoo.com

**Abstract.** Biomass pellet or pellets are bio solid fuel made from compacted sawdust and other bio waste. Pelletized processes consist of three major unit operations, i.e. drying, size reduction (grinding), and densification. Increasing biomass density needs a machine, to produce pellet with best quality. Pellet mill or pelletization machine in Indonesia still expensive because due to import duties, difficult maintenance process, and suitable for small and medium enterprise or industry. The purpose of this study was to design a prototype pellet mill integrated size reduction that encompasses size reduction as well as extrusion using the fixed dies type pelletizing method, and fabricated using locally available construction materials. Main features of the machine are hopper, gearbox, electric motor (2.23 kW, 1 phase, 3 hp rating), axial shaft, roller, and fixed dies. Material used in this research was *A.mangium* bark and the results of this study indicated the pellet mill has been able to produce *pellets* with a throughput capacity of 7.70 kg/h, with rotation speed 68.98 rpm, and load of machine range 0.44-0.51 kW.

## 1. Introduction

In paper and pulp processing, *Acacia mangium Wild* used as main raw material [1],[2]. Before entering pulp and paper processing, *A.mangium* wood was separated from the bark, called debarking process, and produced 8–9 % of *A.mangium* bark from round wood intake by 216.49 ton/day or 74,465 ton/year [3]. This research to explore *A.mangium* bark processing into biomass pellets which used as an alternative energy with increase densification process, because *A.mangium* has high calorific value [4]. According Uasuf and Becker (2011) [5], one of biomass utilization into solid fuel gave many advantages, like: solution of environmental problem and increase value added of waste. The advantages of biomass pellet can produce higher calorific value [6], was a stable alternative energy and competitive price rather than another solid fuel [7].

Pellet mill making was done by pelletization technology [8] or densification [9], Wongsiriamnuay and Tippayawong (2015) [10] was a process which added pressure into sawdust material or in dry and small size to be converted into pellet or cylindrical form [11]. Densification was principally related with extrusion process [12]. According to pellet mill mechanism, there were several pellet mill type, such as: Hendra (2012) [13] was designed hydraulic press type pellet mill with 2.67 kg/h capacity which need 2.55 kW electrical power supply. Ring mold type pellet mill has been developed by Wei et al. (2016) [14], but it needs big energy for 15 hp. Mushiri and Mbohwa (2017) [15] designed fixed dies pellet mill with 900 kg/h capacity and 25 kW power.

If the pellet mill adopted in Indonesia and be applied in small-medium scale enterprises, it will be faced many problems like difficulties in finding spare part, high maintenance cost, and needs skilled technician. In other hands, local pellet mill machine technically and economically proper in Indonesia. To solve the problem, this research was conduct to increase the working performance of pellet mill with rational cost. Pellet mill designed and tested in this research was fixed dies type with stationer dies, and using roller which rotate at shaft [16] with 10 kg/h capacity. This mechanism using combinations force between rotation and centrifugal. Main component in this pellet mill were roller, cutting knife, electrical motor, gearbox, and dies. The purpose of this study was to design a prototype pellet mill integrated size reduction that encompasses size reduction as well as extrusion using the fixed dies type pelletizing method, and fabricated using locally available construction materials.

## 2. Material and Method

### *2.1. Problem identification*

The criteria that should followed in designing a pellet mill were: (1) 6.5 – 7 mm diameter; (2) 25 – 27 mm length; (3) 650 kg.m$^{-3}$ of bulk density; (4) 10 kg/h of production capacity; (5) maximize densification process without using additional heater. Designing machine in agricultural should consider the physical characteristics of material. Biomass processed into pellet was *A.mangium* bark with (< 10 mesh) particle size, 7.352±0.88% of moisture content and 425.93±5.45 kg.m$^{-3}$ of bulk density.

### *2.2. Design concept*

According to problem identification, a design concept to solved the problem was generated. Pellet mill designed in this research was fixed dies type, containing from several parts, such as: hopper, roller, cutting knife, electrical motor, gearbox and fixed dies. Biomass material in dust form collected at hopper for several times. Materials then going down from hopper then drilled and pressed by roller. Rollers rotated centrifugally and produce friction force then pushed the dies. The rotation force worked on fixed dies gave many advantages, such as: produced heat at fixed dies, generated from rollers rotation (rpm) so this pellet mill did not need any additional heater; grinding the material softer and more solid; for give boost and pressed the material entering the dies hole. The sketch of pellet mill and explanation of the main components described at Figure 1.

### *2.3. Functional design analysis*

The main function of this pellet mill was produced pellet with 6.5 – 7 mm diameter, 25 – 27 mm diameter, 650 kg.m$^{-3}$ of bulk density and 10 kg/h of producing capacity. Based on the main function, the detailed sub-function then developed, intended to one component.

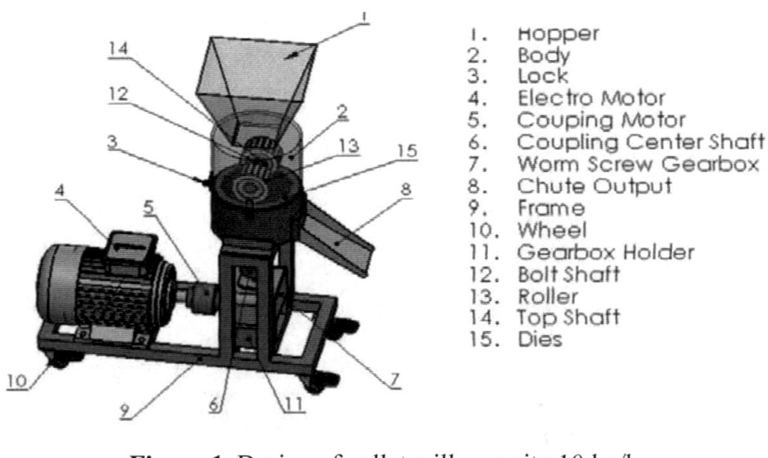


**Figure 1.** Design of pellet mill capacity 10 kg/h.

Component chose from each alternative design based on its lack and specialty, so it will support the main component. Output of functional analysis in pellet mill expected to change opinion that this pellet mill can give best performance. Sub-function, component and mechanism of each component used in designing pellet mill described in Table 1.

### *2.4. Structural design analysis*

Technical perspective was important part in designing pellet mill. Structural analysis includes technical analysis to measure size, form, and construction material in each component. The steps were (1) sketched circuit of each component; (2) gathered mathematical equation used to calculate each process; (3) calculate and determined form, material, size, and performance of each component.
Material inserted to pellet mill collected in hopper to keep the continuity and stability of forming process. Considering the manufacturing process, in this research hopper designed in cone cut-off form. The equation to determine hopper showed in Equation 1.

$$V_h = \frac{1}{3}\, t_h \left(L_1 + L_2 + \sqrt{L_1 \,.\, L_2}\right) \tag{1}$$

According to Xiao et al. (2014) [17], in determining roller diameter based on the size of inserted material used the Duan's half-theory equation [18]. By the Duan's half-theory, we can measure the most accurate method to measure optimum diameter of roller for pellet making proces. Another important component was fixed dies form the pellets. Material pressed using roller and went out through holes in fixed dies [12]. This forming part shaped disc with holes to release pressing result in pellet form. Dies holes measured using pellet diameter standard, 7.5 mm. dies thickness was measure using Equation 2.

$$t_d = k_f \,.\, D_d \,.\, \sqrt{\frac{P_c}{\sigma_y}} \tag{2}$$

Diameter of dies ($D_d$) was measure by calculate all part in dies:

$$D_d = 2 \times \left(w + r_s + r_j + r_b\right) \tag{3}$$

Force ($F$) needed in densification process was measure by Equation 4.

$$F = \tau \times A \tag{4}$$

$$\tau = \frac{m_v}{t} \tag{5}$$

Measured the contact area of roller (A) calculated by Equation:

$$A = w \times I \times D_r \tag{6}$$

**Table 1.** Pellet mill sub-function, components and mechanism.

| Sub-Function | Component | Mechanism |
|---|---|---|
| Material container | Hopper | Material inserted to pellet mill collected to keep the process continuity |
| Pressing | Rollers | Material pressed by two rollers toothed and parallel above dies. Roller rotated at connecting shaft |
| Forming | Fixed dies | Dies was in pellet mill body, roller was in the centre of shaft |
| Dies hole | Normal counter-bore | This mechanism divided dies into rooms with different diameter. The top part was for collecting material and the bottom part was for solidifying process and forming |
| Transmission | Shaft | Transmit power from gearbox to roller |
| Cutter | Rotated knife | Knife was moved with shaft which rotated roller. Knife was below the dies so the output of dies can be cutting |
| Power transformation | Worm screw gearbox | Gearbox used to minimize motor movement into 70 rpm, and used to transform motor momentum from horizontal to vertical on shaft |
| Power supply | Electro motor | Power was very important instrument in forming pellet, it will be converted the movement into vertical using transmission sub-function |

Time ($t$) calculated by equation:

$$v = \frac{S}{t} \tag{7}$$

$$t = \frac{S}{v} \tag{8}$$

According to Equation 5, vertical pressure $F_v$ at fixed dies calculated by:

$$P = \frac{F_v}{A} \tag{9}$$

$$F_v = F \times \text{Cos } 45° \tag{10}$$

$$A = \text{a} \times w \tag{11}$$

*Working stress dies* $\sigma_w$ measure by using equation:

$$\sigma_w = \frac{\sigma_y}{n_k} \tag{12}$$

The basic of choosing electrical motor as power supply was the minimum production of vibration [19], switchable rotation, and high efficiency compared to diesel and Otto machine. Measuring the electrical motor power needs calculated by using equation:

$$p = \frac{I.k_f.K_t.D_r.w.r_d.\pi.n_s}{60} \tag{13}$$

Rotation force produced from electrical motor was transmitted to shaft through worm screw gearbox so the rotation produced was 70 rpm. Rotation from machine transmitted to shaft using coupling, then used to rotated the roller [14] at dies to transform the shape from dust into pellet. The shaft diameter can be measured using equation:

$$T = p \,.\, \frac{60}{2\pi N_s} \tag{14}$$

$$T = \frac{\pi}{16} \,.\, \tau \,.\, {d_s}^3 \tag{15}$$

Empirically, number of pellet produced by pellet mill per hour measure by designed pellet mass by using equation:

$$M_p = V_p \times \rho_e \tag{16}$$

Pellet volume $V_p$ measured using the equation:

$$V_p = \pi r^2 \times H \tag{17}$$

The number of pellet produced in one hour ($P_h$)was measure by using the equation:

$$P_h = \frac{H_p}{M_p} \tag{18}$$

Dies was an important component in pellet mill to forming the pellets, therefore dies component simulated by static pressure using SolidWorks 2012 software simulation. The object became to the simulation analysis was dies strength material. Material used in dies simulation was steel SNI (Indonesia Standard) 07-0601-2006 equivalent S5C, modulus Young 206 kN mm$^{-2}$, yield strength 280 kN mm$^{-2}$, and shear strength 182 kN mm$^{-2}$ [20].

Material flow in forming process generally rotated following the rotation process of roller and shaft, then slowly entering the fixed dies holes. Fixed dies designed so the heating and pressing process can occur. Figure 2 showed interaction process between material in fixed dies and roller rotation movement, contained feed zone shape forming zone or deformation zone, and extrusion zone which optimum densification occur [21].

## 3. Result and Discussion

### *3.1. Hopper*

The capacity of this pellet mill were designed at 10 kg/h, with estimation of material entering the hopper was 4 times/h, therefore needed 2.5 kg of hopper capacity. According to cut-off hopper shape optimization, the height of hopper was 200 mm, top side was 250 mm with square area was 62,500 mm$^2$, bottom side 100 mm with square area of = 10,000 mm$^2$, bulk density of *A.mangium* was 425.93 kg/m$^3$, therefore hopper volume was 2.8 kg, and hopper of pellet mill showed at Figure 3.

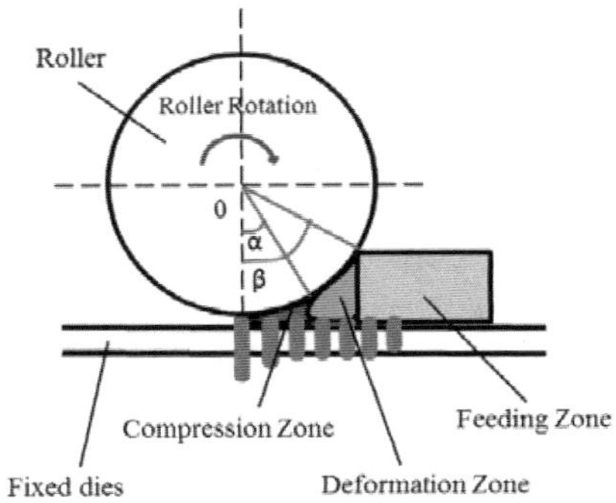


**Figure 2.** Pellet formation process
Source: Gao et al. (2015) [21]

**Figure 3.** Hopper of pellet mill

*3.2. Roller*

Based on Duan's half-theory, can be know the optimum roller diameter for pellet mill. Roller in cylindrical form and toothed at the outer surface, biomass material size that will be processed has ±10 mm diameter, therefore roller diameter needed was 90 mm width of roller measured by square area of fixed dies in the hole groove of dies. Roller specification measure ware material mild steel 43, wight 1.2 kg, width 50.20 mm, diameter 89.34 mm, addendum 3.33 mm, tooth thickness 10.02 mm, and toothed (pieces) 19 mm.

*3.3. Fixed Dies*

Fixed dies was at the bottom side of rollers that has bee hives arrangement and hexagonal shape, showed at Figure 4. This design expected to can reduce the space between holes (space between small holes), so there was no biomass stacked up at the space between holes. Fixed dies hole arrangement showed at Figure 5. Total dies holes was 251 holes with normal counter bore shape has 7.5 mm diameter and 7 mm at the bottom part. Shear stress between roller and dies known by calculating the processed biomass viscosity divided by time. According to Mushiri and Mbohwa (2017) [15], biomass viscosity for pellet processing ($m_v$) was 32 Nsm$^{-2}$ so with time ($t$) of 0.85 second, the shear stress happened was 37.33 N/m$^2$. Contact area of roller ($A$) was $4.48 \times 10^{-3}$ m$^2$ and shear stress (τ) has been known, so the total force was 0.167 N. Width of roller tooth (a) was 10.02 mm, angle of α was 45°, so the pressure produced was $1.74 \times 10^{-4}$ N/m$^2$.
Each pellet has 25 mm of length dimension, 7 mm diameter with $9.6 \times 10^{-7}$ m$^3$. *A.mangium* bark has 425.93 ±5.45 kg.m$^{-3}$ of bulk density, so the pellet mass produced was $6.2 \times 10^{-4}$ kg. If the pellet mill has 10 kg/h capacity, so it produced 15,998.56 pellet/h or 4.44 pellet/s.

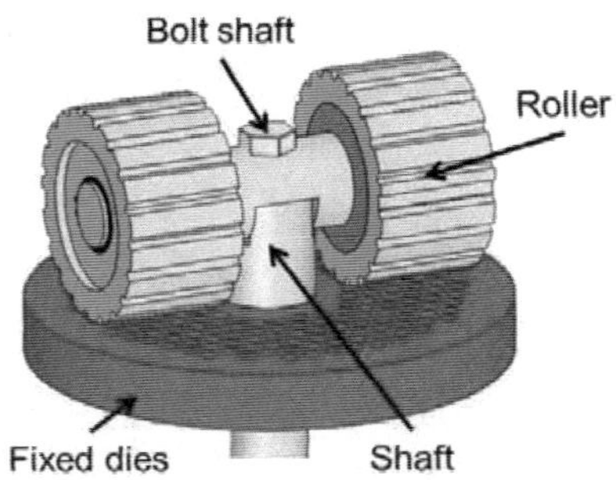


**Figure 4.** Position of fixed dies, shaft, bolt shaft and roller

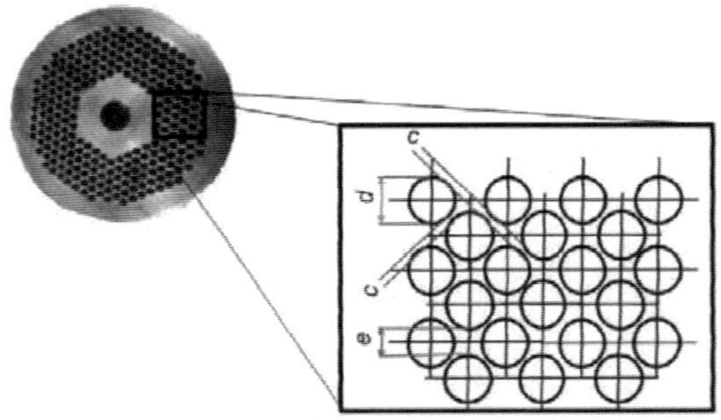


**Figure 5.** Arrangement holes in the fixed dies

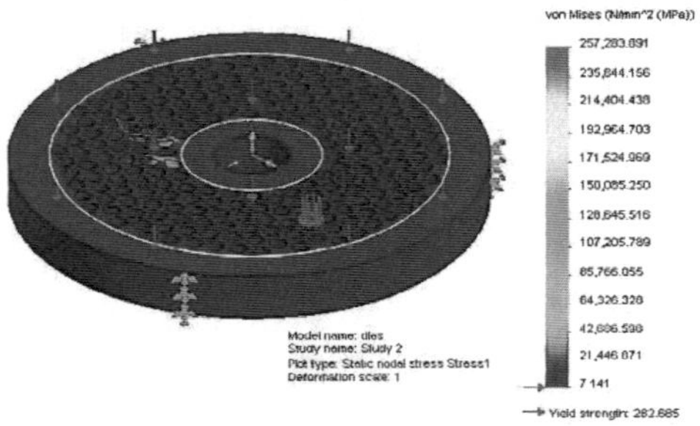


**Figure 6.** Result of simulation dynamic stress in the fixed dies

**Figure 7.** Shaft of pellet mill

Empirically, dies thickness $t_d$ needed was 21.54 mm, to increase the safety of dies, dies designed with 24.30 mm thickness. Diameter of dies *Dd* was 200.10 mm based on all dies component such as shaft radial *rs* 15.87 mm, radius between shaft and roller *rj* was 16.83 mm, radius between roller width to dies body *rb* was 17.14 mm, and roller width was 50.20 mm.

Dies material used was mild steel 43 which has 280 kN.mm$^{-2}$ of yield strength, in material simulation, we used SNI 07-0601-200 steel equivalent to S5C. This selection based on yield strength of both materials was 280 kN.mm$^{-2}$. Safety factor recommended for dies was 6 [25] so dies working stress ($\sigma_w$) was 46.55 N.mm$^{-2}$. The selection of this safety factor based on dies work at pressing and roller shifting process. According to dies working stress ($\sigma_w$) value (46.55 N.mm$^{-2}$), and pressure given by roller ($F_v$) was $1.74 \times 10^{-4}$ N/m$^2$, so dies was safe. Result of static pressure simulation (Figure 6) showed that dies was in blue colored scale, indicated that dies material was very safe to endure the pressure load.

### *3.4. Electro Motor and Gearbox*

According to Equation 13, the pellet mill need 0.93 kW of power and was adequate to process the pellet, but considered the tolerate number which times by two, the power need was 2.23 kW. The aim of increasing tolerate number was to keep the safety of motor when operated. The specification of electrical motor in this research were model and manufaktur by Orsatti O.M.E. Motori Elettrici s.r.l. Via Niccolò Tartaglia, Gussago (BS), Italia, single phase, rate power 3 HP (2.23 kW), rotation speed (r.min$^{-1}$), and Effisiency 88.5%. In order to made pellet mill produced 70 rpm shaft rotation speed so the calculation result of gearbox rotation speed ratio needed was 1:20.

### *3.5. Shaft*

Shaft was the important component to connect the power from electrical motor to move the roller at dies. Shaft diameter was determined by using collected data, such as: maximum shear stress of solid steel VCN 150 was 240 mPa [19], shaft rotation 70 rpm, power given was 2.23 kW. Torsion value *T* given by shaft was $0.304 \times 10^3$ N/m, so the diameter of shaft need was 1.358 mm. The considering importance of shaft diameter in pellet making operational, so the safety factor was 23 times. Shaft diameter used in this pellet mill was 1.25 inch (31.75 mm), and showed at Figure 7.

### *3.6. Trial Performance of pellet mill*

Trial performance of pellet mill showed in Table 2 with average of rotation was 68 rpm. There was no significance change with adhesive variation. Increasing the motor load of pellet mill effected to increase the electrical power needs, caused by adhesive characteristics that make *A.mangium* dust became more solid and clot, so the roller moving force were inhibited.

Throughput capacity increased because *A.mangium* bark material mixed with adhesive so it easier to form at dies holes. It showed at 0%, 5%, and 10% adhesive produced each throughput capacity were 0.651, 3.032 and 5.076 kg.h$^{-1}$. Based on the design of material flow at pellet mill, material rotate centrifugally, matching with shaft rotation, but slowly showing the exit movements form the dies holes. There was low friction between material mixed with adhesive and contact area of dies, reduced the retention of material at dies hole [26], proved has increasing the pellet mill throughput capacity. Accumulation of material at the side of pellet mill body was not as much as without adhesives.

**Table 2.** Pellet mill perform

| Parameter | Percentage of Adhesive | | |
|---|---|---|---|
| | 0% | 5% | 10% |
| Rotation Speed (rpm) | 68.98 | 68.79 | 68.56 |
| Load of Machine (kW) | 0.44 | 0.47 | 0.51 |
| Throughput Capacity (kg.h$^{-1}$) | 0.65 | 3.03 | 7.70 |

## 4. Conclusion

This research was successfully produced pellet mill with 10 kg/h of capacity or produced 44,440 pellets per hour. Features of this pellet mill were hopper, gearbox, electric motor, shaft, roller, and fixed dies. Material used in this research were *A.mangium* bark added with starch adhesive with 0%, 5%, and 10% variations. The results of this study indicated the pellet mill has been able to produce *pellet*s with a throughput capacity of 7.70 kg/h, with rotation speed 68.98 rpm, and load of machine range 0.44-0.51 kW.

## 5. References

[1] Krisnawati H, Kallio M, and Kanninen M 2011 *Acacia mangium Willd.: ecology, silviculture and productivity* (in Indonesian Language) (Bogor: Center for International Forestry Research).

[2] Rustam R, Sucahyono M P, and Salbiah D 2014 Biology of Helopeltis Theivora (Hemiptera : Miridae) on Acasia Mangium Willd. *Int. J. Adv. Sci. Eng. Inf. Technol.* **4** 62–65.

[3] Suprapti S, Santoso E, Djarwanto, and Turjaman M 2012 Utilization of Composted Mangium Bark as Growth Media for Arbuscular Micorrhizal Fungi and Acacia mangium Willd. Seedlings. (in Indonesian Language)," *J. Penelit. Has. Hutan.* **30** 114–123.

[4] Orwa C, Jamnadass R, Anthony R, Kindt S, and Mutua A 2009 Acacia mangium brown salwood: Wild. Fabaceae-Mimosoideae *Agrofor. Database.* **1** 1–5.

[5] Uasuf A and Becker G 2011 Wood pellets production costs and energy consumption under different framework conditions in Northeast Argentina *Biomass and Bioenergy* **35** 1357–66.

[6] Munawar S S and Subiyanto B 2014 Characterization of Biomass Pellet made from Solid Waste Oil Palm Industry *Procedia Environ. Sci.* **20** 336–341.

[7] Nunes L J R, Matias J C O, and Catalão J P S 2016 Wood pellets as a sustainable energy alternative in Portugal *Renew. Energy* **85** 1011–16.

[8] Jackson J, Turner A, Mark T, and Montross M 2016 Densification of biomass using a pilot scale flat ring roller pellet mill *Fuel Process. Technol.* **148** 43–49.

[9] Donghui L, Tabil L G, Decheng W, Guanghui W, and Zhiqin W 2014 Optimization of binder addition and compression load for pelletization of wheat straw using response surface methodology *Int. J. Agric. Biol. Eng.* **7** 67–78.

[10] Wongsiriamnuay T and Tippayawong N 2015 Effect of Densification Parameters on the Properties of Maize Residue Pellets *Biosyst. Eng.* **139**.

[11] Sjöström J and Blomqvist P 2014 Direct measurements of thermal properties of wood pellets : Elevated temperatures, fine fractions and moisture content *Fuel* **134** 460–466.

[12] Oduntan O B, Koya O A, and Faborode M O 2014 Design, fabrication and testing of a cassava pelletizer *Res. Agric. Eng.* **60** 148–154.

[13] Hendra D 2012 Design and Manufacture of Wood Pellets Machine and Testing of its Product. (in Indonesian Language) *J. Penelit. Has. Hutan* **30** 144–154.

[14] Wei G, Tabil L G, Rongfei Z, and Dejun L 2016 Optimized design and experiment on ring mold pelletizer for producing biomass fuel pellets *Int. J. Agric. Biol. Eng.* **9** 57–66.

[15] Mushiri T and Mbohwa C 2017 Design of a sawdust pelleting machine *Proceedings of the International Conference on Industrial Engineering and Operations Management* 1763–76.

[16] Tumuluru J S 2014 Effect of process variables on the density and durability of the pellets made from high moisture corn stover *Biosyst. Eng.* **119** 44–57.

[17] Xiao Q, Li B, and Kang 2014 H The Effect of Fine Grinding Medium Feature on Grinding Results, *Proc. 2nd AASRI Conference on Power and Energy Systems* vol. 7 p 120–125.

[18] Duan X 1991 Theoretical Calculation of Mill Steel Ball Size," *Sci. China*, **34** 220–228.

[19] Okechukwu I K 2012 Mechanized Balling Disc Machine (Pelletizer) for Industrial Development in Nigeria *Proceedings of the World Congress on Engineering and Computer Science* vol 2.

[20] Kim K S, Chen X, Han C, and Lee H W 2002 Estimation methods for fatigue properties of steels under axial and torsional loading *Int. J. Fatigue* **24** 783–793.

[21] Gao W, Tabil L, Zhao R, Liu Q, and Dejun L 2015 Motion simulation analysis and experimental study of a ring mold pelletizer *Proc. CSBE/SCGAB Annual Conference* p 1–10.

[22] García-Maraver A, Popov V, and Zamorano M 2011 Data Bank A review of European standards for pellet quality *Renew. Energy* **36** p 3537–40.

[23] Renjani R A, Wulandani D, Saulia L, and Hermawan W 2016 Design Improvement of Pelletization Machine for Production Solid Fuel from Acacia mangium Bark *Proc. International Conference on the Role of Agricultural Engineering for Sustainable Agriculture Production* vol 1 p 197–202.

[24] Monedero E, Portero H, and Lapuerta M 2015 Pellet blends of poplar and pine sawdust: Effects of material composition, additive, moisture content and compression die on pellet quality *Fuel Process. Technol.* **132** p 15–23.

[25] Juvinall R C and Marshek K M 2012 *Fundamentals of Machine Componenet Design* ed Fifth. (Danvers: John Wiley & Sons).

[26] Mišljenović N, Čolović R, Vukmirović Đ, Brlek T, and Bringas C S 2016 The effects of sugar beet molasses on wheat straw pelleting and pellet quality. A comparative study of pelleting by using a single pellet press and a pilot-scale pellet press *Fuel Process. Technol.* **144** p 220–229.

**Acknowledgments**

This research was fully funded and supported by Stiper Agricultural University (INSTIPER) Yogyakarta. We would also like to thank Department of Mechanical and Biosystem Engineering, Bogor Agricultural University (IPB) which provides the opportunity to conduct this research and publication.

# Exergy Analysis on Pyrolysis Process of Oil Palm Empty Fruit Bunch

**Tiara Etika S[1], Agus Susanto Ginting[2], Radite PA Setiawan[1], Joelianingsih[3], Armansyah H Tambunan[1]**

[1] Department of Mechanical and Biosystem Engineering, Bogor Agricultural University, Dramaga Campus, PO Box 220, Bogor 16680, Bogor, Indonesia.

[2] Polytechnic of Gorontalo, Saptamarga Bone Bolango Street, Gorontalo 96112, Gorontalo, Indonesia.

[3] Department of Chemical Engineering, Indonesia Institute of Technology, Raya Puspitek Serpong Street, Banten 15314, West Tangerang, Indonesia.

**Abstract.** Exergy analysis is useful to evaluate the effectiveness of energy conversion process. This study is aimed to evaluate the effectiveness of oil palm empty fruit bunch conversion by pyrolysis process using the exergy analysis compared to the theoritical of the cellulose pyrolysis. The effect of pyrolysis temperature was evaluated based on the amount of selected syngas produced. The results show that Increasing the temperature of pyrolysis from 473 K until 723 K will increase the concentration of $H_2$ and $CH_4$, but it will reduce the concentration of CO. Based on exergy efficiency, the optimum composition of gases that can be produced in the pyrolysis process of 1 mol cellulose are 3 mol of $H_2$, 5 mol of CO, and 1 mol of $CH_4$ at a temperature of 723K. The maximum amount of exergy efficiency is 96.33% in the reaction 2. Based on Gibbs free energy, the maximum value of hydrogen will be produced start at temperature 628K.

## 1. Introduction

Hydrogen is a chemical element that has great potential as an energy carrier in the future. Hydrogen has the highest energy content compared to other common fuels (about three times more than gasoline). In addition to its high energy content, the source of this element is available abundantly, as in biomass. Many studies noted that hydrogen may be the only alternative fuel that can reduce the country's dependence on foreign oil while significantly reducing greenhouse gases. Chen (2008) states that fuel cell technology can be used as transportation fuel, or primary energy generation, while hydrogen is a vital fuel in the technology. Therefore, any technology to produce hydrogen efficiently with the lowest cost becomes very important and indispensable to be studied.

IOP Conf. Series: Materials Science and Engineering **557** (2019) 012058 doi:10.1088/1757-899X/557/1/012058

Biomass is a renewable energy source that can be converted to hydrogen. Thermochemical reaction, such as pyrolysis is one of the available mechanism. Specific characteristics from biomass will affect its role in producing the hydrogen. The composition of the products can be controlled using process control. Normally, the fraction of hydrogen produced by pyrolysis method is very small compared to other gases. A technology to enrich $H_2$ composition through this pyrolysis process is necessary. This technology can be developed through a thermodynamic analysis of the process. In general, the products produced in a thermochemical process can be predicted through thermodynamic approaches and simulations.

Energy balance is an important concept to analyse the performance of a system. This concept is based on the first law of thermodynamic which is often called as the law of energy conservation. The first law of thermodynamic states that energy cannot be created or destroyed. Therefore, energy balance in a system must be maintained. When the system changes from the initial states to a equilibrium state (final state), the system can absorb or release energy into its environment.

The equilibrium state is a situation where there is no change in the system, or between the system and the environment. At equilibrium, temperature and pressure are uniform througout the system. In pratice, the total energy content in a situation (beginning or end) is very difficult to determine, so the thermodynamic approach is often based on energy difference between the intial and final conditions (Burghardt,1993).

The second law of thermodynamic states that besides having quantity, energy also has quality. A real process will take place in the dicrection of decresing the energy quality. Although there is no loss of energy quantity, the energy quality always decrease during the process. The amount of this energy quality is called exergy. Energy and exergy analysis is useful to evaluate the effectiveness of energy conversion process.

## 2. Methodology

This study is aimed to evaluate the effectiveness of oil palm empty fruit bunch conversion by pyrolysis process using the exergy analysis compared to the theoritical of the cellulose pyrolysis An exergy analysis must consider input and output of the system. In the pyrolysis system, the input that get into its system is cellulose, while the outputs are gas, bio-oil, and biochar products. But in this study, the analysis is focused only on syngas product. In the system, there will always exergy loss, but in this study the value will be ignored.

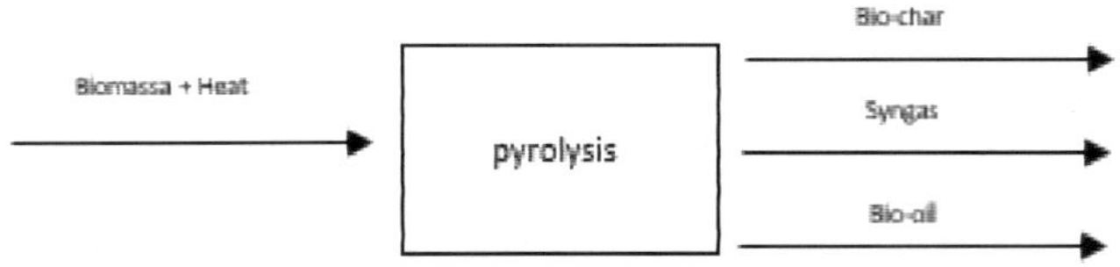


**Figure 1.** Scheme of biomass pyrolysis system

The composition of cellulose pyrolysis products is calculated based on the Gibbs free energy of several possible reactions that can occur during the cellulose pyrolyis process. The reaction with the smallest amount of free energy Gibbs is a reaction that will occur during the process. The composition from that reaction is the maximum gas composition from cellulose pyrolysis. The composition produced for each reaction that may occur during the process will be analyzed based on the exergy efficiency of the syngas. The reaction with the greatest exergy efficiency value is the best reaction. The composition will be compared to the experiment data. The experiment data in this study were carried out using secondary data from [1]. The experiment data used palm oil empty fruit bunch (EFB) as feedstock. EFB contains more cellulose than hemicellulose and lignin. Proximate and ultimate analysis, lignocellulose analysis, and pyrolysis product of palm oil empty fruit bunch shown at table 1, 2, and 3, respectively.

**Table 1.** Proximate and ultimate analysis *

| Analysis | Value |
|---|---|
| **Proximate** | |
| Water content $_{db}$(%) | 1.7 |
| Volatile matter (%) | 75.7 |
| Ash (%) | 7.27 |
| Fixed Carbon (%) | 17.44 |
| **Ultimate** | |
| C (%) | 60.4 |
| H (%) | 7.6 |
| O (%) | 29.8 |
| N (%) | 2.2 |
| S (%) | 0 |
| HHV (MJ/kg) | 18.74 |

**Table 2.** Lignocellulose analysis of palm oil empty fruit bunch [1]

| Component | EFB (%) |
|---|---|
| Cellulose | 56.05 |
| Hemicellulose | 17.63 |
| Lignin | 5.97 |
| Other components | 20.35 |

**Table 3.** Pyrolysis product of palm oil empty fruit bunch [1]

| Temperature (K) | Intial mass (g) | Solid product (g) | Bio-oil (g) | Gas (g) |
|---|---|---|---|---|
| 473 | 152,9 | 60,13 | 10,92 | 81,85 |
| 523 | 152,9 | 65,35 | 11,59 | 75,96 |
| 573 | 152,9 | 54,49 | 16,51 | 81,9 |
| 623 | 152,9 | 62,94 | 14,53 | 75,43 |
| 673 | 152,9 | 50,54 | 23,8 | 78,56 |
| 723 | 152,9 | 50,39 | 14,15 | 88,36 |

The Gibbs free energy value states the spontaneity of a reaction process where the smaller the Gibbs energy value, the more spontaneous the reaction will be. However, to ensure that the reaction can continue, the Gibbs free energy value must exceed its equilibrium limit or be less than 0. In equilibrium, the total Gibbs free energy of a gas emission can be stated as follows:

$$\Delta G = \Delta H_{f,i,T} - T\Delta S \quad (1)$$

where $H_{f,i,T}$ is entalphy formation at T, S is entropy and G is gibbs free energy value.

*2.1 Exergy analysis*

According to Wang et al (2016), exergy does not obey conservation law due to the unavoidable irreversibility of reaction processes. Exergy balance equation can be established with these assumption : pyrolysis is a control volume unit, input exergy is from biomass, output exergy is from product gas, bio-oil, and bio-char, and the irreversibility (internal exergy loss) is in the lost part.

$$Ex_{biomass} = Ex_{gas,T} + Ex_{bio\text{-}oil,T} + Ex_{biochar,T} + Ex_{loss,T} \quad (2)$$

This study only focus on the exergy efficieny of syngas on biomass exergy so that the amount of bio-oil exergy, biochar exergy, and exergy loss will be temporarily ignored. The total exergy value equals to sum of all kinds of exergy value [3]

$$E_x = E_x^{po} + E_x^{ki} + E_x^{ph} + E_x^{ch} \quad (3)$$

where $E_x$ is energy total while $Ex^{ki}$ ,$Ex^{po}$ ,$Ex^{ch}$ , and $Ex^{ph}$ are kinetic exergy, potential exergy, chemical exergy, and physical exergy, respectively. The potential exergy and kinetic exergy values are slight enough so that in this study those values can be neglected. Equation (3) can be equation (4).

$$E_x = E_x^{ph} + E_x^{ch} \quad (4)$$

Physical exergy of pyrolysis gas :

$$Ex^{ph} = m\,((h - ho) - To(s - so)) \quad (5)$$

where m is mass value (kg), h is entalphy (MJ/kg), $h_o$ is standard entalphy of gases at $T_o$ (MJ/kg), $s_o$ is standard entropy of gases at $T_o$ (MJ/kg.K), and $T_o$ is standard temperature (298K). The standard entalphy, and entropy of some gases are shown at table 3.

The specific entalphy and entropy of gases at T can be determined with equation (6) until equation (11)

$$h = h_0 + \int_{T0}^{T} C_p dT \quad (6)$$

$$h - h_0 = \int_{T0}^{T} C_p dT \quad (7)$$

$$\int_{To}^{T} C_p\, dT = aT + \frac{b}{2}T^2 + \frac{c}{3}T^3 + \frac{d}{4}T^4 \quad (8)$$

$$s = s_0 + \int_{T0}^{T} \frac{C_p}{T} dT \quad (9)$$

$$s - s_0 = \int_{T0}^{T} \frac{C_p}{T} dT \quad (10)$$

$$\int_{T_0}^{T} \frac{\Delta C_p}{T} dT = alnT + bT + \frac{1}{2}cT^2 + \frac{1}{3}dT^3 \quad (11)$$

where Cp is specific entalphy of each component, a until d values are coefficient of constant pressure specific heat capacity of some gases, which is shown at table 5.

**Table 4.** Standard entalphy and entropy [2]

| Gas | ho (MJ/kg) | So (MJ/kg.K) |
|---|---|---|
| $H_2$ | 4.23400 | 0.00464 |
| CO | 0.30961 | 0.00706 |
| $CO_2$ | 0.21282 | 0.00763 |
| $CH_4$ | 0.00000 | 0.00665 |
| C | 0.00000 | 0.00020 |
| $H_2O$ | 0.55022 | 0.01048 |

**Table 5.** Coefficients of constant pressure specific heat capacity of some gases [2]

| Gas | A | b ($10^2$) | c ($10^5$) | d ($10^9$) | Range (K) |
|---|---|---|---|---|---|
| $H_2$ | 29.11 | -0.1916 | 0.4003 | -0.8704 | 273-1800 |
| CO | 28.16 | 0.1675 | 0.5327 | -2.222 | 273-1800 |
| $CO_2$ | 22.26 | 5.981 | -3.501 | 7.469 | 273-1800 |
| $CH_4$ | 19.89 | 5.024 | 1.269 | -11.01 | 273-1800 |
| $H_2O$ | 32.22 | 0.192 | 1.055 | -3.593 | 273-1800 |

Chemical exergy of pyrolysis gas :

$$Ex^{ch} = me^{ch} \tag{12}$$

where m is mass of gas (kg), and $e^{ch}$ is standard chemical exergy of gas (MJ/kg) which is shown at table 6.

**Table 6.** Standard chemical exergy of some gases [3]

| Gas | $e^{ch}$ (MJ/kg) |
|---|---|
| CO | 275.1 |
| $H_2$ | 236.1 |
| C | 410 |
| $CH_4$ | 831.65 |
| $H_2O$ | 1.3 |
| $CO_2$ | 19.87 |

HHV value and exergy of biomass can be determined with equation (13) and (14), respectively :

$$e_{DB}^{CH} = 1812.5 + 295.606C + 587.354H + 17.506O + 17.735N + 95.615S - 31.8\,A \tag{13}$$

$$HHV = -1.3675 + 0.3137C + 0.7009H + 0.0318\,(100 - C - H - A) \tag{14}$$

IOP Conf. Series: Materials Science and Engineering **557** (2019) 012058 doi:10.1088/1757-899X/557/1/012058

where is chemical exergy of biomass in dry base (kJ/kg) and HHV is its high heating value (MJ/kg). H, C, O, and N (%) are the value of ultimate analysis of palm oil empty fruit bunch [9].

Exergy efficiency of cellulose gas product can be determined with equation (15) . The gas measured in the secondary data used are only CO, H2, and CH4 so that the other gases are ignored. So, equation (15) can be equation (16)

$$\psi_{gas,T} = \frac{Ex_{CO,T} + Ex_{H2} + Ex_{CH4} + Ex_{H2O} + Ex_{co2}}{Ex_{biomass}} \quad (15)$$

$$\psi_{gas,T} = \frac{Ex_{CO,T} + Ex_{H2} + Ex_{CH4}}{Ex_{biomass}} \quad (16)$$

## 3. Result And Discussion

### *3.1 Composition of maximum syngas produced from cellulose pyrolyis*

Table 7 shows the product result of each reaction that may occur during the process of cellulose pyrolyis. The reactions are based on the stoichiometric equation for pyrolysis of 1 mol cellulose. Based on the table 6, the maximum amount that can be produced for each CO, $H_2$, $CH_4$, $H_2O$, $CO_2$, and C are 5 mol, 2 mol, 2 mol, 1 mol, and 3 mol respectively. The distribution of each products is different for each possible reaction. This study expect maximum hydrogen so, the reaction that is expected to occur should be reaction 1 with the moles of $H_2$, CO, and C are 5 moles, 5 moles, and 1 mol in a row.

**Table 7.** Maximum pyrolysis product of 1 mol cellulose

| Cellulose | | CO (mol) | $H_2$ (mol) | $CH_4$ (mol) | $H_2O$ (mol) | $CO_2$ (mol) | C (mol) |
|---|---|---|---|---|---|---|---|
| Reaction 1 | $C_6H_{10}O_5 \rightarrow 5CO + 5H_2 + C$ | 5 | 5 | 0 | 0 | 0 | 1 |
| Reaction 2 | $C_6H_{10}O_5 \rightarrow 5CO + CH_4 + 3H_2$ | 5 | 3 | 1 | 0 | 0 | 0 |
| Reaction 3 | $C_6H_{10}O_5 \rightarrow 4CO + CH_4 + C + 2H_2 + H_2O$ | 4 | 2 | 1 | 1 | 0 | 1 |
| Reaction 4 | $C_6H_{10}O_5 \rightarrow 3CO + CO_2 + 2CH_4 + H_2$ | 3 | 1 | 2 | 0 | 1 | 0 |
| Reaction 5 | $C_6H_{10}O_5 \rightarrow 3CO + CH_4 + 2C + H_2 + 2H_2O$ | 3 | 1 | 1 | 2 | 0 | 2 |
| Reaction 6 | $C_6H_{10}O_5 \rightarrow 2CO + CO_2 + 2CH_4 + C + H_2O$ | 2 | 0 | 2 | 1 | 1 | 1 |
| Reaction 7 | $C_6H_{10}O_5 \rightarrow 3CO + 3H2 + 3C + 2H_2O$ | 3 | 3 | 0 | 2 | 0 | 3 |
| Reaction 8 | $C_6H_{10}O_5 \rightarrow 2CO + CO2 + 3C + 4H_2 + H_2O$ | 2 | 4 | 0 | 1 | 1 | 3 |

The simulation of each reaction based on the Gibbs free energy value has been conducted at 298 – 950 K. Figure 2 shows the effect of the temperature changes on the Gibbs free energy value of the cellulose pyrolysis process. At 298 -668 K, reaction 6 is the reaction with the smallest Gibbs free energy value but at 668-950 K, the reaction with the smallest Gibbs energy change to reaction 8. Maximum product produced from reaction 8 are 2 mol CO, 1 mol $CO_2$, 3 mol C, 4 mol $H_2$, and 1 mol $H_2O$. It means that based on the Gibbs free energy value, the maximum amount of hydrogen reaches maximum at 4 moles and start at 668 K for 1 mol of cellulose.

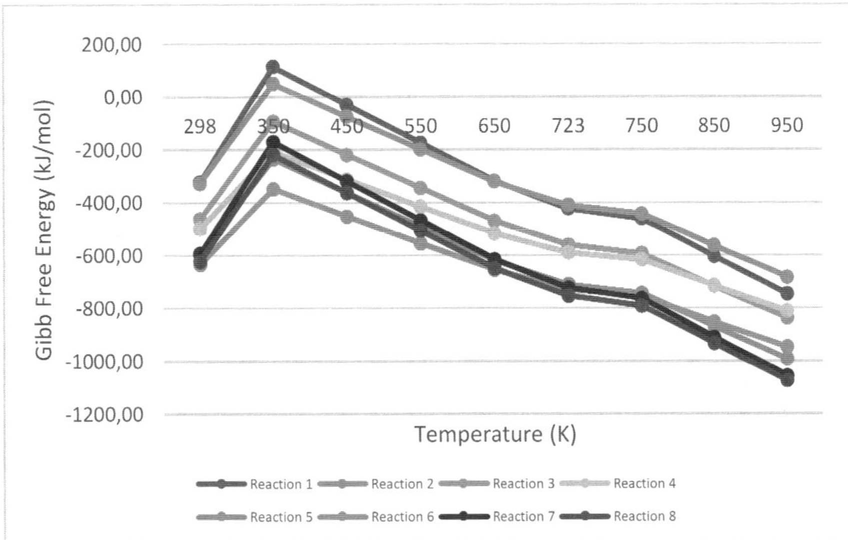


**Figure 2.** Gibbs free energy of cellulose pyrolysis

**Table 8.** Gas product of palm oil empty fruit bunch pyrolysis

| Temperature (K) | CO (mol) | $H_2$ (mol) | $CH_4$ (mol) |
|---|---|---|---|
| 473 | 2.805 | 0.042 | 0.201 |
| 523 | 2.101 | 0.648 | 0.990 |
| 573 | 2.130 | 0.929 | 1.274 |
| 623 | 2.236 | 0.474 | 0.742 |
| 673 | 1.622 | 2.155 | 1.801 |
| 723 | 1.731 | 2.332 | 2.202 |

Table 8 shows the gas product result of palm oil empty fruit bunch pyrolysis that has been conducted by [1] at 473 – 723 K. The result shows that the temperature will increase the amount of $CH_4$ and $H_2$. It is occured because tar decomposes into $CH_4$ and $H_2$ at high temperature. This result is consistent with the result from Sukiran et al (2014) where increase the temperature in pyrolysis process will increase the amount of $CH_4$ and $H_2$ concentrations. As for CO, the concentration decreases with increase the temperature. This is due to characteristic of the CO that will more be released in the decomposition of the hemicellulose component. The cellulose component decomposes at low temperature.

At temperature 723 K, gas composition produced from 56.05% cellulose of palm oil empty fruit bunch are 1.731 mol of CO, 2.332 mol of $H_2$, and 2.332 mol of $CH_4$. Those amounts are still smaller than the simulation result. It shows that the pyrolysis process from [1] is still not optimal.

### *3.2 Maximum hydrogen from pyrolysis cellulose based on exergy analysis*

Table 9 shows the exergy of each gas product can be produced from each reaction that may occur during the pyrolysis process of 1 mol of cellulose at 723K compared to the gas product of oil palm empty fruit bunch from experiment data. The result shows that reaction 8 with the maximum hydrogen produces the smallest amount of gas product exergy, which is 9.6412 MJ/kg. It means that the gas composition of reaction 8 is actually not optimum to be produced in the pyrolysis process compared to the composition of other possible reactions.

**Table 9.** Exergy total of syngas product of 1 mol cellulose pyrolysis at 723 K

| Cellulose | | Exergy (MJ/kg) | | | | | |
|---|---|---|---|---|---|---|---|
| | | CO | H2 | CH4 | H2O | CO2 | Total |
| Reaction 1 | $C_6H_{10}O_5 \rightarrow 5CO + 5H_2 + C$ | 8.6417 | 7.4319 | 0.0000 | 0.0000 | 0.0000 | 16.0736 |
| Reaction 2 | $C_6H_{10}O_5 \rightarrow 5CO + CH_4 + 3H_2$ | 8.6417 | 4.4591 | 5.1834 | 0.0000 | 0.0000 | 18.2842 |
| Reaction 3 | $C_6H_{10}O_5 \rightarrow 4CO + CH_4 + C + 2H_2 + H_2O$ | 6.9134 | 2.9728 | 5.1834 | 0.0438 | 0.0000 | 15.1133 |
| Reaction 4 | $C_6H_{10}O_5 \rightarrow 3CO + CO_2 + 2CH_4 + H_2$ | 5.1850 | 1.4864 | 10.3668 | 0.0000 | 0.1682 | 17.2064 |
| Reaction 5 | $C_6H_{10}O_5 \rightarrow 3CO + CH_4 + 2C + H_2 + 2H_2O$ | 5.1850 | 1.4864 | 5.1834 | 0.0876 | 0.0000 | 11.9424 |
| Reaction 6 | $C_6H_{10}O_5 \rightarrow 2CO + CO_2 + 2CH_4 + C + H_2O$ | 3.4567 | 0.0000 | 10.3668 | 0.0438 | 0.1682 | 14.0355 |
| Reaction 7 | $C_6H_{10}O_5 \rightarrow 3CO + 3H_2 + 3C + 2H_2O$ | 5.1850 | 4.4591 | 0.0000 | 0.0876 | 0.0000 | 9.7318 |
| Reaction 8 | $C_6H_{10}O_5 \rightarrow 2CO + CO_2 + 3C + 4H_2 + H_2O$ | 3.4567 | 5.9455 | 0.0000 | 0.0438 | 0.1682 | 9.6142 |
| Experiment data [1] --> Oil palm empty fruit bunch [1] | | 3.1699 | 3.6723 | 12.0915 | - | - | 18.9337 |

**Table 10.** Exergy efficiency of syngas product of 1 mol cellulose pyrolysis at 723 K

| Cellulose | | Syngas ex (MJ/kg) | Total input ex (MJ/kg) | Ex Efficiency (%) |
|---|---|---|---|---|
| Reaction 1 | $C_6H_{10}O_5 \rightarrow 5CO + 5H_2 + C$ | 16.0736 | 18.9800 | 84.69 |
| Reaction 2 | $C_6H_{10}O_5 \rightarrow 5CO + CH_4 + 3H_2$ | 18.2842 | 18.9800 | 96.33 |
| Reaction 3 | $C_6H_{10}O_5 \rightarrow 4CO + CH_4 + C + 2H_2 + H_2O$ | 15.1133 | 18.9800 | 79.63 |
| Reaction 4 | $C_6H_{10}O_5 \rightarrow 3CO + CO_2 + 2CH_4 + H_2$ | 17.2064 | 18.9800 | 90.66 |
| Reaction 5 | $C_6H_{10}O_5 \rightarrow 3CO + CH_4 + 2C + H_2 + 2H_2O$ | 11.9424 | 18.9800 | 62.92 |
| Reaction 6 | $C_6H_{10}O_5 \rightarrow 2CO + CO_2 + 2CH_4 + C + H_2O$ | 14.0355 | 18.9800 | 73.95 |
| Reaction 7 | $C_6H_{10}O_5 \rightarrow 3CO + 3H_2 + 3C + 2H_2O$ | 9.7318 | 18.9800 | 51.27 |
| Reaction 8 | $C_6H_{10}O_5 \rightarrow 2CO + CO_2 + 3C + 4H_2 + H_2O$ | 9.6142 | 18.9800 | 50.65 |
| Experiment data → Oil palm empty fruit bunch [1] | | 18.9337 | 24.4600 | 77.41 |

Calculation of gas product exergy efficiency has also been done for each possible reaction and experimental data. The amount of theoritical exergy input is derived from the biomass exergy and heat needed to reach a temperature of 723K, whereas the exergy input from experiment data is only from biomass exergy. The heater input energy is not recorded during the process, so the exergy from the heater is ignored. In the experiment data, the gas components produced are only 3 main gases. Those are CO, $H_2$, and $CH_4$ gases. The result shows that the highest exergy efficiency is generated from reaction 2 which is 96.33%. The highest exergy efficiency from reaction 2 shows that 3 mol of hydrogen is the optimum amount which actually must be produced from the pyrolysis process of 1 mol of cellulose, while the other gas composition being 5 mol of CO, and 1 mol of $CH_4$.

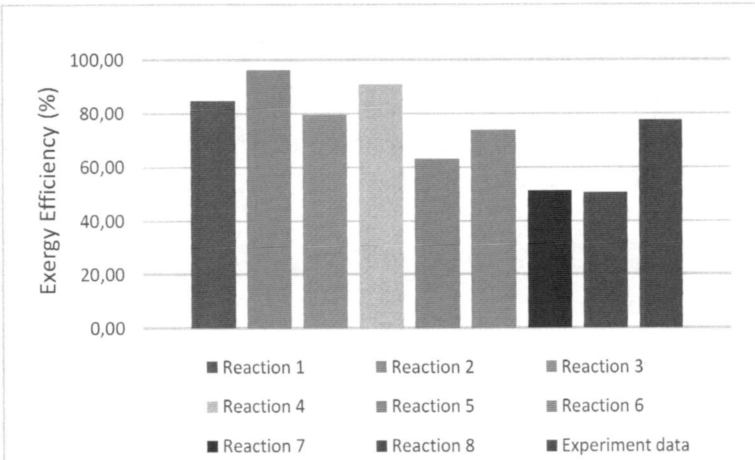


**Figure 3.** Exergy efficiency of syngas product at 723 K

## 4. CONCLUSION

Increasing the temperature of pyrolysis will increase the concentration of $H_2$ and $CH_4$, but it will reduce the concentration of CO. Based on exergy efficiency, the optimum composition of gases that can be produced in the pyrolysis process of 1 mol cellulose are 3 mol of $H_2$, 5 mol of CO, and 1 mol of $CH_4$ at a temperature of 723K. The maximum amount of exergy efficiency is 96.33% in the reaction 2

## REFERENCES

[1] Ginting, AS. 2014. Karakterisasi gas-gas hasil pirolisis tandan kosong kelapa sawit [thesis]. Bogor (ID) : Bogor Agricultural University.

[2]Cengel YA, Boles MA. 2015. Thermodynamics: an engineering approach. Volume 8. New York (US): McGraw-Hill Education

[3] Wang X, Wei L, Li G, Ming Z, Peng D. 2016. Energy and exergy analysis of rice husk high-temperature pyrolysis. *International journal of hydrogen energy*, xxx : 1-10.

[4] Szargut J, Morris DR, Steward FR. 1987. Exergy analysis of thermochemical, and metallurgical processes.

[5] Jacobs L.O, Peregrine C, Akwasi A.B., Charles A.M. 2011. Mass balance and exergy analysis of a fast pyrolysis system. *Proceedings of the ASME 2011 International Mechanical Engineering Congress & Exposition* : USA.

[6] Sukiran, M.A., Loh S.K., Nasrin A.B., Choo Y.M. 2011. Production and characterization of bio-char from the pyrolysis of empty fruit bunches. American Journal of Applied Sciences 8 , 10 :984-988

[7] Zhang Y, Li B, Li H, Liu H. 2011. Thermodynamic evaluation of biomass gasification with air in autothermal gasifiers. *Thermochim Acta,* 519 : 65-72.

[8] Peters JF, Petrakopoulou F, Dufour J. 2015. Exergy analysis of synthetic biofuel production via fast pyrolysis and hydroupgrading. *Energy*. 79 :325-362.

[9] Chen WH, Hsieh TC, Jiang TL. 2008. An experimental study on carbon monoxide conversion and hydrogen generation from water gas shift reaction. *Energy Conversion and Management*. 49: 2801-2808.

IOP Conf. Series: Materials Science and Engineering **557** (2019) 012059 doi:10.1088/1757-899X/557/1/012059

# Mechanical Properties of Concrete Contain Waste Fibers of Plastic Straps

**Hadeel R. Khatab, Samaher J. Mohammed, Laith A. Hameed**

College Of Engineering, Samarra University, Samarra, Iraq

**Abstract.** This study examined the utilization of polyethylene terephthalate (PET) strap waste as fibers to reinforce the concrete. The fibers were obtained by simply cutting the strap. An investigation was carried out on 7 batches containing PET waste fibers as a volumetric fraction of concrete. The parameters were the fibers fraction ratio ranging from (0-0.5) % and aspect ratio (2.5 and 3). Compressive strength, slump test, and splitting tensile strength were the test supplied to concrete. The test results indicate that the PET fibers improve the splitting tensile strength and slightly decrease the compressive strength and workability. So that, the concrete produced can be used in paving, industrial flooring, precast concrete, and shotcrete. Furthermore, a way to reuse plastic waste will be utilized to contribute to saving the environment.

## 1. Introduction

Since ancient times, the idea of adding one substance to another has been applied to improve the property or properties of one of these materials. Straw and horse hair have been added to clay to improve the characteristics of the bricks. Today, fibers are used in various forms to enhance the concrete.

Normal concrete is characterized by its relatively high compressive strength on the one hand, and on the other hand, it has low tensile strength, limited flexural strength and low resistance to the formation and expansion of cracks that appear even before the load is applied as a result of volumetric changes resulting from shrinkage or temperature change. Therefore, many researchers have concluded that the addition of fiber to the normal concrete leads to confrontation these problems and improve the properties of concrete [1][2][3]. There are many types of fibers that are added to concrete, such as iron fibers, synthetic fibers, fiberglass, natural fibers, iron fibers are the most common type. The addition of fiber to concrete is used in various applications such as industrial flooring, concrete paving, precast concrete, shotcrete, and many other applications.[1]

The size of the fiber sample is one of the most influential variables on the properties of fibrous concrete. There are also many characteristics such as the surface area of the fiber, the distance between the fiber and the aspect ratio (the ratio of the length of fiber to diameter).

Plastic, one of the most important discovery of the 20th century, have become an unavoidable part of our lives. A notable increase in the using of plastic annually has been rising steadily. It is used in various applications namely automotive, agriculture, electronics and electrical commodity, packaging, and other uses [4][5][6][7].

In general, Plastics can be categorized into two groups, thermosetting plastic, and thermoplastic. Thermosetting plastics are known as silicone, melamine, phenolic, epoxy resin, unsaturated polyester, and polyurethane, cannot be melted by heating. Otherwise, thermoplastics can melt and harden with heat and cooling, such as polyethylene (PE), polystyrene (PS), polypropylene (PP), polyethylene terephthalate (PET), and high-density polyethylene (HDPE) [8][9].

Because of the plastics benefits, such as lower-cost and suitability of utilizing, a significant demand is ever increasing all over the world. As a result, About 7 billion tons of trash were produced from 2015 until 2017 [10], [11]. Only 9% have been recycled and 12% have been burned while 79% have been accumulated in a landfill or natural environment. Pollution by plastic is one of the most dangerous types of pollution harmful to both the environment and human beings. The non-degradation of plastics means the accumulation and damage to many organisms. Among the plastic wastes, PET is the most common. It is gained in large quantity from containers for liquids and foods, strap for long-term storage and/or transport, etc. So, it is important to reducing and/or re-utilizing PET waste in varied applications.

A different application of plastic waste in concrete have been examined, thus achieve to reusing the waste materials and provide alternative sources of sand in novel concrete mixes. Research has been carried out for utilizing recycled PET as an aggregate in concrete by N. Saikia and J. de Brito[12], who explored the effect of different sized and shaped of PET particles on the fresh and hardened properties of concrete. Three different ratios of natural aggregate replacement, ranging from 5% to 15%, for each shape and size of PET aggregate were tested. It was found that the Compressive strength is inversely proportional to increased replacement ratio. This lack attributed to the lack of interaction between the PET particles and cement mortar.

The effecting of contents and sizes of plastic wastes on the fresh properties and compressive strength of self-compacting concrete were investigated by Hama and Hilal[13]. Three various sized of plastic wastes with six different contents were used as a partial replacement by volume of natural sand. The tests results showed that with the increase in the size and content of the plastic waste parts lead to a decrease in workability properties and compressive strength, but remains possible for use as a fine aggregate in self-compacting concrete.

Islam et. al. [14] reached the same results but using another type of PET as coarse aggregate. PET aggregate was resulted after shredded the used PET bottles into flakes and melted using the oven and then cooled. The Aggregate with a relatively smooth surface and has a rounded shape. Five of replacement ratios ranges from 0% to 50% with three different water-cement (w/c) ratios (0.42, 0.48 and 0.57) were selected to investigate fresh and hardened properties of concrete with PET aggregates.

Nibudey et al.[15] added plastic fibers to the concrete at 0% to 3% of the weight of the cement with aspect ratios of 35 and 50 to study the effects of plastic fibers on the properties of concrete. Test results show that the dry density, compressive strength, and workability decreased with increasing aspect ratios and fiber content.

Oliveira and Castro-gomes [16] investigated the using of (PET) bottle fiber recycled as fiber reinforced renders mortar. Different volumes of fibres, i.e. 0%, 0.5%, 1.0% and 1.5%, were introduced in dry mortar mixes. The results indicate that the incorporation of PET fibers significantly improves the flexural strength of mortars with a major improvement in mortar toughness.

This research will study the physical and mechanical properties of concrete with PET strips; to our knowledge, no study has been done on the properties of PET strips as fiber in concrete. Therefore, the novelty of this study is to provide data to help fill this gap by using a new type of plastic waste in the concrete. For this purpose, two different aspect ratios (2.5 and 3) and three volumetric additions ratio (0.1, 0.25 and 0.5 %) of fiber will be used to obtain compressive and splitting tensile strength and workability of concrete.

## 2. Experimental program and tests

### *2.1 Materials and mix proportions*

#### *2.1.1 Concrete mixture*

Portland Limestone Cement (CEM II/A-L 42.5R) produced by the Lafarge cement company from Iraq having properties shown in Table 1 was used. The results showed that this type is identical to the Iraqi standard specifications (IQS No.5/1984) [17]. Coarse aggregates with 12.5 mm maximum size which are locally available around Samarra city were. Tigris river sand was used as fine aggregates after sieving on a sieve size 4.75 mm. Coarse and fine aggregates had 2.71 and 2.63 specific gravity with

0.5% and 1.78% water absorption, respectively. Table 2 and 3 and Figure 1 and 2 show the results of sieve analysis for coarse and fine aggregate, respectively, and comparison it with Iraqi specifications IQS No.45/1984[18].

**Table 1.** Chemical and physical properties of cement

| Oxides composition | Content % | Limit of Iraqi specifications IQS No.5/1984 |
|---|---|---|
| $Al_2O_3$ | 4.65 | 6.2 - 6.5 |
| $SiO_2$ | 20.1 | 20.7 - 21.5 |
| $Fe_2O_3$ | 3.42 | 2.54 - 2.64 |
| CaO | 60.45 | 62.0 - 65.0 |
| $SO_3$ | 2.13 | Max. (2.8%) |
| MgO | 3.9 | Max. (5%) |
| **Physical properties** | **Test results** | **Limit of Iraqi specifications IQS No.5/1984** |
| Setting time (vacat apparatus), min<br>Initial<br>Final | <br><br>130<br>245 | <br><br>Not less than 45 min.<br>Not more than 10 hrs |
| Compressive strength (MPa)<br>At 3 days<br>7 days | <br>29.1<br>33.7 | <br>Not less than 15 MPa<br>Not less than 23 MPa |

**Table 2.** Sieve analysis for coarse aggregate

| Sieve size (mm) | Cumulative passing % | Limit of IQS No.45/1984 |
|---|---|---|
| 12.5 | 93.98 | 90-100 |
| 9.5 | 85.46 | 85-100 |
| 4.75 | 16.7 | 10-30 |
| 2.36 | 0.164 | 0-10 |

**Table 3.** Sieve analysis for fine aggregate

| Sieve size (mm) | Cumulative passing % | Limit of IQS No.45/1984, zone no.3 |
|---|---|---|
| 4.75 | 94.4 | 90-100 |
| 2.36 | 91.35 | 85-100 |

| | | |
|---|---|---|
| 1.18 | 84.3 | 75-100 |
| 0.6 | 76.32 | 60-79 |
| 0.3 | 34.25 | 12-40 |
| 0.15 | 9.45 | 0-10 |

*2.1.2 Strap plastic waste (SPW) fibers*

Plastic straps that use to fasten and hold effectively boxes, closing shipping containers and papers, made from polyethylene terephthalate (PET) were cut into fibers with 2.5 and 3 aspect ratio, as shown in Figure 3. Varying fractions of fibers from 0% to 0.5% by volume of concrete were mixed. Concrete with 0% of SPW was deemed as reference concrete.

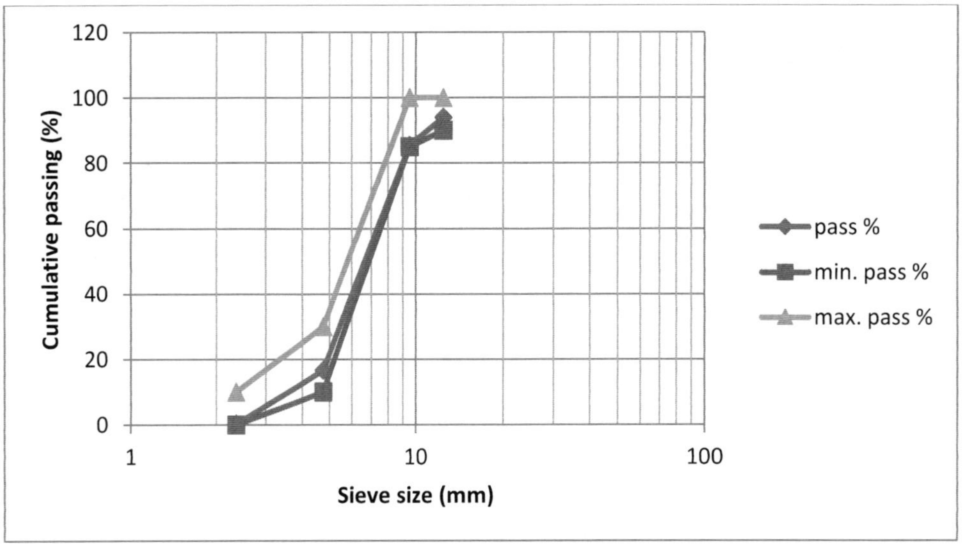


**Figure 1.** Coarse aggregate gradient curve

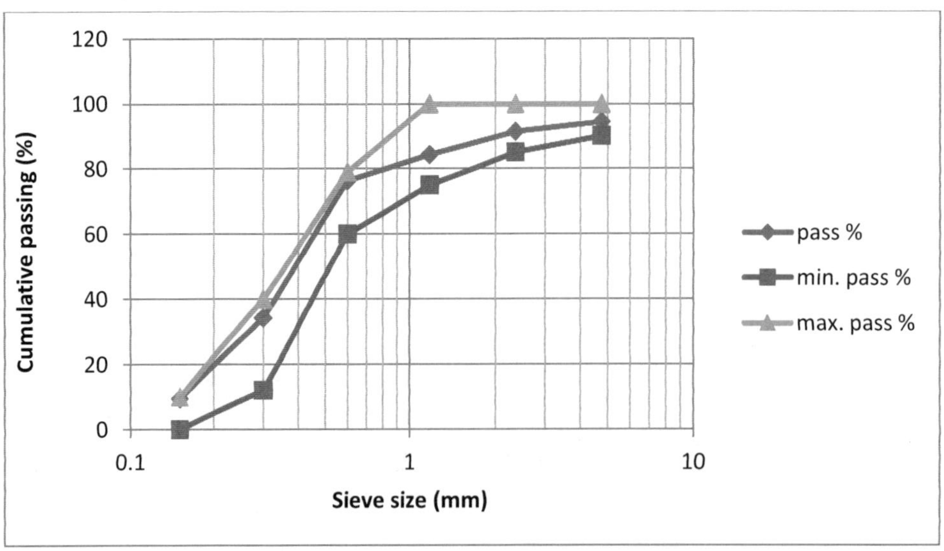


**Figure 2.** Fine aggregate gradient curve

*2.2 Mixture proportions and specimens*

Seven mixtures were mixed and cast. Mix proportions and the details of the mixtures are illustrated in Tables 4 and 5. Before water addition, all of the components had been mixed in dry conditions until the SPW homogenization in the mix after few initial rotations of the mixer. To obtain compressive strength for each mixture, three (150 mm * 150 mm * 150 mm) cubes were prepared and tested. Also, three cylinders (150 mm * 300 mm) were cast in order to split tensile strength test. The effect of the SPW addition of the fresh concrete was investigated by the slump tests. Total 42 specimens were prepared, including 21 cubes and 21 cylinders. Basic methods for the preparation of materials and the batching, mixing and sampling of fresh concrete in the laboratory and curing of the test samples complied with (BS 1881: Part 125:1986) and (BS1881:Part111:1983) [19][20]

a b

**Figure 3.** a) PET strips b) PET fibers

**Table 4.** Mix proportions of concrete

| W/C ratio | Water (kg) | Cement (kg) | Fine agg. (kg) | Coarse agg. (kg) |
|---|---|---|---|---|
| 0.6 | 210 | 350 | 700 | 1400 |

**Table 5.** Batch details of concrete mixes.

| Batch | PET waste fraction | Aspect ratio |
|---|---|---|
| B1 | 0% | 0 |
| B2 | 0.1% | 2.5 |
| B3 | 0.25% | 2.5 |
| B4 | 0.5% | 2.5 |
| B5 | 0.1% | 3 |
| B6 | 0.25% | 3 |
| B7 | 0.5% | 3 |

*2.3 Workability tests on fresh concrete*

The homogeneity of the concrete mixes and its workability were examined by the slump test as shown in Figure 4. The British Standard (B.S 1881: Part 102:1983)[21] was adopted in the test procedure. Slump values for concrete mixtures are shown in Table 6.

**Figure 4.** Slump test of concrete

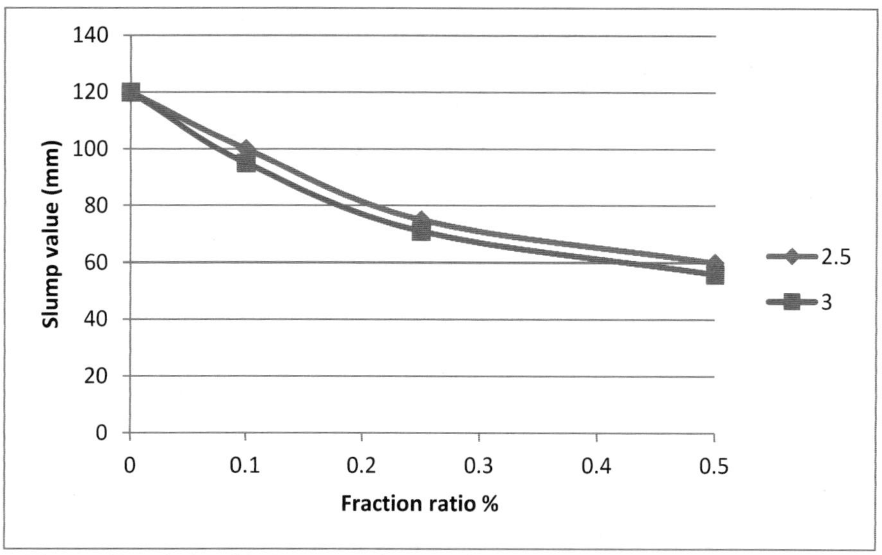


**Figure 5.** Relationship between slump values and PET fiber fraction variation

*2.4 Tests on hardened concrete*

*2.4.1 Compressive strength test*

The compressive strength of the cubes was examined according to the British standard (B.S 1881:Part116:1983)[22]. Compression load was subjected to the cube specimens by using a universal machine of 1000 kN loading capacity with a constant rate of loading till the specimen failure. The tests were executed at age 7 days. By dividing the failure load on the cross-sectional area of a cube, the compressive strength was obtained as stress causing the failure and showing the significant fracture in a specimen.

*2.4.2 Splitting tensile strength*

The indirect tensile strength of concrete was measured by splitting tensile strength. The axial compression load was subjected by using a universal machine of 1000 kN loading capacity along the length of the cylinder specimens at age 7 days. To ensure the uniformity of applying a load, a steel plate was employed as shown in Figure 6. The test complies with British standard (BS:1881:Part117:1983)[23].

## 3. Results and Discussion

Concrete workability was significantly affected by the addition of PET waste fibers. Increasing the fiber fractions were lead to decrease the workability, as shown in Figure 7. Similar response to the addition of plastic waste fibers has been achieved by different researchers.

**Figure 6.** Splitting tensile specimen

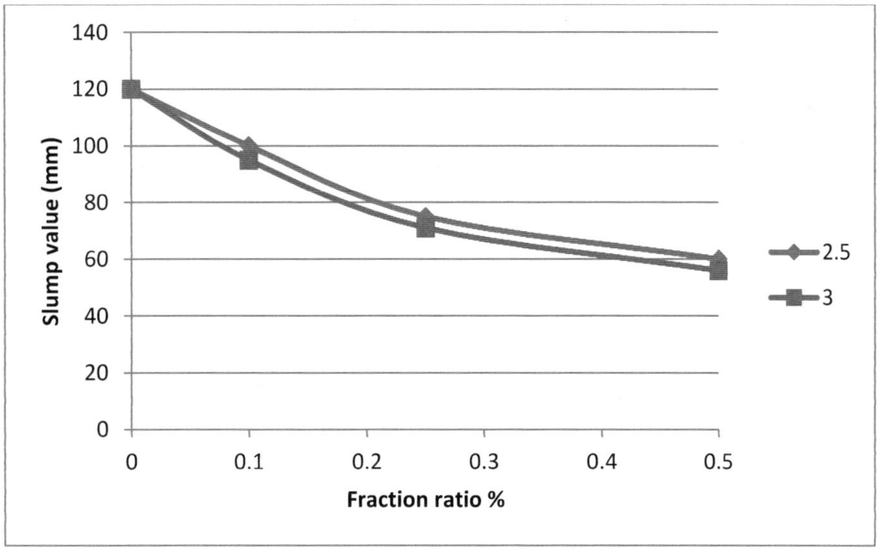


**Figure 7.** Relationship between slump values and PET fiber fraction variation

Concrete containing 2.5 aspect ratio showed a slump reduction of 16%, 37% and 50% for varying fraction 0.1%, 0.25, and 0.5 %. Concrete with 3 aspect ratio has gained slump relatively more than the previous type in the range between 20%, 41%, and 53%. This reduction may be attributed to the increased ability of fibers to restrict the flowability of the fresh concrete, which impedes the workability.

Test results of compressive strength are shown in Figure 8. Compressive strength was reduced with increasing of PET fiber fractions. Whatever aspect ratio, it is noted that the strength reduction was negligible up to a fiber dosage of 0.5% by volume of the mix. A 2.5 aspect ratio showed a 5% reduction in strength while 3 aspect ratio reduces the strength up to 9%.

It was noted that the mechanism of crack-resisting was enhanced and the material's nature was changed from brittle to ductile due to the effects of PET fibers.

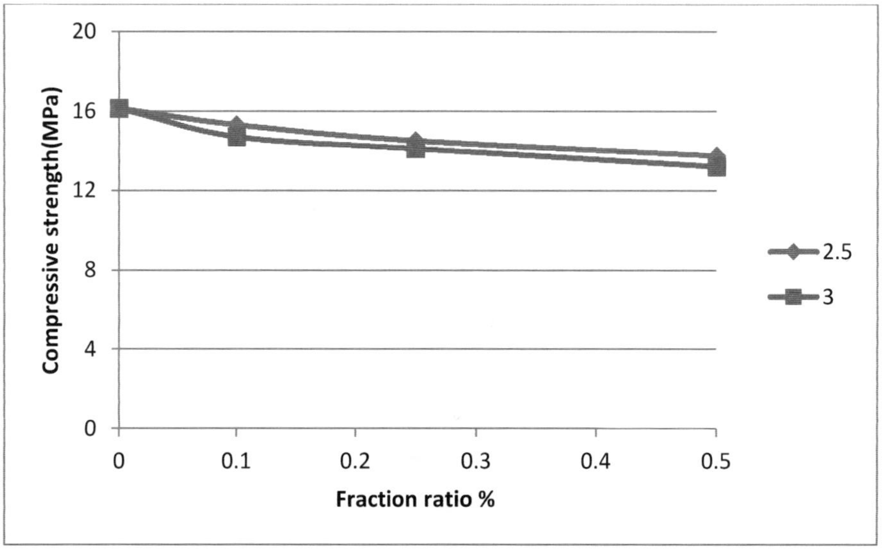


**Figure 8.** Relationship between compressive strength and PET fiber fraction variation

Because of PET fibers addition, there was an enhancement in splitting tensile strength in the range of 2–6% and 4–7% for 2.3 and 3 aspect ratio, respectively. Furthermore, crack development was restricted. Figure 9 show test results of splitting tensile strength.

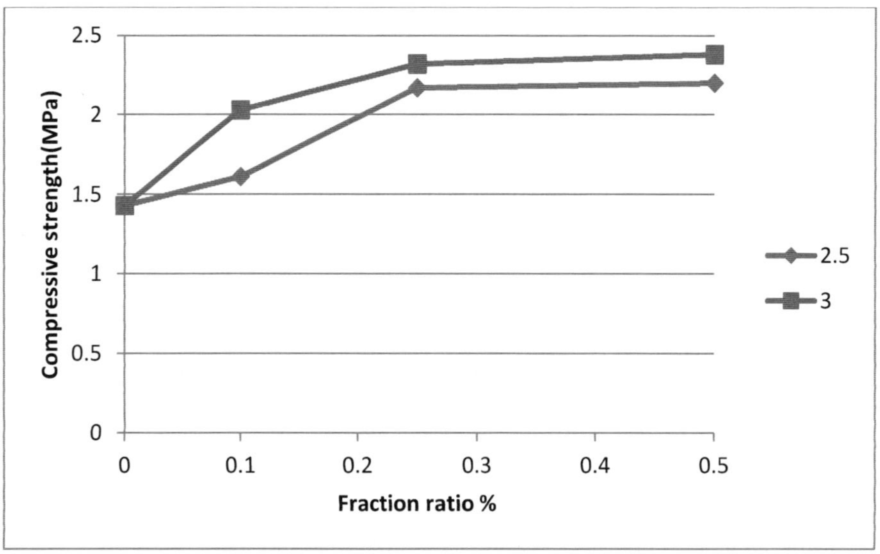


**Figure 9.** Relationship between splitting tensile strength and PET fiber fraction variation

## 4. Conclusion

The addition of plastic waste to the concrete led to changes in the properties of fresh and hardened concrete. The workability is inversely proportional to the ratio of waste addition and aspect ratio. As well as the compressive strength is affected by the addition of waste with a slight decrease with the increase in the ratio of addition and aspect ratio. This decrease is due to the weakness of the structure of the concrete and its low density due to the existence of large void around the fibers. As for splitting tensile strength, it has increased significantly. This increase can be explained by the fact that plastic

fibers work to increase the bonding of concrete components and operate with a principle similar to the reinforcing process and act as a conveyor medium for stresses in the cracking area.
Test results exhibited good strength of crack-resisting and participated in the ductile response of concrete contain PET waste fibers.

**References**

[1] R. Sharma and P. P. Bansal, "Use of Different Forms of Waste Plastic in Concrete – A Review," *J. Clean. Prod.*, 2015.
[2] D. Foti, "Use of recycled waste pet bottles fibers for the reinforcement of concrete," *Compos. Struct.*, vol. 96, pp. 396–404, 2013.
[3] A. C. Bhogayata and N. K. Arora, "Fresh and strength properties of concrete reinforced with metalized plastic waste fibers," *Constr. Build. Mater.*, vol. 146, pp. 455–463, 2017.
[4] K. Balaed, N.Z. Noriman, O. S. Dahham, S. T. Sam, R. Hamzah and M. F. Omar "Characterization and properties of low-linear-density polyethylene/Typha latifolia composites", International Journal of Polymer Analysis and Characterization, vol 21(7), pp. 590-598, 2016.
[5] M. Mohammed, A. R. Rozyanty, A. F. Osman, T. Adam, U. Hashim, A. M. Mohammed, N. Z. Noriman, O. S. Dahham and B. O. Betar, "The Weathering Effect in Natural Environment on KenafBast Filled Unsaturated Polyester Composite and Integration of Nano Zinc particle for Water Repellent" Micro and Nanosystems, vol 9(1), pp. 16-27, 2017.
[6] A. Alnaid, N. Z. Noriman, O. S. Dahham, R. Hamzah, M. N. Al-Samarrai, S. S. Idrus and Z. Shayfull, "Effects of Date Seeds Size and Loading on Properties of Linear Low-Density Polyethylene/Date Seeds Powder Composites" Journal of Physics: Conference Series, vol. 1019, pp. 012060, 2018.
[7] O. S. Dahham, N. Z. Noriman, R. Hamzah, S. Sudin, S. S. Idrus, Z. Shayfull and T. Adam, "The Effect of Cyperus Odoratus Size and Loading on the Properties of Polypropylene/Cyperus Odoratus (PP/CY) Composites" Journal of Physics: Conference Series, vol. 1019,pp. 012056, 2018.
[8] R. Siddique, J. Khatib, and I. Kaur, "Use of recycled plastic in concrete: A review," *Waste Manag.*, vol. 28, no. 10, pp. 1835–1852, 2008.
[9] P. Panyakapo and M. Panyakapo, "Reuse of thermosetting plastic waste for lightweight concrete," vol. 28, pp. 1581–1588, 2008.
[10] R. Geyer, J. R. Jambeck, and K. L. Law, "Production , use , and fate of all plastics ever made," *Sci. Adv.*, vol. 3, no. July, pp. 25–29, 2017.
[11] Plastics Europe, "Plastics – the Facts 2017: An analysis of European plastics production, demand and waste data," *Plast. Eur.*, 2017.
[12] N. Saikia and J. De Brito, "Mechanical properties and abrasion behaviour of concrete containing shredded PET bottle waste as a partial substitution of natural aggregate," *Constr. Build. Mater.*, vol. 52, pp. 236–244, 2014.
[13] S. M. Hama and N. N. Hilal, "Fresh properties of self-compacting concrete with plastic waste as partial replacement of sand," *Int. J. Sustain. Built Environ.*, vol. 6, no. 2, pp. 299–308, 2017.
[14] J. Islam, S. Meherier, and A. K. M. R. Islam, "Effects of waste PET as coarse aggregate on the fresh and harden properties of concrete," *Constr. Build. Mater.*, vol. 125, pp. 946–951, 2016.
[15] R. N. Nibudey, D. K. Parabat, and A. M. Pande, "Strengths Prediction of Plastic fiber Reinforced concrete (M30)," *Int. J. Eng. Res. Appl.*, vol. 3, no. 1, pp. 1818–1825, 2013.
[16] L. A. P. De Oliveira and J. P. Castro-gomes, "Physical and mechanical behaviour of recycled PET fibre reinforced mortar," *Constr. Build. Mater.*, vol. 25, pp. 1712–1717, 2011.
[17] Iraqi Organization of Standards, "IOS 52: 1985" Portland Cement.
[18] Iraqi Organization of Standards."IOS 45: 1984" Natural Aggregate of Concrete.
[19] BS 1881: Part 125, " Methods for mixing and sampling fresh concrete in the laboratory", British Standards Institution, 1986.
[20] BS 1881: Part 111, "Method of normal curing of test specimens (20 "C method) L," British Standards Institution,1983.
[21] BS 1881: Part 102, "Method for determination of slump", British Standards Institution, 1983.
[22] BS 1881: Part 116, "Method for determination of compressive strength of concrete cubes",

British Standards Institution, 1983.
[23] BS 1881: Part 117, "Method for determination of tensile splitting strength", British Standards Institution, 1983.

# Carbon Dioxide Adsorption on MOF-199 Metal-Organic Framework at High Pressure

**Rana Th. A. Alrubaye**[1,2], **Huda M. Kareem**[1]

[1] Chemical Engineering Department, College of Engineering, University of Baghdad, Jadrai, Baghdad, Iraq
[2] Corresponding author. Email : Alrubaye.rana@gmail.com

**Abstract.** MOFs have considered as promising materials for CO2 adsorption due to their excellent chemical stability, large pore volume, and high surface area. In this study, MOF-199 has been synthesized using the solvent-thermal method with focusing on the significant parameters that affecting the structure of the MOF-199 (solvent, synthesis time, and temperature). Produced MOF-199 samples were characterized by electron scan microscope, X-ray diffraction, and microporosimetric analysis. High crystallinity degree and high surface area were found at 1:1 of the H2O: C2H5OH (within the solvents range from 1:1 to 1:2 ratios), whereas the changing of the synthesis duration from 18 to 48 h, and synthesis temperatures from 40 to 140 °C have revealed that the best synthesis conditions are 100 °C for 48 h. The appropriate activation condition has been determined to be 60 °C for 16 h. For CO2 adsorption isotherms on MOF-199 were at a temperature (36, 40, 60, and 70) 0C and pressure range (0-5) bar using a volumetric technique. Temkin, Freundlic and Langmuir isotherm models have been used to analyze the equilibrium data. This research presents the effect of significant parameters on the synthesis of MOF-199 to understand their influence on the formation of MOF structures and the CO2 adsorption capacities.

**Keywords:** MOF-199, $CO_2$ adsorption, synthesis temperature and time

## 1. Introduction

MOFs are new crystalline coordination polymers that have investigated in the past two decades [1]. These new materials are ordered porous solids that are forming by soluble complexes consist of inorganic particles or ion clusters and organic linkers, then self-assemble into frameworks. The formation of MOFs has a significant advantage of designing and tailoring the structure and sizes of pores in the framework, and metal ions and organic ligands selection could achieve that. Different MOFs structures can be obtained by varying the coordination environments of the central metals controlled by the several factors that affecting the synthesis process these factors are the solvent, synthesis temperature and time [2].

Cui [1] experiments revealed that the process of solvation and organize formation for MOFs of different environments for self-assembly could be affected by the water present in a mixed solvent system. Therefore dimensionality of the final structures seemed to be tuned by varying $C_2H_5OH:H_2O$ ratio. The effect of the choice of the solvent has been clearly reflected in the obtained structures.

Solvothermal of MOFs offers advantages compared to other structural methods. The synthesis under solvothermal in a closed system at pressures higher than 1atm and a temperature higher than 100 °C will ensure the high solubility of all the reactants and high-quality crystals can be obtained which led to the high crystallization of MOFs [3].

Sarawade.et reported the influence of reaction temperature and time on size, shape, and morphology of MOF microcrystals formed by microwave synthesis. In addition, the higher dimensional MOFs were produced which indicated, the easier deprotonating of the ligand at higher temperatures [4].
The Metal-Organic Frameworks have recently used for numerous applications as promising alternatives for gas separation due to their ease of manipulation in porous structures and flexibility in composition and chemical properties.
MOF-199 is considered as a highly demanding porous material in gas separation and adsorption due to its large pore volume high, high surface area (generally in a range of 600–1600 $m^2$ $g^{-1}$) and excellent thermal stability. MOF-199 is one of the most widely studied MOF materials as a $CO_2$ capturing agent. Researcher reported that the uptake capacity for $CO_2$ of MOF-199 (at 1 atm and 220 °C) was 69 ml $g^{-1}$ [5], and the capacity of the $CO_2$ uptake at high-pressure ranges at 25 °C from 8 to 11 mmol $g^{-1}$ as examined by [3].
In this study, the synthesis of the highly porous metal-organic framework MOF-199 is reported. The effect of parameters such as solvent volume, synthesis temperature and reaction time were studied. The $CO_2$ storage capacity of the material and static adsorption isotherms on MOF-199 was also investigated.

## 2. Materials and Methods

### *2.1. Materials*
MOF-199 was synthesized using copper (II) nitrate trihydrate [$Cu(NO_3)_2{\cdot}3H_2O$, 99%] and benzene-1,3,5 tricarboxylic acid (BTC, 95%) obtained from Sigma Aldrich. Absolute ethanol ($C_2H_5OH$) from Fisher scientific.

### *2.2. Solvent-thermal synthesis of MOF-199*
MOF-199 was synthesized following Schlichte's method [6] using solvent-thermal synthesis and BTC (0.42 g) dissolved in 24 ml of 1:1 or 2:1 $C_2H_5OH$:$H_2O$ slovent ratio. The prepared mixture was stirred until a clear solution was obtained (10 min). Then $Cu(NO_3)_2{\cdot}3H_2O$ (0.875 g) was added to the mixture and stirred thoroughly for another 10 min. Once completely dissolved reached in the solvent, the resulting blue solution was transferred to a 50 ml stainless steel Teflon lined autoclave and heated to a specific temperature at specific crystallization time (see Table 1). In order to investigate the effect of the operation conditions changing on the solve-thermal synthesis of MOF-199 the heating temperature was changed from 40 to 140 °C and the reaction time was varied from 18 to 48 h. Then, the reactor was cooled to room temperature, and a blue crystalline powder was formed. The powder was then filtered and washed thoroughly with a 60 ml mixture of water and ethanol (1:1). The product was finally activated under vacuum at 60°C for 21 h and stored in closed vials at 60°C in oven for further experiments.

**Table 1.** Summary of the MOF-199 samples synthesized

| Sample | Time h | Temperature °C | $C_2H_5OH$:$H_2O$ ratio |
|---|---|---|---|
| S1 | 18 | 100 | 1:1 |
| S2 | 24 | 100 | 1:1 |
| S3 | 30 | 100 | 1:1 |
| S4 | 48 | 40 | 1:1 |
| S5 | 48 | 85 | 1:1 |
| S6 | 48 | 100 | 1:1 |
| S7 | 48 | 140 | 1:1 |
| S8 | 48 | 100 | 2:1 |

### *2.3. Characterization of MOF-199*
scanning electron microscope (SEM) was carried out using FEI Quanta 200 ESEM equipment with a high voltage mode of 20 kV to obtain MOF-199 synthesized samples morphology. All samples were characterized to confirm the formation of MOF-199 using powder X-ray diffraction (XRD), which

was acheived by a Rigaku Miniflex diffractometer (CuKα radiation, 30 kV, 15 mA, λ = 1.5406 Å) using a step-scan mode (0.03° per step) with a range of 5° < 2θ < 46°. ASAP 2020 Micromeritics instrument was used for N2 adsorption/desorption cycles, which achieved by Microporosimetric analysis at 77 K, the surface area was determined using the Brunauer- Emmett-Teller (BET) method. All synthesized samples were degassed at 180 °C for 3 h prior to characterization.

*2.4. $CO_2$ isotherms adsorption*

The $CO_2$ adsorption capacity of MOF-199 was investigated using pressure swing adsorption (PSA). The necessary steps in the pressure swing adsorption were evaluated using an experiential apparatus, which was designed for wide-range conditions (high pressure, temperature, flow rate, adiabatic, etc.). The apparatus was automated and controlled, with the only manual operations being the analysis of samples and the recording of flow rates and pressure as shown in Figure 1

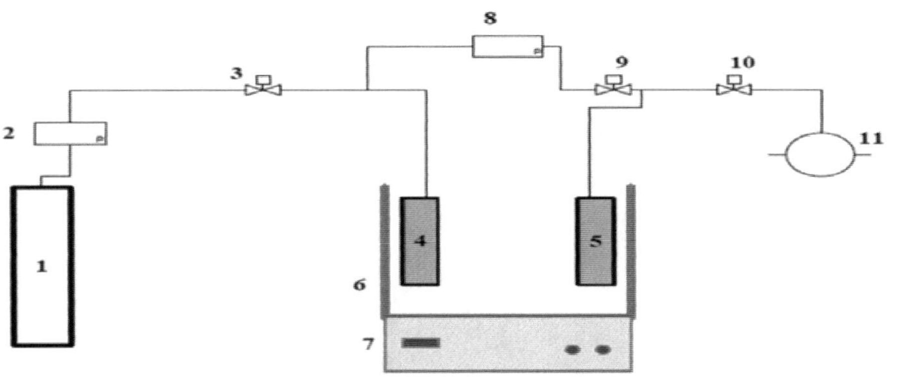


**Figure 1.** Schematic diagram of the apparatus used for adsorption equilibrium measurement

The volume of the whole system was measured manometrically using $N_2$ gas. A weighted sample had been regenerated in the adsorption column by degasifying the sample to a pressure less than 3 bar for an overnight prior to each test. After the MOFs had been regenerated, the heating was on, and a stream of $CO_2$ was passed through the apparatus for about 120 min or until the whole adsorbent bed reaches the specific temperature. Simultaneously, the gage pressure was monitored. The amount of gas adsorbed at equilibrium was determined from Eq(1):

$$q = \frac{(cri - crf)Vr - Crf\ \varepsilon\, Va}{W} \tag{1}$$

q is adsorbent loading at equilibrium (mmol/g), V is volume (l), C is concentration (mmol/l), f is overall bed void fraction , and W is the weight of the adsorbent in the absorber vessel (g), the subscripts r, a, i and refer to reservoir, adsorbed vessel, initial and final condition respectively,Ce (Cri-Crf) concentration equilibrium.TC critical temperature of $CO_2$ (31°C ) , Pc critical pressure of $CO_2$ ( 73 atm), R gas constant (8.314 L.Kpa/ mol. K) , T is Temperature use for adsorption.
Cri and Crf where calculate from equation of real gases :

P V= Z n R T (n/v=C)

$$Cri = \frac{Pi}{Zi\,R\,T} \tag{2}$$

$$Crf = \frac{Pf}{Zf\,R\,T} \tag{3}$$

Z compressibility factor of feed and out gases may be estimated according to the generalized equation :

$$Z = 1 + \frac{BP}{RT} \tag{4}$$

The coefficient of B where calculated by:

$$\frac{BPc}{RTc} = B + \omega\, B' \tag{5}$$

The coefficient of Ḃ and B' may be estimated as follows:

$$B = 0.083 - \frac{0.422}{Tr^{1.6}} \tag{6}$$

$$B' = 0.139 - \frac{0.172}{Tr^{1.2}} \tag{7}$$

Tr= T/T

## 3. Results and discussion

### *3.1. Characterization of MOF-199*

All collected XRD diffraction data of synthesized samples, which are shown in Figures 2 to 4, are consistent with those reported for the typical XRD pattern of MOF-199 (Nobar,2012). Formation of large size crystals can explain the significant high intensity of peaks in the XRD pattern, however, in this study, it could be related to the lower content of water inside the pore of the synthesized MOF. And that because of the sample immediately was inserting inside the analysis chamber after the drying process at 110 °C The XRD pattern of MOF-199 exemplifies its crystalline phase with characteristic peaks at $2\theta \approx 6.5^{o}$, $9.5^{o}$, $11.5^{o}$, and $13.4^{o}$, which were confirming the formation of MOF-199 crystalline phase. The $Cu_2O$ characteristic diffraction peaks were at $2\theta \approx 36.4^{o}$, $42.3^{o}$ and $43.3^{o}$ [7].

The relative crystallinity of metal-organic frameworks of all synthesized samples was calculated by finding the percentage ratio of the sum of the peak intensities for the synthesized MOF-199 samples and the sum of the peak intensities of the standard MOF-199 [8].

Figure 2. show the effect of the changing the reaction time from 18 to 48 hours at constant synthesis temperature (100°C) and 1:1 $C_2H_5OH$:$H_2O$ solvent ratio on the structure of the synthesized samples (S1, S2, S3 and S4). The formation of MOF-199 is indicated by the appearance of characteristic peaks after 18 hours at reaction temperature 100 °C.

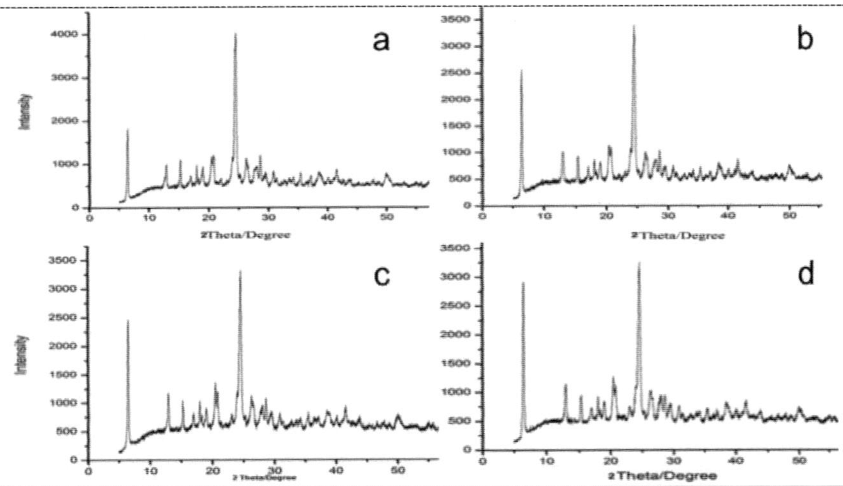


**Figure 2.** XRD pattern MOF-199 at T=100°C for a) 18 h, b) 24h, c) 30h, and d) 48h.

Figure 3 illustrates the effect of increasing ethanol content in solvent volume for samples S6 and S8 in which all the synthesis conditions for both of these samples were practically the same expect the $C_2H_5OH$:$H_2O$ ratio were increased from 1:1 in sample S6 to 2:1 in sample S8. The Diffractogram data of crystals synthesized obtained with different solvent volume show the same typical pattern of MOF-199. However, decreases with the increase in $C_2H_5OH$:$H_2O$ ratio shows decreasing in MOF-199 crystallinity (Table 2) which indicates that the best solvent ratio in at 100 °C was 1:1 $C_2H_5OH$:$H_2O$ ratio.

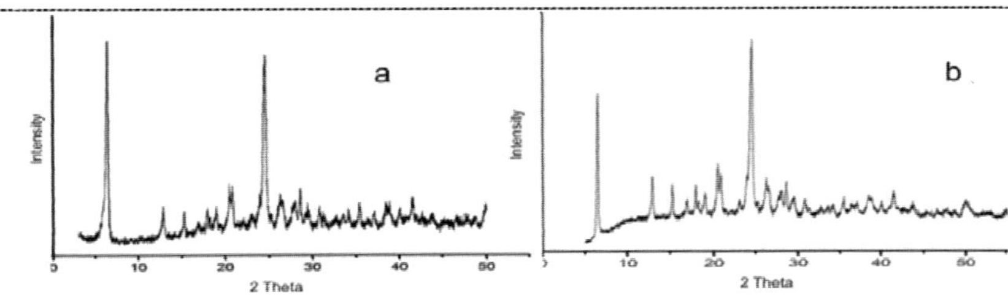


**Figure 3.** XRD pattern of MOF-199 at 100 °C, 48, and ;a) 1:1 C2H5OH: H2O, b) 2:1 C2H5OH: H2O

XRD pattern of MOF-199 synthesized for 48 hours and at different heating temperatures (40, 85, 100, and 140°C) are shown in Figure 4. For 48 h reaction time, the MOF-199 crystals have started to appear at 40°C as indicated by the presence of characteristic peaks of MOF-199 at $2\theta = 6.5°$, and 13.4°. The MOF-199 is fully formed at heating temperatures of 85°C as can be identified by peaks characteristic to MOF-199. As can be shown in Figure 5, the obtained MOFs with reaction times of 85–140 °C have the same XRD pattern to that of typical MOF-199 in the literature [4]. However, the main peaks intensity increased with the increase in heating temperature from 40-100 °C and when increasing the temperature to 140 °C the main peaks intensity were decreased.

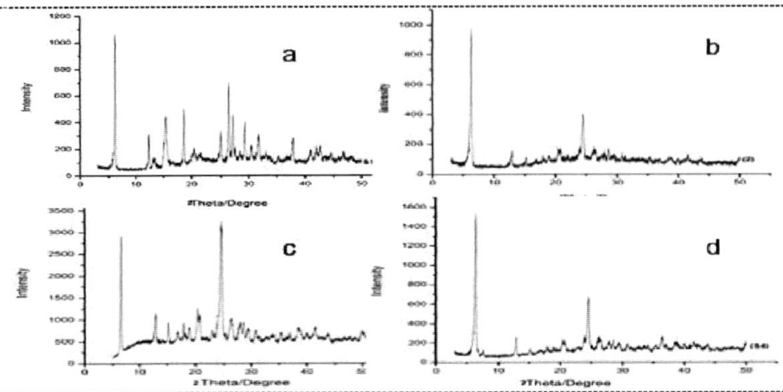


**Figure 4.** XRD pattern of MOF-199 at 48 h and temperatures: a) 40 °C, b) 85 °C, c) 100 °C, and d) 140 °C.

These results have been confirmed by calculating the crystallinity of the MOFs (Table 2), which indicates that the MOF-199 crystallinity decreases with the overheating temperature for that reaction time. The crystallinity results and the BET surface area results are listed in Table 2. Therefore, based on the crystallinity of the obtained crystals and the XRD pattern for all synthesized samples, the crystallinity of samples S5 was the highest crystallinity. Nevertheless, from a comparison of the crystallinity for the XRD of MOF-199 and the BET surface area results, it is seen that the highest surface area (5518 $m^2/g$) with high crystallinity off 103% was for S6 sample. Therefore, the optimum solvothermal for the synthesis of MOF-199 for the chosen operation conditions have chosen to be at 100 °C and 48 h. The solvothermal conditions are chosen based on the XRD patterns of the MOFs follow the pattern of typical XRD of MOF-199 with the high crystallinity and the highest surface. The surface area of prepared MOFs that was measured by physical nitrogen adsorption at liquid nitrogen temperature using the BET method. An increase in the surface area generally increases the catalyst activity [9]. The obtained value of the surface area of prepared MOF-199 in this research, is higher than the value obtained by Jiang (1247$m^2/g$ ,100 °C, 13 h) [8].
Figure 5 a and b shows representative morphology pictures of optimum synthesized MOF-199 for samples synthesized with reaction time 48 h at 85 °C and 100 °C, respectively. SEM characterization shows are in good agreement with the result obtained by Nobar [4].

**Table 2**. Summary of the Surface area and Crystallinity of MOF-199 samples

| Sample | Time (h) | Temperature (°C) | Surface area $m^2/g$ | Crystallinity % |
|---|---|---|---|---|
| S1 | 18 | 100 | 1835 | 102 |
| S2 | 24 | 100 | 2902 | 100 |
| S3 | 30 | 100 | 3635 | 100 |
| S4 | 48 | 40 | NoMOF-199 | NoMOF-199 |
| S5 | 48 | 85 | 2679 | 106 |
| S6 | 48 | 100 | 5518 | 103 |
| S7 | 48 | 140 | 1059 | 63 |
| S8 | 48 | 100 | - | 94 |

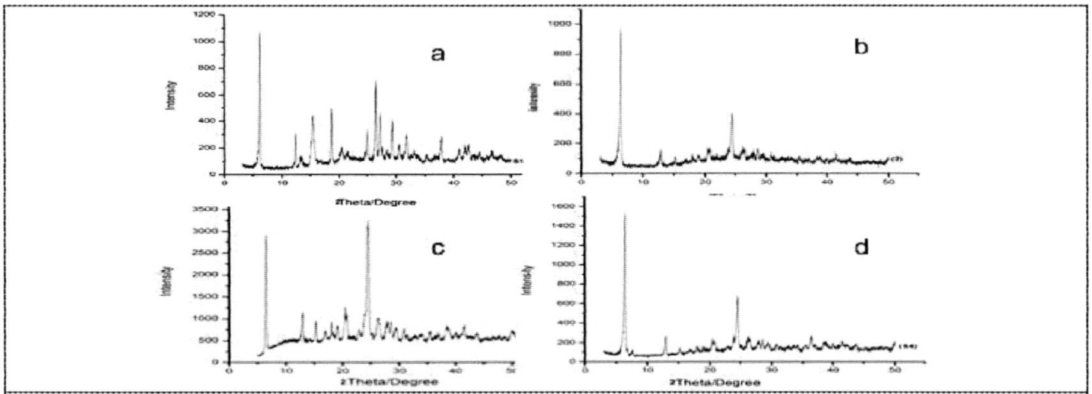


**Figure 5.** SEM images of synthesized MOF-199 samples at a). 24 h and 85°C, and b). 48 h and e 100°C

### 3.2. $CO_2$ adsorption on MOF-199

Carbon dioxide adsorption and desorption isotherms were measured on the optimum sample that was synthesized at 100 °C for 48 h (S6). The BET specific surface areas and pore volumes were calculated as 5518 m2.g−1 and 0.609 cm3.g−1, respectively, Figure 6 presents the relation between amount of $CO_2$ adsorption q (mmol/g) and gas phase concentration equilibrium (Ce)

The experimental adsorption isotherms of $CO_2$ on to MOF-199 was measured overpressure range (0-5 bar) at (36, 40·, 60, 70 °C) as shown in Figure 6.the adsorption capacity on MOF-199 decrease with increase in temperature. In this study obtained a high adsorption capacity equal to 4.9 mmol/g at 5 bar, 36 °C. The effect of temperature on adsorption isotherm for carbon dioxide on MOF-199 (S6). From this Figure, it is seen that by increasing the equilibrium adsorption temperature the total amount of adsorbed on

S6 samples decreasing. With increasing temperature, the $CO_2$ adsorbed on MOF-199 surface becomes unstable because molecular diffusion increased that lead to more desorption of $CO_2$ molecules. This behavior has also been mentioned by [4]. Also the adsorption capacities in this study is in good agreement Zenga, who reported to obtaine 4 mmol/g at 1bar at 25 °C [2].

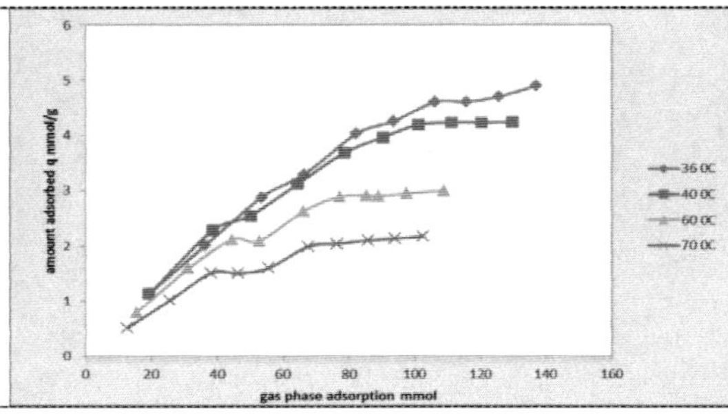


**Figure 6.** Adsorption equilibrium isotherm of carbon dioxide on MOF-19

## 4. Conclusions

From the experimental results, it has been concluded that the MOF-199 has been successfully synthesized at the optimal conditions of 100°C for 48 h. The synthesized sample MOF-199 value of the surface area (BET = 5518 $m^2\ g^{-1}$) porous size of crystal 11.8 Å, the specific volume 0.693 $cm^3\ g^{-1}$ with high crystallinity of 103. The synthesized MOF-199 should be considered as an effective adsorbent for $CO^2$ with dioxide storage capacity of the 4.9 $m^2\ g^{-1}$ at the pressure of 5 bar at 36 °C.

## References

[1] Cui, P., Wu, J., Zhao, X., Sun, D., Zhang, L., Guo, J., and Sun, D. 2011 *Two solvent-dependent zinc (II) supramolecular isomers: rare kgd and lonsdaleite network topologies based on a tripodal flexible ligand.* (Crystal Growth & Design) vol 11(12) p 5182-5187.

[2] Zeng, M., Wang, T., Liu, Y., Long, Z., and Sun, C. 2018 *Research on Planning Method of Microgrid Energy Unit in Intelligent Districts Containing Electric Vehicles Considering CO 2 Emission* (Chemical Engineering Transactions) vol 66 p 1339-1344.

[3] Wang, X. 2017 *Computer Simulation and Process Improvement of CO 2 Capture by Chemical Absorption* (Chemical Engineering Transactions) vol 59 p 547-552.

[4] Nobar S.N., 2012, *Adsorption and diffusion of gases in Cu-BTC*, (PhD Thesis, Sharef University of Technology Tahran, Iran).

[5] Xie, J., Yan, N., Qu, Z., and Yang, S. 2012 *Synthesis, characterization and experimental investigation of Cu-BTC as CO2 adsorbent from flue gas* (Journal of Environmental Sciences) vol 24(4) p 640-644.

[6] Schlichte, K., Kratzke, T., and Kaskel, S. 2004 *Improved synthesis, thermal stability and catalytic properties of the metal-organic framework compound Cu3 (BTC) 2* (Microporous and Mesoporous Materials) vol 73(1-2) p 81-88.

[7] Lin, K. S., Adhikari, A. K., Ku, C. N., Chiang, C. L., and Kuo, H. 2012 *Synthesis and characterization of porous HKUST-1 metal organic frameworks for hydrogen storage* (international journal of hydrogen energy) vol 37(18) p 13865-13871.

[8] Jiang, D., Mallat, T., Krumeich, F. and Baiker, A. 2011 P*olymer-assisted synthesis of nanocrystalline copper-based metal–organic framework for amine oxidation.* (Catalysis Communications) vol 12(7) p 602-605..

[9] Rallan Ch.A., AlRubaye R.Th. and Garforth A.A. 2015 *Generation of Catalytic Films of Alumina and Zeolites on FeCralloy Rods* (Chemical Engineering Transactions) vol 43 p 907-913.

# Synthesis and Characterization of Some New Heterocyclic Compounds Derived from Benzothiazole

**Muhanned A. Mahmoud**
Department of Chemistry/College of Science for Women, University of Baghdad, Baghdad, Iraq.
Corresponding Author E-mail: alnemimu2006@yahoo.com

**Abstract.** The present work involves study of some new 1,3-oxazepine and diazepines derivatives by cycloaddition reaction between Schiff s bases and (Succinic imide, phthalic imide, and 3-nitophthalic) anhydride with dry benzene as a solvent. Oxazepine and diazepines derivatives from 4-chloro-6-nitrobenzo[*d*]thiazol-2-amine were synthesized. These derivatives were synthesized by the reaction of Schiff base with different anhydrides in presence of dry benzene. The starting material was 4-Chloro-6-nitro 2- aminobenzothiazole which prepared from 2-Chloro -4-nitroaniline and ammonium thiocyanate with presence of bromine and sodium acetate. The prepared compounds have been characterized by melting points and some physical properties besides the FT-IR,H-NMR spectra. The purity for these compounds was checked by TLC.
**Keywords**: Schiff bases, aminobenzothiazole, oxazepine and diazepines.

## 1. Introduction
Schiff' s bases possess biological activity such as antibacterial, antifungal, antitumor, anti-HIV and herbicides [1-3]. Heterocyclic compounds are found as building units via some biological molecules [4], mostly are molecules, which have (5,7) membered ring [5-7].

The Schiff bases (SB) had been prepared by (Schiff) for the first time [8] from the condensation reaction of aldehyde or ketones with primary amines by refluxing the mixture in absolute $C_2H_5OH$, $C_6H_6$, or any other suitable solvent for one hour or few hours some times, the reaction may be catalyzed by acid [9-11]. Oxazepine, un-saturated non-homologous (7) membered heterocycle that contains oxygen in 1$^{st}$ position and nitrogen in 3$^{rd}$ position, is prepared by the pericyclic cycloaddition of Sciff bases with (Succinic imide, phthalic imide, and 3-nitophthalic) anhydrides [12]. The importance of 1,3-oxazepine is ascribed to their applications as anticonvulsant.

## 2. Experimental
Melting points in ($^0$C) of all the prepared compounds were recorded. FT-IR spectra had been recorded as KBr pellets, on a FT-IR BRUKER. $^1$H-NMR spectra in DMSO solution had been recorded on the instrument BRUKER. Chemical shifts were reported as δ ppm in relation to TMS as inner standard. The progress of the reaction was observed by TLC with silica gel. Solvents had been of analytical reagent grade and of the top quality available commercially.

### *2.1. Preparing of 4-chloro-6-nitrobenzo[d] Thiazol-2-amine*
This compound had been prepared by the means of ammonium thiocyanate with bromine and sodium acetate. The product obtained was re-crystallized from ethanol [13,14]. The structure was confirmed by FT-IR, $^1$H-NMR and the melting point was checked. 2-chloro-4-nitro aniline was treated with $CH_3COOH$, then a solution ammonium thiocyanate was added to it. The thiourea compound obtained was treated with bromine. After completion of reaction, the reaction mixture was basified with liquor ammonia. A solid product of 4-chloro-6-nitrobenzo[*d*]thiazol-2-amine obtained was filtered, washed, and recrystallized.

### *2.2. General Method for Preparing of Schiff Bases [15-17]: The Schiff s Bases were Prepared by Reacting 2-amino-1,3-benzothiazole with Selected Aldehydes.*
A solution of 4-chloro-6-nitrobenzo[*d*]thiazol-2-amine (3mmol) and appropriate substituted benzaldehyde (3mmol) in 15 ml absolute ethanol with a few drops of $CH_3COOH$ acid was refluxed for (5-6) hrs., the

reaction was observed by TLC, and the maximum solvent had been distilled out .The mixture was cooled to room temperature afterward.
The nucleophilic addition of $-NH_2$ group of hetero to aromatic aldehydes is not so straight forward and easier due to the presence of N in heterocyclic rings [14]. All compounds were purified by re crystallization using absolute ethanol in order to avoid the hydrolysis.

*2.3. Preparing of Oxazepine Derivatives* [15-17]:
A mixture of the prepared Schiff base,( 0.02m) and 3-nitrophthalic anhydride (0.02m) were dissolved in 15ml dry benzene with the mixture were refluxed for (3-5) hours and the excess solvent was distilled . The precipitate obtained was filtered after cooling, re crystallized in ethanol and melting points were determined.

*2.4. Preparing of Diazepines Derivatives* [15-17] *(4, 5): N-(4-Chloro-6-nitrrobenzo[d]thiazol-2-yl)-N-(4-(dimethylamino)benzyl)-N⁴-methylsuccinamide(4), N-(4-Chloro-6-nitrrobenzo[d]thiazol-2-yl)-N-(4-(dimethylamino)benzyl)-N⁴-methylphthalamide(5).*

A mixture of the prepared Schiff base (0.02m) and ((Succinic imide and phthalic imide) anhydrides (0.02m) were dissolved in 15ml dry benzene. The mixture then was refluxed for (5-6) hours . Excess solvent was distilled. The obtained precipitate was then filtered after cooled, re-crystallized by $C_2H_5OH$ and melting points were determined.

**3. Results and Discussion:**
Schiff s bases are prepared through the condensation of benzothiazole with aromatic aldehyde to give the azomethine/imine compounds. The reaction is followed by disappearance of the Carbonyl group absorption band at (1690-1720) $cm^{-1}$ with the appearanc of (C=N) absorption band at (1645 -1649) $cm^{-1}$ (Table 1). Derivatives of oxazepine are prepared by reaction of (Succinic imide, phthalic imide, and 3-nitophthalic), anhydrides with Schiff s base derivative. It was noted disappearance of the azomethine (C=N) absorption band and appearance of the Carbonyl group absorption band at (1670-1730) $cm^{-1}$, and other important FTIR absorptions of compounds [18].

*3.1. Characterization of 4-Chloro-N-(4(dimethylamino) Benzylidene)-6-nitrobenzo[d]thiazol-2-amine.*
Compound (2) has been synthesized from the reacting of compound [1] along with 4-dimethyl amino benzaldehyde in absolute ethanol gave the imine. Condensation reaction of (equimolar quantity) of (4-chloro-6-nitrobenzo[d]thiazol-2-amine) with the 4,4 – dimethyl amino Benzaldehyde is the main technique to prepare Schiff base. The FT-IR Spectrum of compound [2] showed the appearance of (C=N) of stretching band at 1591 $cm^{-1}$ along with (C-H) aliphatic at 2918 $cm^{-1}$ and (C-H) aromatic at (3005) $cm^{-1}$ and (N-H) at 3080 $cm^{-1}$.
$^1$H-NMR spectrum of compound (2) exhibited some signals at δ 3.08 (6H) that were assigned to $2(CH_3)$ protons, signals at δ 3.85 (1H) that can be assigned to (NH) protons, a signal peak at δ 9.51 (1H) that can be assigned to (N=CH) proton .There were also multiplet signals at δ 6.72-7.63 (10 H) that can be assigned to benzene ring protons.

**Table 1:** The IR Characteristic Bands of Compound(2).

| Comp. No. | Changed part(x) | IR, KBr , υ, $cm^{-1}$ | | | | |
|---|---|---|---|---|---|---|
| | | (C-H) Ar. | (C-H) Aliph. | (C=N) | (C=C) | Other Bands |
| 2 | $N(CH_3)_2$ | 3080 | 2918, 2879 | 1645-1649 | 1487, 1525 | $N-CH_3$ 1228,1246 |

1,3-oxazepine compound (3) was synthesized from 1,3-dipolar cycloaddition reaction of Schiff base(2) with 3-nitophthalic anhydride in dry benzene as a solvent.
The mechanism involves the addition of one σ-carbonyl to π –bond (N=C) to give 4-membered cyclic and 5-membered cyclic ring of anhydride in the same transition state [T.S]a, which opens into 3-nitrophthalic anhydride to give 7-membered ring 1,3-oxazepine(3) [19]. The FT-IR spectrum indicated band at (1695-700) $cm^{-1}$ for lactone and (1653-1660) $cm^{-1}$ for lactam. The FT-IR spectra identified the appearance of carbonyl groups at (1742, 1774) $cm^{-1}$ besides the appearance of NH group at (3178) $cm^{-1}$.

IOP Conf. Series: Materials Science and Engineering **557** (2019) 012061 doi:10.1088/1757-899X/557/1/012061

**Table 2:** Physical Properties of the Synthesized Compounds.

| Comp. No. | Molecular Formula | Molecular Weight (g/mole) | Yield (%) | M.P (°C) | Colour | Rf |
|---|---|---|---|---|---|---|
| 1 | $C_7H_4N_3O_2SCl$ | 229.69 | 65 | 92-94 | Yellow | 0.71 |
| 2 | $C_{16}H_{13}N_4O_2SCl$ | 360.87 | 67 | 134-137 | Yellow | 0.89 |
| 3 | $C_{24}C_{16}N_5O_7SCl$ | 553.99 | 77 | 236-238 | Pale yellow | 0.91 |
| 4 | $C_{20}H_{18}N_5O_4SCl$ | 459.952 | 55 | 180-182 | White | 0.92 |
| 5 | $C_{24}H_{18}N_5O_4SCl$ | 508.009 | 59 | 207-209 | Red | 0.93 |

**Table 3:** The IR Characteristic Bands of Compounds (3-5).

| Comp. No | IR, KBr, υ, $cm^{-1}$ | | | | | |
|---|---|---|---|---|---|---|
| | (N-H) | (C-H) Ar. | (C-H) Aliph. | (C=O) lactam | (C=O) lactone | (C-S-C) |
| 3 | - | 3050 | 2921,2898 | 1630 | 1732 | 1097 |
| 4 | 3109 | 3031 | 2960,2890 | 1670 | 1714 | 1033 |
| 5 | 3150 | 3050 | 2946,2890 | 1678 | 1789 | 1034 |

## 4. Conclusion

The research involved the preparation of a number of homogeneous ring compounds containing a derivative benzothiazole as shown in Figure 1.

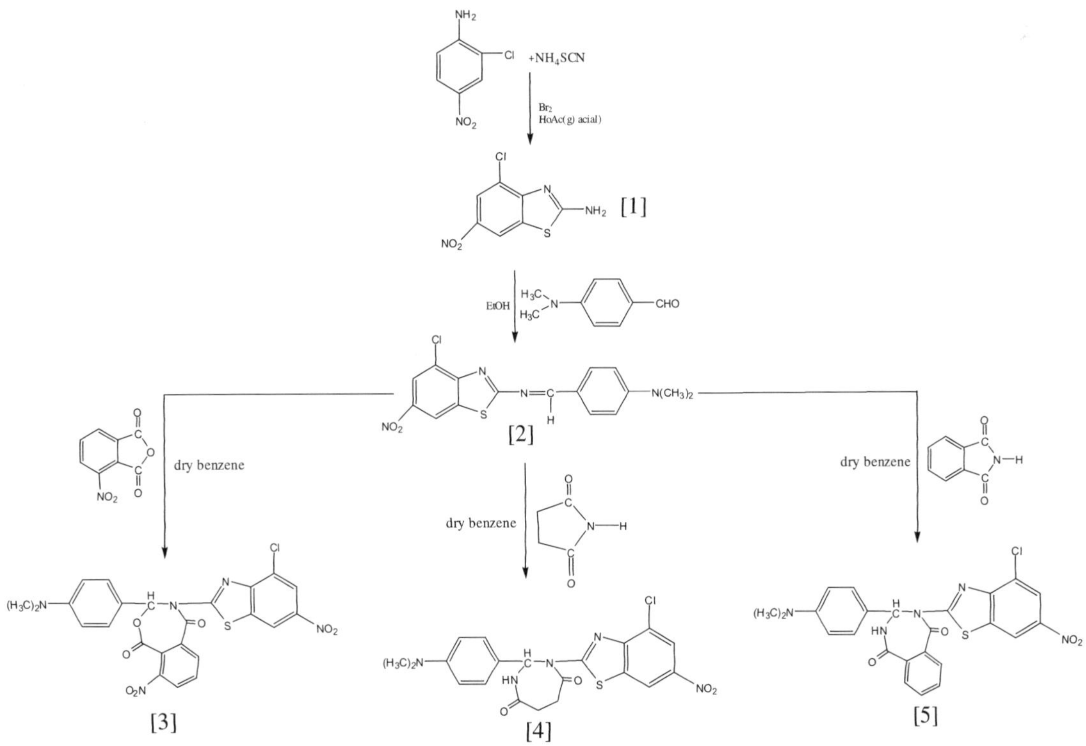


**Figure 1.** Homogeneous ring compounds containing a derivative benzothiazole

**References:**

[1] G.W.Wilkinson, R,D,Gillard and J.A.MC.cleverly, comprehend sivecorrdination chemistry 1st. pergamonpress, Oxrord, Ed, England, 715-535,(1987).
[2] V.Markandan, and R,R.angappan, J.chem.Bio.phy.Sci.sec.A 2(2),677-622,(2012).
[3] K.Krishnankutty, M,M. Basheer and P.S.Sayudev, "J. Argentine.chem.socity" (96),13-12,(2008).
[4] Desai S.B.Desaip.B and Desia K.R.,Hervocyd.Commun. 7-83,(2001).
[5] R.S.chan'Introduction to chemical Nomenclature", Butterworths,London, (1978).
[6] Goodwin, Jill.A Dyer s manual, Palham, 1982,p11.
[7] Geronikaki, A,Theophilidis, G.Eue. J. Med. Chem1992. 27,709.
[8] D.G.Cozzi "Chem .Soci.Rev.,33-410,(2004).
[9] A.M.Ali, Ahmed,T,A.Mohamed and B.H.Mohamed." J.Appl. sci-Res". (32),109-118,2007.
[10] T.Nagamien, A.J anuszko, K. other,p .Kaszynski and Y.Endo" Liq.crys.". (32),8,985,(2005).
[11] H.H,Monfred, O.Pouralimarlim ardan and C.Jamak:Z.Natureforreh; (62)717, (2007).
[12] Huisgen R.,Agew. Chem.Internet,Edit.7,321(1962).
[13] A.G. Awale, Dr. S. B. Gholse, Dr. P.S.Utale, J. of Pharmacy and Biological Sciences(IOSR-JPBS), volume 6, Issue 2 ,2013,1-7.
[14] F.A.Abdiseed, M.M. El-ajaily, International J.of pharm Teeh Rese (USA),1,1097-1103M(2009).
[15] Abdel-Hafez A. and Abdel-Wahab B. (2008); Bioorganic and Medicinal Chemistry, vol. 16(17), pp.:7983-7991
[16] Adnan S., Hassan K. and Thamer H. (2014);. Asian J. Res. Chem., vol. 7(3), pp.:251-255
[17] Serrano-Wu M., Laurent D., Chen Y., Huang S., Lam K., Matson J., Mazzucco C., Stickle T., Tully T., Wong H., Vyas D. and Balasubramanian B. (2002);. Bioorg.Med.Chem.Lett., vol. 12(19), pp.:2757-2760
[18] M.Silverstien and X. Webster, spectrometric Identification of Organic Compounds ,Sixth Edition , 96,(1999).
[19] Z.H.Abood, ph.D. Thesis, University of Babylon, (2007).

IOP Conf. Series: Materials Science and Engineering **557** (2019) 012062 doi:10.1088/1757-899X/557/1/012062

# Characterizations of Fe/Mn binary oxide by nitrogen adsorption

**Samara J Mohammad[1], Fadhil Abd Rasin[2], Abdulkareem M A Alsammarraie[3]**
[1]Physics department /College of Science for women/ University of Baghdad, Iraq
[2] material engineering /University of technology, Iraq
[3]Chemistry Department /College of Science/University of Baghdad, Iraq

**Abstract.** Nanostructure Fe/Mn as a binary oxide with 1:1 molar ratio has successfully synthesized by oxidation raw materials and co-precipitation process. The synthesized Fe/Mn was characterized by $N_2$ adsorption, Atomic Force Microscope (AFM) and X-Ray Diffraction (XRD).The synthesized Fe/Mn demonstrated poorly crystalline structure as an effective mesopores material with high surface area 403.6165 $m^2/g$ and average diameter of particles about 55.66 nm.

**Keywords:** Fe/Mn binary oxide, co-precipitation, Pore size analysis

## 1.Introduction

Development new technologies to synthesize nanosized metal oxides (NMOs) with high internal surface area for a variety of applications has attracted the attention of the scientific community. Metal oxide nanoparticles, including nanosized ferric oxides, manganese oxides etc., provide high internal surface area. Iron oxides with poorly crystalline can be synthesized using a well-known protocol "wet chemical methods" [1]. Some of the synthesis techniques include chemical precipitation, sol-gel, hydrothermal, surfactant mediated-precipitation and microwave assisted hydrothermal technique.
These oxides find applications as catalysts, sorbents, flocculants, coatings, and ion exchangers and for lubrication [1] [2] [3] [4] [5]. Iron oxide nanoparticles have many applications in different areas such as magnetic resonance imaging, magnetic recording, magnetic data storage devices, wastewater treatment, bio separation, and medicine [1]. Because of the unique properties of Iron oxide nanoparticles, they can provide economical solution to the most complex and challenging environmental remediation cleanup problems for example soil, groundwater contamination, sediment, or surface water [1] [6] [7] [8]. Adsorption is a low energy technology[9] [10] [11]. Gao sheng *et al.* in 2014 studied porous Fe-Mn binary as adsorbent in water treatment and environmental remediation [12].
The most important properties of adsorbent, which determines its usage, are the specific surface area and the pore structure [13].$N_2$ adsorption are the first stage in the characterisation of microporous and mesoporous solids because of its size and it poorly interacts with the adsorbent. The adsorption isotherm of nitrogen gas is always important 'fingerprint' and provide a amount of information about BET-area and pore volume. many types of adsorbents such as metal oxides, activated carbons and zeolites have been reported in the literature[14-16].

In this work, a novel Fe/Mn binary oxide sorbent has been successfully synthesized by oxidation and co-precipitation process. The synthesized Fe/Mn was fully characterized by $N_2$ adsorption, AFM and X-Ray Diffraction (XRD). The synthesized Fe/Mn shows as effective mesopores adsorbent with a smaller particle size, high surface area and porous structure.

## 2. Experimental Section

### *2.1. Materials*

Nitric acid $HNO_3$ (PDH, 1%) used to clean all vessels glass of reaction. Iron (II) Sulfate heptahydrate $FeSO_4.7H_2O$ (SCRC China) and Potassium permanganate $KMnO_4$ panreac spain.

### *2.2. Synthesis*

Nanostructure Fe/Mn binary oxides was synthesized at 1:1 molar ratio. The Fe/Mn as a binary oxide was prepared modified from submission by Zhang et al. with 3:1 mol ratio [17].So with different parameters in my work that 0.015 mol Potassium permanganate ($KMnO_4$) and 0.015 mol iron (II) sulfate heptahydrate ($FeSO_4.7H_2O$) were dissolved in 200 ml of deionized water and under magnetic stirring, the sulfate solution has been added into the permanganate solution with 4 M of NaOH drops solution to remain the solution about range 7 – 8 of pH. After addition, the suspension was formed when stirred it for 1 h, aged for 4 h at room temperature, and then washed with deionized water repeatedly. Filtrated suspension and dried at 65 °C with 24 hours. The dry powder was crushed and keep in a desiccator.

### *2.3 Characterization methods*

X-Ray Diffraction (XRD) data of Fe/Mn binary oxide sample was collected on diffractometer (Cu Kα radiation30 mA and40 kV). The surface morphology, root-mean-square (r.m.s) roughness and average diameter were analyzed using Atomic Force Microscope (AFM). Nitrogen adsorption/desorption isotherms of Fe/Mn binary oxide sample at -196.15 °C were obtained by Micrometrics Accelerated Surface Area and Porosimetry (ASAP) 2020 analyser, the isotherms' measurement of sample. Equilibration time was set for each data point of isotherms during the analysis. specific surface areas (SSAs) of materials determine Brunauer–Emmett–Teller (BET) method [18]. Langmuir theory was used to study the collected adsorption data in this study [19]. Langmuir model explains the adsorption by assuming a monolayer adsorption where the interaction between the adsorbate (gas) layer and the solid surface is significantly stronger than the one between gas layers.

## 3. Results and Discussion

XRD of sample in Figure (1) shows the two high broad peaks at 35.51 and 62.72 of d spacing 2.52542 and 1.47460 nm, estimate the average crystallite particle size via using Debye-Scherer formula and spacing of inter-planar between atoms (d-spacing) is calculated by using Law of Bragg

$$D = k\lambda / (\beta \cos\theta) \quad (1)$$

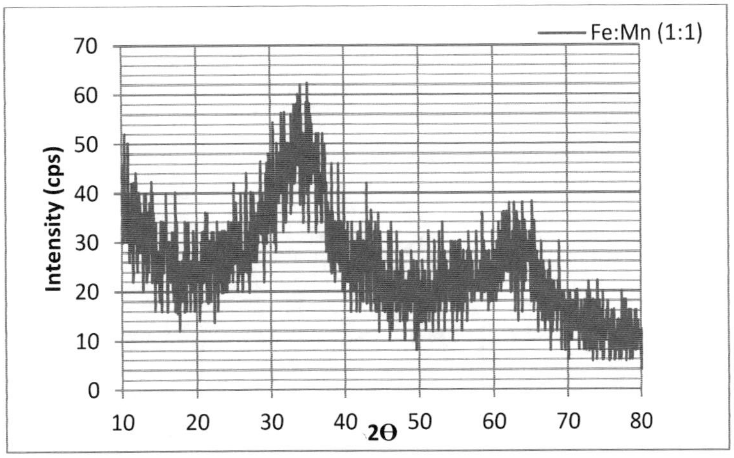


**Figure 1**. X _Ray Diffraction

D mean the crystallite size in nanometers, the wavelength λ (1.54056 Å for $CuK_\alpha$ radiation), which is comparable to the size of atom, k is the geometric factor related to crystallite shape equal to 0.9 and D crystallite size by using Debye-Scherer formula is 2.29 (nm).

Many of broadening in x-ray diffraction lines are done when that particle size is less than 100 nm. The discrete θ dependencies of both effects laid the basis for the separation of size and strain broadening in the analysis of Williamson and Hall. Diffraction pattern showed broadening because of particle size and strain the overall integral breadth of a Bragg peak [20].
2D & 3D view of AFM pictures of surface for Fe/Mn as a binary oxide of nano particle, the general shape is spherical, and quite dispersed, as shown in figure (2) provide more parallel-aligned surface structure, which are suitable for oxygen recognition on cell surfaces and responsible for different color quality. The ferrihydrite morphologys is spherical and does not resemble other iron oxides it occurs only as nano crystals with high specific surface areas [21].
The average grain size, average roughness and (r.m.s) roughness of sample 55.66 nm ,0.57 nm and 0.66nm respectively.

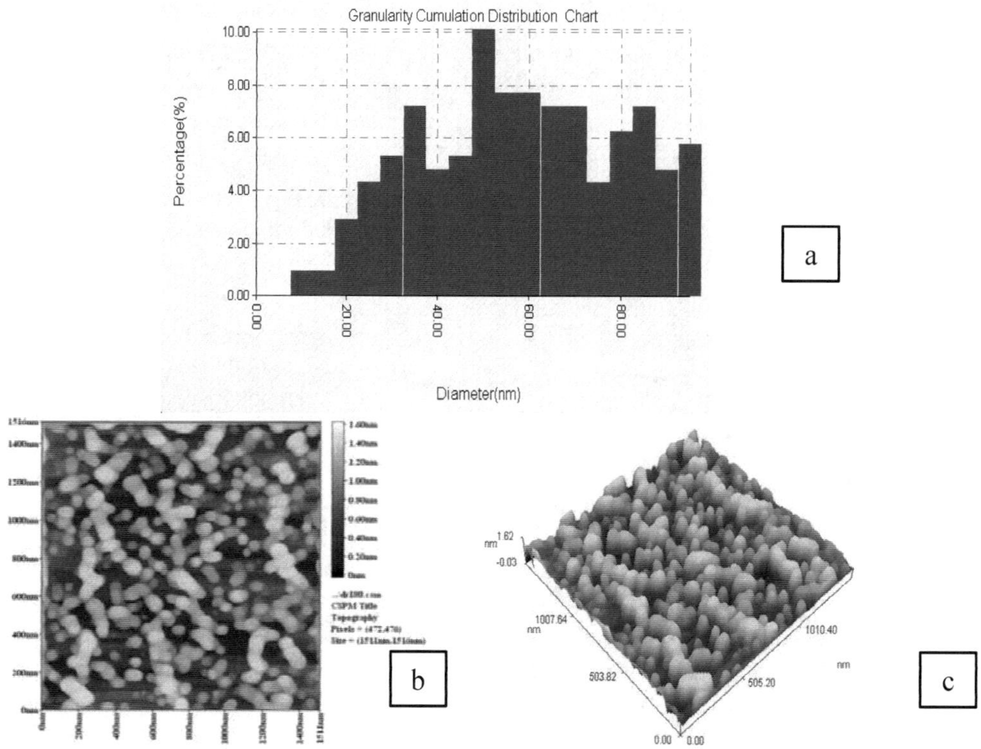


**Figure 2.** (a,b&c) is (a) average particle size, 2Dand 3D view of surface images of sample

Specific surface area (SSA), $N_2$ adsorption−desorption isotherm by BET method [18],measurement of surface area for Fe-Mn as a binary oxide nanoparticles sample give a good surface area( 314.8 $m^2/g$) [22]. The results are recorded in adsorption-desorption isotherms diagrams characteristics with high surface area. The values are a perfect with the specific surface area measurements and the microscopic characteristics.

The adsorption-desorption isotherm of a sample is shown in Figure (3). The formation of a hysteresis loop is visible. This narrow loop can indicate a narrow pore size distribution. Hysteresis loops that discern in the multilayer range of physisorption isotherms related with the stuffing and voiding of mesopores, the desorption branch was favored for mesopore size analysis [23].

IOP Conf. Series: Materials Science and Engineering **557** (2019) 012062 doi:10.1088/1757-899X/557/1/012062

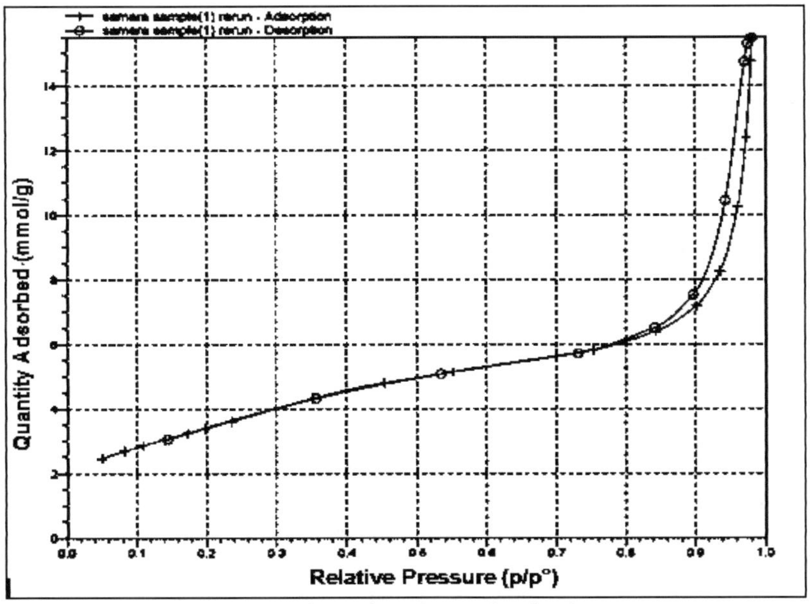


**Figure 3**. adsorption-desorption isotherms

Langmuir's prodigious work on monolayer adsorption resulted in repeated interest in the performance of adsorption data, the Langmuir model called 'ideal localized monolayer adsorption' the amount adsorbed at the stabilization of a Type one I shown in Figure (4) isotherm identification to full monolayer perfusion [23].

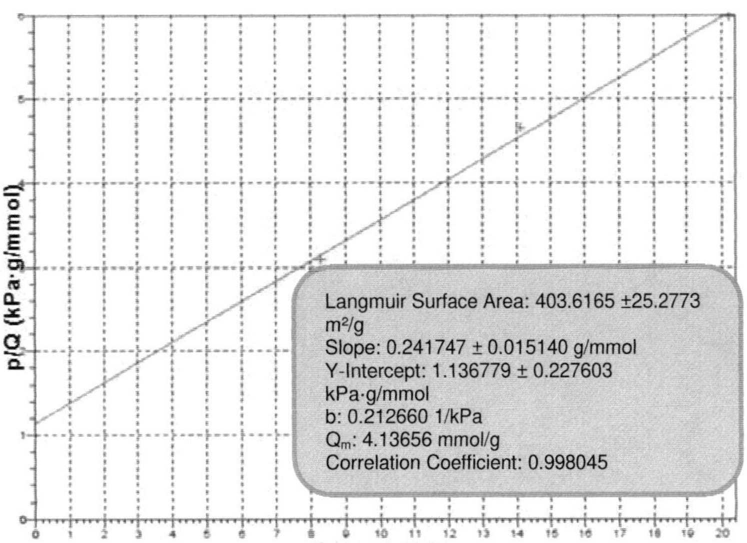


**Figure 4.** Langmuir Surface Area Plot

From desorption information by using(BJH) Barett–Joyner–Halenda model, showed (Figure.5,6 and 7), marking that the sample had mesopores in the nanomaterial see table1 for characterization results of physisorption isotherms.

**Table 1.** Characterization results of physisorption isotherms

| **Pore volume ($cm^3.g^{-1}$)** | |
|---|---|
| Total pore volume of pores of Single point adsorption less than (489.667 Å )radius at p/p° = 0.979823370 | 0.511475 |
| Total pore volume of pores of Single point desorption less than( 401.309 Å ) radius at p/p° =0.975274888 | 0.530074 |
| BJH model of adsorption accumulated volume of pores between (8.500Å) and (1500.000 Å) radius | 0.564989 |
| BJH model of desorption accumulated volume of pores between( 8.500 Å) and (1500.000 Å) radius | 0.562833 |
| **Size of Pore (Å)** | |
| Average pore width of adsorption of BET | 72.7758 |
| Average pore width of desorption of BET | 75.4222 |
| BJH model of Adsorption average pore radius | 35.0150 |

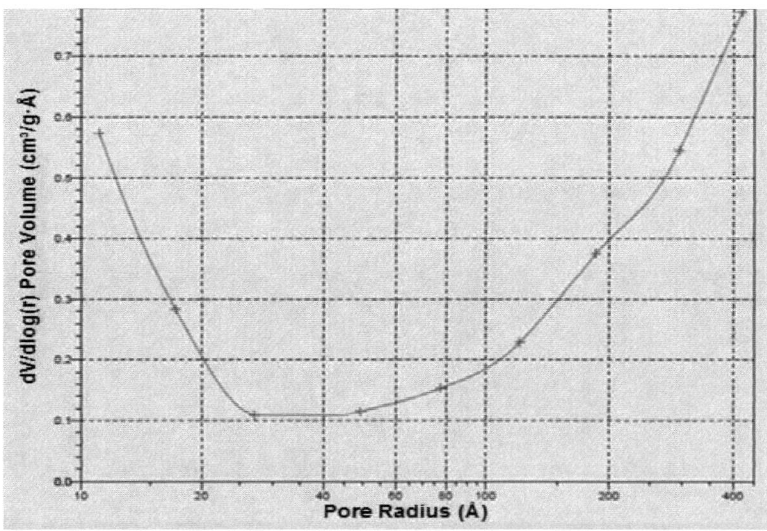


**Figure 5.** BJH Adsorption model ( Pore Volume)

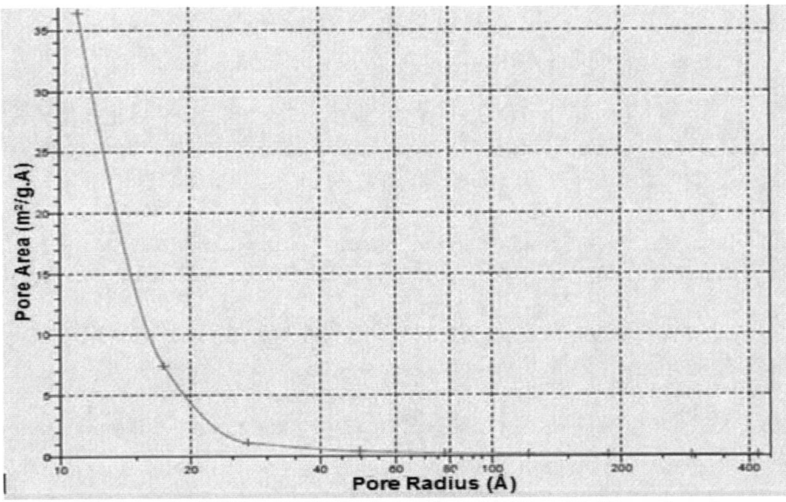


**Figure 6.** BJH model of adsorption dA/dlog(r) (Pore Area)

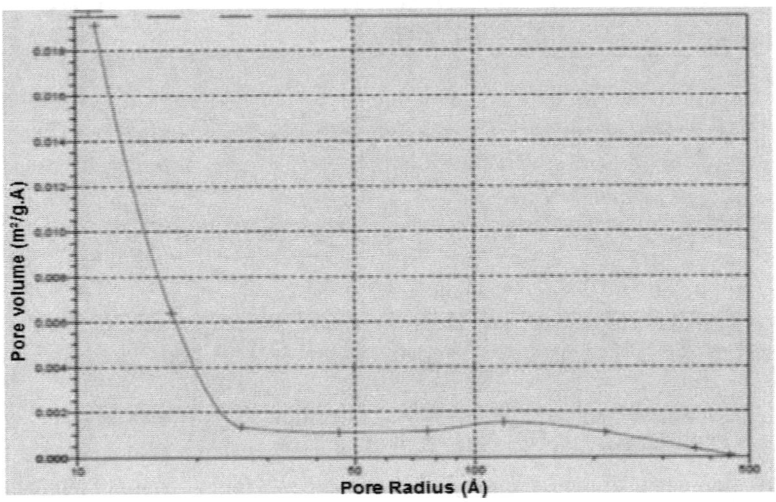


**Figure 7.** BJH desorption dV/dr (Pore Volume)

## 4. Conclusions

Porous Fe-Mn binary as adsorbent nanoparticles was synthesized by oxidation and co-precipitation method. Diffraction pattern showed broadening because of particle size and strain in two peaks at 35.51 and 62.72 with d spacing of 2.52542 and 1.47460 nm. The average grain size, average roughness and (r.m.s) roughness of sample 55.66nm, 0.57nm and 0.66nm respectively by using AFM. Adsorption–desorption isotherms, mesoporous material with high surface area 403.6165 $m^2/g$ in Langmuir Surface Area for characterization results of physisorption isotherms by nitrogen adsorption measurements at 77 K.

## 5.References

[1] M. Mohapatra, and S. Anand, 2011 "Synthesis and applications of nano-structured iron oxides/hydroxides – a review," *Int. J. Eng. Sci. Technol.*, vol. 2, no. 8,.

[2] R. Baxter, N. Hastings, A. Law and E. J. .. Glass, 1978 "Catalytic reduction of aromatic nitro compounds with hydrazine hydrate in the presence of Iron (III) oxide hydroxide," *Synthesis (Stuttg).*, vol. 39, pp. 834–835.

[3] M. P. Sharrock, and R. E. Bodnar, 1985 "Magnetic materials for recording: An overview with special emphasis on particles," *J. Appl. Phys.*, vol. **57**, no. 8, pp. 3919–3924.

[4] C. RICHARD A. CATLOW, JOHN CORISH, JOHN HENNESSY and WILLIAM C. MACKRODT, 1988 "Atomistic Simulation of Defect Structures and Ion Transport in α-$Fe_2O_3$and α-$Cr_2O_3$," *J. Am. Ceram. Soc.*, vol. **71**, no. 1, pp. 42–49.

[5] Kwang Ho Choo, and Suck Ki Kang, 2003 "Removal of residual organic matter from secondary effluent by iron oxides adsorption," *Desalination*, vol. **154**, no. 2, pp. 139–146,.

[6] Glen E. Fryxell, and Guozhong Cao, 2012 *"ENVIRONMENTAL APPLICATIONS OF NANOMATERIALS" Synthesis, Sorbents and Sensors.*

[7] Pragnesh N. Dave, and Lakhan V. Chopda, 2014 "Application of iron oxide nanomaterials for the removal of heavy metals," *J. Nanotechnol.*, vol.

[8] Yidong Zou *et al.*, 2016 "Environmental Remediation and Application of Nanoscale Zero-Valent Iron and Its Composites for the Removal of Heavy Metal Ions: A Review," *Environ. Sci. Technol.*, vol. **50**, no. 14, pp. 7290–7304,.

[9] Thaer M. Al-Jadir, Flor R. Siperstein. 2018 "The influence ofthe pore size in Metal–Organic Frameworks in adsorption and separation of hydrogen sulphide: A molecular simulation study", *Microporous and Mesoporous Materials*

[10] Davide Papurello, Andrea Lanzini. 2018 "SOFC single cells fed by biogas: Experimental tests with trace contaminants", Waste Management,

[11] A. Khorsand Zak, W.H. Abd. Majid, M.E. Abrishami, RaminYousefi. 2011 "X-ray analysis of ZnO nanoparticles byWilliamson–Hall and size–strain plot methods", *Solid State Sciences.*

[12] Gaosheng , Fudong Liu, Huijuan,Jiuhui and Ruiping 2014"Respective Role of Fe and Mn Oxide Contents for Arsenic Sorption in Iron and Manganese Binary Oxide: An X-ray Absorption Spectroscopy Investigation", *Environmental Science & Technology* , American Chemical Society August 5.

[13] Rodriguez-Reinoso F, Linares-Solano A 1989 Chemistry and physics of carbon: a series of advances (P.A.Thrower, Ed.), Marcel Dekker, Inc. New York,.

[14] Hye-Jin Hong, 2011 Removal of organic· Inorganic complex pollutant from water by using metal oxide, Phd.

[15] Maghsoudi H, Soltanieh M, Bozorgzadeh H, Mohamadalizadeh A, 2013 Adsorption isotherms and ideal selectivities of hydrogen sulfide and carbon dioxide over methane for the Si-CHA zeolite: comparison of carbon dioxide and methane adsorption with the all-silica DD3R zeolite, *Adsorption* **19** 1045–1053.

[16] Sigot L, Fontseré Obis M, Benbelkacem H, Germain P, Ducom G, (2016) Comparing the performance of a 13X zeolite and an impregnated activated carbon for $H_2S$ removal from biogas to fuel an SOFC : influence of water, *Sci. Technol. Humanit.* **1** 2–10.

[17] Zhang, G. S., Qu, J. H., Liu, H. J., Liu, R. P. and Wu, R. C. 2007Preparation and evaluation of a novel Fe–Mn binary oxide adsorbent for effective arsenite removal. *Water Res.*, **41**, 1921–1928

[18] Nadeen Al-Janabi, Haoran Deng, Jayane Borges, Xufei Liu, Arthur Garforth, Flor R. Siperstein, Xiaolei Fan. 2016 " A Facile Post- Synthetic Modification Method To Improve Hydrothermal Stability and CO Selectivity ofCuBTC Metal–Organic Framework ", *Industrial & Engineering Chemistry Research*,

[19] Ruthven D. M., 1984 " Principle of Adsorption and Desorption process ", New York, John Wiley and Sons.

[20] Sirdeshmukh, D.B., Sirdeshmukh, L., Subhadra, K.G. 2006 Micro and Macro. Properties of solids: Thermal, Mechanical and Dielectric properties. *Springer*, New York.

[21] Cornell R. M., and Schwertmann U., 2003 Iron oxides. 2nd Edn., Wiley-VCH Verlagsgesellschaft Weinheim.

[22] Xu W, Wang H, Liu R, Zhao X, Qu J 2011 "The mechanism of antimony (III) removal and its reactions on the surfaces of Fe−Mn binary oxide". *J. Colloid Interface Sci.*, **363**, 320

[23] Kenneth Sing. 2001 "The use of nitrogen adsorption for the characterisation of porous materials", Colloids and Surfaces A: Physicochemical and Engineering Aspects,

# Laboratory investigation on the properties of asphalt concrete mixture with GGBFS as filler

**Abbas Al-Hdabi[1], Nibras A AL-Sahaf [1], Layth A Mahdi[1], Zahraa Yesar[2], Haneen Moeid[2], and Mortada Hassan[2]**

[1] Civil Engineering Department, Faculty of Engineering, University of Kufa, Al-Najaf, Iraq,
[2] BSc Graduate, Civil Engineering Department, Faculty of Engineering, Al-Najaf, Iraq.

**Abstract.** Consuming of by product materials in build manufacture such as making of hot asphalt mixtures introduces worthy interest on the way of ecological, economic standpoint. Accordingly, in this experimenter research, it was scrupulous for using Ground Granulated Blast Furnace Slag (GGBFS) in state traditional metal filler in hot asphalt mixtures. The mechanical characteristic was rating by lineal Marshall Stability & Indirect Tensile Strength tests. However, moisture harm and long-range aging were investigated by locate Index of retained Strength (IRS) & Mean Marshall Stability Ratio (MMSR), respectively. The empirical outcomes have presented a considerable refinement in the mechanical properties and a fundamental promotion in durability of the generated mixtures i.e. Asphalt Concrete Mixtures with GGBFS as a mineral filler (GGBFSAC) in rapprochement with the control mixtures that were infectious by ordinary Portland cement (OPC) as a mineral filler (OPCAC). As an outcome of this research, GGBFS can be united in state classical mineral filler in asphalt concrete mixtures especially in region where there is large GGBFS waste.

**Keywords**. Ground Granulated Blast Furnace Slag, OPC, durability, filler

## 1. Introduction

Asphalt concrete common applications in pavement as a result of its excellent service achievement, resistance and water impedance, are mostly a base, binder or surface course contain graded aggregate caught simultaneously by asphalt cement [1, 2]. Asphalt cement is the classical binder applied in build highways and road. Whereas, the aggregate particles work as the constitutional skeleton of the pavement which may represents about 90% of the volume of Hot Mix Asphalt (HMA) [3]. Additionally, filler materials resort to solidify the asphalt cement. Many materials like cement, limestone dust, lime, fine sand and rise husk ash normally applied as metal filler in HMA. As popular that lime, limestone dust and cement are costly and applied effectively for else aims. However, ash fine, sand and GGBFS passing 0.075 mm sieve magnitude are suitable as mineral filler.

The benefit from using the waste fine particles as filler in HMA were inspected by several investigators. Phosphate garbage filler [4], Jordanian oil shale fly ash [5], bag house fines [6], municipal solid waste incineration ash [7], and waste lime [8] were inspected as filler. It was stated that these garbage materials may be applied in asphalt concrete mixtures without any major decline in its effectiveness.

Several studies have been conducted to improve the performance of cold asphalt emulsion mixtures by using different waste materials such as (but not limited to) GGBFS, fly ash, paper sludge ash …etc [9-11].

Many studies proves that effective mineral filler will react with asphalt cement to modify the produced mastic [12]. Also, addition of mineral filler can improve the elastic modulus of HMA. But, high amount of filler might be increase the demand of asphalt cement [13, 14].

GGBFS (Ground Granulated Blast Furnace Slag) is produced from the ferrous making manufacture. GGBFS was gained by mashing the muffle molten runny iron brusqueness. GGBFS is a granulated item that has so restricted crystal fashioning, with quite cementitious characteristic when take place cement fineness. About 2.2 million tons of GGBFS are generated per year in the UK and applied as texture aggregates or like a cementitious binder in the shape of ground granulated blast furnace slag (GGBFS). The BF slag is shaped in an uninterrupted operation by the incorporation of limestone (and/or dolomite) and other inflow with the dust from the carbon origin (coke) and the non- mineral combination of the iron ore. The slag buoys on the face of the slushy iron and is then pulled off and pliable to phlegmatic to produce a semi-dense pored crystal item (lightweight aggregate) recognized as air cooled breath oven slag.

It has two ingredient; 80-100% interactive elements fundamentally from the galenites ($2CaO.Al2O_3.SiO_2$) and akermanite ($2CaO.MgO.2SiO_2$), whereas the non-reactive elements are the crystalline metals gehlenite, akermanite, diopside ($2CaO.MgO.2SiO_2$) and merwinite ($3Cao.MgO.2SiO_2$).

The popular pozzolanic elements from biomass and manufacture by output such as RHA, fly ash and GGBFS are turn into effective areas of research due to the favourable ecological effects in increment to the economic matter [8].

For the time being, there is a growing benefit in the profiteering of garbage elements that is one of the fundamental goal sketches for the environmentally cordial operations. In the situation of build manufacture, there was a growing proclivity to the evolution and consuming of garbage as additional binding material. The popular pozzolanic elements from biomass and industry by products is attractive to many researchers due to the encouraging economic and environmental effects [15-17].

In this research, GGBFS has been used as mineral filler in HMA to inspect the durability and mechanical properties of the produced mixtures. To achieve this aim two types of mixtures have been prepared i.e. with GGBS and OPC and compared with the standards which are adopted by the State Commission of Roads and Bridges (SCRB/R9) 2003 [18].

## 2. Methodology

### *2.1 Selected Materials*

#### *2.1.1 Aggregates*

Crushed quartz which was collected from Al-Nibaa Quarry was used as coarse aggregate. Tables 1 and 2 show the chemical and physical properties, respectively. While natural sand was used as fine aggregate. In accordance to (SCRB/R9) 2003 [18], type III surface course gradation with 12.5 mm maximum size was adopted, Fig. 1.

#### *2.1.2 Asphalt cement*

40-50 penetration grade asphalt cement collected from local refinery was used as a binder. Its physical properties are presented in Table 3.

#### *2.1.3 Filler*

Conventional mineral filler (OPC) and GGBFA were used as mineral filler to produce the traditional and modified HMAs. Table 4 presents the physical properties of OPC.

On the other hand, GGBFS that can be generated from the iron making industry under organised process by means of temperature and duration or to get iron sections. Table 5 shows the physical properties of GGBFS. The specific gravity of the used GGBFS is 2.2, while Fig. 2 displays the used GGBFS material photo. The produced mixtures are nominated as OPCAC and GGBFSAC.

*2.2 Test Methods*

*2.2.1 Marshall Stability and flow*

ASTM D6927 has been adopted to test the specimens for specify MS and flow. From this test MS represent the maximum load reistance to the plastic flow while the flow corresponds to the strain value at the maximum load record.

*2.2.2 Volumetric properties*

ASTM D2726 has been adopted to determine the dry bulk density. While, ASTM D3203 was followed to determine Air Voids (AV), Voids Filled with Asphalt (VFA) and Voids in Mineral Aggregate (VMA).

*2.2.3 Indirect tensile strength test*

ASTM D4123 was conducted to indicate the Indirect Tensile Strength (ITS) [19]. The maximum tensile strength ($\sigma_t$) determined in accordance to Eq. (1).

$$\sigma_t = \frac{2 \times P_{max}}{\pi H D} \quad (1)$$

Where: $P_{max}$ , H and D are the maximum applied load (kN), the specimen height (m) and specimen diameter (m), respectively.

*2.2.4 Durability*

*2.2.4.1 Water damage*

Water sensitivity for treated and untreated mixtures was evaluated by determine the Index of Retained Strength (IRS). ASTM D 6927 was used to determine IRS which is adopted by SCRB specifications and its recommend that its value must be more than 70% for surface course mixtures. Two set of samples are needed, the first set represent the standard samples (dry samples) i.e. without curing. While the other set (wet samples) was immersed for 24 hours at 60°C before testing for MS at 60°C, Eq. (2).

$$IRS = \frac{S_2}{S_1} \times 100\% \quad (2)$$

Where: IRS = Index of Retained Strength, %

$S_1$= Marshall Stability of the dry specimens, kN.

$S_2$= Marshall Stability of the wet specimens, kN.

*2.2.4.2 Long-term aging*

The Strategy Highway Research Program (SHRP) A-003A has been adopted in this study to evaluate the Long Term Ageing (LTA). This procedure recommend that curing of the samples for 2 or 5 days at 85°C represent age hardening in the field for 5 or 10 years, respectively [20]. The later was conducted in this study to evaluate age hardening after 10 years. MS results were indicated and Mean Marshall Stability Ratio (MMSR) was calculated, which is the ratio between MS after and before ageing.

## 3. Results and discussion

First, the optimum binder content for OPCAC and GGBFSAC mixtures were determined in accordance to Marshall mix design method (ASTM D6927). Five binder contents were selected with 0.5% increment i.e. 4, 4.5, 5, 5.5 and 6%, by mass of aggregate. ASTM D2726 and ASTM D2041 were adopted to indicate the bulk specific gravity and theoretical maximum specific gravity. On the other hand, ASTM D3203 have been followed to determine percent air voids. afterwards, Marshall Stability test were implemented for each sample by Marshall apparatus. Accordingly, the optimum binder content for the two types of mixtures were determined and nominated as 5.5% by mass of aggregate.

### *3.1 Influence of GGBFS on Marshall Test Results*

Figs. 3–5 show Marshall test results for the two types of mixtures i.e. OPCAC and GGBFSAC. Also, the requirements in accordance to SCRB Standards are presented in in these figures. As shown in Fig 3, there is a significant enhancement in MS when OPC has been replaced by GGBFS, as the increment is about 40% in comparison with the control mixtures. The same enhancement was observed in Marshall Stiffness results that is shown in Fig. 5. It is worthy to say that both of these mixtures comply with SCRB specifications.

This enhancement in MS of GGBFSAC mixtures can be attributed to the reinforcing of the binder by GGBFS particles and increasing the stiffness and cohesion of the new mastic.

### *3.2 Influence of GGBFS on the volumetric properties*

Figs. 6–9 show these parameters for OPCAC and GGBFSAC. it can be indicated the specific gravity of GGBFSAC mixtures were improved in comparison with the control mixtures. Also, air voids and VMA remain comply with the Iraqi specifications. This performance can be attributed to the increment of the stiffness of the binder when adding GGBFS particles.

### *3.3 Influence of GGBFS on Indirect Tensile Strength*

ITS results for the two mixtures (OPCAC and GGGBFSAC) are presented in Fig. 10. GGBFSAC has higher value in comparison with OPCAC, almost 11% increment can be observed.

### *3.4 Influence of GGBFS on the durability of asphalt mixtures*

#### *3.4.1 Moisture damage*

Fig. 11 shows IRS results for treated and untreated mixtures. A significant improvement was reported when traditional mineral filler was replaced with GGBFS as IRS increased from 73% to 88%.

#### *3.4.2 Long term aging*

Long Term Aging results are shown in Fig. 12 for OPCAC and GGBFSAC. A significant improvement is observed when using GGBFS as filler because MS after aging increased by 28% in comparison with OPCAC mixtures. However there is a considerable increase in the results for OPCAC mixtures with and without aging (MMSR=129%).

**Table 1.** Selected coarse aggregate Chemical composition.

| **Chemical compound** | **Results, %** |
|---|---|
| $SiO_2$ | 82.52 |
| MgO | 0.78 |
| $SO_3$ | 2.7 |
| $Fe_2O_3$ | 0.69 |
| $Al_2O_3$ | 0.48 |
| CaO | 5.37 |
| L.O.I. | 6.55 |
| **Mineral composition** | |
| Quartz | 80.03 |
| Calcite | 10.92 |

**Table 2.** Physical properties of aggregates

| **Property** | **ASTM Designation** | **Test results** | **SCRB specifications** |
|---|---|---|---|
| **Coarse aggregate** | | | |
| Apparent specific gravity | C 127 | 2.695 | …. |
| Bulk specific gravity | C 127 | 2.64 | …. |
| Percent wear by Los Angeles abrasion , % | C131 | 22.7 | 30 Max. |
| Flat and elongated particles ,% | C 4791 | 5 | 10 Max. |
| Soundness loss by sodium sulphate solution,% | C88 | 3.4 | 12 Max. |
| Degree of crushing, % | D5821 | 96 | 90 Min. |
| **Fine aggregate** | | | |
| Apparent specific gravity | C127 | 2.701 | …. |
| Bulk specific gravity | C127 | 2.67 | …. |
| Clay lumps and friable particles, % | C142 | 1.85 | 3 Max. |
| Angularity ,% | C1252 | 54 | …. |

**Table 3.** Asphalt cement properties

| Property | Portland Cement |
|---|---|
| Specific gravity (ASTMC188-95) | 3.05 |
| Passing sieve No. 200, (%) | 94.76 |

**Table 4.** Portland cement physical properties

| Property | ASTM Designation | Test Results | Requirements |
|---|---|---|---|
| Penetration at 25 °C, 0.10 mm | D5 | 46 | 40-50 |
| Ductility at 25 °C , cm | D113 | 115 | >100 |
| Specific gravity at 25 °C | D70 | 1.03 | ------ |
| Solubility in trichloroethylene, % wt | D2042 | 99.31 | >99 |
| Flash point , °C | D92 | 273 | >232 |
| Residue from thin –film oven test | D1754 | | |
| -Ductility at 25 °C, cm | D113 | 55 | >25 |
| - Retained penetration , % of original | D5 | 69 | >55 |

**Table 5.** GGBFS properties

| S. NO | Property | Result |
|---|---|---|
| 1 | Specific Gravity | 2.2 |
| 2 | Bulk density | 570kg/m$^3$ |
| 3 | Size | 75μ |
| 4 | Surface area | 200m$^2$/kg |
| 5 | $SiO_2$ | (90-96)% |
| 6 | $Al_2O_3$ | (0.5-0.8)% |

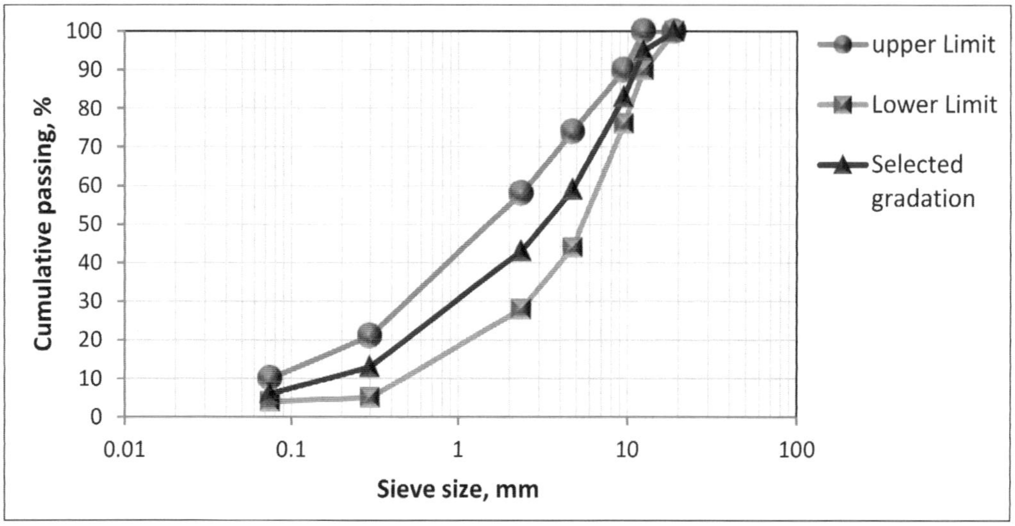


**Figure 1.** Type IIIA surface course hot asphalt gradation

**Figure 2.** GGBFS material photo

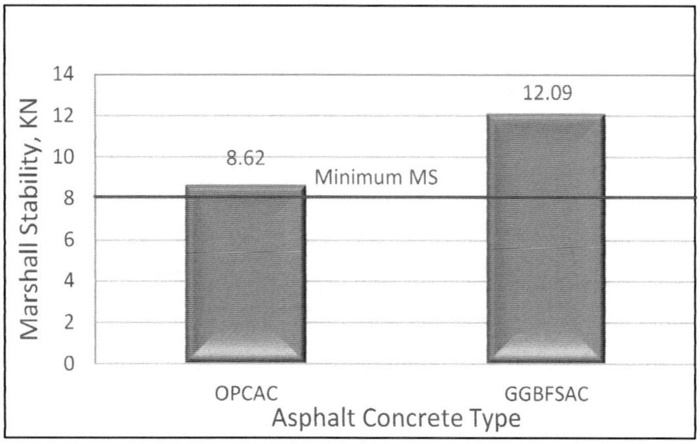


**Figure 3.** Marshall Stability results for treated and untreated mixtures

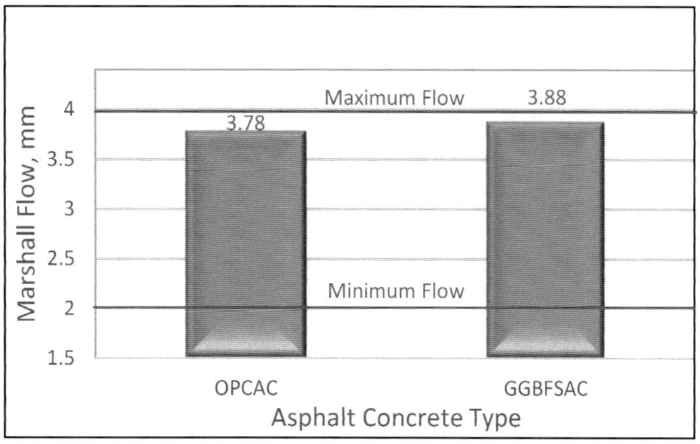


**Figure 4.** Marshall Flow results for treated and untreated mixtures

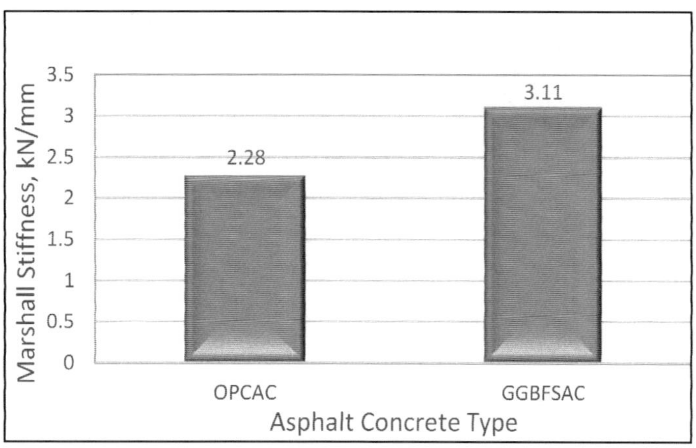


**Figure 5.** Marshall Stiffness results for treated and untreated mixtures

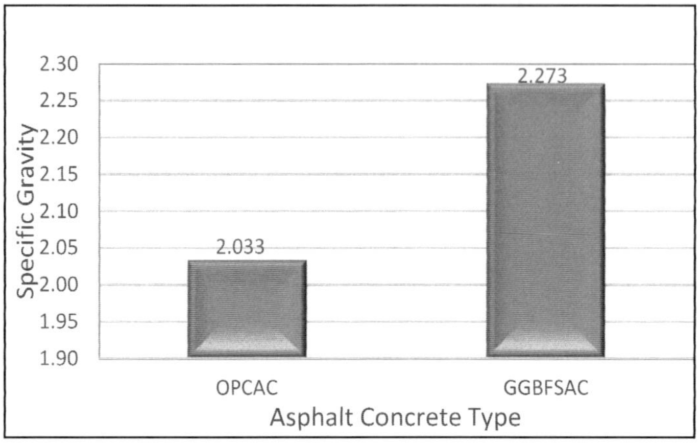


**Figure 6.** Specific Gravity results for treated and untreated mixtures

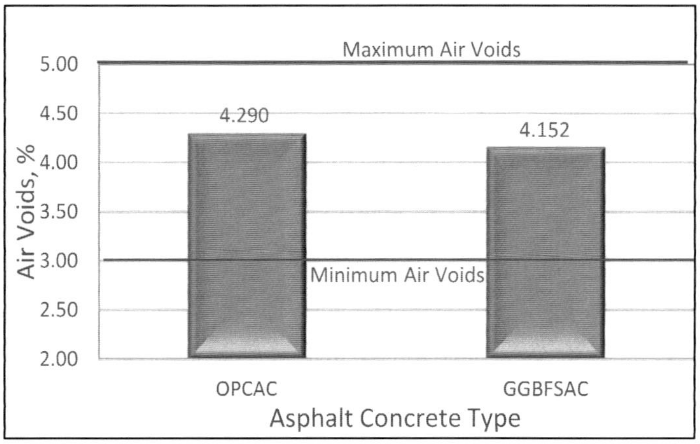


**Figure 7.** Air Void results for treated and untreated mixtures

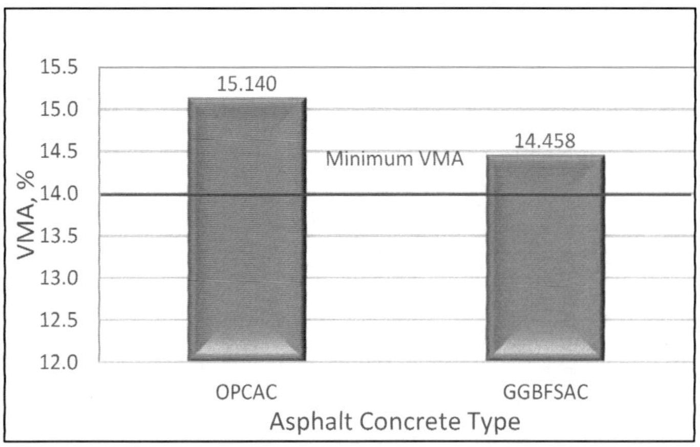


**Figure 8.** Voids in Mineral Aggregate for treated and untreated mixtures

IOP Conf. Series: Materials Science and Engineering **557** (2019) 012063 doi:10.1088/1757-899X/557/1/012063

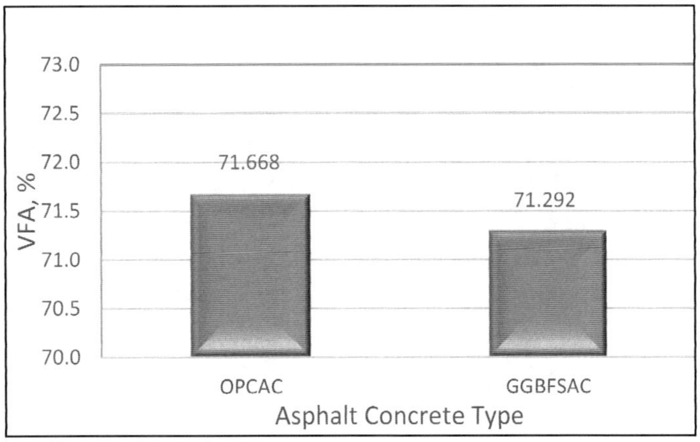


**Figure 9.** Voids Filled with Asphalt for treated and untreated mixtures

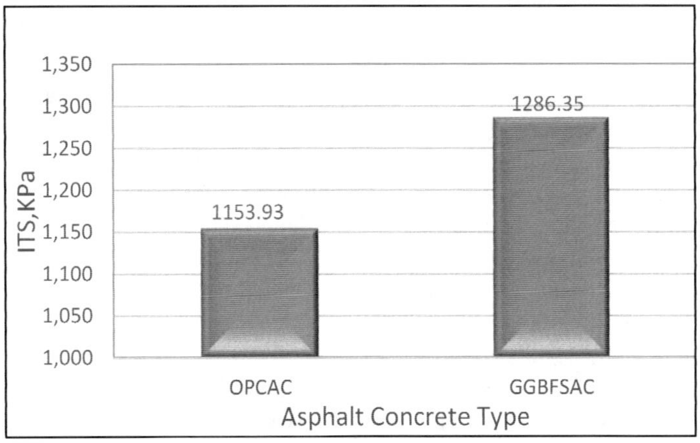


**Figure 10.** Indirect Tensile Strength results for treated and untreated mixtures

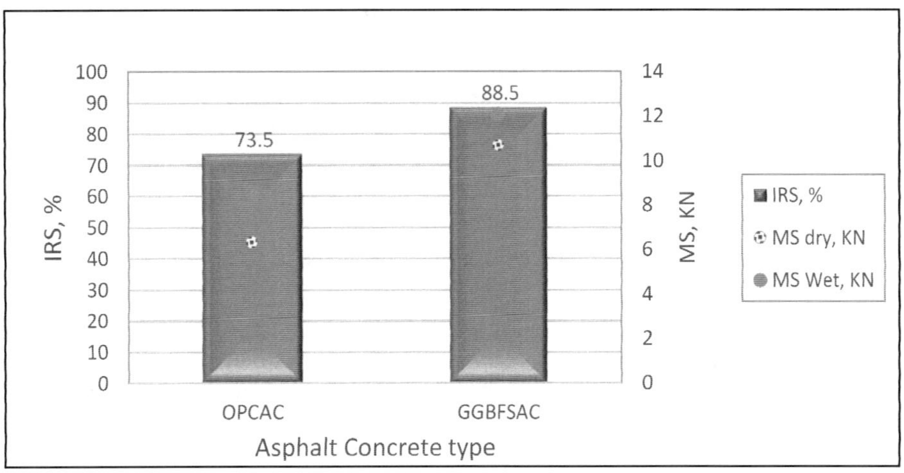


**Figure 11.** Index of Retained Strength for treated and untreated mixtures

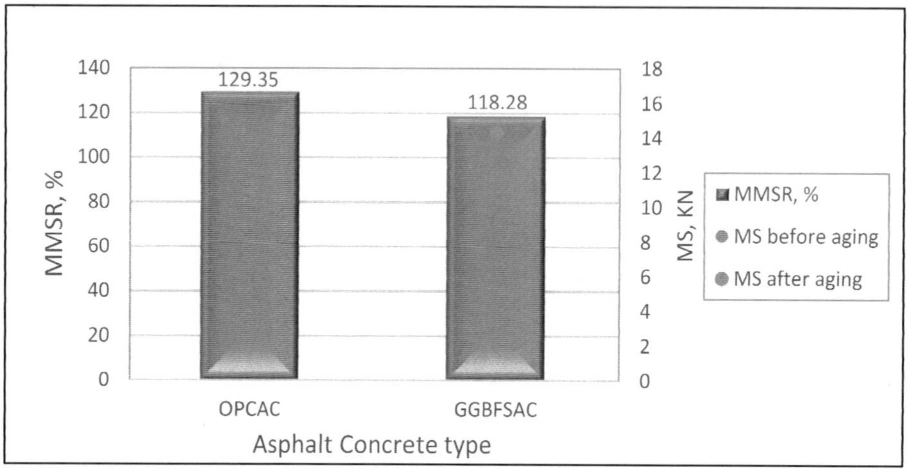


**Figure 12.** Effect of long-term aging on treated and untreated mixtures

## 4. Conclusions

Ground Granulated Blast Furnace Slag (GGBFS), which is generated from iron making industry has been used in this study as a replacement to the conventional mineral filler i.e. OPC to produce the novel hot mixture. The mechanical properties for GGBFSAC and OPCAC were evaluated by Marshall stability and indirect tensile strength tests. Water sensitivity and long term aging were used to assess the durability of the produced mixtures. Marshall stability of GGBFSAC improved substantially in comparison with OPCAC. Which is increased about 40% more than those for the control mixtures. Also, GGBFSAC mixtures implement better than the traditional hot mixtures by means of ITS results. Although there is a decrease in air voids of GGBFSAC mixtures, the results fulfil the Iraqi specifications for surface course mixtures. On the other hand, IRS increased for GGBFSAC mixtures in comparison with control mixtures and there mechanical properties were increased after aging.

## 5. References

[1] Uzun, I. and Terzi, S. (2012). "Evaluation of andesite waste as mineral filler in asphaltic concrete mixture." Construction and Building Materials, 31:284-8.
[2] Yilmaz, M., Kök, B. and Kuloglu, N. (2011). "Effects of using asphaltite as filler on mechanical properties of hot mix asphalt." Construction and Building Materials, 25:4279-86.
[3] Read, J. and Whiteoak, D. (2003). "The Shell Bitumen Handbook." 1 Heron Quay , London: Thomas Telford Publishing.
[4] Katamine, N. (2000). "Phosphate waste in mixtures to improve their deformation." Journal of Transportation Engineering, 126:382-9.
[5] Ibrahim, A. and Abdullah, A. (2005). "Effect of Jordanian oil shale fly ash on asphalt mixes." Journal of Materials in Civil Engineering, 17:553-9.
[6] Deng-Fong, L., Jyh-Dong, L. and Shun-Hsing, C. (2006). "The application of baghouse fines in Taiwan." Journal of Resources Conservation and Recycling, 46:281-301.
[7] Yongjie, X., Haobo, H., Shujing, Z. and Jin, Z. (2009). "Utilization of municipal solid waste incineration ash in stone mastic asphalt mixture: pavement performance and environmental impact." Journal of Construction and Building Materials, 23:989-96.
[8] Hwang, S., Park, H. and Rhee, S. (2008). "A study on engineering characteristics of asphalt concrete using filler with recycled waste lime." Journal of Waste Management, 28:191-9.
[9] Al-Hdabi, A., Al Nageim, H., Ruddock, F. & Seton, L. 2013. A novel Cold Rolled Asphalt mixtures for heavy trafficked surface course. Journal of Construction and Building Materials, 49, 598–603.

[10] Al-Hdabi, A., Al Nageim, H., & Seton, L. 2014. Superior cold rolled asphalt mixtures using supplementary cementations materials. Journal of Construction and Building Materials, 64, 95–102.
[11] Dulaimi, A., Al Nageim, H., Ruddock, F. & Seton, L. 2016. New developments with Cold Asphalt Concrete Binder Course Mixtures Containing Binary Blended Cementitous Filler (BBCF), 124, 414–423.
[12] Anderson, D. "Guidelines for use of dust in hot mix asphalt concrete mixtures." Proceeding Association of Asphalt Paving Technologists, 1987. Association of Asphalt Paving Technologists, 492-516.
[13] Elliot, R., Ford, M., Ghanim, M. and Tu, Y. (1991). "Effect of aggregate gradation variation on asphalt concrete mix properties." Transportation Research Record. Washington, D.C.: National Research Council.
[14] Kandhal, P., Lynn, C. and Parker, F. (1998). "Characterization tests for mineral fillers related to performance of asphalt paving mixtures." In: 98-2, R. N. (ed.) National Centre for Asphalt Technology.
[15] Givi, A., Rashid, S., Aziz, F. and Salleh, M. (2010). "Assessment of the effects of rice husk ash particle size on strength, water permeability and workability of binary blended concrete." Journal of Construction and Building Materials, 24:2145-50.
[16] Chindaprasirt, P. and Rukzon, S. (2008). "Strength, porosity and corrosion resistance of ternary blend Portland cement, rice husk ash and fly ash mortar." Constr Build Mater, 22(8):1601-6.
[17] Bhanumathidas, N. and Mehta, P. (2004). "Concrete mixtures made with ternary blended cements containing fly ash and rice husk ash." International conference proceeding seventh CANMET-Chennai, . India.
[18] State Commission of Roads and Bridges (SCRB/R9) (2003). "General Specification for Roads and Bridges." Republic of Iraq, Ministry of Housing and Construction, Department of Planning and Studies, Baghdad.
[19] ASTM Standards (2004). "Roads and Paving Materials. Annual Book of the American Society for Testing and Materials Standards, Section 4, Vol. 03-04.". Washinton, USA.
[20] KLIEWER, J., BELL, C. and SOSNOVSKE, D. (1995). "Investigation of the Relationship Between Field Performance and Laboratory Ageing Properties of Asphalt Mixtures (SHRPA - 003A ), Engineering Properties of Asphalt Mixtures and the Relationship to their Performance, ASTM STP 1265Gerals A.". Philadelphia: Hubber and Dale S. Decker, Eds. American Society for Testing and Materials.

# Magnetic and structural Behavior of Fe-CoO Nanocomposites Mechanically Milled

**Abderrahmane YOUNES[1,a,*], Mohamed Khorchef[1,b], Amirouche Bouamer[1,c] and Hichem Amar[1,d]**

[1] Research Center in Industrial Technologies (CRTI) P.O.Box 64, Cheraga 16014 Algiers, Algeria

[a]younesabdo11@gmail.com, [b]m.khorchef@crti.dz, [c]a.bouamer@crti.dz, [d]h.amar@crti.dz

**Abstract.** The $Fe_{60}(CoO)_{40}$ nanostructured alloys have been prepared from pure iron and cobalt oxide powders by mechanical alloying technique within a high energy planetary ball-mill. Morphology, microstructural and magnetic properties of this powder were investigated by a Scanning Electron Microscope (SEM), X-ray diffraction (XRD) and Vibrating sample magnetometer (VSM). The effect of time of milling on magnetic behaviour of Fe(CoO) nanostructured composite has been investigated. Apparition of new phase polycrystalline sample having a size in the range of 12 and 26 nm, it is confirmed by X-ray diffraction testing. The enhanced magnetic properties and structural behaviour of the nanoparticle are due by the diminution of size of crystallite. After 40 hours of milling, the appearance of spinel structure of CoFe2O4. The reduction in particle size leads to a significant increase in magnetic hardening, the coercive field at room temperature increases from 6 Oe to 208 Oe

## 1. Introduction

A magnetic powder material is among the first areas of research in recent time for material researchers in accordance to its emerging potential applications in various technological fields and as well as new structural and magnetic properties at the essential level [1]. The CoO is used extensively in the ceramics industry and transition metal oxides, the CoO have higher magneto-crystalline anisotropy [2-3]. In other way, iron nanostructured particles have attracted significant attention for the reason of their attractive magnetic properties and applications in high density magnetic recording media [4]. The main applications of cobalt iron oxide nanoparticles are as a nanoscale magnetic material with applications in molecular imaging and drug delivery [5-6]. In this paper, we are studying the magnetic properties in the Fe/CoO nanostructured alloy. The effect of crystallite size in the hysteresis loops has been observed by vibrating sample magnetometer.

## 2. Experimental procedure

The particle size and purity of elemental powders are 80 μm, 99.5% for Fe and 100 μm, 99.9% for CoO. The mixture of Fe, CoO powders was grinded in planetary ball mill (PM400) under controlled atmosphere used a hard chromium grinding balls. The initial powder mixtures are contains of 60% wt of Fe and 40% wt of (CoO). The mass ratio of powder to balls was approximately 1:30.The time of milling was varied from 5 to 40 hours. The mechanical alloying method was used to elaborate $Fe_{60}$ $(CoO)_{40}$ powder alloys and to decrease the particles sizes after severe plastic deformation.

Morphology, particle size and particle distribution of powders alloy formed after mechanical milling was characterized by scanning electron microscopy (Gemini SEM 300) attached with Energy Dispersive X-ray unit (EDX). Crystallite size, lattice strain and lattice parameters were determined by X-ray diffraction analysis (XPERT PRO) using Co Kα radiation. Magnetic measurement properties were determined by Vibrating Sample Magnetometer (VSM).

## 3. Results and Discussion

### *3.1. Powder morphology*

Morphology of the $Fe_{60}(CoO)_{40}$ powder mixture milled for different times was shown in figure 1. During the milling time, $Fe_{60}(CoO)_{40}$ powder having uniform particle size distribution, flattened and irregular shapes, this morphology is caused by severe plastic deformation, welding process and fracture process [12-13]. Cold welding and fracture process lead to considerable changes in the morphology of the powder particles during mechanical alloying. The mean particle size of $Fe_{60}(CoO)_{40}$ powder alloy after 40 hours of milling (figure 1.e) is smaller than 5, 10,15 and 25 hours of milling, indicating that the milling effect speed up the reduction of particle size.

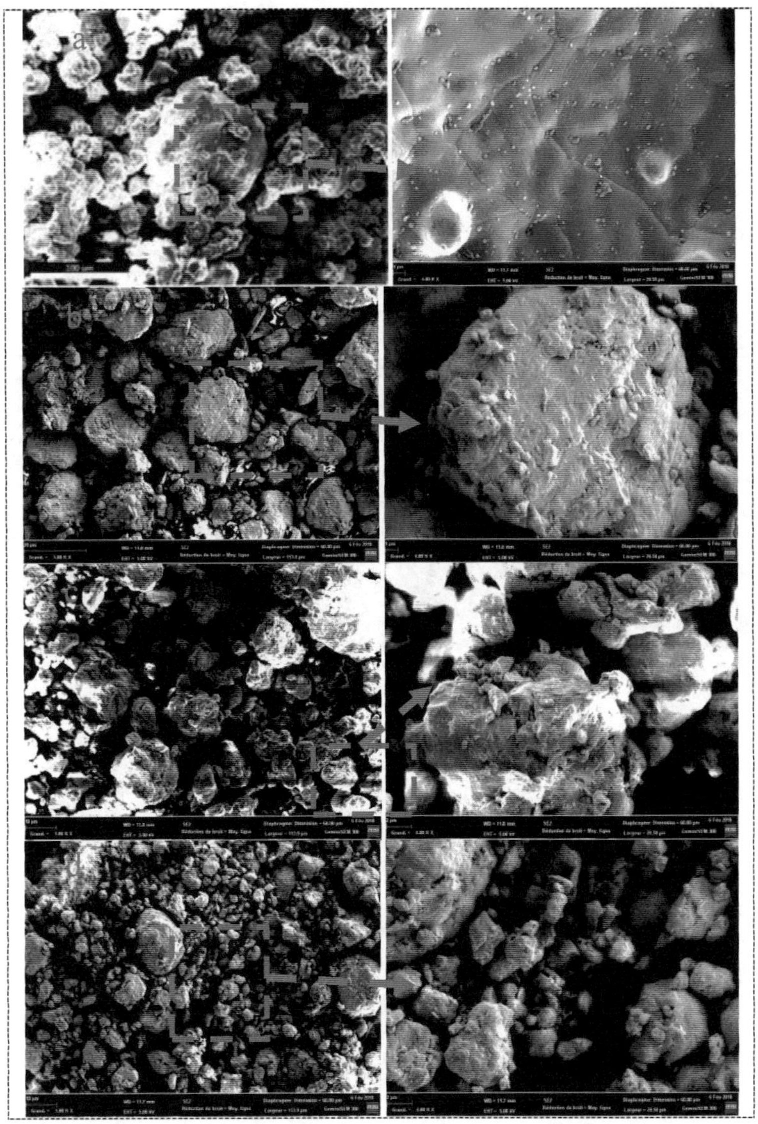

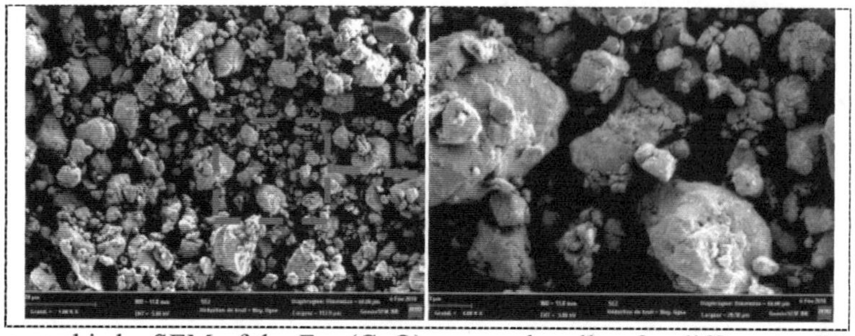

**Figure 1.** Micrographie by SEM of the $Fe_{60}(CoO)_{40}$ powder alloy for different milling time: a(5 h), b(10 h), c(15 h), e(25 h) and d(40h).

Figure 2 show the repartition of chemical elements determined using Energy Differential Scanning, the EDS analysis; indicate the existence and the same percentage of chemical elements in the FeCoO. We have found a relationship between the composition of initial powder and the milling powder studied in this work. The spectrum of existence element of FeCoO alloy powder shows the presence of different element Fe, Co and O.

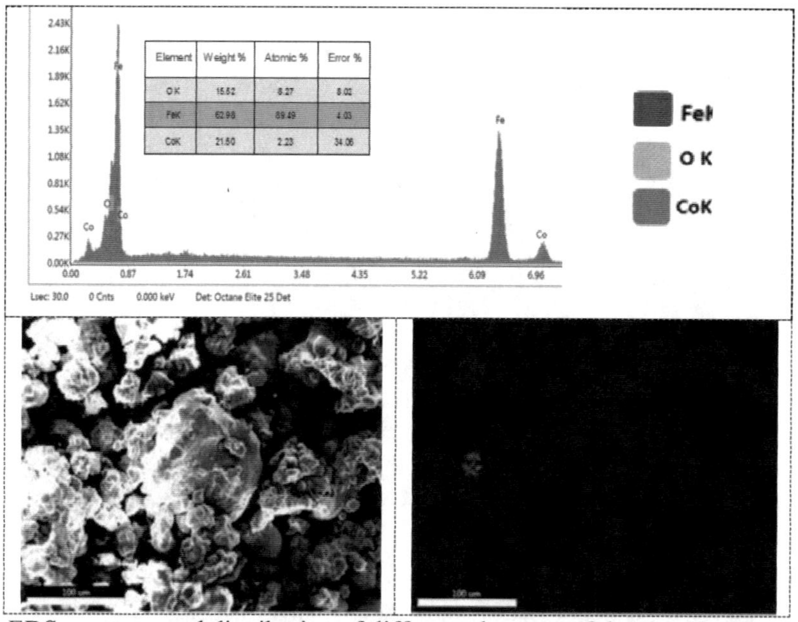


**Figure 2.a.** EDS spectrum and distribution of different elements of the $Fe_{60}(CoO)_{40}$ powder alloy milled to 5h

The repartition of different elements Fe, Co and O are homogeneously distributed on $Fe_{60}(CoO)_{40}$ alloy powder, this distribution are caused by mechanical milling parameters, concerning the value of mass percentages of Fe, CoO determined by EDS analysis are 62.98 %wt, 37.02%wt respectively (figure 2a).

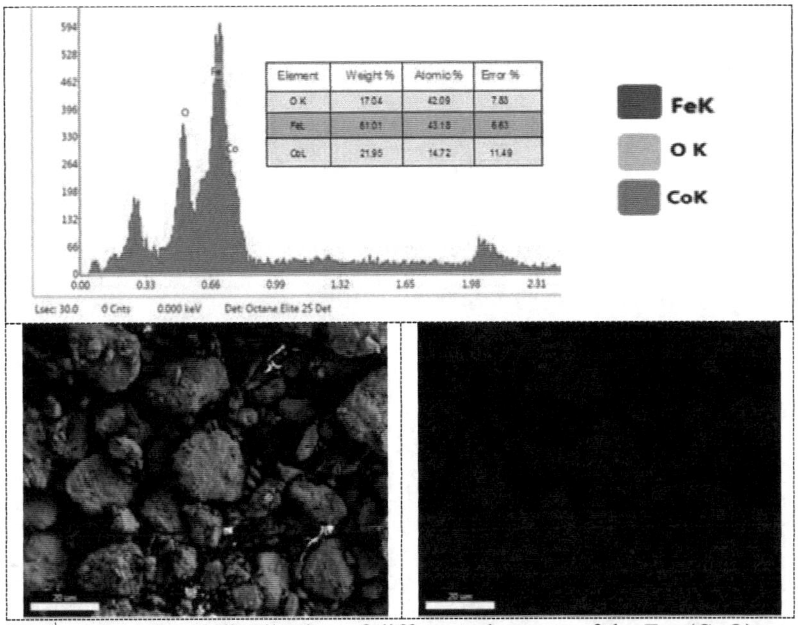


**Figure 2b.** EDS spectrum and distribution of different elements of the $Fe_{60}(CoO)_{40}$ powder alloy milled to 10h

Figure 2b shows the uniformly distributed of Fe, Co and O in the $Fe_{60}(CoO)_{40}$ powder milled to 10h, and the mass percentage of Fe and CoO is 61.01 % Wt and 38.99%Wt respectively determined by EDS analysis.

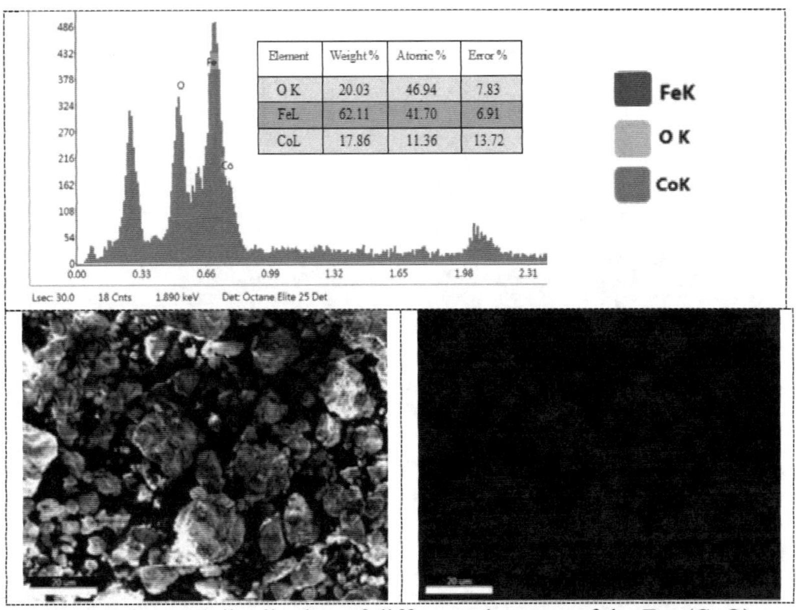


**Figure 2c.** EDS spectrum and distribution of different elements of the $Fe_{60}(CoO)_{40}$ powder alloy milled to 15h.

The cartography of different element presented in figure 2c shows that the reparation of Fe, Co and O are homogeneous and the mass percentage of Fe and CoO is 62.11 % wt and 37.89%wt respectively determined by EDS analysis

IOP Conf. Series: Materials Science and Engineering **557** (2019) 012064 doi:10.1088/1757-899X/557/1/012064

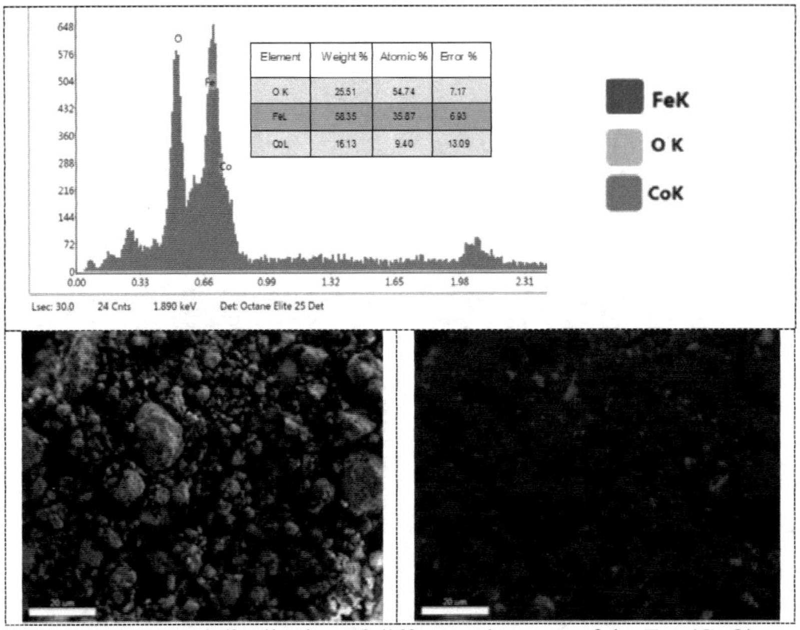


**Figure 2d.** EDS spectrum and distribution of different elements of the $Fe_{60}(CoO)_{40}$ powder alloy milled to 15h.

The distribution of Fe, Co and O in the Fe60 (CoO)40 milled to 25 hours presented in figure 2d shows that have a homogeneity of this elements in $Fe_{60}(CoO)_{40}$ powder milled to 25h, and the mass percentage of Fe and CoO is 58.35 % wt and 41.65%wt respectively determined by EDS analysis.

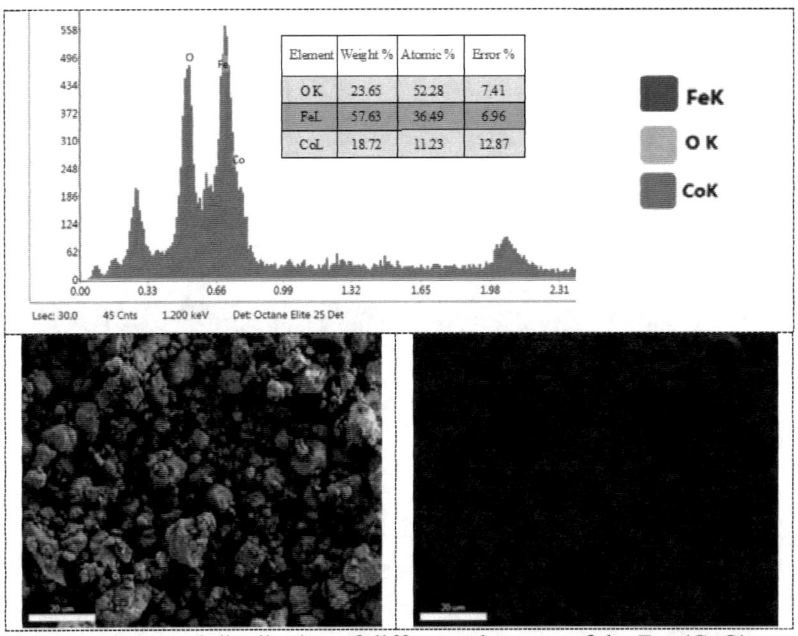


**Figure 2e.** EDS spectrum and distribution of different elements of the $Fe_{60}(CoO)_{40}$ powder alloy milled to 40h.

The EDS analysis presented in figure 2e shows the presence of Fe and CoO in $Fe_{60}(CoO)_{40}$ powder milled to 40h with mass percentage 57.63 % Wt and 42.37%Wt respectively. The repartition of these elements (Fe, Co and O) presented in figure 2d shows a good homogeneity in the powder alloy.

### *3.2. Structural characterization of $Fe_{60}(CoO)_{40}$*

Figure 3 shows the progression of the crystallites size and lattice strain, kinetics of formation, evolution of lattice parameters determined by X-ray diffraction analysis of $Fe_{60}(CoO)_{40}$ powder alloy during mechanical milling time to 5 until 40hours.

Figure 3 has shown the presence of Fe, CoO and CoFe2O4 after milling time, the corresponding peaks in iron decrease in intensity. In parallel, the displacement of peaks of iron for a new position of Bragg angle is caused by the phenomenon of formation of alloy. After 40 hours of milling appeared a new peak of $CoFe_2O_4$.

X-Ray diffraction measurements are used to determine the lattice parameter of the different phases presented in $Fe_{60}(CoO)_{40}$ powder alloy in function of milling time, the principle (hkl) plane (110) were used to determined the lattice parameter according to equation (Eq. 1) [14]:

$$\frac{1}{d^2} = \frac{h^2 + k^2 + l^2}{a^2} \quad (1)$$

The crystallite size (D) was calculated from the width at half maximum (FWHM) using Scherer equation according to equation (Eq. 2)

$$D = \frac{0.9 * \lambda}{FWHM * \tan\theta} \quad (2)$$

Where $\lambda = 1.78901$ (Co radiation), $\theta$ is the peak angle (radians).

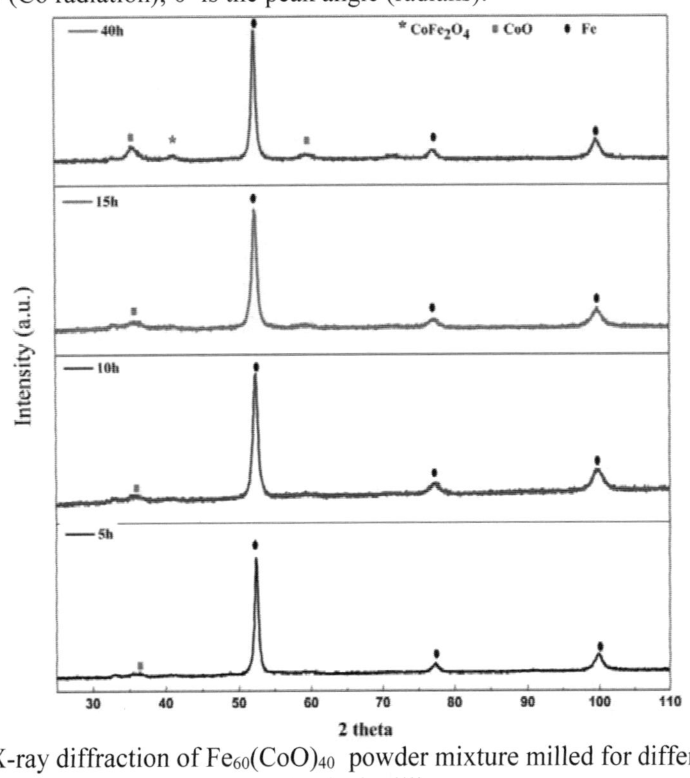


**Figure 3** X-ray diffraction of $Fe_{60}(CoO)_{40}$ powder mixture milled for different time of mechanical milling.

Figure 4 shows the variation of average crystallite size and lattice strain of $Fe_{60}(CoO)_{40}$ powder alloy during mechanical milling time, the reduction of the size of crystallite is due of the increase of energy stocked inside particles during plastic deformation. The increase of lattice strain and the decrease of the size of crystallite in function of mechanical milling time are caused by distortion effect in the crystallite lattice with a high density of dislocations.

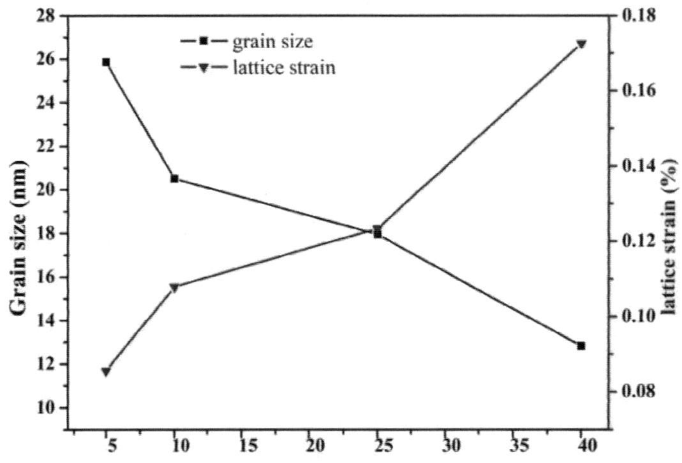


**Figure 4.** Change of average crystallite size and lattice strain of $Fe_{60}(CoO)_{40}$ powder alloy during mechanical milling time.

Figure 5 shows the Change of the lattice parameter of $Fe_{60}(CoO)_{40}$ powder alloy during of mechanical milling time, the increase in the lattice parameter during milling time has been noticed. This variation is easily explained in the first step by the dislocation introduced during mechanical alloying. The decrease in the next step can be explained by a dynamic equilibrium between the mechanical alloying effects and the restoration of particle powder.

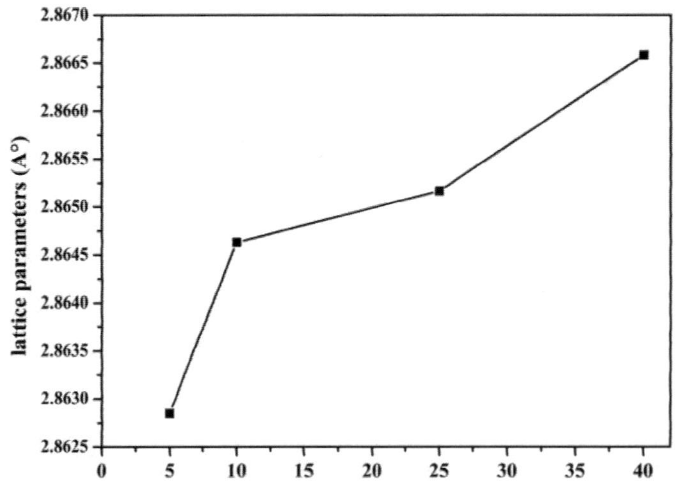


**Figure 5.** Change of the lattice parameter of $Fe_{60}(CoO)_{40}$ powder alloy during of mechanical milling time.

IOP Conf. Series: Materials Science and Engineering **557** (2019) 012064 doi:10.1088/1757-899X/557/1/012064

*3.3. Magnetic measurements*

Figure 6 show the hysteresis loops for $Fe_{60}(CoO)_{40}$ powder alloy milled during mechanical alloying. The hysteresis loops determined seem as soft magnetic materials with the magnetic properties varied versus mechanical milling time. The variation of coercive force (Hc), magnetization saturation (Ms), and residual magnetization (Mr) during mechanical alloying is shown in figure 6.

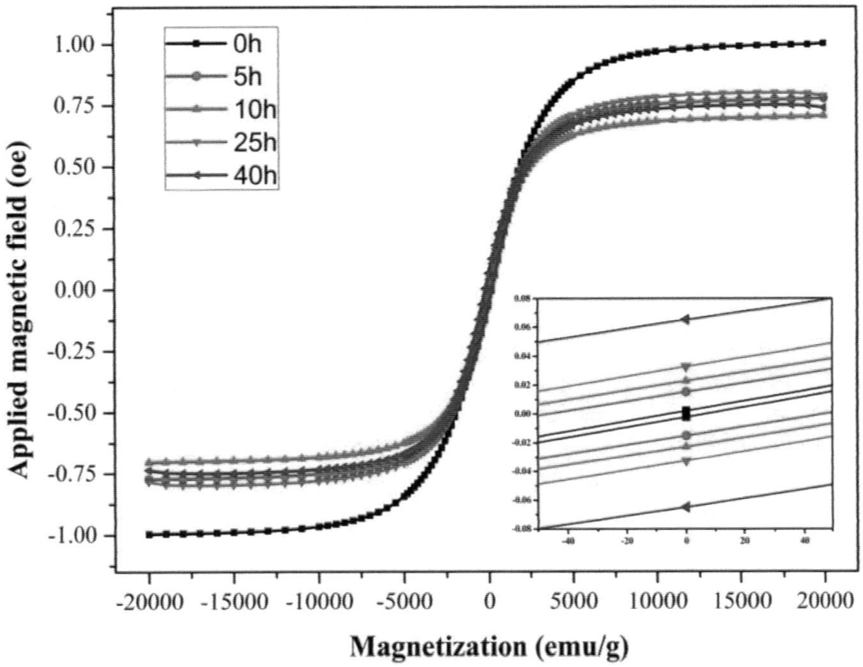


**Figure 6.** Hysteresis curve of $Fe_{60}$ $(CoO)_{40}$ powders alloy for different milling time.

Figure 7 shows the variation of coercivity, remanence and area of hysteresis loops in function of milling time of $Fe_{60}$ $(CoO)_{40}$ powder alloy, These variation of coercivity, remanence and area of hysteresis loop have the same evolution, we can explain this rise by many factors such as reduction of crystallite size, internal stress, dislocation density and impurities which are introduced during the mechanical alloying process (figure .7).

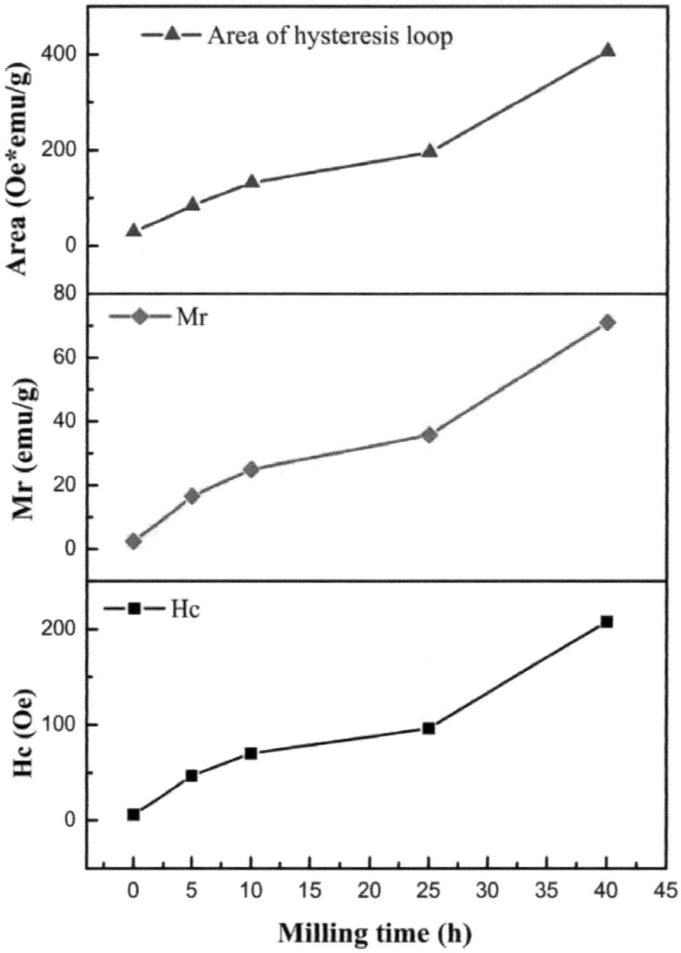


**Figure 7.** Evolution of area of hysteresis loop, Mr and Hc of $Fe_{60}$ $(CoO)_{40}$ powder alloy for different milling time.

Figure 8 shows the evolution of susceptibility, permeability and Ms of $Fe_{60}$ $(CoO)_{40}$ powder alloy for different milling time, The variations of saturation magnetization, susceptibility and permeability during milling time of the powder alloy are shown in figure 8. The reductions of particle size decrease the magnetocrystalline anisotropy; this reduction is due to the magnetic effect of nanoparticles that are oriented in an arbitrary manner, these parameters are highly dependent on the chemical composition of $Fe_{60}(CoO)_{40}$ powder alloy

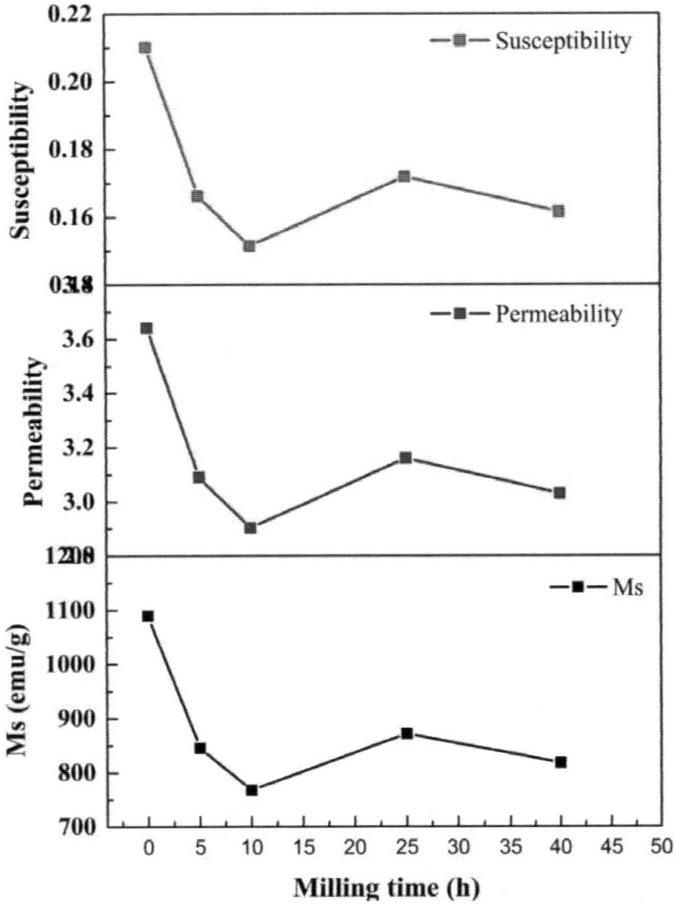


**Figure 8.** Evolution of susceptibility, permeability and Ms of $Fe_{60}$ $(CoO)_{40}$ powder alloy for different milling time

## 4. Conclusion

The $Fe_{60}(CoO)_{40}$ alloy powders were successfully elaborated by mechanical milling technique using PM400. The magnetic behavior, morphological of powder and structural parameters of $Fe_{60}(CoO)_{40}$ powder alloy could be determined. The structural parameters such as the crystallite size modify the magnetic behavior of $Fe_{60}(CoO)_{40}$ powder alloy.

The mappings of the powders alloy are distributed in a homogeneous way and kept the same proportion with the initial chemical composition. The coercivity of the $Fe_{60}(CoO)_{40}$ alloy highly depends on the structural parameters, the saturation magnetization is highly dependent on the chemical composition of powder alloy.

## Reference

[1] S.D. Bader, Colloquium: opportunities in nanomagnetism, Reviews of Modern Physics 78:1, (2006)

[2] N.R. Panda, S.P. Pati, A. Das, D. Das, Annealing temperature induced phase evolution and exchange bias properties of Fe/CoO nanocomposites, App. Surf. Sc. 449 (2018) 654-659.

[3] A. J. Sievers, III, and M. Tinkham, Far Infrared Antiferromagnetic Resonance in MnO and NiO Phys. Rev. 129 (1963) 1566-1571.

[4] J.L. Dormann, , D. Fiorani, Magnetic properties of fine particles, North-Holland delta series, Amsterdam, (1992) 323

[5] J. Kudr, , Y. Haddad, , L. Richtera, , Z. Heger, , M. Cernak, , V. Adam, and O. Zitka, , Magnetic Nanoparticles: From Design and Synthesis to Real World Applications, Nanomaterials, 7(9) (2017) 243.
[6] S. P. Pati, D. Das, Inter-particle and collective states of interactions in mechanically milled Fe(CoO) nanocomposites, J. Nanopart. Res. 16:2278 (2014) 1-12.
[7] G. C. Lavorato, E. Lima Jr, D. Tobia, D. Fiorani, H. E. Troiani, R.D. Zysler, E.L.Winkler, Size effects in bimagnetic $CoO/CoFe_2O_4$ core/shell nanoparticles, Nanotechnology 25:355704 (2014) 1-9
[8] S. P. Pati, A. Roychowdhury, S. Kumar, and D. Das, Signature of exchange bias and spin-glass like phenomena in Fe/CoO nanocomposite, Journal of applied physics 113 (2013) 17D708(1)-17D708(3).
[9] R. Abrudan, J. Miguel, M. Bernien, C. Tieg, M. Piantek, J. Kirschner, and W. Kuch, Structural and magnetic properties of epitaxial Fe/CoO bilayers on Ag(001), Physical review B 77 (2008) 014411(1)- 014411(7).
[10] E. Młynczak, , B. Matlak, , A. Kozio-Rachwa, , N. Gurgu, Spiridis, and J. Korecki, Fe/CoO(001) and Fe/CoO(111) bilayers: Effect of crystal orientation on the exchange bias, Physical Review B 88 (2013) 085442(1)- 085442(10).
[11] S.R. Mishra, I. Dubenko, J. Losby, S. Roy, N. Ali, and K. Marasinghe, Magnetic Behavior of Mechanically Milled FeNi-CoO Nanocomposites, .IEEE Transactions on magnetics, 40(4) , (2004) 2716 - 2720.
[12] A. Younes, N. E. Bacha, M. Zergoug, N. Dilmi, Structural and magnetic properties of Fe–$Co/Al_2O_3$ nanocomposite powder produced by mechanical alloying, Powder Metallurgy and Metal Ceramics, 56 (3) (2017)148-157.
[13] A. Younes, N.E. Bacha, M. Zergoug, M. Gousmine, H. Dehdouh, and A. Bouamer, Effect of Grain Size of Nano Composite on Raman and Magnetic Proprieties, Applied Mechanics, Behavior of Materials, and Engineering Systems , 32(1) (2017) 425-433.
[14] A. Younes, N. Dilmi, M. Khorchef, A. Bouamer, N-E. Bacha, and M. Zergoug, Structural and magnetic properties of FeCuNi nanostructured produced by mechanical alloying, Applied Surface Science 446 (2018) 258–265.

# Studying structural properties of thin film nanostructure of pentoxide vanadium Prepared by reactive DC magnetron sputtering

**M Sh Muhammed[1], M K Kalaf [2], S J Mohammed[3]**

[1] Education Mission. Education Department. Salah al-Din. Iraq
[2] Ministry of Sciences and Technology. Baghdad, Republic of Iraq
[3] Tikrit University. Tikrit. Republic of Iraq
E-mail: muhshr@gmail.com

**Abstract.** In this work, the structural properties of the monocrystalline vanadium pentoxide have been presented. Vanadium pentoxide ($V_2O_5$) films were deposited by using a DC reactive magnetron sputtering system at working pressure of $8.5x10^{-2}$mbar. The sputtered vanadium atoms were sputtered and oxidized in presence $O_2$:Ar gas mixture by (5/95,10/90,15/85,20/80,30/70,50/50). Employment of magnetron results in the formation of $V_2O_5$ in the final samples according to the XRD analysis, increase the roughness and hence surface area of the produced $V_2O_5$nanostructures. The results of X-rays are shown to us, the deposited films were formed by nanoparticles with average grain size in the range of (52.11nm to 98.03) nm and roughness Ave (nm) in the range of (1.04nm to 8.88nm) . The deposited films are identified to be polycrystalline nature with a cubic structure along ((001), (111)) and ((200)) orientation also MonoV2O5, Cub VO were found as deposited. The texture of the films was observed using SEM and AFM, it was observed that the grain size was increased with increased the $O_2$ percentage. These improvements in the structural properties of the produced vanadium pentoxide make these nanostructures good candidates for specific applications, such as photo detectors, solar cells, electro chromic smart window and gas sensor.

## 1. Introduction

Vanadium creates wide range of composites with oxygen which have variation structural and different valence states and founds in a many phases of oxide comprise of vanadium oxide, There are at least 15 variation vanadium oxides communally to now, for example; VO, $V_2O_3$, $VO_2$, $V_2O_5$,$V_6O_{13}$ and so on[1]. V2O5 are known to be the most stable compound form for vanadium metal , but other non-stable stoichiometry can be constrain under particular oxidation conditions.

Different phases of vanadium oxides can be obtained by controlling the deposition method parameters, or by proportion process manipulate, e.g., additional thermal annealing .properties of these phases of vanadium oxides depend on their structure, which determines other properties which exhibit a complex stoichiometric composition[2]. Thin-walled oxide nanoparticles belong to this smart inventory list, and these thin-layer oxide structures have been accelerated for practical applications such as electronics and optoelectronics. Most of these vanadium oxides are characters for the transfer of semiconductor phases to metal to reversible [3], which is accompanied by large changes in infrared optical properties and their electrical resistance.[4][5]. Vanadium oxide (V/O) structures are a well-known catalyst among various metal oxides, and so many fundamental studies have been developed wide-spreading centering on catalytic oxidation[6]. They show reversible metal-semiconductor transition through a wide band of temperatures depending on the ratio of O/V[7][8], Which results in sudden changes that are accompanied by large changes in infrared optical properties and electrical resistance [7]. The thin layer of vanadium / oxide with defined stoichiometry with the appropriate selection of the network simulation subsystem and the adjustment of many parameters including sedimentation, precipitate rate, pressure and temperature. However, the difficulty of precisely controlling all the parameters and preventing the emergence of defect results in several cases in the stoics is impossible and challenges the synthesis of these oxides as an additional task. Due to these problems, vanadium oxide is mixed with valentine for complex synthesis and research has been reduced. However, vanadium oxide mixed Usually in the process of oxidation / reduction VO2 – V2O5[9]. Vanadium pentoxide (V2O5) has been broadly studied because of its highest oxidation state in the V / O system, a wide band gap, a better stability and its electro thermal effects it is useful for device applications[10]. Except for the pure values of vanadium 0, +2, +3, +4 and +5, respectively, are related to phases V, VO, V2O3, VO2 and V2O5. Two sets of compost properties are identified as stable phases. These series VnO2n-1 and VnO2n + 1 are defined as magnetism. [11][12] , Wesley phases, respectively, and they connect the three most common phases, $V_2O_3$, $VO_2$ and $V_2O_5$[11][13][14]. The working mechanism of electro chromic materials consists of electrolyte, intercalated mobile ions which performs intercalation/de intercalation while a potential difference is applied. For this purpose, layered structure of V2O5 is a promising material. The most commonly used phases, which found in wide range of applications due to their novel properties, are the VO, VO2, V2O3, and V2O5 oxide phases[15].

## 2. Experimental details

Vanadium pure was chosen as the primary material to study for this project. Sputtering processes normally uses argon gas because it does not react with the target material allowing thin films to be formed. The addition of a gas that reacts with the target material, such as oxygen, will form compounds of the material and the reactant. To determine how vanadium oxide is formed using reactive magnetron sputtering a reactive gas test was performed. Because the sputtering system uses a DC power supply that allows both voltage and current to float while maintaining constant power, the target voltage can be easily measured and used to infer changes in the impedance of the sputtering plasma. The gun in which the target is mounted is designed to provide a narrow range of impedance to the power supply. With the addition of a reactive gas the impedance of the target changes. Using the fixed conditions of 660V of DC power and a chamber pressure of 0.08 mb while introducing increased amounts of oxygen the voltages were recorded. Figure 1 illustrated the outcome of this experiment

IOP Conf. Series: Materials Science and Engineering **557** (2019) 012065 doi:10.1088/1757-899X/557/1/012065

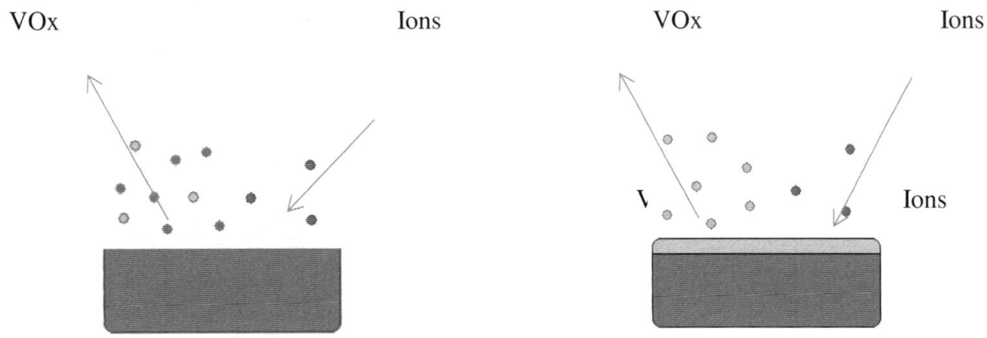


**Figure 1.** (a) Fully Oxidized Phase. (b) Metallic Phase

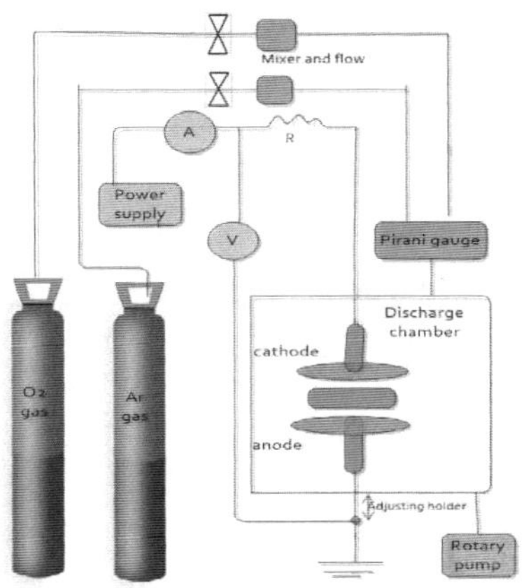


**Figure 2. The main experimental set-up used in this work.**

## 3. Results

Figure 3 additionally shows that there are at145, 198, 287, 308, 413, 491, 535 and 1002 cm-1 bands of crystalline V2O5 nanoparticles. [16] The formation of nanoparticles for catalysts with a V V7.7 nm-2 load or more, which is related to the surface coating of a layer (~ 8 V nm-2). [17] The supported V2O5 bands at 491 and 535 cm 1 are wide, while the 705 cm-1 sample does not exist for the V2O5 crystal, suggesting that this V2O5 NP support is not large and well-ordered. All supported AMV catalysts have a Raman group of 1035 cm-1, which extends to Vagabond (VVO) from the surface of the oligomer, an exo VO4 [18,19,20-21] Sharp tapes at 140, 194, 283, 302, 404, 482, 528, 705, and 998 cm-1 are formed from crystalline V2O5 nanoparticles. The peaks at284 and 401 cm-1 indicate that they have vibrational vibrational results V = O. Vibrational vibrations of V-O-V bonds and 3D overlapping V3-O bonds are obtained in 487 [22]. These bands are asymmetric and symmetric, respectively, to the vibrational vibrational states V-O, V-O-V and V = O. The FT-IR spectrum was packed with 100 watts of RF power at 477.21 cm, 1 506.17 cm-1 and 617.16 cm-1. These are vibration banding vibrations of deformation of the O-V gradient and the band of 1046.65 cm-1 relative to the V = O bond [23].

IOP Conf. Series: Materials Science and Engineering **557** (2019) 012065 doi:10.1088/1757-899X/557/1/012065

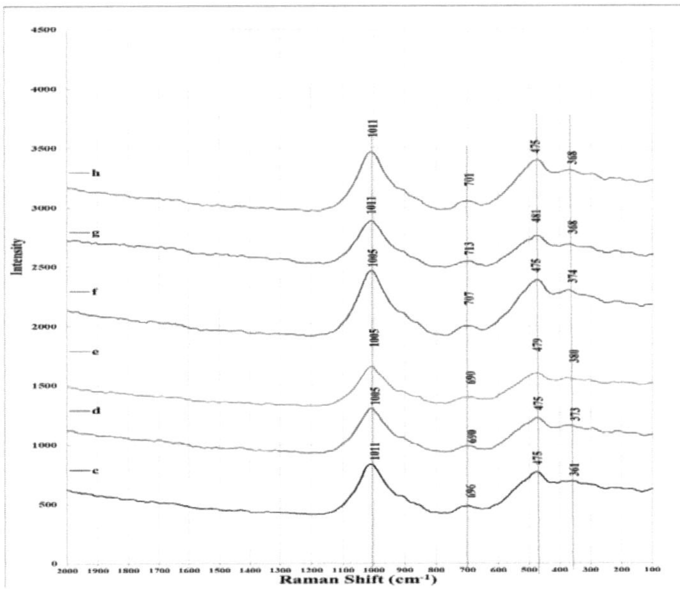


**Figure 3.** shows the Raman spectrum of vanadium oxide films produced by the O2: Ar ratio.

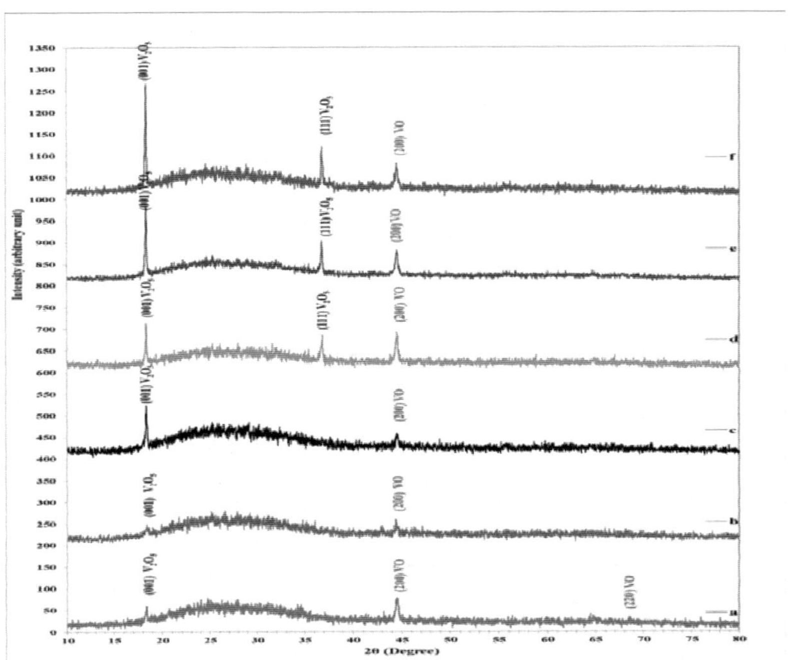


**Figure 4.** X-ray diffraction patterns for a thin sheet of vanadium deposited with different ratios of $O_2$: Ar.

**Table 1.** X-ray diffraction patterns for a thin sheet of vanadium covered with different ratios of $O_2$: Ar

| | 2θ (Deg.) | FWHM (Deg.) | $d_{hkl}$ Exp.(Å) | G.S (nm) | $d_{hkl}$ Std.(Å) | hkl | Phase | card No. |
|---|---|---|---|---|---|---|---|---|
| a | 18.3105 | 0.2826 | 4.8413 | 28.5 | 4.3830 | (001) | Mono. $V_2O_5$ | 96-901-2222 |
| | 44.5100 | 0.4037 | 2.0339 | 21.3 | 2.0310 | (200) | Cub. VO | 96-900-8767 |
| b | 18.4316 | 0.3633 | 4.8098 | 22.2 | 4.3830 | (001) | Mono. $V_2O_5$ | 96-901-2222 |
| | 44.3889 | 0.5248 | 2.0392 | 16.3 | 2.0310 | (200) | Cub. VO | 96-900-8767 |
| c | 18.3509 | 0.2422 | 4.8307 | 33.2 | 4.3830 | (001) | Mono. $V_2O_5$ | 96-901-2222 |
| | 44.4697 | 0.2825 | 2.0357 | 30.4 | 2.0310 | (200) | Cub. VO | 96-900-8767 |
| | 18.3105 | 0.1615 | 4.8413 | 49.8 | 4.3830 | (001) | Mono. $V_2O_5$ | 96-901-2222 |
| d | 36.8051 | 0.2825 | 2.4400 | 29.6 | 2.4434 | (111) | Mono. $V_2O_5$ | 96-901-2222 |
| | 44.4697 | 0.3633 | 2.0357 | 23.6 | 2.0310 | (200) | Cub. VO | 96-900-8767 |
| | 18.3509 | 0.2018 | 4.8307 | 39.9 | 4.3830 | (001) | Mono. $V_2O_5$ | 96-901-2222 |
| e | 36.6840 | 0.2422 | 2.4478 | 34.6 | 2.4434 | (111) | Mono. $V_2O_5$ | 96-901-2222 |
| | 44.4697 | 0.3229 | 2.0357 | 26.6 | 2.0310 | (200) | Cub. VO | 96-900-8767 |
| | 18.3509 | 0.2018 | 4.8307 | 39.9 | 4.3830 | (001) | Mono. $V_2O_5$ | 96-901-2222 |
| f | 36.6840 | 0.2018 | 2.4478 | 41.5 | 2.4434 | (111) | Mono. $V_2O_5$ | 96-901-2222 |
| | 44.4697 | 0.3633 | 2.0357 | 23.6 | 2.0310 | (200) | Cub. VO | 96-900-8767 |

The X-ray diffractions of the Vanadium target is also displayed . The target material exhibits peaks corresponding to the mono V2O5 phase at 2θ = 18.3105°, 18.4316°,18.3509°, 36.8051°, 36.64840°, and 36.6840°, direction identify with standard peaks[ card. No(96-901-2222)]. It are belongs to Pmmn space group having lattice parameters as a = 11.544 A°, b = 3.571 A° and c = 4.383 A°. The XRD pattern illustrates that the (V2O5) films have polycrystalline (111), (001) and The X-ray diffractions of the Vanadium target is also displayed cub(VO) . The target material exhibits peaks corresponding to the cub VO phase at 2θ=44.5100°,44.3889°,44.4697°, direction identify with standard peaks[card96-900-8767]. It are belongs to Pmmn space group having lattice parameters as a=4.0620A°. The XRD pattern illustrates that the (V2O5) films have polycrystalline (200). also, note that increases the different $O_2$:Ar ratio ., leads to peak intensity increases (i.e. increase films crystallinity). This might be due to a change in its phase or crystal structure as deposition rate varied. Also, increasing RF power making increases in grain size, as shown in Table 1, this may be due to the enhancement of films crystallinity. So, improved the crystalline of the films by increasing grain size to decrease the number of grain boundaries. A significant amount of line broadening which is a characteristic of nanoparticles. All the XRD patterns show same behavior where the FWHM decreases with increase of the different $O_2$:Ar ratio .with increasing the crystallite size. [180,181].

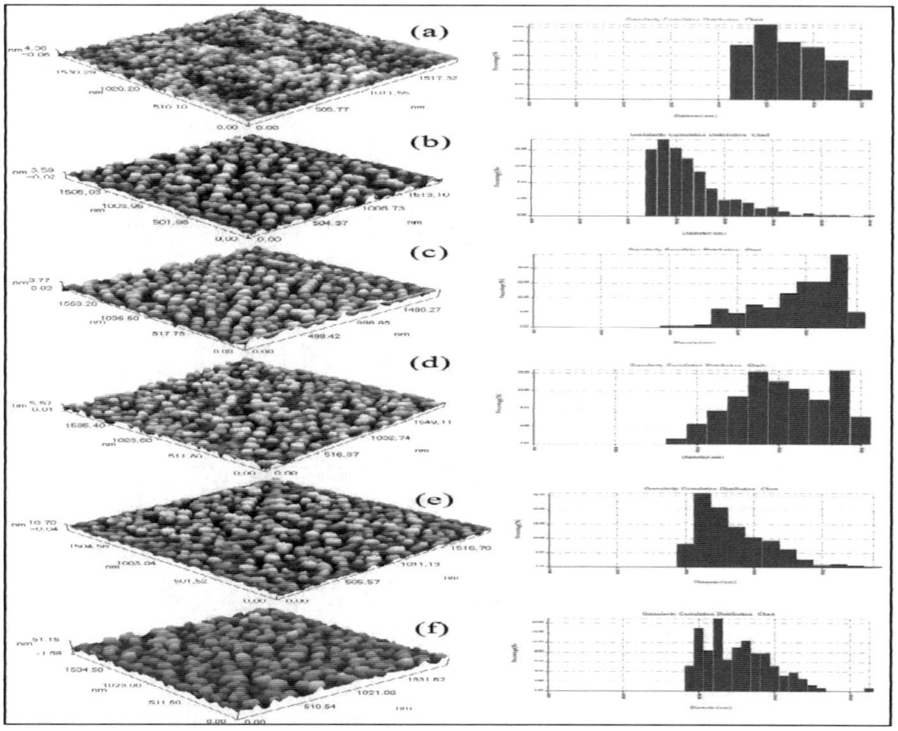

**Figure 5.** 3D AFM image and their granularity accumulation distribution for vanadium thin films at different $O_2$:Ar ratio.

The values of grain sizes for Vanadium oxide thin films are given in Table 2. This table increases grain size between (52.11-98.03) nm and the RMS roughness between (1.04 and 8.88) nm with increases $O_2$/Ar ratio. The images at the low pressure confirm that the films are uniform and the substrate surface is well covered with grains that are nearly uniformly distributed. When increasing the gas pressure. The possibility of some small grains agglomerated to form greater grains as a consequence. The deposited film exhibited small nicely combined with coalescence of some columnar grains at few places. The decrease in surface roughness of the deposited films with higher pressures is due to the growth of grains with favorable orientation dictated by surface and grain boundary diffusivity, ad-atom mobility, film thickness and induced thermal stress. This behavior is similar and agrees with the results obtained in the reference.

**Table 2.** of AFM 3D image and its gravitational accumulation distribution for vanadium thin sheets with different O2: Ar ratios

| $O_2$:Ar ratio | Average Diameter (nm) | RMS roughness (nm) | Roughness Ave.(nm) |
|---|---|---|---|
| 5/95 | 52.11 | 0.881 | 1.04 |
| 10/90 | 63.57 | 0.902 | 1.04 |
| 15/85 | 75.97 | 0.845 | 0.991 |
| 20/80 | 58.52 | 1.31 | 1.53 |
| 30/70 | 51.47 | 2.42 | 2.85 |
| 50/50 | 98.03 | 7.37 | 8.88 |

**4. Conclusions:**
In this paper, an analysis of the structural properties of vanadium powder oxide produced by reactive spectrometry has been observed. An increase in the surface roughness of the samples was observed with an increase in the ratio of the mixture of gases. This allows for the use of samples prepared in the use of gas sensors.

**References**

[1] Schneider K 2015 *Structural and optical properties of VOx thin films. Archives of Metallurgy and Materials* **60.2** 957-961
[2] Golan G et al 2004 *Investigation of phase transition mechanism in vanadium oxide thin films. Journal of Optoelectronics and Advanced Materials* **6.1** 189-195
[3] M E A Warwick and R. Binions *Advances in thermochromic vanadium dioxide films J. Mater. Chem.* **2. 10** 3275–3292
[4] Donev Eugenii U 2008 *Metal-semiconductor transitions in nanoscale vanadium dioxide-thin films, subwavelength holes, and nanoparticles* PhD Thesis
[5] Liu Yueyan et al.*2016 Effect of annealing temperature on the structure and properties of vanadium oxide films. Optical Materials Express* **6.5** 1552-1560
[6] Zou Mengyang *2015 Deposition Methods and Thermoresistive Properties of Vanadium Oxide and Amorphous Silicon Thin Films* PhD Thesis. University of Dayton.
[7] M Farahmandjou *2017 Chemical Synthesis of Vanadium Oxide (V2O5) Nanoparticles Prepared by Sodium Metavanadate," J. Nanomedicine Res* **5.1** 2–5
[8] Lamsal Chiranjivi Ravindran M 2013 *Optical properties of vanadium oxides-an analysis. Journal of materials science* **48.18** 6341-6351
[9] Urena Begara, Ferran Crunteanu, Aurelian Raskin, Jean-Pierre 2017 *Raman and XPS characterization of vanadium oxide thin films with temperature. Applied Surface Science* **403** 717-727
[10] Vijayakumar Yelsani et al 2015 *Influence of the substrate temperature on the structural, optical and thermoelectric properties of sprayed V2O5 thin films. Material in technologies* **49.2** 17
[11] Ramplberg Geert et al 2015 *In situ X-ray diffraction study of the controlled oxidation and reduction in the V–O system for the synthesis of VO 2 and V 2 O 3 thin films. Journal of Materials Chemistry* C3 **43** 11357-11365
[12] Jeong Jaewoo et al 2013 *Suppression of metal-insulator transition in VO2 by electric field–induced oxygen vacancy formation Science* **339.6126** 1402-1405
[13] Masina Bathusile N et al2015 Optimizing the synthesis of vanadium–oxygen nanostructures by plasma plume dynamics using optical imaging. *Optical Engineering* **54.3** 037106

[14] Fay Alexander 2005 *Engineering in vernetzten, offenen, durchgängigen Systemen (Engineering in Linked and Open Systems). at-Automatisierungstechnik* **53.4-5** 205-210

[15] Yimam Elias 2015 *Fabrication of Vanadium Oxide Nanoparticles by Pulsed Laser Ablation*

[16] Carrero Carlos A et al 2013 *Anomalous reactivity of supported V 2 O 5 nanoparticles for propane oxidative dehydrogenation: influence of the vanadium oxide precursor. Dalton Transactions* 42.35 12644-12653

[17] Deo Goutam Wachs Israel E 1994 *Reactivity of supported vanadium oxide catalysts: partial oxidation of methanol. Journal of catalysis* **146.2** 323-334

[18] Tian Hanjing Ross Elizabeth I 2 Wachs Israel *E2006 Quantitative determination of the speciation of surface vanadium oxides and their catalytic activity. The Journal of Physical Chemistry B* **110.19** 9593-9600

[19] Argyle Morris D et al 2002 *Effect of catalyst structure on oxidative dehydrogenation of ethane and propane on alumina-supported vanadia. Journal of Catalysis* **208.1** 139-149

[20] Went Gregory T Oyama S Ted Bell Alexis T *1990 Laser Raman spectroscopy of supported vanadium oxide catalysts. Journal of physical chemistry* **94.10** 4240-4246

[21] Olthof Bryan et al 2000 *Effects of support composition and pretreatment conditions on the structure of vanadia dispersed on SiO2, Al2O3, TiO2, ZrO2, and HfO2. The Journal of Physical Chemistry B* **104.7** 1516-1528

[22] Lee Se-Hee et al 2003 *Raman spectroscopic studies of amorphous vanadium oxide thin films. Solid State Ionics* **165.1-4** 111-116

[23] Murgia Viviana et al 2006 *Sol–gel synthesis of V2O5–SiO2 catalyst in the oxidative dehydrogenation of n-butane. Applied Catalysis A: General* **312** 134-143

# Metal Oxide and Activated Carbon as Photocatalyst for Waste Water Treatment

**Suhaibatul Aslamiyah Abdul Shukor[1], Rosniza Hamzah[1], Mohamad Abu Bakar[2], N Z Noriman[1], Awad A. Al-rashdi[3], Z M Razlan[4], Shahriman A B[4], I Zunaidi[5] and Wan Khairunizam[4]**

[1] Center of Excellence Geopolymer and Green Technology (CEGeoGTech), Faculty of Engineering Technology (FETech), Universiti Malaysia Perlis (UniMAP), UniCITI Alam Campus, 02100, Perlis, Malaysia.
[2] Nanoscience Research Laboratory, School of Chemical Sciences, Universiti Sains Malaysia, 11800 Minden, Penang, Malaysia.
[3]Chemistry Department, Umm Al-Qura University, Al-qunfudah University College, Al-qunfudah Center for Scientific Research (QCSR), Saudi Arabia
[4]School of Mechatronic Engineering, Universiti Malaysia Perlis (UniMAP), Pauh Putra Campus, 02600 Arau, Perlis, Malaysia
[5]Faculty of Technology, University of Sunderland, St Peter's Campus, Sunderland, SR6 0DD, United Kingdom

**Abstract.** Nanoparticles of metal oxide have great potential in the photocatalytic system for removal of heavy metals in wastewater. The usage of titanium dioxide (TiO2) as photocatalyst has gained much attention because of its unique properties compared to others metal oxide. The usage of activated carbon (AC) as adsorbent from agricultural wastes such as coconut shell, cotton stalks and rice husk (RH) has recently used nowadays due to it is characteristic and low cost. Besides, it can reduce the abundant of wastes from open burning occurs and also can minimizing air pollution. The combination of photocatalyst-absorption process has been introduced in photocatalytic activity via dip-coating techniques without involving the filtration process. This technique has seeking more attention of researchers for removal of heavy metals such as cadmium (Cd), chromium Cr) and copper (Cu) in industrial wastewater treatment.

## 1. Introduction

Environmental pollution has been occurring because of the developmental activities such as growing in manufacturing, transportation and construction that produce a large amount of wastes. These wastes may lead to pollution of soil, water, air and oceans that affects our environment and human being. In this review, we focus on method of treatments specifically for industrial wastewater to overcome the water pollution issues. The water pollution occurs due the present contaminants and its amount not suitable in daily life used such as cooking, drinking, bathing and other uses that usually influenced by human activities [1]. Thus water pollution limits the fresh water resources in the whole world due to climate change, gaining in population and urbanization where demands for water increased [2].

Wastewater treatment is the important process to treat the polluted water into the fresh or safe water that can be used for living life purposes. Nowadays, the rapid development in paper, electronics [3] and dye industries [4] as well as metal plating facilities [5] affects the water resources quality mainly for drinking supply. Typically, the wastewater from these industries are highly toxic due to the presence of heavy metals which were directly or indirectly discharged into the environment [6]. Consequently, contaminated industrial wastewater with heavy metals risk human health and aquatic ecosystems as well as serious environmental problems [5]. For example, in Saudi Arabia, the major heavy metal contaminants from industries are Arsenite (As), Cadmium (Cd), Cobalt (Co), Chromium (Cr), Lead (Pb) and Zinc (Zn). These heavy metals are very dangerous even exposure at very low concentrations because it can cause cancer, brain damage, accumulative poisoning and destroy liver. The Cr itself is highly toxic, carcinogenic and mutagenic. The mutagenic of Cr inhibits mitotic process of plants [2]. Mostly Cr are found in tannery industrial wastewater. Then, Pb, Nickel (Ni) and Copper (Cu) [8] also can cause harmful for kidney, sensory system, liver, brain, reproductive system, sterility, capillary damage, stillbirths and causes neonatal deaths [7-9].

Thus to protect human life and prevent any consequence due to contaminated wastewater with heavy metals, the specific standards of water has been introduced in the many of developed countries. Even though high end treatment of industrial wastewater is expensive [6] but many technologies have been developed such as photocatalytic, adsorption, membrane filtration, precipitation, electrochemical and ion-exchanged techniques [7]. Various models of photocatalytic cell have been developed such as photocatalytic reactor using CFD (Computational fluid dynamics) simulation, electrospinning of mesoporous zinc oxide (ZnO) nanoporous for photocatalytic activity [10] and immobilized titanium dioxide ($TiO_2$) nanofilm on the substrate [11]. The model of photocatalytic cell that was proposed in this paper is immobilizing of $TiO_2$ and adsorbent via dip-coating technique. Mainly the advantages of this models are both of adsorption and photocatalysis process will occur simultaneously during treating the wastewater without involves the filtration process [12-13]. In general, photocatalytic methods were widely applied for water purification, air production, antibacterial activity and degradation of organic pollutants in wastewater.

## 2. Metal Oxide as Photocatalyst

Typically, photocatalytic cell consists of photocatalyst, adsorbent, adhesive and substrate. Generally, a photocatalytic cell is able to perform photocatalysis activity which involves capabilities of metal oxide or metal oxide nanoparticles such as titanium oxide that used to remove heavy metals from wastewater [14]. In photocatalysis, both light and catalyst were used simultaneously to help and rapidly increase a chemical reaction where it can decompose organic pollutants. There are two types of photocatalyst which are homogeneous or heterogenous. The homogenous photocatalyst are having similar phase of matter between photocatalyst and targeted medium that involves complexation process between catalyst and heavy metals. While heterogenous photocatalyst involves different phase of matter between photocatalyst and targeted medium. Thus degradation of pollutants occurs via photocatalysis process through irradiation of UV light [15]. Different type of metal oxides and adsorbents were variously studied in heterogenous photocatalyst. Many of nanomaterials adsorbents was developed due to the ability of removing contaminants from wastewater by enhancing the efficiency and capacity of adsorption. There are variety of membrane that acts as adsorbents in photocatalytic cell. The combination of both metal oxide and adsorbent shown high synergistic impact to clear the contaminated water [12].

Last few decades, the catalyst from metal oxide such as silica [7], ZnO, $TiO_2$ and zirconium dioxide ($ZrO_2$) has been used in the photocatalytic process for wastewater treatment. These metal has been proposed for hazardous pollutants removal like heavy metals such as Cd [15] due to environmental-friendly materials, efficient costly and have the capability for pollutants adsorption in low concentration (ppb) [14]. The average size of nano-particles which contains structure components with one dimension is less than 100 nm. The properties of these metal oxides are high surface area, reduce solubility and no secondary pollutants.

As known ZnO is as photocatalyst for heavy metals removal in the aqueous solutions [16]. ZnO is a potential photocatalyst in removal of heavy metals such as Cd [17] and Pb. For example, ZnO was act as photocatalyst to remove Cd from aqueous solution by using experimental-laboratory batch study [17]. ZnO features nanostructure with high sufficient surface area and its surface consists of many functional groups such as hydroxyl groups that can be active sites for adsorption activity and uptake positively charge ions [16]. ZnO possesses variety of nanostructures such as porous thin film, nanorods, nanosphere and nanowired [10]. However, nanorods structures is widely used as photocatalyst in removal heavy metals such as Cd and Pb. There are varieties of techniques to synthesis of ZnO such as hydrothermal synthesis, homogeneous precipitate and sol-gel method. But, the sol-gel method was usually proposed due to it has good crystalline structure of particles and has a narrow size distribution. Then, this method also was required simple operation and mild conditions.

$TiO_2$ is the common metal oxide in photocatalyst. $TiO_2$ features unique properties such as high stability, low cost, high surface activity, high refractive index and have strong catalyzing ability. $TiO_2$ is a strong photocatalyst due to its chemical stability and has high ability reactivity under ultraviolet (UV) light that is wavelength lower than 390 nm. The simple mechanism of the working of photocatalysis depends on the photoexcitation of electron in the catalyst that the holes ($h^+$) and released electrons ($e^-$) was created by irradiation from light (UV) in the conduction band [14]. Instead of heavy metals removal, $TiO_2$ can be combined to remove organic or inorganic pollutants in water and air. $TiO_2$ structure can be amorphous or crystalline that consists of anatase, rutile and brookite. But, the most stable phase of $TiO_2$ is the rutile structure compare to others.

Typically, $TiO_2$ particles is used heavy metal removal of wastewater. Thus $TiO_2$ can be synthesize via sol-gel technique, hydrolysis or hydrothermal. The synthesis of $TiO_2$ via sol-gel technique were carried out using titanium precursor in either acidic or alkaline condition to form the $TiO_2$ sol. The preparation of $TiO_2$ nanofilm is similar to $TiO_2$ particles but it undergoes a specific heat treatment such as the Ag/$TiO_2$ nanofilm was heat treated at 400 °C to formed the nanoparticles of Ag that was dispersed on the $TiO_2$ surface [11]. Other than that, anodic oxidation technique was introduced from researchers that the temperature required is $< 600$ °C to obtain a thicker film at low voltage oxidation and obtaining the smooth surfaces.

Recently, $TiO_2$ also has been used as composites compound with adsorbent such as the powder activated carbon (PAC) and $TiO_2$ that used in treatment of heavy metals wastewater. Thus $TiO_2$ and PAC will reduce overall cost, biologically and chemically inert, and increase it turn over frequency (TOF) [12].

## 3. Adsorbents

In removal of heavy metals such as Pb, Cu, iron (Fe) and Zn from electroplating industrial wastewater are using activated carbon (AC) produced from coconut shell as an adsorbent. Carbon nanotubes (CNT) also have been used as an adsorbent to remove Cd in wastewater due to its characteristics which are have large specific areas and high chemical stability but CNT requires high cost [9].

Adsorption process is a technique that purified wastewater for removing an organic and inorganic pollutants from industrial wastewater. This method also more economical to remove metal ions from wastewater [9]. Recently, AC is most commonly used as adsorbent that suitable for removal of pollutants due to high surface area that allows adsorption. Other than that, this absorbent was also used to obtain the treated water because of its properties like a good porosity structure that can remove pollutants in water [7].

AC have made from agricultural wastes such as rice husk (RH) [8], bamboo [18], cotton stalk [19] and coffee husk due to the microporous of plant structures and have long dispersion path through solid particle of adsorbents [8]. It is also can have commercially from supplier such as PAC. For example, AC from cotton stalks were used to remove heavy metals such as Cd, Pb and Cu in wastewater via mixed batch reactor. This method was proposed due to its good properties such as low cost and have high carbon contain beside it can reduce air pollution from burning the agricultural waste [19]. Then, the AC from bamboo was used to remove the heavy metals which are Pb and Cd due to it has renewable raw materials and less costly [18].

In most of Asian countries, the production of AC from RH has been proposed to reduce environmental pollution from abundant of agricultural waste due to low cost and it is effectively accessible materials. Thus, AC from RH mainly used as dye removal like methylene blue [4], metals removal such as Pb, Cu [8] Cd and Cr [2]. The RH content also can be sources of carbon based on containing high cellulose and lignin. This method was obtained the porosity of surface area of RH as an absorbent. Other than that, RH also is the suitable materials as an adsorbent that containing low ash, high carbon and silica contents.

There are some of researchers was used commercially AC in the photocatalytic studies such as PAC. For example, the magnetic PAC particles has been approached in magnetic field to be treated the wastewater. This process has occurred without involving filtration process, but there are limitations which are not sufficient in surface areas and porosity compared to be AC. Other than that, PAC also have used as adsorbent and $TiO_2$ as a catalyst without blending together because it will make the reducing of porosity. However, this method has disadvantage such as highly cost that mainly used on a large scale [13].

## 4. Adhesive

In most studies, the composite polymers were formed membrane for removal pollutants. It undergoes filtration process. The usage of polymer as adhesive used to form another layer after substrate for dip-coating method. Photocatalytic film has been proposed by using $Ag_2O/TiO_2$ modified chitosan-based photocatalytic film. This method showed the removal of organic pollutants due to the good adsorption capacity by modified chitosan [20]. Next, the other example of adhesive is epoxidized natural rubber (ENR) that was used as a polymer in the adsorption-photocatalyst process for dip-coated method. This method showed the ENR was blend together with PAC to coating the substrate [12-13]. This polymer widely used because its good properties such as low glass transition temperature, high strength and solubility parameter [21].

The other adhesive that has been proposed in photocatalytic activity is polyamide-amine-epichlorohydrin (PAE) which is this polymer was used with microfibrillated cellulose and nanoparticles of $TiO_2$ for photocatalytic activity via filtration process [22]. This paper will focus on photocatalytic activity that combining with adsorbent in the wastewater treatment for removal heavy metals without involving filtration process. This method also will use dip-coating techniques layer-by-layer of materials for removal of heavy metals such as Cu, Cd and Cr.

## 5. Substrate

Substrate typically used are quartz and glass slides that have formed self-assembled film. This film has good characteristics such as stable, has unique physical and chemical properties, particularly dense and electron transport properties. For example, both of the glycine betaine and cetyl-trimethyl ammonium bromide (CTAB) was using to be formed this film due to it is potential to absorb the Pb ion in wastewater [23].

In addition, the other substrate was determined is borosilicate glass slide. This substrate was used for removal COD efficiencies by using $TiO_2$. The synthesis $TiO_2$ via sol-gel process was using to be coated the borosilicate glass slide to form the film before undergo the photocatalytic activity [24].

## 6. Conclusion

To conclude, photocatalytic process is widely used to be treated the contaminated wastewater through metal oxides and absorbance. There are varieties of pathways to shows the physico-chemical parameters for metal oxide nanoparticles and activated carbon such as size, shape, porosity and surface area. This requires a good design of a photocatalytic cell.

**References**

[1] F. W. Owa, "Water pollution : sources , effects , control and management," *International Letters of Natural Sciences.*, vol. 3, pp. 1–6, 2014.

[2] J. H. Al-Baidhani and S. T. Al-Salihy, "Removal of Heavy Metals from Aqueous Solution by Using Low Cost Rice Husk in Batch and Continuous Fluidized Experiments," *International Journal of Chemical Engineering and Applications.*, vol. 7, no. 1 February 2016, pp. 6–10, 2016.

[3] K. Le Van and T. T. Luong Thi, "Activated carbon derived from rice husk by NaOH activation and its application in supercapacitor," *Progress in Natural Science: Materials International*, vol. 24(3), pp. 191–198, 2014.

[4] Riyanto, R. Astuti and B. I. Mukti, "Simple Preparation of Rice Husk Activated Carbon (RHAC) and Applications for Laundry and Methylene Blue Wastewater Treatment," *AIP Conference Proceedings*, vol. 1911, pp. 020033-1-020033-9, 2017.

[5] T.-K. Kim, T. Kim, W.-S. Choe, M.-K. Kim, Y.-J. Jung and K.-D. Zoh, "Removal of heavy metals in electroplating wastewater by powdered activated carbon (PAC) and sodium diethyldithiocarbamate-modified PAC," *Environmental Engineering Research*, vol. 23, no. 3, pp. 301–308, 2018.

[6] A. Catenacci, "Heavy metal removal from water: characterization and applicability of unconventional media," no. March 2014, pp. 1–123, 2014.

[7] M. Karnib, A. Kabbani, H. Holail and Z. Olama, "Heavy Metals Removal Using Activated Carbon, Silica and Silica Activated Carbon Composite," *Energy Procedia*, vol. 50, pp. 113–120, 2014.

[8] M. Zakir, "Adsorption of lead (II) and copper (II) ions on rice husk activated carbon under sonication," *International Symposium on Chemical and Bioprocess Engineering*, (June), pp. 25–28, 2013.

[9] F. A. Al-Khaldi, B. Abu-Sharkh, A. M. Abulkibash and M. A. Atieh, "Cadmium removal by activated carbon, carbon nanotubes, carbon nanofibers, and carbon fly ash: a comparative study," *Desalination and Water Treatment*, vol. 53, no. 5, pp. 1417–1429, 2015.

[10] P. Singh, K. Mondal and A. Sharma, "Reusable electrospun mesoporous ZnO nanofiber mats for photocatalytic degradation of polycyclic aromatic hydrocarbon dyes in wastewater," *Journal of Colloid and Interface Science*, vol. 394, no. 1, pp. 208–215, 2013.

[11] W. Zhang, Y. Liu, D. Zhou, J. Wen, W. Liang and F. Yang, "Photocatalytic activity of Ag nanoparticle-dispersed N-TiO2 nanofilms by magnetron sputtering," *RSC Advances*, vol. 5, 2015.

[12] N. N. Bahrudin and M.A. Nawi, M. A., "Immobilized titanium dioxide/powdered activated carbon system for the photocatalytic adsorptive removal of phenol," *Korean Journal of Chemical Engineering*, vol. 35, no. 7, pp. 1532–1541, 2018.

[13] N. N. Bahrudin and M.A. Nawi, "Fabrication of immobilized powdered activated carbon as a sub-layer of $TiO_2$ for the photocatalytic-adsorptive removal of phenol," *Reaction Kinetics, Mechanisms and Catalysis*, 2017.

[14] M. Anjum, R. Miandad, M. Waqas, F. Gehany and M.A. Barakat, "Remediation of wastewater using various nano-materials," *Arabian Journal of Chemistry*, October 2016.

[15] R. Ameta, M.S. Solanki, S. Benjamin and S.C. Ameta, "Photocatalysis," Advanced Oxidation Processes for Wastewater Treatment: Emerging Green Chemical Technology, 2018.

[16] L. Khezami, K.K. Taha, E. Amami, I. Ghiloufi and L. El Mir, "Removal of cadmium (II) from aqueous solution by zinc oxide nanoparticles: Kinetic and thermodynamic studies," *Desalination and Water Treatment*, vol. 62, pp. 1-9, 2016.

[17] Salmani M.H, Zarei S, Ehrampoush M.H, and Danaie S., "Evaluations of pH and High Ionic Strength Solution Effect in Cadmium Removal by Zinc Oxide Nanoparticles," *J. Appl. Sci. Environ. Manage*, vol. 17, no. 4, pp. 583–593, 2013.
[18] N.U. Udeh and J.C. Agunwamba, "Removal of Heavy Metals from Aqueous Solution using Bamboo Based Activated Carbon," *International Journal of Engineering Inventions*, vol. 6, no. 2, pp. 1–12, 2017.
[19] J. Li, D. Jia, C. Li and B. Yu, "Adsorption Removal of Copper and Nickel Ions from Wastewater by Ammonia Modified Cotton Stalks," *Advanced Materials Research*, vol. 955–959, pp. 2440–2443, 2014.
[20] Y. Zhao, C. Tao, G. Xiao and H. Su, "Controlled synthesis and wastewater treatment of Ag2O/TiO2modified chitosan-based photocatalytic film," *RSC Advances*, vol. 7, no. 18, pp. 11211–11221, 2017.
[21] M. Sasitaran, S. Manroshan, C.S. Lim, B.N.K. Veni, S.K. Ong and R. Gunasunderi, "Preparation and Characterisation of Crosslinked Natural Rubber ( SMR CV 60 ) and Epoxidised Natural Rubber ( ENR-50 ) Blends," *ASEAN J. Sci. Technol. Dev.*, vol. 34, no. 2, pp. 106–118, 2016.
[22] U.M. Garusinghe, V.S. Raghuwanshi, W. Batchelor and G. Garnier, "Water Resistant Cellulose-Titanium Dioxide Composites for Photocatalysis," *Scientific Reports*, vol. 8, no. 1, pp. 1–13, 2018.
[23] C. Lin, B. Fan, J.X. Zhang, X. Yang, and H. Zhang, "Study on lead ion wastewater treatment of self-assembled film," *Desalination and Water Treatment*, vol. 57, no. 45, pp. 21627–21633, 2016.
[24] O. Rojviroon, T. Rojviroon and S. Sirivithayapakorn, "Study of COD Removal Efficiency from Synthetic Wastewater by Photocatalytic Process," *Environmental Engineering Research*, vol. 19, no. 3, pp. 255–259, 2014.

IOP Conf. Series: Materials Science and Engineering **557** (2019) 012067 doi:10.1088/1757-899X/557/1/012067

# Physical properties of plasticized PLA/HNTs bionanocomposites: effects of plasticizer type and content

**Abdulkader M. Alakrach[1], N Z Noriman [1*], Omar S Dahham[1], Awad A. Al-rashdi[2], I. Johari[3], Z M Razlan[4], Shahriman A B[4], I Zunaidi[5] and Wan Khairunizam[4]**

[1]Center of Excellence Geopolymer and Green Technology (CEGeoGTech), Faculty of Engineering Technology (FETech), Universiti Malaysia Perlis (UniMAP), 01000, P.O.Box, D/A Pejabat Pos Besar, Kangar, Perlis, Malaysia
[2]Chemistry Department, Umm Al-Qura University, Al-qunfudah University College, Al-qunfudah Center for Scientific Research (QCSR), Saudi Arabia
[3]School of Civil Engineering, Engineering Campus, Universiti Sains Malaysia, Penang, Malaysia
[4]School of Mechatronic Engineering, Universiti Malaysia Perlis (UniMAP), Pauh Putra Campus, 02600 Arau, Perlis, Malaysia
[5]Faculty of Technology, University of Sunderland, St Peter's Campus, Sunderland, SR6 0DD, United Kingdom

**Abstract.** Halloysite nanotubes (HNTs) reinforced polylactic acid (PLA) nanocomposite films were utilized for different applications, such as packaging, drug recover and other applications. The incorporation of plasticizer into polymer nanocomposites modifies some of their functional and physical properties, such as increasing flexibility, moisture sensitivity, in addition to other functional properties. However, the effects of Polyethylen glycol (PEG) and sesame oil (SO) on selected physical properties (moisture content (MC), contact angle (CA) and water vapor permeability (WVP)) of PLA)/HNTs bionanocomposite films were examined. The plasticized PLA/HNTs (5 wt % HNTs loading) bionanocomposite films were prepared using the solution casting method at room temperature. The concentrations of each plasticizer that used indivisually were (0, 10, 20 and 30 wt %). Results show that the increasing of PEG content led to increase in moisture content and water vapor permeability and decrease in contact angle of the films. On the contrary, the increasing of SO levels led to decrease in moisture content and water vapor permeability and increase in contact angle of the films. Differences in measured physical properties of films with plasticizer type and concentration may be attributed to differences in the hydrophilic and hydrophobic properties of the plasticizers. SO was the plasticizer that showed the most interested effect (low moisture content and water vapor permeability) on PLA/HNTs films for food packaging applications compared to PEG.

## 1. Introduction

Considerable attention has been given to biodegradable polymers as potential candidates for packaging due to the aforementioned challenges posed by non-degradable polymers. Biodegradable polymers are either obtained from biological resources (renewable resources) or from petroleum feed-stocks (non-renewable resources) and offer means of keeping a sustainable development of a technology that is both ecologically and economically attractive. To date, biodegradable polymers are considered as potential materials that can reduce waste disposal of packaging materials and greenhouse effect, as less carbon dioxide will be released into the atmosphere [1].

The development of environmentally-friendly and biodegradable polymeric materials has attracted extensive interest in materials science. Within biodegradable polymeric materials, Polylactic acid (PLA) is one of the best candidates particularly suitable for different application including of biomedicine, pharmaceutical and food packaging. Since it has a biodegradable composition, it is primarily used in developing bioplastic products such as disposable tableware, compost bags and loose-fill packaging. PLA is used in food-packaging industry either for hot drinks and beverages where they shape the lining of the disposable (yet compostable) cups or for whole cups of cold drinks, both should be biodegradable. A shorter polymer of lactic acid known as OLA or Oligolactic acid had been in use within the pharmaceutical industry to act as a surfactant for liquids [2].

Polymer nanocomposites (PNC) are a new class of materials which has promising potential future applications such as high-performance materials. The main feature of the polymer nanocomposite basically consists of a host polymer that is reinforced with nanosized inorganic fillers. The nanofiller will significantly impact the overall macroscopic properties of the host polymer. The interfacial interaction between organoclay and matrix as opposed to conventional composites as well as the nanometre scale pertaining to dispersed fillers both offer the reason for the enhanced properties of nanocomposites [1-3].

Halloysite nanotubes (HNTs) reinforced polymer nanocomposites are gaining extensive popularity both from academic and industrial sectors due to their improved mechanical, thermal, electrical and fire-retardant properties. The unique features of HNTs such as high length to diameter ratio (L/D), nanoscale lumens, and high-temperature resistance are some of the reasons behind its widespread usage in various application sectors. Moreover, HNTs are abundantly available from nature, environmentally friendly material and biocompatible and hence find applications in the biomedical field, alone as well as with various polymer matrices [4].

However, the brittleness problem is still one of the inherent limitations which inhibits the extensive applications of PLA/HNTs nanocomposites, this problem can be solved by addition the plasticizers [5]. the aim of this work was to investigate the effect of polyethylene glycol and Seasem oil on some physical properties of PLA/HNTs bionanocomposites films, such as moisture content (MC), contact angle (CA) and water vapor permeability (WVP).

## 2. Experimental

### *2.1. Materials*
PLA (grade 4032D) was kindly supplied by NatureWorks Ingeo™, USA. Halloysites Nanotubes (HNTs) used bto reinforce PLA was puechased from Sigma-Aldrich chemicals company. Sigma Aldrich provided (PEG) having an average molecular weight of 400g/mole in liquid state. The SO, was purchased from a local grocery store KL, Malaysia.

### *2.2. Preparation of plasticized PLA/HNTs bionanocomposites*
Plasticized polylactic acid/HNTs bionanocomposites with two different plasticizers (hydrophilic plasticizer polyethylene glycol (PEG) and hydrophobic sesame oil (SO) have been prepared via solution casting method at room temperature (23 ± 2 °C). The method used in this study has been utilized by Al-Mulla et al with some modifications [6]. Table 1 lists the amounts of UHP HNTs and PLA/plasticizer that had been used in this research.

**Table 1.** Materials compositions and designations of unplasticized and plasticized PLA/HNTs bionanocomposites by PEG and SO plasticizers.

| Plasticizer loading (wt.%) | Sample Identity | Weight of PLA (g) | Weight of plasticizer (g) | Weight of HNTs (g) |
|---|---|---|---|---|
| 0 | PLA/HNTs | 4.75 | 0 | 0.25 |
| 10 | PLA/HNTs-PEG10<br>PLA/HNTs-SO10 | 4.25 | 0.5 | 0.25 |
| 20 | PLA/HNTs-PEG20<br>PLA/HNTs-SO20 | 3.75 | 1 | 0.25 |
| 30 | PLA/HNTs-PEG30<br>PLA/HNTs-SO30 | 3.25 | 1.5 | 0.25 |

### *2.3. Film moisture content*
A digital scale was used to weigh the film samples and the weights were recorded as Wi. The samples were weighed again after drying for a whole day below 50 °C, and the second weight was recorded as Wf. Eq. 1 was applied to assess the moisture content for the whole set of film samples. This process was repeated in triplicate in order to measure an average moisture content for each film.

$$\text{Moisture content}=[(Wi-Wf)/Wi] \times 100 \quad (1)$$

Where Wi and Wf symbolize initial and final weights of film sample, respectively. This method was used by Sanyang et al., (2015) with some modifications [7].

### *2.4. Contact angle*

Contact angle (θ) measurements of the PLA/HNTs and plasticized PLA/HNTs films were carried out using a goniometer (Rame - Hart model 250 - Standard Goniometer- Tensiometer equipment) under room temperature. The medium was water and the droplet size was 4 μL. A total of 10 measurements at different places on the films were taken for each concentration.

### 2.5. *Water vapour permeability*

According to ASTM E96-95 and including a few modifications, WVP or water vapor permeability test was applied within this research. The test included using 20 g of silica gel within cylindrical cups. The open ends of the cylinders were sealed using circular films. The test included measuring the cups' weights prior to keeping them in a relative humidity chamber having a relative humidity of 75 ± 2% and a temperature of 25 °C. To get this level of humidity, the test used a saturated solution of NaCl kept at 25 °C. A periodic set of measurements was implemented for the cups until the state of equilibrium was begotten. All increments in weight were observed and noted down, the WVP was obtained through the following calculations:

$$WVP = (m \times d) / (A \times t \times P) \quad (2)$$

Where m(g) represents the increment in weight pertaining to the test cup, d(m) represents the thickness of the film, A ($m^2$) represents the area pertaining to the exposed film, t(s) represents time duration needed for permeation and P(Pa) represents the partial pressure across the films (of the water vapor). All results had been expressed in terms of g/s.m.Pa.

## 3. Results and Discussions

### 3.1. *Film moisture content (MC)*

Packaging films should maintain moisture levels within the packaged product. Therefore, the knowledge moisture content of the film is very important for food packaging applications. The results for moisture content (MC) for the PLA/HNTs bionanocomposite films plasticized by PEG and SO are shown in Figure 2.

Two opposite ways were detected for moisture content in PLA/HNTs bionanocomposites films plasticized by EPG and SO. In the case of plasticization by EPG, PLA/HNTs behaves like a hydrophilic polymer. Because of two factors. Firstly, the addition of HNTs to the PLA increases the moisture absorption of resulting nanocomposite films, probably due to the hydrophilic character of the HNTs. Furthermore, MC values have been increased by the addition of PEG due to its hydrophilic nature, these results were similar to the results of Ozkoc and Kemaloglu, which they studied the morphology, biodegradability, mechanical, and thermal Properties of bionanocomposite films based on PLA and Plasticized PLA by PEG [8].

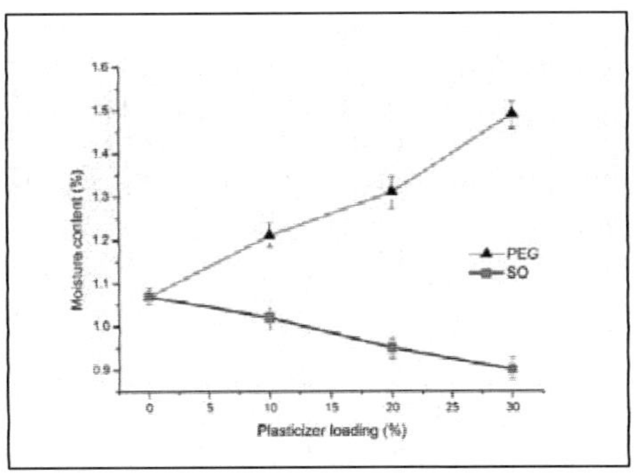


**Figure 1.** Moisture content of PLA/HNTs bionanocomposites plasticized by PEG and SO plasticizers.

On the other case, which PLA/HNTs plasticized by SO, The MC values of films were significantly lower than the corresponding to the non-plasticized PLA/HNTs film, that was due to the inclusion of sesame oil. Moreover, the decrement of MC values was in parallel of the increase of SO loading. More than one factor could affect the MC values in presence of SO as plasticizer. First, the hydrophobicity nature of sesame oil is a main reason to prevent water to penetrate inside PLA/HNTs films. The small ratio that was found could be attributed to the hydrophilicity nature of HNTs, which it is able to form hydrogen bonding with water by the side groups of O-H. in addition, the diminished availability of OH groups of HNTs and carboxylic groups of PLA to make hydrogen bonding with the carboxylate groups of fatty acids of sesame oil cause limited access of water molecules to PLA/HNTs films plasticized by SO, thus resulting in a decrease of moisture content of composite films. Similar results were observed for pistachio globulin protein and fatty acid emulsified films [9]. Partial protein–water interactions replaced with protein–olive oil interactions may account for the MC reduction occurred in the emulsified films.

*3.2. Contact Angle (CA)*
Water contact angle values are good indicators of the hydrophilic or hydrophobic character of films, being higher when hydrophilicity is lower, so thus the final state of the water drop on the film surface can be taken as an indication of surface wettability [10]. To understand the effect of the hydrophilic and hydrophobic plasticizers on the film wettability, contact angles were investigated for films with different contents of the PEG and SO and were related to the control film PLA/HNTs, as it is shown in Figure 3.

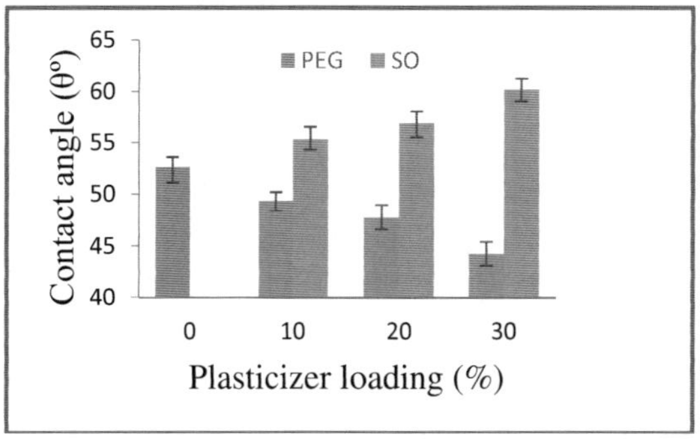


**Figure 2**. Contact angles of PLA/HNTs bionanocomposites plasticized by PEG and SO as plasticizers.

It can be seen that as the loading of PEG increases, the water contact angle gradually decreases. This could be attributed to hydrophilic nature of PEG which enhanced the surface hydrophilicity of the films. PEG enrichment at the surface observed in the morphology section earlier is in consistent with this theory. Hence, excellent hydrophilic nature led to increase in hydrophilicity of the films and reduced their water contact angle. These results were confirmed by FE-SEM results, which exhibited an increasing of the surface smoothing with the increase of PEG loading. Enhancement in hydrophilicity of bionanocomposite films due to addition of PEG has been reported in literature previously [11].

According to Figure *3*, increasing sesame oil concentration caused a slight increase of contact angle values, from approximately 52 for 0% SO to 60◦ for 30% SO, which is due to the hydrophobic nature of the added lipid. Other authors found also similar behaviors, for example Ojagh et al. (2010) reported that the incorporation of cinnamon essential oils into chitosan films resulted in decreasing hydrophilicity of the composite films, which was attributed to the loss of free functional groups (amino and hydroxyl groups) [12].

*3.3. Water vapor permeability (WVP)*
As shown in Figure 4, the WVP value of unplasticized PLA/HNTs film was $1.43 \times 10^{-10}$ g.s$^{-1}$. m$^{-1}$ Pa$^{-1}$. At the same HNTs contents, the WVP values of PLA/HNTs films plasticized by PEG increased with plasticizer content (10 – 30 wt.%). Which increased from $2.720 \times 10^{-10}$ to $5.37 \times 10^{-10}$ g. s$^{-1}$. m$^{-1}$ Pa$^{-1}$. Meanwhile, the plasticization by SO resulted in decreasing the WVP, which the increasing of SO concentration from 10 %-30% decreased the WVP from $1.03 \times 10^{-10}$ to $0.62 \times 10^{-10}$ g. s$^{-1}$. m$^{-1}$. Pa$^{-1}$. Similar findings have been reported by Yu et al., (2008) [13].

The PLA/HNTs plasticized by PEG behaves like a hydrophilic material. It has a high WVP due to the presence of hydrophilic groups in the PEG component of the blend. In addition, there was significant difference in WVP between the PLA/HNTs plasticized by PEG and PLA/HNTs plasticized by SO films. Incorporation of SO decreased WVP of the PLA/HNTs films due to the hydrophobic nature of the SO. Furthermore, the more SO contents were, the less the values of WVP, due to the moisture barrier of hydrophobic plasticizer. On other hand, with the increasing of PEG contents, the values of WVP got increase, due to the increase of free volume around the polymer [14-15].

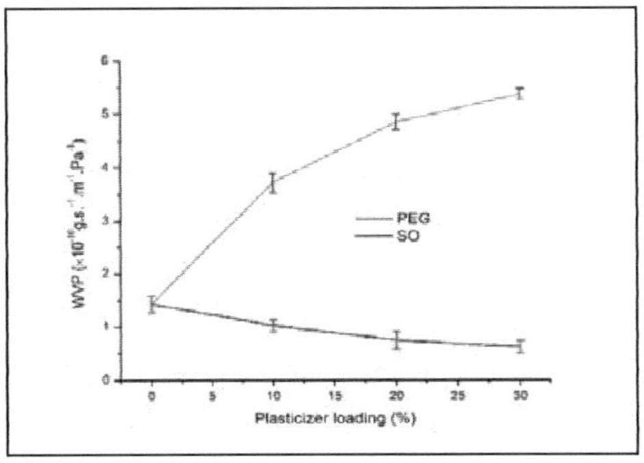


**Figure 3**. Water vapor permeability of PLA/HNTs films plasticized by PEG and SO with different plasticizer loadings.

## *4.* Summary

Without plasticizers, PLA/HNTs films are brittle with many visible cracks and not easily peeled from the casting surface. Hence, the introduction of plasticizers helped to overcome brittleness and, enhance flexibility and peelability of PLA/HNTs films. Two plasticizer types with different concentrations were exploited to investigate their effect on physical properties of PLA/HNTs films. The results demonstrated that plasticizer type and concentration influences film moisture content, wettability and water vapor permeability. Gradually, increasing the SO plasticizer concentration from 10 to 30 % decreases the moisture content and water vapor permeability, but increases the film contact angle. PEG plasticized films exhibited higher moisture content, water vapor permeability but lower contact angle than SO plasticized films. Overall, SO plasticized PLA/HNTs films showed the best performance with respect to physical properties.

## References

[1] Alakrach, A. M., N. Z. Noriman, Omar S. Dahham, R. Hamzah, Mohammed A. Alsaadi, Z. Shayfull, and SZ Syed Idrus. J. Phys. Conf. Ser, **1019**, 1 (2018) 012066. IOP Publishing.
[2] Abdallah, W., Mirzadeh, A., Tan, V., and Kamal, M. J NANOMATER, **9** (2019) 29.
[3] Osman, A. F., Alakrach, A. M., Kalo, H., Dahham, O. S., and Al Bakri Abdullah, M. M.. AMA **754** (2015) 24-28. Trans Tech Publications.
[4] Alakrach, A. M., Osman, A. F., Noriman, N. Z., Betar, B. O., and Dahham, O. S., MATEC Web Conf. **78** (2016) 01074) EDP Sciences.
[5] Mekonnen, T., Mussone, P., Khalil, H., and Bressler, D., J. Mater. Chem. A, **1** (2013), 13379-13398.
[6] Al-Mulla, E. A. J., Yunus, W. M. Z. W., Ibrahim, N. A. B., and Rahman, M. Z. A., J. Mater. Sci., **45** (2010) 1942-1946.
[7] Sanyang, M. L., Sapuan, S. M., Jawaid, M., Ishak, M. R., and Sahari, J., Polymers, **7** (2015) 1106-1124.
[8] Ozkoc, G., & Kemaloglu, S., J. Appl. Polym. Sci., **114** (2009) 2481-2487.
[9] Pereda, M., Dufresne, A., Aranguren, M. I., and Marcovich, N. E., CARBOHYD POLYM., **101** (2014) 1018-1026.
[10] Alakrach, A. M., Noriman, N. Z., Dahham, O. S., Hamzah, R., Alsaadi, M. A., Shayfull, Z., and Idrus, S. S., J. Phys. Conf. Ser, **1019**, 1 (2018) 012065. IOP Publishing.
[11] Serra, T., Planell, J. A., and Navarro, M., Acta Biomater., **9** (2013) 5521-5530.
[12] Ojagh, S. M., Rezaei, M., Razavi, S. H., and Hosseini, S. M. H., Food Chem., **122** (2010) 161-166.
[13] Yu, J. H., Wang, J. L., Wu, X., and Zhu, P. X. Starch - Stärke, **60** (2008) 257-262.
[14] Dahham, O. S., Noriman, N. Z., Sam, S. T., Omar, M. F., and Alakrach, A., AMA (2015).
[15] Alakrach, A. M., Hamzah, R., Noriman, N. Z., Dahham, O. S., Shayfull, Z., Idrus, S. S., and Sudin, S., J. Phys. Conf. Ser, **1019** (2018) 012068. IOP Publishing.

# Slow Release Fertilizer from Treated Rice Straw/Urea Beads Coated With Natural Rubber: FTIR and UV-Vis Analysis

**Aini A K[1], R Hamzah[1], N Z Noriman[1], Awad A. Al-rashdi[2], I. Johari[3], Z M Razlan[4], Shahriman A B[4], I Zunaidi[5] and Wan Khairunizam[4]**

[1] Center of Excellence Geopolymer and Green Technology (CEGeoGTech), Faculty of Engineering Technology (FETech), Universiti Malaysia Perlis (UniMAP), UniCITI Alam Campus, 02100, Perlis, Malaysia.
[2]Chemistry Department, Umm Al-Qura University, Al-qunfudah University College, Al-qunfudah Center for Scientific Research (QCSR), Saudi Arabia
[3]School of Civil Engineering, Engineering Campus, Universiti Sains Malaysia, Penang, Malaysia
[4]School of Mechatronic Engineering, Universiti Malaysia Perlis (UniMAP), Pauh Putra Campus, 02600 Arau, Perlis, Malaysia
[5]Faculty of Technology, University of Sunderland, St Peter's Campus, Sunderland, SR6 0DD, United Kingdom

**Abstract.** Slow release fertilizer (SRF) that consist of rice straw (RS) and urea coated with natural rubber (NR) and salt abbreviated as RS/Urea beads coated with NR/salt composites were prepared. RS was sieved in two different sizes which were 500 µm and 1 mm. Both of the RS sizes were treated with sodium hydroxide to expand the fibre capacity in order to store more urea. The RS/urea beads were then coated with NR/salt composite. Two concentrations of the salts prepared which were 5 % and 10 % towards the weight of the NR while 0 % salt act as a control. The role of salt is to introduce pore to the NR which allow urea to pass through its coating when soaked in the water. All SRF samples were analysed using FTIR and UV-Vis technique.

## 1. Introduction

Nowadays, food demand keeps increasing from all over the world [1] [2] and there are increment in population growth per year from all over the world. However, continuous reduction in cultivated land per capita to provide food supply and steady intensification of fertilizer application will accelerate fast increase in food production of land per hectare. Then, increasing in population growth will burden the agronomic production which need to be enhanced with same limited resources of land and water.

In order to fulfil food supply in limited working land, the sufficient and targeted delivery fertilizer is required [3][4]. Uncontrolled intensification of fertilizer application will lead to decline in land quality due to high solubility of fertilizers. Normally, about 40 – 75 % of fertilizers are leached out and affects loss towards environmental issues, plant growth and health problems. A normal fertilizer contains 80 – 90 % of phosphorus, 50 – 70 % of potassium and 40 – 70 % of nitrogen. Since this fertilizer is lost to environment and cannot be absorbed by plants, it will cause a huge wastage of fertilizers as well as economic implication [5].

All of the above issues must be take seriously and a challenge to change the agronomic practices by designing a new green technology fertilizer. Recently, the usage of slow release fertilizers (SRF) is a potential approach to save minimize environmental problems and fertilizer consumption. The conventional fertilizer releases fertilizer without any control rate will cause high solubility of fertilizer inside the water while SRF also used to release fertilizer over stimulated period of time. This can be achieved by coating with biodegradable polymer [6]. So, fertilizer releases gradually and permits the maximum uptake of the nutrient to the plant. At the same time, this minimized fertilizer losses due to leaching, volatilization, or excessive growth to the plant [7].

SRF defined as releasing active fertilizer nutrients in a control and delayed manner that synchronize with the requirement of plant. These fertilizers can be made physically by coating granules of conventional fertilizers with various material that reduce dissolution rate [5]. The release and dissolution rates of water-soluble fertilizers depend on the coating materials. Chitosan is one of the suitable coating material and been classified as carbohydrate polymer. Chitosan-coated NPK compound is one of the product that use slow release fertilizer concept. The product not only has slow release property but also can absorb water and preserve the soil moisture at the same time.

## 2. Experimental

### *2.1. Materials*

Rice straw as the raw material that used to store the urea as its structure was obtained at paddy field from Ninamaju Sdn. Bhd. at Jitra, Kedah. The natural rubber as coating material for urea fertilizer was supplied by Tiong Huat Rubber Factory, Sungai Petani, Kedah. Urea fertilizer was supplie by Petronas Fertilizer (Kedah) Sdn. Bhd. while sodium hydroxide supplied by Shama Fertilizer Chemical Company, Kulim, Kedah. Others chemical which is sodium chloride, acetone, methanol, chloroform and tetrahydrofuran were supplied by Kumpulan Saintific F. E. Sdn. Bhd., Selangor.

### *2.2. Rice Straw Preparation*

Firstly, rice straw (RS) was collected about 500 g and washed with tap water, followed by distilled water to remove impurities like dust. The clean RS was dried by using air dryer to remove moisture. RS was sieved at two different size, 500 μm and 1 mm. Hermatic plastic containers was used to put the sieved RS to avoind microbial attact like fungi before it was use in subsequent treatment. RS was undergo with alkaline treatment by using sodium hydroxide (NaOH) solution to improve it performance and wettability. It was soaked in 1 % of NaOH solution for 24 hours at room temperature. Then, treated RS was washed using distilled water several times until it become neutral at ph 7. Apart from that, centrifuge was used during washing process at 2000 rpm. Finally, RS was dried in oven to remove excessive moisture. Another sample of RS was prepared simultaneously without treatment process as a control.

### *2.3. Preparation of RS / Urea Beads*

RS / urea beads was prepared about 100 g of concentration urea being dissolved in 100 ml of water. Urea was poured into the beaker until it saturated and concentration was calculate using formula weight / volume. Both of treated and untreated RS was immersed in urea solution for 24 hours. Once the immersion process complete, RS been recoverd by centrifugation or filtration and try it in oven. The same step was repeated for untreated RS and from this process, product of RS / Urea beads will be obtain.

### *2.4. Purification of Natural Rubber*

Natural rubber (NR) was prepared by immersed 20 g of it for 2 days in 400 ml chloroform and continuously stirred for 24 hours at room temperature. Then, NR was filtered through filter cloth to separate low and high molecular weight of NR. Low molecular weight of NR will passed through the filter while high molecular NR remained at filter cloth. To coat RS / Urea beads, NR with low molecular weight was used because it is easy to coat and mold. High molecular weight of NR was more sticky and difficult to used as coating agent. Low molecular weight NR was poured into petri dish and dried for two days in fume hood and form a thin layer.

### *2.5. Preparation of NR / Salt Composite*

Two different beaker was used to mix 0.1 g sodium chloride (NaCl) dissolved in 20 ml of methanol and about 2 g of purified natural rubber dissolve back in 30 ml of chloroform. So, 12 beakers was prepared for different parameter. NR / Salt Composites was produced when both of the prepared solvent was completely mixed. Then, another NR without salt as control was prepared.

*2.6. Preparation of RS/Urea Beads coated with NR / Salt Composites*
RS / Urea beads was poured into petri dish with NR / Salt Composite. Next, 2 g of RS / Urea beads was poured into NR/ salt composite which is known as solvent casting process where NR / Salt Composite coat RS / Urea beads. Ratio for NR : RS was 1 : 1 used and the product was dried in fume cupboard for 2 days until all parts were fully dried. Then, continue dried in vacuum oven for 48 hours at 50 ° C.

*2.7. Release Study*
All product was soaked in known volume of $H_2O$. 10 g of dry product with different salt loading was added into conical flask that contain 250 ml distilled water. This experiment was carried out for 40 days by take out 3 ml of product solution and 3 ml of distilled water was refill again to maintain constant amount of solvent. Then, product was tested using UV-vis in term of time period of release.

## 3. Results And Discussion

*3.1. Charcterization of Untreated Rice Straw and Treated Rice Straw*
The size of RS does not influence its alkaline pre-treatment because both sizes undergoes similar breakage of lignin and provides space for use in the fibrous structure of RS [8]. FTIR analysis in Figure 1shows changes in treated and untreated RS when peaks range 3000 - 3500 $cm^{-1}$ from the FTIR expressed to hydroxyl group (— OH). For untreated RS, both size (c) and (d), the peak 3400.22 $cm^{-1}$ and 3400.18 $cm^{-1}$ were broad as compared to treated RS for both size (a) and (b). The peaks for treated RS seems to be more stretched. This shows that there was reduction in hydroxyl group when RS being treated because the hydroxyl group involve in hydrogen bonding [9].

Apart from that, the band at 2800 - 3000 $cm^{-1}$ is attributed to C — H stretching vibrations of $CH_2$ and $CH_3$. For untreated RS, C — H stretching vibration was broader compared to treated RS dur to the characteristic bands of the C — H usually present in the hemicellulose and cellulose [10]. So, reduction of C — H group can be seen for treated RS due to NaOH treatment [11]. For untreated RS, the peaks at 1320 $cm^{-1}$ that corresponds to lignin can be seen clearly as compared to treated RS which almost disappeared. These results proved that alkali pre-treatment can removes the lignin content. The range of peak 1000 - 1100 $cm^{-1}$ linked with C — O, C = O, C — C — O band stretching present in untreated RS as well as the treated RS indicates the band of cellulose [8].

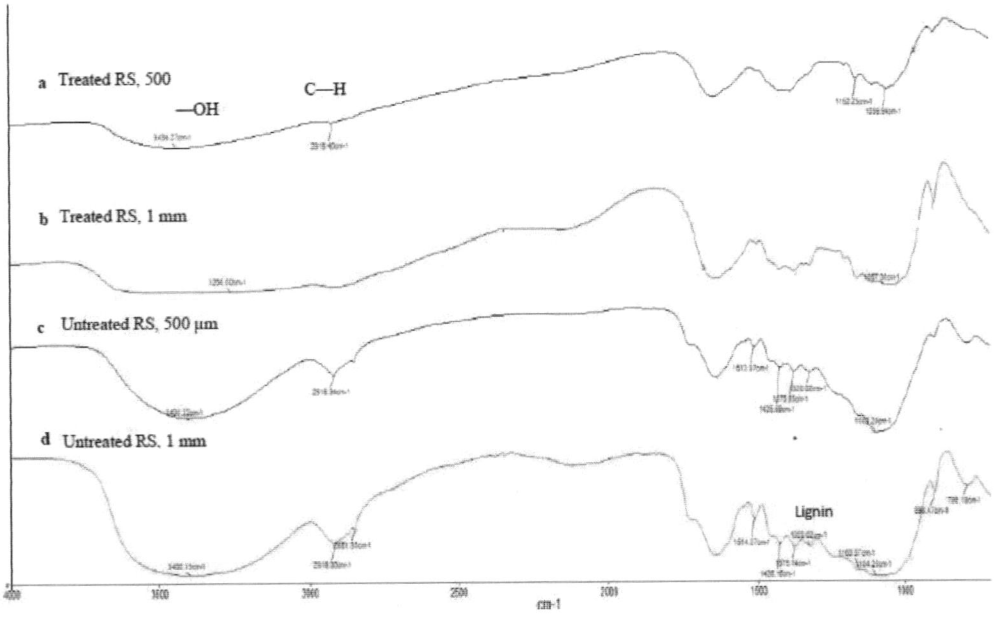


**Figure 1.** FTIR scpectra of treated and untreated Rice Straw with size for treated Rice Straw (a) 500µm (b) 1 mm and untreated Rice Straw with size (c) 500 µm (d) 1 mm

### *3.2. Charcterization of Treated and Untreated RS/Urea Beads*

The FTIR result in Figure 2 illustrated about treated and untreated RS/Urea Beads. Treated RS with size (a) 1 mm and (b) 500µm, hydroxyl group can be seen at range 3000 - 3500 $cm^{-1}$ with stretching vibration while for untreated with size (c) 1 mm and (d) 500µm, broader peak can be observed due to the treatment with NaOH which the hydroxyl group form hydrogen bonds. For treated fibre, pores were created by dissolution of hemicellulose and lignin content which will provide surface area for the urea to retain in the fibrous structure of RS [8]. Therefore, the peaks of C = O (amide I) can be observed at 1634 $cm^{-1}$ and 1637 $cm^{-1}$ for both sizes respectively. However, urea peaks for untreated RS for both size can be observed but the intensity of peaks was lower than treated RS which were 1575 $cm^{-1}$ and 1571 $cm^{-1}$ respectively. This shows that treated urea can store more urea compared untreated RS especially for size 500 µm compared 1 mm. Other researcher reported, when diameter of fibre being reduced, there will be increase in surface area. Surface area of fibres will clearly have an effect on the ability of particle to diffuse and adsorb [12].

IOP Conf. Series: Materials Science and Engineering **557** (2019) 012068 doi:10.1088/1757-899X/557/1/012068

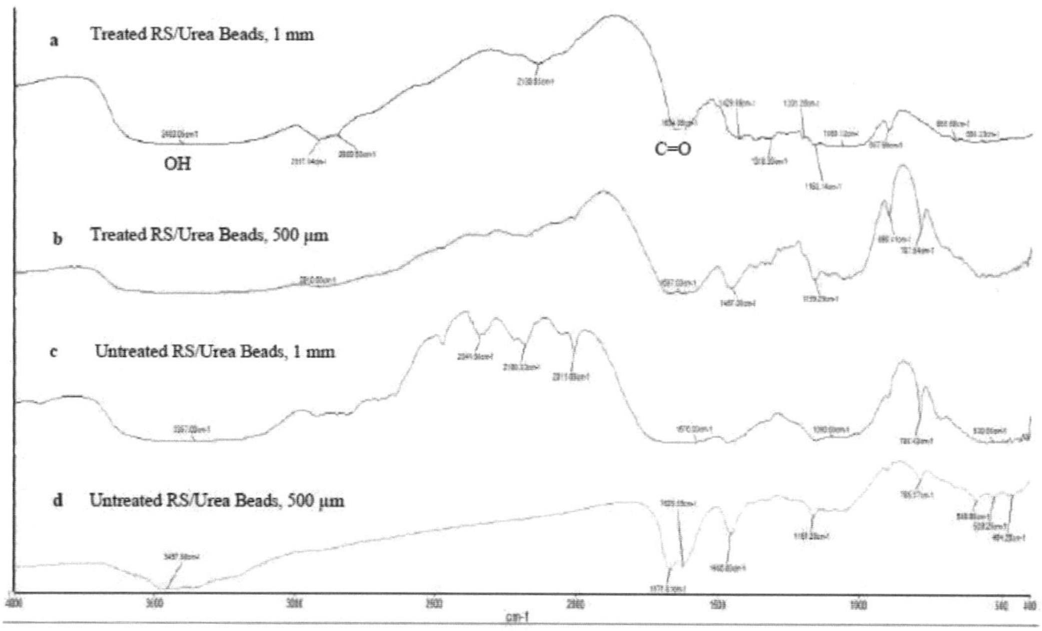


**Figure 2.** FTIR result for treated and untreated / urea beads for both sizes, treated RS (a) 1 mm and (b) 500µm and untreated RS (c) 1 mm and (d) 500µm

*3.3. Release Study*

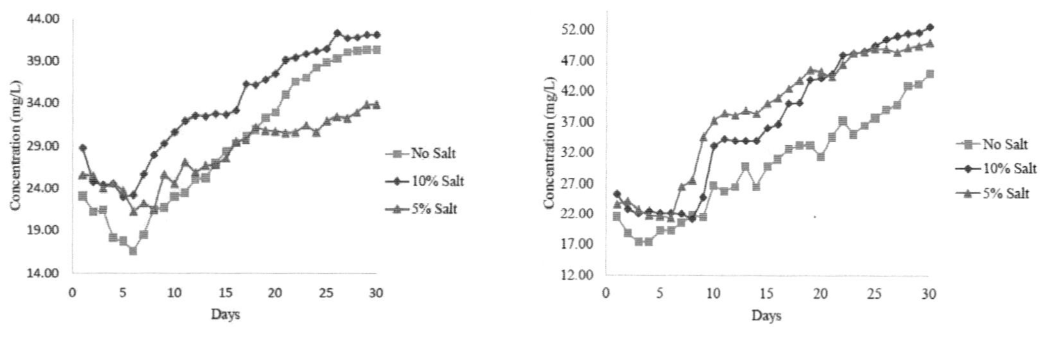


(a) (b)

**Figure 3.** (a) Concentration of urea release for untreated RS with size 1 mm and (b) Concentration of urea release for untreated RS with size 500µm

Figure 3 shows the result of slow release for two different size which is Figure 3a is the concentration of urea release for untreated RS with size 1 mm and Figure 3b Concentration of urea release for untreated RS with size 500 µm. Based on trend shows in Figure 3a, 10 % salt has higher concentration of urea release compared to 5 % salt. This is due to 10 % salt introduce more pore to the coating surface thus allow the urea from the core to diffuse out. This is due to the release path of the urea through the coating is totally depend on natural voids and holes of the polymer chain in NR. While Figure 3b represent the untreated RS with size 500 µm had similar trend with untreated RS with size 1 mm. 10 % salt loadings had higher concentration of urea release as compared to 0 % and 5 %.

Figure 4 shows the concentration of urea release for treated RS (a) size 1 mm and (b) size 500 μm. It can be observed that 10 % salt release more urea as compared to 5% salt. For untreated RS, the concentration of urea release from day 1 until 6 decreased whereas for treated the concentration of urea increased constantly because treated RS undergoes treatment with NaOH where the lignin of fibrous structure RS was broken thus allowing urea to be occupied.at the expand structure. For Figure 4b, 10 % salt loading has higher concentration of urea release than 5 %. So, treated RS can hold up more urea compared to untreated as the releasing rate of urea increases constantly and does not stop at day 30.

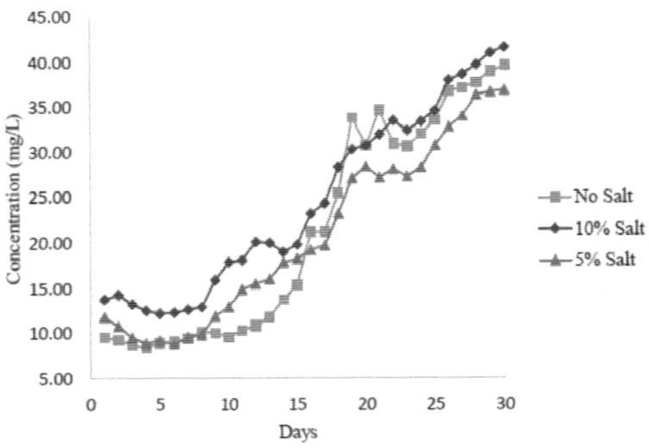


(a)

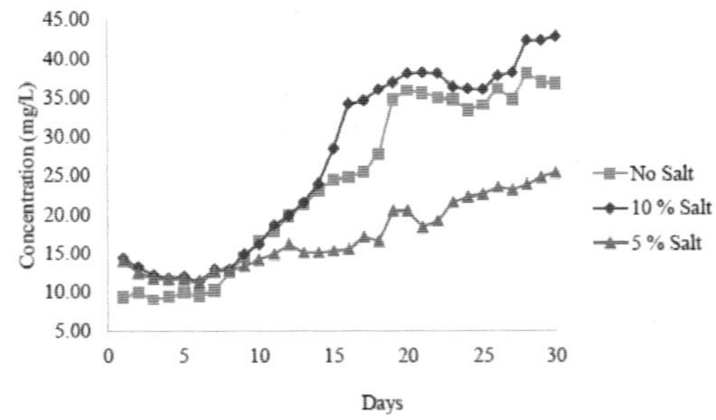


(b)

**Figure 4.** (a) Concentration of urea release for treated RS with size 1 mm and (b) Concentration of urea release for treated RS with size 500 μm

## 4. Conclusion

In conclusion, RS / Urea result shows that treated RS can retain more urea compared with untreated RS. However in term of sizes, 500 μm is more preferable due to 500 μm occupies larger surface area than 1mm. Then, RS / Urea beads was successfully coated by NR / salt composite and percentage of 10 % give broader peak for untreated RS for both sizes whereas for treated RS, 5 % salt give broader peak. The performance of the finished product which is RS / Urea beads coated with NR / salt composites determined by using UV-Vis analysis shows untreated RS with size 500 μm has higher concentration of urea release as compared to 1 mm for the two salt loadings.

## 5. References

[1] Can We Meet the World's Growing Demand for Food? (n.d.). Retrieved August 31, 2018, from https://www.agmrc.org/renewable-energy/renewable-energy/can-we-meet-the-worlds-growing-demand-for-food

[2] Schierhorn, M. E. (2016, April 07). Global Demand for Food Is Rising. Can We Meet It? Retrieved October 31, 2018, from https://hbr.org/2016/04/global-demand-for-food-is-rising-can-we-meet-it

[3] Vedhika, S., Mishra, V. N., and Purnendra, K. S. (2017). Soil test based fertilizer recommendation for targeted yield of crops: A review, *International Journal of Chemical Studies 5*(5): 1298-1303.

[4] Liu, X., Wang, H., Zhou, J., Hu, F., Zhu, D., Chen, Z., & Liu, Y. (2016). Effect of N Fertilization Pattern on Rice Yield, N Use Efficiency and Fertilizer–N Fate in the Yangtze River Basin, China. *Plos One,11*(11).

[5] Wu, L., & Liu, M. (2008). Preparation and properties of chitosan-coated NPK compound fertilizer with controlled-release and water-retention. *Carbohydrate Polymers,72*(2), 240-247.

[6] Lateef, A., Nazir, R., Jamil, N., Alam, S., Shah, R., Khan, M. N., & Saleem, M. (2016). Synthesis and characterization of zeolite based nano–composite: An environment friendly slow release

fertilizer. *Microporous and Mesoporous Materials, 232*, 174-183.

[7] Azeem, B., KuShaari, K., Man, Z. B., Basit, A., & Thanh, T. H. (2014). Review on materials & methods to produce controlled release coated urea fertilizer. *Journal of Controlled Release, 181*, 11-21.

[8] Asghar, U., Irfan, M., Iram, M., Huma, Z., Nelofer, R., Nadeem, M., & Syed, Q. (2015). Effect of alkaline pretreatment on delignification of wheat straw. *Natural product research, 29*(2), 125-131.

[9] Oushabi, A., Sair, S., Hassani, F. O., Abboud, Y., Tanane, O., & El Bouari, A. (2017). The effect of alkali treatment on mechanical, morphological and thermal properties of date palm fibers (DPFs): Study of the interface of DPF–Polyurethane composite. *South African Journal of Chemical Engineering, 23*, 116-123.

[10] Sheng, J., Ji, D., Yu, F., Cui, L., Zeng, Q., Ai, N., & Ji, J. (2014). Influence of chemical treatment on rice straw pyrolysis by TG-FTIR. *IERI Procedia, 8*, 30-34.

[11] Boonterm, M., Sunyadeth, S., Dedpakdee, S., Athichalinthorn, P., Patcharaphun, S., Mungkung, R., & Techapiesancharoenkij, R. (2016). Characterization and comparison of cellulose fiber extraction from rice straw by chemical treatment and thermal steam explosion. *Journal of Cleaner Production, 134, Part B*, 592-599.

[12] Eichhorn, S., & Sampson, W. (2010). Relationships between specific surface area and pore size in electrospun polymer fibre networks. *Journal of The Royal Society Interface, 7*(45), 641-649.

# The Influences of Different Bamboo Filler Loading on Tensile Properties and Impact Strength of RHDPE/BF Composites

**Aini Asifa Ahmad Kamal[1], N.Z. Noriman[1], S.T. Sam[2], Awad A. Al-rashdi[3], I. Johari[4], Z M Razlan[5], Shahriman A B[5], I Zunaidi[6] and Wan Khairunizam[5]**

[1]Center of Excellence Geopolymer and Green Technology (CEGeoGTech), Faculty of Engineering Technology (FETech), Universiti Malaysia Perlis (UniMAP), UniCITI Alam Campus, 02100, Perlis, Malaysia.
[2]School of Bioprocess Engineering, Universiti Malaysia Perlis, Kompleks Pengajian Jejawi 3, 02600 Arau, Perlis, Malaysia
[3]Chemistry Department, Umm Al-Qura University, Al-qunfudah University College, Al-qunfudah Center for Scientific Research (QCSR), Saudi Arabia
[4]School of Civil Engineering, Engineering Campus, Universiti Sains Malaysia, Penang, Malaysia
[5]School of Mechatronic Engineering, Universiti Malaysia Perlis (UniMAP), Pauh Putra Campus, 02600 Arau, Perlis, Malaysia
[6]Faculty of Technology, University of Sunderland, St Peter's Campus, Sunderland, SR6 0DD, United Kingdom

**Abstract.** Recycled high-density polyethylene and high-density polyethylene reinforced with bamboo filler were compounded with twin-screw extruder and injection molding process. The main objective of this study is to investigate the effect of different loading of bamboo filler reinforced with recycled high-density polyethylene composites via injection molding. The fillers loading were reinforced with plastic for different loading which is 0 wt.%, 5 wt.%, 10 wt.%, 15 wt.% and 20 wt.% for two different matrixes, which is recycled high-density polyethylene and high-density polyethylene. Mechanical measurements will show the presence of different loading filler will insignificantly effect in the composites tensile strength and also impact. The result of the mechanical analysis carried out showed that the presence of natural filler in composites will improve the properties of the material.

## 1. Introduction

In recent years, there has been increasing attention to the application of biodegradable composites reinforced with natural fibers with the view towards economic incentives and environmental concerns [1] [2]. The advantages of plant-based fibers over traditional synthetic fillers are their lightweight, low cost, specific mechanical properties and excellent biodegradable properties. There is evidence that shows the presence of natural fiber in polymeric materials can enhance the performance of virgin materials [3].

IOP Conf. Series: Materials Science and Engineering **557** (2019) 012069 doi:10.1088/1757-899X/557/1/012069

Among many types of such materials, the bamboo fiber itself is a unidirectional reinforcing composite consisting of long and parallel cellulose fibers embedded in an amorphous matrix of lignin and hemicellulose [4], and then is recognized as one of the most attractive reinforcement fillers because of its excellent inherent properties [5] [6]. Furthermore, the inexhaustible supply and natural abundance of bamboo make it an excellent candidate for developing natural composites. Thus, the use of bamboo fibers as a reinforcing material in structural composites has gained popularity in the building and automotive industries [7] [8].

Plastics injection molding filled the need for quality of the product with the latest developments which made the industries to find for new methods for manufacturing and production of components with low cost. It is probably the most widely used cyclic process for manufacturing derived of polymers including fiber reinforced thermoplastics, elastomers, thermoplastics materials and thermosetting plastics. This processing technology is possible to fabricate excellent surface finish of complex geometric components with good dimensional and specification properties. [9] [10] [11].

In these research, recycled high-density polyethylene (RHDPE) and virgin high-density polyethylene (RHDPE) will act as matrix and bamboo fiber particle as the reinforcing filler in order to prepare a particle-reinforced composite. It is used to determine the mechanical properties of polymer composite according to filler content and to examine the possibility of using waste materials as reinforcing filler.

## 2. Experimental Procedures

### *2.1. Matrix*

A matrix is a binder material that is used to hold fibers in position and transfer external loads to internal reinforcements [12]. The recycled high-density polyethylene (RHDPE) as the matrix will be supplied by Zarm Scientific and Supplies Sdn. Bhd. in solid pallet form with the melt flow rate of 0.236 g / 10 min at 200 ° C, 5 kg load.

### *2.2. Fiber Preparation*

The bamboo will be obtained abundantly around Perlis, Malaysia area. In this study, automatic hammer continuous grinder (HK-08B) will be used to reduce the particle size of bamboo. Then, the bamboo flour will be sieved based on the required size using a vibratory sieve shaker AS 200 controls before undergoing the particle size analyzer (Malvern Sciro 2000 Mastersizer) for the particle size confirmation.

### *2.3. Composites Preparation*

Different contents of bamboo flour will be used in this study according to the formulation in Table 1. The composites were prepared by compounding using the twin screw extruder with a barrel temperature of 170 ° C. RHDPE mixed with bamboo flour with different ratio will be loaded into the feeder and run through the twin screw extruder. Then, the pallets of composites will be produced. The composites pallet will undergo with injection molding process for the final product. The samples of the final product will be cut into shape according to the requirements of test conduct in this study.

**Table 1** The filler loading of the composites

| Parameters | Materials | Formulation |
|---|---|---|
| Filler loading | RHDPE | RHDPE |
| | RHDPE / 5 wt.% BF | 95 wt.% RHDPE / 5 wt.% BF |
| | RHDPE / 10 wt.% BF | 90 wt.% RHDPE / 10 wt.% BF |
| | RHDPE / 15 wt.% BF | 85 wt.% RHDPE / 15 wt.% BF |
| | RHDPE / 20 wt.% BF | 80 wt.% RHDPE / 20 wt.% BF |
| | HDPE | HDPE |
| | HDPE / 5 wt.% BF | 95 wt.% HDPE / 5 wt.% BF |
| | HDPE / 10 wt.% BF | 90 wt.% HDPE / 10 wt.% BF |
| | HDPE / 15 wt.% BF | 85 wt.% HDPE / 15 wt.% BF |
| | HDPE / 20 wt.% BF | 80 wt.% HDPE / 20 wt.% BF |

### *2.4. Tensile Test*
Tensile testing was performed using Axial-Torsion Universal Testing Machine (INSTRON-5982) at room temperature according to ASTM D638. In this study, all tensile specimens will be pulled under a constant crosshead speed of 50 mm/min and 50 mm gauge length. Before the test, the thickness of specimens will be measured using Vernier caliper. Five specimens will be conducted and the tensile strength values will be averaged to obtain a mean value.

### *2.5. Impact Test*
Charpy impact test was conducted according to ASTM 790 at room temperature by using Compact Charpy Impact Tester (Model MAT 23). From this test, the fracture energy (Joule) will be determined. Five specimens will be conducted and the impact energy values will be averaged to obtain a mean value.

## 3. Results and Discussion

### *3.1. Tensile Properties*
Figure 1 shows the effect of fiber loading on the tensile strength of RHDPE / BF composites and HDPE / BF composites. The results obtained from tensile strength indicates that increasing of fiber loading will increase the tensile strength from 5 wt.% until 10 wt.% and drop when 15 wt.% and 20 wt.% of fiber loading used. 10 wt.% of filler loading for both RHDPE and HDPE polymer give the highest result which is RHDPE / BF, 24.749 MPa while HDPE / BF, 23.944 MPa.

A great interface between RHDPE and HDPE matrix with BF fiber will improve tensile strength due to encourages of perfect stress transfer. A great tensile strength mostly depends on effective and uniform stress distribution [13]. Tensile strength starts to decrease at higher BF fiber loading respectively for both of different matrix used. The BF particles tend to agglomerate which will lead to insufficient wetting of filler by the small amount of matrix. So, fiber will give a negative effect which limits the tensile strength without a great fiber-matrix interfacial [14]. Apart from that, the decreasing of tensile strength with 15 wt.% and 20 wt.% BF fiber was possibly due to the presence of voids getting trapped in the composite during processing. So, a specific technique while preparation process of composites is required when dealing with high filler loading composite to make sure it has an excellent dispersion of fibers and great interfacial bonding [15].

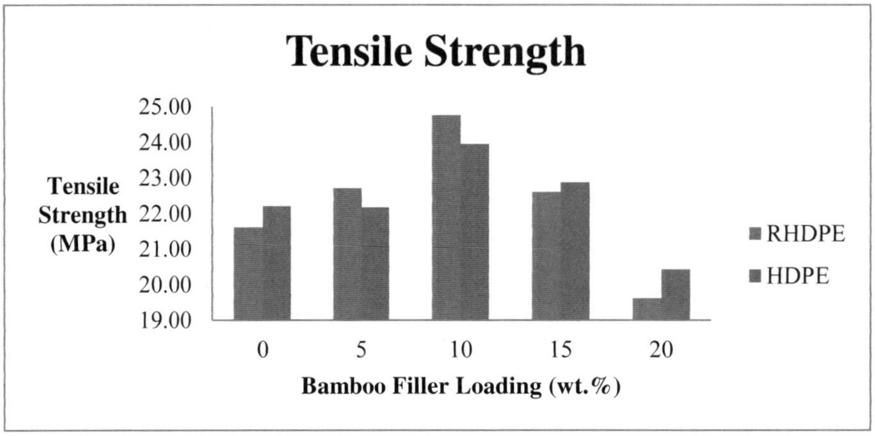


**Figure 1.** Effect of different filler loading on tensile strength.

Figure 2 shows the elongation at break of RHDPE / BF composites and HDPE / BF composites with different fiber loading. The elongation of break results decreased with increasing amount of BF fibers regards of the brittle nature by BF in the composites that indicated a decrease polymer matrix ductility with the presence of BF fibers [14] [16]. Thus, it is proved that a strong but brittle RHDPE / BF and HDPE / BF wood polymer composites are produced due to lower values of elongation at break which is the cracks travel through weaker interfacial regions [15] [17].

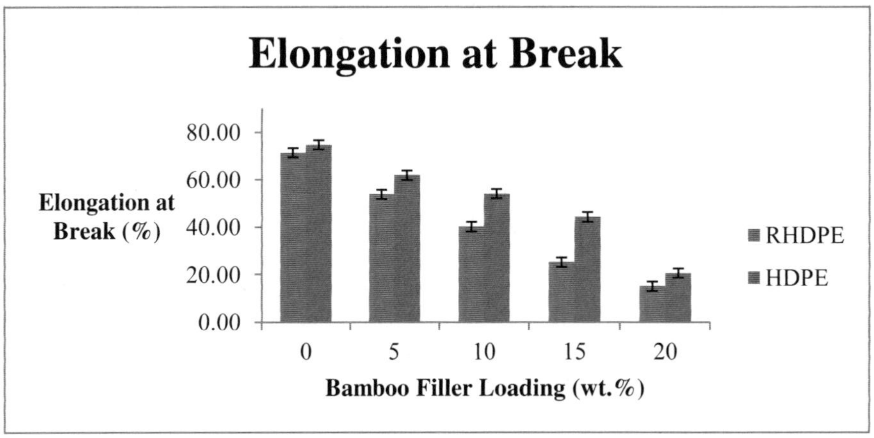


**Figure 2.** Effect of different filler loading on elongation at break.

The Young's modulus of RHDPE / BF composites and HDPE / BF composites with different fiber loading is shown in Figure 3. The results showed that increasing of BF fiber will give positive influence of Young's modulus for composites, compared with the pure polymer, RHDPE and HDPE. Modulus of composites was enhanced with the addition of BF fiber due to increasing restrictions on the movement of polymer molecules. These results indicate that higher fiber loadings will increase Young's modulus by the intrinsic property of high stiffness of BF [17]. In this study, the values of Young's modulus decrease at 5 wt.% of BF filler and gradually increase with highest values obtained at 20 % of BF for both of the RHDPE / BF and HDPE / BF composites.

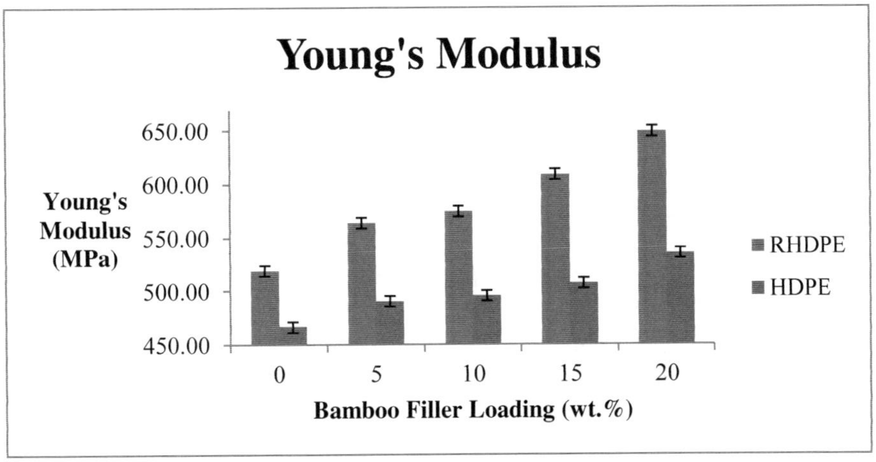


**Figure 3.** Effect of different filler loading on Young's modulus.

### 3.2. *Impact Properties*

Impact properties will determine the ability of a material to withstand an impact load, which ability depends on materials toughness [18]. The effect of increasing filler loading on the notched impact of RHDPE / BF and HDPE / BF composites is shown in Figure 4. The highest impact strength for both of material at 10 wt.% filler content and then it keeps decreasing with increasing of filler content. The lowest impact strength is 20 wt.% of filler loading for both composites. The strength of composites increases with filler loading up to 10 wt.% due to the ability of composite to absorb energy favorable entanglement of filler and matrix [19]. This strength slumps with excessive increasing of filler loading from 15 wt.% to 20 wt.% due to the reduction of elasticity of material when filler was added. So, it will reduce the deformability and ability of the matrix to absorb energy which reducing the toughness [20].

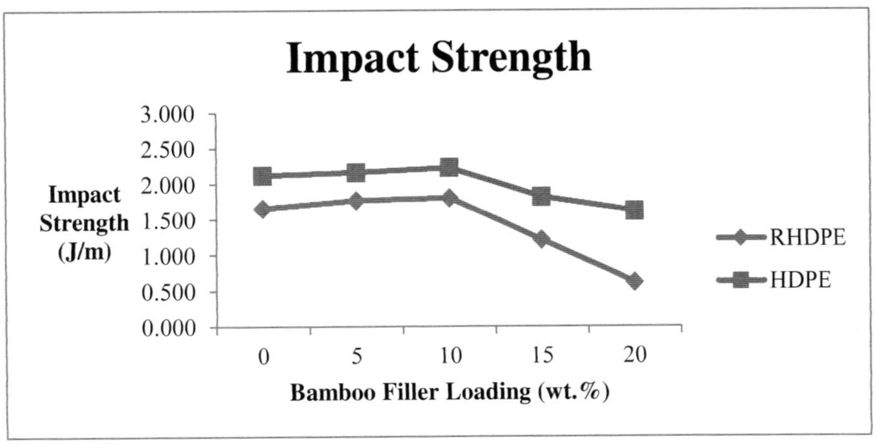


**Figure 4.** Effect of different filler loading on impact strength.

## 4. Conclusions

In conclusion, it has revealed that bamboo act as natural filler to produce polymer composites and create new composites materials with great properties and save cost with the ability to reduce environmental problems. The additional bamboo filler in composite increase the tensile strength and tensile Young's modulus. The tensile elongation at break declines with increasing of filler loading. The impact strength also shows improvement with the optimum result at 10 wt.% of filler loading.

## References

[1] Joshi, S. V., Drzal, L. T., Mohanty, A. K., & Arora, S. (2004). Are Natural Fiber Composites

Environmentally Superior to Glass Fiber Reinforced Composites? *Compos. A, 35,* 371-376.

[2] Abdul Khalil, H. P. S., Bhat, A. H., & Ireana Yusra, A. F. (2012). Green Composites from Sustainable Cellulose Nanofibrails: A Review. *Carbohyd. Polym, 87,* 963-979.

[3] Manshor, M. R., Anuar, H., Nur Aimi, M. N., Ahmad Fitrie, M. I., Wan Nazri, W. B., Sapuan, S. M., El-Shekeil, Y. A., & Wahit, M. U. (2014). Mechanical, Thermal and Morphological Properties of Durian Skin Fibre Reinforced PLA Biocomposites. *Mater. Des*, 59, 279–286.

[4] Rao, K. M. M., & Rao, K. M. (2007). Extraction and Tensile Properties of Natural Fibers: Vakka, Date and Bamboo. *Compos. Struct. 77,* 288–295.

[5] Ray, A. K., Das, S. K., Mondal, S., & Ramachandrarao, P. (2004). Microstructural Characterization of Bamboo. *J. Mater. Sci., 39,* 1055–1060.

[6] Wang, F., Shao, J. X., Keer, L. M., Li, L., & Zhang, J. Q. (2015). The Effect of Elementary Fibre Variability on Bamboo Fibre Strength. *Mater. Des., 75,*136–142.

[7] Lo, T. Y., Cui, H. Z., Tang, P. W. C., & Leung, H. C. (2008). Strength Analysis of Bamboo by Microscopic Investigation of Bamboo Fibre. *Construct. Build. Mater.*, 22, 1532–1535.

[8] Abdul Khalil, H. P. S., Bhat, I. U. H., Jawaid, M., Zaidon, A., Hermanwan, D., & Hadi, Y. S. (2012). Bamboo Fibre Reinforced Biocomposites: A review. *Mater. Des., 42,* 353–368.

[9] Kavade, M. V. (2012). Parameter Optimization of Injection Molding of Polypropylene by using Taguchi Methodology. *IOSR Journal of Mechanical and Civil Engineering,4*(4), 49-58.

[10] Radzi, M. K., Muhamad, N., Sulong, A. B., & Razak, Z. (2017). Optimizing Injection Parameters of Kenaf Filler Polypropylene Composite by Taguchi Method. *Materials Science Forum, 894*, 81-84.

[11] Reddy, B. S., Arun, K., & Dr. M. Chandra Sekhara Reddy. (2017). Optimization of Process Parameters on Injection Moulding Machine with LLDPE Reinforced Flyash using Taguchi. *Nctet-2K17.*

[12] Kabir, M., Wang, H., Lau, K., & Cardona, F. (2012). Chemical treatments on plant-based natural fibre reinforced polymer composites: An overview. *Composites Part B: Engineering, 43*(7), 2883-2892.

[13] El-Shekeil, Y., Sapuan, S., Abdan, K., & Zainudin, E. (2012). Influence of fiber content on the mechanical and thermal properties of Kenaf fiber reinforced thermoplastic polyurethane composites. *Materials & Design, 40*, 299-303.

[14] Nourbakhsh, A., Baghlani, F. F., & Ashori, A. (2011). Nano-SiO2 filled rice husk/polypropylene composites: Physico-mechanical properties. *Industrial Crops and Products, 33*(1), 183-187.

[15] Chen, R. S., Ahmad, S., Ghani, M. H., & Salleh, M. N. (2014). Optimization of high filler loading on tensile properties of recycled HDPE/PET blends filled with rice husk. *AIP Publishing LLC, 1416*, 46-51.

[16] Choudhury, A., Kumar, S., & Adhikari, B. (2007). Recycled milk pouch and virgin low-density polyethylene/linear low-density polyethylene based coir fiber composites. *Journal of Applied Polymer Science, 106*(2), 775-785.

[17] Ayswarya, E., Francis, K. V., Renju, V., & Thachil, E. T. (2012). Rice husk ash – A valuable reinforcement for high density polyethylene. *Materials & Design, 41*, 1-7.

[18] Siregar, J., Sapuan, S. M., Rahman, M., & Zaman, H. (2010). The Effect of Alkali Treatment on the Mechanical Properties of Short Pineapple Leaf Fibre (PALF) Reinforced High Impact Polystyrene (HIPS) Composites. *Journal of Food, Agriculture & Environment, 8*(2), 1103-1108.

[19] Elanchezhian, C., Vijaya Ramnaht, B., Ramakrishnan, G., Rajendrakumar, M., Naveenkumar., V. & Saravanakumar, M. K. (2018). Review on mechanical properties of natural fiber composites, *Materials Today: Proceedings 5,* 1785–1790.

[20] Mohammed R. (2013). Study of Some Mechanical Properties of Unsaturated Polyester with the Seed Shells of Sunflower and Water-Melon. *Journal of Babylon University: Engineering Sciences, 4*(21), 1270-1277.

# Analysis of Spatial Distribution of Soil Moisture Content for Different Soil Layers in Mango Greenhouse

**F A Azizan[1], F N Mohamad Zalani[1], A Nagarajan[1], A A Aznan[1] and R Ruslan[1]**

[1]School of Bioprosess Engineering, Universiti Malaysia Perlis, 02600 Arau, Perlis, Malaysia.

Corresponding author: fathin@unimap.edu.my

**Abstract.** Water content in soil or generally known as soil moisture plays an important role in plant growth. In moving towards a better farm management in precision agriculture concept, it is vital to maintain the uniform growth of plants by preserving the same amount of moisture in the soil. In this study, spatial distribution analysis of soil moisture content was conducted to observe the variability of topsoil and subsoil. A total of 80 soil samples were collected randomly and analysed using the gravimetric method. The data were used to produce soil moisture distribution maps using ArcGIS software. The results show subsoil layer has higher moisture content compared to the topsoil layer with a mean value of 26.17% and 22.84%, respectively. Even though the maps revealed two different patterns of moisture content, both layers fall in high moisture class (21.01% – 28%) with less variation. Thus, this indicates both layers have sufficient amount of moisture content for mango growth and irrigation water supply was adequate.

## 1. Introduction

Soil is one of natural resources that can be found on earth's surface. The soil generally contain mineral components, air, water, and organic matter. Soil plays an important role in supporting growth of plants by holding roots firmly and supplying mineral and water. Therefore, soil should be able to hold right amount of water as per plant water requirement. Water content or moisture content in soil normally differ according to soil type, weather condition and water supply.

In a greenhouse, irrigation system installed is a sole water supply. The irrigation systems can be in a type of sprinkler, micro, drip and others. Among these types, drip irrigation is the most suitable and best irrigation system for mango plant grown in greenhouse because the systems greatly reduces water used. The water supply through this system has a direct relationship to soil moisture content. As water supply is important for plant growth, it is necessary to ensure the soil moisture is at sufficient amount needed by plant.

In precision farming concept, it is important to maintain the uniformity of plant growth for better management practice. Therefore, it is an urge in reducing any variability that might occur. One of the most important input parameter is soil. Several researchers have agreed that the best way to access soil variation is through soil mapping. Brevik et al. [1] stated that, soil mapping is the key in advancement of knowledge of soil.

Even though, many soil mapping method was produced and used since 19th century, nonetheless it still could not meet additional requirement needed in soil mapping. Starting 21st century, many researcher has started to added other elements in soil mapping in parallel with geospatial revolution to improve the accuracies in presenting the data. This includes geographic positioning systems, geographic information systems and recently remote sensing ([2]-[4]). All techniques have its function to collect related data, analysis the data and predict information related to soils to create the soil mapping.

Among all types of soil mapping, soil moisture and soil nutrient mapping are very decisive. Lei Yang et al., [5] in their study stated that, soil moisture and its spatial variation are closely related with plant growth. According to Zhang C. et al. [6] stated that the based on spatial distribution map of soil moisture is useful for managing the precision irrigation in time, analyse the water requirement of plant and predict the variation of soil.

As water supply is important for plant growth, this study aim to map the soil moisture variation at two soil layers; topsoil and subsoil. Mapping these two layers may indicate the adequateness of water supply as subsoil layer is where the active root zone is located. In previous research [7], it is said that soil moisture present in topsoil horizon layer has shown maximum variation compared to subsoil horizon layer in open space area. Despite the fact that there is quite similar research that has been done in this area, however, none has been investigated on the variation of this parameter in high density planting system greenhouse with less environmentally depended.

## 2. Materials and methods

The study was carried out in one of 50 greenhouses at Sustainable Agrotechnology Institute (InSAT), Universiti Malaysia Perlis (UniMAP), Sg. Chuchuh, Padang Besar, Perlis. The mean temperature of the area is about 32°C at daytime and the relative humidity is about 80.7%. The average elevation of the area is about 53m above the level sea level. The selected greenhouse (GH19) was planted with a total of 212 Harumanis mango. The greenhouse has a dimension 24.2 m wide and 84 m length.

### *2.1. Soil Sampling*

40 plant locations were selected using random grid sampling technique for soil sample collection. The samples were taken after an hour of irrigation. The greenhouse was irrigated on once-daily basis as the plants are at flowering stage. Overall, there are 80 samples, with 40 samples collected for topsoil (at depth 0-15cm) and another 40 samples for subsoil (at depth 15-30cm). Each soil sample weight around 50g.

Soil moisture content of each sample was measured using gravimetric method. This method uses hot oven to dry the sample. This standard method is commonly and widely used because it gives high accuracy, more simple procedure, reliability results. The moisture content was calculated by the following equation:

$$\text{Moisture content (\%)} = \frac{W2-W3}{W3-W1}\ X\ 100 \qquad (1)$$

Where W1 is weight of empty can (g), W2 is weight of moist soil + can (g) and W3 is weight of dried soil + can (g).

### *2.2. Interpolation and Mapping Method*

To create a mapping in any of geographic information system (GIS) tool, Global Positioning System (GPS) data were collected at plant sample locations. A data of latitude, longitude and elevation were gathered and received using Topcon GR-5 instrument. The GPS and soil information data were imported in ArcGIS software to map the soil moisture. An interpolation method Inverse Distance Weighted (IDW) was adapted as it is simple and accurate method since known data were quite close to one another. This method calculates the unknown values based on a power of the Euclidean distance between the measured location and unknown location [8].

## 3. Results and discussion

### *3.1 Soil Moisture Data*

Table 1 shows basic statistical value of soil moisture data for topsoil and subsoil layers. The soil moisture varies from 23.80 to 28.00% and from 21.06 to 24.50% for topsoil and subsoil layer, respectively. It can be seen that; topsoil layer has higher mean value compared to subsoil. However, the difference in mean value of both layers is very small (3.33%).

**Table 1.** Basic Statistical Values of the Soil Moisture Data

| Parameter | Topsoil | Subsoil |
|---|---|---|
| Count (n) | 40 | 40 |
| Minimum (%) | 23.80 | 21.06 |
| Maximum (%) | 28.00 | 24.50 |
| Mean (%) | 26.17 | 22.84 |
| Standard Deviation | 0.980 | 1.060 |
| Skewness | 0.170 | 0.020 |
| 1$^{st}$ Quartile (%) | 25.34 | 22.19 |
| Median (%) | 25.91 | 22.63 |
| 3$^{rd}$ Quartile (%) | 26.88 | 23.73 |

In view of results obtained, the standard deviations were low for both layers. This shown that, there are less variations in soil moisture data. In addition, the skewness values for topsoil and subsoil layers were near to 0. Hence, it can be interpreted that the data were normal distributed and approximately symmetrical.

### *3.2 Spatial Distribution of Soil Moisture*

Soil moisture variation or distribution maps were created using ArcGIS and classified into several classes based on the percentage of soil moisture. By refereeing to several literatures, the highest moisture percentage that the soil of a type clay can hold is around 35 to 39%. For that reason, five classes of range have been setup covering from 0 to 35% moisture. The classes are; very low, low, moderate, high and very high (see Table 2).

Nevertheless, this range cannot clearly present the variation of soil moisture as all 80 soil moisture data fit in only one class, which is high (21.01% to 28.00%). Consequently, the subclasses were created by diving the high class (21.01% to 28.00%) to five subclasses of High I, High II, High III, High IV and High V as in Table 3.

**Table 2.** Classification of soil moisture

| Classes | Soil Moisture Content (%) |
|---|---|
| Very Low | 0 - 7 |
| Low | 7.01 - 14 |
| Moderate | 14.01 - 21 |
| High | 21.01 - 28 |
| Very High | 28.01 - 35 |

**Table 3.** Subclasses range for 'High' class

| Classes | Soil Moisture Content (%) |
|---|---|
| High I | 21.01 – 22.40 |
| High II | 22.41 – 23.80 |
| High III | 23.81 – 25.20 |
| High IV | 25.21 – 26.60 |
| High V | 26.61 – 28.00 |

Figure 1 shows two soil moisture maps for topsoil and subsoil. The variations or distributions of soil moisture content in greenhouse were discussed based on this figure. Selected color fills were chosen to indicate the subclasses; grey (High I), green (High II), yellow (High III), blue (High IV) and red (High V).

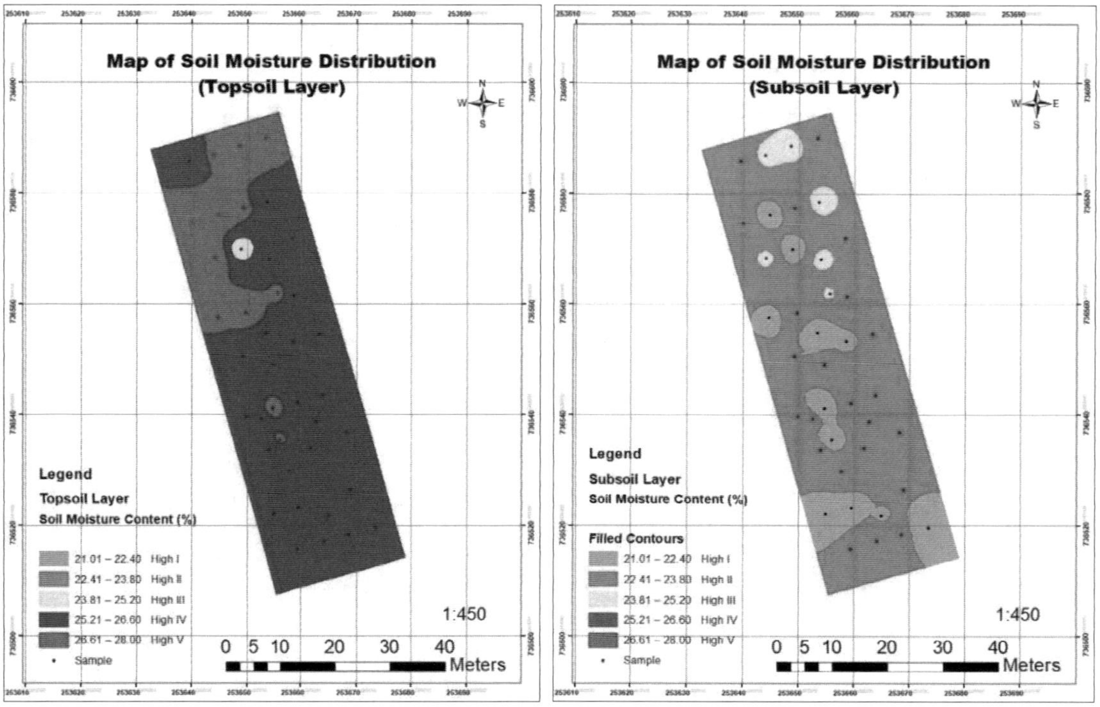


**Figure 1.** Soil Moisture Distribution Map of Topsoil (a) and Subsoil Layers (b).

The soil moisture distribution map shows difference color fill and pattern for both layers. For topsoil layer, it was filled with colors of yellow, blue and red which ranging from High III to High IV. This layer was dominated by blue color, however the red color did exist and has concentrated at northwest of the greenhouse. It is predicted that, high moisture for red region was because the area was slightly lower in elevation although the soil was leveled before greenhouse was built. This has been crossed validate with elevation data collected during GPS data captured. Indirectly, it can be said that certain fraction of irrigation water might seep and run-off towards lower elevation area before it infiltrate into soil. This was in agreement with research study [9]. Soil moisture and elevation also understood to have an inverse relationship to one another [10].

On the other hand, the subsoil layer was filled with grey, green and yellow colors from High 1 to High III class. The green color vastly occupied the layer with seven gray and five yellow spots. The spots were scattered in that layer. Again, they yellow spots with a higher moisture value was seen on northwest and this may be related to above reason in topsoil layer. As the topsoil layer possess more water, more water will be assumed to infiltrate.

As whole, it can be said that, topsoil and subsoil layer can be represented by blue (High IV) and green (High II) class, respectively as the color spread widely in that particular layer. Clearly, as expected, topsoil layer has higher soil moisture content compared to subsoil layer. This may be correlated to the infiltration process. The movement of water downwards or called infiltration process from topsoil to subsoil is happened respectively to time and textural of a soil. There are few more factors that may influence infiltration time or rate such as soil porosity, soil texture, and vegetation ([11] – [12]). On the other hand, some water or moisture in topsoil may be lost before it may infiltrate. This may be due to runoff and evaporation process. According to Li Rong et al., [10] it is said that higher soil water evaporation and vegetation cover transpiration in the topsoil and subsoil layers plays a responsible role in soil moisture content.

## 4. Conclusion

In this study, the soil moisture mapping of both layers have represented a better visual picture of moisture variation. Both maps have shown different spatial variation and distribution of moisture content even though all data fall in high moisture content class. There is less variation of moisture content throughout each soil layers can be seen. Therefore, it can conclude that; spatial distribution of the soil moisture content may be affected by several factors and soil depth has an inverse relationship with soil moisture content. The maps may help the farmers to reduce the variability (i.e supplying more water to lower soil moisture spots) and uncertainties that might occur in future.

## Acknowledgments

The author would like to acknowledge the support from the Fundamental Research Grant Scheme (FRGS) under a grant number of FRGS/1/2016/WAB01/UNIMAP/03/6 from the Ministry of Higher Education Malaysia.

## References

[1] Brevik, E. C., Calzolari, C., Miller, B. A., Pereira, P., Kabala, C., Baumgarten, A., & Jordán, A., 2016. Soil mapping, classification, and pedologic modeling: History and future directions. Geoderma, 264, 256-274.

[2] Lagacherie, P., 2008. Digital soil mapping: a state of the art. In: Hartemink, A.E., McBratney, A., Mendonça-Santos, M.L. (Eds.), Digital Soil Mapping with Limited Data. Springer, Dordrecht, The Netherlands, pp. 3–14. http://dx.doi.org/10.1007/ 978-1-4020-8592-5_1.

[3] McBratney, A.B., Mendonça Santos, M.L. & Minasny, B., 2003. On digital soil mapping. Geoderma 117, 3–52.

[4] Scull, P., Franklin, J., Chadwick, O.A. & McArthur, D., 2003. Predictive soil mapping: a review. Prog. Phys. Geogr. 27, 171–197.

[5] Lei Yang, L. C., 2015. Effects of vegetation restoration on the spatial distribution of soil moisture at the hillslope scale in semi-arid regions. In CATENA, 138-146.

[6] Zhang C., Liu S., Fang J., &Tan K., 2008. Research on the Spatial Variability of Soil Moisture Based on GIS. The International Federation for Information Processing, 719-727

[7] Liding Chen, Z. H., 2007. The effect of land cover/vegetation on soil water dynamic in the hilly area of the loess plateau, China. In CATENA, 200-208

[8] Chen, H., Fan, L., Wu, W. et al. Environ Monit Assess, 2017. 189: 525.

[9] Grayson, R.B., Western, A.W. & Chiew, F.H.S. 1997. Preferred states in spatial soil moisture patterns, local and nonlocal controls. Water Resour. Res. 33, 2897–2908

[10] Li Rong, X. D. (2017). Soil Moisture Variation in a Farmed Dry-Hot Valley Catchment Evaluated by a Redundancy. WATER, 92.

[11] Lipiec, J., Kuś, J., Słowińska-Jurkiewicz A., & Nosalewicz, A. 2006. Soil porosity and water infiltration as influenced by tillage methods. Soil and Tillage Research, 89(2), 210-220

[12] N.B. English, J.F. Weltzin, A. Fravolini, L. Thomas, D.G. Williams, 2005. The influence of soil texture and vegetation on soil moisture under rainout shelters in a semi-desert grassland. Journal of Arid Environments, 63 (1), 324-343

IOP Conf. Series: Materials Science and Engineering **557** (2019) 012071 doi:10.1088/1757-899X/557/1/012071

# Prediction of Ethylene Production Rate of Barangan Banana Using Kinetic Enzymatic Michaelis-Menten

**A F Rahman[1], E Darmawaty[2] and U Ahmad[2]**

1. Major Post-Harvest Technology, Post-Graduated School, Bogor Agriculture University. Dramaga-Bogor Indonesia
2. Mechanical Engineering and Biosystem Department, Bogor Agriculture University. Dramaga-Bogor, Indonesia

**Abstract.** MM is a model that describes the rate of reaction of product ethylene towards the substrate concentration and MMC is an enzymatic kinetics model to see the effect of the increase of $O_2$ on the rate of reaction. The objective of the study was to estimate ethylene production rate in bananas by applying the Michaelis-Menten enzymatic kinetics equation. Barangan Banana with harvest age 11 and 10 weeks were measured respiration rate with continuous gas analyzer in temperature 10, 20, and 28 °C with 3 replications along measurement of ethylene production rate using ethylene meter obtained from a closed system method and represented by Michael-Menten equation. The temperature dependence was represented base on Arrhenius law. The model parameters ($Km_{C_2H_4}$ dan $r_{C_2H_4\,max}$) are calculated using Lineweaver-Burk linier graph method. MM and MMC applications can be used to estimate ethylene production rate of Barangan banana. MM equation resulted that 11-week harvest age obtaining $R^2$ was higher than 10 weeks 0.95 while the 10-week harvested $R^2$ 0.94. MMC equation, the 11-week harvest age $R^2$ 0.95 was higher compared to 10 weeks $R^2$ 0.94. It could be predicted that MM and MMC equations can predicted the rate of ethylene production of Barangan banana

**Key words**: Barangan Banana, Respirations, Ethylene rate, Arrhenius, Michaelis-Menten Kinetic

## 1. Introduction

Ethylene has properties that can be detrimental because it can shorten the shelf life of horticultural products but on the other hand is beneficial because it can trigger and homogenize the ripening process (Genard M, 2005)

The Michaelis-Menten (MM) and Michaelis-Menten Cooperative (MMC) models are mathematical models used to estimate the rate of ethylene production based on enzymatic reactions. MM is a model that describes the reaction rate of ethylene based on substrate concentration (Copeland 2000). Binding of $O_2$ to improve the performance of active compounds (ACC oxide enzyme) in forming ethylene (De Wild *et al.* 1999). The activity of this enzyme will work optimally until the peak of the ethylene production and is directly proportional to the increase in $O_2$ levels. MMC equations, which indicates the performance of enzyme cooperativity in binding the substrate during the reaction is called the degree of cooperativity (h) (Sanders and De Wild 2003).

The MM and MMC models have been investigated to estimate the rate of ethylene production in fruits and vegetables (De Wild *et al* 1999; Sanders and De Wild. 2003; Mendoza *et al* 2016).. Diego *et al.* (2016) used the MMC model on estimating the rate of ethylene production of Cavendish bananas with 3 storage temperatures. The temperature of 15 ° C produces $R^2$ 0.98, 20 °C had a value $R^2$ 0.97 and the temperature of 25 °C produces $R^2$ 0.97. This model is also based on Michaelis kinetics-Menten. Mendoza *et al.* (2016) model ethylene in Cavendish bananas with the Michaelis-Menten enzymatic kinetics equation on a MAP (*Modified Atmosphere Packaging) system* with different temperatures, with the results of the MM model having a more predictive value than the distribution model. The MM model as a model for estimating the rate of ethylene production using enzymatic kinetics can only be used to predict ethylene production to the climacteric peak. (Diego *et al* 2016; Mendoza *et al.* 2016).

Estimation of ethylene production is very useful for carrying out proper postharvest handling both for inhibition and maturity regulation of bananas, especially Barangan banana, which is a local banana that has been designated as a superior fruit in the area of North Sumatra and has the potential as an export fruit. The aim of the study was to estimate the rate of ethylene production in Barangan bananas using the approach of the Michaelis-Menten enzymatic kinetics equation

## 2. Materials and Methods

### *2.1 Sample preparation*

Barangan banana is obtained from PT. Perkebunan Nusantara VIII Parakansalak. Bananas used are harvested 10 and 11 weeks. After being washed with alum water (the concentration according to the SOP of PT PN VIII), Barangan bananas are packaged and on take it to the Agriculture and Food Product Engineering Laboratory of the Bogor Agriculture University. After arriving, It is re-sorted from mechanical damage. Bananas selected as the experiment sample were weighed and measured of volume using the Archimedes method. The volume of the material as a base determines the free volume in the *chamber.* Samples were prepared for the measurement of respiration and ethylene at three temperatures of 10, 20 and room temperature (27-29) °C with 3 replications.

### *2.2 Measurement of Respiration Rate ($O_2$)*

Barangan respiration rate measurement uses a closed system (Hasbullah 2007). There are inserted into the chamber. The chamber is closed with a cover that has been equipped with two flexible plastic pipes as air intake and intake lines. The distance between the chamber and the lid is closed with plasticine to prevent air from entering the *chamber*. Data obtained in the measurement of respiration rate in the form of changes in $O_2$ gas concentration. The concentration of gas $O_2$ is measured using *Continuous Gas Analyzer (Shimadzu IRA-107)* with data collection intervals every 3 hours for 4 days due to peak respiration rate on day 4. Measurements are made at three different temperatures (temperatures of 10, 20, 28) °C. The rate of gas production $O_2$ (ml /kg-hour) during respiration in confined spaces is calculated by the equation of respiration (Fonseca 2002):

$$r_{O_2}(t) = \left(\frac{v}{w}\right)\,(y_{O2\,t} - y_{O2\,t+1})/\Delta t \tag{1}$$

Where $r_{O_2}$ is consumption rate of $O_2$ each time (ml $kg^{-1}h^{-1}$), V is free volume (ml), w is sample weight (kg), $y_{O2}$ is $O_2$ mole fraction (it is the initial mole fraction, and t + 1 is the value of the mole fraction during the measurement) and $\Delta t$ is the changing of time (hour)

### *2.3 Measuring Ethylene Production Rate*

Along with the measurement of the concentration of $O_2$, the measurement of ethylene concentration in Barangan banana harvested (10 and 11) weeks at different temperatures (10, 20, 28) °C with ethylene meter (cosmos) for 4 days with 3 replications. The rate of ethylene production can be calculated by equations (Diego *et al.* 2016):

$$r_{C_2H_4} = \left(\frac{V}{W}\right)(y_{C_2H_4t} - y_{C_2H_{4+1}})/\Delta t \quad (2)$$

Where $r_{C_2H_4}$ is ethylene production rate (ml kg$^{-1}$h$^{-1}$), V is free volume (ml), W is sample weight (kg), $y_{C_2H_4}$ is ethylene fraction mole, and $\Delta t$ is the changing of time (hour)

*2.4 Arrhenius*

The rate of respiration and ethylene production is changed which according coefficient of to the storage temperature along respiration rate in enzymatic kinetics is the storage temperature (Wang *et al,* 2009). So that the effect of storage temperature on Michaelis-Menten kinetics formulation ($r_{C_2H_4}$) (4) and (5) stated in Arrhenius law (Diego *et al.* 2016):

$$k = \text{ko}\, e^{-Ea/RT} \quad (3)$$

where $k$ is oxygen consumption rate (ml kg$^{-1}$h$^{-1)}$, Ea is energy activation (kJ/mol), Ko is constants (not depend on temperature), R is gas constants (8,341 J/mol), *T* is temperature of storage (°K)

*2.5 Application of Michaelis-Menten Equations at Ethylene Production Rate*

The equations of MM enzymatic kinetics (De wild *et al 1999)* and MMC (Mendoza *et al.* 2016) in estimating the rate of ethylene production is based on the reaction between enzymes and $O_2$ of ethylene production with positive cooperativity (Sander and De Wild 2003).

$$r_{C_2H_4} = \frac{r_{C_2H_4} Y_{O_2}}{(Km_{C_2H_4}) + Y_{O_2}} \quad (4)$$

$$r_{C_2H_4} = \frac{r_{C_2H_4\,max} (Y_{O_2})^h}{(Km_{C_2H_4})^h + (Y_{O_2})^h} \quad (5)$$

Where $r_{C_2H_4}$ ethylene production rate (ml kg$^{-1}$h$^{-1}$), h degree of cooperation (Hill coefficient), $r_{C_2H_4\,max}$ the rate of ethylene production in the maximum state (parameter), $Km_{C_2H_4}$ Michaelis-Menten constant$Y_{O_2}$ $O_2$ mole fraction

Determination of value $r_{C_2H_4\,max}$ and $Km_{C_2H_4}$ (4) and (5) using a Lineweaver-Burk linear curve (Coopeland, 2000). That will form the equation y=ax+b, value of $r_{C_2H_4\,max}$calculated by the value of b, whereas $Km_{C_2H_4}$calculated by the value of a. Hill coefficient (h) (5) is calculated using equation (Mendoza *et al.* 2016):

$$\frac{1}{r_{C_2H_4}} = \frac{1}{r_{C_2H_4\,max}} \left(Km_{C_2H_4}(y_{O_2})^h + 1\right) \quad (6)$$

Equation 6 is a linear form of Equation (5) (Mendoza *et al.* 2016) making it easier to calculate Hill coefficient (h).

*2.6 Model Validation*

Validation in the Michaelis-Menten equation is done by calculating the coefficient of determinant ($R^2$) and the error in the equation used. $R^2$ is calculated by linear regression analysis (Abdullah *et al.* 2015) to explain the ability of independent variables in explaining the dependent variable and error was calculated by Sum

Square of Error analysis (Bhande *et al.* 2008). SSE (*Sum Square Error*) is one of the statistical methods to measure the difference the total of the true value of the value achieved by the model. Data validation uses replication of the data (1 data replication for validation and 2 data replications for model). SSE is calculated by equation (Pratama and Harjoko, 2015):

$$\mathrm{SSE} = \sum_{i=1}^{n} (X_i - Y_i)^2 \quad (7)$$

Where X actual value , Y value achieved n data

## 3. **Results and Discussion**

### *3.1 Respiration rate*

In this study, the measured respiration rate was consumption of $O_2$ Barangan bananas based on the temperature and time of harvest as show on in Fig 1. Harvesting 10 and 11 weeks which an increasing of temperature causes increase of the rate of consumption of $O_2$. This was due to the higher storage temperatures can lead to increased respiration rate of a horticultural products. A chemical reaction will take place faster by increasing the temperature, with the temperature being increased, the kinetic energy of the reacting molecules are increases, so that the reaction is faster than the low temperature (Genard M, 2005).

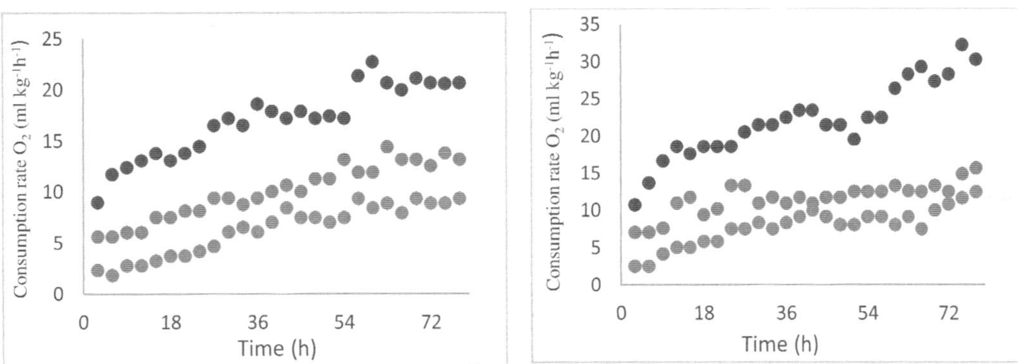


(a) (b)

**Figure 1.** $O_2$ consumption rate of Barangan banana (a) week 10 (b) week 11 at temperature (10, 20, 28) °C

At the 10 weeks harvest, maximum rate of consumption of $O_2$ temperature (°C) 28,20,10 are 20.68, 13.16, 9.35 (ml k g $^{-1}$ h $^{-1)}$, while the harvesting of 11 weeks, maximum consumption rate of $O_2$ is 30.30, 15.66, 12:52 ml k g $^{-1}$ h $^{-1}$). This result consistent with the research by De Wild *et al* (2003); Diego *et al.* (2016) in Cavendish bananas that the consumption of $O_2$ in bananas will increase when the storage temperature is increased, the rate of consumption of $O_2$ at low temperatures is smaller than the high temperature. Barry and Giovvanoni (2007) explains that the rate of consumption of $O_2$ is influenced by the temperature and age of harvesting, the higher storage temperature and the physiological of harvest of horticulture products, the consumption rate $O_2$ will be even greater.

### *3.2 Ethylene Production*

Ethylene production of Barangan banana at 10 and 11 weeks at 3 different temperatures (10, 20, 28) °C show on as Figure 2.

IOP Conf. Series: Materials Science and Engineering **557** (2019) 012071 doi:10.1088/1757-899X/557/1/012071

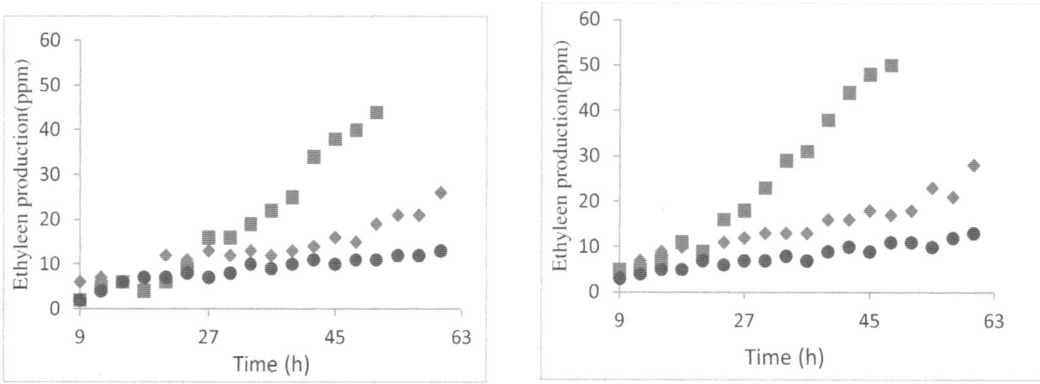


(a) (b)

F**Figure 2.** Ethylene production of Barangan banana (a) 10 week, (b) 11 week at temperature ■ 28 °C ◆ 20 °C, ●10 °C

Production of ethylene measurement show that the ethylene production will increase during the increasing of temperature and due the physiological 11 weeks production the ethylene higher than 10 week. 10 weeks produce the ethylene maximum (13, 27, and 44) ppm while 11 weeks produce (16, 29 and 50) ppm ad the maximum production. When the storage temperature going up, the metabolism of bananas will progress faster and ethylene produced will be high (Genard M 2005). This result consistent with study by Nurjannah (2009) where the ethylene production will increase with increasing storage temperatures and physiological age of horticulture product.

*3.3 Arrhenius*

The Arrhenius model application is preceded by the calculation of the value of k using Equation 5 to produce an equation calculating the value of k for each harvest age as show in Table 1.

**Table 1**. Activations energy using Arrhenius equations

| weeks | Ea (kJ/mol) | k0 | The equations of k |
|---|---|---|---|
| 10 minggu | 98.12 | $2,66776x10^{15}$ | $k = 4.68902x10^{15^{98.12}/_{R.T}}$ |
| 11 minggu | 81.49 | $2,43262x10^{12}$ | $k = 2,43262x10^{12^{81.49}/_{R.T}}$ |

The equation of k (Table 1) is obtained from finding the k value for each temperature (10, 20, 28) °C so that the graph 1 / T vs ln K. will be used to find the Ea and k0 values. There are entered into the Arrhenius equation (5) so that they get the k value equation in Table 1. The low of Activation Energy (Ea) indicate a product has a higher reaction speed (Khatir *et al* 2014). The results of this study were 10 weeks of harvesting have Ea 98.12 kJ/mol and 11 weeks of harvesting have Ea 81.49 kJ/mol so that the ethylene production speed of Barangan banana 11 weeks higher than 10 weeks as in Figure 2. Diego et al. (2016) explained that the lower Ea on the ethylene reaction will run faster so the ethylene production will be higher. Positive activated energy indicated that the effect of temperature has linier effect to ethylene production of Barangan banana (Diego *et ai.* 2016).

IOP Conf. Series: Materials Science and Engineering **557** (2019) 012071 doi:10.1088/1757-899X/557/1/012071

### *3.4 Estimate of Ethylene Production Rate of Barangan Banana*

Estimating the rate of ethylene production at 3 different temperatures and two different harvest ages show on (Figures 3 and 4) based on the mole fraction using MMC and MM equations

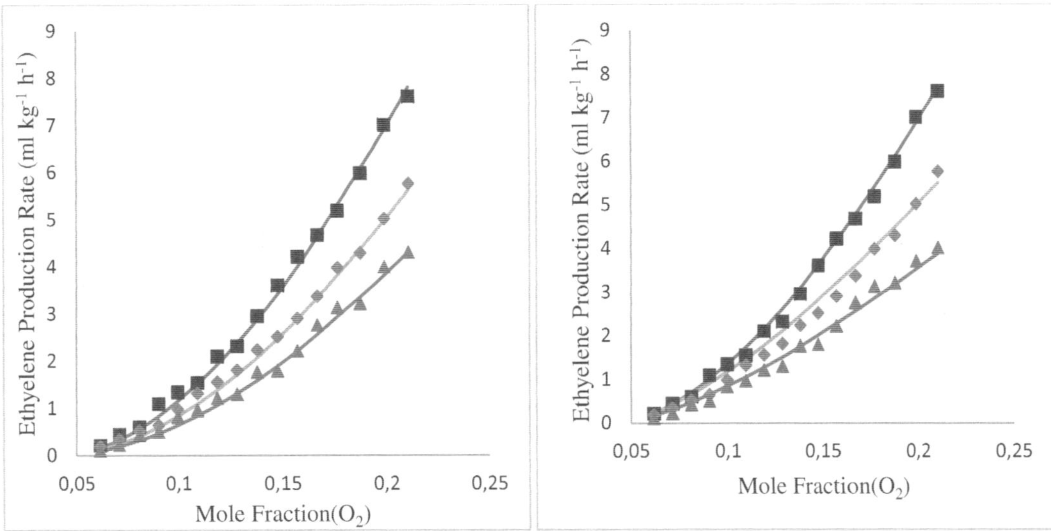


(a) (b)

**Figure 3.** The result of measurement and result of ethylene production rate of Barangan banana using Michaelis-Menten Equations 10 weeks of harvested. a (MM),b (MMC) temperature (28, 20, 10) oC.
■ 28 oC, ♦ 20 oC, ▲ 10oC

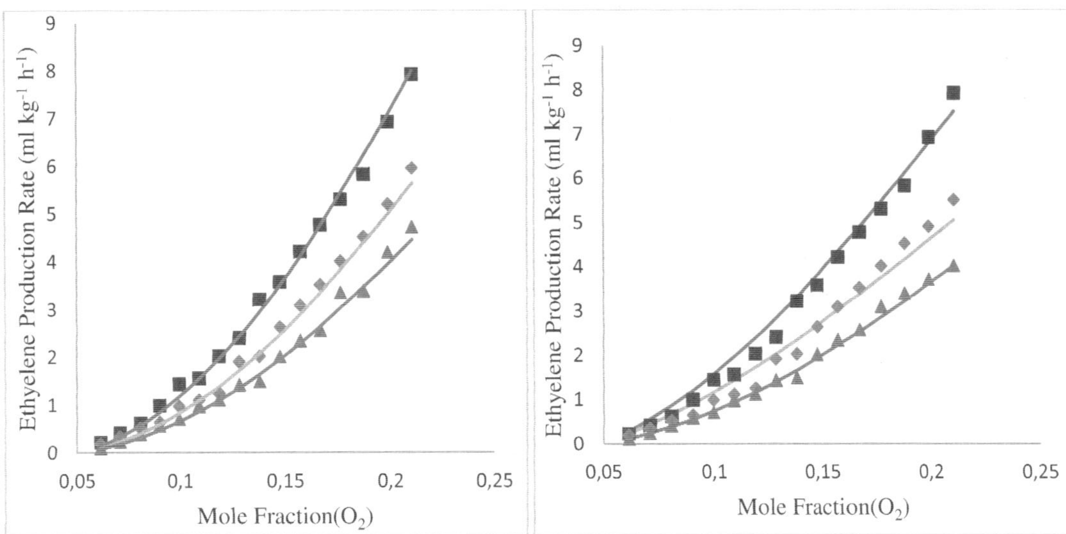


(a) (b)

**Figure 4.** The result of measurement and result of ethylene production rate of Barangan banana Using Michaelis-Menten Equations 11 weeks of harvested. a (MM),b (MMC) temperature (28, 20, 10) oC.
■ 28 oC, ♦ 20 oC, ▲ 10 oC

The differences in the MM and MMC models can be seen in the model parameters produced as show in (Table 2).

**Table 2**. Parameter of equations MMC dan MM in estimated productions of ethylene rate

| | 10 weeks | | | 11 weeks | | |
|---|---|---|---|---|---|---|
| | 28 °C | 20 °C | 10 °C | 28 °C | 20 °C | 10 °C |
| MM | | | | | | |
| $Km_{C_2H_4}$ | 4.56±8.32x10$^{-3}$ | 4.40±8.35x10$^{-3}$ | 4.28±3.47x10$^{-3}$ | 4.51±6.32x10$^{-3}$ | 4.39±3.68x10$^{-3}$ | 4.30±3.11x10$^{-3}$ |
| $r_{C_2H_4\,max}$ (ml.kg/h) | 5.61±6.32x10$^{-3}$ | 4.00±7.23x10$^{-3}$ | 2.01±4.32x10$^{-3}$ | 5.90±7.44x10$^{-3}$ | 4.10±5.54x10$^{-3}$ | 2.27±6.32x10$^{-3}$ |
| $R^2$ | 0.94 | 0.93 | 0.93 | 0.95 | 0.95 | 0.92 |
| SSE | 1.15 | 1.11 | 1.19 | 0.97 | 1.11 | 1.23 |
| MMC | | | | | | |
| $Km_{C_2H_4}$ | 4.56±8.32x10$^{-3}$ | 4.40±8.35x10$^{-3}$ | 4.28±3.47x10$^{-3}$ | 4.51±6.32x10$^{-3}$ | 4.39±3.68x10$^{-3}$ | 4.30±3.11x10$^{-3}$ |
| $r_{C_2H_4\,max}$ (ml.kg/h) | 5.61±6.32x10$^{-3}$ | 4.00±7.23x10$^{-3}$ | 2.01±4.32x10$^{-3}$ | 5.90±7.44x10$^{-3}$ | 4.10±5.54x10$^{-3}$ | 2.27±6.32x10$^{-3}$ |
| h | 0.93 | 0.81 | 0.72 | 0.82 | 0.79 | 0.70 |
| $R^2$ | 0.94 | 0.93 | 0.90 | 0.95 | 0.93 | 0.91 |
| SSE | 0.98 | 1.02 | 1.27 | 0.95 | 0.97 | 1.09 |

Based on model parameters produced, the harvest of 10 weeks and 11 weeks, the MMC and MM equation has a different $R^2$ . The MM model is more appropriate at the age of 11 weeks because it has a higher $R^2$ 0.96. De Wild *et al* (1999) shortly explain that the differences in physiological age harvest will affect the results of the model ($R^2$) where the physiological harvest age is older, the value of $R^2$ produced will be better. Likewise with MMC models, harvesting age 11 weeks resulted in a value of $R^2$ is larger than the 10 weeks. This result is consistent with those reported by Sanders and De Wild (2003); (Diego *et al.* 2016) where the model MMC that older harvesting produce $R^2$ is larger

Based on Table 2, the value of the parameter h (coefficient hill) at 11 weeks harvest smaller than the harvest age of 10 weeks. The value of h illustrates the magnitude of the effect of $O_2$ during ethylene production. Sander and De Wild (2003) ; (Mendoza *et al.* 2016) describes the parameter values are affected by age physiological from the sample, the high of it make h value would be diminished. Copeland (2002) ; Castellanos (2016) explains that the rate of consumption of $O_2$ during high ethylene production, the value of h will be low. 11 weeks, the consumption rate of $O_2$ is greater than 10 weeks, so 11 weeks has a small value (h). The value of h for each temperature is different where the temperature of storage is high, the value of h will be higher.

The parameter values of Km every temperature is dissimilar. The km values in the MM and MMC equations describe the activities of enzyme in the ethylene process based on the level of consumption of $O_2$ (Segel 1993). Due the parameters as show in Table 2, the temperature is rising then the value $Km_{C_2H_4}$ at MM and MMC will be large, and low temperature the value of $Km_{C_2H_4}$will be small. This result consistent with those reported by Sanders and De Wild (2003); Wang *et al* (2009) ; Mendoza *et al.* (2016) that parameter values $Km_{C_2H_4}$both MM and MMC will decrease according to the decrease in storage temperature used. This is because the enzyme performance will experience inhibition at low temperatures and the performance of the enzyme have been effective at high temperatures (Copeland 2002). Consumption of $O_2$ at high temperatures will be greater so that the value $Km_{C_2H_4}$both the MM and MMC equations will be even greater (Sanders and De Wild 2003)

The smallest SSE indicates the model is closer to direct measurement (Sari *et al.* 2014). As show in Table 2, SSE at 11 weeks smaller than the age of 10 weeks. This illustrates the MM and MMC equations can be used to predict the rate of ethylene production of Barangan banana. De wild *et al* (2003), the physiological harvest age of bananas will influence the results of the models used where the harvest age is older, the SSE produced will be smaller.

The rate of ethylene production will increase along the increasing of temperature storage, the higher the storage temperature, can involve the rate of ethylene production. Rashid *et al.* (2012) explained that an increase of storage temperature is directly proportional to the increase in the rate of ethylene production where high temperatures will cause the enzyme working reaction (ACC Synthase) to increase, pushing the metabolic reaction in bananas will be going up, so the rate of ethylene production will increase.

## 4. Conclusion

The application of the MM and MMC equations can be used to estimating the rate of ethylene production of Barangan banana. The MM model, 11 weeks of harvest produced the highest $R^2$ value compared of 10 weeks harvesting with the value of $R^2$ 0.96 while the age of 10 weeks the highest $R^2$ 0.95. The MMC, harvest age 11 weeks produce the highest $R^2$ value of 0.94 at while 10 weeks harvested produces the highest $R^2$ 0.94. The conditions of MM and MMC are more appropriate to use at 28 °C to predict the rate of ethylene production Barangan banana. Value $Km_{C_2H_4}$ on MM and MMC equations will going up while decreasing of storage temperature. MMC equation is more appropriately used at 11 weeks harvested where the value of h on 11 weeks low than 10 weeks of harvested.

## 5. References

[1] Abdullah M, Alhamdan, Diealdin OE and Kheled AE 2015 Modeling OF Respiration Rate OF Date Fruits (Barhi Cultivar) Under Aerobic Conditions. *J Advanced Agricultural Technologies*. **2** 120

[2] Bhande SD, Ravinda MR and Goswani TK 2008 Respiration rete of banana fruit under aerobic conditions at different storage temperature. *J Food and Engineering*. **87** 116

[3] Barry CS and Giovannoni JJ 2007 Ethylene and fruit ripening. *J of Plant Growth Regulation*. 26: 143–159

[4] Castellanos DA, Cerisuelo JP, Hernandez-Munoz P, Herrera AO and Gavara R 2016 Modelling the evolution of $O_2$ and $CO_2$ concentrations in MAP of a fresh product: application to tomato. *J Food Engineering*. **168** 84

[5] Copeland RA 2000 *Enzymes: A Practical Introduction to Structure, Mechanism, and Data Analysis.* Wiley-VCH, Inc. New York.

[6] De Wild HPJ, Woltering EJ and Peppelenbos HW 1999 Carbon dioxide and 1-MCP inhibit ethylene production and respiration of pear fruit by different mechanisms. *J Experimental Botany*. **50** 837

[7] De Wild, Otma EC and Peppelenbos HW 2003 Carbon dioxide action on ethylene biosynthesis of preclimacteric and climacteric pear fruit. *J Experimental Botany*. **54** 1537

[8] Diego AC, Rolando M, Rafael G and Anibal OH 2016 Respiration and Ethylene Generation Modeling of 'Hass' Avocado and Feijoa Fruits and Application in Modified Atmosphere Packaging. *J Food Properties*. **1** 1

[9] Fonseca SC, Oliveira FAO and Brecht JK 2002 Modeling Respiration Rate of Fresh Fruits and Vegetables For Modified Atmosphere Packages. *J of Food Engineering*. **52** 99.

[10] Genard M and Barbara Gouble 2005 Theory of Fruit Climacteric Ethylene Emission. *J Plant Physiology*. **139** 531

[11] Heydari A, Shayesteh K, Eghbalivam N, Bordari H and Falahatpiseh S 2010 Studies On The Respiration Rate Of Banana Fruit Based On Enzyme Kinetics. *International Journal Agriculture Bioloy*. **12** 145.

[12] Mendoza RAC, Juan CG and Julio CV 2016. Original Article Ethylene Production Respiration and Gas Exchange Modelling in Modified Atmosphere Packaging for Banana Fruits. *J Food Properties* **51** 777
[13] Nurjannah S 2002 Kajian Laju Respirasi dan Produksi Etilen Sebagai Dasar Penentuan Waktu Simpan Sayuran dan Buah-buahan. *Jurnal Bionatura*. **4** 148
[14] Pratama A and Harjoko A. 2015. Aplications of Algorithms of Invasive Weed Optimnization for determinations the centre of claster K-Means. *IJCCS*. **9** 55
[15] Rasyid A, Naflaa A, Fahad A, Majeed A, Annamalai and Adel A 2012 Effect Of Storage Conditions on Physico-Chemical Attributes and Physiological Response of 'Milk' (Musa spp., ABB) during Ripening. *Postharvest Technology and Innovation*. **2** 370
[16] Sanders MG, De Wild HPJ. 2003. The relation between in vivo ethylene production and oxygen partial pressure. *J Postharvest Biology and Technology*.**30** 143
[17] Sari IP, Andrizal and Devianto D 2014 Design of human face recognitions system using camera Web with *Summary Squared Error* (*SSE*) method. *IJCCS*. **9** 65
[18] Segel IH 1993 *Ezyme Kinetics. Behavior and Analysis of Rapid Equilibrium and Steady-state Enzyme Systems*. New York: 945.
[19] Wang ZW, Duan HW, and Hu CY 2009 Modelling the respiration rate of guava (Psidium guajava *L.*) fruit using enzyme kinetics, chemical kinetics and artificial neural network. *J Food Res. Technol.* **229** 495

# Single packaging design of papaya (*Carica Papaya* L.) variety IPB 9 (calina) for transportation and distribution

**P L Ismaya[1], E Darmawati[1] and Setyadjit[2]**

[1] Postharvest Technology, Departement of Mechanical and Biosystem Engineering, Bogor Agricultural University, Bogor, Indonesia
[2] Agricultural Research center and Postharvest Development, Bogor, Indonesia

**Abstract**. Papaya Calina (Carica Papaya L) is one of the favorite fruits. The thin skin and soft flesh of the papaya, easily cause the damaged of papaya mainly due to mechanical damage at the distribution process. The distribution of papaya generally carried out in bulk and arranged on the vehicles without packaging. Improvements in packaging for distribution that combined with retail packaging will reduce mechanical damage and increase consumer appeal. The research objective is to design retail and distribution packaging. Retail packaging is designed for individual fruits made from polypropylene plastic bags and corrugated board. Distribution packaging are made from corrugated board. The weight of papaya is selected in the range of 700 g -1000 g with a diameter of 90.5 mm - 110 mm and a length of 225 mm -240 mm. Based on these data, retail packaging is designed in the form of fruit units of 250x125x 250 mm for plastic bags and 90x90x140 for corrugated. The distribution packaging is designed with half telescopic container type using the flute BC with two different positions, both of vertical and horizontal positions. The packaging with the horizontal position contains of 4 fruit per packaging, the dimensions are 410x250x115 mm (KS1) and for vertical contains of 6 fruit per packaging with dimensions 330x230x70 mm (KS2). Corrugated board packaging dimensions is designed by considering the efficiency of pallets or generally vehicles that used in product distribution. The efficiency of the packaging that arranged on the pallet with the size of 1200x1100 mm are 85.41% for KS1 and pallet with the size of 1067x1067are 80% for KS2, while for M5 vehicle with the size of 2200 x 1585 mm the efficiency is 82.56% for KS1 and M2vehicle with the size 2630x1460 mm the efficiency is 90,92% for KS2.

## 1. Introduction

Papaya (Carica Papaya L.) is one tropical fruit that has a high economic value. The Center for Tropical Horticulture Studies (LPPM-IPB) has produces superior varieties of products produced by research innovations in papaya commodities based on market demand criteria. The criteria for papaya desired by consumers for fresh consumption include sweet taste, oval shape, weight range from 500-1000 g, red orange fruit flesh, and long shelf life. Currently, the most desirable varieties of papaya are papaya IPB9 or papaya calina [1].

Papaya calina has some properties such as perishable and rapid weight loss due to its thin fruit skin and soft flesh. In addition, papaya also easily suvjected mechanical damage caused by transportation and distribution activities [2]. Papaya is generally distributed without packaging. Fruit taken from farmers, directly arranged on pickup truck. However the distribution is carried out for supermarkets, fruit shops and other cities, papaya is packaged in bulk using plastic containers or boxes wrapped in newspaper. This packaging makes papaya in the hands of consumers experiencing a low quality and consumer under expectations .

The transportation process also results in quality degradation during postharvest. The vibration occur during transportation on the highway or railroad can cause weight loss and shorten self life. Damage during transportation occurs mostly in bulk conditions or improper use of packaging, resulting in damage to the product when arriving at the destination reaches 30-50% [3]. In addition, mechanical damage that occurs in the process and distribution will accelerate physiological damage when the papaya is stored [2].

One way to reduce mechanical damage to papaya fruit during transportation and distribution is packaging use. Packaging has an important role, among others, to improve the appearance of the product, help prevent or reduce damage, and protect the fruit from contamination and other physical disorders. Good packaging is not only minimize damage and maintain fruit quality, but also reduce transportation and distribution costs [4].

Research conducted in an effort to reduce mechanical damage to papaya using corrugated packaging for distribution has been carried out by Tawakal [2]. Tawakal said the provision of corrugated cardboard packaging with foam net combination and wrapping during transportation is the best treatment because it is able to maintain the quality of papaya before sold. Packaging that is commonly used is bulk packaging, while single packaging that is ready to display (display) has not been widely done. In addition to better protecting the fruit during transportation, single packaging is used for retail that are readily to be displayed when they arrive at fruit shops / fruit outlets. Therefore in this research, a single packaging (retail) made from PP plastic and corrugated cardboard will be conducted. Retail packaging is designed with regard to the convenience of consumers in carrying papaya fruit such as handholes for plastic packaging and ropes for cardboard packaging. In the process of transportation and distribution, retail packaging is packed again in a carton box made of corrugated cardboard.

Based on the description above, the main objectives of this study are: 1) designing single / retail packaging made of PP plastic and corrugated cardboard and distribution packaging made from corrugated cardboard for papaya; 2) Analyzing the strength of packaging and; 3) simulate the preparation of the distribution package design for various transport sizes and pallets to get the best efficient arrangement.

## 2. Material and Methods

### 2.1. *Material*

The materials used in this study were papaya with 1 tier maturity level (papaya weight 700 - 1000g), individual retail packaging made from flute wave C (double face) and BC (double wall) flute type packaging, and polypropylene plastic packaging 0.08 mm. The equipment used consists of scales, meters, sealers, measuring ruler, and UTM (Universal Testing Machine).

### 2.2. *Methods*

#### 2.2.1 *Preparation of fruit samples includes sorting based on weight.*

The weight of the papaya chosen was 700-1000 g based on the heavy information demanded by the export market. This information wasobtained from exporters who have made export sales of papaya fruit. The results of fruit sorting based on the weight range measured fruit dimensions, namely length and diameter to determine the inner dimensions of thedesigned package.

The weight of one papaya is measured using a digital scale (Mettler PM-4800) with accuracy of 2 numbers behind the comma. Measuring the diameter of papaya fruit was done using a measuring tape by measuring the circumference of the width and the height of the papaya using equation 1.

$$D = 2 \times (K / (2 \times \pi)) \quad (1)$$

Information
K = circumference of circle (mm)
D = circle diameter (mm)

2.2.2. *Designing the dimensions of retail packaging*

2.2.2.1. *Polypropylene plastic packaging*

Plastic packaging for papaya fruit that will be designed is a polypropylene (PP) plastic packaging 250x250 mm with a thickness of 0.08 mm. The number of fruits in one plastic package is 1 fruit with the position of the fruit sleeping horizontally. The package is given a round perforation hole with a width of 5 mm. The number of holes perforated as many as 20 (twenty) holes. Determination of retail packaging dimensions is calculated by equations 2 and 3.

$$Pp = Pb + 1 \quad (2)$$

$$Tp = Tb + 1 \quad (3)$$

Information :

Pp = Plastic package length (mm)

Pb = length of papaya fruit (mm)

Tp = plastic packaging height (mm)

Tb = height papaya fruit (mm)

2.2.2.2. *Corrugated cardboard packaging*

Corrugated cardboard packaging for papaya fruit that will be designed is made from double face using flute C 0.3 cm thickness type RSC (Regular Selotted Carton) which is given a rope for ease of carrying. The number of pieces in one carton package is 1 fruit with vertical standing fruit position Packaging is given a round vent with an area of 2% of the packaging area. The number of ventilation is 2 (two holes). The stages for designing cardboard primary packaging are determination of inner dimensions, design dimensions and outer dimensions.

**Table 1**. Calculation of corrugated cardboard pacakaging design

| Information | Length (mm) | Wide (mm) | Height (mm) |
|---|---|---|---|
| Inner dimension | $DP_b$ | $DL_b$ | $T_b$ |
| Design dimensiom | $P_{kd} + T_f$ | $L_{kd} + T_f$ | $T_{kd} + (2xT_f)$ |
| Outer dimension | $P_{kdes} + (2xT_f)$ | $L_{kdes} + (2xT_f)$ | $T_{kdes} + (2xT_f)$ |

information :

D = diameter

Pb = length of papaya (mm)

Lb = width of papaya (mm)

Tb = height of papaya (mm)

$P_{kd}$ = length of inner dimension (mm)

$L_{kd}$ = width of design dimension (mm)

$T_{kd}$ = Height of outer dimension (mm)

$P_{kdes}$ = length of packaging design (mm)

$L_{kdes}$ = width of packaging design (mm)

$T_{kdes}$ = high of packaging design (mm)

$T_f$ = thick *flute* C

2.2.3. *Designing the packages for distribution*

Packaging distribution for papaya fruit designed was made from double wall flute corrugated cardboard BC type RSC (Regulated Slotted Container) and HTC (Half Telescope Container). For one horizontal distribution package, 4 packets of papaya were packaged in PP plastic retail packaging with horizontal ground position, while vertical distribution packages packed 6 papaya packaged in carton packaging with vertical standing position.

**Tabel 2**. Calculation of distribution packaging design horizontal position

| Information | Length (mm) | Width (mm) | Height (mm) |
|---|---|---|---|
| Inner dimension | $N_P \times P_{lr}$ | $N_L \times L_{lr}$ | $T_{lr}$ |
| Design dimension | $P_{kd} + T_f$ | $L_{kd} + T_f$ | $T_{kd} + T_f$ |
| Outer dimension | $P_{kdes} + (2xT_f)$ | $L_{kdes} + (2xT_f)$ | $T_{kdes} + (2xT_f)$ |

**Table 3.** Calculation of distribution packaging design vertical position

| Vertical position | information | Outer dimension | Design dimension | Outer dimension |
|---|---|---|---|---|
| *Body* | Length (mm) | $N_P$ x $P_{lr}$ | $P_{kd} + T_f$ | $P_{kdes} + (2 x T_f)$ |
| | Width (mm) | $N_L$ x $L_{lr}$ | $L_{kd} + T_f$ | $L_{kdes} + (2 x T_f)$ |
| | Height (mm) | $T_{lr}$ | $T_{kd} + T_f$ | $T_{kdes} + (2 x T_f)$ |
| *Cover* | length (mm) | - | $P_{Kdesbody} + (2 x T_f)$ | $P_{Kldescover} + (4 \times T_f)$ |
| | width (mm) | - | $L_{Kdesbody} + (2xT_f)$ | $L_{Kldescover} + (4 \times T_f)$ |
| | Height (mm) | - | $^1/_3 \times (T_{Kdes\ body} - T_f)$ | $T_{Klcover} + T_f$ |

Keterangan
$n_p$ = length number of retail packaging (mm)
$P_{lr}$ = outer retail packaging length (mm)
$n_L$ = wide number of retail packaging (mm)
$L_{lr}$ = outer retail packaging width (mm)
$T_f$ = thickness *flute* BC (mm)
$P_{kd}$ = inner packaging length (mm)
$L_{kd}$ = inner packaging width (mm)
$T_{kd}$ = inner packaging height (mm)
$P_{kdes}$ = packaging design length (mm)
$L_{kdes}$ = packaging design width (mm)
$T_{kdes}$ = packaging design height (mm)
$P_{Kl\ cover}$ = outer packaging length cover (mm)
$L_{Kl\ cover}$ = outer packaging width cover (mm)
$T_{Kl\ cover}$ = outer packagin height cover (mm)

2.2.4. *Perform packaging strength tests and analyze the strength of packaging stacking*

Calculation of the number of packing stacks is done with equations 4 and 5

$$SF = \frac{P}{f} \quad (4)$$

$$\text{Number of stack packs} = \frac{SF}{\text{Berat total box}} \quad (5)$$

Information
SF = maximum safe load
P = compressive strength (N / m2)
f = safety coefficient value

2.2.5. *Simulating the preparation of packaging on the pallet and like a transport car (pick up)*

2.2.5.1. *Packaging arrangement efficiency on pallet*

The packaging design results are structured simulations on a pallet. The size of the pallet with the most efficient packaging arrangement is the best pallet size, namely with the most packaging arrangement. Determination of packaging arrangement efficiency on a pallet using equation 6.

$$E_k = \frac{\quad}{\quad} \quad (6)$$

Information
Ek = Efficiency of packaging on a pallet
Pp = Compilation pattern
Lk = The area of the packaging floor
Lb = Pallet area

2.2.5.2. *Efficiency of packaging arrangement in transport car (pick up)*

The efficiency of packaging arrangement efficiency above the conveyance was intended to determine the freight charge that can affect the quantity of fruit and shipping costs. Determination of the efficiency of packaging arrangement above transport car using equation 7.

$$E_b = — \quad (7)$$

Information
Eb = Body efficiency
Lk = packaging area
Lb = Tub area

## 3. **Result and Discussion**

### 3.1. *Measurement of Dimensions of Papaya Fruit*

The dimensions of papaya measurements included fruit weight, fruit diameter and fruit height. Papaya fruit obtained in this research was varying size because it comes from several different trees. The results of measurements of 30 samples of Calina papaya fruit were obtained as shown in table 4.

Tabel 4. Measurement results of dimension papaya

| Packaging | Part | Average |
|---|---|---|
| Plastic | weight (g) | 883,54±74,53 |
| | Diameter (mm) | 29,31±0,52 |
| | Height (mm) | 23,70±1,23 |
| Corrugated cardboard | Weight (g) | 808,63±66,74 |
| | Diameter (mm) | 26,35±1,66 |
| | Height (mm) | 22,85±0,57 |

The results obtained from the dimensions of papaya measurements were used to determine the dimension retail and distribution packaging design.

### 3.2. *Retail packaging*

3.2.1. *Polypropylene plastic packaging*

Based on the results of measurements of the weight and dimensions of papaya fruit, plastic packaging design were obtained (table 5 and figure 1 and 2).

**Tabel 5**. Dimensions of designed plastic packaging

| Parameter | Hasil rancangan |
|---|---|
| Plastic thickness (mm) | 0.08 |
| Number of perforation holes | 20 |
| Area of perforation hole (mm) | 5 |
| Percentage perforation (%) | 1.6 |
| Plastic length (mm) | 25o |
| Plastic width (mm) | 125 |
| Plastic height (mm) | 250 |
| Packaging area ($cm^2$) | 2187.5 |
| Handhole length (mm) | 100 |
| Handhole height (mm) | 20 |

The use of perforated plastic packaging is to prevent condensation from occurring so as to minimize moisture condensation in the package. The use of plastic packaging without perforation on fruits and vegetables can cause decay and anaerobic reactions that cause unpleasant odor decay occur, so it is necessary to have a perforation hole that can get rid of respiration through the perforation hole [5].

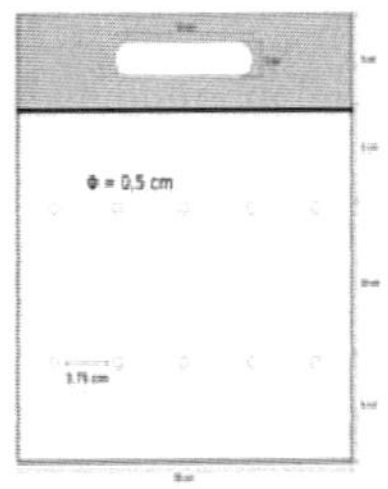


**Figure 1.** Design dimension plastic packaging

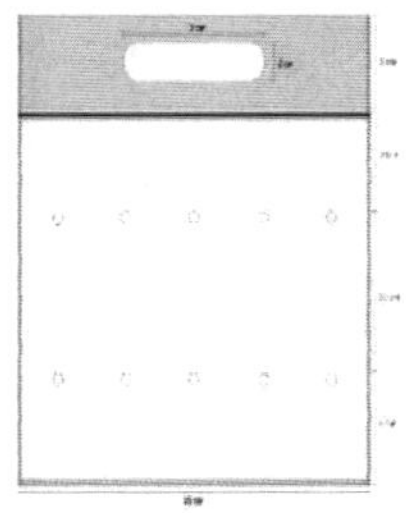

**Figure 2.** Design plastic packaging

3.2.2. *Corrugated cardboard packaging*

Based on the results of measurements of the weight and dimensions of papaya fruit, corrugated cardboard design were obtained (table 6 and figure 3 and 4).

**Tabel 6**. Dimension of corrugated cardboard packaging

| Parameter | Dimension packaging | | |
|---|---|---|---|
| | Inner dimension | Design dimension | Outer dimension |
| Length (mm) | 81 | 84 | 90 |
| Widht (mm) | 81 | 84 | 90 |
| Height (mm) | 231 | 234 | 240 |
| Packaging area (mm$^2$) | 656.1 | 705.6 | 810 |
| Packaging volume (mm$^{3)}$ | 15155.9 | 16511 | 19440 |

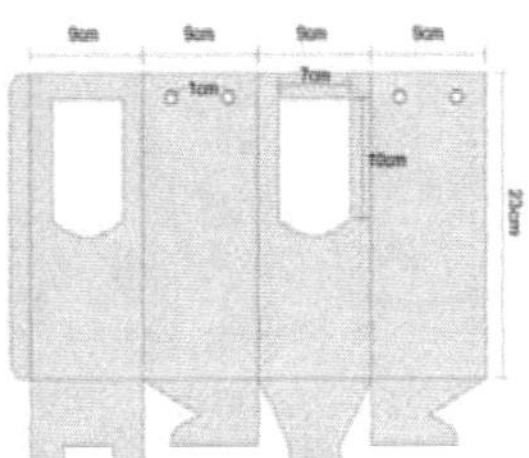

**Figure 3.**dimension corrugated cardboard packaging design

**Figure 4.** corrugated cardboard packaging design

Corrugated cardboard was a type of packaging that was often used to package products such as fruit, vegetables and industrial goods. The use of corrugated cardboard packaging was choosen because, easy to formed, easy to obtained, can be recycled [6]. Corrugated cardboard packaging was also very efficient for shipping, and the use of this packaging had been used in general for distribution, transportation and storage [7].

3.3. *Design distribution packaging*

The type of packaging designed for this distribution package is a double wall flute BC corrugated cardboard with RSC (Regular Slotted Container) and HTC packaging type (Half Telescope Container) which consists of body parts and cover. This material was chosen because it could reduce vibration well and has a smooth surface. Horticultural product packaging has several properties that are more profitable than wood, including having a smooth surface, easy to assemble, and the most important is having good vibration-reducing properties [8]. The use of BC flute typecorrugated cardboard packaging was chosen because it has the best flat compressive resistance so that it is expected to maintain the quality of fruit during distribution [9].

Each distribution package was ventilated. Ventilation on horizontal packaging (KS1) is 12 ventilation and for vertical packaging (KS2) there are 8 ventilation, where the hole diameter of each ventilation is 2 cm. Ventilation serves to maintain adequate air distribution according to product needs so that the heat transfer of the product to cold air will be optimal, this can reduce product damage during transportation [10]. Ventilation was circle shaped, because ventilation of this circle results in having air circulation and good temperature distribution [11].

Packaging distribution of design results for horizontal packaging (KS1) is 41x25x11.5 cm and for vertical packaging (KS2) is 33x23x7 cm. The results of calculation of inner dimensions, design dimensions and external dimensions could be seen in tables 7 and table 8. Packaging design and secondary packaging design results in Figures 5,6,7,8 and 9.

**Tabel 7**. Horizontal packaging design (KS1)

| Parameter | Dimension packaging | | |
|---|---|---|---|
| | Inner dimension | Design dimension | Outer dimension |
| Length (mm) | 399 | 396 | 410 |
| Width (mm) | 229 | 236 | 250 |
| Height (mm) | 94 | 100.1 | 115 |
| Packaging area (mm$^2$) | 9137.1 | 9345.6 | 10250 |
| Packaging volume (mm$^{3)}$ | 85888,7 | 93549.4 | 117875 |

**Tabel 8**. Vertical packaging design (KS2)

| Parameter | Part of packaging | |
|---|---|---|
| | *Bottom (Body)* | *Top (Cover)* |
| Inner dimension | | |
| Length (mm) | 289 | |
| Width (mm) | 189 | |
| Height (mm) | 229 | |
| Design dimension | | |
| Length (mm) | 296 | 316 |
| Width (mm) | 196 | 216 |
| Height (mm) | 236 | 56 |
| Outer dimension | | |
| Length (mm) | 310 | 330 |
| Width (mm) | 210 | 230 |
| Height (mm) | 250 | 70 |
| Packaging area (mm$^2$) | 6510 | 7590 |
| Packaging volume (mm$^3$) | 162750 | 53130 |

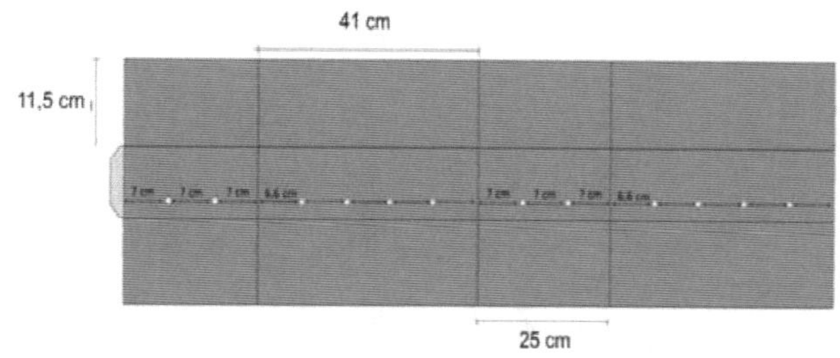


Figure 5. Dimension horizontal packaging (KS1) design

Figure 6. horizontal packaging (KS1) design

Figure 7. Vertical packaging design (KS2)

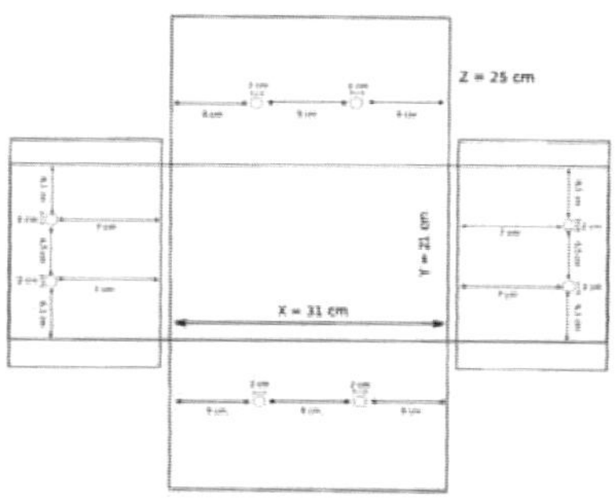


Figure 8. vertical packaging (KS2) design body

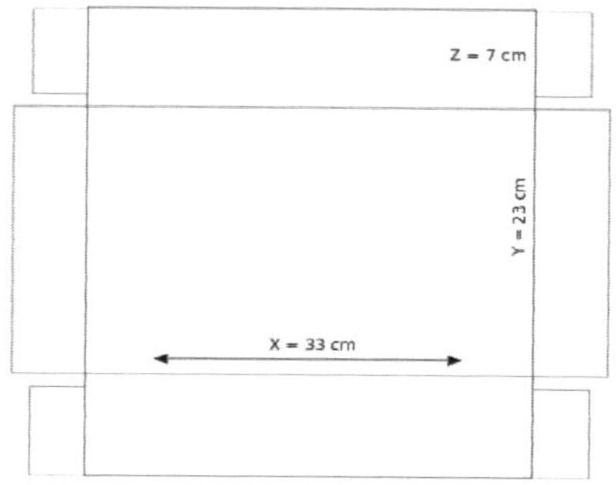


Figure 9. vertical packaging (KS2) design cover.

3.4. *Packaging Strength and Number of Stacks*

The compressive strength test is very important to do on corrugated cardboard packaging to find out the maximum strength of the carton when stacked during distribution using containers and when storing in a warehouse. This piles causes pressure from top to bottom, commonly called top to bottom compression. In addition, the compressive strength value could be used to calculate the maximum number of piles that can be used as a consideration for choosing the appropriate packaging design for papaya [2]. In the real condition in the field the stack height is arranged so that the packaging in the lowest position does not suffer damage caused by the static load from the packaging above [12]. The results of the calculation of the packaging strength test and the maximum number of stack packs in table 9.

**Table 9**. Packaging Strength and Number of Stacks

| Design packaging | Dimension (PxLxT) cm | Fruit capacity | Compressive strength (N/m$^2$) | Number of stacks |
|---|---|---|---|---|
| KS1 | 41 x 25 x 11,5 | 4 | 16623, 99 | 23 |
| KS2 | 31 x 21 x 25 | 6 | 6673,97 | 7 |

3.5. *Packaging arrangement efficiency for transportation packaging*
3.5.1. *Packaging arrangement efficiency on pallet*

The packaging arrangement on the pallet is intended for export so that it is known that the pallet size matches the design package. In the papaya export-import trade the efficiency of product distribution would be maximized if the producing country and importing countries use the same pallet size [2].

The optimal efficiency of pallet used consists of pallet load efficiency of 90%, namely good fit, 80% efficiency, namely average fit and 70% efficiency, namely poor fit. The highest efficiency value for the package designed is a pallet with dimensions of 1200x1000 mm with efficiency of 85.41% for horizontal packaging (KS1) and dimensions of 1067x1067 mm with 80% efficiency for vertical packaging (KS2). According to [14], the presence of residual space on the pallet can provide air circulation space between packages and facilitate the preparation and dismantling during distribution.

The results of calculating the efficiency of using various sizes of pallet based on ISO pallet standards for packaging design results are presented in table 10. The composition of distribution packaging on the pallet with the best efficiency can be seen in Figure 10 and 11.

**Table 10**. Value of efficiency of use of pallets for design packages

| Pallet size (mm) | Country | Pallet eficiency (%) | |
|---|---|---|---|
| | | Horizontal packaging | Vertical packaging |
| 1200 x 800 | Europe, Singapore, Chinese | 74,73 | 71.15 |
| 1140 x 1140 | Several european country, Chinese, Germany, Netherlands, Taiwan, | 78.87 | 70.08 |
| 1200 x 1000 | Singapore, Thailand, Chinese, Indonesia USA, Chinese | 85.41 | 75.9 |
| 1219 x 1016 | USA and Canada | 82.76 | 73.54 |
| 1067 x 1067 | Japan, Taiwan, Korea, | 72.02 | 80 |
| 1100 x 1100 | Singapore, Thailand | 84.71 | 75.15 |

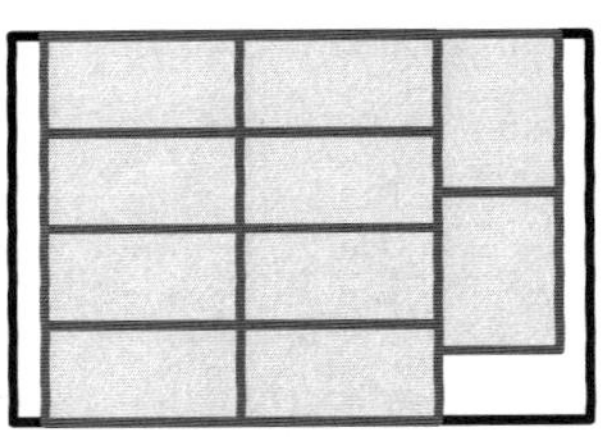

1200x1000

**Figure 10.** efficiency on pallet for horizontal packaging (KS1)

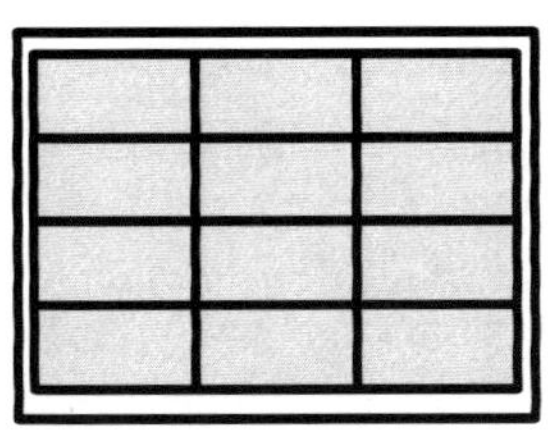

1067x1067

**Figure 11.** efficiency on pallet for vertical packaging (KS2)

3.5.2. *Efficiency of packaging arrangement in transport cars*

The efficiency of packaging arrangement on transport cars is intended to find out shipments that could affect fruit quantity and shipping costs. Some of the transport vehicles that were often used for transportation could be seen in table 11 and in Figure 12.

**Table 11**. Efficiency of packaging arrangement horizontal packaging (KS1) and vertical packaging (KS2) in various types of transport vehicles

| Type of packaging | Types of transport vehicles | Size transport cars (mm) | | Large transport cars (mm$^{2)}$ | Number of packages | Packaging area (mm$^2$) | Eficiency (%) |
|---|---|---|---|---|---|---|---|
| | | Length | Width | | | | |
| Horizontal packaging | M1 | 2200 | 1480 | 3256000 | 25 | 2562500 | 78,70 |
| | M2 | 2630 | 1460 | 3839800 | 30 | 3075000 | 80,08 |
| | M3 | 2350 | 1585 | 3724750 | 30 | 3075000 | 82,56 |
| | M4 | 2425 | 1600 | 3880000 | 33 | 3385200 | 87,17 |
| | M5 | 2200 | 1585 | 3487000 | 30 | 3075000 | 88,18 |
| Vertical packaging | M1 | 2200 | 1480 | 3256000 | 36 | 2732400 | 83,91 |
| | M2 | 2630 | 1460 | 3839800 | 48 | 3643200 | 94,87 |
| | M3 | 2350 | 1585 | 3724750 | 42 | 3187800 | 85,58 |
| | M4 | 2425 | 1600 | 3880000 | 42 | 3187800 | 82,16 |
| | M5 | 2200 | 1585 | 3487000 | 36 | 2732400 | 78,36 |

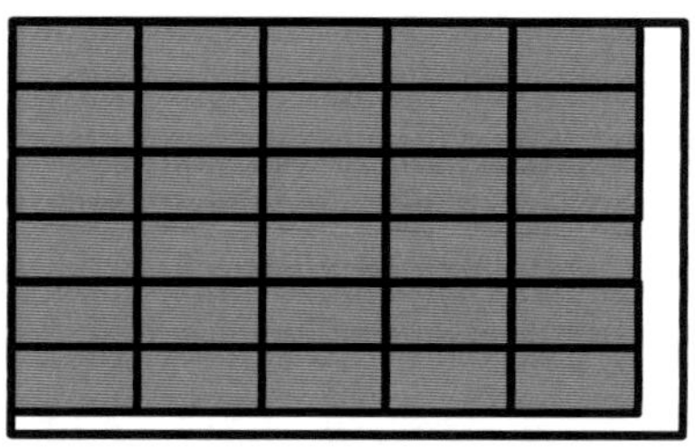

**Figure 12.** efficiency on pallet for horizontal packaging (KS1)

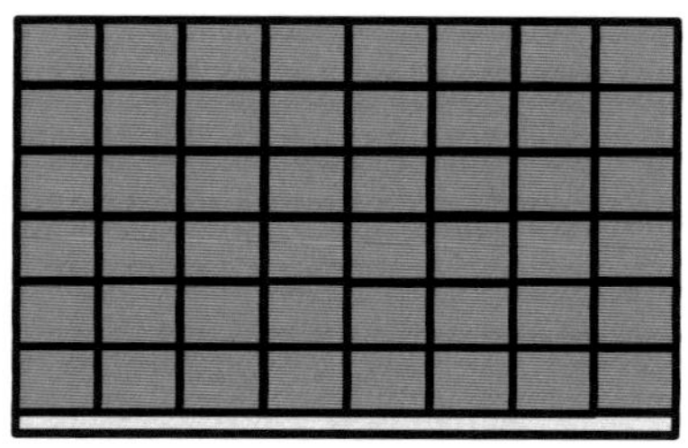

**Figure 13.** efficiency on pallet for vertical packaging (KS2)

Product distribution efficiency will be maximized if horizontal packaging (KS1) uses the type of M5 transport vehicle that is for 88.18% and for vertical packaging (KS2) using M2 type of car that is for 94.87%.

## 4. Conclusion

The average dimension of one papaya was the average weight of 808.63-883.54 g, the average diameter is 26.35-29.31 cm and the average height is 22.85-23.70 cm. The PP plastic packaging was 25x12.5x25 cm and carton packaging was 9x9x24 cm. Horizontal packaging was 41x25x11.5, while vertical packaging was 33x23x7 for top /cover packaging, and 31x21x25 for bottom / body packaging. The packaging strength test results, for horizontal packaging amounted to 16623. 99 N / m2 with a maximum number of packing stacks of 23 piles, and for vertical packaging of 6673.97 N / m2, with a maximum number of packing piles, namely 7 stacks. The efficiency of packaging preparation will be better using a pallet size of 1200x1000 mm (85.41%) for horizontal packaging and pallet sizes of 1067x 1067 mm (80%) for vertical packaging. For horizontal packaging, the efficiency of packaging preparation on a transport vehicle would be better to use a size of 220x158.5 cm or M5 (88.18%) and 263x146 cm or M2 (90.92) for vertical packaging.

**References**

[1] Suketi K, Suhartono M R, Fariyanti A, Harti H, Gunawan E, Darma K 2014 *Prosiding Seminar Nasional PERHORTI. Malang, Indonesia*. 136-142.
[2] Tawakal M I 2017 *Thesis Institut Pertanian Bogor*
[3] Sutrisno, Seesar Y A and Sugiyono 2009 *Makalah Bidang Teknik Produk Pertanian ; Seminar dan Gelar Teknologi FERTETA*. Indonesia.
[4] Qantyah and Ambarsari A 2011 *J.Litbang Pertanian* **30** 8-15
[5] Asgar A 2017 *J.Horti* **27** 127-131
[6] Thompson J F, Park J M and Singh R P *Applied Engginering in Agriculture* **26** 919-924
[7] Han J and Park J M 2009 *Packaging Technology and Sciency* **26** 919-924.
[8] Wasala W C M B, Dharmasena D A N, Dissayanke T M R and Thilakarathe B M K S 2015 *Tropical Agriculture Research* **26** 355-367
[9] Nelinda 2017 *Thesis Institut Pertanian Bogor*
[10] Opara U L and Fadiji T 2018 *J. Scienta* **227** 154-161.
[11] Darmawati E, Sakti G A W 2011 *Kajian Teknik Pasca Panen dan Proses Hasil Pertanian* 227-235.
[12] Yulianti N L, Sutrinso and Darmawati E 2010 *Jurnal Keteknikan Pertanian* **24** 55-60
[13] Iswahyudi , Darmawati E and Sutrisno. 2014 *Jurnal Keteknikan Pertanian*. **3**

# Effects of slurry concentration and rotational speeds of drum drier to the characteristics of pregelatinized breadfruit (*Artocarpus altilis*) flours

**T C Sunarti[1], W S L Br. Sinaga[2], and C Pandji[1]**

[1]Department of Agroindustrial Technology, Faculty of Agricultural Engineering and Technology, Bogor Agricultural University, Indonesia
[2]Program Study of Food Science, Faculty of Science and Technology, University of Pelita Harapan, Tangerang, Indonesia

**Abstract.** Breadfruit is one alternative carbohydrate source in Indonesia. However, breadfruit is seasonal and climacteric crops; it should be processed immediately to produce flour. Native breadfruit flour has many limitations since it is difficult to be solubilized and swollen in cold water. The flour is modified physically by pregelatinization to improve its properties. Pragelatinized breadfruit flour can be used as instant food, porridge, and baby food. This research is aimed to investigate the effect of slurry concentration (20 and 30%) of native breadfruit flour and rotational speeds (4, 6 and 8 rpm) of double drum drier to the chemical composition and physico-chemical properties of the pregelatinized flours. The results showed that pregelatinization did not influence to the chemical composition. Pregelatinization of the flour improved the water solubility and swelling power on room (30°C) and hot (70°C) solutions, reduced the freeze-thaw stability, caused partial gelatinization which is monitored by losing the birefringence of starch granule, and darkened the flour. The pregelatinization also reduced the gelatinization temperature and had low final viscosity. Low concentration of slurry and low rotation speed of drum increased the gelatinized part of the flour, which is influenced to the cold and hot water solubility.
**Keywords:** breadfruit flour, drum drying, physico-chemical properties, pregelatinization

## 1. Introduction

Breadfruit (*Artocarpus altilis)* or sukun (local name) belongs to the Moraceae a family of about 50 general and over 1000 species. The species originated in Oceania, and is grown across the Pacific region. It is now found in many countries in tropical zones across the world, including in Africa, Australia, South America, South and Southeast Asia [1]. It has long been an important staple crop and a primary component of traditional agroforestry systems in Oceania [2].

The plant *Artocarpus altilis*, seedless variety, is a fruit-producing plant. In Indonesia, breadfruit is harvested twice a year. The fruit can be eaten in all stage of maturity, since it is a good source for carbohydrate, vitamin and minerals [2]. Breadfruit is a large, round, starchy fruit borne by a tree. It is classified as climacteric fruit, which has fast respiration process [3]; therefore it has short shelf-life. In order to anticipate abundant breadfruit during the peak harvest session and to prolong its shelf life, the fruits should be processed immediately.

Breadfruit is consumed after cooking by boiling, steaming, roasting, baking and frying. Breadfruit has been processed into many forms for utilization in the food industry. It can been processed into starches [4] into flour [5,6] and other modified products [7]. Native breadfruit flour has some limitation in application as food products. It has low solubility and swellability in room temperature. Studies on the modification of breadfruit starch involving heat-moisture-treatment and annealing, as forms of physical modification collectively referred to as hydrothermal treatment which entail modification of temperature and moisture content have been carried out and reported by Adebowale et al. [8]; and Ojifolokun and Adeniran [7]reported the study explored the potential of substituting breadfruit into cassava to obtain gari analoque.

Pregelatinization is one of the physical methods used to modify starch by using drum dryer, spray dryer or extruder [9]. This modification method affects physicochemical and functional properties of flour significantly [10]. Due to starch granule disruption, pregelatinized flour can absorb water and increase viscosity immediately even with cold water [11,12]. For powder production from fruit or root and tuber, drum drying is considered a special method for numerous materials with many benefits such as good characteristics of product and economic efficiency of the process. Pregelatinized starch is primarily used as a thickener in many instant products, such as baby food, soups and desserts, due to its ability to form pastes and dissolve in cold water. The use of pregelatinized starch is preferred in sensible foods because it does not require heating to form a paste [13]. The four variables which govern the operation of drum are steam pressure, rotational speed, and film thickness and feed material characteristics. The steam pressure or heating medium temperature will regulate the drums temperature. The rotational speed of the drum determines the all performance of dryer [14]. This research is aimed to investigate the effect of slurry concentration of native breadfruit flour and rotational speeds of double drum drier to the chemical composition and physico-chemical properties of the pregelatinized flours.

## 2. Methods

### *2.1. Materials*
The breadfruit used in this study was collected from local farmer in Jawa Barat. They were on the optimum green maturity, and taken in polypropylene plastic bags to the laboratory to use and process immediately.

### *2.2. Processing of Breadfruit into Flour*
Fresh fruits were peeled and washed with clean water, and then sliced, blanched with hot steam for one minute, and dried at 60°C for 6-8 h. Dried chips were milled into flour, and the flour was screened by 60-mesh standard sieve.

### *2.3. Production of Pregelatinized Flour*
The slurry (A1-20% and A2-30%) was prepared by mixing the flour with distilled water, and then dried in double drum drier with different rotational speeds (R1-4 rpm, R2-6 rpm, and R3-8 rpm). The drier has two stainless steel drums with 12 in outer diameter and 8 in cylinder lenght, and 80±5°C surface drier plate temperature. Then the dried flour was milled and screened by 60-mesh standard sieve. Samples were kept in polyethylene bag, sealed and stored at room temperature until used.

### *2.4. Flour Characteristics*
The flour was analyzed its proximate components (moisture, ash, crude fiber, crude protein, crude fat, and carbohydrate *by difference* contents) according to AOAC methods [15], starch content (Luff Schorl method [15], the whiteness of flour by Colortec, shape and size of starch granules by using polarized light microscope.

The viscoamylographical properties of each flour was studied by using Brabender Viscoamylographer. Flour solubility and swelling power at 30°C and 70°C, paste clarity, and freeze-thaw stability were analyzed by using modified Perez *et al.* [16]. Viscosity of 5% flour paste was analyzed by using Brookfield viscosimeter.

## 3. Results and Discussion
Breadfruit should be harvested at the stage of maturity that is required for the method of preparation for consumption. In this research, the breadfruit is harvested as "green mature". At this stage the fruit has completed its development and has reached the dimensions of a mature fruit but has not begun to ripen and is still very firm. The flesh is very starchy and not sweet to the taste.

Native breadfruit flour (A0R0) and six pregelatinized flour were analyzed for their chemical composition (Table 1). It has shown that all flour contained relative low minor components such as fat (1.004-3.671%) and protein (3.694-5.028%), but high in carbohydrate and fiber. The starch is the main carbohydrate in breadfruit flour. The starch content in native flour was determined as 31.6%. Similar results also reported by Wooton and Tumaalii [5], besides the starch the flour also contained lots of total sugar with fructose is predominant sugar found in less mature fruits. The protein and fat contents showed decreasing values compared to native flour. The major cause of this significant decrease in the crude protein content may be the pre-gelatinization process, which allows elution and dissolution of water-soluble proteins during steam treatment [17]. In food application, this composition is a key to achieve desirable product quality as well as nutrition value and even the lipid content in flours was low, it could affect to the flour properties dramatically such as swelling and pasting properties [18].

**Table 1** Proximate components of breadfruit flour

| Treatment | Chemical components*(%) | | | | | |
|---|---|---|---|---|---|---|
| | Moisture | Ash | Crude Protein | Crude Fat | Crude Fiber | Carbohydrate (*by difference*) |
| A0R0 | 3.537 | 0.429 | 5.028 | 3.671 | 5.044 | 82.286 |
| A1R1 | 2.237$^{a}$ | 0.332$^{a}$ | 3.694$^{a}$ | 1.004$^{d}$ | 11.037$^{a}$ | 82.953$^{a}$ |
| A1R2 | 1.876$^{a}$ | 0.402$^{a}$ | 4.060$^{a}$ | 2.249$^{bc}$ | 8.536$^{a}$ | 82.547$^{a}$ |
| A1R3 | 2.721$^{a}$ | 0.38$^{a}$ | 3.977$^{a}$ | 2.576$^{a}$ | 8.060$^{a}$ | 82.493$^{a}$ |
| A2R1 | 2.438$^{a}$ | 0.384$^{a}$ | 4.824$^{b}$ | 2.070$^{c}$ | 10.200$^{a}$ | 80.501$^{a}$ |
| A2R2 | 2.048$^{a}$ | 0.382$^{a}$ | 4.019$^{b}$ | 2.789$^{a}$ | 10.613$^{a}$ | 79.464$^{a}$ |
| A2R3 | 2.844$^{a}$ | 0.510$^{a}$ | 4.315$^{b}$ | 2.494$^{abc}$ | 7.091$^{a}$ | 85.377$^{a}$ |

*All calculated in dry-basis except for moisture content; $^{a}$Values shown in same column with the same letters are not significantly different ($P<0.05$)

Pregelatinized flour is flour that has undergone a cooking process until complete or partial gelatinization and a simultaneous drying process. Since the process is physical process, it did not affect to the chemical composition, as shown in Table 1 for different slurry concentration and rotational speeds of drum dryer. The main consequence of this treatment is the destruction of the granular structure, resulting in complete granular fragmentation, and the absence of birefringence properties as shown in Figure 1.

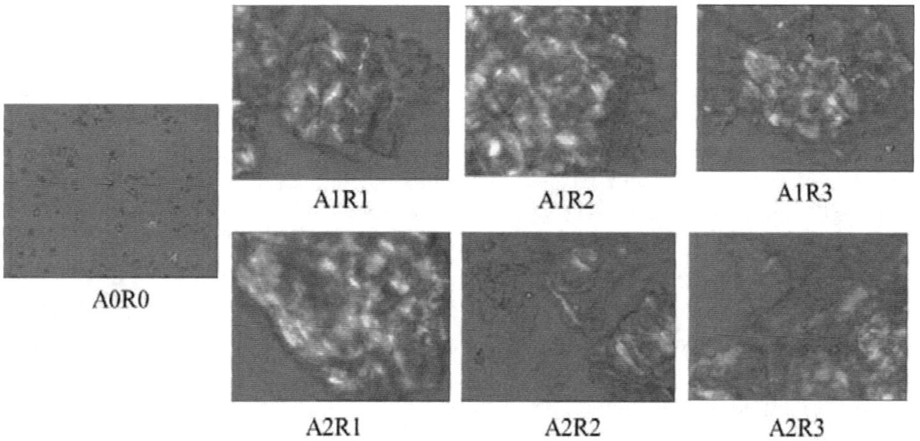


**Figure 1.** Morphology of breadflour in water by using polarized light microscope (magnification 200 times)

Figure 1 (A0R0) showed the small size granule of breadfruit native starch. Rincon and Padilla [4] reported the breadfruit granules size are around 3-8 µm. It expressed the common granules that could not swell in room temperature, but vice versa the pregelatinized flour have tendency to cold swelling properties. The principal properties of pregelatinization flour are an increase in swelling capacity, solubility and cold water dispersion. Table 2 showed that pregelatinization slightly increased the solubility and swelling power at 30°C and 70°C, paste viscosity and clarity. Murayama et al. [17] reported that the solubility of the native flours showed slight changes between 50 and 70°C, and significant increases at 80°C and above. The treatments (slurry concentration and rotational speed of drum dryer) did not significantly affect to the flour properties, but it caused darkening of the flour as shown as low degree of whiteness (35-42%).

The native breadfruit flour has high tendency to retrogradation as shown in high freeze-thaw stability and pasting properties (Figure 2 and Table 3 for A0R0). Similar pattern for breadfruit starch also reported by Rincon and Padilla [4]. Pregelatinization effectively affected to the cold swelling flour with low (30°C) initial gelatinization temperature. Higher slurry concentration and higher rotational speed caused the increasing of maximum viscosity and reduced the tendency to retrogradation.

**Table 2.** Physico-chemical properties of breadfruit flour

| Characteristics | Flour Treatment | | | | | | |
|---|---|---|---|---|---|---|---|
| | A0R0 | A1R1 | A1R2 | A1R3 | A2R1 | A2R2 | A2R3 |
| Solubility at 30°C (%) | 23.08 | 31.28$^{b}$ | 29.39$^{b}$ | 33.12$^{a}$ | 28.92$^{b}$ | 31.50$^{b}$ | 33.40$^{a}$ |
| Solubility at 70°C (%) | 37.43 | 41.53$^{a}$ | 43.16$^{a}$ | 43.82$^{a}$ | 40.62$^{a}$ | 41.80$^{a}$ | 37.51$^{a}$ |
| Swelling Power at 30°C (%) | 36.97 | 44.65$^{a}$ | 51.06$^{a}$ | 44.41$^{a}$ | 41.74$^{a}$ | 42.57$^{a}$ | 44.37$^{a}$ |
| Swelling Power at 70°C (%) | 44.94 | 50.93$^{a}$ | 50.81$^{a}$ | 53.52$^{a}$ | 58.99$^{a}$ | 50.32$^{a}$ | 63.40$^{a}$ |
| 0.1% Paste Clarity | 90.55 | 83.55$^{a}$ | 93.90$^{a}$ | 94.60$^{a}$ | 89.85$^{a}$ | 94.05$^{a}$ | 93.75$^{a}$ |
| Freeze-thaw stability (% Sineresis) | 80.00 | 85.71$^{a}$ | 84.06$^{b}$ | 84.29$^{b}$ | 90.71$^{a}$ | 84.29$^{b}$ | 84.29$^{b}$ |
| Paste viscosity (cP) | 5.50 | 7.00$^{a}$ | 9.25$^{a}$ | 9.25$^{a}$ | 8.35$^{a}$ | 6.65$^{a}$ | 9.10$^{a}$ |
| Degree of whiteness (%) | 54.77 | 40.77$^{a}$ | 42.41$^{a}$ | 35.82$^{a}$ | 38.73$^{b}$ | 37.41$^{b}$ | 38.27$^{b}$ |
| Amylase digestibility (%) | 44.27 | 32.72$^{a}$ | 42.76$^{a}$ | 45.31$^{a}$ | 47.71$^{a}$ | 42.33$^{a}$ | 37.88$^{a}$ |

[a]Values shown in same row with the same letters are not significantly different (P<0.05)

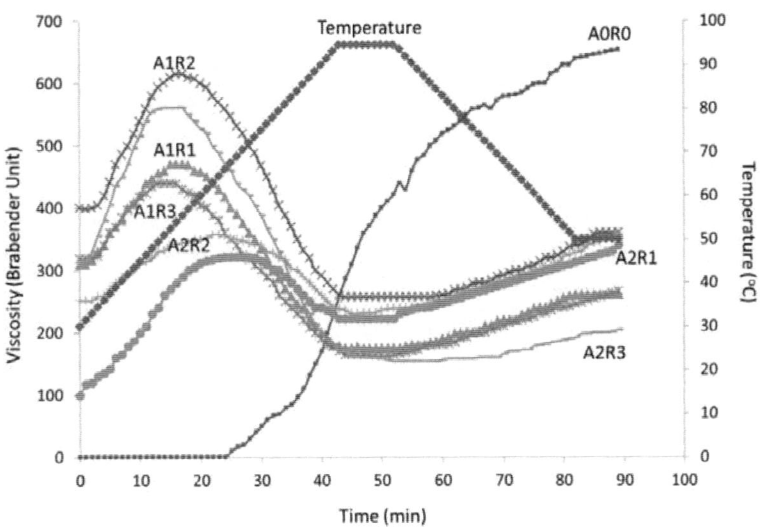


**Figure 2.** Pasting properties of breadfruit flour by Brabender Viscoamylographer

## 4. Conclusion

Breadfruit flour contained high amount of carbohydrate mainly starch, and low minor components such as protein and fat, but the pregelatinization process did not influence to the chemical composition. Pregelatinization of the flour improved the water solubility and swelling power on room and hot solutions, reduced the freeze-thaw stability, caused partial gelatinization which is monitored by losing the birefringence of starch granule, and darkened the flour. The pregelatinization also reduced the initial gelatinization temperature and had low final viscosity. Low concentration of slurry and low rotation speed of drum increased the gelatinized part of the flour, which is influenced to the cold and hot water solubility.

**Table 3** Parameters of Brabender Viscoamylography of Breadfruit Flour

| Treatment | Initial Gelatinization Temperature (°C) | Maximum Viscosity (BU) | Breakdown Viscosity (BU) | Set Back Viscosity (BU) | Final Viscosity (BU) |
|---|---|---|---|---|---|
| A0R0 | 67.5 | nd* | nd | nd | 654 |
| A1R1 | 30 | 470 | 295 | 85 | 260 |
| A1R2 | 30 | 615 | 358 | 103 | 360 |
| A1R3 | 30 | 440 | 278 | 104 | 266 |
| A2R1 | 30 | 322 | 100 | 118 | 340 |
| A2R2 | 30 | 358 | 126 | 127 | 359 |
| A2R3 | 30 | 562 | 408 | 50 | 204 |

*nd: not detected

## References

[1] FAO 2018 Breadfruit http://www.fao.org/traditional-crops/breadfruit/en/
[2] Ravinder A, Waghray K , Subbarao and Evangeline S J 2018 *J. Food Nutr. Populat Health* **2**(1) 8
[3] Thompson A K, Bean B O, and Perking C 1974 *Trop. Agric.* **511** 701
[4] Rincon A M and Padilla FC 2004 *Archivos Latinoamericanos de Nutrición* **54**(4) 449
[5] Wootton M and Tumaalii F 1984 *Food Sci.* **49**(5) 1396
[6] Olaoye O A, Onilude A A and Oladoye C O 2007 *Afr. J Food Sci.* **October** 20
[7] Ajifolokun O M and Adeniran H A 2018 *J. Nutr. Food Sci.* **8** 658
[8] Adebowale K O, Olu-Owolabi B I, Olawumi E K, and Lawal O S 2005 *Indust. Crops and Products* **21**(3) 3
[9] Colonna P, Doublier J L, Melcion J P, and Mercier C 1984. *Cereal Chem.* **61** 543
[10] Lai H M 2001. *Food Chem.* **72**: 455
[11] Glicksman M 1969 *Gum Technology in Food Industry* (New York: Academic Press)
[12] Doublier J L, Colonna P, and Mercier C 1986. *Cereal Chem.* **63**: 240
[13] Majzoobi M, Radi M, Farahnaky A, Jamalian J, Tongdang T, and Mesbahi Gh 2011 *J. Agr. Sci. Tech.* **13** 193
[14] Valous N A, Gavrielidou M A, Karapantsios TD, and Kostoglou M 2002 *J. Food Eng.* **51**(3): 171
[15] AOAC 1995 *Official Methods of Analysis of the Association of Official of Analitycal Chemist* (Washingtong DC: AOAC Int.)
[16] Bello-Pe´rez L A, Agama-Acevedo E, Sa´nchez-Herna´ndez L, and Paredes-Lo´pez O 1999 *J. Agric. Food Chem.* **47** 854
[17] Murayama D, Kasano M, Santiago D M, Yamauchi H, and Koaze H 2014 *Food Sci Technol Res* 20(6): 1131
[18] Eliasson A -C and Gudmundsson M 1996 Starch: physicochemical and functional aspects *Carbohydrates in Food* ed A -C. Eliasson (New York: Marcel Dekker, Inc) pp. 431-503

## Acknowledgements

Thank you very much to Forest Research and Development Agency for providing the raw material (breadfruit) for this research.

# Determination of Mung Bean Seed Viability Change in Vacuum Packaging during Storage in Different Temperatures

**M. Faisal[1], U. Ahmad[2*], and D. Wulandani[2]**

[1]Postharvest Technology Master Course Program, Graduate School, Bogor Agricultural University, Indonesia
[2]Dept. of Mechanical and Bio-system Engineering, IPB, Bogor Agricultural University, Indonesia

e-mail: uahmad2010@gmail.com

**Abstract.** Mung bean seeds decreased their viability very fast during storage, especially in tropical area without proper packaging. The purpose of this study is to observethe decrease in viability of mung bean seeds during storage and to build a mathematical model for shelf life prediction. Dried mung bean seeds were vacuum packed and stored in Accelerated Shelf-life Testing chamber at five different temperatures of 30° C, 35° C, 40° C, 45° C, 50° C with approximately 80% in relative humidity, to facilitate rapid deterioration with different rate of physical and chemical change in the seed. The results show that there was an increase in moisture content and the germination decreased in different rates. Changes in values for some parametersof the seeds affected the viabilty of the seeds during rapid deterioration. The value of constant (k)for viability decreasing was calculated and used to build the mathematical model for predicting the shelf life of mung bean seeds during storage at any temperature. Using the developed model, it was observed that the shelf life of the Sriti variety of mung bean seeds in vacuum packaging could maintain 90% of germination level for 421 days or 14.3 months if stored at20 °C, and for 173.37 days or 5.78 months if stored at 25 °C.

## 1. Introduction

Mung bean (*Vigna radiata (l.)* Wilczek*)* is the third important legumes crop in Indonesia after soybeans and peanuts. However, the production of mung bean in Indonesia is not stable and small if compared to its consumption, and Indonesia needs to import the rest from other caountry. During the ten years of 2006-2015, average mung beans production was 288,199 ton/year, ranging from 204,670 ton/year to 342,342 ton/year **[1]**. Until now, many efforts to increase mung bean production in Indonesia have been introduced, but the result is still under expectation.

One important aspect fora success mung beans production is the use of good quality seeds.The availability of high quality seeds is very important for a success business in agriculture, including cultivation and production of mung beans **[2]**. Quality seeds of superior varieties are one key factor for high productivity in mung beans production. According to **[3]** production of seeds is usually carried out some time before the planting season begins, so that the seeds will experience chemical changeduring storage. Therefore, the seed must be stored properly so that it will experience minimum chemical change so that the seeds would have optimal viability when they were planted. It needs a long several steps processes to produce good quality seeds, from selection of cultivating resourcesincluding seeds, and a good facility to store seeds in packaging. Since the seeds reach physiological maturity stage until they are planted, the seeds will be in the storage period. During this storage period the seeds will experience decline in viability because of chemical change.

The decrease in the quality of mung bean seeds is usually caused by the change of relatively high protein and fat content in the seeds, increasing of moisture content of the seeds if the temperature and humidity of the storage environment are relatively high. This decrease in quality is a gradual and cumulative process of deterioration in viability, and cannot be reversed due to physiological and biochemical changes **[4-5]**. The process of decreasing seed viability cannot be stopped, but might be slow down to get a longer shelf life. The rate of chemical change in seeds during the storage period is influenced by two factors, internal and external factors. The internal factor is related to the seeds characteristics that influence germination process, a genetic trait called physiological decline. The external factor isrelated to seed packaging, storage temperature and humidity which is called physical factors, and biological causing damage to the seeds such as bacteria, fungi, insects and mouse **[6].**

There are two methods to determine shelf life of storing food or agricultural products, namely conventional methods and ASLT (Accelerated Shelf-life Testing) method **[7]**. Conventional method is carried out by storing products in normal conditions as they should be. This method is accurate but it takes a long time and is expensive **[8].** To facilitate a faster chemical change, the ASLT method is applied by storing the product in a modified environment so that the stored product will undergoa quick chemical change under the spesific condition of temperature and humidity of the modified space. Under the condition of storage temperature and humidity, the acceleration method with the Arrhenius model is considered suitable for estimating the shelf life of stored seeds. Therefore, seed shelf life can rich their limit for a relatively short time, without having to wait for the seed shelf life in the actual storage conditions **[9]**. The purpose of this study is to examine the parameters of viability of mung bean seeds during storage under vacuum packaging and estimate the shelf life of mung bean seeds at several storage temperatures using the ASLT (Accelerated Shelf-life Testing) method.

## 2. Research Methods

### *2.1. Time and Research Place*
The study was conducted from February 2018 - September 2018, at the Food and Agricultural Products Processing Engineering Laboratory, Mechanical and Biosystem Engineering Department, Faculty of Agricultural Engineering and Technology, and Seed Science and Technology Laboratory, Faculty of Agriculture, Bogor Agricultural University.

### *2.2. Materials and tools*
The mung bean seeds used as the material in this study are Sriti varieties. As much as 40 kg mung beans are used, packed in a vacuum pack each contains 200 g. The material used for packaging was vacuum plastic packaging (hermetic), aquades, paper, and chemicals for analysis of free fatty acids. The equipment used was Eyela Environmental Chamber, an Oven of 2120 Isuzu Seisakusho, a Vacuum sealer, germinator,scissor, scales, petri dish, desiccator, soxhlet device, camera, container (plastic tray).

### *2.3. Material Preparation and Seed Packaging*
Mung bean grainsare obtained from aseeds producer in Bolo District of Bima Region in West Nusa Tenggara (NTB) Province. The grains were sorted manually from separated from dirtand broken grains, and then sundried until itsmoisture content reached 10% wet base or lower.Then the mung bean seeds were packed in a vacuum plastic package (hermetic,13 x 21 cmin size, and thickness of 0.08 mm)according to the mung bean seed capacity of 200 g with a

packaging, then sealed using a vacuum sealer. Mung bean seeds that have been packed were put in the Eyela Environmental Chamber with different temperatures, namely temperature settings for each treatment (30°C, 35 °C, 40 °C, 45 °C, 50 °C) and 80% relative humidity.The number of samples needed with two replications as much as two packs for each observation at each temperature treatment, so that the sample needed were 10 packs of each temperature parameter. Since the observations are made once before storage and four times during the storage period, then total sample needed for the whole experiment was 50 packs of each temperature treatment. Storage period was determined based on the estimated storage time until the quality of the seeds of mung beans decreases significantly.

### *2.4. Moisture Content Analysis*

Moisture content measurement was carried out using the oven method **[10]**, with a petri dish as a container that was dried in an oven 150ºC for 1 hour before using. The petri dish was cooled in the desiccator for 15 minutes ,then it was weighed using an analytical balance. A sample of 5 g was added to the cup and then weighed and put the sample into the oven, drying with a temperature of 105ºC for 24 hours. After that the sample was taken out form the oven, cooled again in a desiccator, then weighed. Drying is carried out until the weight of the material is constantand moisture content was calculated using a formula :

$$m = \frac{a-b}{a} x\ 100\% \qquad (1)$$

where :

m = moisture content (% wb)
a = sample initial weight (g)
b = final sample weight (g)

### *2.5. Germination Analysis of Between Paper Test Methods*

The germination level test used in this study was based on the ISTA standard [**11**], a between paper test method. Mung bean seeds were germinated in moist substrate conditions for a certain period of time with the purpose of distinguishing between normal sprouts (*seedling*) and abnormal ones. The measurement procedure is that the seeds were prepared as many as 400 grains taken randomly from the components of pure seeds as a result of physical purity analysis. Next, the paper media to be used is moistened with aquades and 25 seeds were sprinkled repeatedly (the seeds were arranged in a zigzag position). Seeds that have been sown thenclosed with the paper that has been moistened and and rolled neatly. The paper roll containing the seed was put into plasticbag, given a name or experimental code on each roll of paper. The paper rolls, then put into the *germinator* and then observed. Observations were carried out according to the schedule set by [**11**], namely for the first count mung bean seed was done on the $5^{th}$ day and the final count was held on the $7^{th}$ day. Percentage of germination can be calculated by formula:

$$V = \frac{NSp}{NSd} x\ 100\% \qquad (2)$$

where :
V= viability level (%)
NSp =number of sprouting seeds
NSd = number of seed

*2.6. Mung bean Seed shelf-life estimation with ASLT Method*

Germination becomes a critical parameter in determination of mung bean seeds shelf life estimation using the Arrhenius model of the *Accelerated Shelf-Life Testing* (ASLT) method **[12]**. ASLT was carried out using five different temperature treatments namely 30 °C, 35 °C, 40 °C, 45 °C, 50 °C. The percentage of germination during storage was plotted on a semi logarithmic chart. The degradation kinetics can be calculated based on zero order reactions as follow:

$$A_t - A_0 = -kt \tag{3}$$

where $A_0$ and $A_t$ were the initial condition (germination level) and condition at time t, and k is the constant decrease in viability, and t is time. When referring to a first-order reaction it can be replaced with another equation as follow:

$$\ln A_t - \ln A_0 = -kt \tag{4}$$

Correlation between storage temperatures and seed viability degradation was calculated by the Arrhenius method, by plotting $^1/_T$ (temperature in the Kelvin scale) on ln k (reaction constants that have been converted to ln forms). From this new linear regression equation curve the value of constants and $^{Ea}/_R$ can be used to find the Arrhenius constant (k). From the curve obtained the following equation:

$$\ln k = \ln k_0 - \frac{Ea}{RT} \tag{5}$$

where $k_0$ is the initial constant, R is the ideal gas constant (1,986 cal/mol.K), T is the temperature (K) and Ea is the activation energy (kal/mol.K).

In this research, the shelf life prediction was determined based on minimum germination level of 90% which was then calculated at each storage temperature using the next equation:

$$t = \frac{A_0 - A_t}{k} \tag{6}$$

where t is time (day). This equation is used for zero-order reactions, whereas for first-order reactions, the initial and final conditions need to be converted into ln forms, so that equation 7 is obtained as follows:

$$t = \frac{\ln A_0 - \ln A_t}{k} \tag{7}$$

## 3. Results and Discussion

*3.1. Moisture Content*

In this research, moisture content was measured at various storage temperatures, from five treatments in the experiment. The moisture content was believed not to directly influence the shelf life of mung bean seeds during the storage process, however this parameter is believed having contribution in such away to the viability of seeds. The moisture content of the mung bean seeds during storage related to predetermined seed quality standards. The graph of changes in moisture content during storage is shown in Figure 1.

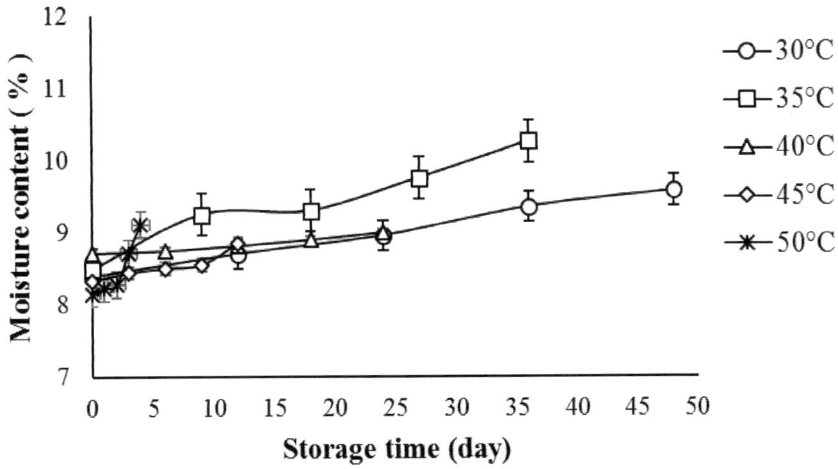


**Figure 1.** Changes in mung bean seed moisture content at five temperatures during storage period (30 °C, 35 °C, 40 °C, 45 °C and 50°C).

Moisture content for each temperature treatment slightly increased day by day. Allegedly the higher temperature during storage results in higher increasing in moisture content of mung bean seeds. This prooved that mung bean seeds are hygroscopic so they decrease (deterioration) in viability depending on the relative humidity of the air and temperature of the environment in which the seeds were stored. One factor that causes seeds to absorb water from the environment is the chemical composition of seeds, including protein. According to **[4]**, proteins are hygroscopic so they will absorb more water if the seeds are stored in humid conditions, and respiration, even slow enough, is also a factor that can cause a decrease (deterioration) in seeds. Respiration increases with increasing seed moisture content and temperature **[2]**. If the seeds are stored at low temperatures, the enzyme activity can be minimized so that respiration will be slow down. When stored at high temperatures, enzyme activity takes place faster so that respiration is faster, which causes the process of reforming food reserves faster. In this research, the results show that during storage of mung bean seeds at various temperature, the percentage of seed moisture content has increased but still within the limits of seed quality requirements. The quality requirements of mung bean seeds is a moisture content not higher than 11% **[13]**. Therefore the level of moisture content during storage is very important to minimize the reduction of mung bean seeds. Previous research has examined the effect of various water levels and storage temperature as stated in **[14]** that mung bean seeds with moisture content 9.4, 10.6 and 12.9 stored at 29 with 75% humidity for three months has a different vigor index, namely for water content 9.4 (23.46), 10.6 (21.55) and 12.9 (20.35).

*3.2. Germination*

In this research, changes in germination of mung bean seeds during the storage at five different temperatures of 30 °C, 35 °C, 40 °C, 45 °C and 50 °C are shown in Figure 2. In general, Figure 2 shows a decrease in the percentage of germination running faster at high temperatures than at low temperatures. This is similar to [**15**] that the seed has decreased the percentage of germination below 80% after being stored at temperatures above 50 °C for one month. While according to **[16]** seeds when stored at room temperature 30 °C the germination of seeds was still relatively high, which was above 80% after two months.

Another possibility for the reason lower germination level due to storage at high temperatures is the increase in mung bean seed respiration activity so that the food reserves in the seeds run out quickly. This phenomena is almost the same as the opinion expressed by [**5**] that storage of seeds at high temperatures can accelerate the activity of respiration enzymes.

Remodeling food reserves in the form of carbohydrates, protein and fat due to the process of respiration can produce metabolites. Some of these metabolites are inhibiting or poisoning other metabolism resulting in lost of germination before the energy supply in the seeds runs out.

The quality of germinating mung bean seeds in all temperatures setting after storage was still relatively good. Germination level was still above the average minimum threshold for seed quality requirements. Seed size was relatively small, protein and fat content was lower than average soybean seeds and the water content that ws at the safe limit, combined with vacuum packaging (hermetic) for storage were thought to be some factors that maintains the shelf life of the quality of mung bean seeds during storage. This fact is according to [**17**], with study which examined the effect of seed size on water absorption and germination of lentil seeds (*Lens cullinaris* Medik.), states that large-sized seeds absorb water more and faster than medium and small-sized seeds. According to [**18**] Boiyd and Van Acker (2004) the imbibition speed is influenced by the size of the seed and the diffusivity of water into the seed. With the storage of vacuum packaging (hermetic), it is suspected as important factor that maintains the germination level of mung beans was quite high. According to [**19**] Shu *et al.* (2018) vacuum packaging can inhibit changes in the morphology of mung bean seeds and maintain the quality of germination and combination of vacuum packaging with low temperatures further increasing the maintenance of germination level. This is because low temperatures decreases enzymes activities especially those contributes in respiration, can be suppressed. Meristematic cell death and decreased food reserves and degradation of enzymes can be slowed so that seed viability is higher.

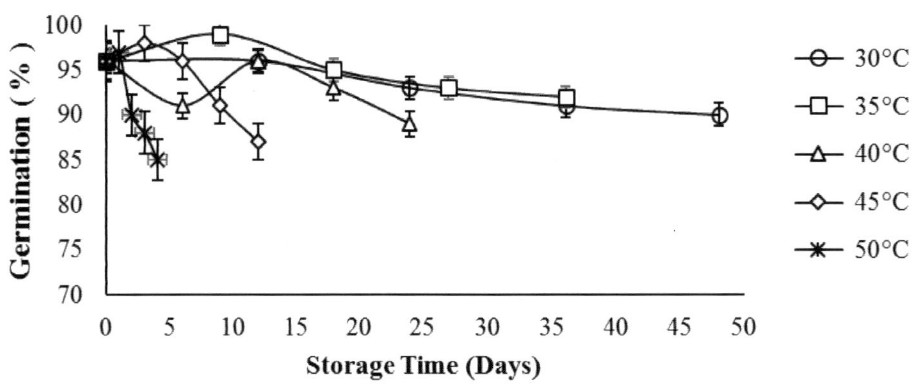


**Figure 2.** Changes in germination of mung bean seeds at five temperatures during storage time (30 ° C, 35 ° C, 40 ° C, 45 ° C and 50 ° C).

Figure 2 also shows that at 35 °C, 40 °C, 45 °C and 50 °C there is a slight change (fluctuation) of germination level, but after that it kept decreasing if compared to the one stored at room temperature (30 °C) that had no change (fluctuation) during storage. Allegedly mung bean seeds have a dormancy stage, but they can absorb water during the dormancy stage. However it will experience inhibition mechanism in the embryo causing the radicle to not appear and an imperfect growth or immature embryo and obstruction of water entering the seed so that causing seeds to fail to germinate.

### *3.3. Decrease in Germination*

Long storage life seeds is very important for farmers and seed growers, nad they often having trouble in determining the expiration date or shelf life of seeds they stored. Almost similar to agricultural products such as fruit and vegetables which is quickly damaged and decay when stored at room temperature, but the decay in the seeds tend to happen much slower and take months until they really damaged. According to [**12**] shelf life estimation on storage condition, can be predicted based on material degradation which depends on the reaction order, reaction constant and activation energy. The kinetic model or order of reaction parameters is selected based on the regression equation with the highest coefficient of determination ($R^2$). The zero order indicates a quality change that occurs in a linear with storage time, whereas in order one the changes are exponential. Reaction order, rate of decline in quality and the coefficient of determination of the germination power of various types of mung bean seeds is shown in Table 1.

The slope values for all temperatures setting during storage show that the higher the temperature treatment results in the greater reaction rate, which means that the quality of mung bean seed quality decreases faster at higher storage temperatures (Table 1). It can also be seen that the coefficient of determination ($R^2$) for decreasing of the quality of mung bean seeds in all temperature treatments, for the first order reactions were higher than that of zero order reactions. Thus the kinetics reaction that decrease the quality of mung bean seeds during storage can be found, following the order pattern of the first order. The higher the coefficient of determination, the more accurate results of the data analysis. Declining quality of seeds tends to follow a first order reaction pattern [**9**].

**Table 1.** Order of reactions, rate of deterioration, and regression coefficient of germination for mung bean seeds after storage at five different temperatures.

| **Parameters** | **Temperature (°C)** | **Order 0** | | **Order 1** | |
|---|---|---|---|---|---|
| | | **Slope** | **$R^2$** | **Slope** | **$R^2$** |
| | 30 | 0.1417 | 0.9383 | 0.0015 | 0.9395 |
| | 35 | 0.1556 | 0.6533 | 0.0016 | 0.6620 |
| Germination | 40 | 0.2000 | 0.3789 | 0.0022 | 0.3805 |
| | 45 | 0.8333 | 0.7697 | 0.0090 | 0.7707 |
| | 50 | 3.1000 | 0.8998 | 0.0341 | 0.9049 |

After determining the reaction order, the first order reaction was chosen for the next step in the effort to determine shelf life by transforming the graph, plotting between 1/T as the absis and ln k as the ordinat so that a new equation can be developed, as shown in Figure 3. Figure 3 shows that the correlation value ($R^2$) for ln k and 1/T was 0.8505. This indicates that the variable reaction constant (y) was influenced by the storage temperature for germination level above 80%. The value of $R^2$ is approaching one ($R^{2=}1$) shows a very strong correlation between the constant reaction of mung bean seeds and storage temperature. Therefore the effect of temperature to provide changes in germination level parameters is obvious. Besides that the activation energy relationship tends to be inversely proportional to the rate of reaction. The smaller the activation energy, the faster the reaction rate because the minimum energy for the occurrence of a reaction gets smaller. On the other hand, the greater the activation energy, the reaction rate will be slower because the minimum energy for the reaction to occur is greater. Activation energy value can be calculated from the slope of the curve on the graph for relation of ln k and 1/T. The amount of activation energy obtained was 30769.098 kal/mol. Activation energy means that to reduce the germination of mung bean seeds during storage requires a minimal energy molecule of 30769.098 kal/mol. In this case it refers to the collision theory and movement of molecules of a material due to activation energy capable of initiating a reaction on the material (Rahmi *et al.* 2016).

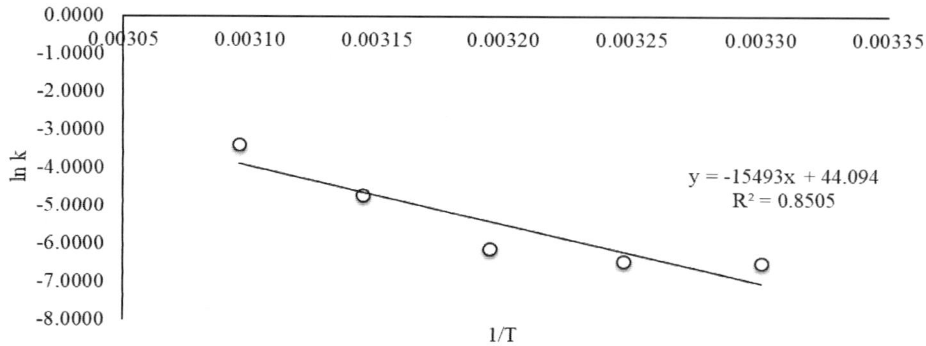


**Figure 3.** Arrhenius plot of order 1 reaction based on the relationship between the rate constant of germination and storage temperature.

*3.4. Estimation of Mung Bean Seeds Shelf Life*

One parameter that is able to describe seed viability directly is germination, not to mention mung bean seeds. The shelf life test of the Arrhenius method was chosen because the germination of mung bean seeds was sensitive to extreme temperature changes in a short time, with the equation for sheld life predicition was ln k = 44.094 + 15493/T. It is assumed, changes in the quality factor of mung bean seeds are only caused by one type of reaction and other factors do not result in significant quality changes, and the shelf life of mung bean seeds is determined from the critical limit of the germination level of 90%. Mung bean seeds germination at two storage conditions can be seen in Figure 4.

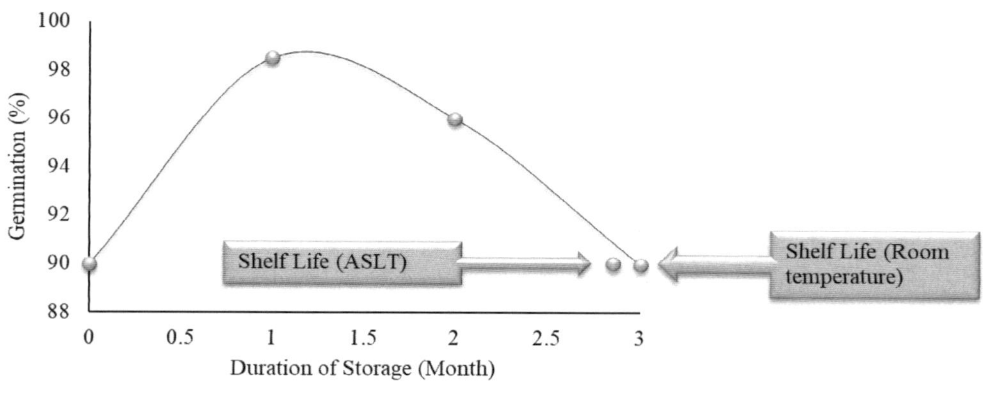


**Figure 4.** Germination of mung bean seeds in two storage conditions at room temperature 29 °C [**15**].

Figure 4 shows that mung bean seeds that stored naturally at room temperature 29 °C with RH 80% were able to maintain germination above 90% for 3 months at 29 °C [**15**]. In this experiment, the shelf life estimation results using the Accelerated Shelf-life Testing (ASLT) method with 90% germination level was predicted up to 2.9 months. According to [**18**] that seeds stored at 30 °C within 6 months of germination of mung bean seeds are above 90%. The same in the study of [**20**] that the storage of mung bean seeds at room temperature 30 °C with 75-80% RH, was able to maintain germination above 90% for 6 months. Besides that, if mung bean seeds are stored at room temperature 27 °C and RH 44% can only maintain 80% germination [**21**]. Allegedly different types of varieties can be seen in terms of seed physical size including medium and small seeds. So the smaller the seeds, the better the storability will be. Estimation of mung bean seeds shelf life results in various treatments at several storage temperatures with tolerance of viability levels above 90% are shown in Table 2.

**Table 2.** Estimation of the shelf life of mung bean seeds as a result of various treatments at several storage temperatures with tolerance of viability levels above 90%.

| Parameters | Temperature (°C) | Shelf Life (Days) | Shelf Life (Month) |
|---|---|---|---|
| Germination level ≥ 90% | 20 | 421.00 | 14.03 |
| | 25 | 173.37 | 5.78 |
| | 30 | 73.52 | 2.45 |
| | 35 | 32.06 | 1.07 |
| | 40 | 14.35 | 0.48 |
| | 45 | 6.59 | 0.22 |
| | 50 | 3.10 | 0.10 |

Table 2 shows that the higher the storage temperature of seeds, the shorter the shelf life of mung bean seeds. The higher the storage temperature treatment results in a greater reaction speed which means a decrease in the quality of mung bean seeds occurs faster at temperatures higher than low temperatures. According to [**9**] that rapid method using high temperatures storage can reduce enzyme activity at embryo and cotyledon and reduce the amount of protein dissolved simultaneously within great shorter period. Decreasing enzyme activity and the amount of dissolved protein can increase the accumulation of peroxide which can inhibit the appearance of the radicle due to the formation of hydroxyl radicals. If there is a continuous increase in lipid peroxide, the seed germination and vigor will decrease over time.

## 4. Conclusion

The results of the study showed that during the storage moisture content of the seeds increased while their germination level were found decreased but still above the 80% as minimum limit of seed quality requirements. Activation energy value (Ea) to reduce the germination of mung bean seeds was found to be 30769.098 kal/mol, and this value can be used to calculate shelf life with the Arrhenius equation $\ln k = 44.094 + 15493/T$. using the developed equation, the estimation of the shelf life of the mung beans of the Sriti variety was found able to maintain 90% of germination level for 421 days or 14.03 months when stored at 20 °C, and for 173.37 days or 5.78 months when stored at 25 °C.

## References

[1] Center of Statistical Berau. 2015. Mung Bean Production by Province. Available online at www.bps.go.id. [Accessed on 8 December 2018]

[2] Justice OL, Bass LN. 2002. Principle and Practice of Seed Storage. Roesli R, translator. Jakarta (ID): Grafindo Persada. Translation from: Principles and Practices of Seed Storage. 446 p.

[3] Indartono. 2011. Assessment of storage room temperature and packaging techniques for soybean seed quality. *Echo technology* 16 (3): 158-163.

[4] Tatipata A. 2008. Effect of initial water content, packaging and shelf life on membrane proteins in soybean mitochondria. *Bul. Agron*. 36 (1): 8-16.

[5] Purwanti S. 2004. Study of room temperature on the quality of black and yellow soybean seeds .Journal of Agricultural Sciences 11 (1): 22-23.

[6] Copeland LO, Mc Donald MB. 1985. *Principles of Seed Science and Technology*. New York (US): Burgees Publishing Company. 369 p.

[7] Floros JD, Ghanansekharan D. 1993. Shelf life prediction of packaged foods: chemical, biological, physical, and nutritional aspects. G. Chlaraslambous (Ed). *J Elsevier Pub . London.*

[8] Kusnandar, F., Adawiyah, D.R., dan Fitria, M. 2010. Estimation of shelf life of biscuit products using the acceleration method based on the critical moisture content approach. J. Food Technology and Industry 21 (2): 1-6.

[9] Rahmi S, Ahmad U, Wulandari D. 2016. Estimation of soybean seed shelf life using *Accelerated Shelf-life Testing (*ASLT)*.Journal of Agricultural Engineering.*

[10] AOAC - Association of Official Analytical Chemists. 2005. *Official Methods of Analysis,* 16thed. 45:5-6. Washington DC (US).

[11] ISTA - International Seed Testing Association. 2013. *International Rules For Seed Testing*. Bassersdorf (CH): International Seed Testing Association.

[12] Asiah N, David W, Cempaka L. 2018. Practical Guidelines for Estimating the shelf life Food Products. Jakarta (ID): UB Press.

[13] Ministry of Agriculture. 2009. Ministry of Agriculture Regulation Number 55/ PERMENTAN / SR.120 / 12/2009 about seed production guideline for soybean, peanut, mung bean, cassava, and sweet potato.

[14] Wafit, D. 2010. The Effect of Moisture Content and Containers on the Viability of Mung Bean Seeds and Mung Bean Beetle Powder Pest (*CallosobruchusChinensis L. )* Populations. *AgriSains Journal* 1 (1): 2086-7719.

[15] Shaumiyah F, Damanhuri, Basuki N. 2014. Pengaruh pengeringan terhadap kualitas benih kedelai (*Glycine max* (L.) Merr). J Prod Tanaman. 2 (5): 388-394.

[16] Abu, A.K, K.U. Sarker, M. M. Haque, M. T. Rubayet& I. H. Mian. 2018. Storage Container, Seed Moisture Level and Storage Condition Effects on Germination and Prevalence of Seed-Borne Fungi of Onion Seed. Global Journal of Science Frontier Research. 18 (3): 2249-4626.

[16] Abdillah M, S Purwanti, Supriyanta. 2016. Savings of Mung Bean Seeds *(Vigna radiata (L.) R. Wilczek)* Intercropping Results with Sweet Corn (*ZeamaysL. Saccharata)* in the Row.*Vegetalika.* 5 (1): 1-12

[17] Al-Karaki. G.N. 1998. Seed size and water potential effects on water uptake, germination and growth of lentil. J. Agron. Crop Sci. 181(4) :237-242.

[18] Boyd, N. S., & Van Acker, R. C. 2004. Imbibition response of green foxtail, canola, wild mustard, and wild oat seeds to different osmotic potentials. Canadian Journal of Botany, 82 (6), 801–806.

[19] Shu J.Z, T. T. Hua, H. K. Liub, Y. Y. Chena, X. Panga, L.L. Zhenga, S.M. Changa, Y.F. Kanga. 2018. Moderate vacuum packing and low temperature effects on

[20] Babariya, C.A.,Patel, J.B., Ribadiya, K.H. Sondarva Jyoti and Bhatiya, V.J.2016. Permance of neem products on the storability of mungbean (*Vigna radiata* (L.) Wilczek) seeds. Indian J. Agric. Res.,50 (6) : 573-578.

[21] Mutungi C.M., Affognon H., Njoroge A.W., Baributsa D., Murdock L.L. 2014. Storage of mung bean (*Vigna radiata* [L.] Wilczek) and pigeonpea grains (*Cajanus cajan* [L.] Millsp) in hermetic triple-layer bags stops losses caused by *Callosobruchus maculatus* (F.) (Coleoptera: Bruchidae) J. Stored Prod. Res. 2014;27:239–243.

# Maintenance Cost Reduction of Paddy Seed Production Machinery by Implementing Preventive Maintenance System

**S Pertiwi *,[1], W Hermawan[1] and E Prahmawati[2]**

[1] Department of Mechanical and Biosystem Engineering, Bogor Agricultural University, Bogor, Indonesia.
[2] Graduate School, Mechanical Engineering for Agriculture and Food Industry, Bogor Agricultural University (alumnae), Bogor, Indonesia.

* *e-mail: pertiwi@ipb.ac.id*

**Abstract.** Production of paddy seed in seed industry involves several different functions and operations of seed production machinery. Maintenance of those machinery has a fundamental role in the production system through its impact on productivity as well as cost management. This paper will discuss the case study on analysis of possible cost reduction on maintenance of paddy seed production machinery through changes from the current practiced corrective maintenance system into preventive maintenance system. The study was mainly based on twenty four years of company's maintenance records of its seed production machinery which covered the failures history and actions undertaken. The methods were including the identification of critical components using Equipment Criticality Rating (ECR) and optimization of replacement schedule of critical components. The results indicated that eleven (11) components out of total 294 components in 8 specific machines have been identified as having considerable criticality rating and the changes from corrective maintenance system into preventive maintenance system may reduce about 39.4% of direct machinery maintenance cost.

## 1. Introduction

Seed is the foundation of any rice crop. In rice production, seed must be grown, harvested, and processed correctly for best yield and quality results. Certified paddy seed use is highly recommended. However, Indonesian certified rice seed production can only fulfil half of the total domestic needs of the seeds. According to BPS, seed requirements in 2015 reaching 374 thousand tons while the total seed production only reached 184 thousand tons. The factor that causes the lack of certified seeds production, among others, is the poor condition of seed production machinery. Machinery condition greatly affects the quality of the seeds produced. Maintenance of those machinery has a fundamental role in the production system through its impact on productivity as well as cost management. Therefore, machinery maintenance activities need to be undertaken to maintain the condition of the machinery in its optimal state.

Prahmawati, Pertiwi, and Hermawan [1] investigated the machinery maintenance system in the largest seed production industry in Indonesia, namely PT SHS. The company produces approximately 20% of the domestic seed needs. Seed production activities sometimes were stopped due to engine failure. Systematic efforts to overcome engine disruption need to be done through improvement of machinery maintenance system.

In general, there are two machinery maintenance approaches, namely corrective (or reactive) machinery maintenance and preventive machinery maintenance. Corrective maintenance is maintenance which is carried out after the occurrence of failure to restore the failed equipment, machine, or facilities to an operational condition. On the other hand, preventive maintenance is the care and servicing by qualified personnel to maintain equipment, machine and facilities in satisfactory operating condition by providing systematic inspection, detection, and correction of incipient failures either before they occur or before they develop into major defects. The machinery maintenance activities in PT SHS is carried out correctively. Impacts of corrective maintenance include unscheduled downtime resulting in damage to seed products, cessation of operations under certain conditions of damage, idled labour time, and an increasing in machine start-up costs. To improve the situation, implementation of preventive maintenance is highly recommended. However, the implementation of a preventive maintenance can be time consuming and costly. This paper will discuss the case study on analysis of possible cost reduction on maintenance of paddy seed production machinery through changes from the current practiced corrective maintenance system into preventive maintenance system.

## 2. Method

Production of paddy seed in seed industry involves several different functions and operations of seed production machinery. In PT SHS, the machinery consist of a set of machines such as intake, chain conveyor, buckets elevators, pre-cleaners, hopper, dryer, lockers bin, air screen separators, sealer, bet conveyor, cyclone and silo as shown in figure 1.

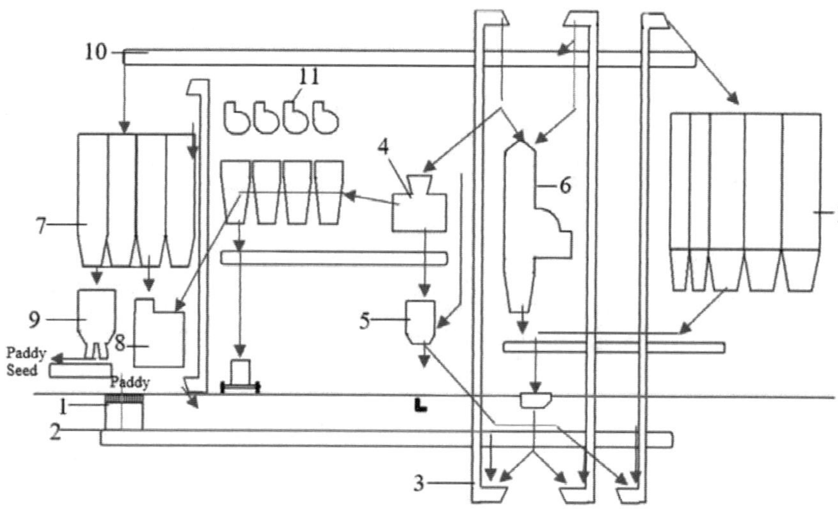


| | | | |
|---|---|---|---|
| 1. | Seed Intake | 7. | Locker Bin |
| 2. | Chain Conveyor | 8. | Air screen Separator |
| 3. | Bucket Elevator | 9. | Sealer (packaging) |
| 4. | Pre-cleaner | 10. | Belt Conveyor |
| 5. | Hopper | 11. | Cyclone |
| 6. | Dryer | 12 | Silo |

**Figure 1.** Paddy seed production machinery in PT SHS.

Given the complexity of the machinery, the analysis is limited to the critical machines and/or components of the machinery system. Hence the analysis was carried out in three steps, namely 1) determining the equipment criticality rating, 2) optimizing the replacement schedule of components with respect to the existing routine maintenance schedule, and 3) calculating the direct maintenance costs.

Suryadi and Setyanta [2] proposed a combination model of serial and parallel criteria to assess Equipment Criticality Rating or ECR as shown in figure 2. The principle used is if equipment is not functioning, how big it will affect to company as a whole. The method has been adopted by Cahyati, Pramudya, Pertiwi and Herodian [3] in the study of machinery maintenance in sugarcane industry. It was found that the method was very helpful in determining the critical machines and/or components in the factory that enabled the researchers to be more focus on determining maintenance strategy. Results of ECR assessment was also in line with the real conditions in the field.

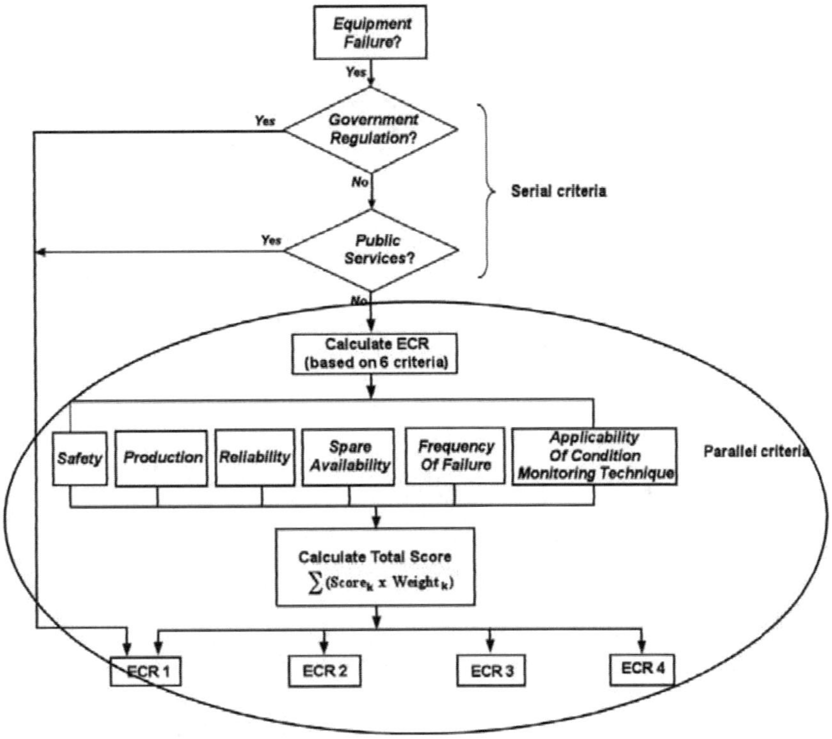


**Figure 2.** Combination model of serial and parallel criteria for ECR assessment.

The assessment of ECR in paddy seed production machinery was undertaken by adopting only the parallel part of the criteria (in the ellipse) as the machinery are clearly comply to government regulation and are not used for public service. The weight of the criteria was determined by using pairwise comparison [4] made by resource persons in PT. SHS. Table 1 exhibits the formulas and weights involved in the assessment. At the end the ECR values were classified as very critical (ECR1**:** ECR > 60), critical (ECR2**:** 40< ECR≤ 60), slightly (less) critical (ECR3**:** 20< ECR ≤40) and not critical (ECR4**:** ECR≤20) to describe the level of machine criticality. From each of machines identified as very critical and critical the optimum schedule for replacement of critical components were then determined.

Optimization of the components replacement schedule was carried out by combining approaches suggested by Jardine and Tsang [5] with respect to the existing routine maintenance schedule, i.e. optimization of replacement time interval (tr) that minimize total cost per unit time C(tr), optimization of replacement time interval (tp) that minimize total cost C(tp), optimization of replacement interval (tp) between preventive replacements to minimize total downtime per unit time, and optimization of preventive replacement age (tp) to minimize total downtime per unit time. The optimum times to replace the components were determined as the weighted values of the four optimum replacement times. Figure 3 shows the approaches while table 2 indicates formulas and weights involved in the optimization process.

The direct maintenance costs were calculated for both, corrective and suggested preventive maintenance system, to be compared. They were involving number of required components, component prices (IDR), production loss (IDR) due to unscheduled down time, wages of idle paid labour (IDR), wages of overhauling for each scheduled downtime (IDR), and additional start-up cost (IDR) due to unscheduled down time. All prices were based on year 2016 market prices.

**Table 1.** Formulas and weights involved in the ECR assessment.

| Criteria | Weight | Formula | |
|---|---|---|---|
| Safety | 0.251 | Rating on safety risk : 20, 40, 60, 80, 100 | |
| Effect on Production | 0.201 | $EfP = \frac{PL}{SC} \times 100$ | (1) |
| Reliability | 0.167 | $URF = \left[1 - \left(\frac{MDT - (SDT + USDT)}{MDT - SDT}\right)\right] \times 100$ | (2) |
| Spare Availability | 0.143 | $SAF = \left(1 - \left(\frac{SUC}{RUC}\right)\right) \times 100$ | (3) |
| Frequency of Failure | 0.126 | $FoF = \left[\frac{NoF}{NoY}\right] \times 100$ | (4) |
| Applicability of Monitoring Technique | 0.112 | Rating on applicability of monitoring technique: 0, 10, 25, 50, 100 | |

Note:
PL: production loss
SC: sustainable capacity of equipment
MDT: machinery working time
SDT: scheduled downtime
USDT: unscheduled downtime
SUC: standby unit capacity
RUC: running unit capacity
NoF: number of machine failure
NoY: time span of all failures occurrences (year)

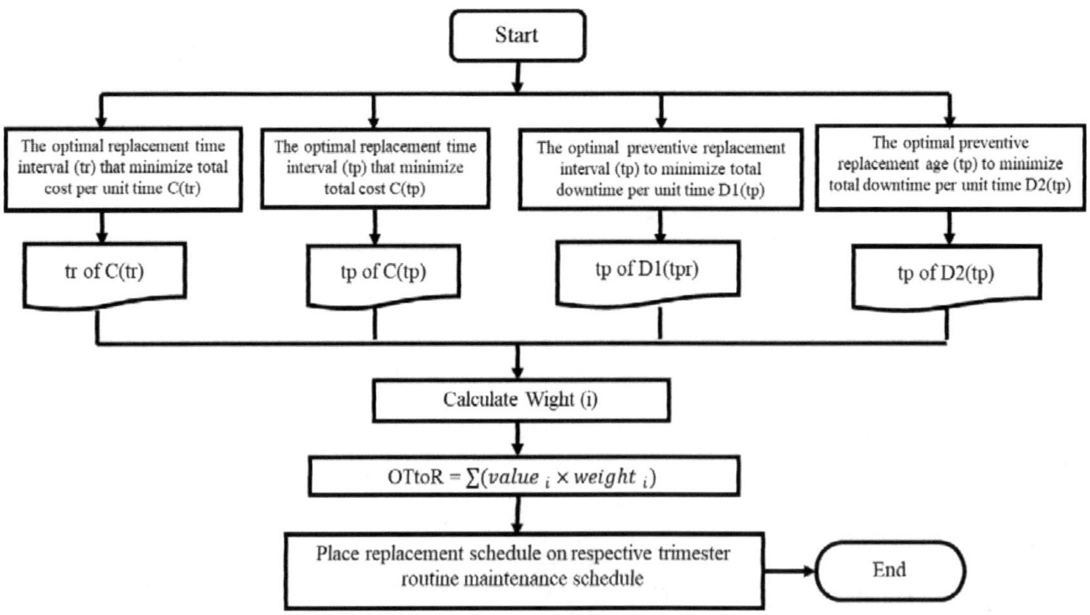


**Figure 3.** Combination of approaches for optimizing component replacement schedule.

**Table 2.** Formulas and weights involved in the optimization of component replacement schedule

| Parameter | Weight | Formula | |
|---|---|---|---|
| tr of C(tr) | 0.315 | $C(t_r) = \frac{1}{tr}\int_0^{tr} [FC - VC \times e^{-kt}]dt + Cr$ | (5) |
| tp of C(tp) | 0.262 | $C(t_p) = \frac{Cp \times R(tp) + Cf \times [1 - R(tp)]}{tp \times R(tp) + \int_{-\infty}^{tp} tf(t)dt}$ | (6) |
| tp of D1(tp) | 0.226 | $D1(tp) = \frac{H(tp)Tf + Tp}{tp + \left(\frac{Tp}{8670}\right)}$ | (7) |
| tp of D2(tp) | 0.197 | $D2(tp) = \frac{TpR(tp) + Tf[1 - R(tp)]}{(tp + Tp)R(tp) + [M(tp) + Tf][1 - R(tp)]}$ | (8) |

Note:
FC: fixed cost
VC: variable cost
k: constant, rate of deterioration
Cr: component replacement cost
Cp: cost of preventive maintenance
Cf: cost of failure
R(tp): reliability function in interval tp
f(t): probability density function of item's failure time
H(tp): expected number of failures in interval tp
Tp: mean time of annual preventive replacement
Tf: mean time for annual replacement due to failure
M(tp): mean time to failure

## 3. Results and Discussion

Currently the company carries out reactive maintenance system, in which repair and replacement of machine components was undertaken after the occurrence of machine breakdown. Routine maintenance were undertaken every four-month (trimester) during off season, i.e. after the completion of harvest season. Activities during routine maintenance were including overall checking, replacement of broken components, and factory cleaning.

The records of maintenance history from 1992-2016 (twenty four years) indicated a number of machine breakdown and the handling activities carried out to repair or to replace machine component. There were several components that experience recurring breakdown every year such as bearings, cables, v-belts, bucket elevators, chain, and gear. In addition, breakdown also occurred in the bottom of the hoper funnel and the box dryer floor plate so it must be replaced frequently. Impact of breakdown varies greatly depending on which machine components were breakdown. It may cause physical damage to seeds, unpredictable machine downtime, and even may lead to cessation of overall operation. The downtime occurrences during 18 years were 72 times of four monthly routine downtime and 34 times of unscheduled downtime that occurs when the machine is operating. The duration of downtime was varied depending on the availability of spare parts and time to repair. Figure 4 shows the distribution of unscheduled breakdown times.

The company severely limits the number of spare parts. This policy leads to problem if parts from broken components are not available since it may cause the stopping of production process for a certain period. Stopping the production process may cause prolonged production process, damage to the seed produced, as well as additional startup costs. For example there is damage to the bearing on the bucket elevator which causes the drying operation to stop, while the seeds are located the dryer has reached 15% moisture content and has to be stopped drying. The bearing replacement process takes 2-3 hours so the seed being dried has to be delayed. The water content of the seed will be increased again the presence of water vapor absorbed back into the seed. As a result, drying will last longer resulting in costs additional electricity. Another impact of the cessation of engine operation is the idle time of laborers, especially those of production staff. Therefore, the maintenance costs that occur consist of the cost of replacing the components coupled with the cost of paid idle laborers and cost of production loss due to engine breakdown during the process.

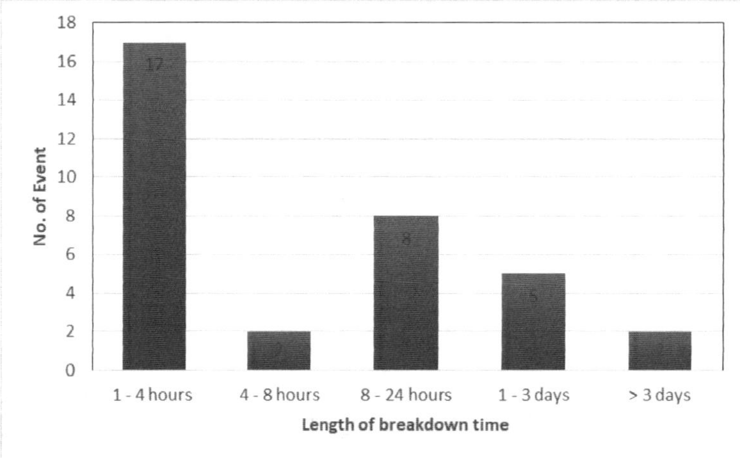


**Figure 4.** The distribution of unscheduled breakdown times.

Assessment on criticality ratings of the involved machines resulted in table 3.There were no machine categorized into very critical. Four machines were categorized into critical, i.e. chain conveyor, dryer, sealer, and belt conveyor, from which chain conveyor had the highest criticality rating. With this results, further analysis was extended into those categorized as slightly critical machines. Based on maintenance and spare parts replacement record, optimization of replacement schedule on eleven (11) out of 294 spare parts frequently broke down in the critical and slightly critical machines was undertaken. The optimum time for replacing the components varies, ranging from 9 months to 3 years. It is very uneconomical to implement the replacement schedule exactly as the optimization results since it will cause very frequent scheduled machinery breakdown time. Therefore, by taking the longest time for replacement, i.e. 3 years as the preventive maintenance cycle, the replacement schedules were rounded-up or rounded-down into the closest on going four-month routine maintenance schedule. The results are shown in table 4.

**Table 3.** Results of equipment criticality rating.

| | Safety (0.251) | PF (0.201) | URF (0.167) | SAF (0.143) | FoF (0.126) | ACMT (0.112) | ECR Score | Category |
|---|---|---|---|---|---|---|---|---|
| Seed Intake | 20 | 0 | 0 | 10 | 10 | 0 | 7.71 | Not Critical |
| Chain Conveyor | 40 | 60 | 30 | 60 | 80 | 80 | 54.73 | Critical |
| Bucket Elevator | 20 | 10 | 20 | 40 | 60 | 50 | 29.25 | Slightly Critical |
| Pre-Cleaner | 20 | 30 | 20 | 10 | 40 | 20 | 23.10 | Slightly Critical |
| Hopper with weighing scale | 20 | 0 | 10 | 10 | 10 | 20 | 11.62 | Not Critical |
| Dryer | 40 | 80 | 30 | 30 | 50 | 50 | 47.32 | Critical |
| Locker Bin | 20 | 15 | 40 | 40 | 30 | 20 | 26.45 | Slightly Critical |
| Air screen separator | 20 | 10 | 20 | 10 | 50 | 40 | 22.58 | Slightly Critical |
| Sealer | 40 | 60 | 40 | 60 | 50 | 20 | 45.90 | Critical |
| Belt Conveyor | 40 | 60 | 30 | 30 | 50 | 50 | 43.30 | Critical |
| Cyclone | 20 | 30 | 20 | 10 | 40 | 20 | 23.10 | Slightly Critical |
| Silo | 20 | 0 | 10 | 10 | 70 | 0 | 16.94 | Not Critical |

**Table 4.** Result of components replacement scheduling.

| Machine | Spare Parts | tr of C(tr) | tp of C(tp) | tp of D1(tp) | Tp of D2(tp) | Optimum Replacement Time (Year) | Replacement Schedule*) |
|---|---|---|---|---|---|---|---|
| Chain conveyor | Chain | 0.33 | 0.5 | 1.07 | 1.87 | 0.85 | 1, 3, 6 |
| Belt conveyor | NYY HY Cable | 4 | 7 | 0.60 | 0.34 | 3.03 | 1, (+1) |
| Air screen separator | Bearing 6202 | 1 | 1 | 1.2 | 1.2 | 1.08 | 1, 4, 7 |
| Cyclone | Bearing 2211 | 0.83 | 1 | 1.03 | 1.13 | 0.98 | 1, 4, 7 |
| | V belt B118 | 1 | 1 | 1.2 | 1.2 | 1.08 | 1, 4, 7 |
| Dryer | Plate screen | 4 | 4 | 1.05 | 0.60 | 2.66 | 1, 7 |
| | Burner NYY Cable | 6 | 7 | 1.2 | 0.46 | 4.09 | 1, (+4) |
| | Acer Plate | 4 | 4 | 1.05 | 0.60 | 2.66 | 1, 7 |
| Sealer | Heating element | 0.72 | 0.92 | 0.66 | 0.60 | 0.74 | 1, 3, 6 |
| Bucket Elevator | Gear elevator Z31 | 2 | 3 | 0.73 | 0.53 | 1.69 | 1, 5, 9 |
| Seed cleaner | Screen | 1.1 | 1.5 | 0.87 | 0.83 | 1.10 | 1, 4, 7 |

*) Replacement schedule on 3-year cycle, 4-month routine maintenance schedule; plus sign means next cycle

Having preventive replacement schedule of those ten major components, the direct costs of preventive maintenance system was calculated and it was compared to those of the actual corrective maintenance system. The results are shown in table 5. The table indicated that applying preventive machinery maintenance with periodic replacement of the spare parts may reduce direct maintenance cost to about 59.6% of the existing practiced corrective machinery maintenance. Seeing its nominal value, the cost reduction may be less attractive for the decision makers to change the machinery maintenance approach into preventive machinery maintenance. Inaccurate calculation of yield loss due to unscheduled machinery breakdown highly affected the cost calculation result. However, it should be noted that the nominal value will become greater if some more spare parts/components involved are analysed.

**Table 5.** Direct cost of corrective maintenance and preventive maintenance (based on 11 spare parts).

| Machine | Part/Component | Unit Price (IDR) | Corrective Maintenance (24 years) | | Preventive Maintenance (3-year cycle) | |
|---|---|---|---|---|---|---|
| | | | Number of Parts | Direct Cost (IDR) | Number of Parts | Direct Cost (IDR) |
| Air Screen Separator | Bearing 6202 | 28,000 | 26 | 728,000 | 3 | 84,000 |
| Box Dryer | Plate Screen | 45,000 | 6 | 270,000 | 2 | 90,000 |
| Box Dryer | Burner NYY Cable | 180,000 | 5 | 900,000 | 1 | 180,000 |
| Box Dryer | Acer Plate | 170,000 | 8 | 1,360,000 | 2 | 340,000 |
| Belt Conveyor | NYY HY Cable | 180,000 | 8 | 1,440,000 | 2 | 360,000 |
| Chain Conveyor | Chain | 50,000 | 24 | 1,200,000 | 3 | 150,000 |
| Bucket Elevator | Gear Elevator Z31 | 15,000 | 17 | 1,513,000 | 3 | 45,000 |
| Sealer | Heating element | 15,000 | 84 | 1,260,000 | 12 | 180,000 |
| Cyclone | Bearing 2211 | 30,000 | 26 | 390,000 | 3 | 90,000 |
| Cyclone | V Belt B118 | 55,000 | 18 | 990,000 | 3 | 165,000 |
| Seed Cleaner | Screen | 147,000 | 21 | 3,087,000 | 3 | 441,000 |
| Paid idle labor (due to unscheduled breakdown time) | | | | 7,236,000 | | - |
| Cost of machinery start-up | | | | 1,630,200 | | - |
| Production lost due to unscheduled machinery breakdown | | | | 6,536,250 | | - |
| | | Total direct maintenance cost (24 years) | | 28,540,450 | | - |
| | | **Average direct maintenance cost (3-year cycle)** | | **3,567,556** | | **2,125,000** |

IOP Conf. Series: Materials Science and Engineering 557 (2019) 012075 doi:10.1088/1757-899X/557/1/012075

## 4. Conclusions

A case study on analysis of possible cost reduction on maintenance of paddy seed production machinery through changes from the current practiced corrective maintenance system into preventive maintenance system has undertaken with positive results. Based on criteria of component criticality, four (4) out of 12 machinery involved in paddy seed production were rated as critical, i.e. chain conveyor, belt conveyor, dryer, and sealer; four (4) were rated as slightly critical (bucket elevator, pre-cleaner, locker bin, and air screen separator). The rest were not critical. Analysis on eleven (11) components out of 294 involved components in the machinery showed that the optimum time for replacing the components/spare parts varies, ranging from 9 months to 3 years. Applying the preventive machinery maintenance with periodic replacement of the components may reduce maintenance cost to about 59.6% of the existing practiced corrective machinery maintenance. Therefore, preventive machinery maintenance is recommended to be implemented instead of corrective machinery maintenance

## References

[1] Prahmawati E, Pertiwi S and Hermawan W 2018 Optimization of Decision Support System for Maintenance of Machineries Production (in Bahasa Indonesia) J Keteknikan Pertanian Vol 6 No 2 p 115-122.
[2] Suryadi K and Setyanta H 2009 Equipment Criticality Classification Model Based on AHP Proc. of the International Symposium on the Analytic Hierarchy Process (ISAHP 2009) University of Pittsburgh, Pittsburgh, Pennsylvania, USA July 28 – August 1 2009 p 1-17.
[3] Cahyati S, Pramudya B, Pertiwi S and Herodian S 2011 The Identification of Component Criticality at Production Line of Sugar Fabrication Using ECR Method (in Bahasa Indonesia) J Keteknikan Pertanian Vol 25 No 1 p 73-78.
[4] Saaty T L 2008 Decision making with the analytic hierarchy process International Journal Services Sciences 1(1) p 83-98.
[5] Jardine AKS and Tsang AHC 2013 Maintenance, Replacement, and Reliability, Theory and Applications, Second Edition CRC Press Taylor and Francis Group p 332.

IOP Conf. Series: Materials Science and Engineering **557** (2019) 012076 doi:10.1088/1757-899X/557/1/012076

# Synthesis and characterization of aloe vera-based edible film incorporated with shellac resin and hydrocolloids

**Nanik Purwanti*, Fajri Yunus, Emmy Darmawati**

Departement of Mechanical and Biosystem Engineering, Bogor Agricultural University, Dramaga IPB Campus, PO BOX 220, Bogor 16002 West Java, Indonesia
* Corresponding author

**Abstract.** Aloe vera-based edible film usually has weak structure, thus it has low strength. Incorporation of shellac resin and gelatine/guar gum might create composite edible film with preferable properties. This research aimed to synthesis edible film of aloe vera gel that was incorporated with shellac resin and gelatine or and guar gum. The initial concentration of shellac resin was kept constant at 6% w/w; meanwhile the initial concentrations of gelatine were 4% and 6% w/w and of guar gum were 1% and 2% w/w. The ratio between SHGE/SHGG and aloe vera gel in the suspension (ALSHGE/ALSHGG) was 1:1. The suspension was kept in the fridge for 4 days and the film was prepared from this suspension at day 0, 2 and 4. The suspension was characterized in terms of its viscosity, sedimentation and microbial activity during storage. The film was characterized in terms of its thickness, solubility, moisture content and water vapor transmission rate (WVTR). The results showed that ALSHGE and ALSHGG suspension were pseudoplastic, depending on their concentrations, with sedimentation occurred in ALSHGG suspension after 4 days. Microbial growth in the suspension was inevitable in both concentrations of ALSHGG, but there was no microbial growth observed in both concentrations of ALSHGE. After 20 g of suspension was casted into film, ALSHGE4% and 6% resulted in thicker films, with lower moisture content, than ALSHGG1% and 2%. These could be the reason for high solubility of ALSHGG films. The WVTR of the films varied, independent of storage time of the suspension, with the least WVTR was observed in the film from ALSHGE.

## 1. Introduction

Horticultural products such as fruits and vegetables are perishable; therefore, they have relatively short shelf life. The basic methods to extend shelf life of fresh horticultural products are by retarding their metabolisms, including respiration, transpiration, evaporation, and preventing microorganism contamination [1]. This might be achieved by applying, one amongst others, an edible film to fresh horticultural products. Aloe vera gel has great potential to be applied as edible film in Indonesia because aloe vera is widely cultivated in the country and it is cheap [2]. Aloe vera gel is rich of polysaccharide compounds which would make the film has good barrier properties against $O_2$ at low to moderate humidity condition. The film would also have ability to maintain moisture content of coated products by controlling evaporation and exchange of water-soluble compounds with environment [3]. However, this particular ability also becomes the drawback of aloe vera-based edible film, i.e., it adsorbs water from its environment due to its hydrophilicity [4]. Imbalance water activity between the coated product, the film, and their environment might lead unexpected water transport to the products that leads to spoilage. Therefore, incorporation of other materials that improve the drawback of film properties shall be considered. Shellac resin was then selected in this research to add certain hydrophobicity to aloe vera-based edible film. Nevertheless, shellac alone has poor mechanical property and unstable; therefore, a hydrocolloid, i.e., guar gum or gelatine was added as a stabilizer.

This research aimed to synthesize aloe vera-based edible film which was incorporated with shellac resin and a hydrocolloid. The suspension containing those materials was characterized in terms of viscosity, sedimentation and its microbial activity during storage. Further, the resulted composite film was characterized in terms of its physical properties, i.e., thickness, solubility, moisture content and water vapor transmission rate (WVTR).

## 2. Materials and Methods

### *2.1. Materials*

Aloe vera leaves which were the main material to prepare aloe vera gel were cultivated in Bogor and harvested at 8-10 month after planting. Other materials used were yellow shellac resin, obtained from Lubuk Linggau, South Sumatera, guar gum (Premium Guar Gum, Bob's Red Mill), gelatine (Halal Gelatine Powder, Naturich), citric acid, ascorbic acid, alcohol 96%, and aquadest (local chemical stores around Bogor).

*2.2. Methods*

*2.2.1. Preparation of aloe vera gel.* Aloe vera leaves were cleaned under running tap water and then trimmed. The pulp was filleted, cleaned with water and then pulverized using a blender for 10 s. The pulverized pulp was further homogenized at 10,000 rpm for 30 min at 5 °C. The homogenized pulp was filtered, pasteurized at 70-75 °C for 15 min, rapidly cooled, and then citric acid and ascorbic acid were added into the homogenized pulp. The pulp exposed to these treatments was then referred as aloe vera gel.

*2.2.2. Preparation of composite suspension.* The composite suspension consisted of aloe vera gel, shellac resin solution and guar gum or gelatine solution. Solution of shellac resin was prepared by dispersing shellac flakes into alcohol with concentration of 6%w/w. The dispersion was stirred for 24 h and then centrifuged at 10,000 rpm, 5 °C, for 20 min. The supernatant was harvested as shellac solution. Guar gum solution was prepared by dispersing guar gum powder in aquadest at 1% and 2% w/w and then stirring at 650 rpm was performed until the powder was completely dissolved. Gelatine solution was prepared in similar way as guar gum solution but heating at 50 °C was applied during stirring to speed up solubilization of gelatine. Concentrations of the solution were 4% and 6% w/w. The composite suspension was prepared as follows, guar gum or gelatine solution (40%) was mixed with shellac resin solution (60%) for 3 h. Aloe vera gel was added into the mixed solution with a ratio of 1:1 and then homogenized at 10,000 rpm for 1 min. This composite suspension containing guar gum was coded as ALSHGG and the one containing gelatine was coded as ALSHGE. The percentage following the codes represented concentration of the guar gum or gelatine.

*2.2.3. Characterization of the composite suspension.* The suspension was kept for 4 days in the fridge and then, its viscosity, as well as its flow behaviour, and number of its microbial contaminants were measured at day 0, 2 and 4. Viscosity and flow behaviour of the suspension was measured using a rotational test (rheometer MCR301, Anton Paar, Austria) at shear rate range of 0.1 – 500 $s^{-1}$ for up-ramp shear rate sweep and vice versa for down-ramp shear rate sweep, at a temperature of 30 °C. Each sample was measured in triplicate. Number of microbes that contaminated the suspension was examined by performing total plate count (TPC) method [5]. The composite suspension was diluted 100-fold and 1000-fold, spread over the surface of agar plates, and then the plates were incubated at 35 – 37 °C for 48 h. Number microbe colonies appeared on the surface was counted to indicate extent of microbial growth in the composite suspension. Number of colonies per ml of the suspension was calculated using Equation (1). Sedimentation that occurred in the suspension was observed every day during storage. The suspension was placed in sample tubes and the suspension height was between 5.5 cm and 6.6 cm. If sedimentation took place, the height of the sediment was measured.

$$Number\ of\ colonies\ per\ ml\ suspension = \frac{Number\ of\ colonies\ on\ the\ plate}{dilution\ factor} \quad (1)$$

*2.2.4. Preparation of the film and its characterizations.* Aloe vera-based edible film was prepared by casting 20 g of the composite suspension, which was stored for 0, 2 and 4 days, onto plastic plates of diameter 85 mm. The casted solution was dried at room temperature for 2 days and then the resulting film was peeled off of the plate. Thickness of the film was measured using a thickness gauge with 0.005 mm precision. Ten points on the film was measured and the results were presented as an average thickness. The film solubility was measured according to the method of Khoshgozaran-Abras et al. [6].

The film was cut into pieces with a dimension of 1x4 $cm^2$ and its dry weight was measured. The film was then immersed in 50 mL of aquadest for 24 h. After 24 h, the soaked film was filtered and then dried in an oven at 105 °C for 24 h before measuring its final dry weight. Solubility of the film is defined in Equation (2). Moisture content of the film was measured according to Costa et al. [7]. About 50 mg of film was weighed and then it was dried in an oven at 105 °C for 24 h. The dried film was weighed, and the moisture content was calculated using Equation (3). Gravimetric method according to Warkoyo et al. [8] was applied to measured WVTR of the film. The film with a designated area was weighted and then, it was placed exactly on the mouth of sample container. The sample container with the film on it was kept in a desiccator at 25 °C and 75% relative humidity for 24 h. To maintain a 75% RH gradient across the film, anhydrous calcium chloride (0% RH) was placed inside the sample contained; meanwhile saturated solution of sodium chloride (75% RH) was placed at the bottom of desiccator. In this way, RH inside the sample container was always lower than that outside of the container. WVTR was determined from the weight gain by the film [9] and calculated with Equation (4).

$$Solubility\ (\%) = \frac{W_0 - W_t}{W_0} \times 100\% \quad (2)$$

where $W_0$ is the initial dry weight (g) and $W_t$ (g) is the final dry weight of the film.

$$Moisture\ content\ (\%) = \frac{m_i - m_f}{m_i} \times 100\% \quad (3)$$

where $m_i$ is the initial weight (g) and $m_f$ (g) is the final weight of the film after drying.

$$WVTR\ (g/m^2h) = \frac{b}{A} \times 100\% \quad (4)$$

where $b$ is the rate of weight gain (g/h) and A is the film area ($m^2$).

## 3. Results and Discussion

The composite suspension containing aloe vera gel, shellac resin and guar gum/gelatine was fairly stable during storage. ALSHGE at 4% and 6% did not show any precipitation during 4-day storage but ALSHGG at both concentrations of 1% and 2% w/w started to precipitate at the fourth day of storage. Nevertheless, the maximum thickness of the sediment was only about 4% of the suspension's initial height. Low concentration of ALSHGG and ALSHGE was prone to Newtonian-like behaviour. At the highest concentration tested, both ALSHGG and ALSHGE suspension were pseudoplastic (Figure 1a). After 4-day storage, these suspensions were Newtonian-like behaviour similar as those ALSHGG and ALSHGE at low concentrations (Figure 1b). Pseudoplastic behaviour of the suspensions at high concentrations was probably contributed by guar gum, aloe vera, and into certain extent by gelatine and shellac. Guar gum as low as 1% w/v showed Pseudoplastic behaviour measured at 25 °C. At low shear rate, the apparent viscosities slightly reduced during storage up to 37 days, but overall flow behaviour remained Pseudoplastic [10]. The mucilaginous fresh aloe vera gel was also reported to be Pseudoplastic behaviour, regardless harvesting time of the leaves, irrigation status of the leaves in the field, or gel treatment [11-12]. On the other hand, gelatine solution was often reported as Newtonian-like behaviour, even at high concentration [13-14]. Nevertheless, Pseudoplastic behaviour of gelatine solution could also be found depending on gelatine sources [14-15]. In this research, the gelatine was originated from bovine and its solution has Newtonian behaviour. Flow behaviour of shellac solution measured at the range shear rates tested in this research would be Newtonian behaviour as has been shown by Gao et al. [16]. At shear rates between 1 $s^{-1}$ and 100 $s^{-1}$, sodium shellac solution at 0.1-15% w/w was Newtonian and deviation from Newtonian occurred at lower and higher shear rates than 1 $s^{-1}$ and 100 $s^{-1}$, respectively. Considering flow behaviour of each component in the suspension, significant change on flow behaviour of ALSHGG2% and ALSHGE6% after 4-day storage was most probably contributed by aloe vera gel. Composition within aloe vera gel underwent significant changes during storage for 48 h at 40 °C that led to Newtonian behaviour [11]. Glucose content in the gel dropped up to 75% of its original content; meanwhile, mannose content increased up to 44% of its original content during storage. Overall, polysaccharide content of aloe vera gel tended to decrease during storage. In addition, imperceptible precipitation of remaining fibre in the gel might contribute to the change of Pseudoplastic to Newtonian.

Microbial contamination was inevitable in the suspension containing guar gum of both concentrations (ALSHGG1% and ALSHGG2%). The contamination took place already after suspension preparation and increased during storage, except for ALSHGG1% at day 2 and 4 but the numbers are not significantly different (Figure 2). Number of colonies per ml suspension could be suppressed with increasing guar gum concentration. Contrarily to ALSHGG suspension, ALSHGE suspension at both concentrations did not show significant contamination of microorganism. Guar gum has been proven as a good media to replace agar for microbial culture media [17]. This gum was shown to act as prebiotics for the growth of *Lactobacillus reuteri* at 37 °C for 16 h [18]. It is also the case for gelatine because protein is a good source for microbial growth. Various bacteria such as *Staphylococcus aureus* and *Listeria monocytogenes* grew well on gelatine-based film [19]. On the other hand, shellac was not reported to have certain support towards microbial growth nor against it. Therefore, application of shellac-based film for retarding microbial activity was often in combination with other antimicrobial agents such as lemongrass oil [20]. The only component that owns antimicrobial activity in the suspension was aloe vera gel. Various methods have been tested to screen antimicrobial properties of aloe vera gel and most of the methods indicated that aloe vera gel is capable of inhibiting gram-positive and gram-negative bacteria to grow [21]. In this research, microbial activity remained occur in ALSHGG suspensions. The reason might be the fact that guar gum is better food source for microbes than gelatine; therefore, the existence of aloe vera gel in the suspension is not enough to suppress their activities.

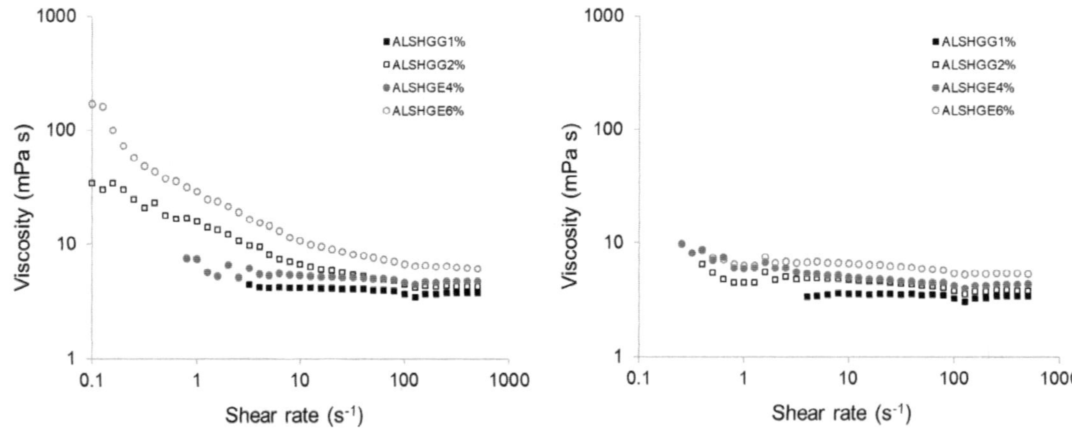


a. Flow behaviour at day 0 b. Flow behaviour after 4-day storage

**Figure 1.** Flow behaviour and apparent viscosity values of ALSHGG and ALSHGE suspension during 4-day storage.

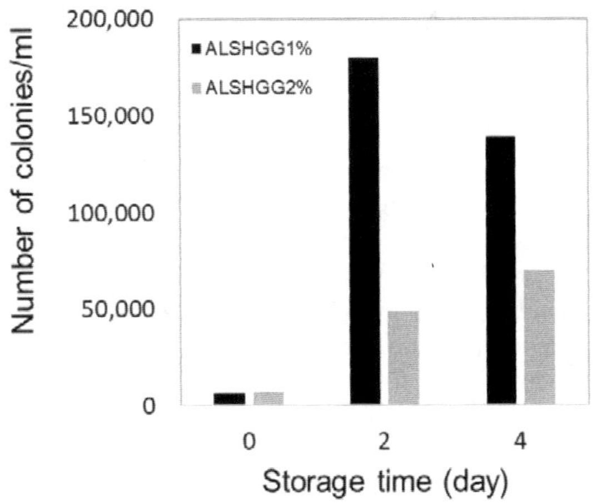


**Figure 2.** Microbial growth in ALSHGG suspension during storage.

After casting the suspension into films, different film colours were obtained. Originally, film made of aloe vera gel was transparent as shown in Figure 3a. Incorporation of shellac resin turned all film colours with certain yellow tone. Shellac flakes had yellow golden colour initially. As the results, ALSHGE films (Figure 3d and e) had bold yellow colour without significant difference between the two concentrations (colour data and its statistical analysis are not shown). ALSHGG films (Figure 3b and c) had less yellow tone with significant difference between the two concentrations.

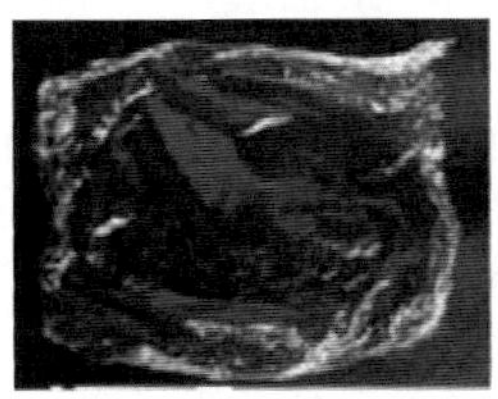

a. Film of aloe vera gel

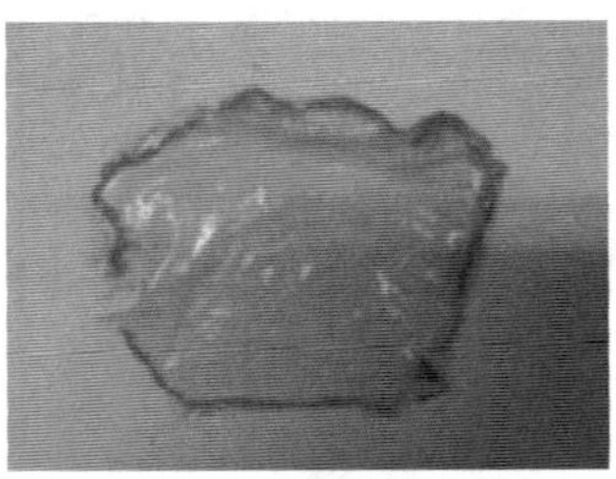

b. ALSHGG1%

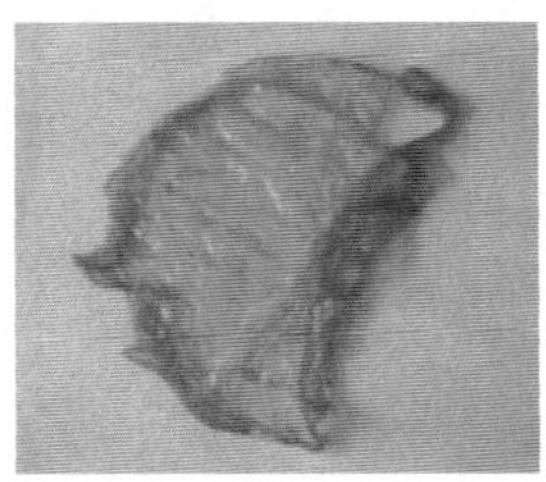

c. ALSHGG2%

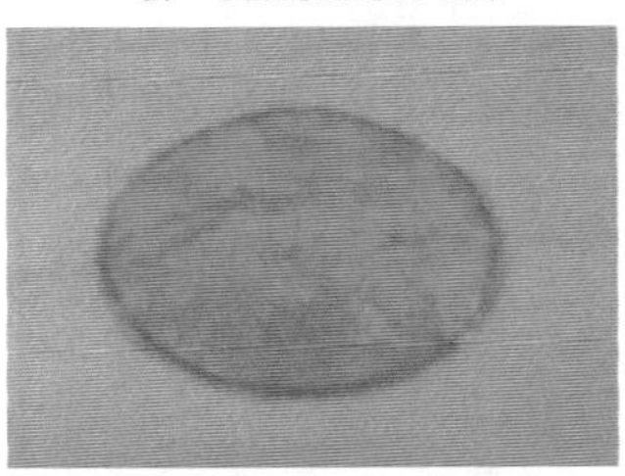

d. ALSHGE4%

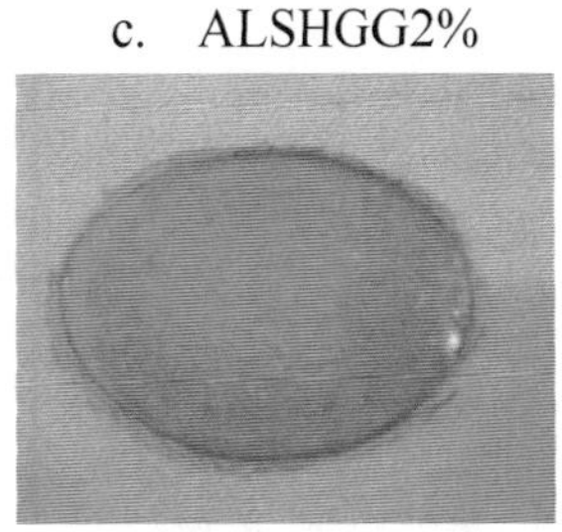

e. ALSHGE6%

**Figure 3.** Appearance of ALSHGG and ALSHGE films. Film of aloe vera gel was depicted as the reference.

Various film thickness was also obtained. ALSHGE suspension at both concentrations of gelatine resulted in thicker films than ALSHGG suspension (Table 1). Higher viscosity of ALSHGE suspension than ALSHGG suspension, as shown in Figure 1, could be the reason. At the same weight, casted ALSHGE suspension would result in thicker layer than ALSHGG suspension, leaving thicker film when it was dried. This thicker film took longer time to dissolve in water and had low WVTR as shown in Table 1. Some researches indicated anomalous relationship between film thickness and WVTR. WVTR of a hydrophilic film increased linearly with the thickness [22-23]; however, other research suggested differently. Relatively constant WVTR of film was found over a thickness range of 0.046 - 0.061 mm (methyl cellulose film) and a thickness range of 0.023 – 0.14 mm (hydroxypropyl cellulose film) [24]. Molecular weight of the film components and their hydrophilicity dictated degree of WVTR. In ideal polymer film, thickness does not have any effect on WVTR [23]. Degree of WVTR might vary depending on the film materials, hence, film solubility might change in accordance with the WVTR. ALSHGG and ALSHGE films had similar basic materials, i.e. aloe vera which is highly hydrophilic and shellac resin which is highly hydrophobic, while, both guar gum and gelatine are hydrophilic materials. If WVTR is related to molecular weight of the film material as stated by Park et al. [24], thus, results of this research are inline with that theory. Molecular weight of guar gum is much higher than that of gelatine, therefore, WVTR of ALSHGG films is higher than that of ALSHGE films.

**Table 1.** Physical characteristics of various aloe vera-based edible films.

| Film | Thickness (mm) | Moisture content (%) | Solubility (%) | | | WVTR |
|---|---|---|---|---|---|---|
| | | | day 0 | day 2 | day 4 | ($g/m^2$ h) |
| ALSHGG1% | 0.04 (cv. 6%) | 19.1 (cv. 2%) | 31.5 | 37.2 | 42.5 | 8.3 |
| ALSHGG2% | 0.06 (cv. 3%) | 17.4 (cv. 6%) | 27.8 | 41.4 | 32.0 | 7.7 |
| ALSHGE4% | 0.13 (cv.1%) | 10.1 (cv. 3%) | 23.6 | 25.3 | 25.6 | 4.6 |
| ALSHGE6% | 0.14 (cv. 2%) | 11.0 (cv. 5%) | 23.0 | 24.3 | 27.2 | 6.6 |

## 4. Conclusion

This research showed that aloe vera-based edible film was possible to be formed by incorporating shellac resin and guar gum or gelatine. Suspension of the composite film containing gelatine (ALSHGE) seemed to be a better option for forming the film because the suspension withstood sedimentation process and the microbial growth during 4-day storage. This would allow longer time of film preparation because new and freshly prepared suspension might not be necessary. Further, higher viscosity of the ALSHGE suspension allowed formation of thicker film, which consequently, resulted in low solubility and low WVTR of the film. Selection upon the best gelatine concentration to be incorporated into the composite suspension subjects to further statistical analysis.

## References

[1] Muchtadi T, Sugiyono and Ayustaningwarno F 2010 *Ilmu Pengetahuan Bahan Pangan* (Bandung, Indonesia: Alfabeta)

[2] Mardjan S S 2015 *Peran Teknologi Penyimpanan dalam Penanganan Produk Segar Buah Tropika Indonesia* (Bogor, Indonesia: IPB Press)

[3] Dweck A C and Reynolds T 1999 *Aloe vera* L. leaf gel: A review update *J. Enthnophamacol* **68** 3-37

[4] Kester J J and Fennema O R 1986 Edible films and coatings: A review *Food Technology* **40**(12) 47-59

[5] Yunita M, Hendrawan Y and Yulianingsih R 2015 Analisis kuantitatif mikrobiologi pada makanan penerbangan (*Aerofood ACS*) Garuda Indonesia berdasarkan TPC (*Total Plate*

*Count*) dengan metode pour plate *Jurnal Keteknikan Pertanian Tropis dan Biosistem* **3**(3) 237–48
[6] Khoshgozaran-Abrasa S, Azizia M H, Hamidya Z and Bagheripoor-Fallah N 2012 Mechanical, physicochemical and color properties of chitosan based-films as a function of Aloe vera gel incorporation *J. Food Technology* **3**(87) 2058–62
[7] Costa N, Alves V and Isabel M 2005 Barrier properties of biodegradable composite *films* based on kappa-carrageenan/pectin blends and mica flakes *Carbohydr Polym* **79**(2) 269–76
[8] Warkoyo, Rahardjo B, Marseno D W and Karyadi J N W 2014 Sifat fisik, mekanik dan *barrier edible film* berbasis pati umbi kimpul (*Xanthosoma sagittifolium*) yang diinkorporasi dengan kalium sorbat *Agritech* **34**(1) 72–81
[9] Garcia A, Martino M and Mali Z 2004 Barrier. mechanical and optical properties of plasticized yam starch films *Carbohydr Polym* **56**(2) 129–35
[10] Chenlo F, Moreira R and Silva C 2010 Rheological behaviour of aqueous systems of tragacanth and guar gums with storage time *J. Food Eng* **96** 107–113
[11] Yaron A 1993 Characterization of aloe vera gel before and after autodegradation, and stabilization of the natural fresh gel *Phytother Res* **7**(7) S11–13
[12] Opazo-Navarrete M, Tabilo-Munizaga G, Vega-Gálvez A, Miranda M and Pérez-Won M 2012 Effects of high hydrostatic pressure (HHP) on the rheological properties of *Aloe vera* suspensions (*Aloe barbadensis* Miller) *Innov Food Sci Emerg Technol* **16** 243–50
[13] Marcotte M, Hoshahili A R T and Ramaswamy H S 2001 Rheological properties of selected hydrocolloids as a function of concentration and temperature *Food Res Int* **34** 695–703
[14] Rivero S, García, M A and Pinotti A 2009 Composite and bi-layer films based on gelatin and chitosan *J. Food Eng* **90**(4) 531–39
[15] Binsi P K, Shamasundar B A, Dileep A O, Badii F and Howell N K 2009 Rheological and functional properties of gelatin from the skin of Bigeye snapper (*Priacanthus hamrur*) fish: Influence of gelatin on the gel-forming ability of fish mince *Food Hydrocoll* **23** 132–45
[16] Gao J, Li K, Xu J, Zhang W, Ma J, Liu L, Sun Y, Zhang H and Li K 2018 Unexpected rheological behavior of hydrophonic associative shellac-based oligomeric food thickener *J. Agric Food Chem* **66**(26) 6799–805
[17] Jain R, Anjaiah V and Babbar S B 2005 Guar gum: a cheap substitute for agar in microbial culture media *Lett Appl Microbiol* **41**(4) 345–49
[18] Karlton-Senaye B D and Ibrahim S A 2013 Impacts of gums on the growth of probiotics *Agro Food Ind Hi Tech* **24**(4) 10–14
[19] López-Carballo G, Hernández-Muñoz P, Gavara R and Ocio M J 2008 Photoactivated chlorophyllin-based gelatin films and coating to prevent microbial contamination of food products *Int J Food Microbiol* **126**(1-2) 65–70
[20] Jo W -S, Song H -Y, Song N -B, Lee J -H, Min S C and Song K B 2014 Quality of microbial safety of 'Fuji' apples coated with carnauba-shellac wax containing lemongrass oil *LWT-Food Sci Technol* **55**(2) 490–97
[21] Habeeb F, Shakir E, Bradbury F, Cameron P, Taravati M R, Drummond A J, Gray A I and Ferro V A 2007 Screening methods used to determine the anti-microbial properties of aloe vera inner gel *Methods* **42** 315–20
[22] Banker G S, Gore A Y and Swarbrick J 1966 Water vapour transmission properties of free polymer films. *J. Pharm. Pharmacol.* **18**(7) 457–66
[23] McHugh T H, Avena-Bustillos R and Krochta J M 1993 Hydrophilic edible films: modified procedure for water vapor permeability and explanation of thickness effects *J Food Sci* **58**(4) 899–903
[24] Park H J, Weller C L, Vergano P J and Testin R F 1993 Permeability and mechanical properties of cellulose-based edible films *J Food Sci* **58**(6) 1361–64

IOP Conf. Series: Materials Science and Engineering **557** (2019) 012077 doi:10.1088/1757-899X/557/1/012077

# Prediction Shelf Life of Arabica Java Preanger Coffee Beans under Hermetic Packaging Using Arrhenius Method

**S Mardjan[1*] and F R Hakim[1]**

[1] Departement of Mechanical and Biosystem Engineering, Faculty of Agricultural Engineering and Technology, Bogor Agricultural University, Bogor, Indonesia
*Corresponding author: kensutrisno@yahoo.com

**Abstract.** Java Preanger Arabica coffee is a plantation crop with 2-3 months harvest period annually, therefore the storage system is required to keep the product available throughout the year. Postharvest handling, storage, and packaging particularly are the main factors to be considered so that the coffee beans can be stored for a long time. Hermetic packaging is identified as an effective and safe technology to preserve the quality of grain during storage. Specialty coffee beans cannot be accepted if the total quality value is below 80 according the SCAA protocol, therefore an estimation of the coffee beans shelf life is required. The objective of this research was to analyze and compare the difference between the quality changes of Preanger Arabica Coffee during storage period using hermetic packaging and non-hermetic packaging, and to estimate the shelf life of dried coffee beans based on taste value using ASLT (Accelerated Shelf-Life Testing) method by using Arrhenius approach. Storage simulation was conducted within temperature range of 40°C, 50°C, and 60°C, with 80% relative humidity using both hermetic and non-hermetic packaging. The degradation of quality rate (K value) which was obtained based on the simulation results, was used to predict coffee beans shelf-life by putting it on Arrhenius equation. Coffee packaged with hermetic plastic has a slower rate of quality decline compared to coffee packed with non-hermetic plastic. The predicted shelf-life of coffee beans with hermetic and non-hermetic packaging were 232 days and 45 days (10°C), 113 days and 26 days (15°C), and 57 days and 16 days (20°C), 29 days and 10 days (25°C) respectively.

## 1. Introduction

Indonesia is the fourth coffee producer in the world after Brazil, Vietnam and Colombia [7] [3]. Arabica coffee has a distinctive taste and is much in demand by other countries, which is a source of increasing foreign exchange. Likewise, coffee from West Java, known as Java Preanger, productivity continues to increase with a market share that also increases from year to year. However, coffee is a plantation crop with a harvest period once a year with a harvest period of 2 -3 months, while market demand occurs throughout the year, so coffee beans must be stored so that they are available at all times.

Therefore, postharvest handling, especially storage methods, is the main factor that needs to be considered so that the quality of coffee is maintained during storage. In addition to temperature and storage time, one of the important factors in storing coffee is the type of packaging [1] [2]. Hermetic plastic packaging is a package made from special materials and techniques to create a tight environment from outside influences [5] [6], especially on water and gas including $CO_2$, $O_2$ and $N_2$ [1]. The basic rw materials of hermetic packaging consist of plastic High Density Polyethylene (HDPE) which functions as a barrier to oxygen, while Polyvinyl Chloride (PVC) plastic serves to give strength to the packaging [9]. This packaging is suitable for certain products that require manipulation of atmospheric conditions, with the aim of preventing pest attacks, reducing product respiration [6], and maintaining sensory quality, especially the aroma of the packaged material [12].

During storage, coffee beans decline in quality and the length of shelf life is affected, among others, by temperature, relative humidity and the type of packaging. According to the Specialty Coffee Association of America (SCAA), specialty coffee beans cannot be accepted if the taste value is less than 80, so a shelf life estimation is needed to determine the quality changes of coffee during the storage period. One method of estimating shelf life is Accelerated Shelf-Life Testing (ASLT) with the Arrhenius approach [4]. Quality reduction analysis with this method requires several observations, namely parameters measured quantitatively that reflect quality changes that occur during storage [13]. The objectives of this study are 1) to analyze the quality degradation of Preanger Arabica coffee beans during the storage period with hermetic packaging and compare them with non-hermetic plastic packaging, and 2) to estimate the shelf life of Preanger Arabica coffee beans based on flavor value parameters using the ASLT method with Arrhenius approach.

## 2. Methodology

*2.1 Tools and Materials*

The tools used in testing the shelf life of this coffee bean product are digital scales, Grainpro SGB PREMIUM-15RZ hermetic plastic packaging with high strength polyethylene material with barrier, non-hermetic plastic packaging in the form of low density polyethylene (LDPE) plastic, coffee roasting machines W600 type direct flame drum, and KCL-2000 Eyela incubator. The main material used in this study was the Preanger Arabica coffee (greenbean) from the Frinsa plantation in the village of Weninggalih, West Bandung regency.

*2.2 Research Method*

The study was conducted in 4 stages, namely sample preparation, determination of the taste value of coffee beans, simulations accelerating the decline in the quality of coffee beans during storage, and estimation of shelf life with the Arrhenius approach.

*2.2.1 Sample Preparation*

500 grams of dried coffee beans were packed into 12 hermetic plastic packages and 12 non-hermetic plastic packages, so the total samples were 24 packs of coffee beans @ 500 grams. The samples were put into an incubator that has been set with RH 80% and three temperature levels (40°C, 50°C, and 60°C) as a storage simulation.

*2.2.2 Determination of the Taste Value of Coffee Beans Using the Cupping Method*

Determination of the taste value of coffee was done by cupping method which refers to the SCAA cupping protocol [7], which consists of 3 main parts, namely sample preparation, sensory evaluation, and assessment. The sample preparation process consists of roasting, grinding and brewing. The next step was sensory evaluation carried out by 3 panelists or Q-graders. Coffee that had been brewed for 3-5 minutes was stirred using a cupping spoon, an assessment was based on the attributes of flavor, aroma, flavor, aftertaste, acidity, body, uniformity, balance, clean cup, sweetness and overall. Based on these 10 attributes, coffee was classified based on the final total score according to the SCAA as in Table 1, where coffee is not suitable for consumption if the final score reaches 30 or less.

**Table 1.** Coffee qualifications based on total taste values

| *Total Score* | *Quality* | *Classification* |
|---|---|---|
| 90-100 | *Outstanding* | *Specialty* |
| 85-89.99 | *Excellent* | |
| 80-84.99 | *Very Good* | |
| <80.0 | *Below Specialty Quality* | *Not Specialty* |

*2.2.3 Prediction of Shelf-life of Coffee Beans during Storage*

Estimation of shelf life was carried out using the accelerated shelf life testing (ASLT) method with the Arrhenius approach, which was able to predict the actual shelf life by simulating quality degradation data obtained during observation. Samples were stored in extreme conditions, namely at high temperatures of 40, 50 and 60°C, and measured parameters that reflect changes in quality during storage [8], [10], [11], [13]. The Arrhenius equation shows the dependence of the deterioration reaction rate on temperature which is formulated as follows:

$$K = K_o.e^{-Ea/RT} \quad (1)$$

$$\ln K = \ln Ko^{-(Ea/RT)} \quad (2)$$

$$\ln K = \ln Ko^{-\{(Ea/R).(1/T)\}} \quad (3)$$

where, K = reaction rate constant at temperature T, Ko = pre-exponential constant, Ea = Activation energy (cal/mol), T = Absolute temperature (°K), R = Gas constant (1,986 cal/mol°K). The activation energy (Ea) in this equation shows the magnitude of the effect of storage temperature on the reaction that occurs in food. The activation energy value is obtained from the slope of the straight line graph ln K to (1/T). Large activation energy indicates that a slight increase in storage temperature will make the ln K value increase significantly.

## 3. Results and Discussions

*3.1 Determination of the Initial Taste Value of Coffee Beans*

The coffee sample in the initial condition received a total final score of 85.75 so that it was included in the specialty category (above 80), as can be seen in Figure 1.

| Attribute | Score |
|---|---|
| Fragrance | 8 |
| Flavor | 8 |
| Aftertaste | 7.5 |
| Acidity | 8.25 |
| Body | 8 |
| Balance | 8 |
| Uniformity | 10 |
| Clean Cup | 10 |
| Sweetness | 10 |
| Overall | 8 |
| **Total score** | 85.75 |
| **Defect** | 0 |
| **Final score** | 85.75 |

**Figure 1.** Test results for the taste value of coffee beans without treatment

The value of 10 on uniformity attributes shows that this coffee flavor has good consistency, while the value of 10 on the clean cup attribute and the value of 0 defect shows that the coffee processing is done cleanly so that no dirty taste or aroma is found in coffee.

*3.2 Simulation of the Acceleration of Decreasing Quality and Analysis of Declining Value of Coffee Flavors during Storage*

The simulation results of storing packaged coffee at various temperatures with decreasing quality are presented in Table 2.

**Table 2.** Data of decreasing the value of coffee taste during storage

| Days | Total Score of Cupping Test | | | | | |
|---|---|---|---|---|---|---|
| | 40°C | | 50°C | | 60°C | |
| | Hermetic | Non-Hermetic | Hermetic | Non-Hermetic | Hermetic | Non-Hermetic |
| 0 | 85.75 | 85.75 | 85.75 | 85.75 | 85.75 | 85.75 |
| 2 | 85.5 | 85.5 | 85.25 | 84.5 | 56 | 55.5 |
| 4 | 85.25 | 85.25 | 84 | 83.5 | 41.75 | 39.5 |
| 6 | 85 | 76.75 | 83 | 71 | 36.25 | 34.75 |
| 8 | 78.5 | 73.25 | 70.5 | 69.5 | 25.25 | 25 |

After obtaining data on quality degradation for each storage temperature and packaging materials (Table 2), then the graph of the linear regression equation was calculated as follows:

| | | |
|---|---|---|
| Hermetic 40°C : | $Y = -1.5x + 88.5$ | $R^2 = 0.5902$ |
| Hermetic 50°C : | $Y = -3.275x + 91.525$ | $R^2 = 0.6644$ |
| Hermetic 60°C : | $Y = -14.075x + 91.225$ | $R^2 = 0.9093$ |
| Non-Hermetic 40°C : | $Y = -3.375x + 91.425$ | $R^2 = 0.8221$ |
| Non-Hermetic 50°C : | $Y = -4.6x + 92.65$ | $R^2 = 0.8457$ |
| Non-Hermetic 60°C : | $Y = -14.225x + 90.775$ | $R^2 = 0.8961$ |

The value of K for each temperature and type of packaging wa obtained from the value of the slope of each graph. Then the ln K value was plotted into a semi logarithmic graph (Arrhenius plot), where ln K as ordinate and 1/T (temperature in Kelvin) as abscissa to get the value of –E/R for each type of package. The Arrhenius plot for hermetic and non-hermetic packaging can be seen in Figure 2.

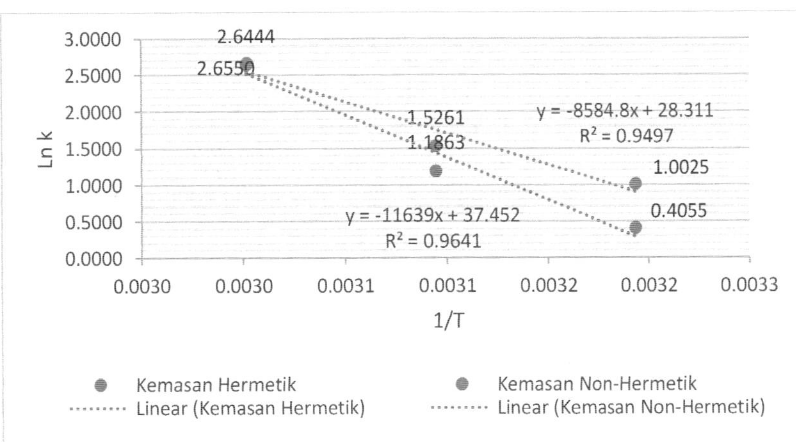


**Figure 2.** Relationship between the value of ln K with 1/T coffee beans using hermetic and non-hermetic plastic packaging

Based on Figure 2, the relationship of the straight line equation between the values of ln K and the value of 1/T for each packaging was obtained, as follows:

| | |
|---|---|
| Hermetic packaging | : $Y = -11639x + 37.452$ $R^2 = 0.9641$ |
| Non-hermetik packaging | : $Y = -8584.8x + 28.311$ $R^2 = 0.9497$ |

The slope value of the equation is the value of -E/R for each equation. The value of R is the gas constant (1,986 cal/mol°K) and the activation energy value is the value of E in the equation, so that the activation energy for each packaging can be obtained as shown in Table 3.

**Table 3.** The value of -E/R for each type of packaging

| | -E/R | R (cal/mol°K) | E (cal/mol°K) |
|---|---|---|---|
| Hermetic packaging | -11639 | 1.986 | 23115.054 |
| Non-hermetic packaging | -8584.8 | 1.986 | 17049.4128 |

The intercept value in the semi logarithmic graph is the value of ln Ko in the Arrhenius equation, so the results of ln Ko and Ko were obtained as shown in Table 4.

**Table 4.** Ln Ko and Ko values for each type of packaging

| | Ln Ko | Ko |
|---|---|---|
| Hermetic packaging | 37.452 | 1.842 |
| Non-hermetic packaging | 28.311 | $1.974 \times 10^{12}$ |

Arrhenius equation can be arranged based on the value of -E/R and Ko that has been obtained. After the Arrhenius equation for each type of packaging is obtained, then the rate of declining of coffee bean taste can be calculated then called K value based on storage temperature and type of packaging as shown in Table 5. K values obtained from calculations with Equation 6, and estimation of shelf life of coffee beans obtained by entering the rate of deterioration (K) in Equation 7, i.e. :

$$K = Ko\ e^{-E/RT} \tag{6}$$

$$t = \frac{A_O - A}{K_O . e^{-E/RT}} \tag{7}$$

**Table 5.** K values and estimation of shelf life for each storage temperature and type of packaging

| Type of Packaging | Storage Temperature | K | t (days) |
|---|---|---|---|
| Hermetic | 40 °C or 313.15 K | 1.329 | 4.5 |
| | 50 °C or 323.15 K | 4.198 | 1.4 |
| | 60 °C or 333.15 K | 12.376 | 0.5 |
| Non Hermetic | 40 °C or 313.15 K | 2.451 | 2.5 |
| | 50 °C or 323.15 K | 5.726 | 1 |
| | 60 °C or 333.15 K | 12.711 | 0.5 |

The initial coffee taste value of 85.75 was then denoted by the letter A0 in Equation 7. Coffee is considered not included in the specialty category if the taste value is below 80 or has reached the value of 79.75, and this value is hereafter referred to as critical taste value (A). Estimation results of coffee shelf life at storage temperatures of 40°C, 50°C, and 60°C were estimated at 4.5 days, 1.4 days and 0.5 days for hermetic packaging, and 2.5 days, 1 day and 0.5 days for non-hermetic packaging, respectivelly. According to the reaction rate theory, increasing temperature will cause an increase in reaction rate because rising temperatures will result in an increase in kinetic energy in the particles. This condition will accelerate the movement of molecules and increase the number of collisions that occur between particles, so that the faster the reaction occurs. This also occurs in coffee beans in the packaging, where reactions can occur from within coffee, such as fat oxidation; coffee reaction with packaging where migration of packaging substances is absorbed in coffee; and also the reaction of coffee with substances outside the environment that can penetrate the packaging such as absorption of moisture content, and the growth of bacteria or fungi.

According to [1] in his book entitled "Handbook of Coffee Post-Harvest Technology", raw or green coffee beans should be stored at temperatures between 10-17 °C, or store at room temperature (20-30°C) with proper packaging. Therefore a shelf life estimation was done at temperatures of 10 to 30°C, the results are presented in Table 6 and Figure 8.

**Table 6.** K values and estimation of shelf life for each storage temperature and type of packaging

| Type of packaging | Storage Temperature | °K | t (day) |
|---|---|---|---|
| Hermetic | 10 °C or 283.15 K | 0.026 | 232 |
| | 15 °C or 288.15 K | 0.053 | 113 |
| | 20 °C or 293.15 K | 0.105 | 57 |
| | 25 °C or 298.15 K | 0.205 | 29 |
| | 30 °C or 303.15 K | 0.390 | 15 |
| Non Hermetic | 10 °C or 283.15 K | 0.134 | 45 |
| | 15 °C or 288.15 K | 0.227 | 26 |
| | 20 °C or 293.15 K | 0.378 | 16 |
| | 25 °C or 298.15 K | 0.617 | 10 |
| | 30 °C or 303.15 K | 0.992 | 6 |

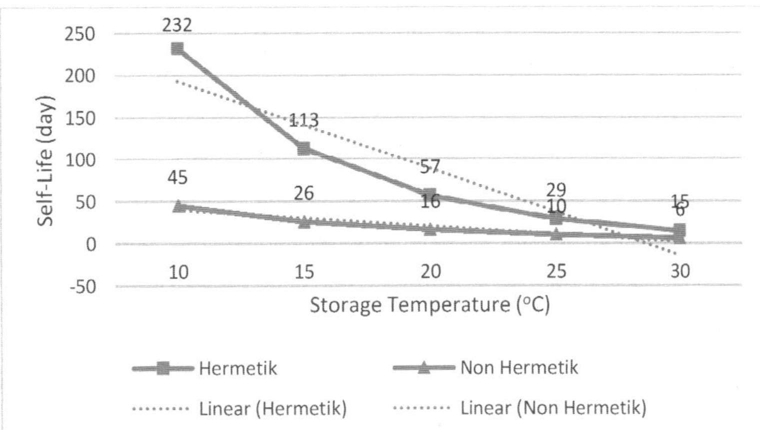


**Figure 3.** Relationship between storage temperature and estimated shelf life

As can be seen in Figure 8 that hermetic plastic packaging has a better ability to maintain coffee quality than non-hermetic plastic packaging. At storage temperatures of 10°C to 30°C coffee beans can last up to 15 to 232 days for hermetic packaging, and 6 to 45 days for non-hermetic packaging. Specialty coffee shelf life is strongly influenced by the rate of decline in the final score of taste value during storage, the lower the rate of declining taste, the longer the shelf life.

## 4. Conclusions and Suggestions

Coffee packaged with hermetic plastic has a slower rate of decline compared to coffee packed with non-hermetic plastic. The shelf life estimation using the ASLT method with the Arrhenius approach shows coffee shelf life of 232 days with hermetic packaging at 10°C, 113 days at 15°C, 57 days at 20°C, 29 days at 25°C, and 15 days at 30°C. With non-hermetic packaging the shelf life is shorter, which is 45 days at 10°C, 26 days at 15°C, 16 days at 20°C, 10 days at 25°C, and 6 days at 30°C.
Suggestions for research on shelf life estimation in agricultural products using the ASLT method and Arrhenius approach, critical parameters used must be able to represent quality in general, including organoleptic testing.

IOP Conf. Series: Materials Science and Engineering **557** (2019) 012077 doi:10.1088/1757-899X/557/1/012077

## 5. References

[1] Borem FM, Riberio FC, Figueiredo LP, Giomo GS, Fortunato VA, Isquerdo EP. 2013. Evaluation of the sensory and color quality of coffee beans stored in hermetic packaging. *J of stored product research* 52:1 – 6.

[2] Bhumiratana, N., K. Adhikari, and E. Chambers. 2011. Evaluation of sensory aroma attributes from coffee beans to brewed coffee. *LWT Food Science and Technology* 44: 2185-2192.

[3] Direktorat Jendral Perkebunan. 2015. Statistik Perkebunan Indonesia (Kopi) 2014-2016. Jakarta (ID).

[4] Ellis, MJ. 1994. The methodology of shelf life determination *in* Shelf Life Evaluation of Foods. Ed. CMD Man and AA Jones. Blackie Academic and Professional. Glasgow.

[5] Emblem A. 2012. Plastics properties for packaging materials. Woodhead Publishing UK.

[6] Guenha R, Salvador BDV, Rickman J, Goulao LF, Muocha IM, Carvalho MO. 2014. Hermetic storage with plastic sealing to reduce insect infestat ion andsecure paddy seed quality: A powerful strategy for rice farmers *in* Hine, DJ. 1997. Modern Packaging, Packaging, and Distribution System for Food. Blackie. London.

[7] ICO (2014). Statistik Perdagangan. www.ico.org, London: International Coffee Organization.

[8] Labuza, TP 1982. Open shelf-life Dating of Foods. Food Science and Nutrition. Press Inc., Westport, Connecticut.

[9] Marsh K, Bugusu B. 2007. Food Packaging Roles, Materials, and Environmental Issues. *J Food Sci.* 72:39-55.

[10] Morten C. Meilgaard, B. Thomas Carr, Gail Vance Civille. 2006. Sensory Evaluation Technique. *Detroit. United States. CRC Press.*

[11] Nurcahyanti, A. 2005. Pengemasan dan Pendugaan Umur Simpan Keripik Talas dalam Kemasan Plastik PP, OPP/VMCPP, and PET/DL/VMPET/SPE dengan Metode Akselerasi [skripsi]. Bogor (ID): Institut Pertanian Bogor.

[12] Ribeiro FC, Borem FM, Giomo GS, Lima RRD, Malta MR, Figueiredo LP. 2011. Storage of green coffee in hermetic packaging injected with $CO_2$. *J of stored product research* 47:341 – 348.

[13] Syarif R and H. Halid. 1993. Teknologi Penyimpanan Pangan. Pusat Studi Antar Universitas. IPB. Bogor.

# Studies on The Steel Reinforcement Performance in Presence of NaCl: Effects of Salt Concentration and Temperature

**Sabah hatam obaid[1], JabbarAbbs Jaber[2], Hadi Raheem Ibrahim[3]**

[1, 2, 3]Al-Mussaib Technical Institute, Al-Furat Al-Awsat Technical University, 51009 Babylon, Iraq
Email: inm.sbh@atu.edu.iq[1], inm.jbr@atu.edu.iq[2], inm.hadi@atu.edu.iq[3]

**Abstract.** The present study is to evaluate the performance of steel reinforcement under NaCl concentration chloride exposure. The loss of weight, yield stress and ultimate stress were tested to evaluate the performance of steel reinforcement before and after exposure. The exposure was also repeated under a variable temperature of (50, 60 and 70 $C^0$). The steel was treated to improve its properties by using hardening in the oxy-acetylene torch, epoxy coating and aluminium spread coating. Results showed that the reduction in reinforcement properties was reduced significantly by using the hardening method and the maximum reduction in the performance of steel reinforcement was in the higher temperature chloride exposure. The increasing of concentration (0.1, 0.2, 0.3 and 0.4 NaCl %) was due to high corrosion in steel wire. The results showed that the treatment of reinforcement was significantly improved.

**Keywords:** Steel reinforcement, the loss of weight, the treatment of reinforcement

## 1. Introduction

Most metals are extracted from raw material by applying a specific amount of energy, and according to the second law of thermodynamics, where the last state is returning to lowest probable level of energy i.e. union form that is called corrosion process [1]. The corrosion occurs in the surface metal that is exposed to atmospher. The Process is accompanied by a reduction in the metallic thickness and mechanical properties [2]. The corrosion leads to impair metallic material and neutralize from its function.To eliminator prevent this state and maintain the performance of the materials, two ways can be achieved: Using cheap alloys with corrosion inhibitors using a high coast alloy which did not trend to erode [3].

The corrosion depends on the atmospheric type that surrounds with the metal part and oxygen amount available with the water vapour and amount of salt in the soil in which metal parts are submerged [4]. There are different types of corrosion: corrosion due to dissimilar metal corrosion, corrosion due to differential aeration cells, corrosion due to metal oxide cells corrosion, corrosion due to deferential stresses cells, corrosion due to differential, concentration of salts, and corrosion due to stray current as well as atmospheric corrosion [5]. For all types of corrosion mentioned above Oxygen, Water vapour and salt solution must be present for process occurrence to protect metallic material from corrosion. The first way is by the approach depending on isolation of the metallic part from corrosion of atmosphere by coating it with a temporary layer organically or inorganically sustainable ormetallic. The second way is by electrochemical approach that depends on the application of anodic or cathodic protection or protection by sacrifice. The third way is by approach depending on inhibitors of erosion processes by adding materials to erosion surrounding to reduce corrosion rate. These approaches contribute to anodic and cathodic interactions, therefore they are called "anodic and catholic inhibitors, and there are inhibitors of adsorption [6].

## 2. Methodology

### *2.1 Preparation of specimens and the procedure of testing*

To achieve the aim of this study, Fourteen samples of Ukrainian steel were tested to measure their yield and ultimate stress according to the (ASTM). The samples were in diameter (10mm) and length of (50mm).The tests were carried out in the construction laboratory of civil engineering department of Al – Mussaib technician Institute. The mineral ingredient ratios of steel used from the produced company are listed in table (1).

**Table 1**: Mineral ingredients and ratio of steel used

| %C | %Mn | %Mg | %Cr | %Ni | %Si |
|---|---|---|---|---|---|
| 0.45 | 0.75 | 0.05 | 0.08 | 0.07 | 0.2 |

The Figure 1 illustrates the samples and tensile tester while the properties of steel tested are listed in Table 2.

**Figure 1:** illustrates: (a) Wire steel reinforcement sample epoxy coating, (b) Tensile tester, (c) Wire steel reinforcement after test

**Table 2:** The results of standard sample test

| Dbefore test(mm) | Ls (mm) | Weight(gm) | Yield stress(Mpa) | Ultimate stress(Mpa) | Dafter test(mm) | Actual elongation $(\ell_f-\ell_i)/\ell_{eff}$ | Equipment elongation |
|---|---|---|---|---|---|---|---|
| 10 | 496 | 294 | 664.5 | 729 | 6 | 5% | 11% |

Notes: effective length = 220 mm; D = sample diameter; Ls = sample length

Three samples of steel were immersed in salty soil solution of (3%) concentration of NaCl for (30,60,90) Days to monitor and evaluate the adverse effect of chloride salt according to the influence of chemical corrosion on mechanical properties of reinforcement steel used in construction processes .The corrosion rate depends on the following equation [7,8,9,10]
Corrosion ratio (mmpy) = 534 (w/DAT):
Where: mmpy: mm penetration per year; W = weight loss (mg); D = density of sample (g / $cm^3$); A = area of sample in ($in^2$); T = time of exposure of the metal sample in hours.
Also Yield and Ultimate stresses were measured as well as the percentage of elongation .The measured results are listed as in Table 3 also as shown in Figures 2 and 3.

**Table 3:** The results of three samples in different intervals

| Sample no. | Immersion time (hurs) | Yield stress (Mpa) | Ultimate stress (Mpa) | Actual elongation % | Equipment elongation % | mmpy (mm/year) |
|---|---|---|---|---|---|---|
| 1 | 720 | 574.8 | 647.3 | 6.8 | 12.8 | 22.50 |
| 2 | 1440 | 532.7 | 590.5 | 8.1 | 13.8 | 23.60 |
| 3 | 2160 | 518.0 | 565.8 | 9.2 | 14.3 | 23.75 |

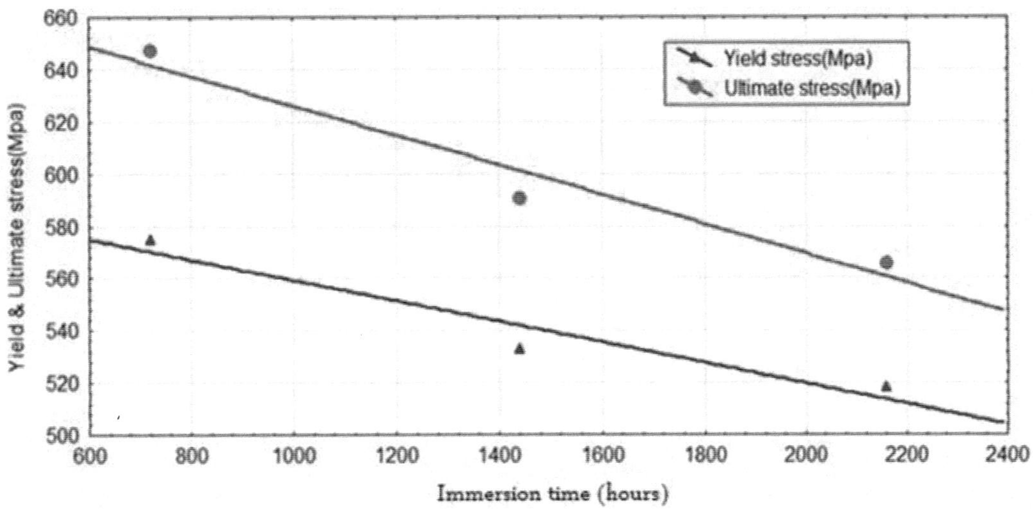


**Figure 2:** The relationship between the yield and ultimate stress of reinforcement steel Corresponding to Immersion time

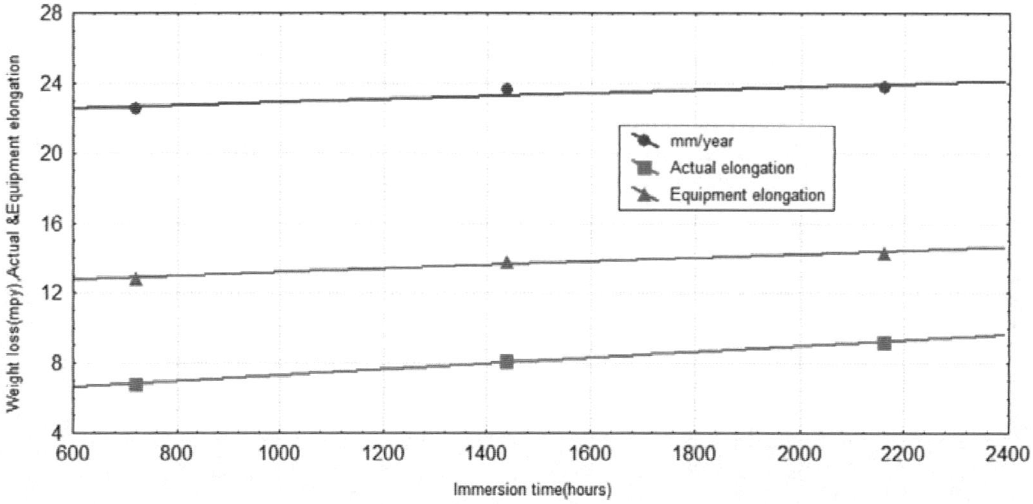


**Figure 3:** The relationship between the weight loss, actual and equipment elongation Corresponding to Immersion time

*2.2 The treatment approaches of wires steel reinforcement*

In this research situational treatments were used three reinforced steel samples with the same physical and mechanical properties and then these samples were immersed in salty solution with the same mentioned previous for interval (2160 hours) for all samples, and weighting the samples to know weight loss and after that samples were subjected to tensile test to produce ultimate and yielding stress values and to compare with untreated samples. The treated ways were as follow:

1 Surface hardening process by Oxy-acetylene torch for one sample and then cooling by solution (water and oil with ratio of 50% for each part).
2 Coating process by epoxy material with hardening material .That are used in ceramic works.
3 Aluminum coating by spread way within the electric oven and by recrystallization temperature for aluminum element after the surrounding the sample by element powder.

All the treatments for the samples have done in the laboratory of Almusaib technical institute and after that immersion the samples in salt solution for interval (2160 hours) and the results were listed in table (4) as following:

**Table 4:** The results of wire steel reinforcement treated

| Samples | Immersion time (hrs) | Yield stress (Mpa) | Ultimate stress (Mpa) | Actual elongation % | Equipment elongation % | mmpy (mm/yr) |
|---|---|---|---|---|---|---|
| Hardening with Oxy-acytelineflam | 2160 | 660 | 709 | 3 | 11 | 10.8 |
| Coating with Epoxy material | 2160 | 662 | 715 | 6 | 11.3 | 9.6 |
| Coating by aluminum | 2160 | --- | 690.7 | 5 | 13.4 | 16.5 |

*2.3 Increasing of temperature and (NaCl) concentration influence*

The temperature in Iraq environment is increasing at high rates in summer season may be Reach to (60 $^0$C) especially in soil which is directly exposed to sun ray and Conceivable for next year's due to thermal retention, there for the temperature is an important variable in this research . (3- samples) was immersed in solution with Concentricity of (3%NaCl) for different temperature (50,60 and 70C$^0$) and for intervals of (700,800,900 and 1000 hours) and then measuring the loss percent ( μg ) in sample Weight relative to unit area .The results are illustrate in figure (4).

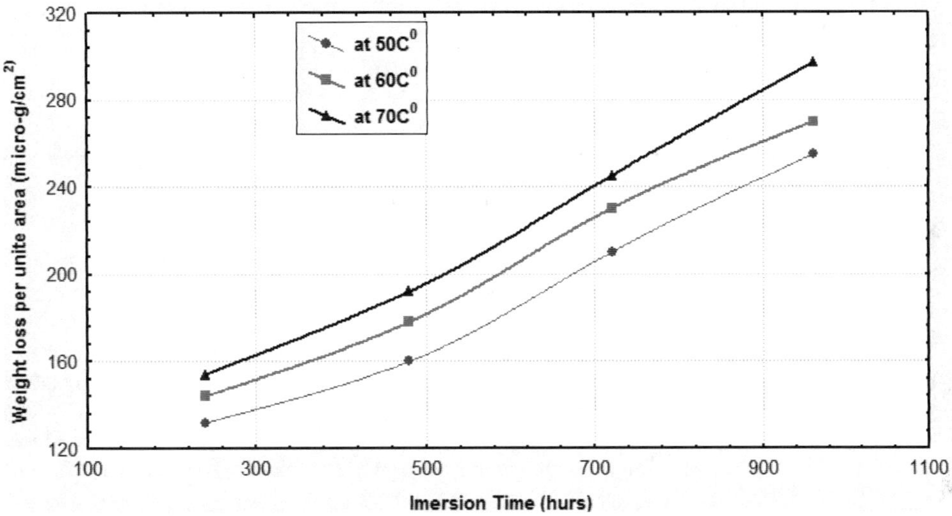


**Figure 4:** The relationship between weight losses from whole sample area corresponding to different immersion times for different temperatures.

The variation of salt concentration from one soil to another especially relative to soil of Iraq where the soil of southern areas are more saltiness than northern regions and middle , there for the variation of (NaCl ) concentration is an important factor that effects on construction steel corrosion which is used in concrete reinforcement works. Four concentration of (0.1, 0.2, 0.3, 0.4 M/NaCl) were used and at high temperature of (70C$^0$) and was calculated the loss percentage in weight relative to the total area of sample (cm$^2$) and for different intervals (in hours) for all the samples which are previously mentioned. The results are showed in figure (5).

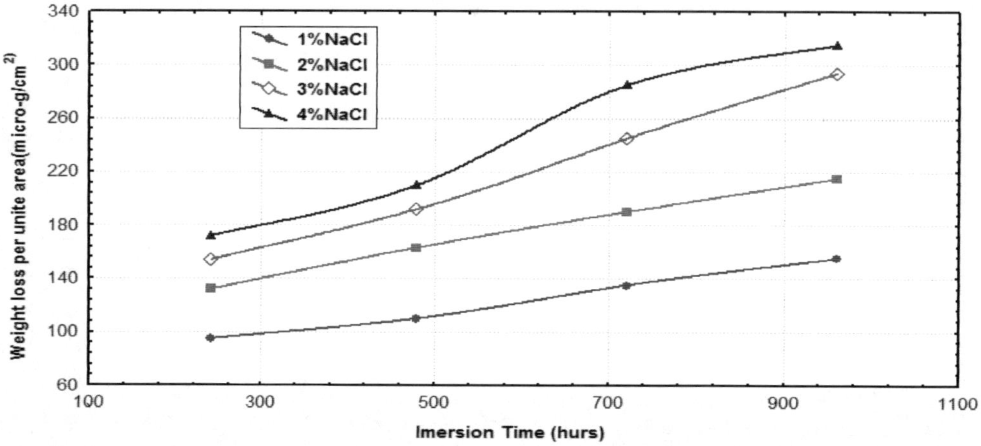


**Figure 5:** The relationship between weight losses corresponding to different immersion period at different ratios of (NaCl) concentration

The treatment was proceeded for one construction steel sample by heating at steel recrystallization temperature and after that cooling the sample by mixture of ( oil and water) in cooling process and immersion the sample in salt solution with concentration (4%NaCl) at temperature of (60 $C^0$) for interval of (720 hours) . The sample was tested by tensile test to know heat treated construction steel sample ability to resist increase soil temperature with increase percentage of (NaCl) .The results in table 5.

**Table 5:** The results of tension test for heat treated sample exposure to (60 C & 4 % NaCl )

| Immersion period (hours) | $\sigma$ Yield (Mpa) | $\sigma$ ultimate (Mpa) | mmpy |
|---|---|---|---|
| 720 | 620 | 693 | 11.2 |

**3. Discussion of Results**

For Figure 2 which contain different intervals of time in hours for samples immersion process in salt solution .The relationship was inverse between immersion time and yield stress, also the same relationship was with ultimate stress for failure, i .e. the effect of salt solution (3% NaCl) on reinforced steel with over time was high and this effect on resistance of concert blocks in foundation casting and thus on constructions set up on it.
The results are taken from table (3).

For figure 3 shows corrosion rate (mmpy) is directly proportional to immersion interval in salt solution and the range between (23.75 – 22.5) for Different intervals. These values are undesired to avoid steel wires failure within the blocks concrete.

When comparing between standard sample readings immersed samples in salt solution for the yield stress value of drawback difference shall be about (150MPa) also, for the ultimate stress shall be about (163 MPa) . These illustrate negative impact of salt property solution on values of failure steel resistance during time interval.

For table 4 shows that the yield stress for treated sample by epoxy coating, it was observed that tensile equipment didn't give any reading , and this refer to that the ductility has reduced to a level that the material was close to brittleness and it is probably that the reason of chemical interaction between coating material and Hardener with salt had an effect on the increase of carbon ratio in steel making it exceed Eutectic area in (iron-carbon) chart ,there for the ductility is reduced and the brittleness is increased in cementite region .

For figure 4 shows that the relationship between weight loss due to temperature increase of soil with concentration of solution (3%NaCl) and between immersion time in (hours) is directly proportional and it can be represented as a second degree equation with a high inclination, as a result of temperature increase of soil especially in Iraq through the summer season has significant influence on lost ratio in construction steel which is applied in reinforcement.

For figure 5 shows that the relationship between weight loss and immersion time in solutions with different concentrations of (NaCl) was directly proportional and have a slope less inclined than temperature increase effected, but when the Concentration is increases up to (3% NaCl) and when time exceeds (50 hours) , it can be noticed that the inclination has changed with a larger angle and also for the concentration ratio ( 4% NaCl) . At time (500 hours) where the straight line turning into curve , this refer to that the high concentration and increment of immersion time will inversely effect on construction steel.

In the case of using the sample that has been thermally treated which was tested by tensile equipment, results in table ( 5 ) show that the thermal treatment at recrystallization temperature gives good results and approaches to standard sample despite the increase of temperature to (60 $C^0$ ) with increase of salt concentration .

## 4. Conclusions

Based on the experimental works,

1. It has been proved through applications that the effect of salt solution (3%NaCl) on increase of corrosion rate for flooded steel reinforcement in solution in side block of concrete , and this effect is Undesired according to values for long period of time ,where possibly leads to failure in concrete foundations.

2. The treatment was conducted on the samples of reinforcement steel wires and for limited interval of time (2160 hours) which has proved the ability of epoxy coating in keeping steel flooding process in salt solution.

3. The treatment of reinforcement steel wire by surface hardening will be economical approach also, maintains on yield stress and ultimate stress for sample asymptotic well for values of standard sample before flooding process. It is possible to apply this approach in the site for treating steel wires reinforcement process and pour foundations.

## References

[1] Cole I., Muster T. and Lau D. (2004), Some recent trends in corrosion science and their application to conservation, Proceedings of metal 2004. National Museum of Australia Canberra.
[2] Sharma A. (2012), Corrosion of metals and polymers: Avital concern, National conference Science media, Organized by YMCA university.
[3] Zohdy K .M. (2015),Surface Protection of carbon steel in Acidic solution using Ethylene diamante tabetic Disodium salt. Int.J. Electrochemical.Sci.

[4] ZHANG X.G. (2011), GALVANIC CORROSION. Teck Metals Ltd. Mississauga, Ontario, Canada.
[5]Wika.S. F. (2012), Pitting and crevice corrosion of stainless steel under offshore conditions, Master thesis, Norwegian University.

[6] Scully J. R. (2013), Investigation of the corrosion propagation characteristics of new metallic reinforcing bars.

[7] ZarrokH., OuddaH., ZarroukA., SalghiR., HammoutiB., BouchrineM.(2011), Weight loss measurement and theoretical study of new Pyridoxine compound as corrosion inhibitor for $C_{38}$Steel in hydrochloric acid solution. Der Pharma chemical.

[8] AfolabiA.S., MuhirwaA.C., Abdulkareem A.S. and Muzenda E.(2014), Weight loss and Microstructural studies of stressed mild steel in apple juice., Int. J. Electrochem. Sci.

[9] GivehhiR. (2016), Filtration of NaCl and Wox Nanoparticles using wire Screens and Nano fibrous filter, Athesis presented to the university of Waterloo, Canada , ©Raheleh Givehchi.

[10] Long Y., Feng C., Peng B.,XieY., Wang J.(2017), Corrosion Behavior of $T_{92}$ steel in NaCl Solution Int.J. Eletrochem. Sci.

# Experimental investigation for solar thermal hybrid installs to inverter air conditioning vapour-compression cycle system

**Ahmed Al-Rubaye** [1, 2]

[1] Department of Mechanical Engineering, University of Al-Qadisiyah, Al-Qadisiyah, Iraq
[2] Corresponding author. Email : ahmed.alrubaye@qu.edu.iq

**Abstract.** This study is designed a new development and integration for the solar hybrid system connected to an inverter air conditioning system. This experimental system will provide an essential addition to the renewable energy future. The device is supplying the same cooling load with significantly less electricity demand. Apparatus system has the solar unit and solar collector with DC compressor that was used to compress the refrigerant in an air conditioning system, and that will effectively reduce the air conditioning electricity consumption. The solar collector unit is installed between the compressor and condenser, which provide part of compression pressure and heating by further superheating the refrigerant. The higher pressure and more substantial temperature difference enhance the condensation process in the condenser, resulting in the high-pressure liquid refrigerant. This configuration realty reduces energy consumption by reducing the load on the electric compressor.

**Keywords:** Compressor Solar Assisted, DC Compressor, Inverter Air Conditioning, Solar Collector, Vapour-Compression Cycle.

## 1. Introduction

The increasing demand for electric power by the consumer has led to higher tax rates for electricity during the hot summer season in Iraq [1] & [2]. In Iraq, the renewable energies that available from solar energy with wind energy have not effectively invested due to excessive heat that reaches 55°C [3] & [4]. This heat leads to the reduction of solar cell capacity to 45% from real capacity processed. There is an actual need to air-conditioning using the compression systems with high power consumption of electricity [5] & [6]. Many solutions and applications use solar energy in air conditioning [5] & [7]. In Iraq, the maximum and minimum value of solar radiation yearly average is 5596 Wh/m2/day and 4266 Wh/m2/day [8]. Solid Desiccant, reduce moisture content in pure ambient air wheel rotates sluggishly, air flows through by separated into two sectors to be dehumidified and other sectors to renew the wheel [9]. Liquid Desiccant, same behave solid desiccants, the water vapor pressure is a role of temperature with moisture content [10] & [11] Most of the applications of air conditioning are derived from renewable energy, which has not achieved as an alternative to traditional compression cycleswith the effect on the coefficient of performance [12] and the cooling of temperatures at increasing temperatures in the summer, and with the exceptional situations of heat such as Iraq.

The aim of this experimentally to improve performance coefficient and reduce the power consumption of the air conditioner to less than of the original value by using a solar heat exchanger with the system and using of an air conditioner split inverter device (12000Btu/h). In this work, a solar thermal radiation has been used to make up the energy provided to the system, that is the same value required (12000 Btu/h), after lowering the energy required from the compressor that connected to regular power supply to minimum speed (3105Btu/h). The solar thermal radiation has been achieved by the heat exchanger placed after compressor DC.

## 2. The apparatus described

Experimentation was completed and installed on the Laboratory Unit at the Faculty of Engineering / University of Qadisiyah, Iraq [13]. Used vapor-compression cycle compressor DC (wall type split unit). External heat exchanger immersed in water path connected between DC compressor and condenser. Control the cooling fluid flow by valve between in and out heat exchanger, and bypass valve to closed recycle to pass flow before heat exchanger see Table 1. The LM35 temperature sensor was used as a semiconductor with very high accuracy (± 0.5) where the linear relationship between voltage and temperature was assumed. Table 2 demonstrated the eight sensors distribution that use micro-shape processing unit (Arduino), calibration of sensors that a connection with the computer and output data as an Excel file for recording readings. Measuring power consumption and the value of voltages and current are shown in 'figure 1', 'figure 2' and'figure 3'.

**Table 1.** Shows the specifications of the device and parts experimental

| Category | Unit |
|---|---|
| Min. Cooling Capacity | 3105Btu/h |
| Max. Cooling Capacity | 13000Btu/h |
| Starting Current | 1.25 A |
| Running Cooling | 5A |
| Refrigerant / Charge | R410A /1.2kg |
| Power Compressor/type Structure | DC Power / Twin Rotary |
| Heat Exchanger | Pipe Coil coper 10 m 1/4" |
| Valve Pipe 1/4" | 5 Parts |
| water path | heat water (65 90) °C |

**Table 2.** Shows the name of the sensor with its location in the device.

| Site | Label temp. sensors |
|---|---|
| $T_1$ | Before the compressor. Discharge pipe |
| $T_2$ | After compressor charging tube |
| $T_3$ | After the condenser. Before expansion |
| $T_4$ | After expansion. Before the evaporator |
| $T_x=T_5$ | After the heat exchanger before the condenser |
| $T_a=T_6$ | Ambient temperatures |
| $T_s=T_7$ | The processed temperatures of the condenser |
| $T_w=T_8$ | Heat the heated water in the water bath |

IOP Conf. Series: Materials Science and Engineering **557** (2019) 012079 doi:10.1088/1757-899X/557/1/012079

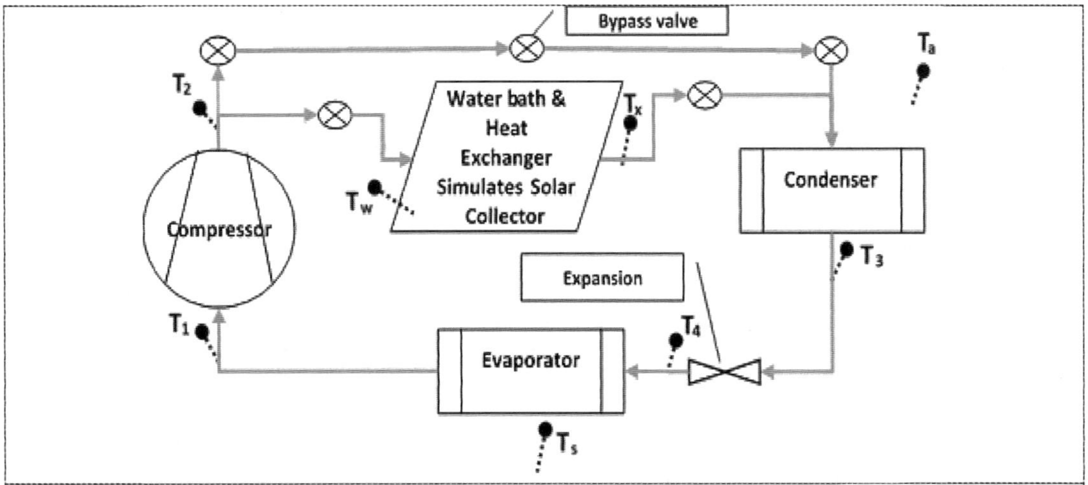


**Figure 1.** Scheme device experimental.

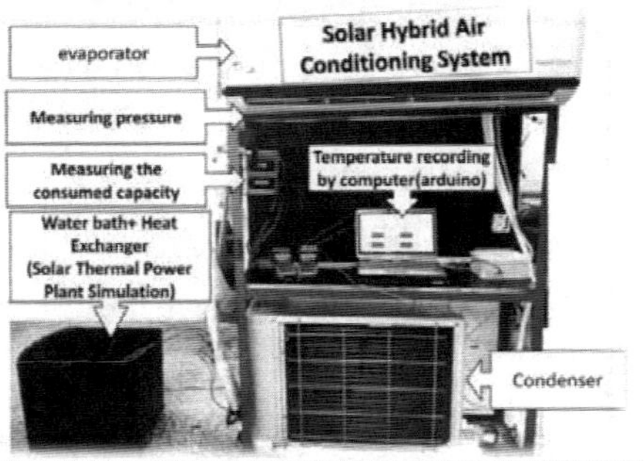


**Figure 2.** Actual structure cycle.

**Figure 3.** Connect the pipe structure with the valve.

## 3. Experimental procedure

The experiment was conducted by recording readings of temperatures from all sensors in standard laboratory. The first case is to completely closure of the cycle by open the bypass valve and close in-out valves in the external heat exchanger. The second case closes the bypass valve and completely closure of the cycle through open in-out valves in the external heat exchanger.

### *3.1. Measurements*

The readings and record were taken in two stages, with and without a water bath. That records the temperature, high and low pressures for cycle cooling, and power consumption. The recording temperature time from unstable primary state to a stable state to show two state 'figure 4' and 'figure 5'.

The text of your paper should be formatted as follows:

IOP Conf. Series: Materials Science and Engineering **557** (2019) 012079 doi:10.1088/1757-899X/557/1/012079

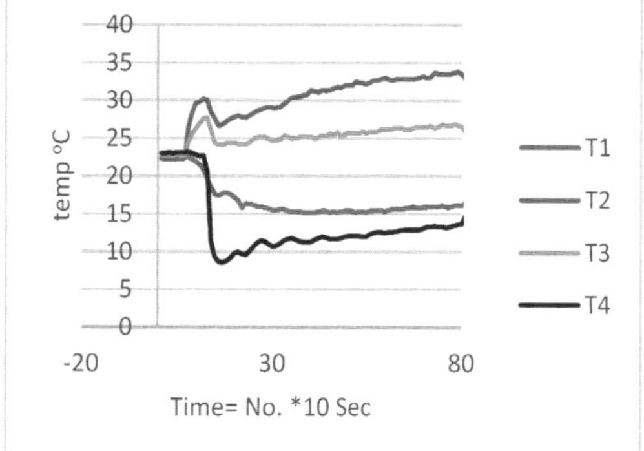


**Figure 4.** It shows the temperature readings with time, where each number represents 10 seconds without the thermal solar collector enters with automatic mode of operation.

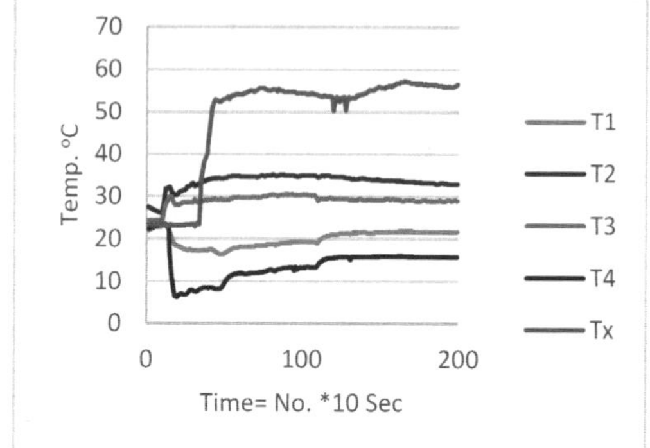


**Figure 5.** It shows the temperature readings with time, where each number represents 10 seconds with the thermal solar collector enters with automatic mode of operation.

## 4. Calculations and results

The experimental procedure showed the effect on the cooling cycle and the performance factor with the consumption in the processed capacity. Used for calculations engineering equation solver (EES) software [14].

### *4.1. Calculations without Compressor Solar Assisted in standard comfort conditions*

The voltage of the compressor with the current used (voltage = 130 volt & current = 1.87 A). The enthalpy values were calculated from the intersection of high and low pressure values used equation (1-5) with the recorded temperatures before and after each part of the compressive system. The diagram was then drawn and the results obtained for the completed work were calculated from the plan, taking into account the efficiency of the compressor and the calculation of the real work extracted from the plan. Calculation of Performance Factor Cooling from Freon Cooling Scheme shows in the 'figure 6'.

Case 1:
$P_1 = 12(\text{bar})$
$T_1 = 16(°C)$
$h_1 = h_{R410a}, T = T_1, P = P_1$
Case 2:
$P_2 = 23(\text{bar})$
$T_2 = 55(°C)$
$h_2 = h_{R410a}, T = T_2, P = P_2$
$$W_{comp} = h_2 - h_1 = 23.2 \text{kj/kg} \quad (1)$$
Case 3:
$P_3 = 24(\text{bar})$
$T_3 = 26(°C)$
$h_4 = h_{R410a}, T = T_3, P = P_3$
Case 4:
$P_4 = 12(\text{bar})$
$T_4 = 15°C$
$h_4 = h_3$
$\eta = 0.85$

$$W_{comp/Real} = \frac{W_{Comp}}{\eta} = 27.29 \text{ kJ/kg} \quad (2)$$
$$Q_{Con} = h_2 - h_3 = 210 \text{ kj/kg} \quad (3)$$
$$Q_{Evap} = h_2 - h_4 = 186.8 \text{ kj/kg} \quad (4)$$
$$COP_R = \frac{Q_{Evap}}{W_{Comp.Real}} = 6.844 \quad (5)$$

IOP Conf. Series: Materials Science and Engineering **557** (2019) 012079 doi:10.1088/1757-899X/557/1/012079

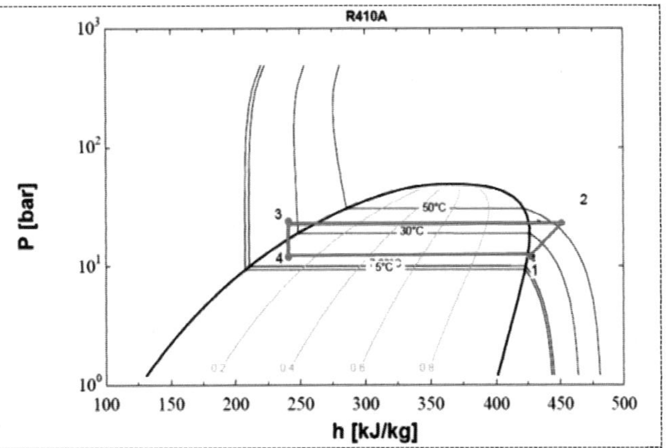


**Figure 6.** P-h the R410a diagram of the vapor compression refrigeration cycle without the presence of solar assisted.

### *4.2. Calculations with Compressor Solar Assisted in standard comfort conditions*

The voltage of the compressor with the current used (voltage = 104 volt & current = 1.178A). The enthalpy values were calculated from the intersection of high and low pressure values used equation (6-13) with the recorded temperatures before and after each part of the compressive system with the solar assisted. The diagram was then drawn and the results obtained for the completed work were drawn from the plan, taking into account the efficiency of the compressor and the calculation of the real work extracted from the plan and the calculation The solar assisted from the solar collector, the networking calculation, the recording voltage and the current consumed by the compressor, and the calculation of the cooling coefficient of the cooling Freon show the 'figure 7'.

Case 1:

$P_1 = 12(\text{bar})$

$T_1 = 16(°\text{C})$

$h_1 = h_{R410a}, T = T_1, P = P_1$

Case x:

$P_x = 23(\text{bar})$

$T_x = 55(°\text{C})$

$h_x = h_{R410a}, T = T_x, P = P_x$

$$W_{Comp/X} = h_X - h_2 = 16.02 \text{ kj/kg} \quad (6)$$

Case 2:

$P_2 = 21(\text{bar})$

$T_2 = 40(°\text{C})$

$h_2 = h_{R410a}, T = T_2, P = P_2$

$$W_{Comp} = h_2 - h_1 = 7.175 \text{ kj/kg} \quad (7)$$

Case 3:

$P_3 = 24(\text{bar})$

$T_3 = 26(°\text{C})$

$h_3 = h_{R410a}, T = T_3, P = P_3$

Case 4:

$P_4 = 12(\text{bar})$

$T_4 = 15°\text{C}$

$h_4 = h_3$

$\eta = 0.85$

$$W_{Comp/Real} = \frac{W_{comp}}{\eta} = 8.442 \text{ kJ/kg} \quad (8)$$

$$Q_{Con} = h_2 - h_3 = 194 \text{ kj/kg} \quad (9)$$

$$Q_{Con/x} = h_x - h_3 = 210 \text{ kj/kg} \quad (10)$$

$$Q_{Evap} = h_1 - h_3 = 186.8 \text{ kj/kg} \quad (11)$$
$$W_{Net} = W_{Comp'Real} + W_{comp'X} = 24.46 \text{ kj/kg} \quad (12)$$

$$\mathbf{COP_R} = \frac{\mathbf{Q_{Evap}}}{\mathbf{W_{Net}}} = \mathbf{7.635} \quad \mathbf{(13)}$$

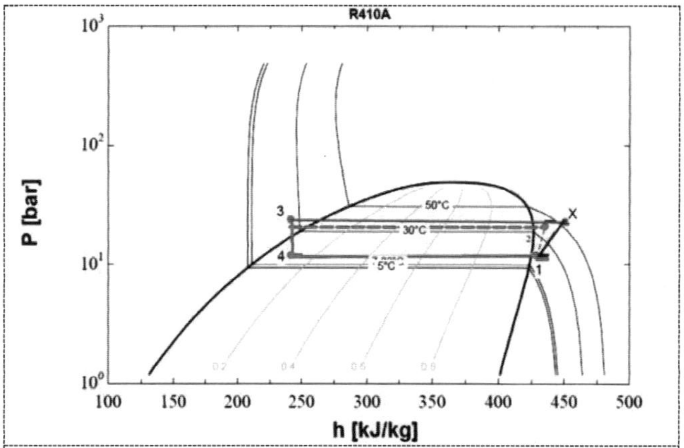


**Figure 7.** P-h the R410a diagram of the vapor compression refrigeration cycle with the presence of solar assisted.

The thermal performance coefficient values were obtained in both cases with and without the solar assisted when the comparison was made. The high coefficient of thermal performance in the case of running the solar assisted and the low pull of the current consumed by the compressor that reduced compressor consumption from the planned cooling cycle for Freon R410a as shown in Table 3.

**Table 3.** The final results with and without the compressor Solar Assisted show the thermal performance values, the processed capacity, and the conventional compressor.

| State | current I (A) | Voltage V (V) | $W_{compReal}$ (kJ/kg) | $Q_{Evap}$ (kJ/kg) | $COP_R$ From chart R410a |
|---|---|---|---|---|---|
| **Without Solar** | 1.87 | 130 | 27.29 | 186.8 | 6.844 |
| **With Solar** | 1.178 | 104 | 8.442 | 186.8 | 7.635 |

## 5. Conclusions

The solar air conditioning system is a great innovation that drives solar energy leaps while enhancing the positive effect on the environment. This project provides a comprehensive energy-saving comparison that can be achieved for various air-conditioner capacities ranging from Btu / h 2388 Btu / h to 24,000 Btu /h. The system is capable of achieving up to 45% energy saving during the day and thus dramatically reduces the peak load of electricity during the day. At the same time, the DC compressor will contribute up to 25% energy saving overnight. The main modifications needed are the compressor (DC). In addition, thermal storage and photoelectric system proposed to provide future cycle performance.

## References

[1] Al-Salihi, A. M., Kadum, M. M., & Mohammed, A. J. 2010 *Estimation of global solar radiation on horizontal surface using routine meteorological measurements for different cities in Iraq*. (Asian Journal of Scientific Research) vol 3 (4) p 240-248.

[2] Al-Rubaye R T A, Al-Rubaye A T A and Al-Khuzaie M M 2018 *Optimal Design of Hybrid Renewable Energy System off grid in Al-Diwaniyah,* (Iraq IOP Conf. Ser. Mater. Sci. Eng 454 012103)

[3] Al-Abidi, A. A., Mat, S. B., Sopian, K., Sulaiman, M. Y., Lim, C. H., and Th, A. 2012 *Review of thermal energy storage for air conditioning systems*. (Renewable and Sustainable energy reviews) No vol 16 (8) p 5802-5819.

[4] Patil, M., Jahagirdar, R., & Deore, E. 2012 *Experimental investigation of heat loss from hemispherical solar concentrator receiver* (Frontiers in Heat and Mass Transfer) 3(3).

[5] Daut, I., Adzrie, M., Irwanto, M., Ibrahim, P., and Fitra, M. 2013 *Solar powered air conditioning system* (Energy Procedia) vol 36 p 444-453.

[6] Boulahia, Z., Wakif, A., Chamkha, A. J., & Sehaqui, R 2017 *Numerical study of natural and mixed convection in a square cavity filled by a Cu–water nanofluid with circular heating and cooling cylinders* (Mechanics & Industry) vol 18 (5) p 502.

[7] Al-chlaihawi K, Al-farhany K and Al-Rubaye A 2018 *Experimental Investigation of Solar Thermal Collector on the Open Parabolic Trough* (Journal of Engineering and Applied Sciences) vol 13 no 16 p 6873-6878.

[8] Rehman, S., Alam, M. M., Meyer, J. P., & Al-Hadhrami, L. M. 2012 *Feasibility study of a wind–pv–diesel hybrid power system for a village* (Renewable energy) vol 38 (1) p 258-268.

[9] Hussein, A. K. 2015 *Applications of nanotechnology in renewable energies—A comprehensive overview and understanding* (Renewable and Sustainable Energy Reviews) vol 42 p 460-476.

[10] Maouassi, A., Baghidja, A., Douad, S. and Zeraibi, N. 2018 *Heat exchanges intensification through a flat plat solar collector by using nanofluids as working fluid* (Frontiers in Heat and Mass Transfer) 10.

[11] Assadi, M. K., Gilani, S. I., & Yen, T. J. 2016 *design a solar hybrid air conditioning compressor system* (In MATEC Web of Conferences) Vol. 38 p. 02001 EDP Sciences.

[12] Al-Rubaye A, Al-farhany K and Al-chlaihawi K 2018 *performance of a portable thermoelectric water cooling* (International Journal of Mechanical Engineering & Technology) vol 9 p277–85.

[13] Al-Rubaye, A., Abas, K., Jaleel, Z., Nadhim, A., Khulaibs, G. 2017 *Design of solar hybrid air conditioning system* (Graduation Project in Department of Mechanical Engineering, University of Al-Qadisiyah).

[14] Yunus, C. A., & Afshin, J. G. 2011 *Heat and mass transfer: fundamentals and applications* (Tata McGraw-Hill, New Delhi, India).

## Acknowledgments

We are thankful to all supported by Faculty of Engineering at University of Qadisiyah. We have to express out appreciation to the Dr. Khaled Al-Farhany for sharing their pearls of wisdom with us during the research. We are also immensely grateful to Zain Alabdien Jaleel, Karrar Abas, Ahmed Nadhim, and Ghufran Khulaibs for their help. We are also grateful to Dr Rana Alrubaye who moderated this paper and in that line improved the manuscript significantly.

IOP Conf. Series: Materials Science and Engineering **557** (2019) 012080 doi:10.1088/1757-899X/557/1/012080

# Remediation of Heavy Metals Using Selected Agricultural Waste: Sunflower Husk and Durian Leaves

**Manal M Abood, F Istiaque and N N Azhari**

Civil Engineering Department, Faculty of Engineering, Infrastructure University Kuala Lumpur (IUKL), 4300, Kajang, Selangor, Malaysia

**Abstract.** This study revealed the behaviour of sunflower husk (SFH) and durian leaves (DL) as an adsorbent that could treat the aqueous solution to be free of heavy metal ions. The batch method was utilized. The influence of pH, contact time and initial metal concentration were described. The impact of pH on the uptake levels of the metal particles by both selected adsorbents were delineated between pH 4 to pH11. It was observed that the best value of pH for copper and zinc removal was 8 in case of SFH and it was 9 for DL. The process tends to be rapid the initial 10 minutes for SFH and 5 minutes for DL and an equilibrium time of two hours was needed for the copper and zinc particles adsorption onto SFH and an equilibrium time of one hour was needed for the copper and zinc adsorption onto DL. The Freundlich and Langmuir isotherm models are used to decide the adsorption parameters. The Freundlich isotherm model tends to fit the equilibrium data better than Langmuir isotherm model. The outcomes demonstrated that SFH and DL hold great potential to be utilized as low cost dry biomass adsorbent in the order SFH < DL.

## 1. Introduction

Wastewater Wastewater refers to the release from assembling plants, rural exercises, household, business and commercial. Or it might be a surface water and storm water. Most recent couple of decades, industrial wastewater has moved toward becoming one of the primary sources of water contamination. The purification of surface water has been influenced by the appearance of heavy metal as the significant contaminations for the aquatic conditions through the release of industrial wastes, storm water runoff, and human exercises. Although numerous heavy metals are essential for people, plants and animals in follow sums, however, the utilization of bigger sums prompts intense and perpetual harmfulness which connected to learning incapacities, cancers, and sometimes death. [1].

Over the most recent couple of years numerous inquiries about have been led to opined the removal of heavy metals by utilizing agricultural materials such as Sawdust [2], Pomegranate [3], Micro particles of dry plants [4], waste tea leaves [5], saraca indica leaf [6], tobacco stems [7], neem leaf powder [8]. [9] Studied the removal of iron utilizing the shell of almond as adsorbent, and [10] utilized the Tilapia Mossambica as an Adsorbed, while [11] describe the utilization of rice husk as adsorbent for the removal of iron particle from the solution.

The existing technology, in generally, dissolvable extraction, precipitation, liquid film and ion exchange are presently being utilized to remove heavy metals from wastewaters. However, these regular materials have a proficiency to be applied as adsorbents with minimal cost as they serve as unused resources, are for the most part naturally well-disposed and are likewise within reach [12]. Normal adsorbents or all the more delicately known as bio sorbents that were previously used for research of utilization of heavy metal such as pomegranate peel, tea leaves, sawdust, , maize bran, wood ash.. In this research, Sunflower Husk (SFH) and Durian Leaf (DL) are being used as a low cost adsorbent.

In the South East Asia areas, Durian (Durio Zibethinus) is an outlandish organic product which is notable. Numerous tropical nations devour it in an enormous sum, along these lines durian estates covers a colossal regions in this locale. The estimation of hectare in repeating crops was 75,713.1 hectares in 2013, while the yearly creation added up to 373,087 million tons. It was observed to be the biggest organic product cultivate inside Malaysia as indicated by the Agriculture Department Fruit Crop Statistics.

As an organic product creation, durian estates can be viewed as one of the biggest waste producing natural product area. The durian leaves has been chosen to be studied as an adsorbent for the utilization of heavy metals from industrial waste released since it is of no utilization for any commercial purposes.

The yearly making of sunflower was 9,273,000 toms as shown by the oil world section, it is seen to be the greatest bloom home in Malaysia. It is being devoured enormously all over tropical region, and thusly, huge zones are being put something aside for sunflower estate.

This study is a part of full research has been conducted to predict the utilization of SFH and DL as adsorbent with a minimal cost for the remediation of zinc and copper particles from aqueous solution and to determine the impact of pH, initial concentration of Zn (II) and Cu (II) particles and contact time, on the uptake of Zn (II) and Cu (II) particles on the surfaces of SFH and DL.

## 2. Methodology

### *2.1 Preparation of adsorbent materials*

SFH was collected from the local market. It was washed softly using the distilled water to remove dust, then it was ground using laboratory mill, sieved to 0.5- 0.8 mm before use. From local house yard in Kajang, Selangor Malaysia, DL were collected. To remove dust and different pollutions, each leaf was washed thoroughly by utilizing tap water. The DL were dried under daylight for 7 days until the point when the leaves transformed into darker shading. . The leaves were additionally dried in a broiler at 105°C for 1 day, at that point grounded into powder shape by utilizing blender. To ensure no contact of atmospheric moisture, the DL was put away in impermeable compartment and seal the ground powder.

### *2.2 Reagents and chemicals*

Solutions of various concentrations were gotten by diluting the stock solution with distilled water. Stock solutions of Zinc sulfate, Copper sulfate 1000 mg/l were utilized as adsorbate, zinc and copper concentrations were controlled by spectrophotometer. All experiments were completed in 500 ml glass bottles at the research centre temperature of 25 ± 2°C, and every one of the synthetic concoctions utilized were of diagnostic review reagent.

### *2.3 Experimental work*

Exploratory work utilizing batch adsorption strategy were performed by mixing a series of jugs containing different measures of SFH and DL and heavy metal particles independently at ideal pH. 1.5 g of the two adsorbents with 300 ml of copper in form of copper sulphate mixed in a 500 ml beaker, similarly, zinc in form of zinc sulphate had been mixed in a 500 ml beaker. The pH was adjusted to a desire value using 0.1M of $H_2SO_4$ and 0.1M of NaOH and the pH was balanced, a jar test machine was used at 150 rpm mixing rate and keep running for 2 hours of contact time until equilibrium is come to. At that point, toward the finish of mixing the adsorbent particles were isolated from the suspensions by filtration through 0.40 μm filter paper. The filtrate is then sampled to measure the concentration of copper and zinc ions using a spectrophotometer. All samples are done under standard condition and the normal outcomes are figured.

## 3. Results and Discussion

### *3.1 Viability of pH*

By modifying the pH of the solution within a range of 3 – 11, the result of pH of the selected heavy metal solutions upon the adsorption of SFH and DL effectiveness was displayed. The outcomes got as shown in Fig. 1(a) and (b) demonstrated that zinc and copper ions were adequately adsorbed in pH range (4-9), and the adsorption rate expanded toward the start from pH 5 to 8. At pH 4 which has the highest acidity value, the concentration of zinc and copper particles removed was at a minimum. This is due to fact that the $H^+$ ions will be greater at low pH the concentration .Hydrogen particles have a tendency to contend with metal particles for adsorption site of the adsorbent. This credits at higher grouping of H+, the adsorbent turns out to be all the more emphatically charged. This obstructs the electrostatic fascination between the metal and the adsorbent. Most extreme adsorption happened for zinc and copper particles by utilizing DL take place at pH 7 while the greatest adsorption of zinc and copper utilizing SFH take place at pH6. Presently, the gathering of hydrogen particles would be in any event. This is in concurrence with the outcomes acquired by [5] for adsorption of heavy metals onto tea leaves.

### *3.2 Initial concentration impact*

The impact of initial concentration on Copper and Zinc remediation was investigated and completed at 25 ±2 ° C utilizing various initial metal ion concentrations (10, 20, 40, 60, 80 and 100 mg/l) at rpm 150 and optimum ph.

The outcomes are shown in Figure 2 (a) and (b), which demonstrated that the percentage uptake decreases with the increments in primary stage of metal ion concentration. This is on account of there were no more adsorption locales on the adsorption surface of the adsorbent material.

The outcomes showed that the copper uptake for the initial 20 minutes was 85% by utilizing DL and 70 % by utilizing SFH, while the zinc uptake following 20 minutes was 90% utilizing DL and 77% utilizing SFH.

It was discovered that the level of remediation of metal ions decreases with the increments of primary concentration. Rapid metal uptake is observed to be when the concentration of the metal solution is 10 mg/L.

### *3.3 Contact time impact*

The impact of contact time on the remediation efficiency of SFH and DL for copper and zinc particles was considered: the outcomes are shown in Figure 3(a) and (b). The uptake of metal particles in the initial 20 minutes, utilizing SFH was 63% for zinc and 48% for copper at equilibrium 81% of Cu particles and 70% of Zn particles are remediate. While by utilizing DL, the percentage uptake reach to 70% of Cu and 78% of Zn following 20 minutes and reach to 83% for Cu and 93% for Zn at equilibrium following 2 hours.

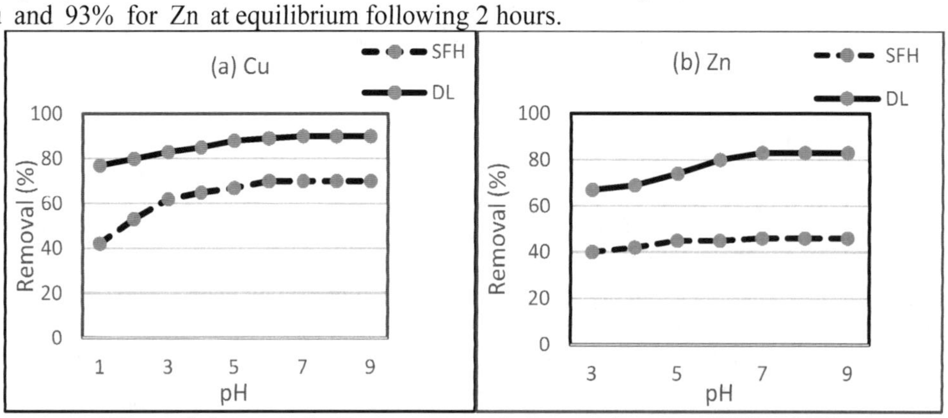


**Figure 1**. Impact of pH on copper and zinc uptake using SFH and DL

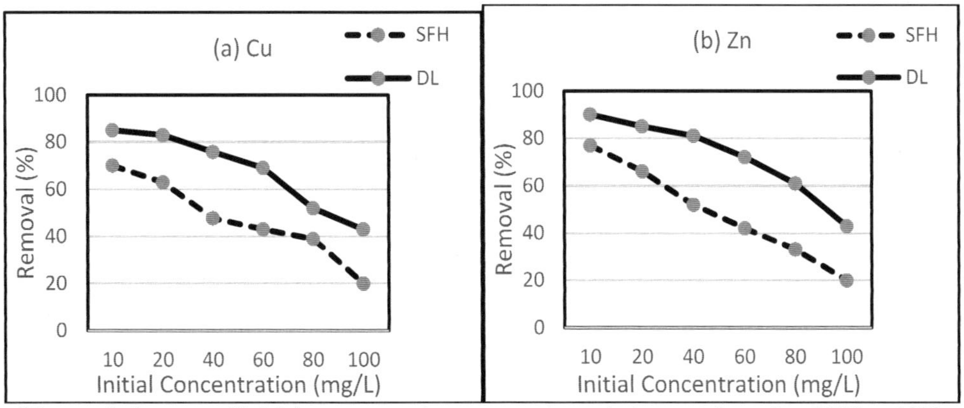


**Figure 2.** Impact of initial concentration on copper and zinc uptake using SFH and DL

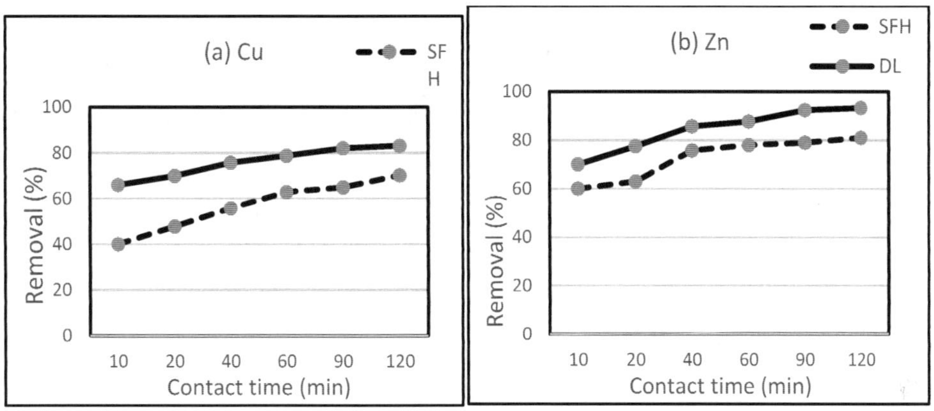


**Figure 3.** Impact of contact time on copper and zinc uptake using SFH and DL

*3.4 Freundlich isotherm model*

The Freundlich Model, an adsorption isotherm, is an observational connection between the concentration of the solute in the fluid and the concentration of a solute on the surface of an adsorbent. This relation is to gives an expression incorporating the surface heterogeneity and the exponential distribution of active sites and their energies. The Freundlich isotherm is defined as:

$$Qe = KC_e^{1/n} \tag{1}$$

Ce is the equilibrium concentration in mg/l, qe = amount of adsorbate adsorbed per unit weight of adsorbent (mg/g). ''k'' is a parameter identified to the temperature and ''n'' is a constant for the adsorption framework under investigation, The plots of log Qe versus log Ce are shown in Figure 4(a) and (b); the adsorption of Zinc and copper particles onto the SFH and DL gave a straight line with $R^2$ equivalent to 0.93 for copper and equivalent to 0.98 for zinc in DL and gave $R^2$ equivalent to 0.96 for copper and equivalent to 0.85 for zinc in SFH ; estimation of ''n'' somewhere in the range of 0.9 and 1.2 demonstrate a good adsorption. The constants and $R^2$for Freundlich isotherm model are listed in table 1.

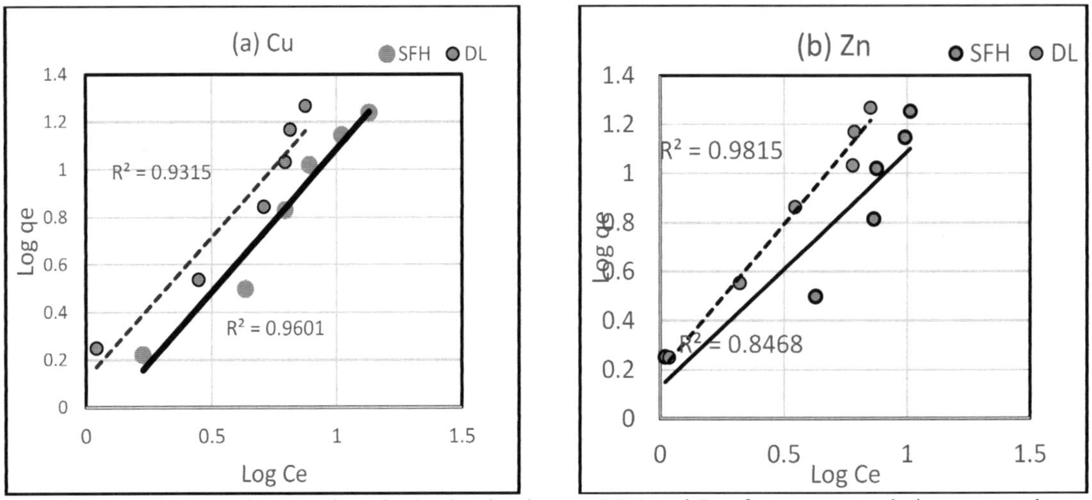


**Figure 4.** Plot of Freundlich for both adsorbents, SFH and DL for copper and zinc removal

**Table 1.** Constant for uptake of copper and zinc onto SFH and DL (Freundlich Model)

| Heavy Metal | Adsorbent | Freundlich constant | | $R^2$ |
|---|---|---|---|---|
| | | k | n | |
| Cu | SFH | 0.12 | 1.2 | 0.96 |
| | DL | 0.12 | 1.19 | 0.93 |
| Zn | SFH | 0.13 | 0.96 | 0.85 |
| | DL | 0.19 | 1.2 | 0.98 |

*3.5 Langmuir isotherm model*

The Langmuir equation depends on the assumptions that greatest adsorption relates to an immersed mono-layer of adsorbate particles on the adsorbent surface, that the vitality of adsorption is steady, and that there is no transmigration of adsorbate in the plane of the surface

The Langmuir isotherm is defined as:

$$Qe = (K_L Q_m Ce)(1 + K_L Ce) \quad (2)$$

Where $Q_m$ and $K_L$ are Langmuir constants identified with the sorption limit, and sorption vitality, respectively, Ce is the equilibrium concentration in mg/l, and qe is the amount of adsorbate adsorbed per unit weight of adsorbent (mg/g). The plots of Ce/qe versus Ce are shown in Figure 5(a) and (b); the adsorption of copper and Zinc particles on SFH and DL gave a straight line with correlation coefficient, $R^2$ equivalent to 0.97 for copper and 0.88 zinc with DL, and $R^2$ equivalent to 0.94 for copper and 0.87 for zinc with SFH It can be concluded that exploratory information were better fitted to the Freundlich equation than to the Langmuir equation, and hence it is more reasonable for the examination of kinetics particularly for DL. The constants and correlation coefficients for Langmuir isotherm model are recorded in table 2.

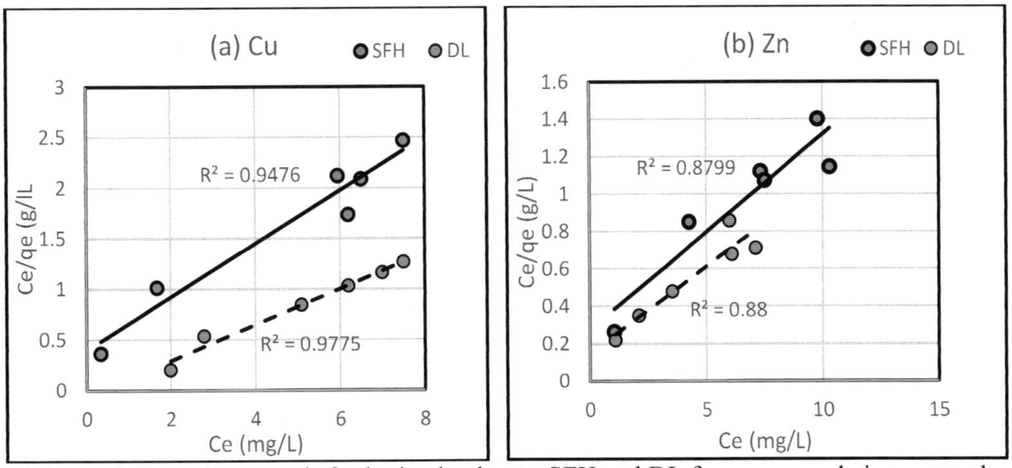


**Figure 5.** Plot of Langmuir for both adsorbents, SFH and DL for copper and zinc removal

**Table 2.** Constant for sorption of copper and zinc onto SFH and DL (Langmuir Model)

| Heavy Metal | Adsorbent | Langmuir constant | | $R^2$ |
|---|---|---|---|---|
| | | b | $q_m$ | |
| Cu | SFH | 0.20 | 0.39 | 0.94 |
| | DL | 0.18 | 0.06 | 0.97 |
| Zn | SFH | 0.12 | 0.27 | 0.87 |
| | DL | 0.09 | 0.15 | 0.88 |

## 4. Conclusion

The exploratory technique was utilized; parameters for example, pH, contact time and metal concentration were studied at an surrounding temperature 25± 2° C, the optimum pH comparing to the most extreme adsorption of copper and zinc remediation was 7–9. Copper and Zinc particles were adsorbed onto the adsorbents very quickly within the initial 20 min, while balance was accomplished within 2h for copper and zinc particles. It very well may be said that the minimal-cost adsorbent like sunflower husk (SFH) and Durian leaves (DL) is successful for the remediation of copper and zinc particles from aqueous solutions. The Freundlich isotherm better fitted the experimental data since the correlation coefficient for the Freundlich isotherm was higher than that of the Langmuir isotherm for the two metals.

## 5. Acknowledgments

Financial support from Department of Civil Engineering, Faculty of Engineering and Technology Infrastructure, Infrastructure University Kuala Lumpur (IUKL) and collaboration with research management centre (RMC) is highly appreciated to complete this research.

## 6. References

[1] Chaturvedi S and Dave P N 2012 *Desalination.* **303** 1
[2] Yasemin B and Zeki T.2007. *Journal of Environmental Sciences.* **19** 160
[3] El- Ashtoukhy E S, Amin N and Abdelwahab O *Desalination* **223** 162
[4] Benhima H Chiban M Sinan F Seta P and Persin M 2008. *Colloids and Surfaces B: Biointerfaces* **61** 10
[5] Ahluwalia S and Goya D 2005. *Engineering in Life Sciences* **5** 2
[6] Goyal P Sharma P, Srivastava S, and Srivastava M 2008. *Environmental Science Technology* **5** 1, 27.
[7] Li W, Zhang L, Peng J, Li N, Zhang S and Guo S 2008. *Industrial Crops and Products* **28** 295

[8] Bhattacharyya K G., and Sharma A 2004. *Journal of Hazardous Materials* B**113**, 97
[9] Anusha D G and Raja Murugadoss, DJ 2014. *International Journal of Scientific Research* **3** 10.
[10] Zayadi N and Othman N 2013. *International Journal of Integrated Engineering*, **5** 23.
[11] Kadir A , Abdullah A and Wai L 2013. *University Hussein Onn.*
[12] AbdelGhani N, Hefny M. and El-Chaghaby G. 2007 *Int. J. Environ. Sci. Tech*, **4** 1 67.

IOP Conf. Series: Materials Science and Engineering **557** (2019) 012081 doi:10.1088/1757-899X/557/1/012081

# One-step chemical bath deposition of $Cu_2O$ flowers grown on ITO seed layer coated glass substrate

**Forat H Alsultany**

Al-Mustaqbal University College, Department of Medical Physics ,Iraq

**Abstract.** Here, the growth of $Cu_2O$ flowers on a glass substrate were successfully fabricated using one-step chemical bath deposition method at low growth temperature. These flowers were grown on ITO seed layers that were prepared on glass substrate by magnetron-frequency RF sputtering. The structure and surface morphology of the products of $Cu_2O$ flowers were characterized in detail using X-ray diffraction and microscopy of field emission scanning electron. Optical properties were examined through the ultraviolet/visible spectro-scopy. The results showed that the $Cu_2O$ flowers were uniformly formed on ITO seeds-coated glass substrate, grow in the cubic structure with film preferential (111) plane orientation, high transmittance, and good crystallinity. The value of band gap is approximately 2.53 eV. This study helps to be the basis for more research on the growth of Cu2O at low-temperature and low cost-effective substrate.

## 1. Introduction

Cuprous oxide (Cu2O) is an attractive semi-conductor material, a cubic crystal structure, and p-type bandgap energy of 2.0-2.6 eV [1]. In addition, its superior physical, chemical, and photo-electronic properties, non-toxic material, natural abundance, and simple setup procedures. Cu2O nanostructures has produced importance in different uses, such as photovoltaics [2], photo-chemical hydrolysis into H2 and O2 [3], catalysis [4], change of solar energy to energy of chemical or electrical [5], batteries [6], photo degradation of dye molecules [7], hetero-junction in solar cells [8], biosensing [9], and electrode materials in batteries of ion-lithium [10]. Requires success in different applications the control and definition of the size, dimension, and geometry of the nanostructured materials [11]. So, tailoring the growth $Cu_2O$ nanostructure is important for specific production orientations, sizes, and surface densities. Preparation of $Cu_2O$ has been carried out by various chemical and physical methods such as electro deposition, sintering, vacuum deposition, spray pyrolysis and deposition of chemical bath (CBD). Among these methods, CBD is inexpensive, simple, cost-effectiveness, and suitable chemical method (wet method) which has been used for preparation of a multiplicity of nanostructured materials [12].

The layer of ITO seeds is very important in promoting the growth of $Cu_2O$ nanostructures. Consequently, the kind and relationship of seed layer with nanostructure is very important because the properties of nanostructures depend mainly on seeds properties, like grain size, morphology, crystalline density, and roughness. So, making a high-quality seed layer is one of the important roles toward developing the nanostructures performance, which can help expand the improvement of novel devices and possible applications [13]. The aim of this research is to use a low-cost substrate in a deposition of the $Cu_2O$ nanostructures on a glass substrate by using chemical bath deposition method at low growth temperature.

## 2. Experiment details

Prior to the growth process of $Cu_2O$ flowers, ITO seeds were prepared on glass substrates by magnetron-frequency RF sputtering with power of 100W and chamber pressure at $10^{-5}$ mbar. The thickness of ITO seed layer was always saved at 75 ± 5 nm. The $Cu_2O$ flowers were grown using CBD employing a double distilled water (100 ml) was mixed with copper (II) chloride dehydrate (CuCl2. 2H2O) (1.705 g) to obtain copper chloride solution (0.1 M). The solution was moved for 1 h at room temperature to get a magnetic stirrer for a well-dissolved and transparent solution. With 15 min boiling, the solution was prepared to boil at temperatures of 70 °C. In order to taken out loosely and larger bonded particles, the ITO/glass substrate was taken away from the bath and rinsed in double distilled water using ultrasonic bath for about 10 min.

The thickness of the $Cu_2O$ film was obtained using an Filmetrics F20 (optical reflectometer). The optical properties, surface morphology, and crystal structure of the flowers nanostructures were investigated using UV–Vis spectrophotometry, X-ray diffraction (XRD), and field emission scanning electron microscopy (FESEM).

## 3. Results and discussion

The morphologies of $Cu_2O$ flowers on ITO seeds/glass substrate by CBD at low growth temperature of 70 °C were measured by FESEM with low and high magnification as shown in Figure 1 (a, b). Clearly, the substrate surface were densely covered with the spherical clusters of the $Cu_2O$ that contain the high-density with size of smaller particle which provide a closer view of typical flower morphology. The dimension of flower particle sizes ranges 85 nm width and length.

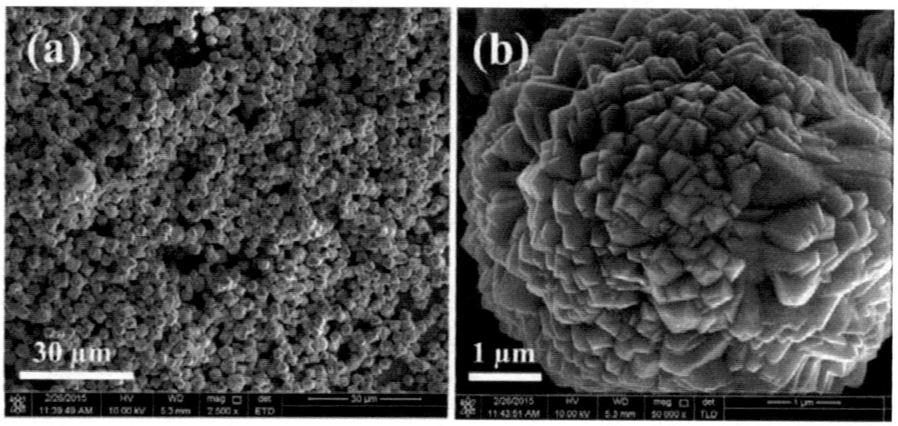


**Figure 1.**Low and high-magnification FESEM image of $Cu_2O$ flowers on ITO seeds coated glass substrate.

The ratio of O to Cu is appear in Figure 2. EDX revealed that $Cu_2O$ flowers has 65.8 and 34.2 At% of Cu and O, respectively; and these values represents the best stoichiometry for the Cu2O.

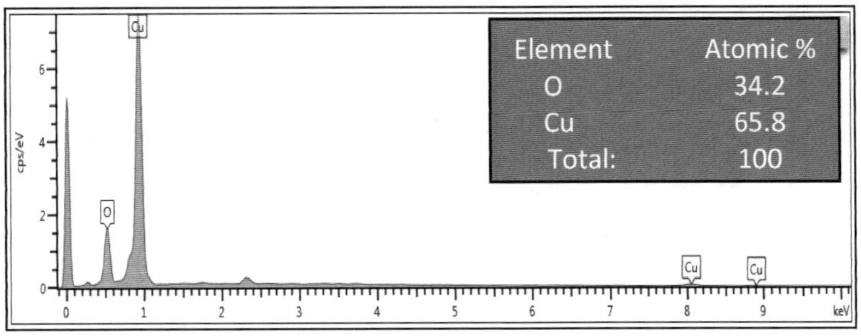


**Figure 2.** EDX spectra of the $Cu_2O$ flowers on ITO seed layer coated glass substrate.

The quality of crystal and orientation of the $Cu_2O$ flowers grown on ITO/glass substrate via CBD were examined by XRD. Figure 3 shows the located of diffraction peaks were at $2\theta = 29.4°$, $36.3°$, $42.2°$, $61.3°$ and $73.4°$, that corresponded to the (110), (111), (200), (220) and (311) diffraction of the $Cu_2O$. The plots of XRD were matched with the $Cu_2O$ reference (JCPDS- 05/0667), having a cubic crystal structure.

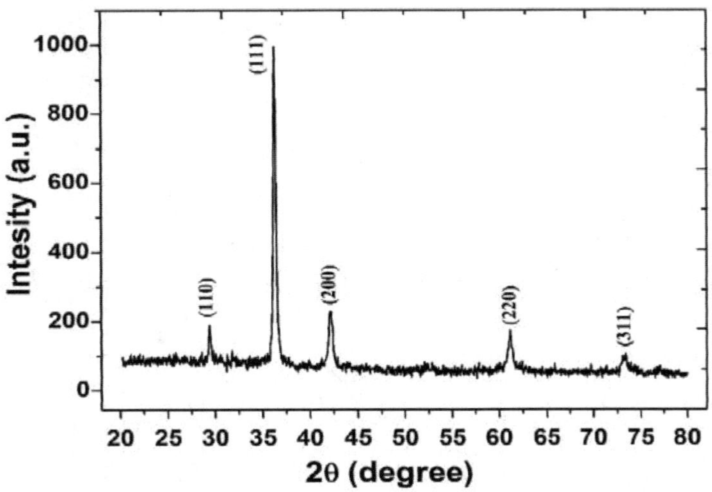


**Figure 3.** The XRD pattern of $Cu_2O$ flowers on ITO seeds coated glass substrate.

Fig. 4 shows the UV–vis transmission of the $Cu_2O$ flowers on ITO seed layer coated glass substrate in a wavelength range of 220–800 nm and the absorption. The Cu2O flowers are transparent in the visible region (Fig. 4). Up to nearly 500 nm, the flowers nanostructures are absorbent. The transmittance of the $Cu_2O$ film increases strongly between 500 and 1800 nm. At long wavelengths, the transmittance values is estimated to be 67%. The transmittance spectrum displays an interference pattern with a sharp fall of transmittance at the band edge . In the visible light range, the high optical transmittance of Cu2O flowers indicates the low-surface roughness and high-homogeneity [14]. The inset displays the typical variation of $(\alpha h\upsilon)^2$ as a function of photon energy ($h\upsilon$) for the $Cu_2O$ flowers.

In order to calculate the energy gap (Eg) of the Cu2O flowers, the absorption coefficient is estimated using the relationship [15]:

$$\alpha = \frac{1}{D} \ln \left(\frac{1}{T}\right) \qquad (1)$$

where D is thickness, T is transmittance, and α is absorption coefficient of the Cu2O flowers.
Eg is given for direct transition using [16]:

$$\alpha\, hv = A\, (hv - E_g)^{1/2} \quad (2)$$

where $hv$ is the energy of incident photon, and A being a constant,.
A Cu2O flowers absorbs the light below a threshold wavelength (λc) i.e $[hv \geq Eg]$ or:

$$hv = \frac{1240}{\lambda} \quad (3)$$

The Eg of the film is resultant using planning the Tauc formula and taking the extrapolation of the linear portion from the plot of $(\alpha h\upsilon)^2$ versus the photon energy ($h\upsilon$). The $Eg$ of the $Cu_2O$ flowers is 2.53 eV, as shown inset of Figure 4. The reduction Eg value comparison with bulk value is caused by different contributions of the excitonic emissions and their phonon replicas

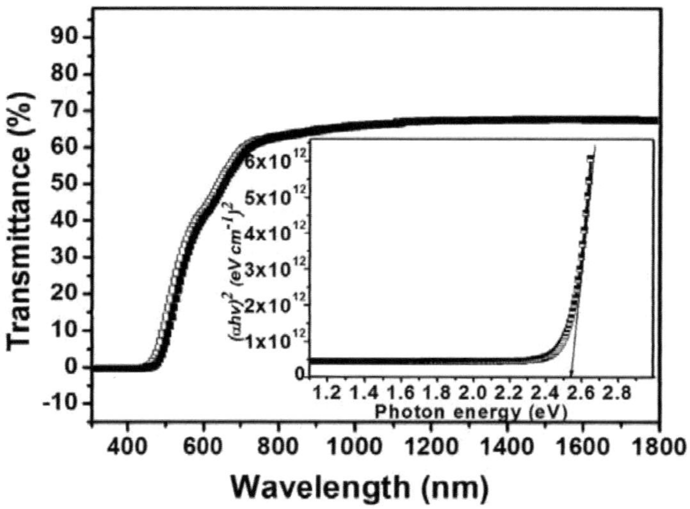


**Figure 4.**UV–vis transmission of the $Cu_2O$ flowers. The typical variation of $(\alpha h\upsilon)^2$ as a photon energy function ($h\upsilon$) of $Cu_2O$ flowers is shown in the inset.

## 4. Conclusion

The growth of $Cu_2O$ flowers was successfully conducted on ITO seeds coated glass substrate using one-step chemical bath deposition method at low growth temperature of 70 °C with 15 min boiling. ITO seeds was prepared on glass substrates by magnetron-frequency RF sputtering with thickness of approximately 75 ± 5 nm. This approach will helps to be the basis for more research on the growth of Cu2O at low-temperature and low cost-effective substrate which can be advantageous in numerous solid-state and opto-electronic device.

## 5. Acknowledgments

"This work was supported by LRGS grant (203/CINOR/6720013) and Universiti Sains Malaysia".

## 6. References

[1] Hwang, Y., Ahn, H., Kang, M. and Um, Y. 2015 Current Applied Physics 15 pp 89-94.
[2] Sears, W.M. and Fortin, E. 1984 Solar Energy Materials 10 pp 93-103.
[3] de Jongh, P.E., Vanmaekelbergh, D. and Kelly, J.J. 1999 Chemical Communications 12 pp 1069-1070.
[4] El Kasmi, A., Tian, Z.Y., Vieker, H., Beyer, A. and Chafik, T. 2016 Applied Catalysis B: Environmental 186 pp 10-18.
[5] Xu, H., Dong, J. and Chen, C. 2014 Materials Chemistry and Physics 143 pp 713-719.
[6] Ni, S., Lv, X., Li, T., Yang, X. and Zhang, L. 2013 Electrochimica Acta 109 pp 419-425.

[7] Xu, H., Wang, W. and Zhu, W. 2006 The Journal of Physical Chemistry B 110 pp 13829-13834.
[8] Musa, A.O., Akomolafe, T. and Carter, M.J. 1998 Solar Energy Materials and Solar Cells 51 pp 305-316.
[9] Huang, Q., Lin, X., Lin, C., Zhang, Y., Hu, S. and Wei, C. 2015 Rsc Advances, 5 pp 54102-54108.
[10] Park, J.C., Kim, J., Kwon, H. and Song, H. 2009 Advanced Materials 21 pp 803-807.
[11] Alsultany, F.H., Hassan, Z. and Ahmed, N.M. 2016 Superlattices and Microstructures 92 pp 68-79.
[12] Zou, X., Fan, H., Tian, Y., Zhang, M. and Yan, X. 2015 RSC Advances 5 pp 23401-23409.
[13] Alsultany, F.H., Hassan, Z. and Ahmed, N.M. 2016 Optical Materials 60 pp 30-37.
[14] Alsultany, F.H., Hassan, Z., Ahmed, N.M., Elafadill, N.G. and Abd, H.R. 2018 Optics & Laser Technology 98 pp 344-353.
[15] Benramache, S. and Benhaoua, B. 2012 Superlattices and Microstructures 52 pp 1062-1070.
[16] Selman, A.M. and Hassan, Z. 2013 Superlattices and Microstructures 64 pp 27-36.

# Detection of XRCC1 Expression and (8-OHdG) Levels as a marker of Oxidative DNA Damage in Individuals Exposed to Low Dose of Gamma Rays

**Rafed Abbas Kadhum*[1], Wathiq Abbas Aldrghi[2]**

[1,2] Institute of Genetic Engineering And Biotechnology for Post Graduate Studies, University of Baghdad, Baghdad, Iraq.

* rafedstudent@gmail.com

**Abstract.** To study ionizing radiation (IR) effects on biological systems there several techniques are used to identify molecular features of IR exposure, urinary detection of 8-ohdG by ELISA as a marker of DNA oxidative stress is one of the techniques was used in this research , when IR absorbs in a cell so that causing DNA damage and releases 8-ohdG through correction damage .The detection of XRCC1 expression by QRT- PCR is else technique in this study. The genotoxic of IR can be observed even at a low doses 5cGy, According to RPC (Radiation Protection Center) limited place in Baghdad as a source of low-dose IR( Europium 152) before it is removed, some of workers and slums are exposed to low dose of γ-rays ( 5cGy to 10 cGy ) .The study groups included the (G1)fifty workers in metal melting are working closely with ( IR source) , (G2)twenty five individuals from slums near the IR source, and(G3) twenty five volunteers individuals of male in different age, didn't exposure to diagnostic x-ray examination or other radiotherapy . Peripheral blood are collected and urine from all individuals . The results indicate that the level of 8-OHDG factor in workers higher than slums individuals and negatively related with XRCC1 expression , also the expression of XRCC1 decreased in individuals exposed for more than 5cGy ,and increasing in slums exposed to 5cGy . The low dose of γ-rays causing genotoxic effects due to a combination of DNA-damaging effects and reduced capacity of DNA repair.

## 1. Introduction:

Radiation can be classified two main categories : ionizing (IR) and others non-ionizing(NIR). The effect of IR with bimolecular more than NIR because of the ability of ionizing to stimulate atom ionization. Radioisotopes ( radio nuclides) are unstable that be main source of IR, these radioisotopes emit particles have high energy to displace atomic electrons , also to easy happen the chain reaction .There are main types of IR include particles as (alpha α , beta β ),rays(X-rays, and gamma - γ-rays) , the particles(α ,β) stopped at barriers of an aluminum or sheet of paper , whereas the most penetrating are the rays (X ,γ ) , environmental exposure to particles (α ,β) less than the rays (X ,γ ) ,therefore; consider these rays the main reason of biological damage with greater degree than the particles( α ,β) , these types used for therapeutic purposes successfully and also lead to cellular damage [1] In this study the Europium 152 was the source of low dose gamma rays ( 5cGy equivalent to 50 mSv, to 10 cGy equivalent to 100 mSv) . The non-IR formed largely type of Environmental radiation, included all region the spectrum such (UV) rays. The effect of non-IR like UV rays on biological molecular is now well-studied [2-3], ionizing radiation most widely employed in research centers and therapies are from sources of man-made like as Co-60 , Cs-137 and Europium 152 . this study focused on the interactions of ionizing radiation with biological molecules [4]. radioactive waste as result of the using radioisotopes in industry, science and medicine . This waste needs to be managed in a way to keep the environment safe and people over long periods of time, as well known that the Ionizing radiation may be occur either naturally or artificially in our environment , these types have same the effects on healthy human. In 1902 were recorded first case of cancers induced by Radiation [5] other cases registered in (1917-1926) included painters deal with radium element [6] as response to some environmental factors or stress such as chemical reagent or physical factors like gamma rays ,the cell activated mechanisms of DNA repair , single strand break repair (SSBR) ,base expression repair (BER) and other mechanisms as apoptosis and cell cycle control all the body parts effected by ionizing radiation but the response depending on the dose and period of exposure. The tissue , such as blood effected by IR and causes DNA damage , cells of mammalian have several types of DNA repair systems that it has main role to correct different types of DNA damage , The gamma rays lead to damage of DNA either directly or indirectly and causes usually DNA base damage that is repaired by base excision repair (BER) [7]. X-ray repair cross complementing group 1 (XRCC1) and others such as (XRCC2 ,XRCC3) are necessary proteins in the BER pathway. The location of XRCC1 gene on chromosome 19q13.2–13.3. XRCC1 protein coordinate with other proteins that play a essential role as scaffold for the DNA repair complex [8] .The primary role for this gene XRCC1 assemble proteins of SSBR in position of DNA damage in order to support the efficiently repair .This operation have controlled by PARP1-XRCC1 scaffold , the mechanisms of repair include the temporary assembly complexes of multi-protein the assembly of which rely upon the kind of the initiating lesion, for more information about the chemistry mediate in mechanisms of repair and post-translational modifications of the constitutive pathway members see Svilar et al.[9-10].

As accordingly, the oxidative stresses play the main effect on certain pathologies, especially in carcinogenesis. The reactive oxygen species (ROS) can be induced deferent of damaged and contamination to DNA, included oxidative bases then be repaired and release in urine the 8-ohdG, can be measured the oxidized in urine or blood with various methods. 8-ohdG urinary measurement usually used in many studies period subjects with occupational exposure. Indeed the factors of chemical and physical can be increased the oxidative stress [11]. The repaired single strand breaks or base excision repair by XRCC1 proteins that be associated with produce 8-ohdG to determined the level of DNA damage used firstly 8-ohdG as marker . The objective of this research is to identify the association between DNA repair gene expression(XRCC1) and low dose of gamma ray then analysis the association of the XRCC1 and level of 8-ohdG in urine with DNA damage from exposed individuals and control group also study the correlation among the age ,smokers and urinary 8-ohdG levels .

## 2.Materials and Methods

### *2.1.Blood and Urine sample collection:*

Blood samples were obtained from 100 male individuals before removed the source of ionizing radiation, and distributed into three groups 50 from workers, 25 from slums who living near the gamma rays source and 25 from healthy donors with no history of radiotherapy , 30- 65 years old with no alcohol or medicine consumption . Blood samples were drawn in tubes.

**Table 1.** Distribution of samples study according to age groups.

| Age group | No | % |
|---|---|---|
| **Less than 40** | 39 | 37.50 |
| **40-50** | 44 | 42.31 |
| **More than 50** | 21 | 20.19 |
| **P-value** | ---- | 0.0083 ** |

**** (P<0.01).**

For the gene expression study, transformed 250 µl blood directly to tube containing Trizol reagent to protect the RNA from analysis .Whole blood from each tube were processed immediately .Urine samples were obtained from same individuals and drawn in free heparin tubes for the 8-ohdG ELISA test. In this study used urinary 8-ohdG test because the level of 8-ohdG in urine more sensitive than monocytes that exposure to same gamma dose. The 8-ohdG levels in monocytes were not increased at doses lower than 10,000 cGy from gamma rays [12].

*2.2.Detection of Urinary 8-hydroxydeoxyguanosine (8-ohdG) Levels by ELISA:*

To measured the Level of 8-hydroxydeoxyguanosine (8-ohdG) in the urine as the indicator of oxidative stress can be used enzyme-linked immunesorbent assay (ELISA) ,this marker lead to know the level of oxidative stress of each person then make other test for the same individuals that appear high result in level of 8-ohdG [13] Kasai et al. show that the DNA damaged by oxidative stress can be seen by the measured concentration of 8 hydroxydeoxyguanosine (8-ohdG)as biomarker for cellular oxidative stress during carcinogenesis[14]. Briefly, Urinary 8-ohdG level was assessment by an ELISA according to the manufacturer's instructions( Germany ), fresh urine samples centrifuged at 2,000 x g for 10 minute and was stored at -20C immediately after collected and before used . The level of 8-ohdG in urine were measured by using a competitive enzyme linked with immune sorbent assay kit, 50µL from standard and samples was prepared then added in triplicate to appropriate plate well that precoated with antibody specific for 8-ohdG and incubated at 37 °C the plate was Covered at room temperature and incubated for 1 hour then the plate washed four time by 1x wash Buffer ,and the plate was covered and developed in the dark for 30 minutes at room temperature.

1- The stop solution (100 µL) was added to each well. The absorbance was measured of each well on the reader of plate at 450 nm .
2- Then Plot the standard curve then determine the concentrations of sample .The result was expressed as 8-ohdG ng /ml [13].

RNA isolation and reverse in to cDNA RNA was extraction according to recommendation of kits .The kits allowed an average yield of up to 30 µg of total mRNA. Total RNAwas reversely transcribed to complementary DNA(cDNA) using WizScript™RT FDmix(Hexamer ) Kit. The procedure was carried out in a reaction volume of 20 µl according to the manufacturer's The total RNA volume to be reversely transcribed was (15µl).A volume of 15 µl of total RNA was added into a tube containing RT FDmix(Hexamer) and pipetted up and down two times to mix The tube was briefly centrifuged to spin down the contents and toeliminate any air bubbles. The tube was placed on ice, the thermal cycler was programmed using the conditions shown in Table 2.

**Table 2**. Show thermal cycler program.

| | **Step 1** | **Step 2** | **Step 3** | **Step 4** |
|---|---|---|---|---|
| **Temperature** | 25 °C | 42°C | 85 °C | **4 °C** |
| **Time** | 10 min | 30 min | 5 min | ∞ |

*2.3.Analysis of XRCC1 m RNA expression by Quantitative real time polymerase chain reaction (QRT-PCR):*

Primer sequence for XRCC1 were 5- GAGGATGAGAGCGCCAACTC-3Forword and 5-TCACTCAGGACCCACGTTGTC-3 Reverse **,** primers sequences for *genes* were prepared according to[15] synthesized by Alpha DNA Ltd (Canada) ,the final volume of PCR reaction was 20 µl, 10µl from SYBR green , 1µl forword primer ,1µl revers primer ,2µl c DNA and completed to 20 µl by the free nucleus water .

The samples were amplified for 40 cycles at 95 °C for 5 min. , 95 °C for 30 sec. and the annealing 62°C for 60 sec , then analysis the Ct value for XRCC1 to calculate the folding times of expression for each group ,the expression levels of XRCC1 were normalized to housekeeping expression .

### *2.4.Statistical Analysis*

The Statistical Analysis System- SAS (2012) program was used affect of difference factors in study parameters Least significant difference –LSD test (ANOVA) was used to significant compare between means Estimate of correlation coefficient between variables in this study [16].

## 3.Results

### *3.1. 8-OHDG Levels in Urine*

Levels of 8-ohdG in IR-exposed workers and slums individuals in urine samples are presente in Table 3. The levels of 8-ohdG in both workers and slums were in the different ranged, There was significantly differeence in the levels of 8-OHdG among three groups , The level of urine 8-OHdG was different between the two groups (workers and slums ).It found that different in urinary 8-OHdG/creatinine excretion between the exposure individuals participated in present study and their matched controls. Exposure to ionizing radiation more than 500 cGy did not increase the levels of 8-ohdG in urine or blood.

**Table 3.** levels of 8-ohdG in gamma-exposed workers and slums individuals in urine samples

| Group | Mean ± SE |
|---|---|
| | 8-ohdG ng/ml |
| **G1: Workers** | 48.81 ± 0.67 a |
| **G2: Slums** | 36.27 ± 0.82 b |
| **G3: Control** | 14.03 ± 0.83 c |
| **LSD-value** | 2.324 ** |
| **P-value** | 0.0001 |

**** (P<0.01).**
**" Means having with the different letters in same column differed significantly"**

### *3.2.Effect of smoking in 8-ohdG level*

In other hand , it observed that significant increased in 8-ohdg levels in smokers control group and significant results in smokers individuals in groups ( worker , slums ) Table 4.

**Table 4 .** Effect of smoking in 8-OHDG level.

| Smoking | Mean ± SE |
|---|---|
| | 8-ohdG ng/ml |
| **Yes** | 45.16 ± 1.55 a |
| **No** | 34.81 ± 1.93 b |
| **LSD-value** | 6.491 ** |
| **P-value** | 0.0021 |

**" ** (P<0.01) , NS: Non-Significant ".**
**" Means having with the different letters in same column differed significantly "**

### *3.3.Effect of age in 8-OHDG level*

Also, it observed that no significant increases in 8-ohdg levels for age less than 40 when compared with (40-50)age group and more than 50, the increased was found but non significant , and significant results in 40-50 age group individuals when compared with age group more than 50( worker , slums and control) Table 5.

**Table 5.** Effect of age in 8-OHDG level.

| **Age Group (year)** | **Mean ± SE** |
|---|---|
| | 8-ohdG ng/ml |
| **Less than 40** | 38.84 ± 2.59 ab |
| **40-50** | 34.65 ± 2.41 b |
| **More than 50** | 43.13 ± 2.72 a |
| **LSD-value** | 7.518 * |
| **P-value** | 0.0497 |

*** (P<0.05) , NS: Non-Significant.**
**" Means having with the different letters in same column differed significantly "**

Urinary 8-ohdG a highly sensitive of oxidative stress, also smoking [17-18] was reported high affect on 8-ohdG levels. The data has analyzed in effect concentration on smoking of 8-ohdG levels . Though experimental study of animals showed that urinary 8-OHdG levels not appear any affect caused by content of nucleic acid diet [19]

*3.4. Influence LD ionizing radiation of XRCC1 DNA repair Gene:*

The expression of XRCC1 gene was determined in whole blood that isolated from different exposure groups. The results show that decreased XRCC1 mRNA expression levels in workers groups(Gamma rays doses higher than slums ) and increased of XRCC1 mRNA expression levels in slums( low gamma rays doses ) as presented in **Table 6** , significantly the decreased of XRCC1 gene expression has been observed at 20 cGy (po0:05) [20]. The expression levels of XRCC1 were negatively correlated with levels of 8-ohdG concentration level.

**Table 6.**Fold of XRCC1 expression depending on $2^{-\Delta Ct}$Method

| **Groups** | Means Ct of XRCC1 | Means Ct of GAPDH | ΔCt (Means Ct of XRCC1 - Means Ct of GAPDH) | $2^{-\Delta Ct}$ | experimental group/ Control group | Fold of gene expression |
|---|---|---|---|---|---|---|
| **Group1** | 25.35 | 21.12 | 4.23 | 0.0532 | 0.0532/0.0125 | 4.2 |
| **Group2** | 24.81 | 21.15 | 3.66 | 0.0791 | 0.0791/0.0125 | 6.32 |
| **Control** | 27.33 | 21.02 | 6.32 | 0.0125 | 0.0125/0.0125 | 1 |

## 4. Discussion

In this study, the determination of urinary 8-ohdG stimulated with ionizing radiation have been performed in both the workers and slums individuals exposure to different doses ranging from 5 cGy to 17 cGy. There significantly changed of the 8-ohdG levels in two groups samples . 8-ohdG levels in workers with (5cGy to 10 cGy) are (48.81 ± 0.67 ng /ml ) in concentration and slums with (less than 5 cGy) was (36.27 ± 0.82 ng /ml ) .The low doses of ionizing radiation were significantly increasing 8-ohdG levels above the standard , because of the fact that ionizing radiation induced single and double strand breakes as a result generation of (ROS) [21] , free radicals or (ROS) involved as share DNA mutation lead to cancer , diabetes, and neurodegenerative diseases [22]. The endogenous ROS are generated by influenced of various metabolic and other biochemical reactions.
The exposured to low dose -LET radiation due to the death cases of 1% from exposure individuals to period 10 years after exposure [23]. The damaging bases was occurred even less than 5 cGy , 8-ohdG as biomarker for BER and SSBR was released during the reaper, that is mean increasing the level of 8-ohdG with increase of gamma rays doses so the level of 8-ohdG in workers higher than slums ,but exposure to doses more than 500 cGy of gamma radiation did not significantly increase the levels of

8-ohdG [20]**.** The range level of 8-ohdG in workers and slums effects with many other factors such as age , gander and smoking [17],chemical reagent and metabolic rate .Most these factors were observed in workers may be lead to increase the 8-ohdag in urine. In this study ,found that damaged DNA oxidative, can be expressed by levels of urinary 8-ohdG/creatinine , the increasing in smoking individuals (45.16 ± 1.55) and (34.81 ± 1.93 ) in non- smoking exposure individuals the urinary 8-OHdG/creatinine levels significant observably in different age, Less than 40 years about ( 38.84 ± 2.59 ng/ml) non significantly increased when compared with age group between 40- 50 years (34.65 ± 2.41 ng/ml) because of the ages are close , more than 50 years were significantly increased urinary 8-ohdG level (43.13 ± 2.72 ng/ml) when compared with the first and second age as result of the aging , smoking and they were working more hours than all age groups above besides affect with the low dose of gamma rays . The levels of urinary 8-ohdG was reported by Tagesson and et al. found that high in patients treated by irradiation for lymphoma patients for bone marrow transplantation of leukemia [24] Plummer reported that measurement level of 8-ohdG may be limited use for dosimetry of radiation when used with urine DNA, because it the measurement only at doses higher than occupational exposures[25].

The mRNA expression levels of XRCC1 gene, , that shown decreased as response to IR . A significantly decreased expression level of XRCC1 gene was observed in workers exposure to 10 cGy,the ct values of XRCC1 shown in Figure 1 , The decreases of XRCC1 expression prove that ionizing radiation not only lead to DNA damage but also influence on the expression. Monika and et.al reported that the level of XRCC1 gene expression in exposure individuals were less than expression in individuals exposure to low dose of IR[26].

Some of researchers found that level XRCC1 protein expression lower in patients treated by radiotherapy vs control group [27] .These results lead to explain the significant increasing XRCC1 expression in slums ( 5cGy ) as comparison with higher doses of gamma rays in workers ( 5cGy to 10 cGy). In other studies was the exposed to high dose lead to increased the XRCC1 expression level,

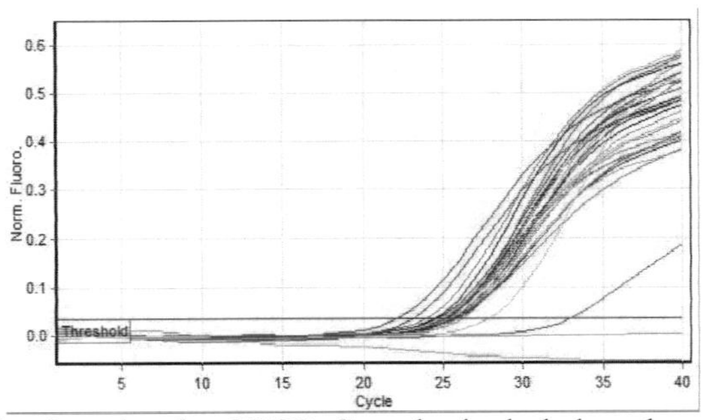


**Figure 1**. XRCC1 amplification plots by QPCR. Samples included workers groups .Ct values ranged from 22.16 to 27.9. The photograph was taken directly from QPCR machine.

inverse the slums individuals the mRNA expression levels of XRCC1 repair gene, was found to be increased in response to gamma radiation , the Ct values as shown in Figure 2 . A significantly increased expression level of XRCC1 gene was observed in individuals exposure to lower than 5 cGy. The observed increasing in XRCC1 expression proved that low IR don't effect clearly to the normal expression levels of XRCC1 repair gene .Table 6 .

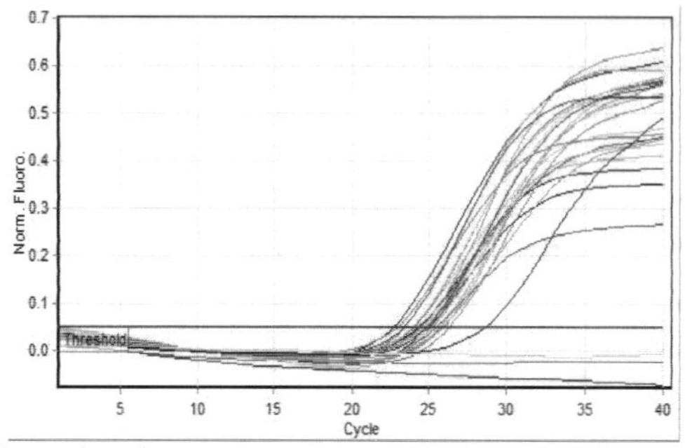


**Figure 2**. XRCC1 amplification plots by QPCR. Samples included slums individuals group .Ct values ranged from 22.75 to 28.64. The photograph was taken directly from QPCR machine

The results indicated is a possibility that there of using the 8-OHDG and gene expression as molecular assay for the assessment of DNA damage in the human peripheral blood and urine of population exposed to low dose of ionizing radiation.

## 5. Conclusion

In present study ,the γ-ray " at a dose as low as 5 cGy " for a long time can cause DNA strand breaks then increased 8-ohdG level in urine . This finding is significant since the genotoxic effects of γ-rays can be observed even at a low dose between 5 cGy – 10 cGy and less than 5 cGy , levels of XRCC1 gene expression " were significantly decreased in workers by a dose of 10 cGy of γ-rays compared to those from lower doses less than 5 cGy" . " These findings suggest that the genotoxic effects of γ-ray may be due to a combination of DNA damage and reduced DNA repair capacity. It is well known that the biological effects of IR do not follow a linear dose–response but a quadratic dose–response pattern, therefore; The results provide " that used 8-OHDG as a marker to DNA damaged and XRCC1 gene also good indication for effecting by γ-ray and some of chemical materials, Mutation of the XRCC1 gene causes defective DNA repair.

## References

[1] Lawrence T, and Rosenberg S 2008 (Eds.)." *Cancer: Principles and Practice of Oncology. Philadelphia,* (PA: Lippincott Williams and Wilkins)"

[2] Karagas M R, Stannard VA, Mott LA, Slattery M J, Spencer S K, and Weinstock M A *J.Natl Cancer Inst.***94** 224–226

[3] Zhang M, Qureshi A A, Geller AC, Frazier L, Hunter DJ, and Han J.2012 *J. ClinOncol* **30** 1588–1593

[4] " Reisz J A, Bansal N, Qian J, Zhao W, and Furdui C M 2014 *J.Antio. Red. Sign.***21**(2) 260–292 "

[5] Frieben A, Cancroid des rechten handrückens 1902 *Deutsche Medicinische Wochenschrift* **28** p335

[6] Gunderman RB, Gonda AS and girls R 2015 *J.Radio.* **274** 314–318

[7] Sak A, Fegers I, Groneberg M and Stuschke M 2008 *J.Lines. Cell Prolif.* **41** 660–670.

[8] Hoeijmakers J H 2001*Nature* **411** 366–374

[9] Svilar D, Goellner EM, Almeida KH and Sobol RW 2011*J.Antio. Red. Sign.*) **14** 2491–2507.

[10] Almeida K H and Sobol RW , 2007 *J.DNA Repair* **6** 695–711

[11] Sajous L, Botta A and MinodierI S 2008 *J.Ann Biol Clin (Paris)* **66** 19-29

[12] Pouget J P , Ravanat J L Douki, T, Richard M J, CadetJ *1999Int. J. Radiat.Biol.* **75** 51–58

[13] Zhang XH, Xuan Z, Xu CW, Li FJ, Zhang PY, Cai XJ, Qing C, Xiau BR, Jian ZC, Qiang W, Yi MZ. Chronic 2011*J.Health* **11**:224

[14] Kasai H, Hayaami H, Yamaizumi Z, Saito H, Nishimura S. 1984 *Detection and identification* of **12** p2127–36

[15] William R , Langer E, Davies S and Robison L 2004*J.Cancer* **100** 411-417.

[16] SAS 2012 Statistical Analysis System, User's Guide. Statistical. Version 9.1th ed. SAS. Inst. Inc. Cary. N.C. USA
[17] Loft S , Vistisen K, Ewertz M , Tjonneland A , Overvad K and Poulsen H E 1992 *Carcinogenesis* **13** 2241-2247.
[18] Halliwell B 1996 *Free Radie. Res.* **25** 57-74
[19] Shigenaga M K, Gimeno C J, Ames BN, 1989 *Proc. Nati. Acad. Sei.* USA **86** p.9697-9701"
[20] Wanwisa Sudprasert, Panida Navasumrit, Mathuros Ruchirawat 2006*Int. J. Hyg. Environ.-Health* **209** 503–511
[21] Hall EJ, 2000 *In: Radiobiology for the Radiologist*, **fifth ed**.( Philadelphia :Lippincott Williams &Wilkins) p. 5–16
[22] Wu L L, Chiou C C, Chang PY and Wu JT 2004 *Clin. Chim.* Acta **339** 1–9.
[23] CRP 2012 ICRP Statement on Tissue Reactions and Early and Late Effects of Radiation in Normal Tissues and Organs-threshold Doses for Tissue Reactions in a Radiation Protection Context, Publication **118**, *Ann. ICRP 41* (1/2) 1–322
[24] Tagesson C, Kallberg M , Klintenberg C, Starkhammar, H 1995 *Eur J.Cancer* **31**A 934-940
[25] Plummer S M, Pheasant A E, Johnson R, Faux, S P ,Chipman, J K and Hulten, M A 1994 *Hereditas* **121** 139–145
[26] Hanova M , Stetina R , VodickovaL , Vaclavikova R , Hlavac P and Smerhovsky Z 2010 *J.Toxic. and Appl. Pharm.* **248** 194–200
[27] Batar B , Guven G , Eroz S , Bese N S , Guven M. 2016 *J.Gene* 5pages

# Association of Hyperhomocysteinemia with Acute Myocardial Infarction in Kurdish Patients

**Shwan Othman Amen[1], Soza Tharwat Baban[2]**

[1]Consultant Interventional Cardiologist, Surgical Specialty Hospital - Cardiac Center, Erbil, Iraq, German Board (Facharzt) in Interventional Cardiology, German Board (Facharzt) in Internal Medicine.
[2]Biomedical scientist, Medical laboratory, Surgical Specialty Hospital- Cardiac center, Erbil, Iraq.

**Abstract.** Cardiovascular disease (CVD) is the main cause of death worldwide. Among multiple cardiovascular risk factors, such as age, blood pressure, diabetes, smoking and hyperlipidemia, hyperhomocysteinemia has recently emerged to have an important role in increasing risk of CVD by impairing oxidative stress-induced endothelial dysfunction and inducing atherothrombosis. However, the association between high circulating level of homocysteine and increasing the risk of CVD in human remains controversial and has not been consistently observed. OBJECTIVES: The aim of this study is to evaluate the association between serum total homocysteine (tHcy) concentrations with increased risk of CAD in patients with acute myocardial infarction (MI) in Kurdish population. In addition, correlations of classical risk factors including hypertension, smoking, diabetes and hyperlipidemia with hyperhomocysteinemia in MI patients were studied. Its role in in development of CVD could be pivotal. METHODS: In a case-control study, a total of 74 patients (51 male; 23 female; aged 22-80 years) who had diagnosed with acute CAD on angiography were enrolled in this study and 74 gender and age matched non-CAD individuals were included as the control group. The serum tHcy level was measured by enzymatic immunoassay. Hyperhomocysteinemia (HHcy) was defined as Hcy>15 μmol/l. RESULTS: the mean ± standard deviation (SD) of serum total homocysteine (tHcy) level in acute myocardial infarction patients was 22.8 ± 1.30 μmol/l, which was significantly higher compared to that of control group being 15.1 ± 1.33 μmol/l ($P < 0.001$), indicating that a high homocysteine level may promote pathogenesis of atherosclerosis. 51 (68.9%) MI cases had hyperhomocysteinemia; of these 35 MI patients (47.3%) had mild hyperhomocysteinemia (15-30μmol/L), and 16 MI patients (21.6%) had moderate levels of hyperhomocysteinemia (30-100 μmol/L). No significant correlation between tHcy levels and MI risk factors such as hypertension, hyperlipidemia, and diabetes mellitus, was observed. The tHcy levels were significantly higher in smoker patients than in non-smokers. CONCLUSIONS: These findings confirm that in patients with acute myocardial infarction, elevated homocysteine levels are associated with development of MI independent of other risk factors. A strong association between elevated levels of serum tHcy in MI patients was observed in Kurdish population. This study concludes that Hcy is considered as a strong emerging cardiovascular risk factor for development of myocardial infarction. Moreover, the measurement of Homocysteine levels might act as a surrogate laboratory marker and as one of the effective ways to diagnose patients with myocardial infarction in hospitals.

**Keywords**: Coronary Artery Disease, Hyperhomocysteinemia, Acute Myocardial Infarction.

IOP Conf. Series: Materials Science and Engineering **557** (2019) 012083 doi:10.1088/1757-899X/557/1/012083

## 1. Introduction

Atherosclerotic coronary artery disease (CAD) continue to be the major cause of death and loss of quality of life worldwide, and its prevention and effective treatment are key strategies in reducing the rate of mortality and morbidity [1]. Myocardial infarction (MI) is the manifestation of CAD. MI is a disease of the vessel that feeds the cardiac muscle, called the coronary artery. Plaque rupture with thrombosis is a well-established critical factor in the pathogenesis of MI. Although detailed mechanisms of plaque rupture are unknown, inflammation is thought to play an important role in its pathogenesis (Ross 1999). The pathological milieu and mechanisms leading to the development of CAD and MI remain incompletely understood although a number of risk factors have been identified over the past several decades [2]. It is widely recognized that CAD is causally attributed to 'classical' atherogenic risk factors including abnormal level of circulating cholesterol with elevated levels of Low Density Lipoprotein (LDL) and reduced level of High Density Lipoprotein (HDL), hypertension, smoking, diabetes mellitus, advancing age, gender, and a positive family history of premature vascular disease [3]. Recent observational studies have found that 50% of the variation in the etiology and progression of atherosclerotic cardiovascular disease still remains unexplained by traditional risk factors. In addition, up to 30% of patients presenting with MI have none of these risk factors. There is an increasing evidence that atherosclerosis may, in part, be an inflammatory disease and circulating inflammatory biomarkers in the blood may be predictors of CVD [4]. This has prompted the focus on identification of other emerging predisposing factors –or cardiovascular risk markers, including homocysteine, to improve CAD risk assessment. Homocysteine is a sulfur containing amino acid which is produced as an intermediate product during the metabolism of methionine via demethylation of methionine [5].

This amino acid has been recognized recently as an important emerging atherogenic risk factor and it has been found to be toxic to the endothelium and prothrombotic. These properties of homocysteine were clearly evident in patients with homocystinuria genetic disorder where elevation of plasm homocysteine causes atherosclosrisis at younger age [6]. More recently, homocysteine has received greater attention as an important risk factor for CAD, although a direct causative role of this emerging cardiovascular risk marker and the exact mechanism in pathology and development of CAD has not been fully elucidated [7]. Elevated level of homocysteine has emerged as an important risk factor that is associated with the development of CAD. There is an evidence that excess homocysteine in the blood stream may cause endothelial cell injuries, promotes oxidative damage of LDL and stimulates smooth muscle cell proliferation in the vessels, and result in inflammation, and plaque formation in the arterial vessels [8]. A study conducted by Ganguly [9] has found that homocysteine is released from damaged tissue, leading to an increased plasma homocysteine concentration, which thus contributes to recruitment of leukocytes. Thus, in this case, an increased homocysteine level might not be a cause but a consequence of vascular injury in the acute phase, but a cause of increasing inflammatory activity in the chronic course. The aim of the present study was to determine the association between elevated homocysteine level and AMI in Kurdish population.

## 2. Materials and Methods

### *2.1. Study population*

The study population comprised two hundred ninety six consecutive patients with acute MI, admitted to the causality department of the Surgical Specialty Hospital – Cardiac center within 12 hours of clinical signs and symptoms onset during the period from March to May 2018. All patients between 20 to 90 years old were included in this study. The control group consisted of 240 age-and gender-matched individuals without Ischemic Heart Disease. Diagnostic criteria of acute MI were followed according to the guidelines of European Society of Cardiology, such as when the patient is presented with typical chest pain, diagnostic ECG changes, and a significant elevation of cardiac enzymes. Exclusion criteria for patients and controls included pregnancy, anemia, and kidney failure. This study was approved by the Ethics Committee of Hawler Medical University, and written informed consent was obtained from all patients and control subjects. Cases and controls filled in a standard questionnaire about their personal histories, major risk factors, history of ischemic heart disease, and provided blood samples for laboratory analysis.

### *2.2. Laboratory analysis*

The level of total homocysteine in human serum was quantitatively determined by Homocysteine Enzymatic cycling immunoassay method using Cobas Integra system. The Venous blood samples were collected in EDTA tubes from patients with MI within the first 12 hour of myocardial infarction. Blood was transferred into EDTA tubes. Following collection, the samples were centrifuged immediately at 4000g for 10 min. The serum was separated and immediately stored at -20C until further analysis. Other laboratory biochemical parameters such as Lipid profile test including total cholesterol (mg/dL), High-density lipoprotein cholesterol (HDL [mg/dL]), Low-density lipoprotein cholesterol (LDL [mg/dL]), Triglyceride (mg/dL) and Hemoglobin A1C test for diabetes, were measured by enzymatic colorimetric method using Roche Cobas system.

## 3. Results

A total of 74 consecutive patients presenting with AMI of case group, of which 23 (31.1%) female and 51 (68.9%) male, was studied. The mean age of AMI patients in this group was 55.5 ±12.47 ranged between 22 to 83 years. The control group consisted of totally 74 healthy individuals, of which 54 (72.9%) female and 20 (27.03%) male. The mean age of healthy individuals was 49.6±16.55 ranged between 21 to 83 years. There was a significant difference in relation to age (P=<0.05) between the case and control groups. This study observed that AMI patients in case group had significantly higher level of serum homocysteine concentration than in healthy individuals in the control group (22.8±12.97μmol/L versus 15.1±13.31μmol/L, respectively, $P$= 0.0001), as shown in Figure 1. There was about two-fold rise in the mean level of serum homocysteien in AMI patients compared to the control subjects.

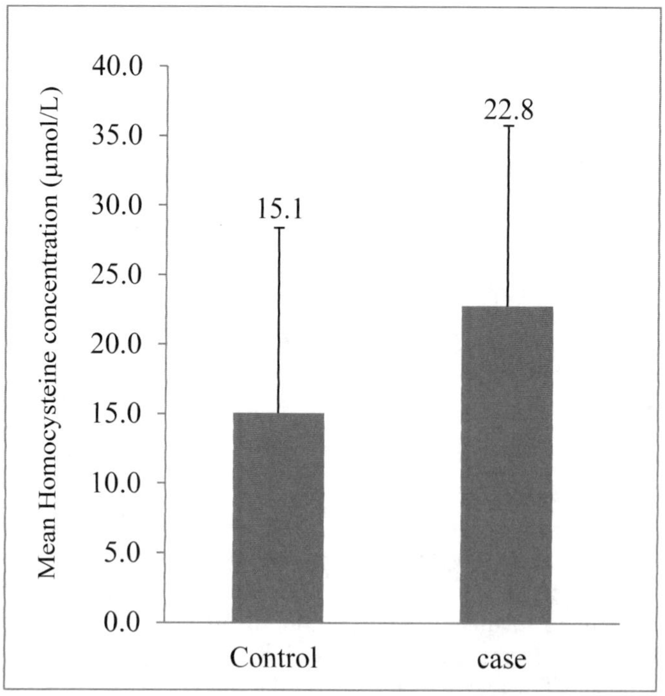


**Figure 1.** Comparison of plasma homocysteine concentration between cases and control subjects.

Moreover, results showed that the mean concentration of homocysteine in male patients (26.8±13.8μmol/L μmol/L) was higher than in female patients (19.3 ±13.2μmol/L) in the case group), as shown in Table 1. This difference was statistically significant (p<0.05).

**Table 1.** Mean serum concentration of homocysteine level in relation to gender.

| Gender | Case | Control | *P* value |
|---|---|---|---|
| **Female** | 19.2±13.7 | 11.6±9.1 | <0.05 |
| **Male** | 26.8±13.8 | 19.3 ±13.2 | <0.01 |

Twenty three patients with AMI (31.1%) in case group had normal homocysteine level, whereas the majority of healthy participants (70.3%) had normal homocysteine level. Interestingly, 47.3% of AMI patients were detected with mild hyperhomocysteinemia, relative to 15% of healthy individuals in control group. This difference was statistically significant (P<0.05). Moreover, moderate homocysteinemia was found in 16 AMI patients (21.6%) in the case group and only 7 (9.5%) of control healthy individuals (P<0.05). Severe hyperhomocysteinemia (>100μmol/L) was not detected in neither case nor in control group, as shown in Table 2.

**Table 2.** Mild, moderate and severe hyperhomocysteinemia in case and control groups.

| **Variable** | Case (n=74) | | Control (n=74) | |
|---|---|---|---|---|
| | **Number (n)** | **Percentage (%)** | **Number (n)** | **Percentage (%)** |
| Normal homocysteine (≤15μmol/L) | 23 | 31.1 | 52 | 70.3 |
| Mild hyperhomocysteinemia (16-30μmol/L) | 35 | 47.3 | 15 | 20.3 |
| Moderate hyperhomocysteinemia (31-100 μmol/L) | 16 | 21.6 | 7 | 9.5 |
| Severe hyperhomocysteinemia (>100 μmol/L) | 0 | 0.0 | 0 | 0.0 |

As described in Figure 2. Incidence of Mild and moderate hyperhomocysteinemia levels were significantly higher in male patients compared to female.

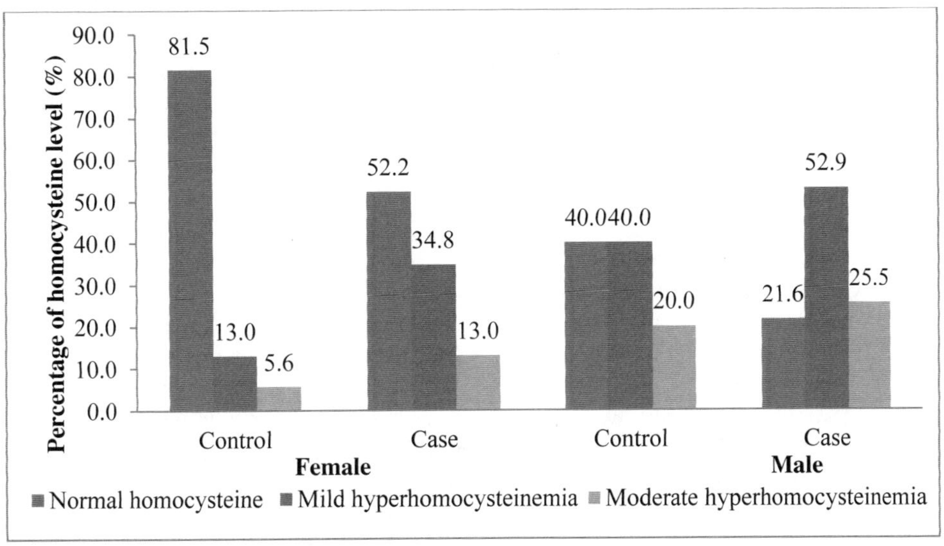


**Figure 2.** Incidence of homocysteine levels in AMI patients compared to healthy controls in relation to gender.

IOP Conf. Series: Materials Science and Engineering **557** (2019) 012083 doi:10.1088/1757-899X/557/1/012083

In addition, elevated level of homocysteine (>15µmol/L) was observed significantly in 68.9% of AMI patients compared to 29.7% of healthy controls, as shown in Table 3. This difference was statistically significant (*P*=<0.05). These findings clearly demonstrate that elevated level of homocysteine was associated with myocardial infarction.

**Table 3.** Incidence of normal Homocysteine level and hyperhomocysteinemia in AMI patients compared to control group.

| Variable | Case (n=74) | | Control (n=74) | |
|---|---|---|---|---|
| | Number (n) | Percentage (%) | Number (n) | Percentage (%) |
| **Normal homocysteine (≤15µmol/L)** | 23 | 31.1 | 52 | 70.3 |
| **Hyperhomocysteinemia (≥15µmol/L)** | 51 | 68.9 | 22 | 29.7 |

The European Concerted Action group reported that when hyperhomocysteinemia coexist with conventional cardiac risk factors such as smoking, diabetes mellitus, hypertension, hyperlipidemia will lead to increase in risk of cardiovascular development. This study examined the correlation between homocysteine, conventional risk factors and CAD risk. Among case group, 35 (47.3%) of AMI patients were found smokers and 29 (39.2%) were nonsmokers and 10 (13.5%) who quick the smoking habit. Hyperhomocysteinemia level was more highly prevalent in smoker patients (56.9%) compared to non-smokers (33.3%) and patients who quit smoking habit (9.8%), as shown in Table 4. Furthermore, male smoking patients (43.1%) were observed with significantly higher rate of hyperhomocysteinemia compared to female smokers (13.7%). High incidence rate of Hyperhomocysteinemia was observed in the smoking patients in the age group of 60-69 years (19.6%).56.9% of smoker patients were observed with hyperhomocysteinemia compared to 26.1% of those with normal homocysteine level. This difference was statistically significantly different (P<0.0001). Next we examined the association between homocysteine and hypertension in these patients. Hypertension was found in 62.5% patients with elevated homocysteine level as compared to those with normal homocysteine level (37.5%). Elevated homocysteine level was detected in 25 (49%) and 21 (41%) of AMI patients with low HDL and high triglyceride levels. However, the incidence of hyperhomocysteinemia was observed in 24 (47%) of AMI patients with high LDL and 14 (27%) with high total cholesterol. These results suggest that hyperhomocysteinemia is not affected by high cholesterol level to increase the risk of ACD in the patients. The association of elevated homocysteine level with diabetes mellitus and CAD risk was investigated in this study. The mean glycated hemoglobin level was 9% in hyperhomocysteinemia diabetic patients, whereas it was 10.2% in diabetic patients with normal homocysteine level. High incidence of Hyperhomocysteinemia was observed in the diabetic patients in the age group of 50-59 years (46.2%). Moreover, we found high incidence of hyperhomocysteinemia in male diabetic patients (69.2%) compared to female diabetic patients (30.8%). The incidence rate of hyperhomocysteinemia in diabetic AMI patients (25.55) was significantly lower compared to non-diabetic patients (74.55%), as shown in Table 4. In addition, only 5 (9.8%) AMI male patients were detected as hypertensive and had high serum homocysteine level. It was observed that elevated level of homocysteine was highly predominant in 32 (62.7%) of AMI patients with stress as compared to 19 (37.3%) without stress. Furthermore, significant higher incidence of Hyperhomocysteinemia was observed in 27 (52.9%) patients with CAD positive family history as compared to 24 (47.1%) of patients with no CAD family history.

**Table 4.** Frequency of hyperhomocysteinemia in diabetic and non-diabetic AMI patients in relation to glycated hemoglobin (HbA1C).

| CAD risk factors | Hyperhomocysteinemia | Normal Homocysteine | Total | *P*-value |
|---|---|---|---|---|
| Smoker | 29 (56.9%) | 6 (26.1%) | 35 (47.3%) | 0.009 |
| Non-smoker | 17 (33.3%) | 12 (52.2%) | 29 (39.2%) | 0.0009 |
| History of smoking | 5 (9.8%) | 5 (21.7%) | 10 (13.5%) | 0.0006 |
| Low HDL | 25.00 (49.0%) | 7.00 (30.4%) | 32.0 (43.2%) | 0.0002 |
| High LDL | 24.00 (47.1%) | 13.00 (56.5%) | 37.0 (50.0%) | 0.0002 |
| High Triglyceride | 21.00 (41.2%) | 9.00 (39.1%) | 30.0 (40.5%) | 0.004 |
| High Total Cholesterol | 14.00 (27.5%) | 10.00 (43.5%) | 24.0 (32.4%) | 0.01 |
| Hypertension | 5 (9.8%) | 3 (13.0%) | 8 (10.8%) | 0.02 |
| Diabetic | 13 (25.5%) | 12(52.2%) | 25(33.8%) | 0.0001 |
| Nondiabetic | 38(74.5%) | 11(47.8%) | 49(66.2%) | 0.0001 |
| Stress (Yes) | 32 (62.7%) | 12 (52.2%) | 44 (59.4%) | <0.05 |
| Stress (No) | 19 (37.3%) | 11 (47.8%) | 30 (40.5%) | <0.05 |
| Family history (Yes) | 27 (52.9%) | 12 (52.2%) | 39 (52.7%) | <0.05 |
| Family history (No) | 24 (47.1%) | 11 (47.8%) | 35 (47.2%) | <0.05 |

## 4. Discussion

The present study aims to define the role of homocysteine in cardiovascular disease. In the last decades, in spite of great efforts have been done to reduce mortality rate of cardiovascular diseases the effective prevention and treatment techniques remains unsatisfactory, particularly in high risk group of patients. Therefore, recent studies focused on the identification of new risk factors for CAD and its prevention. Atherosclerosis is the most common pathological process that leads to cardiovascular diseases such as myocardial infarction (MI), heart failure, and stroke. Arteriosclerosis is a continuous inflammatory damage to the arterial intima with increased permeability to plasma, deposition of plasma lipids in plaques and fibrosis and calcification of plaques. The association between elevated homocysteine concentration level and atherosclerotic disease was first proposed more than 40 years ago due to its atherothrombotic mechanism leading to endothelial dysfunction [10]. Despite that, controversy still exists about the role of elevated homocysteine concentration level as a risk factor for coronary artery disease and myocardial infarction [11]. An observation had pointed towards a clear association between homocysteine concentration and CAD [12-14]. Several studies reported that 10% of CVD risk in patients and overall mortality are attributed to the elevated homocysteine level [15, 16]. The major finding of the present study confirms that hyperhomocysteinemia is an independent emerging cardiovascular risk factor that contributes to the development of CAD and myocardial infarction in Kurdish population. Our results showed that the homocysteine concentration level was significantly higher in patients with myocardial infarction as compared to control subjects, indicating that there is a strong association between hyperhomocysteinemia and cardiovascular disease and its complications such as myocardial infarction. These findings are consistent with a study which reported high homocysteine level in patients with acute myocardial infarction and it was postulated as a risk factor for coronary heart disease. Moreover, consistent relationship between homocysteine concentration and CAD risk were found both in the Schaffer [17] and Al-Obaidi [18]. There is a strong evidence that hyperhomocysteinemia leads to endothelial cell damage, reduction in the flexibility of vessels, and alteration in process of haemostasis. Some studies suggest that this emerging atherogenic risk factor may be secondary to preclinical vascular diseases and it was demonstrated that elevated homocysteine level is a causal risk factor for CAD not the effect. In other words, elevated homocysteine level may enhance the effect of conventional risk factors, such as hyperlipidemia, smoking, hypertension and inflammation on CAD risk. A study by Graham *et al.* [19] demonstrated that elevated level of homocysteine showed more effect on CAD risk in smoker, patients with hypertension and hyperlipidemia compared to normal population. Another study [20] reported that hyperhomocysteinemia was not correlated to CAD in patients without hypertension, diabetes and dyslipidemia.

On the contrary, No any association between hyperhomocysteinemia and CAD risk was found in the Americal multiple risk factor intervention trial (MRFIT) [21]. In addition, we have determined the incidence of hyperhomocysteinemia in AMI patients with conventional risk factors in the case group in the present study. The incidence rate of hyperhomocysteinemia in smokers was higher than that in non-smoker patients. This finding was in agreement with a study by Shengfang *et al.* which found the same results [22]. High incidence rate of hyperhomocysteinemia was found in patients with hyperlipidemia compared to normal homocysteine level in these patients of case group. Similar trend of high incidence of hyperhomocysteinemia condition was found more in hypertensive patients compared to patients without hypertension. However, lower incidence rate of hyperhomocysteinemia was observed in diabetic patients compared to non-diabetic patients. Although these results were statistically significantly different when comparing incidence rates of high homocysteine level to normal level in these conditions in patients with AMI disease in the case group alone, no correlation was shown between elevated homocysteine level and traditional risk factors by multiple regression analysis. This study excludes that hyperhomocysteinemia may be secondary to preclinical vascular disease. These finding implies a strong association between hyperhomocysteinemia and CAD even after adjustment of other coronary risk factors was found. Similar result was found in a heart study conducted by Welch *et al.* [23]. It is important to note that differences in design, statistical approach may partly explain variations in results of different studies but other causes of unsatisfactory results may be nutritional and large ethnic genetic variations. There are very few studies assessing association between serum total homocysteine level and CAD recently in Middle East especially in Kurdistan region and Iraq and its effect on other cardiovascular risk factors was not identified. One of the most important determinants of homocysteine level are genetic factors, vitamin B6 and folic acid uptake. In the present study, interestingly we found no correlation between hyperhomocysteinemia and vitamin B6 or folic acid deficiency in AMI patients (data not shown). AMI patients with hyperhomocysteinemia had normal vitamin B6 and folic acid concentration levels. Moreover, Genotyping of Methylene tetrahydrofolate reductase (MTHFR) enzyme for missense mutations (C677T and A1298C) results showed no correlation between genetic polymorphism, hyperhomocysteinemia and CAD risk. This result was in agreement with a study by Al-Gazally et al. [24] conducted among Iraqi population and showed that ischemic stroke and myocardial groups had significantly higher level of homocysteine and homocysteine was considered as an independent risk factor for CAD. Several studies in Iran [25, 26] showed that the increased level of homocysteine is an independent risk factor for increasing of CAD risk as higher levels of serum homocysteine in patients with abnormal angiography was observed as compared to that in control group. It is postulated that geographical variation in dietary folate intake and frequency of genetic polymorphism and different ethnic group may be important in impact of hyperlipidemia on CAD risk. In this study, we showed that elevated level of homocysteine is significantly associated with CAD risk independent of other risk factors in Kurdish population. In conclusion, the present study concludes that there is a strong association between hyperhomocysteinemia and development of CAD risk in Kurdish population independent of other conventional CAD risk factors. Given these findings, the next step is to establish a direct association between hyperhomocysteinemia and risk of cardiovascular disease morbidity and mortality.

## 5. References

[1] Smith SC. Current and future directions of cardiovascular risk prediction. Am J Cardiol 2006; 97(2A): 28A-32A.

[2] Mangge H., Becker K., Fuchs D., Gostner J.M. Antioxidants, inflammation, and cardiovascular disease. World J Cardiol. 2014;6:462–477.

[3] Wilson PW. Assessing coronary heart disease risk with traditional and noel risk factors. Clin Cardiol 2004; 27:III7-11.

[4] Selhub J. Homocysteine metabolism. Annu Rev Med. 1999;19:217–246.

[5] Lentz S.R. Mechanisms of homocysteine-induced atherothrom-bosis. J, Thromb, Haemost. 2005;3:1646–1654.

[6] Knekt P, Reunanen A, Alfthan G, Heliövaara M, Rissanen H, Marniemi J, et al. Hyperhomocystinemia: a risk factor or a consequence of coronary heart disease? Arch Intern Med 2001; 161: 1589-1594.

[7] Fallest-Strobl PC, Koch DD, Stein JH, McBride PE. Homocysteine: a new risk factor for atherosclerosis. Am Fam Phys. 1997; 56(6):1607- 1612.

[8] Paul Ganguly and Sreyoshi Fatima Alam Role of homocysteine in the development of cardiovascular disease, Nutr J. 2015;14: 6.

[9] McCully KS. Vascular pathology of homocysteinemia: implications for the pathogenesis of arteriosclerosis. Am J Pathol 1969;56:111–28.

[10] Veeranna V, Zalawadiya SK, Niraj A, Pradhan J, Ference B, Burack RC, et al. Homocysteine and reclassification of cardiovascular disease risk. J Am Coll Cardiol 2011;30:1025–33.

[11] Okura T, Miyoshi K, Irita J, Enomoto D, Nagao T, Kukida M, Tanino A, Kudo K, Pei Z, Higaki J Hyperhomocysteinemia is one of the risk factors associated with cerebrovascular stiffness in hypertensive patients, especially elderly males. Sci Rep. 2014 Jul 11; 4():5663.

[12] Paul Ganguly and Sreyoshi Fatima Alam Role of homocysteine in the development of cardiovascular disease Nutr J. 2015; 14: 6.

[13] Fallest-Strobl PC, Koch DD, Stein JH, McBride PE. Homocysteine: a new risk factor for atherosclerosis. Am Fam Phys. 1997; 56(6):1607- 1612.

[14] Nygard O, Vollset SE, Refsum H, et al. Total plasma homocysteine and cardiovascular risk profile: the Hordaland Homocysteine Study. JAMA. 1995;274:1526–33.

[15] Boushey CJ, Beresford SA, Omenn GS, Motulsky AG. A quantitative assessment of plasma homocysteine as a risk factor for vascular disease. Probable benefits of increasing folic acid intakes. JAMA 1995; 274(13): 1049-57.

[16] Akram Saleh, Nabil Basher, Serum Homocysteine Levels in Patients with Acute Myocardial Infarction in Jordan J Med J 2010; Vol. 44 (2):193-197

[17] Schaffer A, Verdoia M, Cassetti E, Marino P, Suryapranata H, De Luca G; Novara Atherosclerosis Study Group (NAS). Relationship between homocysteine and coronary artery disease. Results from a large prospective cohort study. Thromb Res. 2014 Aug;134(2):288-93.

[18] Al-Obaidi MK, Stubbs PJ, Collinson P, Conroy R, Graham I, Noble MI. Elevated homocysteine levels are associated with increased ischemic myocardial injury in acute coronary syndromes. J Am Coll Cardiol. 2000 Oct;36(4):1217-22.

[19] Graham IM, Daly LE, Refsum HM, Robinson K, Brattstrom LE, Ueland PM, et al. Plasma homocysteine as a risk factor for vascular disease. The European Concerted Action Project. JAMA 1997; 277(22): 1775-81.

[20] Donner MG, Klein GK, Mathes PB, Schwandt P, Richter WO. Plasma total homocysteine levels in patients with early-onset coronary heart disease and a low cardiovascular risk profile. Metabolism 1998; 47(3): 273-9.

[21] Evans RW, Shaten BJ, Hempel JD, Cutler JA, Kuller LH. Homocyst(e)ine and risk of cardiovascular disease in the Multiple Risk Factor Intervention Trial. Arterioscler Thromb Vasc Biol 1997; 17(10): 1947-53.

[22] Shengfang Chen, Ping Wu, Lin Zhou, Yuqin Shen, Yunjie Li, andHaoming Song Relationship between increase of serum homocysteine caused by smoking and oxidative damage in elderly patients with cardiovascular disease. Int J Clin Exp Med. 2015; 8(3): 4446–4454.

[23] Welch GN, Loscalzo J. Homocysteine and atherothrombosis. N Engl J Med 1998; 338:1042-1050.

[24] Moaed E. Al-Gazally , Ali H. Al-Saadi , Asmaa H. Radeef , Effect of homocysteine on ischemic stroke and myocardial infarction in Iraqi population, International Journal of PharmTech Research, Vol.8, No.10, pp 139-145, 2015

[25] Kazemi MB, Eshraghian K, Omrani GR, Lankarani KB, Hosseini E. Homocysteine level and coronary artery disease. Angiology 2006; 57(1): 9-14.

[26] Mirdamadi A, Farzamnia H, Varzandeh P, Almasi N, Arasteh M., Association Between Serum Homocysteine Concentration With Coronary Artery Disease In Iranian Patients, Arya Atheroscler. 2011; 7(2):63-7.

IOP Conf. Series: Materials Science and Engineering **557** (2019) 012084 doi:10.1088/1757-899X/557/1/012084

# Oil stove modification to get optimum performance on jatropha oil usage

**S. Endah Agustina* and Rimo Hasnan**

Dept. of Mechanical and Biosystem Engineering, Faculty of Mechanical & Agricultural Technology, Bogor Agriculture University (IPB), Indonesia

*E-mail : endah859@gmail.com

**Abstract**. This research is based on the idea to combine efforts on promoting jatropha oil as one of renewable and green energy resources, and efforts to help "house hold oil stove industries" continuing their production even-though more people change kerosene to LPG as their main fuel for cooking. Regarding the idea, this research aims to modify oil stove which has been sold in the market, which was designed for kerosene fuel, to be able to use with jatropha oil and other oil as its fuel in the good performance. Two approach methods based on the physical oil characteristic has been done. They were modification of wick material, and "oil pre-heating system" implementation. The result shows that no wick material having good performance when implemented for jatropha oil usage due to its high viscosity. Original wick has best performance compare to modified wicks. By using original wick, implementation of oil pre heating system has been successfully increasing fuel rate combustion from 0.43 kg/h to 0.63 kg/h, and increasing the total system efficiency from 9.54% to 16.1%.

## 1. Introduction

Efforts to reach energy mix target which is implemented in many kinds of program, sometimes didn't recognize negative impacts following those program. Promotion LPG to substitute kerosene as main fuel for cooking in house hold and small scale industries, has been successfully decreasing kerosene demand, but in the other side causing negative impact to the stove industries since their market down significantly. In the other hand, substitution fossil energy with renewable and green energy resources also the main target of energy mix program. Therefore, combining efforts on promoting jatropha oil as one of renewable and green energy resources, and efforts to help "house hold oil stove industries" continuing their production activities, should be a great solution. This research aims to modify oil stove which has been sold in the market, which was designed for kerosene fuel, to be able to use with jatropha oil and other pure plant oil as its fuel in the good performance. For the first step, research conducted for jatropha oil usage since other pure plant oil has not popular yet. Therefore, discussion in this paper will be focus on stove modification based on the jatropha oil usage.

## 2. Research Method

Two approach methods based on the physical oil characteristics has been done. There were wick material modification, and implementation of oil pre heating system. Besides of modification on structure and system, modification on stove operation also should be implemented due to the differences of fuel characteristic. The methods described as follows.

IOP Conf. Series: Materials Science and Engineering **557** (2019) 012084 doi:10.1088/1757-899X/557/1/012084

1) Wick material modification: Wick is the main component of the stove, due to its function as fuel distributor which transporting fuel (based on capillarity process) from the fuel tank to the combustion area/surface. The capillarity will be influence wick's capability to transport fuel, or fuel rate consumption. In this research, 3 kind of wick's material has been tested. There are cotton, polyester, and asbestos. Modification also done in order to improve polyester wick performance by inserting a copper wire inside the polyester wick.
2) Structure modification: Modification of stove structure should be done to implement "oil pre-heating system" on the stove. Jatropha oil has higher viscosity compare to kerosene. Higher viscosity will be causing lower capacity of the wick's capillarity and more difficult to combusted. Pre-heating system was needed to decreasing jatropha oil viscosity before its deliver to the combustion zone through the wick. Beside implementation of pre heating system, air inlet for combustion process also must be modified, since the different fuel will be need different amount of primary air for good or perfect combustion.
3) Since the stove has been modified, the operation will be different than before modified (original design). Therefore, a modification of the stove usages should be applied to get optimum stove performance.

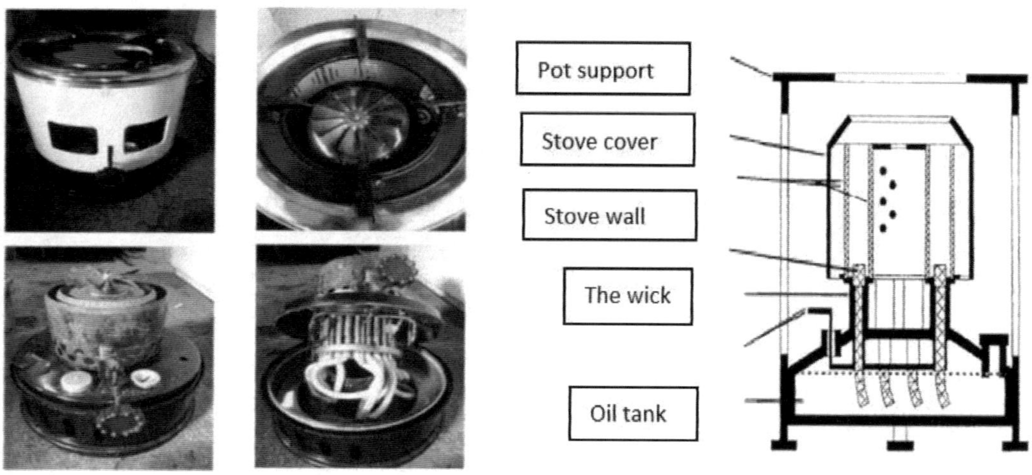


**Figure 1.** Stove components.

Figure 1 shows the original stove and its components, which will be modified in this research. The detail structure inside the stove can be seen in the right picture. The original wick material is polyester. There are 3 (three) layer stove wall surrounding combustion zone. Air inlet holes in the stove wall designed to warranty that amount of air supply (for combustion process) is enough for perfect combustion. Therefore, air inlet holes will be modified based on the new fuel (jatropha). Research stages described in the Figure 2.

Parameters for wick's capillarity are rate of oil level reduction in the measurement glass (cm/minute), and time (duration) of wick's flaming (minutes). Faster oil level reduction means more oil flowing through the wick, meaning the wick having better capillarity. Shorter duration of wick's flaming means the wick easier to burnt out and must be replace in the shorter period. Also, produce more ash that could be influencing combustion process. So, the wick which has longest flaming duration is the best.

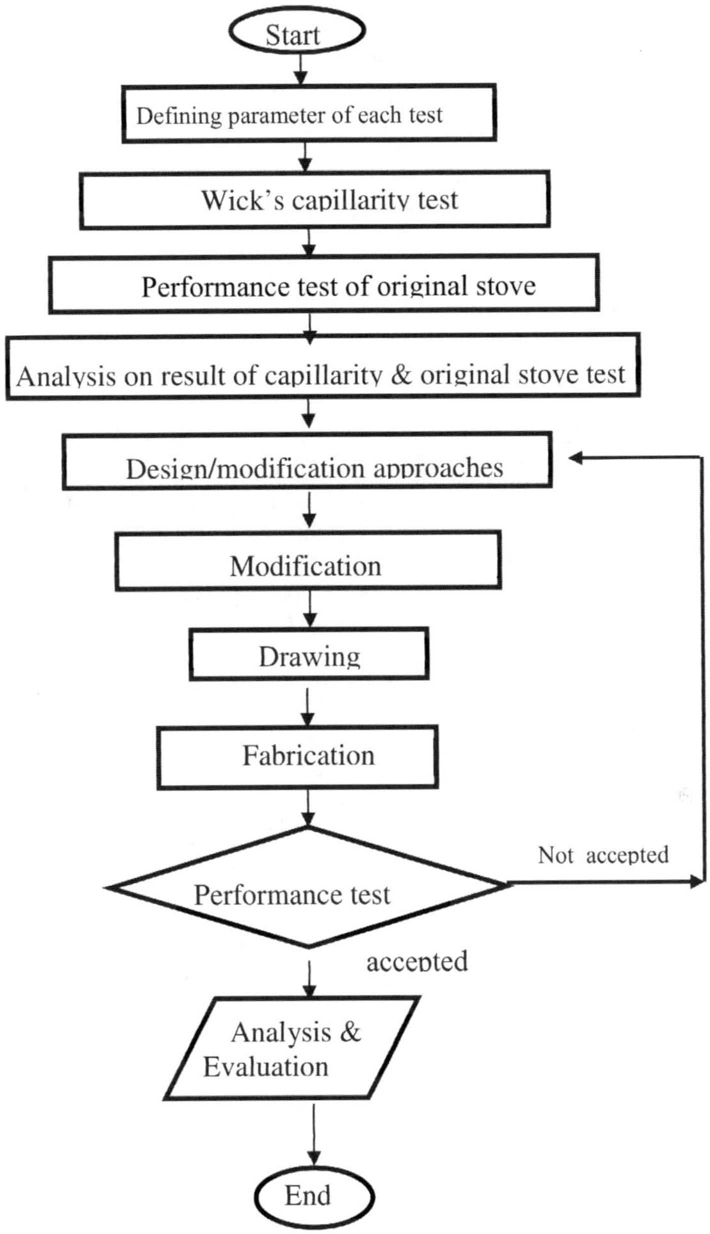


**Figure 2.** Research stages

Stove performance was conducted by using *water boiling test* method. Parameters which has been use are fuel (consumption) rate (kg/hour), time for ignition (minutes), time for turn off fire (minutes), flame's color (range: blue – yellow- red), smoke (range: none – plenty - huge) , stove efficiency (%) and total system efficiency (%) . Figure 3 describes the measurement points on the stove when the performance test conducted.

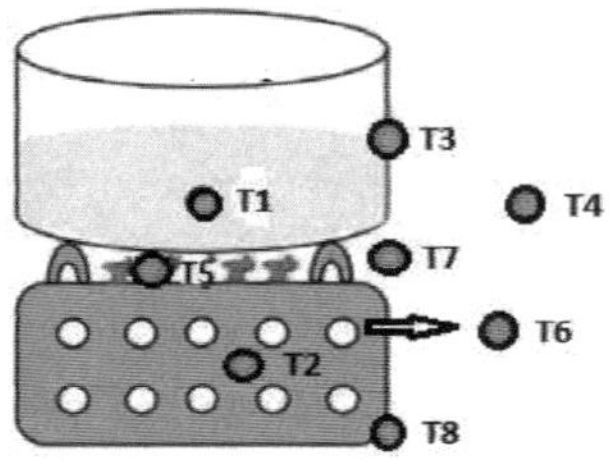


**Figure 3.** Measurement points in the stove performance test.

## 3. Result and Discussion

### *3.1. Wick modification*

First, 3 kind of wick with different material (polyester, cotton, and asbestos) has been tested based on the parameter as mentioned before. Result of the first test showing that the best wick's material is polyester (Table 1). Polyester is the original wick which has been use in the original stove. In fact, result of original stove performance test by using jatropha oil shows that the original wick has very poor performance. The poor performance was indicate by very low rate of oil level reduction (Table 1), meaning the capillarity capacity was poor, since jatropha oil has higher viscosity compare to kerosene. In order to increasing polyester wick's performance, the wick was modified by inserting copper wire inside the polyester wick. The idea is based on the capability of metal (copper) to help transferring heat from the combustion zone to the oil inside the wick. But those polyester wick modification has poor performance. Result of all wick's performance test can be seen in the Table 1 below.

**Table 1.** Result of wick's performance test

| Wick's material | Duration of flaming (minutes) | Rate of oil level reduction (cm/minute) |
|---|---|---|
| Cotton | 25 | 4.2 |
| Polyester | 30 | 5.3 |
| Asbestos | 8 | 1.2 |
| Polyester + 20% copper | 15 | 2.5 |
| Polyester +30% copper | 12 | 2.1 |
| Polyester +50% copper | 10 | 1.6 |
| polyester+80% copper | 10 | 2.0 |

Based on the wick' performance test, we can conclude that we can't use jatropha oil as fuel in the original stove by modification or change the wick material. Jatropha oil available to use in the stove if jatropha oil has same or almost same viscosity like kerosene. Therefore, the stove system and stove structure should be modified by applying pre-heating oil system in order to decrease jatropha's viscosity.

### *3.2. Performance of original stove using jatropha oil as fuel*

Performance of the original stove by using jatropha oil as its fuel should be done to get real data and also to understand how the system work properly. Result of the performance test will be used as basic approach on designing modification. The data performance also will be use as success parameter of the modification. Table 2 shows the result of 3 performance test of the original stove using jatropha oil as its fuel. The data shows that the stove has poor performance, indicate by low fuel rate, low efficiency (highest 10.5%), flame color (red), and produce smoke.

Fuel rate was very low because jatropha oil has higher viscosity, causing difficulty of oil flow through the wick to combustion zone. Flame color was red, indicate poor combustion process, which also causing smoke, and finally produce poor efficiency. Based on those data, pre-heating oil should be implemented to reduce jatropha oil viscosity, so the oil can be easily flowing through the wick to the combustion zone. To improve combustion process, which will (finally) produce higher efficiency, the air inlet (for combustion process) should be designed bigger than the original one, to ensure combustion air supply is enough to support perfect combustion process, produce blue flame, no smoke, and high stove's efficiency.

**Table 2.** Performance test result of original stove using jatropha oil as its fuel

| No | Parameter | Test result | | |
|---|---|---|---|---|
| | | 1 | 2 | 3 |
| 1 | Fuel rate (liter/hour) | 0.48 | 0.34 | 0.48 |
| 2 | Total fuel consumption (kg) | 0.35 | 0.25 | 0.35 |
| 3 | Total energy input (kcal) | 3330.6 | 2379 | 3330.6 |
| 4 | Efficiency of total system (%) | 9.08 | 8.99 | 10.56 |
| 5 | Heat capacity (kJ/minutes) | 190.7 | 192.4 | 189.4 |
| 6 | Duration of stove operation (minutes) | 5.16 | 4.16 | 5.16 |
| 7 | Turn off fire/stove (minutes) | 2.67 | 3.16 | 2.75 |
| 8 | Color of flame | red | red | red |
| 9 | Smoke identification | Produce smoke | Produce smoke | Produce smoke |

### *3.3. Stove modification*

Based on the data shows in Table 2, two kind of modification should be implemented. The first modification aim to decreasing jatropha oil viscosity, done by implementing 'pre-heating oil system'. While second modification aim to improving combustion process, done by modifying air inlet system.

'Pre-heating oil system' designed by added a kind of ditch (the place to put ethanol and burnt it to increase jatropha oil temperature) surrounding stove body (Figure 4). Heat energy produce by ethanol will increasing jatropha oil temperature up to 40 °C, so the oil viscosity will be decrease, and finally make the oil easier to flow through the wick to the combustion zone. Implementation of pre-heating oil system in the stove can be seen in the Figure 4.

**Figure 4.** Applying ditch surround (outside) the stove wall.

Combustion process improvement has been done by addition more air inlet holes around the stove wall, to ensure air supply was enough for perfect combustion process. The air inlet modification based on the AFR (air fuel ratio) calculation for jatropha oil combustion. Based on the calculation, AFR for jatropha oil combustion is 2.08 m3/s. Original stove has air inlet capacity only 0.51 m3/s, with total number of holes is 969 and total area 2.99 m2 (Figure 5), in the actual air flow velocity. To fulfil the air combustion demand, based on AFR (2.08 m3/s), in the actual air flow velocity, area needed (as total holes) is 15.81 m2. According to the number, air inlet holes has been designed as 950 small holes (original holes) and 12 bigger holes at the inner (first layer) wall , and 50 big holes at second layer wall. The result of air inlet modification shows in Figure 6.

**Figure 5.** Air inlet at the stove wall, before modification

**Figure 6.** Air inlet at the stove wall, after modification.

*3.4. Performance test of the modified stove*

By using jatropha oil as fuel, performance of the modified stove is better than the original stove, even though still poor compare to the performance of the stove by using kerosene as original fuel designed for the stove. The performance test result of both stove provided in the Table 3. Based on the performance data showed in Table 3, pre-heating system has been successfully increasing fuel rate from 0.43 kg/hour to 0.63 kg/hour, and reducing time needed for ignition from 4.8 minute to 3.1 minute.

**Table 3.** Stove performance, before and after modified

| No. | Parameter | Performance result | |
|---|---|---|---|
| | | Original design | After modified |
| 1 | Fuel rate (kg/hour) | 0.43 | 0.63 |
| 2 | Total fuel consumption (kg) | 0.32 | 0.48 |
| 3 | Total energy consumption (kcal) | 3013.40 | 3172 |
| 4 | Efficiency of the sistem (%) | 9.54 | 16.17 |
| 5 | Heat stove capacity (kJ/minute) | 190.83 | 202.70 |
| 6 | Time for ignition (minute) | 4.83 | 3.15 |
| 7 | Time of flame turn off (minute) | 2.86 | 2.13 |
| 8 | Flame's color | red | blue |
| 9 | Smoke indication | smoke | none |

In final result, the heat capacity of the stove also increasing due to the fuel rate increasing, indicate by the combustion temperature as shown in the Figure 9 and 10. Modification of air inlet also successfully improving combustion process, indicated by flame color which has been change from red to blue (Figure 11), and decreasing smoke. The better combustion process will be produce higher heat energy, higher combustion temperature, and finally increasing both of stove heat capacity and stove efficiency.

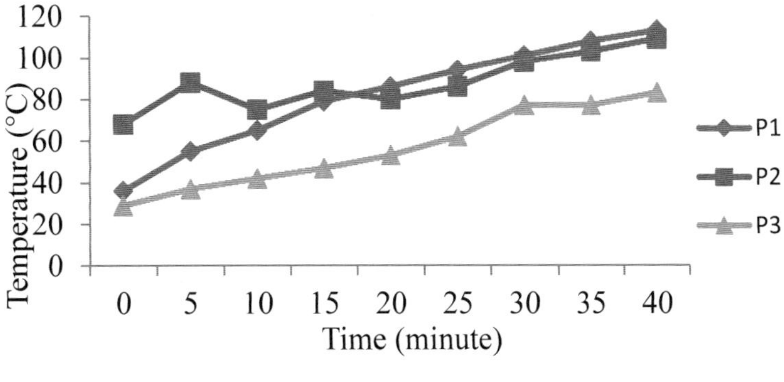


**Figure 7.** Fuel temperature, before stove modification

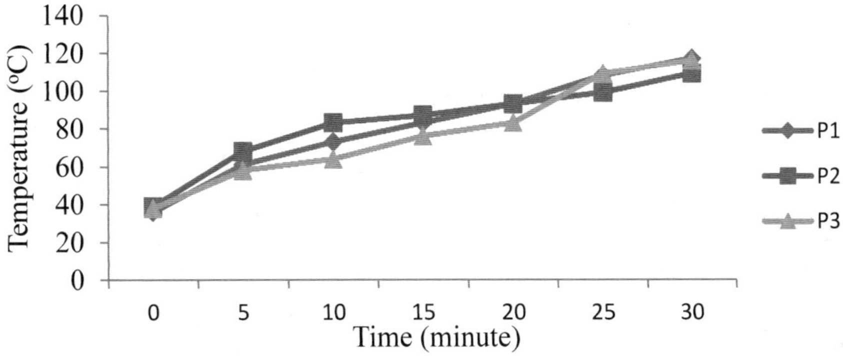


**Figure 8.** Fuel temperature, after stove modification

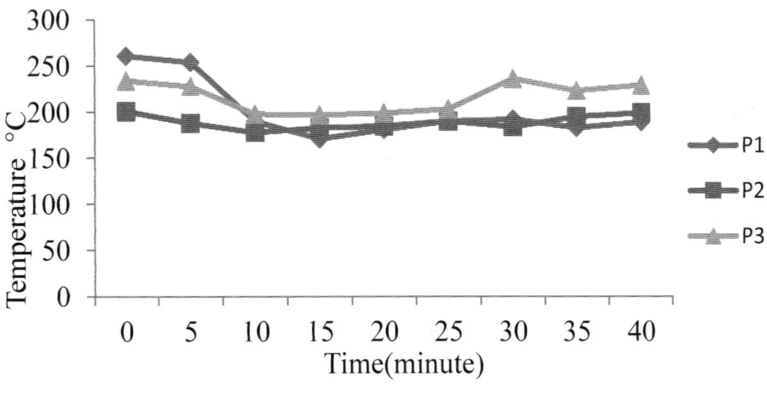


**Figure 9.** Flame temperature, before stove modification

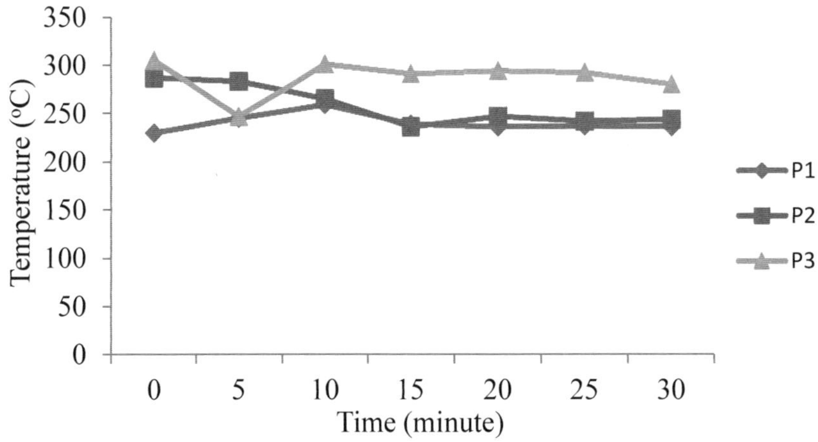


**Figure 10.** Flame temperature, after stove modification

(a) (b)

**Figure 11.** Flame color (a) before modification, and (b) after modification

## 4. Conclusion

Based on the data and observation, we can conclude as follows:

1) Common oil stove which has been sold in the market, which was designed for kerosene, is not available for jatropha oil usage due to its poor performance. To improve stove performance with jatropha as its fuel, modification should be applied to the stove, both of stove structure and stove operation. The modification should be based on the jatropha oil characteristic (oil viscosity and air supply for combustion process)
2) Modification of wick material was not successful increasing stove performance. Observation of wick capillarity and its flaming time shows that the original wick (polyester) is the best wick, even for jatropha oil usages.
3) Implementation of "pre-heating oil system" has been successful decreasing jatropha oil viscosity and made the oil easier flowing through the wick to the combustion zone, indicated by increasing of fuel combustion rate, from 0.43 kg/hour to 0.63 kg/hour. The system was designed as addition a ditch (for ethanol which was burning in order to increase jatropha oil temperature up to 40 °C before flowing through the wick), surround outside of stove wall.
4) Modification of number and size of air inlet holes in order to get amount of air supply based on the ARF calculation, has been successful improving combustion quality, indicate by the blue color of flame and no smoke produce on combustion process.

5) Overall result of all modification has been successful increasing stove performance for jatropha oil usages. The successful result was indicate by increasing of fuel combustion rate (from 0.43 kg/hour to 0.63 kg/hour), shorter ignition time (from 4.8 minute to 3.15 minute), improving combustion quality to produce blue flame, no smoke, and finally, increasing total system efficiency almost 70% (from 9.54% to 16.1%).

**References**

[1] Abdullah K. 1998. Energi dan Elektrifikasi Pertanian. Bogor (ID): IPB Pres
[2] Marlianto T. 2012. Modifikasi dan Unjuk Kerja Kompor Sumbu Tunggal Berbahan Bakar Minyak Bintaro [Thesis]. Bogor (ID): Bogor Agricultural University (IPB)
[3] Prastowo B. 2007. Kompor Berbahan Bakar Minyak Nabati. *Warta Penelitian dan Pengembangan Pertanian* 29(6):10-18.
[4] Raffaella. 2010. Kerosene Wick Stoves. [Internet]. [download on 2017 Jun 27]. http://www.hedon.info/KeroseneWickStoves
[5] Sunandar K. 2010. Kajian Kapilaritas Minyak Nabati pada Kompor Sumbu. [Disertation]. Bogor (ID): Graduate School, Institut Pertanian Bogor.
[6] Windra AM, Tamrin, Agus HB. 2007. Pengaruh Penambahan Serat Tembaga pada Sumbu Kompor terhadap Kinerja Kompor Minyak Jelantah. *Jurnal Keteknikan Pertanian Lampung* 2(3):173-184

IOP Conf. Series: Materials Science and Engineering **557** (2019) 012085 doi:10.1088/1757-899X/557/1/012085

# Analysis of Paddy Productivity Using NDVI and K-means Clustering in Cibarusah Jaya, Bekasi Regency

**Liyantono*, Yusup Almadani, Yusuf Adillah, Muhammad Maulana Yusuf, Muhammad Nyak Reza Mahbub, and Alvin Fatikhunnada**

Department of Mechanical and Biosystem Engineering, Faculty of Agricultural Engineering and Technology (FATETA), Bogor Agriculture University, Bogor, Indonesia.

*E-mail: liyantono@apps.ipb.ac.id

**Abstract.** Information about rice productivity is one of the references for government to maintain food availability. With remote sensing technology, rice productivity can be known faster. This research was conducted using UAV (Unmanned Aerial Vehicle) and Sentinel-2 Satellite. Sentinel-2 NDVI which has a low resolution with high resolution UAV images, both variables have similarity values and regression reaches 0.8. NDVI are grouped into 8 classes using k-means clustering based on the similarity of the waveforms of each data retrieval point. Based on characteristic of k-means classes, field which has earlier planting times and the location closer to the water source, allowing a higher paddy productivity. Further analysis was also carried out to get the best period to estimate paddy productivity using Sentinel-2 imagery. Sentinel-2 was chosen because it has a distance between data as far as 5 days, allowing it to be more accurate. The best time is obtained at 63 DAP (Days After Planting), which is when NDVI reaches its maximum state. The estimation model of rice productivity based on UAV has a high coefficient of determination compared to Sentinel-2 so that the relationship between maximum NDVI UAV and rice productivity is better than Sentinel-2.

## 1. Introduction

Paddy is widely use as a staple food for Indonesian people. The availability of safe rice is very important to achieve stable food security. Information about rice productivity has an important role for government to make national decisions. According to the Central Statistics Agency (2017), rice production in Indonesia is increasing every year. In 2011, Indonesia's rice production amounted to 65.7 million tons and continued to rise to 75.4 million tons in 2015. In estimating rice production in Indonesia, the Agriculture Ministry estimates rice production by considering the parameters of the area of planting / harvesting, the number of seeds spread by farmers and calculation of productivity by utilizing the institutional structure below, namely agricultural extension workers and agricultural staff from the Central Statistics Agency [5]. Most data retrieval is still done manually so it requires more time and effort. Therefore we need a more effective and efficient method of estimating rice productivity, one of which is using remote sensing technology. The development of methods for estimating rice productivity using this technology has been developed in Indonesia.

Remote sensing is science and art in obtaining information about an object area, or phenomenon through analysis of data obtained with a tool without a direct contact [3]. Remote sensing technology has been developed so rapidly and its application is increasingly widespread for various fields including agriculture [4]. Prediction of crop yields can be done by identifying the greenness (vegetation index) of a plant using a ratio between infrared and near infrared bands [1]. The ratio formula is known as the vegetation index which can provide an overview of the greenness of vegetation based on plant biomass and indicates plant health. The formulation that is commonly used as a parameter in determining the vegetation index with the level of crop production is NDVI (Normalized Difference Vegetation Index) [8]. NDVI has a range of values ranging from -1 to 1 where values -1 to 0 indicate objects or non-vegetation objects while values from 0 to 1 indicate the presence of vegetation objects. The value index listed in NDVI determines how much influence vegetated land cover, or not vegetation, and the potential distribution characteristics of an area. The higher the NDVI value, the more plants are in the ready-to-harvest phase, on the contrary the lower the value, the more unproductive plants [6].

Instruments commonly used in remote sensing are UAV (Unmanned Aerial Vehicle) and Satellite such as Sentinel-2. Each instrument has advantages and disadvantages. The UAV has a spatial resolution that is more detailed than Sentinel-2, however Sentinel-2 images are easier to obtain because its available for free every 5 days. In this research, a comparison was made between UAV and Sentinel-2 to estimating rice productivity based on NDVI waveform characteristics and regression from NDVI and paddy productivities.

## 2. Methodology

### *2.1. Data Collection*

The data collected in this research consist of 3 types of data including Sentinel-2 image data, UAV image data and field surveys data. Sentinel-2 image data downloaded from the sentinel official website via http: /scihub.copernicus.eu/. The image taken every 5 days from April 10, 2018 to August 13, 2018 according to one cycle of the second planting season in Cibarusah District, Bekasi Regency. Drone image taking is done every 16 days, from April 17, 2018 to August 8, 2018. Images are taken using Parrot Sequoia multispectral camera. Field surveys were conducted to obtain rice productivity data. Data is collected from farmers through direct interviews in the field

### *2.2. Research Method*

Sentinel-2 image data needs to be processed before further analysis. Data processing consists of several stages, which is summarized in the research methods (Figure 1)

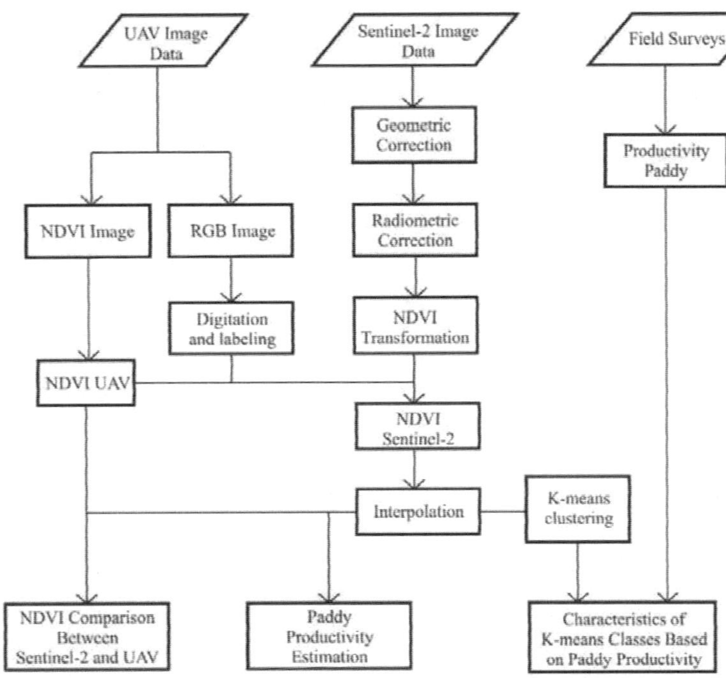


**Figure 1.** Flowchart of research method

Paddy vegetation index calculated using NDVI (Normalized Difference Vegetation Index) with these following equation:

$$NDVI = \frac{[\rho_{NIR} - \rho_{RED}]}{[\rho_{NIR} + \rho_{RED}]} \quad (1)$$

After NDVI obtained, data processed using k-means clustering. The purpose of using k-means is to generalize NDVI waveform. Main principle of k-means is to compile a prototype or center of mass (centroid) from a dimensionless data set n. k-means requires input of the number of classes as much as k and divides the amount of data n into class k based on the similarity with the class centroid [2]. Determinate of the centroid value in k-means using these equation [7]:

$$C_i = \frac{1}{M}\sum_{j=1}^{M} x_j \quad (2)$$

Data that has the shortest distance with the centroid will be entered into the class in the centroid. Calculation to determine the distance between centroid points with each point of the object using following equation:

$$d = \sqrt{(x1 - x2)^2 + (y1 - y2)^2} \quad (3)$$

## 3. Result and Discussion

### *3.1 Comparison Between NDVI Sentinel-2 and UAV*

NDVI UAV is the image that closest to the original conditions in the field. UAV use parrot sequoia camera to capture multispectral images from the field, pixel resolution is capable of reaching 5 cm x 5 cm for RGB and 12 cm x 12 cm for NDVI. Image took using UAV is also carried out at an altitude of 110 m so that the resulting image has minimal atmospheric interference. Comparison of NDVI waves from Sentinel-2 and drone (UAV) as in Figure 2.

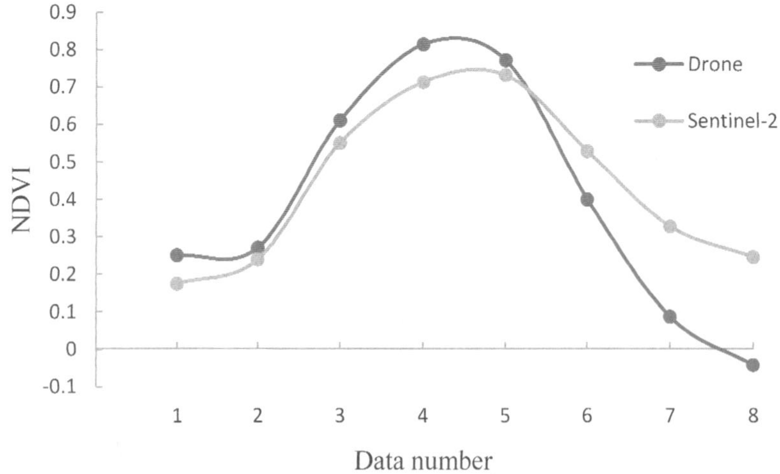


**Figure 2.** Comparison between NDVI Sentinel-2 and UAV

Figure 2 shows that the two images have different characteristics. In the vegetative phase, NDVI Sentinel-2 is lower than the NDVI UAV. Different case in generative phase, NDVI Sentinel-2 is higher than NDVI UAV. In this phase, NDVI UAV is abnormal. NDVI decrease significantly and in the last data number, average NDVI is below 0.1. This value generally occurs in the early phase of rice planting when there is still a pool of water or mud on the land. Validation carried out using RGB drones shows that there were no water body during data collection. This is most likely due to corruptness of the multispectral instrument in parrot sequoia.

**Table 1.** Equation and regression NDVI

| No | Equation of Regression | $R^2$ | No | Equation of Regression | $R^2$ |
|---|---|---|---|---|---|
| 1 | y = 0.448x + 0.0631 | 0.5412 | 5 | y = 0.6118x + 0.2591 | 0.4737 |
| 2 | y = 0.6342x + 0.0681 | 0.7312 | 6 | y = 0.3767x + 0.3773 | 0.8025 |
| 3 | y = 0.6959x + 0.1256 | 0.7909 | 7 | y = 0.3953x + 0.2928 | 0.7376 |
| 4 | y = 0.8015x + 0.0601 | 0.6127 | 8 | y = 0.5086x + 0.2672 | 0.7967 |

From Table 1, the slope value for each take is quite diverse. The highest slope is 0.8015 which is at the 4th take on June 2, 2018. The slope value close to 1 indicates that the value of x approaches the y value. $R^2$ of the equation model has the smallest value on June 22, 2018, which is 0.4737. While the biggest value is on July 7, 2018, which is 0.8025. This shows that the relationship between NDVI drone and NDVI Sentinel-2 has a large percentage of up to 80.25%

*3.2 Paddy Productivity Based on Characteristics of K-means Classes*

NDVI data in one season would be grouping into 8 classes using k-mean clustering. The results of the classification as presented in the figure 3.

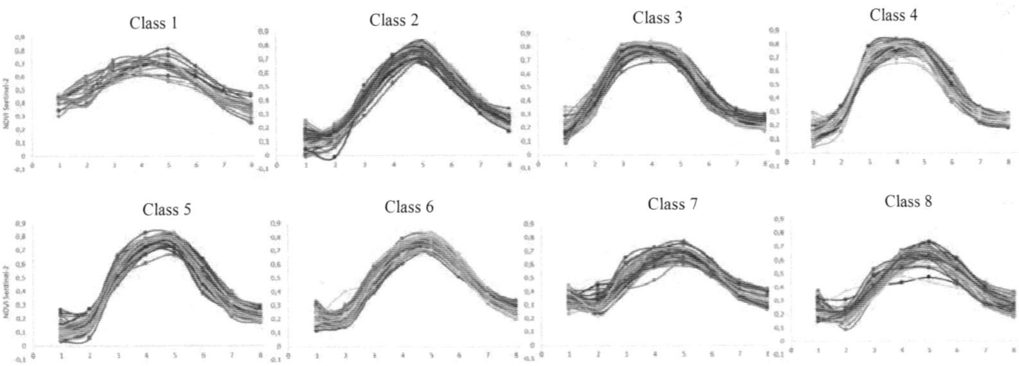


**Figure 3.** Results of NDVI wave grouping based on k-means clustering

Each 8 k-means class has varied characteristics. Class 1 has high NDVI value and does not have a significant change, this class identified as a weed and plots that are disturbed by other vegetation. Class 3 and 4 has the characteristics of NDVI which immediately increase in the first time, this indicates a change from the plot that has been done by processing the land to the plots that have been planted with rice. Class 3 and 4 peaks also reach the peak of NDVI faster because rice growth is better. For other classes, NDVI will go down first after it has increased. There is a change from the plot that has not been carried out till the soil is cultivated, becoming a plot that has been treated by the soil, and then becomes a plot that has been planted with rice. Map those who have not undergone tillage have a higher NDVI because they contain weeds or former planting. NDVI will go down after soil lubrication, higher water content causes NVDI to decrease. After rice is planted, NDVI will increase because there are green plants that grow in the plot.

From the survey result that has been conducted, productivity data are grouped according to the k-means class. Average rice productivity for each class as presented in Table 2.

**Table 2.** Rice productivity for each class

| Classes | Average rice productivity (ton/ha) | Classes | Average rice productivity (ton/ha) |
|---|---|---|---|
| Class 2 | 0.841 | Class 5 | 1.576 |
| Class 6 | 0.919 | Class 3 | 2.122 |
| Class 8 | 0.938 | Class 4 | 3.280 |
| Class 7 | 1.324 | | |

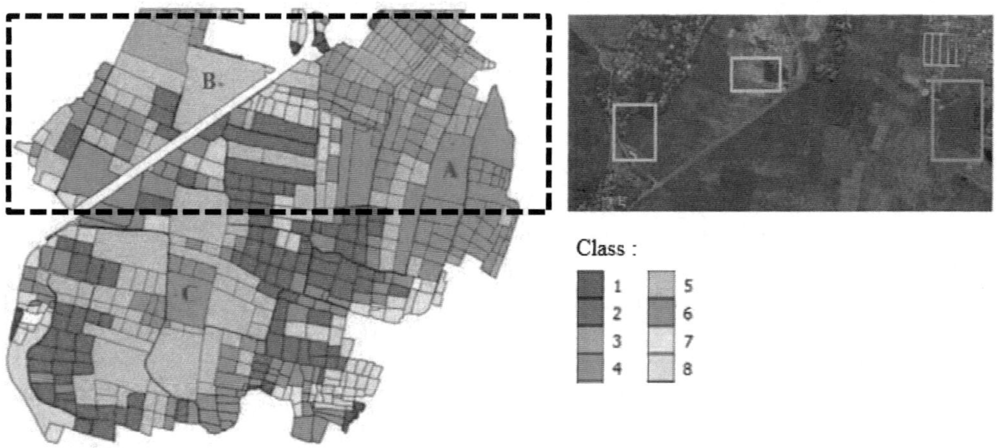


**Figure 4.** Spatial distribution of k-means classes

Figure 4 shows the class distribution in the rice field. Each class is given a different color indicator, the color indicator is adjusted to the level of productivity of k-means class. Class 4 which has the highest productivity is given a green color, class 1 which has the lowest productivity is given a red color, and for other classes the color indicator adjusts. The majority of Class 3 and 4 are in group A which is adjacent to water sources, so that the water needs for rice can be fulfilled. This explains why classes 3 and 4 experience an increase in NDVI faster than other classes which far from water sources.

*3.3 Paddy Productivity Estimation*

The best period to estimate paddy productivity is sought based on the highest regression relationship NDVI with rice productivity. NDVI used from Sentinel-2 images, the data has a range of 5 days so that it has smaller error if interpolated to NDVI each day or DAP (Days After Planting). The regression value between NDVI for each DAP and rice productivity is presented in Figure 5

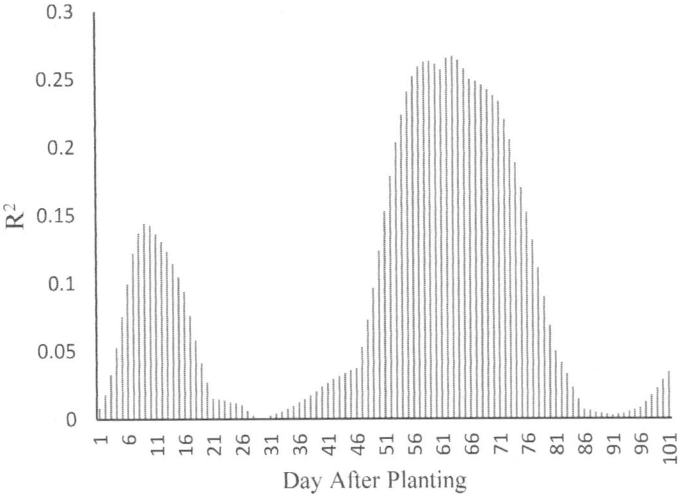


**Figure 5.** Regression paddy productivity and NDVI each DAP

The best period in estimating rice productivity is 63 DAP with a regression value is 0.27. In this period, NDVI Sentinel-2 reaches its maximum state. This is strengthened by the pattern of average NDVI UAV value of paddy with high productivity and paddy with low productivity which is shown in Figure 6. Based on Figure 6, NDVI value between paddy with high productivity and paddy with low productivity are significanly different when it reach maximum state.

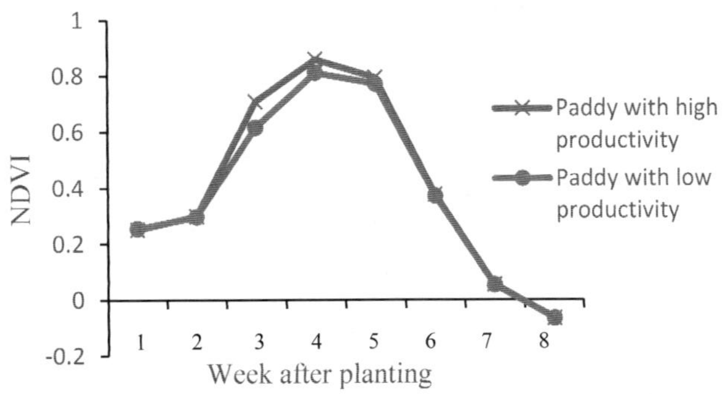


**Figure 6.** Pattern comparison of ndvi value on paddy with high productivity and paddy with low productivity

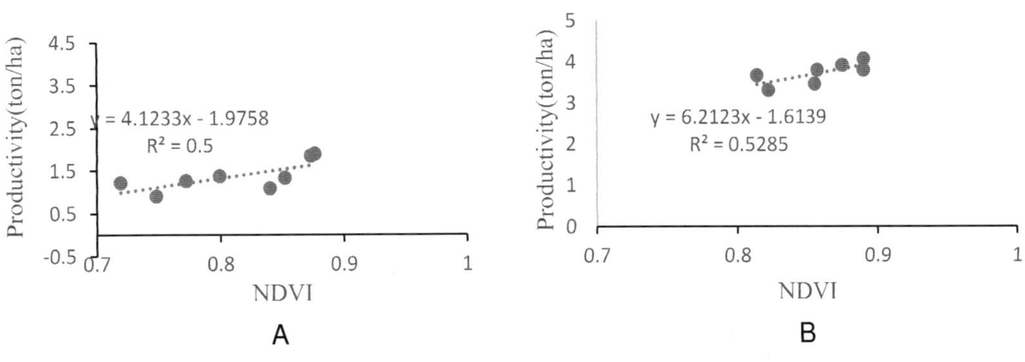


**Figure 7.** Relation between NDVI and productivity, A) low productivity, B) high productivity

Then if the maximum values of NDVI on every K-means classes connected with the productivity, it will make a linear line which is shown at Figure 7. Rice with high productivity has coeficient of determination $R^2 = 0.5$ and $R = 0.71$. Meanwhile, rice with low productivity has coeficient of determination and correlation $R^2 = 0.53$, $R = 0.73$. Based on Figure 6 the productivity for rice with low productivity can be predicted using equation $y = 4.1233x - 1.9758$ and rice with high productivity can be predicted using equation $y = 6.2123x - 1.6139$.

## 4. Conclusion

Sentinel-2 images have NDVI values that approach high-resolution NDVI images such as UAV images. The regression formed between the two variables reaches 0.8 and the linear equation slope value is 0.8. In the analysis of rice productivity, NDVI waveforms are grouped using K-means clustering. Class 3 and 4 have the highest productivity. Based on the waveform and the spatial distribution of k-means class, it is known factors that affect of high paddy productivity are the earlier planting times and the adequacy of water requirements for rice growth. The best period to estimate paddy productivity is 63 DAP (Days After Planting) that NDVI reach its maximum state. The regression formed between productivity and maximum value of NDVI Sentinel-2 image is 0.27, lower than NDVI UAV which has regression 0.53 with estimation model is $y = 6.2123x - 1.6139$.

## References

[1] Affan MJ. 2002. *Assessment of Forest Fire Hazard Levels Based on the Vegetation Index and KBDI* (Bogor : Institut Pertanian Bogor)
[2] Giyanto H. 2008. *Application of the k-means, k-medoid, gath geva clustering algorithm* (Yogyakarta: Thesis in Universitas Gadjah Mada)
[3] Lillesand TM, Kiefer RW, Chipman J 2008 *Remote Sensing and Image Interpretation, 6th ed* (New York : J Willey)
[4] Martono DN. 2008. *Application of Remote Sensing Technology and Validation Test for Detection of Spread of Rice Fields and Land Use / Cover* (Jakarta : PUSDATA LAPAN)
[5] Napitupulu TEM. 1998. *Estimation System for Production Forecast Results in the Context of Safeguarding National Food Production. Workshop on Rice Monitoring and Prediction Systems in Indonesia* (Jakarta : Sari Project – BPPT)
[6] Pradipta D. 2012. *Data Analysis of Time Series NDVI-Spot Vegetation for Rice Plants (case study: Karawang district)* (Bogor : Institut Pertanian Bogor)
[7] Wakhidah N. 2010. Clustering using k-means algorithm *Jurnal Transformatika* 8 (1) 33-39
[8] Wahyunto W, Heryanto B. 2006. Estimating the Productivity of Paddy Crops through Analysis of Satellite Images. *Warta Informatika Pertanian* 15 (2) 853-859

# Prediction of soluble solid content, vitamin C, total acid and firmness in astringent persimmon (*Diospyros kaki* L.) cv. Rendeu using NIR spectroscopy

**Nur Hasnah Ar[1], Yohanes Aris Purwanto[*1,2], I Wayan Budiastra[1], Sobir[2]**

[1] Department of Mechanical and Biosystem Engineering, Bogor Agricultural University
[2] Center of Tropical Horticulture Studies, Bogor Agricultural University

*E-mail: arispurwanto@apps.ipb.ac.id

**Abstract.** Internal content is one of the important quality criteria in fruit. Quantitative analysis is carried out by using a chemical method. This process is destructive, high cost and time consuming, thus not suitable for real-time measurement. The objective of this study was to evaluate the possibility of the use of near-infrared (NIR) spectroscopy as a rapid method of predicting the internal quantitative quality of persimmon fruit such as solid soluble content (SSC), vitamin C, total acid and firmness. The spectra pre-processing and Partial Least Square (PLS) method were utilized to develop a calibration model. The result showed that the best calibration model was SSC with (MSC) multiplicative scatter correction and 17 PLS factors; vitamin C with MSC and 16 PLS factors; total acid with MSC and 12 PLS factors; firmness with n01 (normalization between 0-1) and 12 PLS factors. This study demonstrated the possibility the use of NIR spectroscopy for rapid and non-destructive measurement of SSC and firmness contents in astringent persimmon. The low accuracy in predicting vitamin C and total acid was obtained due to a limitation of those the contents in persimmon.

## 1. Introduction

Persimmon (*Diospyros kaki* L.) is one of the horticultural products grown in the highlands [1]. Persimmon fruit is subtropical fruit originated in China and entered the territory of Indonesia at the beginning of the 20th century.The spread of persimmon in Indonesia includes Sumatra (Tanah Karo, Solok, Brastagi, and Toba) and Java (Cikajang, Indonesia, Bayongbong, Majalengka, Ciloto, Selo, Boyolali, Magetan, Malang-Tirtoyudo, Temanggung, and Junggo) [2]. Persimmon fruit can be consumed as a fresh fruit or made into processed products such as jelly, jam, ice cream and chips.

Persimmon can be classified into two general types, astringent and non-astringent types [2]. Persimmon describes with three rates of aging based on the days after bloom is 150 days, 170 days, and 190 days after bloom [2]. Persimmon shaped round like tomatoes, bright fruit and the fruit skin was red-orange. With the appearance of the beautiful fruit color persimmon have prospects to be developed and become commodities export.

The degree of freshness, hardness, age, size and internal content are the quality criteria. One of the criteria that matter is the quality of the internal content (including soluble solid content (SSC), total acids, vitamin C). Traditionally, determining of SSC, vitamin C, total acid and firmness contents in persimmon fruits are conducted qualitatively based on subjective measurement by looking the fruit appearance. Standard methods for measuring the internal quality of persimmon are destructive and thus the same fruit cannot be continually monitored. This method is not efficient since it is time-consuming and destructive.

Near-infrared (NIR) spectroscopy is an electromagnetic wave which has the wavelength of 780-2500nm that serves to identify the chemical component in the quantitative and qualitative analysis. Most of the NIR spectra dominated by hydrogen bonding causes the hydrogen atom is the more potent. Hydrogen bonds can be either C-H, N-H, S-H, or O-H. Instruments used in NIR spectroscopy are light sources, detectors, optical detectors, sampling and data processing system.

Near Infrared (NIR) spectroscopy is a nondestructive approach [3], accurate, and fast which can replace chemical method [4]. NIR spectroscopy can be used for nondestructive measurements of fruit components, such as water content, acid, carbohydrate, protein and fat [5], starch and SSC [6], moisture content and peel hardness in mangosteen [7], lignin and moisture content in the wood [8].

Near infrared (720-1100mm) is used to determine ripeness for oil palm fresh fruit bunch [9]. The measurements are fast, highly precise, need no particular sample pretreatment, or chemical additives. Several studies have shown the ability of NIR spectroscopy to determine the bioactive component of agricultural products such as a) determining of SSC and acidity of loquats base on FT-NIR spectroscopy using Partial Least Square (PLS) method and first and second derivative, multiple scatter correction (MSC) and standard normal variate (SNV) pre-processing [10], b) determining of firmness and soluble tannin content of Mopashi persimmon using the second derivative based PLS method [11], c) determining of varieties persimmon fruit using PCA method [12].

The objectives of this study were to develop an NIR spectroscopy calibration model to predict SSC, vitamin C, total acid, and firmness in astringent persimmon cv. Rendeu using PLS method with different pre-processing spectral treatment.

## 2. Material and Method

### *2.1. Sample Preparation and Spectra Acquisition*

The fresh astringent persimmon cv. Rendeu were harvested from a farmer persimmon in Garut, West Java, Indonesia. The 147 number of samples were divided randomly into two groups. A 98 sample of fruits were used as calibration set and the rest of the validation set. NIR spectrometer from Buchi NIRFlex N-500 Fiber optics solids which has a gun type probe and wavelength of 1000-2500nm with interval 0.4nm was used for spectra acquisition using NIRware 1.2 (Buchi Labortechnik AG, Flawil, Switzerland) was used to operate the instrument. Measurements of each sample were conducted at three different positions along a transect (near the top, middle, and near the bottom) of each fruit.

### *2.2. Chemical Analysis*

Firmness was measured using hardness tester (FHR-1). SSC was estimated from a single refractometer (PAL-1 Atago, Japan) reading taken from the combined juice extracted from the three penetrometer holes used for firmness measurement (near the top, middle, and near the bottom) of each fruit. Vitamin C and total acid were measured on approximately 10g by fresh fruit weight, in term of the malic acid equivalent. The tissue was macerated and the juice, filtered from the homogenate, titrated using 0.01N Iodium for vitamin C and 0.1N NaOH for total acid. Subsequent analysis of the chemical analysis result used the average value of the three replicate measurements.

### *2.3. Data Analysis*

The data used in this analysis was of the fresh astringent persimmon cv. Rendeu which was still in good (quality) condition, i.e., those acceptable to consumers. The total number of fruit was 147. The two-thirds of the data set was used for calibration, and the remaining one-third used for validation [7].

NIR spectra obtained from NIR instrument contain not only the information of the sample but also the background and noise. Therefore, the spectral pre-processing treatment method is required to get reliable, accurate and stable calibration model [13]. This method was utilized to minimize the contribution of information which was not relevant to spectrum to develop reliable model [14].

This study used both of original and several spectral pre-processing methods. Those were normalization between 0 and 1 (n01), smoothing (sa3), first derivative Savitzky-Golay 9 points (dg1), multiplicative scatter correction (MSC), the combination of n01 with dg1 (n01,dg1) and a combination of dg1 with multiplicative scatter correction (dg1,MSC) [15]. Normalization was used for creating scale sample all of the data were at approximately the same level based on area, mean, hose, maximum, peak, and vector units. Normalization of functioning spectral transforms spectrum into units of length so that the value of reflectance at a longer range. Normalization was used to eliminate multiplicative spectra effects [6]. Savizky-Golay smoothing (sa3) is the method often used for removing noise. Smoothing was also utilized in the optimization of the signal-to-noise rate. In general, this method combined with other data to perform the removal of noise. The algorithms for derivative are direct differentiation and Savitzky-Golay [13]. The first derivatives can be used to eliminate additive baseline effects while second derivatives were used to remove sloped additive baselines [6]. The first derivatives can be used to remove the background and enhance the resolution of spectral processing of derivatives to work separating the spectral that overlaps on the original ones. The resulting reflectances value will be reduced to smaller to clarify each of the peaks and valleys of the spectrum. MSC was used for the correction of scattering light by different particle sizes. These techniques were used to correct for additive and multiplicative effects in the spectral [16].

### *2.4. Calibration Method*

NIR spectra usually contain unselective and widely overlapped bands. Therefore, quantification analysis requires chemometrics analysis such as PLS, principal component regression (PCR) or multiple linear regression (MLR) [17]. The PLS regression method is a very popular method for chemometric analysis [18]. In this study, selection the number of optimum PLS factors was done based on consistency value. Selected PLS factor has consistency value between 80-110%. The number of PLS factors was 1 to 20. The value of consistency was used to ensure that the number of selected PLS factors was optimal [19]. Statistical parameters used to evaluate the resulted model were standard error calibration set (SEC), standard error of prediction set (SEP), coefficient correlation (r), the ratio of prediction to deviation (RPD), and coefficient of variance (CV) [17].

## 3. Results and Discussion

### *3.1. Data of Destructive Persimmon*

Table 1 shows the descriptive statistics summary for SSC, vitamin C, total acid and firmness in the calibration and validation set. Calibration and validation models, based on the correlation between NIR data and destructively determined SSC, vitamin c, total acid and firmness contents, were used to predict. The calibration and validation models were developed using the PLS method. The data employed in this analysis was of that fruit which was still in good (quality) condition, i.e., those acceptable to consumers. The total number of fruit was 147. Two-thirds of the data set was used for calibration, and the remaining one-third used for validation [7].

IOP Conf. Series: Materials Science and Engineering **557** (2019) 012086 doi:10.1088/1757-899X/557/1/012086

**Table 1.** Descriptive SSC, Vitamin C, total acid and firmness data for calibration and validation of NIR by PLS method

| Statistics description | SSC (°Brix) | Vitamin C (mg/100gr) | Total Acid (%) | Firmness Kgf |
|---|---|---|---|---|
| No. Of data | 429 | 369 | 423 | 432 |
| Min(%) | 14.10 | 0.79 | 0.04 | 0.08 |
| Max(%) | 31.33 | 3.64 | 0.32 | 0.95 |
| Avr.(%) | 22.85 | 1.9 | 0.12 | 0.74 |
| Std.dev.(%) | 3.95 | 0.5 | 0.06 | 0.2 |
| CV | 11.26 | 34.05 | 48.34 | 12 |

### *3.2. Analysis of NIR Reflectance*

The original reflectance spectra of persimmon samples (Figure 1) reveals some valley in region 1000-2500nm which indicates high noise. These valleys exist because molecule structures of persimmon contain many hydric groups (such as C-H and O-H). Spectrum data pretreatment by normalization into a 0-1 range was necessary to decrease the error value due to different in particle size and the range of reflectance value. After normalization, the reflectance data spectrum was transformed into absorbance log (1/reflectance). In the absorbance spectrum of persimmon, the peak was in the range of 1160-1200nm, 1140-1460nm, 1920-1940nm which reflect water content, 1160-1200nm, 1780nm, 1920-1940nm which reflect carbohydrate content such as fructose, sucrose, and galactose, 1440-1460nm, 1160-1200nm, 1440-1460nm, 2240-2280nm wich reflects total acid content [13].

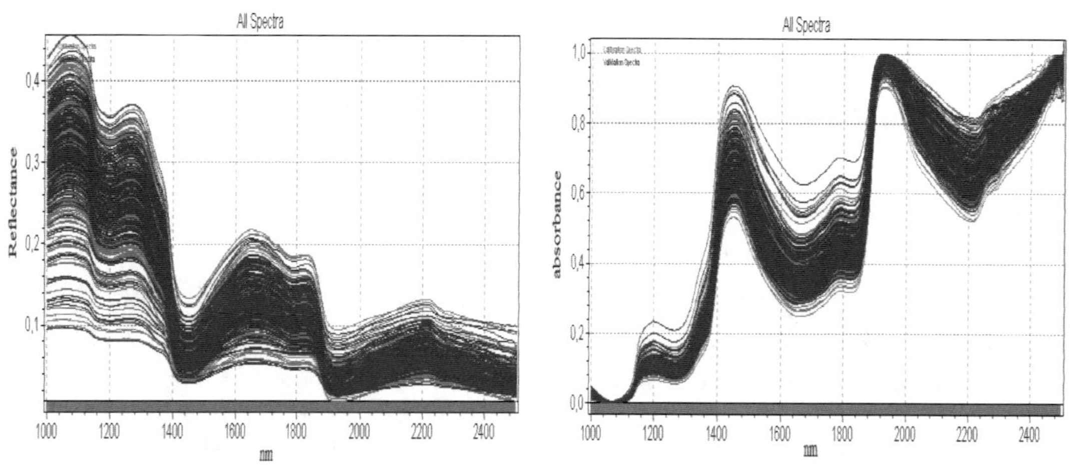


**Figure 1.** The original spectrum of persimmon (a) reflectance; (b) absorbance after normalization

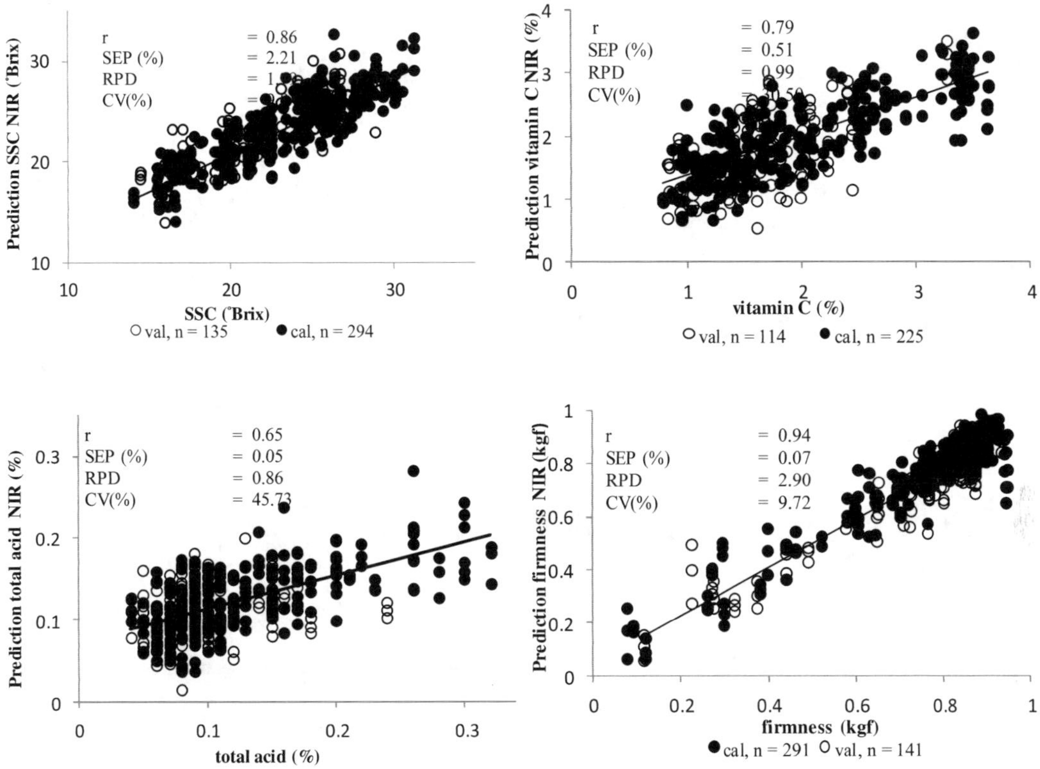


**Figure 2**. Scatter plot of measured the best calibration model to predict persimmon contents: SSC with MSC and 17 PLS factors, vitamin C with MSC and 16 PLS factors, total acid with MSC and 12 PLS factors, firmness with n01 and 12 PLS factors

*C. Prediction of SSC, Vitamin C, Total Acid and Firmness in Persimmon*

PLS is an effective method to reduce the dimension in NIR spectroscopy analysis. The spectra information of sample component is showed of PLS factors is essential to reduce noise and use the whole of spectral information [13]. The performance of parameters of the model using different pre-processing methods to predict SSC, vitamin C, total acid and firmness in persimmon is shown in Table 2.

Pre-processing of data is carried out before modeling will improve the calibration results. Application of NIR data processing and the number of the main components of optimal spectral NIR calibration model PLS applied to obtain the best in predicting the chemical content of material [17]. The data processing results of the spectrum are used in this research (Figure 3).

This pre-processing MSC produce the best calibration to predict SSC with higher accuracy and precision than other pre-processing (r = 0.86, SEP = 2.21%, RPD = 1.79 and CV = 9.84%). It indicated that pre-processing was important before developing a model as it would improve the accuracy and precision of calibration model [20, 21].

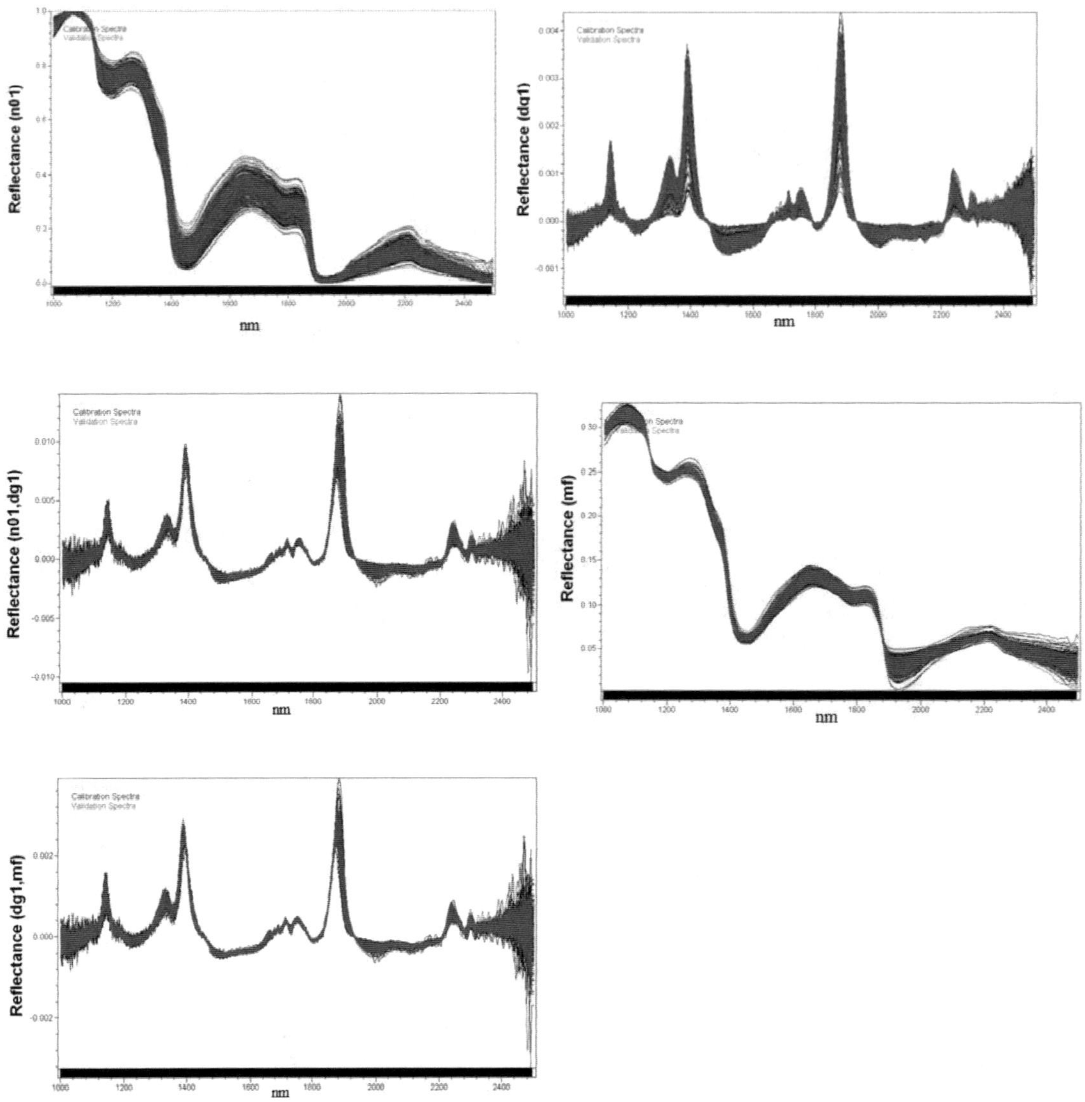


**Figure 3.** Pre-processing data NIR persimmon with (a) normalization (n01), (b) first derivatif Savitzky-golay (dg1), (c) combination n01 and dg1, (d) Multiplicative Scatter Correction (MSC), (e) combination dg1 and Multiplicative Scatter Correction (MSC).

The vitamin C content calibration of persimmon resulted that spectral pre-processing of MSC produced the best calibration model compared to other spectral pre-processing (r = 0.79, SEP = 0.51%, RPD = 0.99, CV = 30.50%). Spectra pre-processing with MSC produced the best calibration model compared to other spectral pre-processing for total acid content calibration. This pre-processing reduced coefficient correlation between combination pre-processing (n01, dg1) model with MSC pre-processing from 0.71 to 0.65 but RPD increased from 0.82 to 0.86. In this study, NIR spectroscopy failed to predict vitamin C and total acid of Rendeu persimmon fruit (RPD < 1.5). The concentration of acids in most fruit and vegetables is smaller than sugar and possibly too small to affect the NIR spectrum significantly. Sugar the dominance of water absorption bands in fresh produce also makes minor contents difficult to measured [22].

IOP Conf. Series: Materials Science and Engineering **557** (2019) 012086 doi:10.1088/1757-899X/557/1/012086

**Table 2.** Parameter of NIR calibration and validation model for SSC, vitamin C, total acid contents and firmness in persimmon

| Content | Pre-processing | Consistency (%) | Factors | Calibration-set | | Validation-set | | |
|---|---|---|---|---|---|---|---|---|
| | | | | r | SEC (%) | SEP (%) | RPD | CV (%) |
| SSC | original | 93.6 | 19 | 0.87 | 2.07 | 2.21 | 1.79 | 9.86 |
| | n01 | 98.4 | 17 | 0.85 | 2.25 | 2.29 | 1.72 | 10.2 |
| | dg1 | 84.4 | 9 | 0.84 | 2.28 | 2.7 | 1.46 | 12 |
| | msc | 97.3 | 17 | 0.86 | 2.15 | 2.21 | 1.79 | 9.84 |
| | n01+dg1 | 86.3 | 9 | 0.85 | 2.23 | 2.58 | 1.53 | 11.5 |
| | dg1+msc | 88 | 9 | 0.85 | 2.22 | 2.52 | 1.56 | 11.3 |
| Vitamin C | original | 89.3 | 17 | 0.79 | 0.5 | 0.56 | 0.89 | 33.8 |
| | n01 | 80.3 | 19 | 0.83 | 0.46 | 0.57 | 0.88 | 34.4 |
| | dg1 | 86.7 | 9 | 0.78 | 0.51 | 0.59 | 0.84 | 35.9 |
| | msc | 99.1 | 16 | 0.79 | 0.5 | 0.51 | 0.99 | 30.5 |
| | n01+dg1 | 91.2 | 9 | 0.78 | 0.51 | 0.56 | 0.89 | 34.1 |
| | dg1+msc | 85.9 | 9 | 0.79 | 0.5 | 0.58 | 0.86 | 35 |
| Total acid | original | 80.9 | 12 | 0.75 | 0.04 | 0.05 | 0.81 | 48.7 |
| | n01 | 95.4 | 13 | 0.67 | 0.05 | 0.05 | 0.86 | 46 |
| | dg1 | 77.9 | 7 | 0.75 | 0.04 | 0.05 | 0.78 | 50.5 |
| | msc | 98 | 12 | 0.65 | 0.05 | 0.05 | 0.86 | 45.7 |
| | n01+dg1 | 84.8 | 7 | 0.72 | 0.04 | 0.05 | 0.82 | 48.3 |
| | dg1+msc | 83.3 | 7 | 0.72 | 0.04 | 0.05 | 0.8 | 49 |
| Firmness | original | 107 | 6 | 0.82 | 0.1 | 0.1 | 2.09 | 13.5 |
| | n01 | 90 | 12 | 0.94 | 0.06 | 0.07 | 2.9 | 9.72 |
| | dg1 | 83.6 | 6 | 0.89 | 0.08 | 0.1 | 2.09 | 13.5 |
| | msc | 84.9 | 12 | 0.94 | 0.06 | 0.07 | 2.74 | 10.3 |
| | n01+dg1 | 84.1 | 6 | 0.92 | 0.07 | 0.09 | 2.35 | 12 |
| | dg1+msc | 74.2 | 6 | 0.9 | 0.08 | 0.1 | 1.94 | 14.5 |

The result of firmness content calibration of persimmon that spectral pre-processing of n01 increased correlation coefficient firmness content between no pre-processing with pre-processing from 0.82 to 0.94 while SEP value decreased from 0.10% to 0.07%. The model with an application of pre-processing n01 method showed higher accuracy and precision than the original model and other pre-processing.

*D. Evaluation of Prediction*

The best calibration model for predicting SSC, vitamin C and total acid contents presented by scatter plot in Fig. 2. The result showed that the best calibration models were SSC with (MSC) multiplicative scatter correction and 17 PLS factors; vitamin C with MSC and 16 PLS factors; total acid with MSC and 12 PLS factors; firmness with n01 (normalization between 0-1) and 12 PLS factors. The model provided a small value of SEC and SEP compared to other models. Since SEP value was not greater than two times of SEC, overfitting can be prevented [17]. The slight different values of SEC and SEP indicated that the model has high precision.

The good model has the high correlation coefficient (r> 0.75). RPD values below 1.5 indicated that the calibration could not be used, between 1.5 and 2.0 is possible to distinguish high and low values [23]. Based on obtaining r, SEC, SEP, RPD and CV values, the resulted model to was possible to predict the persimmon SSC and firmness.

## 4. Conclusion

The calibration and validation model of NIR spectroscopy has been developed to predict SSC, vitamin C, total acid, and firmness in astringent persimmon cv. Reundeu using PLS method with different pre-processing spectra treatment. The best calibration model was obtained for: SSC with MSC and 17 PLS factors and firmness with n01 and 12 PLS factors were applied. It has a little SEC, SEP and CV, high r and RPD. In this study, the accuracy of the prediction of vitamin C and total acid using NIR spectroscopy was little due to a limitation of the content of those parameters in persimmon.

## Acknowledgment

The authors wish to express their appreciation for the financial support from Ministry of Research, Technology and Higher Education of the Republic of Indonesia through Competency Research Grant No. 451/IT3.11/PN/2016.

## References

[1] E.W.M. Verheij, and R.E. Coronel, "Edible fruits and nuts," Plant Resources of South-East Asia (Prosea) No. 2: 298-301, 1991.

[2] Baswarsiati, Suhardi and D. Rahmawati, Potensi dan Wilayah Pengembangan Kesemek Junggo. Buletin Plasma Nuftah, Vol. 12 No. 2, 2006.

[3] L.C. Lee, C.Y. Liong, and A.A. Jemain, "Applying Fourier-transform infrared spectroscopy and self organizing maps for forensic classification of white-copy papers," International Journal on Advanced Science, Engineering and Information Technology Vol. 4 No. 6: 1033-1039, 2016.

[4] R.F. Gallego, J.M.H. Hierro, J.C.R Gonzalo, and M.T.E. Bailón, "Feasibility study on the use of near-infrared spectroscopy to determine flavonol in grape seeds," Talanta 82: 1778-1783, 2010.

[5] C. Pasquini, "Near infrared spectroscopy: fundamentals, practical aspects and analytical applications," J Braz Chem Soc, 14: 198-219, 2003.

[6] Y.A. Purwanto, I.W. Budiastra, E. Darmawati, and N. Arifiya, "Measurement of starch and soluble solid content in papaya using near infrared spectroscopy," Journal of Chemical and Pharmaceutical Research 7(6): 112-116, 2015.

[7] U. Ahmad, Sutrisno, Y.A. Purwanto, I.W. Budiastra, Y. Makino, S. Oshita, Y. Kawagoe, S. Kuroki, and D.D. Novita, "Prediction of hardness development in mangosteen peel using NIR spectroscopy during low-temperature storage," Engineering in Agriculture, Environment and Food 7: 86-90, 2014.

[8] L. Karlinasari, Y.A Purwanto, M. Sabed, I.N J, and Wistara, "Near-infrared (NIR) spectroscopy for estimating the chemical composition of (Acacia mangium wills.) wood," J Indian acad Wood Sci 11(2): 162-167, 2014.

[9] D. Cherie, S. Herodian, U. Ahmad, T. Mandang, and M. Makky, "Optical characteristics of oil palm fresh bunch (FFB) under three spectrum regions influence for harvest decision," International Journal on Advanced Science, Engineering and Information Technology Vol. 5 No. 3: 255-263, 2015.

[10] F.U Xia-ping, L.I Jian-ping, Ying, and Zhou, "Determination of soluble solid content and acidity of loquats base on FT-NIR spectroscopy," Journal of Zhejiang University Science B. ISSN 167-1581: 120-125, 2009.

[11] Z. Peng, Y. Xue, J. Li, X. Feng, and B. Wang, "Research on nondestructive measurement of firmness and soluble tannin content of 'mopanshi' persimmon using vis/nir diffuse reflection spectroscopy," Acta Horticulture 996, 2011.

[12] Z. Shunjuang, J. Dengfei, and Z. Haihong, "NIR spectroscopy identification of persimmon varieties based on PCA-SVM," IFIP Advances in Information and Communication Technology, Vol. 345: 118-123, 2011.

[13] H. Cen, and Y. He, Y, "Theory and application of near infrared reflectance spectroscopy in determination of food quality," Trends in Food Science & Technology 18: 72-83, 2007.

[14] M. Blanco, and I. Villarroya, "NIR Spectroscopy: a rapid-response analytical tool," Trends in analytical chemistry 21: 240-250, 2002.

[15] Y.A. Purwanto, H.P. Sari, and I.W. Budiastra, "Effects of preprocessing techniques in developing a calibration model for soluble solid and acidity in 'Gedong Gincu' mango using NIR spectroscopy,'' International Journal of Engineering Technology Vol. 7 No. 5: 1921-1927, 2015.

[16] Q.S. Chen, J.W. Zhao, S. Caitep, Z.M Guo, "Simultaneous analysis of main catechin content in green tea (Camellia sinensis (L.) by Fourier transform near infrared reflectance (FT-NIR) spectroscopy," Food Chemistry 113: 1272-1277, 2009.

[17] [Andasuryani, Y.A. Purwanto, I.W. Budiastra, and K. Syamsu, "Determination of catechin content in gambir powder from dried gambir leaves quickly using FT NIR PLS model," International Journal on Advanced Science, Engineering and Information Technology Vol. 4 No. 5: 2088-5334, 2014.

[18] H. Buddenbaum, and M. Steffens, "The effects of spectral pre-treatment on chemometric analysis of soil profiles using laboratory imaging spectroscopy," Applied and Environmental Soil Science Vol. 2012: 1-12, 2012.

[19] E. Elfadl, C. Reinbrecht, and W. Claupein, "Development of near infrared reflectance spectroscopy (NIRS) calibration model for estimation of oil content in a worldwide safflower germplasm collection," International Journal of Plant Production 4: 259-270, 2010.

[20] Q. Ouyang, Q. Chen, J. Zhao, and H. Lin, "Determination of amino acid nitrogen in soy sauce using near-infrared spectroscopy combined with characteristic variable selection and extreme learning machine," Food and Bioprocess Technology 6(9): 2486-2493, 2013.

[21] T. Udelhoven, C. Emmerling, and T. Jarmer, "Quantitative analysis of soil chemical properties with diffuse reflectance spectrometry and partial least-square regression: A feasibility study," Plant and Soil 251: 319–329, 2003.

[22] B.M. Nicolai, K. Beullens, E. Bobelyn, A. Peirs, W. Saeys, K.I. Theron, and J. Lammertyn, "Nondestructive measurement of fruit and vegetable quality by means of NIR spectroscopy," A Rev. Postharvest. Biol. Technol 46: 99 -118, 2007.

[23] A.M. Mouazen, W. Saeys, J. Xing, J.D Baerdemaeker, and H. Ramon, "Near infrared spectroscopy for agricultural materials: an instrument comparison," J. Near Infrared Spectroscopy 13: 87-97, 2005.

IOP Conf. Series: Materials Science and Engineering **557** (2019) 012087 doi:10.1088/1757-899X/557/1/012087

# Prediction of chemical content in patchouli oil (*Pogostemon cablin* Benth) by portable NIR spectroscopy

**Vrita Windawati[1], Y A Purwanto[1,2*], I Wayan Budiastra[1], Slamet Widodo[1], Shinichiro Kuroki[3]**

[1]Department of Mechanical and Biosystem Engineering, Bogor Agricultural University, Indonesia
[2]Center of Tropical Horticulture Studies, Bogor Agricultural University, Indonesia
[3]Graduate School of Agricultural Science, Kobe University, Japan

*E-mail: arispurwanto@apps.ipb.ac.id

**Abstract.** The objective of this study was to develop the calibration model for predicting the patchouli alcohol in patchouli oil by portable Near Infrared (NIR) spectroscopy. Patchouli oil samples were obtained from different origin in Indonesia. Portable NIR spectrometer with wavelength of 760-1050nm was used to collect the spectra of the patchouli oils. The calibration model was developed by partial least square regression (PLS). The best model for predicting patchouli alcohol content is y = 0.7209x + 8.2804 with r = 0.86, SEC = 2.04%, SEP = 2.16%, CV = 7.6, RPD = 1.8.

## 1. Introduction

The essential oil is natural extraction from various certain plants. The part of the plant that could be extracted is leaf, flower, wood, and seeds [1]. Currently, there are 150 plants that produce essential oil. The amount of it that being trade internationally is 70 and 40 of it comes from Indonesia. Due to this amount of it, the essential oil is one of Indonesia's predominant export commodities [2].

Patchouli oil is one of the essential oil that belongs to a high selling value export commodity [3]. This oil obtained through steam or water distillation. The main material used can be stem or leaf of patchouli [4]. The quality of oil produced depends on the distillation time and part of the material used [5]. Patchouli alcohol is the major content of patchouli oil that used as quality and price determination. The assessment of its quality usually uses chemical laboratory analysis. This method is accurate yet takes a long time and high cost. Therefore nondestructive method that efficient in the time and cost is needed in order to determine the quality of patchouli oil rapidly. One of technology that can be used is near-infrared spectroscopy (NIRS).

NIRS is a fast analysis method as well as nondestructive that relate to chemometric. The use of this method is appropriate for solid, liquid and biotechnology analysis. On the other hand, NIRS could be implemented in the production process for process monitoring or used in the laboratory for quality control [6]. NIR is enable to predict virgin olive oils. The information about the composition of olive oil given by NIR through real time [7]. FT NIRS can be used to analyze the mixed of crude palm oil and canola oil from the various vendor [8]. The use if NIRS also can classify extra virgin oils based on cultivar [9]. NIRS succeeded to classify patchouli oil based on the origin precisely and accurately [10]. Currently, Portable NIR has been developed so that it easy to use directly in the field. The portable spectrometer is appropriate to use as a daily scanner for customer due to its simplicity, ability to sync and upload measure data wirelessly, and automatically construct data model [11]. Based on the

IOP Conf. Series: Materials Science and Engineering **557** (2019) 012087 doi:10.1088/1757-899X/557/1/012087

explanation above, it is necessary to use NIRS as a method to determine the patchouli oil content that has produced from different region in Indonesia.

## 2. Material and Method

The materials used in this research are 30 samples of patchouli oil that obtained from different regions in Indonesia (Konawe, Kolaka and Masamba from Sulawesi island, Bogor, and Garut from west java and Aceh and Jambi from Sumatra). Instrument used in this study was a portable spectrometer SCiO by Consumer Physics (wave length 760-1050 nm) that used for measuring wave spectrum of patchouli oil samples, *liquid scan accessory* used as a tool for laying the *portable spectrometer* when the measuring in order to avoid the sink of SciO sensor in the oil. GC-MS was utilized to determine the content of clove oil. Furthermore, the software used in this research is *SCiO Lab: Developer Toolkit*, *Octave-4.2.1*, *The Unscrambler X* 10.4, dan *Microsoft Office Excel.* There are five steps of patchouli oil determination nondestructively using *Near Infrared Spectroscopy*; (1) samples preparation, (2)Measurement using GC-MS, (3) wave spectrum data intake using NIR, (4) Pre-treatment data process, and (5) development of calibration and validation model.

**Figure 1.** Spectra acquisition using portable spectrometer

## 3. Results and Discussion

Patchouli oil content was measured using GC-MS method. The results of then used as reference data for prediction model development. The prediction model that was obtained from the correlation between NIR spectrum was shot into patchouli oil with reference data. Hence, the reference data determines the success of predicting quality parameters value of patchouli oil using NIR. GC-MS showed that the content of patchouli oil is patchouli alcohol ($C_{15}H_{26}O$) (Table 1).

**Table 1**. Chemical characteristic of the patchouli oils based on the main chemical component (Patchouli-Alcohol C15H26O)

| Sample | % of the component |
|---|---|
| Jambi | 32.26 |
| Aceh | 30.91 |
| Kolaka | 26.46 |
| Konawe | 29.81 |
| Masamba | 24.15 |
| Garut | 30.98 |
| Bogor | 24.18 |
| average | 28.39 |
| Standard deviation | 3.40 |

The collected spectra from portable spectrometer of 760-1050nm was the normalized transflectant spectra data. Spectrum normalized data is a spectrum that already passed through pre-treatment SNV. Pre-treatment SNV is used for minimizing basic variation and deleting multiplicative interferences from scattering effects of spectrum data [12]. Each peak and the valley of the wave on the spectrum could be interpreted that the materials contain different chemical compound [13]. The spectrum peak of the NIR is affected by the chemical compound in the materials as well as physical characteristic in the materials. The absorption of wavelength on the chemical compound of a material on a transflectan spectrum is showed by the peak. The peaks of absorption occurred due to vibration and strain on atomic bond group O–H, N–H, and C–H that are major components as organic compound forming [14]. The high amount of chemical percentage in a material will cause high absorption. This cause the peak on a transflectan spectrum will get deeper.

The absorption peaks on NIR transflectan of patchouli oil occurred on the 915 and 969 nm wavelength. The deepest absorption peaks occurred on the 915 nm wavelength indicated a high percentage of the chemical compound in patchouli oil. However, the lack of information and reference regarding the chemical compound of material on the <1000 nm wavelength make it difficult to determine the kind of detected chemical compound.

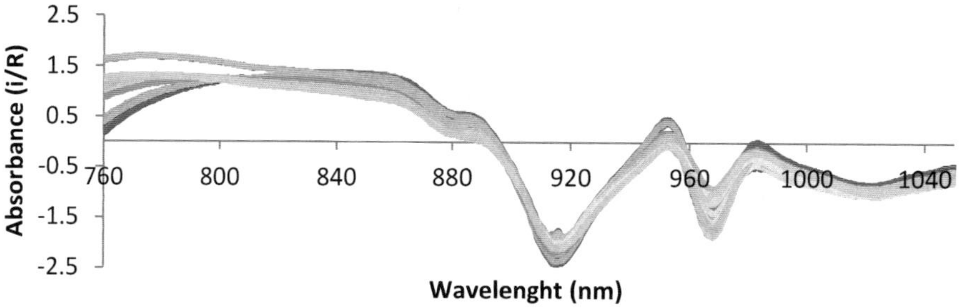


**Figure 2.** Original transflectant spectrum of patchouli alcohol

The NIR's translefctant spectrum data of patchouli oil was processed using PLS. PLS used because of the obtained data as numeric data. PLS is a quantitative analysis method that has been a wide reference for evaluating the quality and suitable for linear analysis. PLS could reduce the data dimension in order to find the relevant factors in predicting and interpreting the data. This method produces a better calibration method than PCR [15].

The prediction model of patchouli alcohol content from patchouli oil samples was developed based on the correlation of NIR transflectant spectrum data with data on the chemical content of GC-MS test results. Ideally, all samples should cover a homogeneous calibration range. Two-thirds of the total sample is used for calibration, while 1/3 of the total sample is used for validation. The spectrum of the same sample must be kept together and put in either the calibration set or the validation set. The spectrum with the lowest concentration (minimum) and highest (maximum) must be selected for the calibration set. The purpose of the calibration is to build a mathematical equation model that can be used to predict the chemical content of the material without having to take direct measurements that damage the material [14].

The NIR transflectant spectrum data obtained contained background and noise which caused the prediction model to be less accurate. This is influenced by the nature of scattering and absorption by a material. The disturbance can be reduced or partially eliminated with the help of proper pre-treatment data during processing [12]. Data pre-treatments were carried out with the aim of improving spectra quality and reducing or eliminating unwanted effects on spectrum data without affecting spectroscopic information needed for predictions [16]. Data pre-treatments are applied before calibration and validation are carried out. Pre-treatments carried out in this study were smoothing (SGs) and de-trending (DT). Smoothing Savizky-Golay (SGs) derives from the idea that around the spectrum measurement points can be installed by low polynomial degrees [17].

**Table 2.** Results of calibration and validation of patchouli alcohol content of patchouli oil with several pre-treatments

| Processing | PLS Factor | Calibration | | Validation | | | Consis-tent (%) |
|---|---|---|---|---|---|---|---|
| | | r | SEC (%) | SEP (%) | RPD | CV (%) | |
| Original | F4 | 0,87 | 2,04 | 2,16 | 1,8 | 2,73 | 94,56 |
| SGs | F4 | 0,86 | 2,05 | 2,16 | 1,8 | 2,94 | 94,94 |
| DT | F3 | 0,85 | 2,15 | 2,29 | 1,7 | 2,62 | 94,22 |

The value of RPD 1.86 states that the estimation model is feasible. The value of RPD will be inversely proportional to the value of the CV so that the higher the RPD, the CV value will be smaller. The best model for predicting patchouli alcohol content is y = 0.7209x + 8.2804, without using pre-treatment (r = 0.87, SEC = 2.04%, SEP = 2.16%, CV = 2.73% , RPD = 1.8). The results of the prediction of the best calibration model for patchouli alcohol content in patchouli oil can be seen in Figure 3.

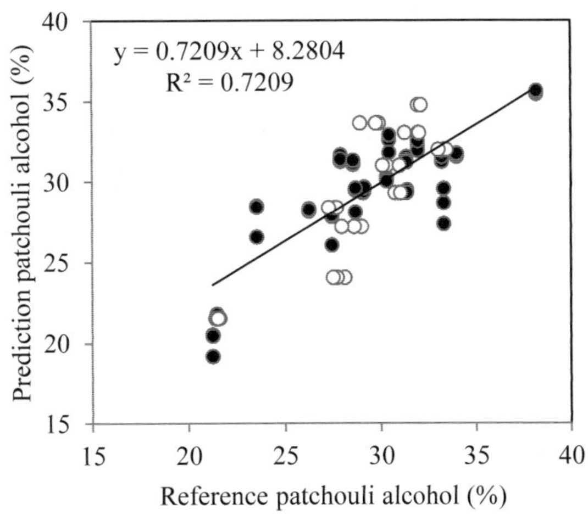


**Figure 3**. Calibration model for patchouli alcohol

## 4. Conclusion

From the results of GC-MS test it can be seen that 30 patchouli oil samples on average contain 28.39% patchouli alcohol as the most dominant chemical content. The absorption peak by the chemical content of patchouli oil occurs at a wavelength of 915 nm as the highest absorption peak and 969 nm. Calibration and validation was done by the Partial Least Squares (PLS) method. The best model for predicting patchouli alcohol content is y = 0.7209x + 8.2804, without pretreatment (r = 0.86, SEC = 2.04%, SEP = 2.16%, CV = 2.73, RPD = 1.8).

**Acknowledgment**

The authors express their appreciation for the financial support from Ministry of Research, Technology and Higher Education of the Republic of Indonesia under the scheme of International Research Collaboration and Scientific Publication Research Grant No. 1571/IT3.11/PN/2018.

**References**

[1] Kemendag RI. 2014. Market brief 2014 minyak atsiri. Berlin (DE): Kementrian Perdagangan Republik Indonesia Atase Perdagangan KBRI Berlin.

[2] ITPC. Peluang ekspor produk minyak esensial (*essential oils*) di Pasar Kanada. Vancouver (CA): Indonesian Trade Promotion Center Vancouver.

[3] Septiana A, Arianata HF, Kumoro AC. 2013. Potensi jeruk nipis (*Citrus aurantifolia*) sebagai bahan pengkelat dalam proses pemurnian minyak nilam (*patchouli oil*) dengan metode kompleksometri. *J Teknologi kimia & Industri* 2 (2): 257-261.

[4] Harimurti N, Soerawidjaja TH, Sumangat D, Risfaheri. 2012. Ekstraksi minyak nilam (*Pogostemon cablin* Benth) dengan teknik hidrodifusi pada tekanan 1-3 bar. *J Pascapanen* 9 (1): 1-10.

[5] Sulaiman A. Harsono D. 2012. Pengaruh lama penyulingan dan komposisi bahan baku terhadap rendemen dan mutu minyak atsiri dari daun dan batang nilam. *J Riset Industri Hasil Hutan* 4 (2).

[6] Roggo Y. 2007. Aerview of near infrared spectroscopy and chemometrics in pharmaceutical technologies. *J of Pharmacetical and Biomedical Analysis.* 44 (3): 683-700.

[7] Inarejos-Garcia AM, Gomes-Alonso, S, Fregapane G, Salvador MD. 2013. Evaluation of minor component, sensory characteristics and quality of virgin olive oil by Near Infrared (NIR) spectroscopy. *J Food Research International. 50: 250-258.*

[8] Mba O, Adewale P, Dumont M, Ngadi M. 2014. Application of near-infrared spectroscopy to characterize binary blend of palm and canola oils. *J Industrial Crops and Products.* 61 : 472-478.

[9] Sinelli. 2010. Application of near (NIR) infrared and mid (MIR infrared spectroscopy as a rapid tool to classy extra virgin olive oil on the basis of fruity attribute intensity. *J Food Research International.* 43 (1) : 369-375.

[10] Diego MCR, Purwanto YA, Sutrisno, Budiastra IW. 2017. Determination of the Characteristics and Classification of Near-Infrared Spectra of Patchouli Oil (*Pogostemon Cablin Benth.*) from Different Origin. IOP Conference Series: Earth and Environmental Science 147.

[11] Thong YJ, Nguyen T, Zhang Q, Karunanithi M, Yu L. 2017. Predicting food nutrition facts using pocket-size near-infrared sensor. 39$^{th}$ Annual International Conference of the IEEE.Seogwipo, Korea Selatan.

[12] Buchi Labortechnik AG. 2013. *Operation manual* NIRCal 5.5. Flawil (CH): BÜCHI Labortechnik AG.

[13] Osborne BG, Fearn T, Hindle PH. 1993. Practical NIR Spectroscopy, with applications in food and beverage analysis 2nd Edition. Longman Scientific and Technical, United Kingdom.

[14] Khalid Z, Meilina H, Syaubari, Munawar AA. 2017. *J Serambi Engineering* 1 (2):60-67.

[15] Kurniasari I, Purwanto YA, Budiastra IW, Ridwani S. 2018. Prediksi tanin dan total padatan tidak terlarut buah kesemek (*Diospyros kaki* L.) menggunakan spektroskopi NIR. *J Keteknikan Pertanian* 5 (3).

[16] Pizarro C, Esteban-Díez I, Nistal AJ, González-Sáiz JM. 2004. Influence of data pre-processing on the quantitative determination of the ash content and lipids in roasted coffee by near infrared spectroscopy. *Analytica Chimica Acta.* 509:217-227.

[17] Ozaki Y, McClure WF, Christy AA. 2007. *Near-infrared spectroscopy in food science and technology.* New Jersey (USA): John Wiley & Sons, Inc.

# Effect of low temperature and period of storage on the quality of Garlic Seeds (*Allium sativum L*)

**Nurmalia[1], Y A Purwanto[1,2,*], Sobir[2,3], Sulassih[2],N Naibaho[2]**

[1]Post Graduate Program of Postharvest Technology, Department of Mechanical and Biosystem Engineering, Bogor Agricultural University.

[2]Center of Tropical Horticulture Studies, Bogor Agricultural University
[3]Department of Agronomy and Horticulture, Bogor Agricultural University

*E-mail: arispurwanto@apps.ipb.ac.id

**Abstract**. The aim the study was to analyze the effect of temperature storage of garlic seeds including weight loss, sprouting and damage during the storage period. The samples of Garlic seeds are local varieties, cv. Lumbu Hijau and cv. Lumbu Kuning were collected from the farmers in Temanggung Regency. The experiments were set at two different temperature conditions, i.e. 5°C and room temperature (29-31°C) as a control. The relative humidity (RH) of storage conditions was 50-70% for cold storage and 70-80% for storage in room temperature. The amout of sampel used was about 47-68 tubers (500 g) for cv. Lumbu Hijau and around 24-32 tubers (500 gr) for cv. Lumbu Kuning , were placed in cold storage at different temperatures and in dark. The period of storage was design for 5 months. All experiments were carried out in three replications. The results showed that the decrease in weight loss for garlic stored at storage temperatures of 5°C and room temperature were 10.48 and 7.46 % for cv. Lumbu Hijau and 15.03 and 11.38 % cv. Lumbu Kuning, respectively. The higest sprouting percentage was found for those seeds garlic cv. Lumbu Kuning at 5°C, i. e, 22.22 %. The highest percentage of damage after 5 months of storage was found in the seeds of garlic cv. Lumbu Hijau stored at room temperature were 2.30 % and cv. Lumbu Kuning at 5°C were 19.04 %.

## 1. Introduction

Garlic is one of the important horticultural commodities and has high economic value. The productivity of garlic in 2016 was 8.79 tons / ha with a harvest area of 2,407 ha and production of 21,150 tons [1]. Garlic is a seasonal horticulture product. On season availability is abundant while off season the availability of garlic is small. This has an impact on the import of garlic to meet national needs. The availability of garlic in Indonesia has been obtained from Chinese and Indian imports. In 1998 Indonesia experienced self-sufficiency in garlic with a total area of 28,000 ha and Indonesia's imports under 10%. However, at present almost 95% of the total national needs which reach 400,000 tons are imported from other countries.

The productivity of garlic is very much determined by the use of high-quality seed bulbs. Garlic is propagated vegetatifly of bulbs. Plant propagation is the proces of growing new plants from a variety of sources; seeds, cuttings, bulbs and other plant parts. Bulbs that are good for planting do not contain disease, are not deformed and are not stored for too long. A good bulbs is a tuber that has broken its dormant period and is healthy and optimal size. Dormancy is a condition of plant organs that cannot grow under conditions below optimum. Quality seeds are seeds that have been declared as high-quality seeds from superior types of plants. High quality seeds have more than 90% germination, growth power or viability, so proper maintenance activities are needed. Improper handling will cause the viability and vigor of the seed to decline. The seeds will experience setbacks caused by humidity and temperature.

Garlic has perishable properties and it is difficult to maintain freshness in a long time. Therefore, proper handling is needed so as not to experience changes in quality at the time of storage. Storage is one of the important ways to overcome the availability of garlic. With good and correct storage, the stock will be controlled continuously with quality in accordance with market demand [2].

Garlic requires a good temperature for storage conditions so that its quality is relatively durable. Temperatures that are too high cause high shrinkage weights and cause the tubers to shrink. While high humidity provides an opportunity for the growth of fungi and molds and stimulates the growth of shoots and roots [3]. One common method is to store fresh agricultural products at low temperatures to extend shelf life and product quality. Storage of garlic seeds at low temperatures can reduce water loss in tubers, maintain respiration rate to be stable and slow the occurrence of metabolism so that it can inhibit the growth of shoots. Various damage that can occur is a decrease in water content, growth of shoots, hollowing, rot and softening (chiling Injury).

Various studies of garlic storage at low temperatures have been carried out. However, there are still not many reports regarding local varieties. It is important to know the changes that occur in storage for the long term through low temperature storage. The purpose of this study was to analyze the effect of low temperature storage on the storage of garlic seeds.

## 2. Material and Method

Sample of local garlic seeds i.e. cv.Lumbu Hijau and cv Lumbu Kuning were harvested from farmers orchad at Temanggung Regency, Central Java. After being harvested, the garlics were cured and dried until the water content reached approximately 60%. The experiments were set at temperature of 5°C (RH 50-70%) and room temperature (29-31°C, RH 70-80%). The weight garlic seeds samples were 500g and consisted of 47-68 bulbs for cv. LumbuHijau and 24-32 bulbs for cv. Lumbu Kuning. The samples were put in the cold storage at a given temperatures under the darkness condition (Figure 1). The storage period was designed for 4 months. Changes in weight loss, percentage of sprouting and damage bulbs were measured during the storage period. All experiments were carried out in three replications.

**Figure 1.** Drying and storage condition of seeds garlic

## 3. Result dan Discussion

The results of this study indicate that weight losses for storage of 5°C and room temperature after 4 months storage were 7.90 and 5.95% for cv. Lumbu Hijau and 10.87 and 9.22% for cv. Lumbu Kuning (Figure 2 and 3). Temperature storage of 5°C can maintain loss of water content of the bulb. Efforts to prevent water loss can be done by adjusting the right temperature and humidity. Because when stored the material will under go a transpiration process where some of the water in the material tissue will evaporate [4]. In addition, changes in weight loss are influenced by the rate of respiration and respiration rate affected by temperature [5].

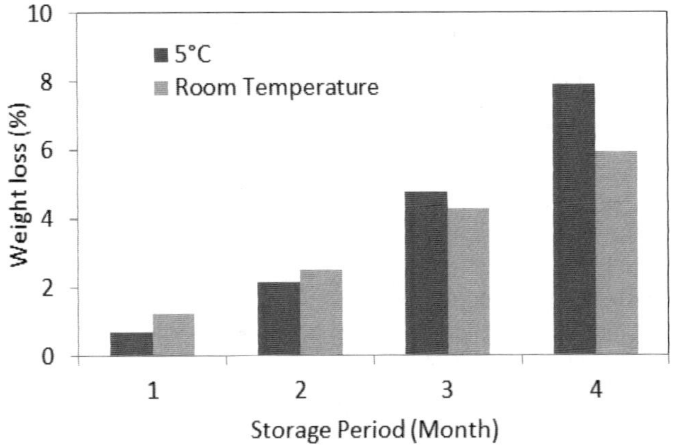


**Figure 2.** Weight loss of cv. Lumbu Hijau at 5 °C and room temperature for 4 months in the darkness condition

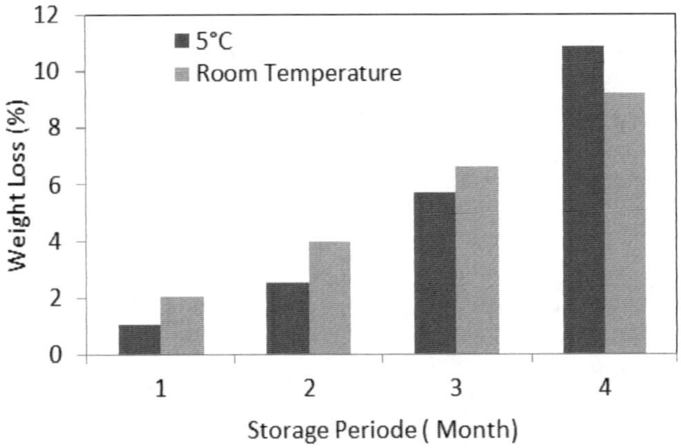


**Figure 3.** Weight loss of cv. Lumbu Kuning at 5 °C and room temperature for 4 months in the darkness condition

After 4 months storage of garlic seeds, sprouting was found at temperature of 5°C of the cv. Lumbu Kuning which was equal to 1.11% and not found in the cv. Lumbu Hijau at all storage temperatures. This shows that each variety gives different responses to the temperature treatment. Storage at various low temperatures for both varieties can be applied to maintain the seeds from sprouting. During storage, sucrose and starch are consumed in high content and parts of carbohydrates that are not consumed cause glucose accumulation from the garlic tissue thereby increasing sprouting. Bulbs stored at 10°C for 30 days storage temperature increase the percentage of sprouting [6]. Storage temperatures of 5 and 30°C can inhibit the growth of onion bulbs [3]. The effect of cold storage depends not only on temperature but also on its duration and depends on cold temperatures which can break its dormancy [6].

Packaging of garlic seed bulbs that produce plants with growth and the best results is the use of plastic nets with a shelf life of 57 days and damage of 9.6% [7]. Garlic cannot be planted immediately after harvest because garlic requires a dormancy period before continuing growth. The dormancy period varies depending on the variation and storage temperature. The optimum storage temperature for garlic buds ranges between 5 and 10°C. Cold storage combination 7°C duration for 30 days has a higher growth percentage [8].

**Table 1.** Sprouting bulbs of seeds garlic cv. Lumbu Hijau and cv. Lumbu Kuning during 4 months storage

| Month | Lumbu Hijau | | Lumbu Kuning | |
|---|---|---|---|---|
| | Sprouting bulb (%) | | Sprouting bulbs (%) | |
| | 5°C | Room Temperature | 5°C | Room Temperature |
| 1 | 0 | 0 | 0 | 0 |
| 2 | 0 | 0 | 0 | 0 |
| 3 | 0 | 0 | 0 | 0 |
| 4 | 0 | 0 | 1 | 0 |

**Table 2.** Damage bulbs of seeds garlic cv. Lumbu Kuning and Cv. Lumbu Hijau during 4 months storage

| Month | Lumbu Hijau | | | | | | | |
|---|---|---|---|---|---|---|---|---|
| | Damage Bulbs | | | | | | | |
| | 5°C | | | | Room Temperature | | | |
| | Empty Bulb | Waxy breakdown | Pest diseases | Rot | Empty Bulb | Waxy breakdown | Pest diseases | Rot |
| 1 | 0 | 0 | 0 | 0 | 1 | 0 | 0 | 0 |
| 2 | 1 | 0 | 0 | 0 | 1 | 0 | 0 | 0 |
| 3 | 1 | 0 | 0 | 0 | 1 | 0 | 0 | 0 |
| 4 | 1 | 0 | 0 | 0 | 1 | 0 | 0 | 0 |
| Month | Lumbu Kuning | | | | | | | |
| | Damage Bulbs | | | | | | | |
| | 5°C | | | | Room Temperature | | | |
| | Empty Bulb | Waxy breakdown | Pest diseases | Rot | Empty Bulb | Waxy breakdown | Pest diseases | Rot |
| 1 | 0 | 0 | 0 | 7 | 0 | 0 | 0 | 0 |
| 2 | 0 | 4 | 0 | 9 | 0 | 0 | 0 | 0 |
| 3 | 0 | 4 | 0 | 9 | 0 | 0 | 5 | 0 |
| 4 | 0 | 4 | 0 | 9 | 0 | 0 | 5 | 0 |

Based on the results of this experiment, there was also some damage to the garlic seeds. Damage that occurs as empty bulb, waxy breakdown, pest diseases and rot. Damage can occur because when storing low temperatures garlic seeds cannot carry out normal metabolic processes. Then a decrease in the fruit's ability for oxidative phosphorylation occurs which will increase further in the presence of low temperatures [4].

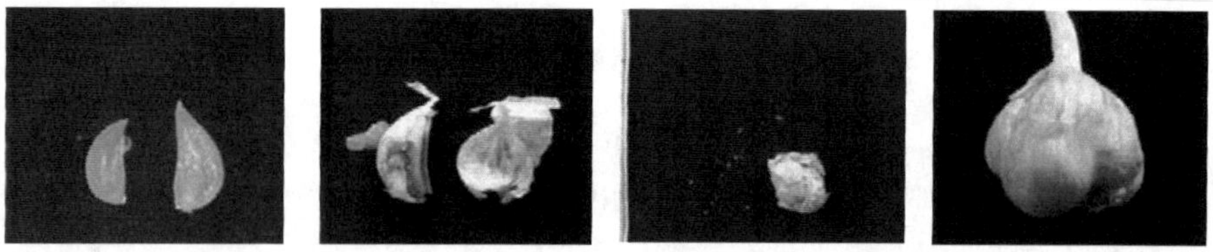

**Figure 4.** The damage of Garlic Seeds cv. Lumbu Hijau (a), cv. Lumbu Kuning (b) at 0 ° C and sprouting of cv. Lumbu Kuning at 5°C

## 4. Conclusion

The temperature of 5ºC RH 50-70% (dark conditions) resulted in the highest percentage of weight loss for seeds garlic of cv. Lumbu Hijau and cv. Lumbu Kuning. The damage of seeds garlic were found in the seeds garlic of lumbu Hijau at a temperature of 5°C and room temperature that occurs as empety bulbs, while in cv. Lumbu Kuning damage that occurs as waxy breakdown and rot at temperature 5°C and pest diseases at room temperatur. The initial sprouting was found at storage period of 4 month in 5ºC cv Lumbu Kuning and did not occur in the cv.Lumbu Hijau.

## 5. Acknowledgment

The authors express their appreciation for the financial support from Ministry of Research, Technology, and Higher Education of Republic Indonesia through National Strategic Research (STRANAS) Fund No. 1560/IT3.11/PN/2018. The first author would like to thank IAARD Ministry of Agriculture of The Republic Indonesia for financial support for the Master study at Bogor Agricultural University.

## Reference

[1] Directorate Jenderal Of Horticulture. 2017. DE. Jakarta. Development of National Garlic. Jakarta. Ministry of Agriculture of the Republic of Indonesia.

[2] Mutia AK, Purwanto YA, Pujantoro L, 2014.Changes In Quality Shallot (*Allium ascalonicum L*) During Storage at Different Temperature and Water Content. *J. Pascapanen* 11 (2) : 108-115.

[3] Miedema. 1994. Bulb dormancy in onion. I. The effect of temperature and cultivar sprouting and rooting. *J. Hort. Sci*. 69 (1) :29-39.

[4] Muchtadi D. 1992. Instructions for Post-Harvest Vegetables and Fruits Physiology Laboratory. Department of Education and Culture. Directorate General and Higher Education Center for Inter-University Food and Nutrition. Bogor Agricultural Institute.

[5] Komar N. Rokhmadiono, S. Kurnia L.2001. Post-Harvest Shallot Storage Techniques in East Java. *J. Agricultural Technology* 2 (2): 79-95.

[6] Woldeyes F, Wtsadik K, Tabor G. 2017. Emergence of Garlic (*Allium sativum* L.) as Influenced by Low Storage Temperature and Gibberellic Acid Treatments. *J.of Agriculture And Ecology Research International* 10 (2) ; 1-7.

[7] Soedomo R.P. 2006. Effect of Packaging Materials on The Keeping Quality of seedbulbs, the growth and yield performance of garlic. J. Hort 16 (4) : 283-289.

[8] Desta B, Woldetsadik K, Mohammed W, Abebie B. 2017. Duration of Low Temperature Storage, Clove Topping and Smoka Water on Garlic Sprouting and Seedling Vigor. *J. Of Agronomy* 16 (3) : 124-130